Type of Problem	Type of Data				
	One Sample	Two Independent Samples	Two Related Samples	Three or More Independent Samples	Three or More Related Samples
Association	Spearman rank correlation coefficient, p. 358 Kendall's tau, p. 365 Confidence interval for tau, p. 377 Olmstead–Tukey test, p. 381 Phi coefficient, p. 401 Yule coefficient, p. 402 Goodman–Kruskal coefficient, p. 404 Cramér's statistic, p. 403 Point biserial coefficient, p. 409	Chi-square test of independence, p. 181		Kendall's coefficient of concordance, p. 386 Chi-square test of independence, p. 181 Partial rank correlation, p. 395	
Regression	Estimation, p. 427 Brown–Mood test, p. 431 Theil test, p. 435 Confidence interval for slope, p. 442	Test for parallelism, p. 444 Confidence interval for difference between two slopes, p. 455			
Miscellaneous	Binomial test, p. 56 Confidence interval for population proportion, p. 61 One-sample runs test, p. 63 Cox–Stuart test for trend, p. 68	Wald–Wolfowitz runs test, p. 113 Hollander test of extreme reactions, p. 116 Fisher exact test, p. 120 Chi-square test of homogeneity, p. 192	McNemar test, p. 163		Cochran's Q, p. 290

APPLIED
NONPARAMETRIC
STATISTICS

PREFACE

I have written *Applied Nonparametric Statistics*, Second Edition, with two purposes in mind.

1. To make available a textbook in nonparametric statistics for a course in which the emphasis is on applications rather than theory.
2. To provide a reference book in nonparametric statistics for the practicing researcher.

The amount of material contained in the book is about right for a one-quarter or one-semester course at the advanced undergraduate or graduate level for students of all disciplines. Most students using the book as a text will have had at least one introductory (nonmathematical) course in classical statistics. Although such preparation is not absolutely necessary, the text does assume a certain mathematical facility—equivalent to that acquired in a course in college algebra.

With the practicing researcher in mind, I have adopted a format intended to make it easy for the researcher to use the book for reference purposes during the planning and analytical phases of an investigation. A separate chapter is devoted to each of the research situations that the researcher is likely to encounter. Thus there are chapters devoted to techniques that are appropriate under the following circumstances:

1. When the data available for analysis consist of observations from a single sample (Chapter 2).
2. When the data available for analysis are from two independent samples (Chapter 3).
3. When the data available for analysis are paired data (data from two related samples) (Chapter 4).

4. When the data consist of frequencies and the researcher is interested in reaching a decision regarding the independence of two criteria of classification or the homogeneity of two or more populations (Chapter 5).

5. When the data available for analysis consist of observations from three or more independent samples (Chapter 6).

6. When the data available for analysis consist of observations from three or more related samples (Chapter 7).

7. When the researcher wants to know whether a single sample has been drawn from a population that is distributed in a specified manner or whether two samples have been drawn from identically distributed populations (Chapter 8).

8. When the data to be analyzed consist of pairs of measurements measured on at least an ordinal scale and the researcher wants to know whether or not the two relevant variables are associated (Chapter 9).

9. When the data conform to the simple linear regression model, but the assumptions underlying parametric inference are suspect (Chapter 10).

If the researcher can determine that one of these situations applies to the problem at hand, he or she can consult the appropriate chapter and then scan the numbered sections to further pinpoint the pertinent procedure. For example, suppose that the problem concerns the difference beween the location parameters of two populations. The researcher should consult Section 3.1. If the problem concerns the dispersion of two independent samples, the researcher should refer to Section 3.2.

Hypothesis-testing procedures are all handled with the same format, which is as follows:

1. *Assumptions:* The assumptions underlying the test are stated.

2. *Hypotheses:* The appropriate null and alternative hypotheses are stated.

3. *Test statistic:* Instructions for computing the test statistic as well as the rationale underlying the test are given.

4. *Decision rule:* The reader is told how to use the appendix tables to decide whether or not to reject each of the possible null hypotheses.

Where appropriate, the procedure for dealing with ties, the large-sample approximation, and the power-efficiency of a given test are discussed in clearly identified paragraphs. The text first presents the methodology for each procedure in general terms and then gives a numerical example.

Each section includes numerous references to the statistical literature. These references can be used in two ways: as a resource for the reader who wishes to explore a particular topic in greater depth and as outside reading assignments for the instructor who wishes to enrich the course. These references have been extensively updated for the second edition.

Wherever possible I have used real data extracted from published research literature. My purpose in doing so is twofold. First, I hope that what might otherwise be a rather pedestrian experience for the reader will take on new life, meaning, and relevance because of these brief glimpses into the real world of research and experimentation. Second, I feel that the student will be more easily convinced of the usefulness of the procedures introduced when he or she sees them applied to data resulting from actual scientific investigations. In analyzing published data, I have in many instances used a statistical procedure different from that employed by the authors. My doing so is not an indication that they are wrong and I am right but merely that their data were also appropriate for illustrating the application of the particular nonparametric procedure. To the researchers whose data I have used, I extend my gratitude.

For the examples and exercises, I have drawn material from a wide variety of disciplines: agriculture, biology, sociology, education, psychology, medicine, business, geology, and anthropology. Again, my purpose was to create as much interest (through variety) in the subject matter as possible. Furthermore, I wanted to demonstrate the wide applicability of nonparametric procedures.

I have been generous with exercises. The instructor may assign them as homework; the nonstudent reader can use them to test his or her comprehension of the techniques involved. So that they can be used for immediate reinforcement of the material covered, I have placed exercises at the end of the discussion of each technique. For review purposes, additional exercises appear at the ends of chapters.

There are so many nonparametric procedures now available that the writer of a textbook on the subject must select only a sample for presentation. My two criteria for selecting the procedures discussed here were *usefulness* and *popularity*. I wanted to include those procedures likely to prove most useful to the researcher, as well as those most likely to be encountered in published research findings.

Among the techniques included in this second edition of *Applied Nonparametric Statistics* that were not in the first edition are the Ansari–Bradley dispersion test, Lehman contrasts for both the one-way and the two-way layout, techniques for comparing all treatments with a control in the one-way and two-way layout, an aligned ranks procedure for the two-way layout, Lilliefors test for normality, and, for analyzing qualitative data, several widely used methods, including the phi coefficient, the Yule coefficient, the Goodman–Kruskal coefficient, Cramér's statistic, and the point biserial coefficient.

The users of nonparametric statistical techniques will be rewarded with the usual advantages of speed, accuracy, and convenience when they use a computer to perform the required computations. With that in mind, I have tried to acquaint the reader with as many sources of computer support as possible. My efforts have resulted in two types of references: (1) published computer programs that have been written for various nonparametric techniques and (2) microcomputer software packages that provide routines for nonparametric statistical techniques. For two reasons I have elected not to include samples of computer printouts in the text. First,

so many software packages are available that choosing one to use for illustrative purposes would be difficult. Second, computer printouts of the results of nonparametric statistical analyses would not contribute a great deal to the pedagogical impact of the text. Printouts for nonparametric techniques tend to be simple, lacking the rich store of information found in printouts for such parametric techniques as analysis of variance and regression analysis.

I would like to express my gratitude to Richard A. Groeneveld, Iowa State University, and S. K. Katti, University of Missouri—Columbia, who read the manuscript for this book. They made many valuable suggestions for improving the text; they are, however, absolved of all responsibility for any deficiencies that remain.

I also want to thank Rickie Domangue, James Madison University, and LeRoy A. Franklin, Indiana State University, who were kind enough to offer many helpful suggestions during the planning stage for this edition of the book.

Wayne W. Daniel
Atlanta, Georgia

CONTENTS

CHAPTER *3*

PROCEDURES THAT UTILIZE DATA FROM TWO INDEPENDENT SAMPLES   82

CHAPTER *4*

PROCEDURES THAT UTILIZE DATA FROM TWO RELATED SAMPLES   144

CHAPTER *5*

CHI-SQUARE TESTS OF INDEPENDENCE AND HOMOGENEITY   178

CHAPTER *6*

PROCEDURES THAT UTILIZE DATA FROM THREE OR MORE INDEPENDENT SAMPLES   220

CHAPTER *7*

PROCEDURES THAT UTILIZE DATA FROM THREE OR MORE RELATED SAMPLES   261

CHAPTER *8*

GOODNESS-OF-FIT TESTS   305

APPLIED
NONPARAMETRIC
STATISTICS

INTRODUCTION AND REVIEW

The subject of statistics encompasses a wide variety of activities, ideas, and results. Practitioners of the science of statistics usually acknowledge that it has two broad subdivisions: *descriptive statistics* and *inferential statistics*. Descriptive statistics is concerned with recording and summarizing, in quantitative terms, the outcomes of events and the characteristics of persons, places, and things. Records of the annual numbers of births, deaths, and marriages are called statistics. So are descriptions of the age, level of education, and ethnic composition of persons living in a given area. Statistical inference, or inferential statistics, involves drawing conclusions from such facts and making decisions based on them.

This book is devoted to the study of inferential statistics. Presumably, most readers have had at least one previous course in statistics. However, a brief review of some important concepts will probably not be wasted. Therefore the first three sections of this chapter are devoted to a quick overview of the subject in general. Section 1.4 discusses measurement scales. Section 1.5 brings us to the basic concepts of *nonparametric* statistics, the subject of this book. Section 1.6 is concerned with the use of computers in nonparametric statistical analysis. Section 1.7 offers a brief preview of Chapters 2 through 10, and Section 1.8 describes the format used to present the statistical techniques in the chapters that follow.

1.1

SOME IMPORTANT TERMINOLOGY

This section defines some terms used in succeeding chapters. These terms are part of the vocabulary of the statistician. Other terms will be defined as they occur later in the book.

Population   The word *population* is used to refer to a collection of persons, places, or things. Just which collection constitutes the population in a given discussion depends on the investigator's sphere of interest. One investigator may want to make statements about all college and university students in the United States. Another may want to make statements about all students at one particular college or university. Each of these investigators considers the population to be the collection of students about which he or she wishes to make statements.

In some contexts we also refer to a *collection of measurements* (sometimes called observations) made on a population of persons, places, or things as a population. For example, if we are interested in the ages of all students at a certain college or university, we refer to a collection of these ages as a population (of ages). More specifically, a population may be defined as *the largest collection of persons, places, or things* (including measurements) *in which we have an interest*. Populations may be either *finite* or *infinite*.

Infinite populations are composed of a limitless number of elements. We may better understand the concept of an infinite population if we consider some never-ending, element-yielding process. Such a process would produce an infinite population of elements. Imagine, for example, a manufacturing process that continues forever. If the results of the process are ball bearings, the process will yield an infinite population of ball bearings. The population of all humans who have ever lived, are now living, and ever will live in the future may, for all practical purposes, be thought of as an infinite population.

When a population is finite, it is possible (though not always practical) to count the elements of which it is composed. Examples of finite populations include the students enrolled at a certain college, all employees of some firm, all items of a particular type produced in a factory on a given day, and the houses located in a given census tract.

Populations may also be either *real* or *hypothetical*. An example of a real population is all students currently enrolled at a given university. An example of a hypothetical population is as follows. Suppose that we designed an experiment to evaluate the effectiveness of three tranquilizing drugs, and randomly assigned subjects to receive one of the three. We could think of each of the three resulting groups as a sample from a population of a large number of subjects to whom the drug in question could be given. While we may imagine such a population, it would be impractical to create it. Such a population, then, is hypothetical rather than real.

Typically in drug research, subjects who today receive an experimental drug are considered to constitute a sample of subjects who now have the disease of interest and who will contract it anytime in the future. These current and future subjects who have or will have the disease are thought of as a hypothetical population. Such a population does not exist; hence it is a *hypothetical* or *potential* population.

Sample   A sample is part of a population. Suppose that a certain population consists of all students at a particular college. Those students enrolled in a statistics course at the college, being a part of the population, would constitute a sample from the population. We could identify samples in many other ways. For example, all students who are majoring in English would be a sample, as would all students who are married or all students who have a car registered for campus parking.

We may, of course, have samples from infinite as well as from finite populations. For example, a sociologist may be interested in some characteristic of all adults who live (in the past, present, and future) in the Southeastern United States. The sociologist would consider this an infinite population. A sample of current adult residents of the Southeastern United States would constitute a sample from this infinite population.

Random Sample   Statistical inference consists of reaching conclusions about a population on the basis of information contained in a sample. When populations are sufficiently large or infinite, it is impractical or impossible to examine every element of the population to gather information on which to base a conclusion about the population as a whole. For this and other reasons, conclusions about a population are usually based on the information contained in a sample that has been drawn from that population.

When statistical inference is used to reach conclusions about populations, just any type of sample is not necessarily appropriate. The validity of results based on statistical inference rests on the assumption that a special type of sample, called a *random sample*, has been employed in the process.

To obtain a random sample of size n, we select it in such a way that the probability of selecting it is known in advance. The simplest type of random sample is the *simple random sample*. A simple random sample of size n is one that is selected in such a way that every random sample of size n that can possibly be selected from the population has the same probability of being selected. Simple random samples are usually selected through the use of a table of random numbers or with the help of a computer. The reader unfamiliar with simple random sample concepts and procedures is referred to an elementary statistics book or a book on sampling techniques. Other kinds of random samples include *stratified random samples* and *cluster samples.*

Samples of Convenience   Readers who are coming fresh from a statistics course in which random sampling in all its purity was the basis of inferential procedures may be shocked by the samples used in much of the research reported in

scientific literature. Instead of random samples drawn with the help of random number tables or the random number generating capabilities of a computer, we find samples consisting of "patients admitted to the stroke clinic during the first three months of the year," or "all students in the first grade at Blank School," or "healthy volunteers." People use such samples because they are available and convenient. How, then, can we rationalize using them to make inferences?

Dunn (1, page 12) and Remington and Schork (2, page 72) suggest that we examine the nature of the population from which such samples might be considered random. For example, if the sample consists of first-grade students in a middle-class, suburban school, perhaps the sample can be considered a random sample of all first-grade students attending similar schools in similar neighborhoods. Armitage (3, pages 99, 100) proposes a slightly different rationale for using samples of convenience. Colton (4, pages 4–7) addresses the same problem when discussing the difference between the *target population* (the population about which we wish to reach a conclusion) and the *sampled population* (the population from which the sample is actually drawn).

Statistic   A *statistic*, which is a function of one or more random variables, is a measure computed from sample data. Statistics familiar to those who have had a course in statistics are the sample mean $\bar{x}$, the sample variance s^2, and the sample correlation coefficient r.

Parameter   A *parameter* is a constant that determines the specific form of a density function. Examples of parameters include the population mean μ, the population variance σ^2, and the population correlation coefficient ρ. Parameters are usually unknown; when they are unknown, we use statistics to estimate them. For example, we may use the sample mean $\bar{x}$ to estimate the unknown mean μ of the population from which the sample was drawn.

In nonparametric statistical analysis, a parameter of considerable interest is the population *median*. This parameter, in nonparametric analysis, frequently replaces the population mean as the preferred measure of location or central tendency.

Random Variable   We usually assume that the numerical data on which we perform statistical analyses are the outcomes of a random sampling procedure or a random experiment. A set of such outcomes is called a *random* (or *chance*) *variable*. In the process of sampling or experimenting, we observe one or more values of the random variable. For example, the time that adult subjects take to react to some stimulus is a random variable. If we apply the stimulus to a randomly selected adult and observe a reaction time of 0.15 second, then 0.15 is a value of this random variable.

Continuous Variable   A random variable is *continuous* if the values that it can assume consist of all real numbers in some interval; that is, a continuous variable

can assume any of the uncountable and infinite number of values within a relevant interval. Time of reaction to some stimulus is an example of a continuous variable.

Discrete Variable   If a random variable can asssume only a finite or countably infinite number of values, it is said to be *discrete*; that is, the number of values may be either finite or infinite, but countable. The values assumed by a discrete variable are characterized by gaps, since such a variable can assume only certain values, rather than all possible values, within an interval. Thus the number of children in a family is a discrete variable, since it can take on only the values 1, 2, 3, 4, 5, and so on. The values that a discrete variable can assume do not have to consist of the natural numbers. A discrete variable might be able to assume values that are fractions or combinations of fractions and whole numbers.

1.2

HYPOTHESIS TESTING

This book is concerned with two types of statistical inference: *hypothesis testing* and *interval estimation*. Hypothesis testing will be treated in this section and interval estimation in the next.

A hypothesis may be defined as *a statement about one or more populations*. A distinction may be made between two general types of hypothesis: the *research hypothesis* and the *statistical hypothesis*. The research hypothesis is one that is formulated by a potential investigator (sample surveyor or experimenter) who is usually not a statistician. A research hypothesis is frequently the result of a hunch or suspicion based on extended observation by the potential investigator.

For example, a teacher may suspect, on the basis of several years of teaching experience, that certain physical conditions in the classroom impede learning. A physician who observes that some patients are short of breath after taking a certain drug may suspect that the drug has adverse side effects, at least in some patients. Such suspicions lead to research hypotheses such as "Third-grade students score higher on arithmetic tests when the room temperature during teaching does not exceed 68°F," and "Shortness of breath following administration of drug A occurs more frequently in patients who have high blood pressure than in patients who do not."

There are two statistical hypotheses: the *null hypothesis* (which we designate H_0) and the *alternative hypothesis* (which we designate H_1).

The null hypothesis is the hypothesis that we test. The null hypothesis is always a statement of no difference, no effect, or the status quo. For example, we may test the null hypothesis that there is no difference in the effects of two drugs when administered to patients for the purpose of curing some illness; we may test the null hypothesis that a certain drug has no effect on the progress of a disease; or we may

test the null hypothesis that one population is identical to another with respect to some characteristic. A null hypothesis is presumed to be true until sufficient evidence to reject it has been amassed.

The test procedure, which is based on information derived from the data of an appropriate sample, results in one of two statistical decisions: (1) a decision to reject the null hypothesis (as false) or (2) a decision *not* to reject the null hypothesis because the sample does not provide sufficient evidence to warrant rejection. When we reject the null hypothesis, we accept the alternative hypothesis as true. We can do so because we state the null hypothesis and the alternative hypothesis in such a way that they are mutually exclusive and complementary. Usually—but not always—the alternative hypothesis and the research hypothesis are the same. Thus the alternative hypothesis in most cases is a statement of what we expect to be able to conclude.

When we reject a null hypothesis, we accept the alternative hypothesis with greater conviction than we would if we "accepted" the null hypothesis because we were unable to reject it. In general, evidence that supports a contention or hypothesis is not as convincing of the truth of the hypothesis as evidence that is incompatible with a hypothesis is convincing of the falsity of that hypothesis.

A hypothesis test may be *two-sided* (nondirectional) or *one-sided* (directional). The following is an example of the statement of the null and alternative hypotheses when the parameters of interest are the means μ_1 and μ_2 of populations 1 and 2, respectively, and the test is two-sided:

$$H_0: \mu_1 = \mu_2, \qquad H_1: \mu_1 \neq \mu_2$$

Alternatively, we may state these hypotheses as

$$H_0: \mu_1 - \mu_2 = 0, \qquad H_1: \mu_1 - \mu_2 \neq 0$$

since, if $\mu_1 = \mu_2$, their difference will be 0, and if $\mu_1 \neq \mu_2$, their difference will be something other than 0. The null hypothesis states that the means of two populations are equal. The alternative states that they are not equal. In this case the investigator is asking, "On the basis of my sample data, can I conclude that the two populations have different means?" An investigator may feel that a more meaningful question would be, "Can I conclude that population 1 has a larger mean than population 2?" In this case, the investigator performs a one-sided test, and the null and alternative hypotheses are

$$H_0: \mu_1 \leq \mu_2, \qquad H_1: \mu_1 > \mu_2$$
$$\text{or} \quad H_0: \mu_1 - \mu_2 \leq 0, \qquad H_1: \mu_1 - \mu_2 > 0$$

The investigator may also state the question leading to a one-sided test in such a way that the statistical hypotheses are

$$H_0: \mu_1 \geq \mu_2, \qquad H_1: \mu_1 < \mu_2$$
$$\text{or} \quad H_0: \mu_1 - \mu_2 \geq 0, \qquad H_1: \mu_1 - \mu_2 < 0$$

A hypothesis of the form $\mu_1 - \mu_2 = 0$ is called a *simple* hypothesis, since a single value only is specified. A hypothesis of the form $\mu_1 - \mu_2 \leq 0$ is referred to as a *composite* hypothesis, since more than one value is specified. Thus we see that a simple hypothesis completely specifies the distribution of the variable of interest. A composite hypothesis does not completely specify the distribution of the variable of interest. In those cases in which the null hypothesis is composite, we perform the test at the point of equality. It can be shown that whatever conclusion we reach from testing at the point of equality is the same conclusion (all other things being equal) that we would reach if we were to perform the test at any other value specified in the composite hypothesis.

To test a null hypothesis, the investigator selects an appropriate *test statistic* and specifies its distribution when H_0 is true. For example, when the hypothesis concerns the difference between two population means, the test statistic is usually

$$t = \frac{(\bar{x}_1 - \bar{x}_2) - D_0}{\sqrt{(s_p^2/n_1) + (s_p^2/n_2)}}$$

Here $\bar{x}_1$ and $\bar{x}_2$ are sample means computed from samples of size n_1 and n_2 drawn, respectively, from population 1 and population 2, D_0 is the hypothesized difference between the population means, and s_p^2 is obtained by pooling the two sample variances. When certain assumptions are met and H_0 is true, t has Student's t distribution with $n_1 + n_2 - 2$ degrees of freedom.

From observed sample data we compute a value of the test statistic and ask ourselves, "Is this a surprisingly extreme (either very large or very small) value to observe when H_0 is true?" In other words, we wonder whether the magnitude of the computed value of the test statistic is sufficiently extreme to cause us to reject the null hypothesis. Before examining the sample data, many investigators formulate a decision rule. This rule says, in effect, that they will reject H_0 if the probability of obtaining a value of the test statistic of a given or more extreme magnitude—when H_0 is true—is equal to or less than some small number α. Most writers of elementary statistics books refer to α as the *level of significance*. Others call α the *size of the test*; for example, Mood, Graybill, and Boes (5) use this term. When people use the decision-rule approach, they usually choose α to be 0.05 or 0.01, or occasionally 0.10.

The *critical value* of the test statistic is the value that is so extreme that the probability of getting it or a more extreme value, when H_0 is true, is equal to α. Alternatively, then, we may state the decision rule in terms of critical values. In a one-sided test, for example, the decision rule instructs us to reject H_0 if the computed value of the test statistic is as extreme as or more extreme (either larger or smaller, depending on the direction of the alternative hypothesis) than the critical value. In two-sided tests there are two critical values. We reject H_0 if the computed value of the test statistic is as extreme as or more extreme than either of two specified critical values. One of the critical values is selected in such a way that if the null hypothesis happens to be true, a value (computed from sample data) of the test statistic as large as or larger than the selected critical value would be considered unusual. The other

FIGURE 1.1   Critical value (1.645) and computed value (1.86) of test statistic z for a one-sided hypothesis test

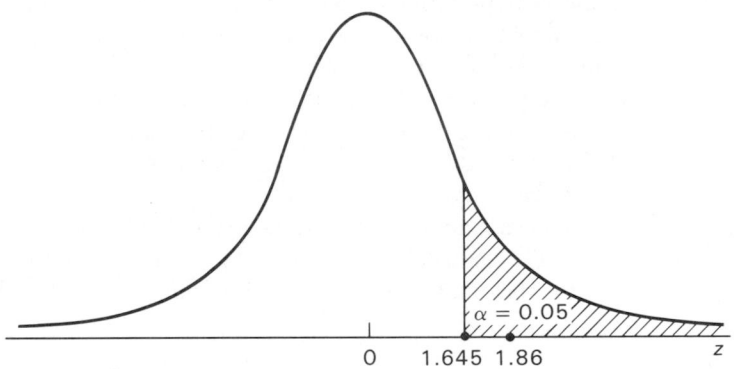

critical value is selected in such a way that, in the same context, a computed value of the test statistic as small as or smaller than this selected critical value would also be considered unusual.

Let us graphically illustrate the use of a decision rule for testing hypotheses. Suppose that the null hypothesis we want to test and its alternative are

$$H_0: \mu_1 \le \mu_2, \qquad H_1: \mu_1 > \mu_2$$

Suppose also that the level of significance is $\alpha = 0.05$, the test statistic has the standard normal distribution, and the computed value of z is 1.86. When we refer to Table A.2, we see that the critical value of the test statistic for $\alpha = 0.05$ and a one-sided test is 1.645. Figure 1.1 shows the distribution of z, its critical value, and its computed value. Since 1.86 is greater than 1.645, we reject H_0.

P Values   Another way to decide whether the sample data cast doubt on the null hypothesis is to determine the probability of observing, when H_0 is true, a value of the test statistic at least as extreme (in the appropriate direction) as the value actually observed. This probability is referred to by a variety of names: critical level, the descriptive level of significance, the prob value, and the associated probability. We shall use the term *P value* to refer to this probability. In doing so, we follow the practice of Gibbons and Pratt (6) in their article on the interpretation and methodology of *P* values. Hodges and Lehmann (7, page 317) suggest that we think of the *P* value (which they call the significance probability) "as giving, in a single convenient number, a measure of the degree of surprise which the experiment should cause a believer of the null hypothesis."

Most writers in the scientific literature report *P* values in such terms as $p > 0.05$, $p < 0.01$, $0.01 < p < 0.05$, or $p = 0.0618$, where p is the *P* value. Thus a *P* value may be reported as an exact value or as an interval, depending on the nature of the

available table of the distribution of the test statistic. Many statistical tables have been constructed (or abridged for inclusion in statistics textbooks) in such a way that they are more convenient for investigators who use a decision rule and pre-select values of α. To use such tables for determining the P value of a test, the investigator must usually be content with an interval rather than an exact value. In the following example we can find an exact P value.

Example 1.1   Suppose that the null and alternative hypotheses are

$$H_0: \mu_1 \leq \mu_2, \qquad H_1: \mu_1 > \mu_2$$

Suppose further that the appropriate test statistic is z and that the computed value is $z = 1.86$. We refer to Table A.2 and see that the area to the right of $z = 1.86$ is $0.5 - 0.4686 = 0.0314$. That is, the probability of observing a value of z as large as or larger than 1.86, when H_0 is true, is 0.0314. Hence the P value is 0.0314. Figure 1.2 illustrates this situation.

FIGURE 1.2   Computed value (1.86) of the test statistic z and the corresponding P value for a one-sided hypothesis test

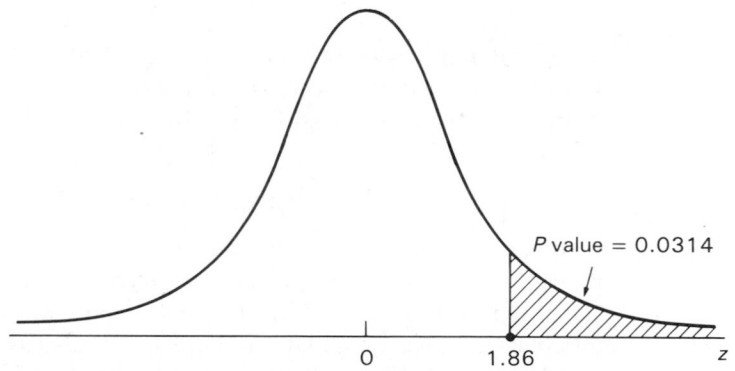

Now consider an example in which we have to report a P value as an interval because the available table is limited.

Example 1.2   Suppose that in our experiment the test statistic follows a chi-square distribution with five degrees of freedom, and that the computed value of the test statistic is 14.665. For sufficiently large values of the test statistic, we will reject H_0. When we enter Table A.11 with five degrees of freedom, we find that 14.665 is between $\chi^2_{0.975} = 12.832$ and $\chi^2_{0.99} = 15.086$. We report the P value, then, as $0.01 < P$ value < 0.025.

Suppose that the sample data had yielded a value of the test statistic of 17.335. Since 17.335 is greater than $\chi^2_{0.995} = 16.750$, we could report P value < 0.005. Figure 1.3 illustrates this example.

FIGURE 1.3   Computed value (14.655) of chi-square test statistic and corresponding P value for a hypothesis test

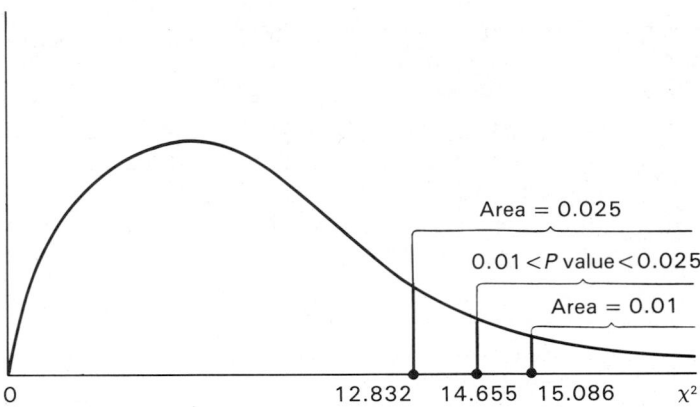

Determining P values in two-sided tests presents a problem. As Gibbons and Pratt (6) point out, the most common practice in a two-sided test is to report the P value as twice the one-sided P value. In Example 1.1, for instance, suppose that the hypotheses had been

$$H_0: \mu_1 = \mu_2, \qquad H_1: \mu_1 \neq \mu_2$$

A computed value of $z = 1.86$ would lead to a P value of $2(0.0314) = 0.0628$. This procedure is satisfactory for cases in which the sampling distribution of the test statistic is symmetric when the null hypothesis is true. Examples of such distributions are the standard normal and Student's t.

If, however, the distribution of the test statistic under H_0 is asymmetric, doubling the one-sided P value to obtain a two-sided P value can lead to a P value that is greater than 1, as well as other absurdities. Gibbons and Pratt (6), who discuss several optional procedures in the two-sided case, favor the practice of reporting the one-tailed P value and stating the direction of the observed departure from the null hypothesis. Suppose we follow this plan when, for example, the hypotheses are

$$H_0: \mu_1 = \mu_2, \qquad H_1: \mu_1 \neq \mu_2$$

and the appropriate test statistic is z. If the computed value of z is 1.86, we may summarize the result in terms of a P value in any one of the following ways. (There are other ways, too.)

1. $P(z \geq 1.86 \,|\, H_0) = 0.0314$

2. $P(z \geq 1.86 \,|\, \mu_1 = \mu_2) = 0.0314$

3. The probability of observing a value of the test statistic *as large as* or *larger than* 1.86 when H_0 is true is equal to 0.0314.

If we were to compute $z = -1.86$ from sample data (and substitute it for 1.86 in 1, 2, and 3), the inequalities in 1 and 2 would be reversed, and the words "large" and "larger" would be replaced with the words "small" and "smaller" in (3), thereby indicating the direction of departure from H_0.

For the computed value of $+1.86$ in this example, we could state that "the P value $= 0.0314$ in favor of a larger value of μ_1." For $z = -1.86$, we could state that "the P value $= 0.0314$ in favor of a larger value of μ_2."

A P value gives us more information than statements such as "the difference was significant at the 0.05 level" or "H_0 can be rejected at the 0.01 level." This is a major argument in favor of reporting P values as a part of research findings. Given a P value, researchers can select their own level of significance, or the level at which they are willing to stop believing that H_0 is true and start believing that H_1 is true.

Some authors distinguish between *hypothesis testing* and *significance testing*. They use "hypothesis testing" when they adopt a decision rule in terms of a preselected α, and "significance testing" when they report the P value. For further discussion of this difference, see Kempthorne and Folks (8) and Lindgren (9).

Since prior exposure to statistics may have acquainted the reader with the decision-rule (preselected α) approach to hypothesis testing rather than the calculation-of-P-values approach, both approaches have been used in the examples and exercises in the first part of this book. Later the practice of preselecting α in favor of the P-value approach is discontinued. By following both approaches, a better understanding of P values (which may be a new concept to the reader) and a better understanding of the relationship between the two approaches may be gained. In examples and exercises in which only P values are calculated, readers may practice arriving at their own significance level in evaluating the results.

When a P value is reported, it may be used as a criterion for rejecting or failing to reject the null hypothesis by persons with different ideas regarding the level of significance at which such action should be taken. If the P value is equal to less than the chosen α, the null hypothesis is rejected. If the P value is greater than the chosen α, the null hypothesis is not rejected. If, for example, the P value for a given test is 0.03, a researcher who had chosen a significance level of 0.05 would reject H_0, but a researcher who had chosen a significance level of 0.01 would not reject H_0.

The reader may find the comments of Bahn (10) and Daniel (11) on the topic of P values of interest.

STATISTICAL SIGNIFICANCE VERSUS PRACTICAL SIGNIFICANCE

In addition to statistical significance, another important concept arises when we attempt to evaluate the results of research—*practical* or *substantive significance*. Unfortunately, the term statistical significance is often used in a way that implies

practical significance instead. When the statistical analysis of research findings uncovers statistically significant results, do not presume that the discovery necessarily has any practical significance. Suppose that we are interested in knowing whether two population means are equal. Sufficiently large samples will reveal any difference, no matter how small, but it may be that only a fairly large difference is of any practical value. Likewise, small samples may fail to detect population differences (or relationships) that are of practical significance.

Because the concepts of statistical significance and practical significance are different, be careful in applying the terms. When speaking of statistical significance, reserve the word "significant" to refer to sample results. Thus, for example, we might say that "there is a significant difference between sample means," meaning that their observed difference led to a P value sufficiently small to cause us to reject the null hypothesis of no difference in population means. Avoid such expressions as "the population means are significantly different," since, if we mean statistical significance, the terminology is incorrect, and if we mean practical significance, we should use a word other than "significantly" to avoid confusion. By all means, avoid such phrases as "the alternative hypothesis is that the two population means are significantly different," since statistical tests of hypotheses do not necessarily determine what is of practical significance. Only the person knowledgeable in the area of investigation is qualified to decide that.

The confusion surrounding the concepts of statistical significance and practical significance has been discussed by Bakan (12), Brewer (13), Cohen (14), Daniel (15), Duggan and Dean (16), Gold (17), Kish (18), and McGinnis (19).

POWER OF HYPOTHESIS TESTS

The *power* of a hypothesis test is the probability of rejecting the null hypothesis when it is false. Power may be defined also as $1 - \beta$, where β is the probability of accepting a false null hypothesis. Recall that accepting a false null hypothesis is referred to as a type II error, and that rejecting a true null hypothesis is a type I error. The probability of a type I error is usually designated α. High power is always a desirable characteristic of a test.

Unfortunately, the estimation of the power of a test is usually not an easy task. The necessary calculations require us to have knowledge of the probability distribution of the test statistic given that the alternative hypothesis is true. When the alternative hypothesis is composite, it is usually desirable to compute more than one power value (one for each of several well-defined alternative hypotheses), thereby adding to the computational burden.

In general, we may increase the power of a test through the manipulation of various components of the hypothesis-testing procedure.

Increasing Power by Increasing Sample Size   It is always possible to increase the power of a test by taking a larger sample. If the price of an individual mea-

surement is high, increasing power by increasing sample size may be an expensive procedure to follow—sometimes prohibitively so.

Increasing Power by Increasing Significance Level   If we are willing to increase the probability of rejecting a true null hypothesis, we can increase the power of a test. In other words, when all other conditions are unchanged and the null hypothesis is false, we will have a more powerful test if $\alpha = 0.10$ than if $\alpha = 0.05$. Similarly, a choice of $\alpha = 0.05$ will result in a more powerful test than will a choice of $\alpha = 0.01$.

Increased Power as a Result of Size of Effect to Be Detected   The size of the effect to be detected by a hypothesis test cannot be manipulated by the investigator. When all other conditions are the same, however, a test will be more powerful the greater the discrepancy between the condition stated in the null hypothesis and the true condition in the sampled population. Suppose, for example, that we wish to test $H_0: \mu = 100$. Suppose further that H_0 is false. If all other conditions are the same, we will have a more powerful test if H_0 is false because the true population mean, μ, is 150 than if H_0 is false because $\mu = 101$; that is, our test will be much more likely to detect a difference of 50 than a difference of 1. When the required assumptions arc mct, many of the classical parametric tests are most powerful for the test at hand. Gibbons (20) points out, however, that there are many cases, especially when the samples are small, in which nonparametric tests are almost as powerful under standard assumptions as their parametric analogues. For most of the tests discussed in succeeding chapters, comments will be made on what we know about their power.

EFFICIENCY OF HYPOTHESIS TESTS

Another criterion for evaluating the performance of a test is *efficiency*. The most frequently encountered index of efficiency for a nonparametric test is its *asymptotic relative efficiency* (usually abbreviated ARE). Since the concept of asymptotic relative efficiency is attributed to Pitman (21), it is frequently referred to as *Pitman efficiency*. High efficiency is a desirable property of a test.

In many practical situations the ARE of a test is a good approximation of its *relative efficiency*. The relative efficiency of test A to test B (for the same H_0, H_1, α, and β) is the ratio n_B/n_A, where n_A and n_B are the sample sizes, respectively, of tests A and B. In general, the asymptotic relative efficiency of a test is the limiting value of relative efficiency as the size of the sample, n, increases without limit. If n_A is smaller than n_B, the efficiency of test A relative to test B is greater than unity, and we say that test A is more efficient than test B. We prefer the test requiring the smaller sample size under the same conditions, since smaller samples generally reduce expenditures of money, time, and other resources. When the alternative hypothesis is composite, with resulting different possible values of β, we may wish to compute the relative efficiency of the test for several of these values.

FURTHER READING

In the chapters that follow, comments are made on the efficiency of most of the tests discussed. Readers who are interested in the mathematical details of ARE, can refer to the articles by Noether (22) and Stuart (23, 24) and the books by Lehmann (25), Fraser (26), Randles and Wolfe (27), and Pratt and Gibbons (28). For other definitions of relative efficiency, see Blyth (29). The article by Smith (30) may also be of interest.

Literature on the various facets of statistical inference is extensive. The following may be of value or interest. Wilson and Miller (31) discuss the "inconclusiveness of accepting the null hypothesis." Edwards (32) discusses the relationship between scientific and statistical hypotheses. Feinberg (33) and Rodger (34, 35) write about type I and type II errors. Edgington (36) discusses nonrandom samples in statistical inference, and Ungerleider and Smith (37) comment on the abuse of statistics.

An article by Brewer and Knowles (38) contains an elementary discussion of power. Additional comments on statistical significance appear in articles by Ahrens (39), Barnard (40), Chandler (41), Labovitz (42), Lykken (43), Krause (44), Morrison and Henkel (45), O'Brien and Shapiro (46), Rozeboom (47), Selvin (48), Skipper (49), Stone (50), Winch and Campbell (51), Zeisel (52), and in an editorial in the *New England Journal of Medicine* (53). Bross (54), Godambe and Sprott (55), Kiefer (56), and Lurie (57) deal with general considerations in statistical inference.

1.3

ESTIMATION

In many instances investigators may want to reach a decision regarding the numerical value of a population parameter instead of (or in addition to) knowing whether they can reject the null hypothesis that it is equal to some specified value. To reach decisions regarding the magnitudes of population parameters based on sample data, we use the process of *estimation*.

There are two types of estimation: *point estimation* and *interval estimation*. In point estimation we compute a single value—called the *estimate*—from sample data and offer it as a candidate for the parameter that we want to estimate. For example, if we wish to estimate the unknown mean of some population, we typically draw a sample from the population, compute the mean of the sample, and use it as the estimate of the population mean.

In most cases the *interval estimate* is a more desirable and more useful estimate. An interval estimate consists of two possible values of the parameter being estimated—a lower value and an upper value. These two values define an interval that enables us to express our degree of confidence that the interval contains the estimated parameter. An interval estimate, therefore, is frequently called a *confidence interval*.

We express our degree of confidence in a confidence interval by using the *confidence coefficient*, which is either a number between 0 and 1 or a percentage. If the confidence coefficient is 0.95 or 95%, for example, we say that we are 95% confident that the interval contains the parameter we are estimating.

The usual method of constructing confidence intervals allows us to interpret them in two ways. The *probabilistic interpretation* is based on the fact that in repeated sampling, $100(1 - \alpha)\%$ of the intervals constructed in the same manner (and with the same sample sizes) contain the parameter being estimated. This interpretation applies to all the confidence intervals that we may construct. In practice we construct only a single interval, and it is to this single interval that we apply the other interpretation, the *practical interpretation*. In expressing the practical interpretation, we say we are $100(1 - \alpha)\%$ confident that the single interval constructed contains the parameter we are estimating. In both interpretations $100(1 - \alpha)\%$ is the confidence coefficient.

We may express a confidence interval for the parameter θ in probabilistic terms as

$$P(L_0 < \theta < U_0) = 1 - \alpha \qquad (1.1)$$

where L_0 and U_0 are random variables satisfying the probability statement. Equation 1.1 is a probability statement. It is a statement of the probability that one of all the possible intervals constructed as just described will contain the unknown parameter θ. Once we have specified the values of L_0 and U_0, say L and U, Expression 1.1 is no longer a chance variable, but a fixed interval; that is, it is the single interval constructed in a practical application. The interval with L_0 and U_0 specified either contains the unknown parameter or it does not. If we have constructed the interval with a high confidence coefficient, however, we have high confidence that the interval does, in fact, contain the parameter we are estimating. We can express the single interval compactly as

$$C(L < \theta < U) = 100(1 - \alpha)\% \qquad (1.2)$$

where C stands for confidence and indicates that the statement is a confidence statement rather than a probability statement, and the interval is called a *confidence interval*. Suppose, for example, that we wish to estimate a population mean μ with a confidence coefficient of 0.95. From the sample results, suppose we are able to compute $L_0 = 60$ and $U_0 = 80$. We may now write

$$C(60 < \mu < 80) = 95\%$$

We read this expression as "We are 95% confident that the unknown population mean μ is somewhere between 60 and 80."

As you may suspect, interval estimation and hypothesis testing are related. Consider the hypotheses

$$H_0: \theta = \theta_0, \qquad H_1: \theta \neq \theta_0$$

where the significance level is α. Possible values of the parameter θ contained in the $100(1 - \alpha)\%$ confidence interval $L < \theta < U$ are those values that are compatible with the null hypothesis. Possible values of θ outside the interval are not compatible

with the null hypothesis. We may therefore test H_0 by means of a confidence interval. If the $100(1 - \alpha)\%$ confidence interval does not contain the hypothesized value of the parameter θ_0, we reject H_0 at the α level of significance. If θ_0 is in the interval, we do not reject H_0 (at the α level of significance). Natrella (58) discusses the relationship between confidence intervals and tests of significance. Also of interest are the chapters on hypothesis testing and confidence intervals in a book of readings by Kirk (59).

1.4

MEASUREMENT SCALES

An objective of nonparametric statistics is to provide procedures whereby we may make statistical inferences on the basis of data that do not conform to the usual assumptions of normality and other assumptions that validate the use of parametric statistics. One of the problems that sends us in search of a nonparametric technique when we are trying to apply a parametric technique has to do with the way in which the variable under consideration is measured. A researcher analyzing numerical data is concerned with the nature of the scale used to make the measurements. Many researchers follow the views on measurement and measurement scales articulated by Stevens (60, 61, 62). Stevens (60) defines measurement as the assignment of numerals to objects or events according to rules, and points out that different rules lead to different kinds of scales and different kinds of measurement. Stevens defines four types of measurement scale: *nominal, ordinal, interval*, and *ratio*.

Nominal Scale   The nominal scale is the weakest of the four measurement scales. As its name implies, the nominal scale distinguishes one object or event from another on the basis of a name. Thus we may classify (name) items coming off an assembly line as defective or nondefective. A newborn infant is male or female. Patients in a psychiatric hospital may be schizophrenic, manic–depressive, psychoneurotic, and so on.

Frequently we use arbitrary numbers, rather than names in the usual sense, to distinguish among objects or events on the basis of some characteristic. For example, we may use the number 1 to designate defective items coming off an assembly line and 0 to designate nondefective items. Usually we use the nominal scale when we are interested in the number of objects falling into each of the various nominal categories. For example, we may want to know how many patients in a psychiatric hospital are diagnosed as schizophrenic, as manic–depressive, as psychoneurotic, and so on. Data of this type are frequently referred to as *count data, frequency data*, or *categorical data*.

Ordinal Scale   The next-most-precise or sophisticated measurement scale is the *ordinal scale*. We distinguish objects or events measured on the ordinal scale from one another on the basis of the relative amounts of some characteristic they possess.

Ordinal measurement makes it possible for objects to be ranked. Salespersons, for example, can be ranked from "poorest" to "best" on the basis of their personalities. Beauty contestants can be ranked from least beautiful to most beautiful. Illnesses can be ranked from least severe to most severe. If we are to rank *n* objects on the basis of some trait, we may assign the number 1 to the object having the least amount of that trait, the number 2 to the object containing the next-smallest amount, and so on to *n*, the object with the largest amount of the trait under consideration. For example, contestants in a track meet may be ranked $1, 2, 3, \ldots$, according to the order in which they cross the finish line. Data of this type are frequently referred to as *rank data*.

The differences between rankings are not necessarily equal. For example, three students taking an examination may be ranked first, second, and third on the basis of the order in which they complete the examination. This does not mean, however, that the time elapsing between completion of the examination by number 1 and by number 2 is the same as that between number 2 and number 3. The student finishing first may, for example, finish five minutes before the second student, who, in turn, may finish eight minutes before the third. If we have only the ranks available for analysis, we do not know the magnitudes of the differences between measurements that are ranked.

Interval Scale   When objects or events can be distinguished one from another and ranked, and when the differences between measurements also have meaning (that is, when there is a fixed unit of measurement), the *interval scale* of measurement is applicable. The true interval scale has a zero point, but it is arbitrary. A familiar example of interval measurement is the measurement of temperature in degrees Fahrenheit or degrees Celsius (centigrade). The zero point on Fahrenheit and Celsius thermometers does *not* indicate an absence of temperature, the trait being measured.

Suppose, for example, that four objects A, B, C, and D are assigned scores of 20, 30, 60, and 70, respectively, where measurement is on the interval scale. Since we used an interval scale, we can say that the difference between 20 and 30 is equal to the difference between 60 and 70; that is, equal distances between the members of each of two pairs of scores indicate equal differences in the amount of the trait being measured. The interval scale, however, does not permit us to speak meaningfully about the ratios of two scores. In our example, we cannot say that a score of 60 for C and a score of 30 for B means that C has twice as much of the trait as B.

Ratio Scale   When measurements have the properties of the first three scales and the additional property that their ratios are meaningful, the scale of measurement is the *ratio scale*. A property of the ratio scale is a true zero, indicating a complete absence of the trait being measured. The familiar measurements of height and weight are examples of measurement on the ratio scale. We can say that a person who weighs 180 pounds weighs 60 pounds more than a person who weighs

120 pounds (as we can with the interval scale). With a ratio scale, we can also say that a 180-pound person weighs twice as much as a 90-pound person. The ratio scale represents the highest level of measurement.

FURTHER READING

Stevens holds that the statistical procedures appropriate for use with empirical data depend on the measurement scale represented by the observations. Many practicing statisticians and researchers follow this philosophy when they perform statistical analyses. Some, however, take issue with Stevens; in particular, they disagree with his contention that the level of measurement determines the nature of the statistical operations that are permissible. Among those who hold opposing views are Anderson (63), Boneau (64), Brown (65), Gaito (66, 67), Lord (68), Mitchell (69), and Townsend and Ashby (70). Khurshid (71) has prepared a bibliography on measurement scales. Other articles of interest include those by Baker et al. (72), Campbell (73), and Gardner (74).

1.5

NONPARAMETRIC STATISTICS

The typical introductory course in statistics examines primarily *parametric statistical procedures*. Recall that these procedures include tests based on Student's *t* distribution, analysis of variance, correlation analysis, and regression analysis. A characteristic of these procedures is the fact that the appropriateness of their use for purposes of inference depends on certain assumptions. Inferential procedures in analysis of variance, for example, assume that samples have been drawn from normally distributed populations with equal variances.

Since populations do not always meet the assumptions underlying parametric tests, we frequently need inferential procedures whose validity does not depend on rigid assumptions. *Nonparametric statistical procedures* fill this need in many instances, since they are valid under very general assumptions. As we shall discuss more fully later, nonparametric procedures also satisfy other needs of the researcher.

By convention, two main types of statistical procedures are treated as nonparametric: (1) truly nonparametric procedures and (2) *distribution-free* procedures. Strictly speaking, the true nonparametric procedures are not concerned with population parameters. For example, in this book we shall discuss tests of goodness of fit and tests for randomness where we are concerned with some characteristic other than the value of a population parameter. As the term suggests, the validity of distribution-free procedures does not depend on the functional form of the population from which the sample has been drawn. It is customary, especially among American writers, to refer to both types of procedure as nonparametric. Kendall and Sundrum (75) discuss the differences between the terms nonparametric and distribution-free.

HISTORY

The first use of what we would now call a nonparametric statistical procedure seems to have been reported in 1710 by John Arbuthnot (76). Uses of such procedures were conspicuously sparse until the 1940s. The word nonparametric appeared for the first time in 1942 in a paper by Wolfowitz (77). Since then, the growth of interest in both the theory and the application of nonparametric statistics has been rapid. Nonparametric statistics is currently one of the most important branches of statistics. The techniques that fall within this category of statistics are used in most, if not all, of the physical, biological, and social sciences. A book edited by Papatoni-Kazakos and Kazakos (78) is devoted to the use of nonparametric methods in communications. Brown and Hayden (79) discuss the clinical applications of nonparametric methods. Their advantages in water quality data analysis are discussed by Helsel (80). Jenkins et al. (81) conclude, on the basis of articles published in a single psychology journal, that nonparametric procedures had not been used as much as parametric procedures; Buckalew (82) argues that nonparametric techniques deserve greater recognition and use in psychology.

Advantages of Nonparametric Statistics The following are some of the advantages of the available nonparametric statistical procedures.

1. Since most nonparametric procedures depend on a minimum of assumptions, the chance of their being improperly used is small.

2. For some nonparametric procedures, the computations can be quickly and easily performed, especially if calculations are done by hand. Thus using them saves computation time. This can be an important consideration if results are needed in a hurry or if high-powered calculation devices are not available.

3. Researchers with minimum preparation in mathematics and statistics usually find the concepts and methods of nonparametric procedures easy to understand.

4. Nonparametric procedures may be applied when the data are measured on a weak measurement scale, as when only count data or rank data are available for analysis.

Disadvantages of Nonparametric Statistics Nonparametric procedures, however, are not without disadvantages. The following are some of the more important disadvantages.

1. Because the calculations needed for most nonparametric procedures are simple and rapid, these procedures are sometimes used when parametric procedures are more appropriate. Such a practice often wastes information.

2. Although nonparametric procedures have a reputation for requiring only simple calculations, the arithmetic in many instances is tedious

and laborious, especially when samples are large and a computer is not handy.

When to Use Nonparametric Procedures The following are some situations in which the use of a nonparametric procedure is appropriate.

1. The hypothesis to be tested does not involve a population parameter.
2. The data have been measured on a scale weaker than that required for the parametric procedure that would otherwise be employed. For example, the data may consist of count data or rank data, thereby precluding the use of some otherwise appropriate parametric procedure.
3. The assumptions necessary for the valid use of a parametric procedure are not met. In many instances, the design of a research project may suggest a certain parametric procedure. Examination of the data, however, may reveal that one or more assumptions underlying the test are grossly violated. In that case, a nonparametric procedure is frequently the only alternative.
4. Results are needed in a hurry, a computer is not readily available, and calculations must be done by hand.

In some instances we may not be able to resort to the remedy of the central limit theorem, because of the excessive skewness or kurtosis of the population from which we must draw our sample.

FURTHER READING

The literature of nonparametric statistics is extensive. In 1962 a bibliography by Savage (83) contained some 3,000 entries. A bibliography of distribution-free methods published in 1979 by Singer (84) runs to 53 pages. Harter (85, 86) is the author of an annotated bibliography of order statistics in two volumes. Volume I, covering the period prior to 1950, contains 942 entries in its 430 pages. Volume II, for the period 1950–1959, is 750 pages in length. Books on nonparametric methods that do not require a strong background in mathematics are those by Bradley (87), Conover (88), Gibbons (89), Hollander and Wolfe (90), Leach (91), Lehmann (92), Marascuilo and McSweeney (93), Maxwell (94), Mosteller and Rourke (95), Noether (96), Pierce (97), Quenouille (98), Runyon (99), Senders (100), Siegel and Castellan (101), Sprent (102), Tate and Clelland (103), and Wilcoxon and Wilcox (104). Books that are more demanding mathematically include those by David (105), Edgington (106), Fraser (107), Gnedenko et al. (108), Gibbons (20), Hájek (109), Hájek and Šidák (110), Hettmansperger (111), Kraft and van Eeden (112), Manoukian (113), Maritz (114), Noether (115), Pratt and Gibbons (28), Puri (116), Randles and Wolfe (117), Sarhan and Greenberg (118), and Walsh (119, 120, 121).

Shorter introductions to nonparametric statistics are to be found in the articles by Fisher (122) and Noether (123). In 1984, issue Number 3, Volume 9, of the *Journal of*

Statistical Planning and Inference was devoted entirely to articles on nonparametric statistics. Likewise, Volume 4 of the *Handbook of Statistics*, edited by Krishnaiah and Sen (124), contains only articles concerned with nonparametric methods.

Fisher (125) discusses graphical methods in nonparametric statistics, and Grimm (126) addresses the issue of the use of transformed variables versus nonparametric techniques.

The articles by Doksum (127), Noether (128), and Ruist (129) contain a good historical review of nonparametric statistics. Hettmansperger and McKean (130) present a graph that is useful in illustrating the connection between nonparametric test statistics and the estimation procedures associated with them. The articles by Buchanan (131), Noether (132) and Scheffé (133) are also of interest.

1.6

AVAILABILITY AND USE OF COMPUTER PROGRAMS IN NONPARAMETRIC STATISTICAL ANALYSIS

In nonparametric statistical analysis, as in statistical analysis as a whole, the computer is a valuable computational tool. Numerous software packages are available for use on both mainframe computers and the ubiquitous microcomputer. Although space does not allow for an exhaustive list, the following are examples of such packages that were available in late 1988. The *BMDP, SAS, SCA*, and *SPSS* are some of the packages that provide nonparametric analysis capabilities for both mainframe and microcomputers. Among the many software packages designed for the microcomputer, *Number Cruncher, STATA, STATGRAPHICS*, and *STATISTIX* are some that possess significant nonparametric analysis capabilities. Woodward et al. (134) have compiled a directory of statistical microcomputer software packages. For each package discussed, the authors provide information on such characteristics as configurations supported, number of current users, program description, bibliography of reviews, documentation, graphics, and the statistical features supported. The directory is a valuable reference for anyone looking for a package with specific statistical capabilities such as nonparametric analysis.

A number of periodicals devoted to statistical computing are available and should not be overlooked by those searching for information on computer programs for a particular nonparametric procedure. Such publications include *Computer Science and Statistics: Proceedings of the Symposium on the Interface; Journal of Statistical Computation and Simulation; COMPSTAT: Proceedings in Computational Statistics; Statistical Software Newsletter; Communications in Statistics, Part B; SIAM Journal on Scientific and Statistical Computing; Computational Statistics and Data Analysis; Computers and Geosciences; Computers in Biology and Medicine; Computers and Biomedical Research; International Journal of Bio-Medical Computing; Computer Programs in Biomedicine*; and *Computers and Medicine*. Several organizations, such as the American Statistical Association and the *SAS* Users

Group, publish proceedings of conferences at which the topic of computer programs for nonparametric statistical analysis is covered. A listing of the contents of many such proceedings is published monthly and cumulated annually in a publication called *Index to Scientific & Technical Proceedings. The American Statistician* regularly publishes information on statistical computing under the headings of "New Developments in Statistical Computing" and "Statistical Computing Software Reviews." Individuals who have compiled extensive software packages for nonparametric statistics include Vegelius (135), Zarnoch [see Kennedy (136)], von Collani (137), and Hannan (138). Information on the availability of these packages is included in the references.

1.7

SCOPE OF THIS BOOK

The emphasis in this book is on the application of nonparametric statistical methods. Whenever possible, the examples and exercises use real data, gleaned primarily from the results of research published in various journals. The use of real situations and real data will, it is hoped, make the book more interesting. Included are problems from as wide a variety of sources as possible to show the broad applicability of the techniques described. Also included are a wide variety of statistical techniques. The techniques discussed are those most likely to prove helpful to the researcher and most likely to appear in the research literature. This text covers not only hypothesis testing, but interval estimation as well.

Here is a brief summary of the topics covered in succeeding chapters.

Chapter 2 *Procedures that utilize data from a single sample*   This chapter covers procedures for estimating and testing hypotheses about location parameters when we are interested in the characteristics of a single population. Also included are a test for randomness and a test for trend.

Chapter 3 *Procedures that utilize data from two independent samples* When the available data consist of independent samples from each of two populations, the procedures of this chapter are appropriate. Included are estimation techniques for the difference between two population parameters and tests for the equality of two dispersion parameters. In addition, a runs test, a test for extreme reactions, and the Fisher exact test are presented.

Chapter 4 *Procedures that utilize data from two related samples*   The procedures presented here are applicable when the data consist of two samples that are related in some way. The observations may be measurements taken on the same subjects before and after some treatment has been applied, or they may be measurements taken on different subjects who have been matched on the basis of one or more criteria. Procedures for estimating intervals and a test for use when the data are frequencies are also discussed.

Chapter 5 *Chi-square tests of independence and homogeneity* Here the chi-square test, perhaps the most widely used of all the nonparametric procedures, is discussed. Two situations are covered. In the first the data come from a single sample of subjects cross-classified on the basis of two criteria, and the objective is to determine whether we should conclude that the two criteria of classification are related. In the other situation we identify two or more populations in advance and draw a sample from each. Our objective in this case is to determine whether we should conclude that the populations are not homogeneous with respect to some characteristic.

Chapter 6 *Procedures that utilize data from three or more independent samples* This chapter includes procedures that are the nonparametric analogues of parametric one-way analysis of variance. It also discusses a multiple-comparisons procedure and a procedure for use when the order of magnitude of the location parameters is specified in the alternative hypothesis.

Chapter 7 *Procedures that utilize data from three or more related samples* This chapter includes tests for use when we want a nonparametric test analogous to the analysis of variance for a randomized block design. It also covers the case of incomplete blocks and a test for use when the alternative hypothesis is ordered.

Chapter 8 *Goodness-of-fit tests* This chapter covers the most frequently used goodness-of-fit tests—the chi-square test and the Kolmogorov–Smirnov tests. It also covers a procedure for constructing a confidence band for a population distribution function.

Chapter 9 *Rank correlation and other measures of association* This chapter discusses the most frequently used tests of association.

Chapter 10 *Simple linear regression analysis* This chapter presents some nonparametric procedures that can be used with the simple linear regression model. The chapter covers tests and confidence interval procedures for the slope coefficient and the intercept parameter. Also included are a test for the parallelism of two regression lines and a confidence-interval procedure for the difference between two slope parameters.

1.8

FORMAT AND ORGANIZATION

To present these statistical procedures, a format designed to facilitate the use of the book has been adopted. Each hypothesis-testing procedure is broken down into four components: (1) assumptions, (2) hypotheses, (3) test statistic, and (4) decision rule.

Thus for a given test the reader can quickly determine the assumptions on which the test is based, the hypotheses that are appropriate, how to compute the test statistic, and how to determine whether to reject the null hypothesis. First these topics are discussed in general, and then an example to illustrate the application of the test is provided.

Where appropriate for a given test, ties, the large-sample approximation, the power efficiency, and—when applicable—the availability of appropriate computer software to do the computations required by the test are discussed. For each procedure references are cited that may be consulted to learn more about the procedure or to pursue further a related topic. Finally exercises for each procedure are provided. These exercises serve two purposes: They illustrate appropriate uses of a test, and they give readers a chance to determine whether they have mastered the computational techniques and learned how to set up the hypotheses and use the applicable decision rule.

In the remaining chapters two types of reference are cited: those that are cited in the body of the text and refer the reader to the statistical literature and those that are cited in the examples and exercises and refer the reader to the research literature. The letter T precedes numbers for references cited in the text, and the letter E precedes those cited in the examples and exercises. The letter A precedes numbers for tables appearing in the Appendix.

REFERENCES

1 Dunn, Olive J., *Basic Statistics: A Primer for the Biomedical Sciences*, second edition, New York: Wiley, 1977.

2 Remington, Richard D., and M. Anthony Schork, *Statistics with Applications to the Biological and Health Sciences*, second edition, Englewood Cliffs, N. J.: Prentice-Hall, 1985.

3 Armitage, P., *Statistical Methods in Medical Research*, Oxford and Edinburgh: Blackwell Scientific Publications, 1971.

4 Colton, Theodore, *Statistics in Medicine*, Boston: Little, Brown, 1974.

5 Mood, Alexander M., Franklin A. Graybill, and Duane C. Boes, *Introduction to the Theory of Statistics*, third edition, New York: McGraw-Hill, 1974.

6 Gibbons, Jean D., and John W. Pratt, "*P*-Values: Interpretation and Methodology," *Amer. Statist.*, 29 (1975), 20–25.

7 Hodges, J. L., Jr., and E. L. Lehmann, *Basic Concepts of Probability and Statistics*, second edition, San Francisco: Holden-Day, 1970.

8 Kempthorne, Oscar, and Leroy Folks, *Probability, Statistics, and Data Analysis*, Ames, Iowa: Iowa State University Press, 1971.

9 Lindgren, Bernard W., *Statistical Theory*, third edition, New York: Macmillan, 1976.

10 Bahn, Anita K., "*P* and the Null Hypothesis," *Ann. Intern. Med.*, 76 (1972), 674.

11 Daniel, Wayne W., "What Are *p*-Values? How Are They Calculated? How Are They Related to Levels of Significance?" *Nursing Res.*, 26 (1977), 304–306.

12 Bakan, David, *On Method*, San Francisco: Jossey-Bass, 1967.

13 Brewer, James K., Letter to the Editor, *Amer. Statist.*, 29 (1975), 171.

14 Cohen, J., *Statistical Power Analysis for the Behavioral Sciences*, revised edition, New York: Academic Press, 1977.

15 Daniel, Wayne W., "Statistical Significance versus Practical Significance," *Sci. Educ.*, 61 (1977), 423–427.

16 Duggan, T. J., and C. W. Dean, "Common Misinterpretation of Significance Levels in Sociology Journals," *Amer. Sociol.*, 3 (1968), 45–46.

17 Gold, David, "Statistical Tests and Substantive Significance," *Amer. Sociol.*, 4 (1969), 42–46.

18 Kish, Leslie, "Some Statistical Problems in Research Design," *Amer. Sociol. Rev.*, 24 (1959), 328–338.

19 McGinnis, R., "Randomization and Inference in Sociological Research," *Amer. Sociol. Rev.*, 23 (1958), 408–414.

20 Gibbons, Jean Dickinson, *Nonparametric Statistical Inference*, second edition, New York: Marcel Dekker, 1985.

21 Pitman, E. J. G., *Lecture Notes on Nonparametric Statistical Inference*, Columbia University, spring 1948, cited in Capon, Jack, "Asymptotic Efficiency of Certain Locally Most Powerful Rank Tests," *Ann. Math. Statist.*, 32 (1961), 88–100.

22 Noether, G. E., "On a Theorem of Pitman," *Ann. Math. Statist.*, 26 (1955), 64–68.

23 Stuart, A., "The Asymptotic Relative Efficiency of Tests and the Derivatives of Their Power Functions," *Skandinavisk Aktuarietidskrift*, 37 (1954), 163–169.

24 Stuart, A., "The Measurement of Estimation and Test Efficiency," *Bull. Int. Statist. Inst.*, Part III, 36 (1956), 79–86.

25 Lehmann, E. L., *Testing Statistical Hypotheses*, New York: Wiley, 1959.

26 Fraser, D. A. S., *Nonparametric Methods in Statistics*, New York: Wiley, 1957.

27 Randles, Ronald H., and Douglas A. Wolfe, *Introduction to the Theory of Nonparametric Statistics*, New York: Wiley, 1979.

28 Pratt, John W., and Jean D. Gibbons, *Concepts of Nonparametric Theory*, New York: Springer-Verlag, 1981.

29 Blyth, C. R., "Note on Relative Efficiency of Tests," *Ann. Math. Statist.*, 29 (1958), 898–903.

30 Smith, K., "Distribution-Free Statistical Methods and the Concept of Power Efficiency," in L. Festinger and D. Katz (eds.), *Research Methods in the Behavioral Sciences*, New York: Dryden, 1953, pp. 536–577.

31 Wilson, Warren R., and Howard Miller, "A Note on the Inconclusiveness of Accepting the Null Hypothesis," *Psychol. Rev.*, 71 (1964), 238–242.

32 Edwards, Ward, "Tactical Note on the Relation between Scientific and Statistical Hypotheses," *Psychol. Bull.*, 63 (1965), 400–402.

33 Feinberg, William E., "Teaching the Type I and Type II Errors: The Judicial Process," *Amer. Statist.*, 25 (June 1971), 30–32.

34 Rodger, R. S., "Type I Errors and Their Decision Basis," *Br. J. Math. Statist. Psychol.*, 20 (1967), 51–62.

35 Rodger, R. S., "Type II Errors and Their Decision Basis," *Br. J. Math. Statist. Psychol.*, 20 (1967), 187–204.

36 Edgington, Eugene S., "Statistical Inference and Nonrandom Samples," *Psychol. Bull.*, 66 (1966), 485–487.

37 Ungerleider, Harry E., and Courtland C. Smith, "Use and Abuse of Statistics," *Geriatrics*, 22 (February 1967), 112–120.

38 Brewer, James K., and Ruth Dailey Knowles, "Some Statistical Considerations in Nursing Research," *Nursing Res.* 23 (1974), 68–70.

39 Ahrens, S. J., "Statistical Tests of Significance: Truth, Paradox, or Folly?" *Res. Quart. Amer. Assoc. Health, Phys. Educ. Rec.*, 42 (1971), 436–440.

40 Barnard, G. A., "The Meaning of a Significance Level," *Biometrika*, 34 (1947), 179–182.

41 Chandler, Robert E., "The Statistical Concepts of Confidence and Significance," *Psychol. Bull.*, 54 (1957), 429–430.

42 Labovitz, Sanford, "Criteria for Selecting a Significance Level: A Note on the Sacredness of 0.05," *Amer. Sociol.*, 3 (1968), 220–222.

43 Lykken, David T., "Statistical Significance in Psychological Research," *Psychol. Bull.*, 70 (1968), 151–159.

44 Krause, Merton S., "Insignificant Differences and Null Explanations," *J. Gen. Psychol.*, 86 (1972), 217–220.

45 Morrison, D. E., and R. E. Henkel, *The Significance Test Controversy—A Reader*, Chicago: Aldine, 1970.

46 O'Brien, Thomas C., and Bernard J. Shapiro, "Statistical Significance—What?" *Math. Teacher*, 61 (1968), 673–676.

47 Rozeboom, W. W., "The Fallacy of the Null Hypothesis Significance Test," *Psychol. Bull.*, 57 (1960), 416–428.

48 Selvin, H. C., "A Critique of Tests of Significance in Survey Research," *Amer. Sociol. Rev.*, 22 (1957), 519–527.

49 Skipper, James K., Jr., Anthony L. Guenther, and Gilbert Nass, "The Sacredness of .05: A Note Concerning the Uses of Statistical Levels of Significance in Social Science," *Amer. Sociol.*, 2 (1967), 16–18.

50 Stone, M., "Role of Significance Testing—Some Data With a Message," *Biometrika*, 56 (1969), 485–493.

51 Winch, R. F., and D. T. Campbell, "Proof? No. Evidence? Yes. The Significance of Tests of Significance," *Amer. Sociol.*, 4 (May 1969), 140–143.

52 Zeisel, H., "The Significance of Insignificant Differences," *Public Opinion Quart.*, 19 (1955), 319–321.

53 Editorial. "Significance of Significant," *N. Eng. J. Med.*, 278 (1968), 1232.

54 Bross, Irwin D. J., "The Role of the Statistician: Scientist or Shoe Clerk," *Amer. Statist.*, 28 (1974), 126–127.

55 Godambe, V. P., and D. A. Sprott, *Foundations of Statistical Inference, A Symposium*, Minneapolis: Winston Press, 1972.

56 Kiefer, J., "Statistical Inference," *Math. Spectrum*, 3 (Fall 1970), 1–11.

57 Lurie, William, "The Impertinent Questioner: The Scientist's Guide to the Statistician's Mind," *Amer. Scientist*, 46 (1958), 57–61.

58 Natrella, Mary G., "The Relation between Confidence Intervals and Tests of Significance," *Amer. Statist.*, 14 (February 1960), 20–22, 38.

59 Kirk, Roger E. (ed.), *Statistical Issues: A Reader for the Behavioral Sciences*, Monterey, Calif.: Brooks/Cole, 1972.

60 Stevens, S. S., "On the Theory of Scales of Measurement," *Science*, 103 (1946), 677–680.

61 Stevens, S. S., "Mathematics, Measurement and Psychophysics," in S. S. Stevens (ed.), *Handbook of Experimental Psychology*, New York: Wiley, 1951.

62 Stevens, S. S., "Measurement Statistics, and the Schemapiric View," *Science*, 161 (1968), 849–856.

63 Anderson, Norman H., "Scales and Statistics: Parametric and Nonparametric," *Psychol. Bull.*, 58 (1961), 305–316.

64 Boneau, C. Alan, "A Note on Measurement Scales and Statistical Tests," *Amer. Psychol.*, 16 (1961), 260–261.

65 Brown, George W., "Counts, Scales, and Scores," *Amer. J. of Diseases of Children*, 139 (1985), 147–151.

66 Gaito, John, "Scale Classification and Statistics," *Psychol. Rev.*, 67 (1960), 277–278.

67 Gaito, John, "Measurement Scales and Statistics: Resurgence of an Old Misconception," *Psychol. Bull.*, 87 (1980), 564–567.

68 Lord, F. M., "On the Statistical Treatment of Football Numbers," *Amer. Psychol.*, 8 (1953), 750–751.

69 Mitchell, Joel, "Measurement Scales and Statistics: A Clash of Paradigms," *Psychol. Bull.*, 100 (1986), 398–407.

70 Townsend, J. T., and F. G. Ashby, "Measurement Scales and Statistics: The Misconception Misconceived," *Psychol. Bull.*, 96 (1984), 394–401.

71 Khurshid, Anwer, "Scales of Measurement: A Selected Bibliography," unpublished manuscript.

72 Baker, Bela O., Curtis D. Hardyck, and Lewis F. Petrinovich, "Weak Measurement vs. Strong Statistics: An Empirical Critique of S. S. Stevens' Proscriptions on Statistics," *Educ. Psychol. Measurement*, 26 (1966), 291–309.

73 Campbell, N. R., "Symposium: Measurement and Its Importance for Philosophy," *Proc. Aristotelian Soc.*, Suppl., 17 (1938), London: Harrison and Sons.

74 Gardner, Paul Leslie, "Scales and Statistics," *Rev. Educ. Res.*, 45 (Winter 1975), 43–57.

75 Kendall, M. G., and R. M. Sundrum, "Distribution-Free Methods and Order Properties," *Rev. Int. Statist. Inst.* 21 (1953), 124–134.

76 Arbuthnot, John, "An Argument for Divine Providence, Taken from the Constant Regularity Observed in the Births of Both Sexes," *Philosophical Transactions*, 27 (1710), 186–190.

77 Wolfowitz, J., "Additive Partition Functions and a Class of Statistical Hypotheses," *Ann. Math. Statist.*, 13 (1942), 247–279.

78 Papatoni-Kazakos, P., and Dimitri Kazakos (eds.), *Nonparametric Methods in Communications*, New York: Marcel Dekker, 1977.

79 Brown, George W., and Gregory F. Hayden, "Nonparametric Methods: Clinical Applications," *Clin. Pediatrics*, 24 (1985), 490–498.

80 Helsel, Dennis R., "Advantages of Nonparametric Procedures for Analysis of Water Quality Data," *Hydrolog. Sci. J.*, 32 (1987), 179–190.

81 Jenkins, Stephen J., Dale R. Fuqua, and Thomas C. Froehle, "A Critical Examination of the Use of Non-Parametric Statistics in the *Journal of Counseling Psychology*," *Perceptual and Motor Skills*, 59 (1984), 31–35.

82 Buckalew, L. W., "Nonparametrics and Psychology: A Revitalized Alliance," *Perceptual and Motor Skills*, 57 (1983), 447–450.

83 Savage, I. R., *Bibliography of Nonparametric Statistics*, Cambridge, Mass.: Harvard University Press, 1962.

84   Singer, Bernard, *Distribution-Free Methods for Non-Parametric Problems: A Classified and Selected Bibliography*, Leicester, England: British Psychological Society, 1979.

85   Harter, H. Leon, *The Chronological Annotated Bibliography of Order Statistics, Volume I: Pre-1950*, Columbus, Ohio: American Sciences Press, 1983.

86   Harter, H. Leon, *The Chronological Annotated Bibliography of Order Statistics, Volume II: 1950–1959*, Columbus, Ohio: American Sciences Press, 1983.

87   Bradley, James V., *Distribution-Free Statistical Tests*, Englewood Cliffs, N. J.: Prentice-Hall, 1968.

88   Conover, W. J., *Practical Nonparametric Statistics*, second edition, New York: Wiley, 1980.

89   Gibbons, Jean Dickinson, *Nonparametric Methods for Quantitative Analysis*, second edition, Columbus, Ohio: American Sciences Press, 1985.

90   Hollander, Myles, and Douglas A. Wolfe, *Nonparametric Statistical Methods*, New York: Wiley, 1973.

91   Leach, Chris, *Introduction to Statistics: A Nonparametric Approach for the Social Sciences*, Chichester, England: Wiley, 1979.

92   Lehmann, E. L., *Nonparametrics: Statistical Methods Based on Ranks*, San Francisco: Holden-Day, 1975.

93   Marascuilo, Leonard A., and Maryellen McSweeney, *Nonparametric and Distribution-Free Methods for the Social Sciences*, Monterey, Calif.: Brooks/Cole, 1977.

94   Maxwell, A. E., *Analyzing Quantitative Data*, New York: Wiley, 1961.

95   Mosteller, F., and R. E. K. Rourke, *Sturdy Statistics*, Reading, Mass.: Addison-Wesley. 1973.

96   Noether, G., *Introduction to Statistics: A Fresh Approach*, Boston: Houghton Mifflin, 1971.

97   Pierce, Albert, *Fundamentals of Nonparametric Statistics*, Belmont, Calif.: Dickenson, 1970.

98   Quenouille, M. H., *Rapid Statistical Calculations. A Collection of Distribution-Free and Easy Methods of Estimation and Testing*, second edition, London: Griffin, 1972.

99   Runyon, Richard P., *Nonparametric Statistics: A Contemporary Approach*, Reading, Mass.: Addison-Wesley, 1977.

100   Senders, V. L., *Measurement and Statistics*, New York: Oxford University Press, 1958.

101   Siegel, Sidney, and N. John Castellan, *Nonparametric Statistics for the Behavioral Sciences*, second edition, New York: McGraw-Hill, 1988.

102   Sprent, Peter, *Quick Statistics: An Introduction to Non-Parametric Methods*, Middlesex, England: Penguin Books, 1981.

103   Tate, Merle W., and Richard C. Clelland, *Nonparametric and Shortcut Statistics in the Social, Biological, and Medical Sciences*, Danville, Ill.: Interstate, 1957.

104   Wilcoxon, Frank, and Roberta A. Wilcox, *Some Rapid Approximate Statistical Procedures*, revised, Pearl River, N. Y.: Lederle Laboratories, 1964.

105   David, H. A., *Order Statistics*, New York: Wiley, 1970.

106   Edgington, Eugene S., *Statistical Inference: The Distribution-Free Approach*, New York: McGraw-Hill, 1969.

107   Fraser, D. A. S., *Nonparametric Methods in Statistics*, New York: Wiley, 1957.

108   Gnedenko, B. V., M. L. Puri, and I. Vincze (eds.), *Nonparametric Statistical Inference*, Amsterdam: North-Holland, 1982.

109   Hájek, Jeroslav, *A Course in Nonparametric Statistics*, San Francisco: Holden-Day, 1969.

110   Hájek, J., and Šidák, Z., *Theory of Rank Tests*, New York: Academic Press, 1967.

111 Hettmansperger, Thomas P., *Statistical Inference Based on Ranks*, New York: Wiley, 1984.

112 Kraft, Charles H., and Constance van Eeden, *A Nonparametric Introduction to Statistics*, New York: Macmillan, 1968.

113 Manoukian, Edward B., *Mathematical Nonparametric Statistics*, New York: Gordon and Breach, 1986.

114 Maritz, J. S., *Distribution-Free Statistical Methods*, London: Chapman and Hall, 1981.

115 Noether, Gottfried E., *Elements of Nonparametric Statistics*, New York: Wiley, 1967.

116 Puri, Mandan Lal (ed.), *Nonparametric Techniques in Statistical Inference*, Cambridge, England: Cambridge University Press, 1970.

117 Randles, Ronald H., and Douglas A. Wolfe, *Introduction to the Theory of Nonparametric Statistics*, New York: Wiley, 1979.

118 Sarhan, S. E., and B. G. Greenberg (eds.), *Contributions to Order Statistics*, New York: Wiley, 1962.

119 Walsh, John E., *Handbook of Nonparametric Statistics*, Princeton, N. J.: D. van Nostrand, 1962.

120 Walsh, John E., *Handbook of Nonparametric Statistics. II*, Princeton, N. J.: D. van Nostrand, 1965.

121 Walsh, John E., *Handbook of Nonparametric Statistics. III*, Princeton, N. J.: D. van Nostrand, 1968.

122 Fisher, N. I., "Non-Parametric Statistics," *Mathematical Scientist*, 7 (1982), 25–47.

123 Noether, Gottfried E., "Elementary Estimates: An Introduction to Nonparametrics," *J. Educ. Statist.*, 10 (1985), 211–221.

124 Krishnaiah, P. R., and P. K. Sen (eds.), *Nonparametric Methods*, Volume 4 of *Handbook of Statistics*, Amsterdam: North-Holland, 1984.

125 Fisher, Nicholas I., "Graphical Methods in Nonparametric Statistics: A Review and Annotated Bibliography," *International Statistical Rev.*, 51 (1983), 25–58.

126 Grimm, H., "Transformation of Variables versus Nonparametrics," in B. V. Gnedenko, M. L. Puri, and I. Vincze (eds.), *Nonparametric Statistical Inference*, Amsterdam: North-Holland, 1982, pp. 351–360.

127 Doksum, K. A., "Some Remarks on the Development of Nonparametric Methods and Robust Statistical Inference," in D. B. Owen (ed.), *On the History of Statistics and Probability*, New York: Marcel Dekker, 1976, pp. 237–263.

128 Noether, Gottfried E., "Nonparametrics. The Early Years—Impressions and Recollections," *Amer. Statist.*, 38 (1984), 173–178.

129 Ruist, E., "Comparison of Tests for Nonparametric Hypotheses," *Ark. Matematik*, 3 (1955), 133–163.

130 Hettmansperger, Thomas P., and Joseph W. McKean, "A Graphical Representation for Nonparametric Inference," *Amer. Statist.*, 28 (1974), 100–102.

131 Buchanan, William, "Nominal and Ordinal Bivariate Statistics: The Practitioner's View," *Amer. J. Polit. Sci.*, 18 (1974), 625–646.

132 Noether, Gottfried E., "The Nonparametric Approach in Elementary Statistics," *Math. Teacher*, 67 (1974), 123–126.

133 Scheffé, H., "Statistical Inference in the Nonparametric Case," *Ann. Math. Statist.*, 14 (1943), 305–332.

134 Woodward, W. A., A. C. Elliott, and H. L. Gray, *Directory of Statistical Microcomputer Software, 1985 Edition*, New York: Marcel Dekker, 1985.

135 Vegelius, Jan, "Siegel, A Fortran IV Program for Nonparametrical Methods," *Educ. & Psychol. Measurement*, 35 (1975), 713–715.

136 Kennedy, William J. (section ed.), "New Developments in Statistical Computing," *Amer. Statist.*, 34 (1980), 115–116.

137 von Collani, Gernot, "NONPARAM: A BASIC Program Package for Nonparametric Procedures," *Behav. Res. Methods & Instrumentation*, 15 (1983), 104.

138 Hannan, Thomas E., "CBASIC Programs for Nonparametric Statistical Analysis," *Behav. Res. Methods, Instruments & Computers*, 18 (1986), 403–404.

2

PROCEDURES THAT
UTILIZE DATA FROM
A SINGLE SAMPLE

In this chapter some nonparametric procedures that utilize data from a single sample are presented. The first section considers estimation and hypothesis-testing procedures that are appropriate when the parameter of interest is a measure of central tendency, or location, as it is sometimes called. Succeeding sections discuss procedures for estimating a population proportion and testing for randomness and the presence of trend.

Wherever possible the following format in presenting the hypothesis-testing procedures will be observed:

1. *Assumptions:* The assumptions necessary for the validity of the test are listed, and the data on which the calculations are based are described.
2. *Hypotheses:* The null hypotheses that may be tested and their alternatives are stated.
3. *Test statistic:* A formula or direction for computing the relevant test statistic is given. When a formula is given, its use in obtaining a numerical value of the test statistic is described.
4. *Decision rule:* The Appendix gives appropriate tables for the distribution of the test statistic. From these tables, we can determine critical

values of the test statistic corresponding to the chosen level of significance. If the computed value of the test statistic is as extreme as or more extreme than a critical value, we reject the null hypothesis and conclude that the alternative hypothesis is true. If we cannot reject the null hypothesis, we conclude that it *may be* true.

As discussed in Chapter 1, we can also use the tables in the Appendix to determine the P value associated with a specific test. If the P value is equal to or less than α, the chosen level of significance, reject H_0.

Each hypothesis-testing procedure and each estimation procedure presented is illustrated with an appropriate example. The inclusion of exercises at the end of each section provides an opportunity, through practice, to become more familiar with these procedures. Exercises at the end of each chapter may be used for review.

2.1

MAKING INFERENCES ABOUT A LOCATION PARAMETER

The two measures of central tendency that are most frequently of interest to the researcher are the *arithmetic mean* (subsequently referred to as the *mean*) and the *median*.

The population mean is the measure of central tendency with which most parametric inferential procedures are concerned. Thus when parametric procedures are appropriate, we may test the *null hypothesis* (designated H_0) that a population mean μ is equal to some hypothesized numerical value μ_0 versus the *alternative hypothesis* (designated H_1) that μ is not equal to μ_0.

At other times, we may be interested in constructing a $100(1 - \alpha)\%$ *confidence interval* for μ, where $1 - \alpha$ is the desired *confidence coefficient*. In most such instances we use the t test based on the Student's t distribution in the hypothesis-testing procedure. Likewise, we usually use the t statistic to construct confidence intervals for a population mean. When sample sizes are large, we may use the central limit theorem to justify the use of the z statistic in constructing confidence intervals for and testing hypotheses about population means.

When we use the t test, we assume that the population from which the sample data have been drawn is normally distributed. Modest departures from this assumption do not seriously affect our conclusions; but when the assumption is grossly violated, we have to seek an alternative method of analysis. One such alternative is a nonparametric procedure.

Several nonparametric procedures are available for making inferences about a location parameter, or population measure of central tendency. A characteristic of these nonparametric procedures is that the median rather than the mean is the location parameter.

Recall from a previous course in statistics that the median is the "middle" value of a set of measurements arranged in order of magnitude. For a continuous distribution, we define the median as the point M for which the probability that a

value selected at random from the distribution is less than M, and the probability that a value selected at random from the distribution is greater than M, are both equal to one-half. When the population from which the sample has been drawn is symmetric, any conclusions about the median are applicable to the mean, since in symmetric distributions the mean and the median coincide.

The distribution of a random variable X is symmetric about the point C if $P(X \geq C + x) = P(X \leq C - x)$ for all values of x. A discrete distribution is symmetric if the left half of the graph of its probability function is a mirror image of the right half. Examples of symmetric discrete distributions are the binomial for $p = 0.5$ and the discrete uniform distribution. The normal distribution is an example of a symmetric continuous distribution.

ONE-SAMPLE SIGN TEST

The *sign test* is perhaps the oldest of all the nonparametric procedures. Its use was reported as early as 1710 by Arbuthnott (T1). It is called the sign test because, as we shall see, we may convert the data for analysis to a series of plus and minus signs. The test statistic, then, consists of either the number of plus signs or the number of minus signs.

Assumptions

  A. The sample available for analysis is a random sample of independent measurements from a population with unknown median M.

  B. The variable of interest is measured on at least an ordinal scale.

  C. The variable of interest is continuous. The n sample measurements are designated by $X_1, X_2, \ldots, X_n$.

Hypotheses

  A. (Two-sided): $H_0: M = M_0,$ $H_1: M \neq M_0$

  B. (One-sided): $H_0: M \leq M_0,$ $H_1: M > M_0$

  C. (One-sided): $H_0: M \geq M_0,$ $H_1: M < M_0$

Select a level of significance α.

Test Statistic

Record the sign of the difference obtained by subtracting the hypothesized median M_0 from each sample value; that is, record the sign of the n differences, $X_i - M_0$, $i = 1, 2, \ldots, n$.

If the null hypothesis is true—that is, if the population median is in fact equal to M_0—we expect a random sample from the population to have about as many plus signs as minus signs when the n differences $X_i - M_0$ have been computed. If we observe a sufficiently small number of either plus or minus signs, we reject null hypothesis A. If we observe a sufficiently small number of minus signs, we

reject null hypothesis B, and if we observe a sufficiently small number of plus signs, we reject null hypothesis C. The test statistic for hypothesis A, then, is the number of plus signs or the number of minus signs, whichever is smaller. The test statistic for hypothesis B is the number of minus signs, and the test statistic for hypothesis C is the number of plus signs.

Decision Rule

The decision rule for each of the possible hypotheses is as follows:

A. Reject H_0 at the α level of significance if the probability, when H_0 is true, of observing as few (or fewer) of the less frequently occurring sign in a random sample of size n is less than or equal to $\alpha/2$.

B. Reject H_0 at the α level if the probability, when H_0 is true, of observing as few (or fewer) minus signs as are actually observed in a random sample of size n is less than or equal to α.

C. Reject H_0 at the α level if the probability, when H_0 is true, of observing as few (or fewer) plus signs as are actually observed is less than or equal to α.

To determine the probability of observing a value as extreme as or more extreme than that actually observed, we must think of our sample of n plus and minus signs as a sample from a population of plus and minus signs. In other words, imagine a population of dichotomous observations from which we have a sample of size n. The sampling distribution of the number of differences yielding the sign of interest under a given hypothesis is the binomial distribution with parameter $p = 0.50$ if H_0 is true. If H_0 is true, the probability that an observation drawn at random from the population yields a plus sign is equal to the probability that it yields a minus sign. In other words, the probability that a difference $X_i - M_0$ yields a plus sign and the probability that $X_i - M_0$ yields a minus sign are both equal to 0.50.

We can obtain the probability of observing a test statistic as extreme as or more extreme than the one actually observed from a table of binomial probabilities such as Table A.1. Suppose that we let K equal the random variable, the number of signs of interest under the given hypothesis, and let k be the observed value of the test statistic. What we seek is

$$P(K \leq k \,|\, n, 0.50) \tag{2.1}$$

We read this as "The probability that K is less than or equal to k, given a random sample of size n from a population in which the proportion of signs of the specified type is 0.50."

Alternatively, we may determine from Table A.1 a critical value of K such that if k is greater than or equal to this value, we reject H_0. Both approaches are illustrated in Example 2.1.

Problem of Zero Differences   As previously mentioned, we assume that the variable of interest is continuous. Then, in theory, no zero differences should occur

when we compute $X_i - M_0$. In practice, however, zero differences *do* occur. The usual procedure in such cases is to discard observations that lead to zero differences and reduce n accordingly. When sample measurements that are equal to the hypothesized median are discarded, each remaining measurement is either greater than the hypothesized median or less than the hypothesized median. In that case the hypotheses may be restated in probability terms. For example, the null hypothesis for the two-sided case may be stated as

$$P(X < M_0) = P(X > M_0) = 0.5$$

Here is an example with medical application.

Example 2.1

In a study of myocardial transit times, Liedtke et al. (E1) measured appearance transit times in a series of subjects with angiographically normal right coronary arteries. The median appearance time for this group was 3.50 seconds. Suppose that another research team repeated the procedure on a sample of 11 patients with significantly occluded right coronary arteries and obtained the results shown in Table 2.1. Could the second team conclude, at the 0.05 level of significance, that the median appearance transit time in the population from which its sample was drawn is different from 3.50 seconds?

TABLE 2.1

Appearance transit times for 11 patients with significantly occluded right coronary arteries

Subject	1	2	3	4	5	6	7	8	9	10	11
Transit time, sec	1.80	3.30	5.65	2.25	2.50	3.50	2.75	3.25	3.10	2.70	3.00

Hypotheses

$$H_0: M = 3.50, \qquad H_1: M \neq 3.50$$

Test Statistic

When we compute the 11 differences $X_i - 3.50$ with the observations in Table 2.1, we find nine negative differences, one positive difference, and one zero difference. Since there are fewer positive differences than negative differences, the value of the test statistic is $k = 1$, the number of differences with a plus sign. We discard the observation yielding a zero difference, which leaves us with a usable sample size of 10.

Decision

We will reject H_0 if the probability of observing one or fewer plus signs when H_0 is true is less than or equal to 0.025. When we refer to Table A.1, we see that this

probability is

$$P(K \leq 1 \mid 10, 0.50) = 0.0108$$

Since 0.0108 is less than 0.025, we reject H_0 and conclude that the population median is not 3.50. Since this was a two-sided test, the P value for this example is $2(0.0108) = 0.0216$, which is less than 0.05.

Another way to test H_0 is to compute a critical value of K. This critical value, which we call K', is a number so small that the probability of observing a value that small or smaller, when H_0 is true, is less than or equal to 0.025 (since $\alpha = 0.05$, and we have a two-sided test). If k is less than or equal to K', we reject H_0 in favor of H_1.

To express it in symbols, we seek a value of K' such that the following statement is true:

$$P(K \leq K' \mid n, 0.50) \leq 0.025$$

From Table A.1, we see that when $n = 10$, the critical value K' of K is 1. Since $k = 1 = K'$, we reject H_0. Since $2[P(K \leq 1 \mid 10, 0.50)] = 0.0216$, we refer to this test as a 0.0216-level test rather than a 0.05-level test.

This example illustrates one of the advantages of reporting P values rather than stating that a result is significant or not significant at some preselected significance level.

Large-Sample Approximation   For samples of size 12 or larger, we may use the normal approximation to the binomial. Since the normal approximation involves approximating a discrete distribution by means of a continuous distribution, we use a continuity correction factor of 0.5. When we do this, we compute

$$z = \frac{(K + 0.5) - 0.5n}{0.5\sqrt{n}} \tag{2.2}$$

which we compare for significance with values of the standard normal distribution (given in Table A.2) for the chosen level of significance.

In another context the formula for the normal approximation might call for subtracting 0.5 from K. The nature of the test statistics in the present situation, however, is such that we will always be seeking the probability, under H_0, of a value less than or equal to K. Consequently the use of the normal approximation requires that we add 0.5 to the computed value of K.

To illustrate the use of the normal approximation, let us use the data of Example 2.1 to evaluate Equation 2.2. After making appropriate substitutions, we have

$$z = \frac{(1 + 0.5) - 0.5(10)}{0.5\sqrt{10}} = -2.21$$

Table A.2 reveals that the probability of observing a value of z this small or smaller is 0.0136 and, as before, we reject H_0; that is, $P(K \leq 1 \mid 10, 0.5) = 0.0108$ is

approximately equal to $P(z \leq -2.21) = 0.0136$. Thus we see that the normal approximation is good even with a sample size as small as 10.

In Example 2.1 the sample measurements were measured on a ratio scale, but such a high level of measurement is not needed for the application of the sign test. We could have used the sign test and obtained the same results if our sample measurements had consisted merely of an indication of whether a given subject's score was above or below the hypothesized median.

FURTHER READING

Dixon and Mood (T2), who discuss the sign test in considerable detail, give critical values of K for the 1, 5, 10, and 25% significance levels and of n from 1 through 100. They also give a table for determining sample size.

More recently, Noether (T3) discusses the problem of sample size determination for the sign test when a certain power level is desired. Nelson (T4) presents an alternative formula for computing the test statistic and a method for determining critical values when the normal approximation is employed. Mantel and Rahe (T5) discuss methods for reducing the conservatism of the sign test by taking into account the rankings of the absolute values. A sign test for correlation is proposed by Nelson (T6). Mackinnon (T7) gives a table of critical values of K for sample sizes up to 1000.

Power-Efficiency Walsh (T8) compared the power functions of the sign test with those of the Student's t test for the case of normal populations, and found the sign test to be approximately 95% efficient for small samples. When sampling from normal populations, he found that the relative efficiency of the sign test decreases as the sample size increases. For samples of size 13, the relative efficiency of the sign test (compared to Student's t test in normal populations) is approximately 75%.

Dixon (T9)—who also compared sign-test power functions with those of the Student's t test for samples from normal populations—reports decreasing power-efficiency for increasing sample size, for increasing level of significance, and for increasing alternative. Cochran (T10), Gibbons (T11), Hodges and Lehman (T12), and MacStewart (T13), among others, have also considered the power-efficiency of the sign test.

EXERCISES

2.1 Lenzer et al. (E2) reported the endurance scores of animals during a 48-hour session of discrimination responding. The median score for animals with electrodes implanted in the hypothalamus was 97.5. Suppose that the experiment was duplicated in another laboratory, except that electrodes were implanted in the forebrain of 12 animals. Assume that investigators observed the endurance scores shown in Table 2.2.

Use the one-sample sign test to see whether the investigators may conclude at the 0.05 level of significance that the median endurance score of animals with electrodes implanted in the forebrain is less than 97.5. What is the P value for this test?

TABLE 2.2

Endurance scores of animals with electrodes implanted in forebrain

| 93.6 | 89.1 | 97.7 | 84.4 | 97.8 | 94.5 | 88.3 | 97.5 | 83.7 | 94.6 | 85.5 | 82.6 |

2.2 Iwamoto (E3) found that the mean weight of a sample of a particular species of adult female monkey from a certain locality was 8.41 kg. Suppose that a sample of adult females of the same species from another locality yielded the weights shown in Table 2.3.

Can we conclude that the median weight of the population from which this second sample was drawn is greater than 8.41 kg? Use the one-sample sign test and a 0.05 level of significance. What is the P value for this test?

TABLE 2.3

Weights of female monkeys, kilograms

| 8.30 | 9.50 | 9.60 | 8.75 | 8.40 | 9.10 | 9.25 | 9.80 | 10.05 | 8.15 | 10.00 | 9.60 | 9.80 | 9.20 | 9.30 |

WILCOXON SIGNED-RANKS TEST

As we have seen, the sign test utilizes only the signs of the differences between observed values and the hypothesized median. For testing $H_0: M = M_0$ there is another procedure that uses the magnitude of differences when these are available. To use the sign test to test a hypothesis about a population median, we need only know whether a sample measurement falls above or below the hypothesized median. In order to use this other procedure, known as the *Wilcoxon* (T14) *signed-ranks test*, we need additional information about each sample measurement. Specifically, we need enough additional information to be able to rank the differences between each sample measurement and the hypothesized median.

To use the Wilcoxon procedure, we first rank the differences in order of absolute size. Then we assign the original signs of the differences to the ranks and compute two sums: the sum of the ranks with negative signs and the sum of the ranks with positive signs. Since the Wilcoxon signed-ranks test uses more information than the sign test, it is often a more powerful test. The Wilcoxon signed-ranks test also assumes that the sampled population is symmetric. When the sampled population meets this assumption, conclusions about the population median also apply to the population mean. When the population is not symmetric, we prefer the sign test over the Wilcoxon test.

Assumptions

    A. The sample available for analysis is a random sample of size n from a population with unknown median M.

    B. The variable of interest is continuous.

    C. The sampled population is symmetric.

    D. The scale of measurement is at least interval.

    E. The observations are independent.

Hypotheses

    A. $H_0: M = M_0$, $H_1: M \neq M_0$

    B. $H_0: M \geq M_0$, $H_1: M < M_0$

    C. $H_0: M \leq M_0$, $H_1: M > M_0$

Select a significance level α.

Test Statistic

To obtain the test statistic, we use the following procedure.

1. Subtract the hypothesized median from each observation; that is, for each observation, find

$$D_i = X_i - M_0 \tag{2.3}$$

   If any observation X_i is equal to the hypothesized median M_0, eliminate it from the calculations and reduce the sample size accordingly.

2. Rank the differences from smallest to largest without regard to their signs. In other words, rank the $|D_i|$, the absolute values of the differences. If two or more $|D_i|$ are equal, assign each tied value the mean of the rank positions occupied by the differences that are tied. For example, if the three smallest differences are all equal, rank them 1, 2, and 3, but assign each a rank of $(1 + 2 + 3)/3 = 6/3 = 2$. There are other ways by which ties may be broken, but none of them has any scientific value over the others. Mathematical simplicity is the guiding principle in choosing a method.

3. Assign to each rank the sign of the difference of which it is the rank.

4. Obtain the sum of the ranks with positive signs; call it T_+. Obtain the sum of the ranks with negative signs; call it T_-. Actually, we need to compute only one of these sums directly. Given one of the sums, we can obtain the other from the relationship $T_+ = [n(n + 1)/2] - T_-$.

If H_0 is true—that is, if the true population median M is equal to the hypothesized median M_0—and if the assumptions are met, the probability of observing a positive difference $D_i = X_i - M_0$ of a given magnitude is equal to the probability of observing a negative difference of the same magnitude. Then, in repeated sampling, when H_0 is true and the assumptions are met, the expected value of T_+ is equal to the expected value of T_-. For a given sample we do not expect T_+ to equal T_-. When H_0 is true, however, we do not expect a great difference in their values. Consequently a sufficiently small value of T_+ or a sufficiently small value of T_- causes us to reject H_0.

Specifically, the test statistic for each hypothesis is as follows.

A. Since we reject $H_0: M = M_0$ for either a sufficiently small value of T_+ or a sufficiently small value of T_-, the test statistic for the hypotheses stated in A is either T_+ or T_-, whichever is smaller. To simplify notation, we call the smaller of the two T.

B. For a sufficiently large sum computed from ranks with negative signs, we reject $H_0: M \geq M_0$, since under this null hypothesis we expect a fairly large sum computed from ranks with positive signs. A sufficiently small value of T_+, then, causes us to reject the null hypothesis specified in B.

C. By a similar line of reasoning, we see that for the hypotheses stated in C, the test statistic is T_-.

Decision Rule

Critical values of the test statistic for the Wilcoxon signed-ranks test appear in Table A.3. Exact probability levels (P) are given to four decimal places for all possible rank totals (T) that yield a different probability level at the fourth decimal place, from 0.0001 up to and including 0.5000. The rank totals (T) are tabulated for all sample sizes from $n = 5$ through $n = 30$.

Decision rules for each set of hypotheses listed in Step 2 above are as follows.

A. We reject H_0 at the α level of significance if the calculated T is smaller than or equal to tabulated T for n and preselected $\alpha/2$. Alternatively, we may enter Table A.3 with n and our calculated value of T to see whether the tabulated P associated with calculated T is less than or equal to our stated level of significance. If so, we may reject H_0.

B. Reject H_0 at the α level of significance if T_+ is less than or equal to tabulated T for n and preselected α.

C. Reject H_0 at the α level of significance if T_- is less than or equal to tabulated T for n and preselected α.

Example 2.2   In a study of drug abuse in a suburban area (E4), investigators found that the median IQ of arrested abusers who were 16 years of age or older was 107. Suppose that a researcher wishes to know whether to conclude that the median IQ of arrested abusers who are 16 or older in another suburban area is different from 107. Table 2.4 shows the IQs of a random sample of 15 persons from the population of interest. What can the researchers conclude? (Let $\alpha = 0.05$.)

TABLE 2.4   **IQs of persons 16 or older arrested for drug abuse in a certain suburban area**

| 99 | 100 | 90 | 94 | 135 | 108 | 107 | 111 | 119 | 104 | 127 | 109 | 117 | 105 | 125 |

Hypotheses

$$H_0: M = 107, \qquad H_1: M \neq 107$$

Test Statistic

The calculations for obtaining the test statistic are summarized in Table 2.5.

TABLE 2.5 **Calculations for obtaining test statistic for Example 2.2**

IQ	$D_i = X_i - M_0$	Rank of $|D_i|$	Signed rank of $|D_i|$
99	−8	7	−7
100	−7	6	−6
90	−17	11	−11
94	−13	10	−10
135	+28	14	+14
108	+1	1	+1
107	0		Eliminate from analysis
111	+4	5	+5
119	+12	9	+9
104	−3	4	−4
127	+20	13	+13
109	+2	2.5	+2.5
117	+10	8	+8
105	−2	2.5	−2.5
125	+18	12	+12
			$T_+ = 64.5$
			$T_- = 40.5$

Since $T_- = 40.5$ is less than $T_+ = 64.5$, the test statistic for our example is $T = 40.5$.

Decision

For a two-sided significance level of $\alpha = 0.05$ and a sample size of $n = 14$, Table A.3 tells us to reject H_0 if T is less than or equal to 21. Since $T = 40.5 > 21$, we cannot reject the null hypothesis. We conclude that the median IQ of the subjects in our population may be 107, since our sample data are compatible with this value as a population median. In other words, the sample data did not provide sufficient evidence to cause us to reject $H_0: M = 107$. Since our calculated value of T of 40.5 is between the tabulated values of 40 and 41, the P value for this test is between $2(0.1384) = 0.2768$ and $2(0.1514) = 0.3028$.

Large-Sample Approximation When n is greater than 30, we cannot use Table A.3 to determine the significance of a computed value of the test statistic. We

can show that for large samples, the statistic

$$T^* = \frac{T - n(n + 1)/4}{\sqrt{n(n + 1)(2n + 1)/24}} \tag{2.4}$$

has approximately the standard normal distribution. For one-sided tests we replace T in Equation 2.4 with T_+ or T_-, as the situation demands.

We can incorporate an adjustment for ties among the nonzero differences in the large-sample approximation given by Equation 2.4 in the following way. Let t be the number of absolute differences tied for a particular nonzero rank. Then the correction factor is

$$\frac{\Sigma t^3 - \Sigma t}{48}$$

and we subtract this factor from the quantity under the radical in Equation 2.4. When we have ties, then, we replace the denominator of the large-sample approximation test statistic by

$$\sqrt{\frac{n(n + 1)(2n + 1)}{24} - \frac{\Sigma t^3 - \Sigma t}{48}}$$

To illustrate the calculation of an adjustment for ties, suppose that we have the following data.

Observation	Rank	t	t^3
3	1.5		
3	1.5	2	8
4	3		
6	5		
6	5	3	27
6	5		
8	7.5		
8	7.5	2	8
9	10.5		
9	10.5		
9	10.5	4	64
9	10.5		
		11	107

The correction for ties, then, is

$$\frac{107 - 11}{48} = 2$$

FURTHER READING

Tables of critical values for the Wilcoxon signed-ranks test have been given by McCornack (T15), Wilcoxon (T14, T16, T17), and Wilcoxon, Katti, and Wilcox

(T18), among others. The latter tables, from which Table A.3 was extracted, give values of rank totals (T) and associated probability levels (P) for values of n through 50.

An alternative procedure for handling zero differences appears in the article by Pratt (T19). Cureton (T20) gives a formula for the normal approximation when following Pratt's procedure for handling zero differences. Buck (T21), Conover (T22), Klotz (T23), and Putter (T24) have also considered the problem of zero differences and ties. Noether (T3) gives a formula for determining sample size when use of the Wilcoxon one-sample test is anticipated. Those interested in APL programs for P values in the Wilcoxon one-sample and two-sample tests may consult the paper by Larsen (T25).

Power-Efficiency The power-efficiency characteristics of the Wilcoxon signed-ranks test have been investigated by Arnold (T26), Klotz (T27, T28), Mood (T29), and Noether (T30), among others. Noether (T30) has shown that the Wilcoxon signed-ranks, one-sample test has an asymptotic relative efficiency of 0.955 relative to the one-sample t test if the D_i are normally distributed, and an efficiency of 1 if the D_i are uniformly distributed. Noether (T30) shows further that the asymptotic relative efficiency of the sign test relative to the Wilcoxon one-sample test is 2/3 if the D_i are normally distributed, 1/3 if uniformly distributed, and 4/3 if the D_i follow the double exponential distribution.

Sampling Distribution of T_+ It may be instructive to show how the sampling distribution of T_+ in the Wilcoxon signed-ranks test is constructed. Suppose we let $n = 4$. Regardless of the absolute values of the differences to be ranked, the ranks are 1, 2, 3, and 4. Under the null hypothesis the differences to be ranked are distributed symmetrically about zero. Consequently each difference to be ranked is just as likely to be positive as negative. In general, there are 2^n possible sets of signs associated with the n ranks. For $n = 4$, then, there are $2^4 = 16$ possible sets of signs associated with the 4 ranks. For example, none (that is, zero) of the ranks may be positive, only rank 1 may be positive, only ranks 1 and 3 may be positive, and so on. For $n = 4$ the following table shows which of the four ranks may be positive and the corresponding value of T_+.

Ranks with positive signs	Value of T_+	Ranks with positive signs	Value of T_+
None	0	2, 3	5
1	1	2, 4	6
2	2	3, 4	7
3	3	1, 2, 3	6
4	4	1, 2, 4	7
1, 2	3	1, 3, 4	8
1, 3	4	2, 3, 4	9
1, 4	5	1, 2, 3, 4	10

Each of these 16 possible sets of ranks with positive signs has an equal chance of occurring. Therefore the probability of any one of them occurring is $1/16 = 0.0625$.

Note that the table contains only 11 distinct values of T_+, since some of the values occurred more than once. The following table shows the distinct values of T_+, the number of times each occurred, and the probability, under H_0, of observing each distinct value.

Value of T_+	Number of times occurred	$P(T_+)$
0	1	0.0625
1	1	0.0625
2	1	0.0625
3	2	0.1250
4	2	0.1250
5	2	0.1250
6	2	0.1250
7	2	0.1250
8	1	0.0625
9	1	0.0625
10	1	0.0625

From this table any cumulative or decumulative probability may be obtained. For example, $P(T_+ \le 1) = 0.0625 + 0.0625 = 0.1250$.

EXERCISES

2.3 Malina (E5) reported the results of a study of weights of football players at the University of Texas at Austin between 1899 and 1970. Suppose that the weights of a random sample of 15 football players during the past 10 years at another large state university are those shown in Table 2.6.

Can we conclude that the median weight of the population from which this sample was drawn is greater than 163.5 pounds? Let $\alpha = 0.05$. What is the P value for this test?

TABLE 2.6   **Weights of football players**

Player	Weight	Player	Weight
1	188.0	9	214.4
2	211.2	10	221.0
3	170.8	11	162.0
4	212.4	12	222.8
5	156.9	13	174.1
6	223.1	14	210.3
7	235.9	15	195.2
8	183.9		

2.4 Moore and Ogletree (E6) investigated the readiness of pupils at the beginning of the first grade. They compared scores on a readiness test of pupils who had attended a Head Start program for a full year with the scores of those who had not. Suppose that a random sample of 20 pupils who had not attended a Head Start program achieved the scores on the readiness test shown in Table 2.7.

Can we conclude that the median score of the population represented by this sample is less than 45.32? What is the *P* value for this test?

TABLE 2.7   **Readiness test scores of pupils who did not attend a Head Start program**

Pupil	Readiness score	Pupil	Readiness score
1	33	11	41
2	19	12	31
3	40	13	46
4	35	14	51
5	51	15	34
6	41	16	37
7	27	17	36
8	23	18	55
9	39	19	52
10	21	20	32

2.5 Palmer (E7) reports that, during a six-month period, new salespersons in an insurance company spent an average of 119 hours per month in the field. Suppose that for a random sample of 16 new salespersons from another company, we have the data shown in Table 2.8. Test the null hypothesis that the median of the population from which this sample was drawn is 119 hours. What is the *P* value?

TABLE 2.8   **Average number of hours per month spent in the field by new insurance salespersons during a six-month period**

136	103	91	122	96	145	140	138	126	120	99	125	91	142	119	137

CONFIDENCE INTERVAL FOR MEDIAN BASED ON SIGN TEST

When we have a random sample of measurements drawn from some population, the median *m* of the sample provides a logical point estimate of the median of the population. The sample median is usually not of particular interest solely as a measure of location for the sample. In most situations it is of interest as an estimate of the median of the population from which the sample was drawn. Point estimates are of limited value, since we cannot attach to them statements regarding the amount of confidence we have that they have estimated the unknown parameter. Of greater value is an *interval estimate*, an estimate about which we can make statements of confidence. For a properly constructed interval estimate of an unknown population median, for example, we may say that we are 95% confident

that the interval includes the unknown median. We are able to make such a statement because of the probability properties of the interval. Suppose, for example, that we wish to construct a 95% confidence interval for an unknown population mean. If we were to select many, many (theoretically, all possible) simple random samples of the same size from the population of interest and, in the usual manner, construct from the data of each sample an interval estimate with a confidence coefficient of 0.95, about 95% of them would include the unknown parameter. The practical importance of this theory is the fact that for a single such interval estimate we can say that we are 95% confident that it contains the unknown population parameter.

In hypothesis testing we begin the statistical analysis with a statement regarding a possible value of the magnitude of the parameter, such as the median, that is of interest. Through the hypothesis testing procedure we determine whether or not the sample data are compatible with the initial statement. In interval estimation, on the other hand, we do not begin the analysis with any stated or preconceived value of the unknown parameter about which we wish to make an inference. We select a sample from the population of interest, and from the sample data we construct an interval estimate of the parameter. Sometimes we are motivated by the results of a hypothesis test to construct an interval estimate. We may, for example, reject a null hypothesis that a population median is equal to 150 and consequently conclude that it is not 150. We then may wish to know just what the likely value of the population median is. The interval estimate satisfies this desire. Two methods for constructing interval estimates of an unknown population median will now be illustrated. The first method is based on the sign test, and the second is based on the Wilcoxon test.

The sample median m is the "middle" value in the ordered array of sample observations. If the sample size is an odd number, the sample median is the middle value in the ordered array. If the sample size is an even number, the sample median is the mean of the two middle values in the ordered array.

A symmetric, two-sided confidence interval for the population median may be obtained by using a procedure that is closely related to the sign test. Thompson (T31) and Savur (T32) describe this procedure, and David (T33) gives the mathematical derivation of the interval.

The $100(1 - \alpha)\%$ confidence interval for M consists of those values of M_0 for which we would not reject the two-sided null hypothesis $H_0: M = M_0$ at the α level of significance. We designate the lower limit of our confidence interval by M_L and the upper limit by M_U. For a $100(1 - \alpha)\%$ confidence interval, we select M_L and M_U as follows: We use Table A.1 to determine the largest number of positive or negative signs (that is, the value of K') such that $P(K \leq K' \mid n, 0.50) \leq \alpha/2$. In other words, we select K' just as we did in the two-sided sign test. When we have arranged the sample values in order of magnitude, the $(K' + 1)$th observation is M_L, the lower limit of the $100(1 - \alpha)\%$ confidence interval. To find M_U, the upper limit of the confidence interval, we count the ordered sample values backward from the largest.

The $(K' + 1)$th observation from the upper end is M_U. Counting from the lower end of the ordered array (that is, beginning with the smallest sample value), we find that M_U is the $(n - K')$th value.

Example 2.3

Agnew et al. (E8), in a study of sleep patterns, reported the data shown in Table 2.9. Let us construct a 95% confidence interval for the median of the population from which these sample data have been drawn. The ordered values are as follows.

0.07, 0.69, 1.74, 1.90, 1.99, 2.41, 3.07,   3.08
3.10, 3.53, 3.71, 4.01, 8.11, 8.23, 9.10, 10.16

TABLE 2.9

Percentage of total sleep time spent in stage 0 sleep by 16 mentally and physically healthy males between the ages of 50 and 60

1.90	3.08	9.10	3.53	1.99	3.10	10.16	0.69
1.74	2.41	4.01	3.71	8.11	8.23	0.07	3.07

Source: Harman W. Agnew, Jr., Wilse W. Webb, and Robert L. Williams, "Sleep Patterns in Late Middle Age Males: An EEG Study," *Electroencephalogr. Clin. Neurophysiol,* 23 (1967), 168–171

The point estimate of the population median is the sample median, which is the mean of the two middle values in the ordered array; that is, the sample median is $(3.08 + 3.10)/2 = 3.09$.

To find the lower and upper limits of the confidence interval, we consult Table A.1 and find that

$$P(K \leq 3 \,|\, 16, 0.50) = 0.0105$$

and

$$P(K \leq 4 \,|\, 16, 0.50) = 0.0383$$

Thus we see that we cannot obtain an exact 95% confidence interval, since $100[1 - 2(0.0105)] = 97.9$, which is larger than 95, and $100[1 - 2(0.0383)] = 92.34$, which is less than 95. This method of constructing a confidence interval for the median does not usually yield intervals with exactly the usual coefficients of 0.90, 0.95, and 0.99. In this example we must decide between (1) the wider interval and the higher confidence and (2) the narrower interval and the lower confidence. Suppose that we choose the latter. Then $K' = 4$ and $K' + 1 = 5$. The fifth value in the ordered array is M_L, the lower limit of our interval. The upper limit M_U is the fifth value from the right or the $(16 - 4) =$ twelfth value from the left. Thus we see that for our example, $M_L = 1.99$ and $M_U = 4.01$. The confidence coefficient is $100[1 - 2(0.0383)] = 92.34$. We say that we are 92.34% confident that the population median is between 1.99 and 4.01. Figure 2.1 illustrates the construction of this confidence interval.

FIGURE 2.1   Construction of approximate 95% confidence interval for Example 2.3

K′	Probabilities (from Table A.1)	Cumulative probabilities
0	0.0000	0.0000
1	0.0002	0.0002
2	0.0018	0.0020
Rank of 3	0.0085	0.0105
lower limit 4 + 1 = 5	0.0278	0.0383
5	0.0667	
6	0.1222	
7	0.1746	
8	0.1964	
9	0.1746	
10	0.1222	
11 Rank of	0.0667	
12 upper limit	0.0278	0.0383
13	0.0085	0.0105
14	0.0018	0.0020
15	0.0002	0.0002
16	0.0000	0.0000

Rank	1	2	3	4	5	6	7	8	9	10	11	12	13	14	15	16
Values	0.07	0.69	1.74	1.90	1.99	2.41	3.07	3.08	3.10	3.53	3.71	4.01	8.11	8.23	9.10	10.16

M_L M_U

Large-Sample Approximation   For large samples we may use the approximation given by Hollander and Wolfe (T34). To employ the approximation, we approximate $K' + 1$ by

$$(K' + 1) \approx (n/2) - z_{\alpha/2}\sqrt{n/4} \tag{2.5}$$

where $z_{\alpha/2}$ is the value of z in Table A.2 corresponding to $\alpha/2$. By Equation 2.5, $(K' + 1)$ is usually not an integer; in that case, we use the closest integer.

Example 2.4   To demonstrate the large-sample approximation, let us apply the normal approximation procedure to the data of Example 2.3. By Equation 2.5 we have, for a confidence coefficient of 0.95 ($\alpha/2 = 0.025$),

$$(K' + 1) \approx (16/2) - 1.96\sqrt{16/4} \approx 4$$

Counting to the fourth observation from each end of the ordered array yields $M_L = 1.90$ and $M_U = 8.11$.

FURTHER READING

Efron (T35) discusses bootstrap intervals for nonparametric situations. Hettmansperger and Sheather (T36) consider the problem of interpolating adjacent

order statistics to form confidence intervals with intermediate values of the confidence coefficients; they provide a simple interpolation formula that they claim works well in most practical situations.

EXERCISES

2.6 Armstrong (E9) studied the daily exposure, in minutes, of 10 North Hawaiian families to the risk of a motor vehicle accident. Suppose that a similar survey in another area yielded the exposure times shown in Table 2.10. Find the point estimate and construct an approximate 95% confidence interval for the population median.

TABLE 2.10 **Exposure times per day, in minutes, of individuals to motor-vehicle accidents**

18.3	45.2	19.1	57.0	63.9	10.3	12.1	35.5	36.6	74.6
10.5	27.8	44.9	40.9	63.7	40.8	59.1	31.5	40.1	8.1

2.7 Abu-Ayyash (E10) found that the median education of heads of households living in mobile homes in a certain area was 11.6 years. Suppose that a similar survey conducted in another area revealed the educational levels of heads of households shown in Table 2.11. Find the point estimate, and construct the approximate 95% confidence interval for the population median.

TABLE 2.11 **Educational levels (years of school completed) of heads of households residing in mobile homes**

13	6	6	12	12	10	9	11	14	8	7	16	15	8	7

2.8 The Agricultural Experiment Station of Auburn University (E11) reported data on ages of farm operators in Alabama in 1950 and 1960. Table 2.12 shows the ages of 16 farm operators in another state. Find the point estimate, and construct the approximate 95% confidence interval for the median.

TABLE 2.12 **Ages of farm operators**

32	42	30	35	57	40	30	52	34	64	55	57	50	45	46	63

CONFIDENCE INTERVAL FOR MEDIAN BASED ON WILCOXON SIGNED-RANKS TEST

We may construct a confidence interval for a population median M by using a procedure based on the Wilcoxon signed-ranks test when we are willing to assume that the population is symmetric. The theory underlying the procedure, which has been attributed to Tukey (T37), is discussed by Noether (T30) and Gibbons (T11).

As noted previously, the $100(1 - \alpha)\%$ confidence interval for M consists of those values of M for which we would not reject the two-sided null hypothesis $H_0: M = M_0$ at the α level of significance.

Arithmetic Procedure   The procedure consists of the following steps:

1. From the sample observations $X_1, X_2, \ldots, X_n$, form all possible averages.

$$u_{ij} = \frac{X_i + X_j}{2}, \qquad 1 \le i \le j \le n \tag{2.6}$$

In other words, compute

$$\frac{X_1 + X_1}{2}, \quad \frac{X_1 + X_2}{2}, \quad \ldots, \quad \frac{X_1 + X_n}{2},$$

$$\frac{X_2 + X_2}{2}, \quad \frac{X_2 + X_3}{2}, \quad \ldots, \quad \frac{X_2 + X_n}{2}, \quad \ldots, \quad \frac{X_{n-1} + X_n}{2},$$

$$\frac{X_n + X_n}{2}$$

There are $n(n - 1)/2 + n$ such averages, distributed symmetrically about the median.

2. Arrange the u_{ij}'s in order of magnitude from smallest to largest.
3. The median of the u_{ij}'s provides a point estimate of the population median.
4. Locate in Table A.3 the sample size and the appropriate value of P as determined by the desired level of confidence. When $(1 - \alpha)$ is the confidence coefficient, $P = \alpha/2$. When the exact value of $\alpha/2$ cannot be found in Table A.3, choose a neighboring value—either the closest value or one that is larger or smaller than $\alpha/2$, depending on whether a slightly wider or slightly narrower interval than desired is more acceptable.
5. The endpoints of the confidence interval are the Kth smallest and the Kth largest values of u_{ij}. $K = T + 1$, where T is the value in the column labeled T corresponding to the value of P selected in Step 4.

Example 2.5   To study personality differences between male and female medical students, Cartwright (E12) analyzed the scores that a series of male medical students made on the responsibility scale of the California Psychological Inventory. Suppose that a random sample of 10 male medical students from a certain section of the country made the scores shown in Table 2.13. Let us obtain a point estimate and construct an interval with a confidence coefficient of approximately 0.95 for the population median.

TABLE 2.13   **Scores made by 10 male medical students on the responsibility scale of the California Psychological Inventory**

28.5	25.2	28.7	41.0	29.1	32.3	37.7	39.9	26.8	28.8

1. Computing all the u_{ij} values, we have the following.

$$\frac{28.5 + 28.5}{2} = 28.50 \qquad \frac{25.2 + 25.2}{2} = 25.20 \qquad \frac{28.7 + 28.7}{2} = 28.70$$

$$\frac{28.5 + 25.2}{2} = 26.85 \qquad \frac{25.2 + 28.7}{2} = 26.95 \qquad \frac{28.7 + 41.0}{2} = 34.85$$

$$\frac{28.5 + 28.7}{2} = 28.60 \qquad \frac{25.2 + 41.0}{2} = 33.10 \qquad \frac{28.7 + 29.1}{2} = 28.90$$

$$\frac{28.5 + 41.0}{2} = 34.75 \qquad \frac{25.2 + 29.1}{2} = 27.15 \qquad \frac{28.7 + 32.3}{2} = 30.50$$

$$\frac{28.5 + 29.1}{2} = 28.80 \qquad \frac{25.2 + 32.3}{2} = 28.75 \qquad \frac{28.7 + 37.7}{2} = 33.20$$

$$\frac{28.5 + 32.3}{2} = 30.40 \qquad \frac{25.2 + 37.7}{2} = 31.45 \qquad \frac{28.7 + 39.9}{2} = 34.30$$

$$\frac{28.5 + 37.7}{2} = 33.10 \qquad \frac{25.2 + 39.9}{2} = 32.55 \qquad \frac{28.7 + 26.8}{2} = 27.75$$

$$\frac{28.5 + 39.9}{2} = 34.20 \qquad \frac{25.2 + 26.8}{2} = 26.00 \qquad \frac{28.7 + 28.8}{2} = 28.75$$

$$\frac{28.5 + 26.8}{2} = 27.65 \qquad \frac{25.2 + 28.8}{2} = 27.00$$

$$\frac{28.5 + 28.8}{2} = 28.65$$

$$\frac{41.0 + 41.0}{2} = 41.00 \qquad \frac{29.1 + 29.1}{2} = 29.10 \qquad \frac{32.3 + 32.3}{2} = 32.30$$

$$\frac{41.0 + 29.1}{2} = 35.05 \qquad \frac{29.1 + 32.3}{2} = 30.70 \qquad \frac{32.3 + 37.7}{2} = 35.00$$

$$\frac{41.0 + 32.3}{2} = 36.65 \qquad \frac{29.1 + 37.7}{2} = 33.40 \qquad \frac{32.3 + 39.9}{2} = 36.10$$

$$\frac{41.0 + 37.7}{2} = 39.35 \qquad \frac{29.1 + 39.9}{2} = 34.50 \qquad \frac{32.3 + 26.8}{2} = 29.55$$

$$\frac{41.0 + 39.9}{2} = 40.45 \qquad \frac{29.1 + 26.8}{2} = 27.95 \qquad \frac{32.3 + 28.8}{2} = 30.55$$

$$\frac{41.0 + 26.8}{2} = 33.90 \qquad \frac{29.1 + 28.8}{2} = 28.95$$

$$\frac{41.0 + 28.8}{2} = 34.90$$

$$\frac{37.7 + 37.7}{2} = 37.70 \qquad \frac{39.9 + 39.9}{2} = 39.90 \qquad \frac{26.8 + 26.8}{2} = 26.80$$

$$\frac{37.7 + 39.9}{2} = 38.80 \qquad \frac{39.9 + 26.8}{2} = 33.35 \qquad \frac{26.8 + 28.8}{2} = 27.80$$

$$\frac{37.7 + 26.8}{2} = 32.25 \qquad \frac{39.9 + 28.8}{2} = 34.35$$

$$\frac{37.7 + 28.8}{2} = 33.25$$

$$\frac{28.8 + 28.8}{2} = 28.80$$

2. The ordered array of these $[10(9)/2] + 10 = 55$ values of u_{ij} appears in Table 2.14.

TABLE 2.14   **Ordered array of u_{ij} values for Example 2.5**

1 25.20	12 28.50	23 29.55	34 33.20	45 34.90
2 26.00	13 28.60	24 30.40	35 33.25	46 35.00
3 26.80	14 28.65	25 30.50	36 33.35	47 35.05
4 26.85	15 28.70	26 30.55	37 33.40	48 36.10
5 26.95	16 28.75	27 30.70	38 33.90	49 36.65
6 27.00	17 28.75	28 31.45	39 34.20	50 37.70
7 27.15	18 28.80	29 32.25	40 34.30	51 38.80
8 27.65	19 28.80	30 32.30	41 34.35	52 39.35
9 27.75	20 28.90	31 32.55	42 34.50	53 39.90
10 27.80	21 28.95	32 33.10	43 34.75	54 40.45
11 27.95	22 29.10	33 33.10	44 34.85	55 41.00

3. The estimate of the population median is the 28th value, which is 31.45.
4. We enter Table A.3 with $n = 10$ and $P = 0.0244$, the value of P closest to $(1 - 0.95)/2 = 0.025$ for $n = 10$.
5. The value of T corresponding to $n = 10$ and $P = 0.0244$ is 8. Consequently, $K = T + 1 = 8 + 1 = 9$. Our confidence interval for the population median is bounded by the ninth smallest and largest values of u_{ij}. The confidence coefficient is $1 - 2(0.0244) = 0.9512$.

For this example, then, the 95.12% confidence interval is bounded by 27.75 and 35.05. In other words, we are 95.12% confident that the median score for the population from which our hypothetical sample was drawn is somewhere between 27.75 and 35.05.

It is obvious that even for moderately sized samples, this method of finding a confidence interval for a population median is rather tedious. We can save some time, however, by ranking the original scores before calculating the u_{ij} values. Then we can compute the K smallest and the K largest averages fairly quickly. Obtaining the point estimate of the population median requires a little more time. If a, the number of averages, is odd, we only have to compute the $(a + 1)/2$ smallest (or largest) averages. If a is even, we only have to compute the $(a/2) + 1$ smallest (or largest) averages in order to find the median.

Graphical Procedure Moses (T38) describes a graphical procedure, based on the Wilcoxon signed-ranks test, for obtaining a confidence interval for the population median. Since this procedure is less tedious than the arithmetic procedure, it is illustrated here with the data of Example 2.5. The steps in this method of solution are as follows. (Figure 2.2 also illustrates these steps.)

FIGURE 2.2 Graphical construction of the 95% confidence interval for the population median in Example 2.5.

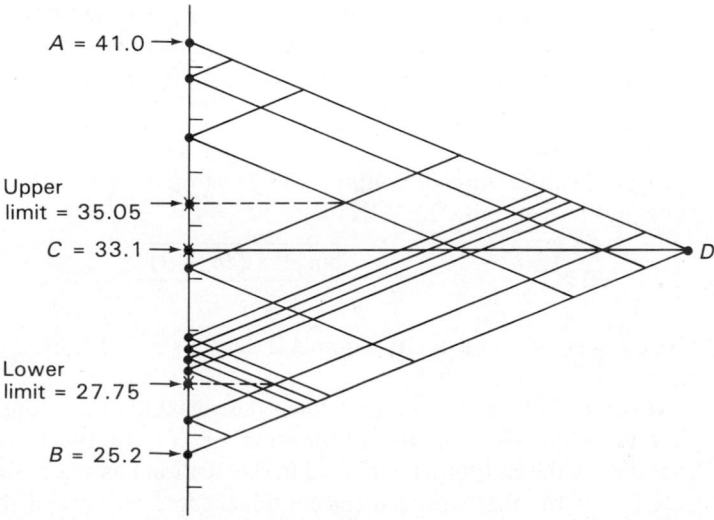

1. Plot the original sample values on the vertical axis of the graph.
2. Label the largest value A and the smallest value B. In our example, $A = 41.0$ and $B = 25.2$.
3. Find the point halfway between A and B and label it C. In our example, $C = 33.1$.
4. Draw through C a line of convenient length, perpendicular to the vertical axis. Label the endpoint of this line D.
5. Connect AD and DB with straight lines to form the triangle ADB.
6. Draw a line parallel to BD from each data point on the vertical axis to line AD.
7. Draw a line parallel to AD from each data point on the vertical axis to line BD.
8. To locate the upper confidence limit, consult Table A.3 to determine $K = T + 1$. Count down from point A to the Kth intersection, including intersections with the vertical axis. Draw a horizontal line from the Kth intersection to the vertical axis. The endpoint of this line is the upper limit of the confidence interval. In this example, with $n = 10$ and $1 - \alpha = 0.9512$, we have $K = 9$. A horizontal line from the ninth intersection from the top intersects the vertical at 35.05.
9. To locate the lower limit of the confidence interval, begin with point B and count up to the Kth intersection from the bottom. Draw a horizontal line from this point to the vertical axis. The endpoint of this line is the lower limit of the confidence interval. In this example we find that the lower limit is 27.75. This lower limit and the upper limit of 30.05 agree with the results obtained by using the arithmetic procedure.

When a value in the original data set occurs more than once, the horizontal lines through the repeated points on the vertical axis coincide. Be careful to count the intersections made by all such lines even though they appear as a single line on the graph.

Large-Sample Approximation　　With samples larger than 30 we cannot use Table A.3 to determine K. When $n > 30$,

$$K \approx \frac{n(n + 1)}{4} - z_{\alpha/2} \sqrt{\frac{n(n + 1)(2n + 1)}{24}} \tag{2.7}$$

where $z_{\alpha/2}$ is the value of z in Table A.2 that has $\alpha/2$ of the area under the standard normal curve to its right.

Noether (T39) has shown that when the variable under consideration is discrete (that is, when we relax the assumption of continuity), the confidence interval bounded by the endpoints described in this section has a confidence coefficient of at least $1 - \alpha$. In other words, if the confidence intervals are of the form $L \leq \theta \leq U$, where θ is the parameter of interest, L is the lower confidence limit, and U is the

upper confidence limit, then we may make a probability statement of the form

$$P(L \leq \theta \leq U) \geq 1 - \alpha$$

where the interval includes the endpoints. If the confidence intervals are open, as $L < \theta < U$, then the confidence coefficients in the discrete case are at most equal to the confidence coefficients in the continuous case, and we can make statements of the form

$$P(L < \theta < U) \leq 1 - \alpha$$

The intervals we construct include the endpoints.

EXERCISES

2.9 Hall et al. (E13) computed the ratio of Lintner Soluble Starch (LSS) to Amylase Azure (AA) values (enzyme units per milliliter) determined from saliva specimens from 11 subjects. The results are shown in Table 2.15. Obtain a point estimate, and construct an approximate 95% confidence interval for the median of the population of similar subjects from which these subjects may be considered a random sample.

TABLE 2.15 **Ratio of Lintner Soluble Starch (LSS) to Amylase Azure (AA), enzyme units per milliliter**

Subject	A	B	C	D	E	F	G	H	I	J	K
Ratio LSS/AA	0.78	0.86	0.82	1.10	0.71	1.00	0.65	0.85	0.64	0.86	0.70

Source: F. F. Hall, C. R. Ratliff, T. Hayakawa, T. W. Culp, and N. C. Hightower, "Substrate Differentiation of Human Pancreatic and Salivary Alpha-Amylases," *Am. J. Dig. Dis.*, 15 (1970), 1031–1038.

2.10 Golde et al. (E14) reported the white blood cell counts taken on seven adult males with some type of leukemia; the counts are shown in Table 2.16.

Obtain a point estimate; construct an approximate 95% confidence interval for the median of the population from which these subjects may be presumed to represent a random sample.

TABLE 2.16 **White blood cell counts in adult males with leukemia**

Subject	1	2	3	4	5	6	7
WBC count	19,000	31,000	1,300	1,500	43,000	14,000	14,000

Source: David W. Golde, Belina Rothman, and Martin J. Cline, "Production of Colony-Stimulating Factor by Malignant Leukocytes," *Blood*, 43 (1974), 749–756; used by permission.

2.11 Silverman et al. (E15) measured forced vital capacity (FVC) in five healthy adult males who were either physicians or medical research workers. The subjects ranged in age from

19 to 30, were nonsmokers, and had no history of chest or cardiovascular disease. The results are shown in Table 2.17.

Obtain the point estimate; construct the approximate 95% confidence interval for the median of the population from which these subjects may be presumed to represent a random sample.

<table>
<tr><td>TABLE 2.17</td><td colspan="6">Forced vital capacity (FVC) in healthy adult males</td></tr>
<tr><td></td><td>Subject</td><td>1</td><td>2</td><td>3</td><td>4</td><td>5</td></tr>
<tr><td></td><td>FVC, milliliters</td><td>5,610</td><td>4,290</td><td>5,555</td><td>5,280</td><td>5,280</td></tr>
</table>

Source: M. Silverman, E. Zeidifard, J. W. Paterson, and S. Godfrey, "The Effect of Isoprenaline on the Cardiac and Respiratory Responses to Exercise," *Quart. J. Exper. Physiol.*, 58 (1973), 7–17.

2.2

MAKING INFERENCES ABOUT A POPULATION PROPORTION

The population proportion is a parameter of frequent interest in research and decision-making activities. A market analyst may wish to know the proportion of families in a certain area who have cable TV. A public health official may be interested in knowing the proportion of school-age children who have been immunized against some childhood disease. A sociologist may want to know the proportion of heads of household in a certain area who are women. A politician may find it useful to know the proportion of residents in some community who are registered Democrats.

When it is impossible or impractical to survey the total population, researchers base decisions regarding population proportions on inferences made by analyzing samples drawn from the population. As usual, inference may take the form of interval estimation or hypothesis testing.

BINOMIAL TEST

First consider the case in which the researcher wishes to test a hypothesis about a population proportion. A test that is appropriate in this case is the *binomial test*. This test uses the binomial formula, which is usually encountered in elementary statistics courses.

Suppose that a population contains only two elements: type A and type B. We let p designate the proportion of type A elements in the population and let $1 - p = q$ denote the proportion of type B elements. If we contemplate drawing a simple random sample of size n from the population, the binomial formula enables us to compute the probability that the sample will contain a specified number of elements of type A (or type B, if we wish) under certain assumptions. If we let r denote the number of type A elements in the sample, then we write the binomial formula for determining the probability $P(r)$ that r is equal to any number greater than or equal

to zero as

$$P(r) = \binom{n}{r} p^r q^{n-r}, \quad \text{where} \quad \binom{n}{r} = \frac{n!}{r!(n-r)!}$$

Suppose that 0.4 of the elements in some population are type A and 0.6 are type B. We may find the probability that a random sample of size $n = 5$ will contain $r = 3$ type A elements, as follows:

$$P(r = 3) = \binom{5}{3}(0.4)^3(0.6)^{5-3}$$

$$= \frac{5!}{3!2!}(0.064)(0.36) = 0.2304$$

We can determine the probability that a random sample of size $n = 5$ will contain 3 or fewer elements by adding individual probabilities $P(r = 3)$, $P(r = 2)$, $P(r = 1)$, and $P(r = 0)$; that is,

$$P(r \leq 3) = P(r = 3) + P(r = 2) + P(r = 1) + P(r = 0)$$

Since

$$P(r = 3) = \frac{5!}{3!2!}(0.4)^3(0.6)^2 = 0.2304$$

$$P(r = 2) = \frac{5!}{2!3!}(0.4)^2(0.6)^3 = 0.3456$$

$$P(r = 1) = \frac{5!}{1!4!}(0.4)^1(0.6)^4 = 0.2592$$

$$P(r = 0) = \frac{5!}{0!5!}(0.4)^0(0.6)^5 = 0.0778$$

we find that

$$P(r \leq 3) = 0.2304 + 0.3456 + 0.2592 + 0.0778 = 0.9130$$

We then find $P(r > 3)$ as follows:

$$P(r > 3) = 1 - P(r \leq 3)$$

$$= 1 - 0.9130 = 0.0870$$

Fortunately, there are tables of these and other probabilities based on the binomial formula, so we do not have to perform the calculations ourselves.

Now suppose, for some population, that p, the proportion of type A elements, is some specified value p_0. Suppose also that we draw a random sample of size n from the population and find that the probability associated with the number of type A elements in the sample is sufficiently small, given n and p. Then we are justified in doubting that $p = p_0$. We formalize these ideas in the following discussion of the binomial test.

Assumptions

A. The data consist of a sample of the outcomes of n repetitions of some process. Each outcome consists of either a "success" or a "failure." These designations are purely arbitrary. The number of successes S is the number of outcomes having a given characteristic. The number of successes in the sample divided by n, the number of trials, yields $\hat{p}$, the sample proportion having the characteristic of interest.

B. The n trials are independent.

C. The probability of a success p remains constant from trial to trial. We use p to designate the proportion of the population having the characteristic of interest.

Hypotheses

We use p_0 to designate the hypothesized population proportion. We may state the hypotheses in such a way that they lead to either a two-sided test or to one of two possible one-sided tests.

A. (Two-sided): $H_0: p = p_0$ $H_1: p \neq p_0$
B. (One-sided): $H_0: p \leq p_0$ $H_1: p > p_0$
C. (One-sided): $H_0: p \geq p_0$ $H_1: p < p_0$

Select a level of significance α.

Test Statistic

Since we are interested in the number of successes S, our test statistic is $S =$ number of successes.

Decision Rule

The decision rule that is applicable in a given situation depends on which of the three null hypotheses we are testing. The three decision rules corresponding to hypotheses A, B, and C are as follows:

A. For either sufficiently large or sufficiently small values of S, we reject $H_0: p = p_0$. Consequently, since this is a two-sided test, we should divide α as nearly in half as possible. Then, to find critical values of the test statistic, we enter Table A.1 with n and p_0 and look for the number s_1 such that $P(r \leq s_1) \approx \alpha/2$, and the number s_2 such that $P(r > s_2) \approx \alpha/2$.

In other words, divide α as nearly in half as possible. Enter Table A.1 with n and p_0; cumulate the probabilities, beginning with zero, until the sum is equal to approximately half of α. Locate the value of r in line with the last probability added into the cumulation and call it s_1. To find s_2, begin with the last probability listed for n and

p_0, and cumulate probabilities until the sum is equal to approximately $\alpha/2$. Locate the value of r in line with the last probability added into the cumulation. Subtract one from this value of r to obtain s_2. Reject H_0 if S is either less than or equal to s_1 or larger than s_2.

B. For sufficiently large values of S, we reject $H_0: p \leq p_0$. Thus we must enter Table A.1 with n and p_0, and find the value of s such that $P(r > s) = \alpha$. We reject H_0 if S is greater than s.

For this one-sided test we find s in the same way that we find s_2 for the two-sided test, except that in the one-sided test we cumulate probabilities until the total is equal to α rather than $\alpha/2$.

If we cannot obtain the exact preselected value of α by cumulating the tabulated probabilities, we cumulate until the total is as close to α as possible.

C. We reject the null hypothesis $H_0: p \geq p_0$ for sufficiently small values of S. To find the critical value of S, we enter Table A.1 with n and p_0 and find the value of s such that $P(r \leq s) = \alpha$. We reject H_0 if S is smaller than or equal to s. Again, if it is not possible to find α exactly in Table A.1, we use an approximate value.

Example 2.6

Cinotti and Patti (E16) found anterior subcapsular vacuoles in the eyes of 11 of 25 diabetic subjects. If these data satisfy the assumptions underlying the binomial test, and if we can consider the subjects a random sample from the population of similar subjects, can we conclude that the population proportion with the condition of interest is greater than 0.27? (Let $\alpha = 0.05$.)

Hypotheses

$H_0: p \leq 0.27, \qquad H_1: p > 0.27$

Test Statistic

Since 11 subjects have the characteristic of interest, $S = 11$

Decision

We enter Table A.1 with $n = 25$ and $P_0 = 0.27$ to find s. Cumulating from the last probability, we find that $r = 11$ is opposite the cumulative total of 0.05. When we subtract 1 from r, we have $s = 11 - 1 = 10$, the critical value of the test statistic. The decision rule requires that we reject H_0 if $S > s$. Since 11 is greater than 10, we reject H_0, and we conclude that the population proportion p is greater than 0.27. Since the cumulative probability from 11 through 25 is 0.05, the probability of observing 11 or more "successes" out of 25 trials when H_0 is true is 0.05. Hence the P value for this example is 0.05.

Large-Sample Approximation   When n is large and p is not too close to 0 or 1, we can obtain the critical value of S by using the following large-sample approximation.

$$s = np_0 + z\sqrt{np_0(1 - p_0)} \qquad (2.8)$$

Table A.2 gives z, the value of the standard normal variate corresponding to α, the chosen level of significance.

To test the null hypothesis specified in A, we obtain s_1 by substituting the negative of z corresponding to $\alpha/2$ into Equation 2.8, and we obtain s_2 by substituting the positive z corresponding to $\alpha/2$. We reject H_0 if S is smaller than or equal to s_1 or larger than s_2. To test H_0 specified in B, we substitute positive z corresponding to α into Equation 2.8, and to test H_0 specified in C, we substitute negative z corresponding to α. In the former case we reject H_0 if S is greater than s; in the latter case we reject H_0 if S is smaller than or equal to s.

Example 2.7

During the 1969–1970 fiscal year, 56% of the felons committed to custody of the State Board of Corrections of Georgia were under 25 years of age (E17). Suppose that 23 felons of a random sample of 50 felons committed to custody of the board of corrections of another state were under 25 years of age. Do such data indicate that the proportion of those in the sampled population who are younger than 25 is less than 0.56?

The appropriate hypotheses are $H_0: p \geq 0.56$ and $H_1: p < 0.56$. Suppose that we let $\alpha = 0.05$. Then, by Equation 2.8, we have

$$s = 50(0.56) + (-1.645)\sqrt{50(0.56)(0.44)} = 22.2$$

Since 23 is not less than 22.2, we cannot reject H_0, and we conclude that the population proportion may be 0.56.

We usually consider that the application of the large-sample approximation is valid if np and $n(1 - p)$ are both greater than 5. Note that the sign test discussed earlier is a special case, in which $p_0 = 0.5$, of the binomial test.

EXERCISES

2.12 Bowman et al. (E18) investigated 15 patients diagnosed as having psittacosis, and found that 11 complained of persistent or recurrent symptoms from one to four years following the initial illness. If we can assume that these 15 constitute a random sample from psittacosis patients, do the data provide sufficient evidence to indicate that the proportion of such patients who suffer from recurrent symptoms is greater than 0.50? Let $\alpha = 0.05$. Compute the P value.

2.13 In a sample of families with incomes between $5,000 and $5,999, Moles (E19) found that 19% were receiving public assistance. Suppose that in a random sample of 20 families in the same income group from another section of the country, 6 are receiving public assistance. On the basis of these data, could we conclude that the population proportion receiving public assistance is greater than 0.19? Let $\alpha = 0.05$. Compute the P value.

2.14 Chin et al. (E20) did indirect fluorescent antibody tests on pretreatment sera against falciparum malaria in 57 successfully treated subjects. They found 38 positives. If this sample satisfies the assumptions of the binomial test, can we conclude from these data that the proportion of positives in the population is greater than 0.50? Let $\alpha = 0.05$.

2.15 In a large group of females between the ages of 15 and 19, Zackler et al. (E21) found 6.9% to be positive for *Neisseria gonorrhoeae*. Suppose that a random sample of 25 females in the same age group is selected from another population, and 4 are positive. Would these data provide sufficient evidence to indicate that the proportion of positives in this population is greater than 0.07? Let $\alpha = 0.01$. Compute the P value.

CONFIDENCE INTERVAL FOR A PROPORTION

When a sample meets the assumptions underlying the binomial test, we can construct a $100(1 - \alpha)\%$ confidence interval for the population proportion p. Table A.4 gives lower and upper confidence limits for confidence coefficients of 0.90, 0.95, 0.99, and for values of n from 1 through 30. The entries in this table, which is adapted from the one given by Crow (T40), were calculated by modifying the method proposed by Sterne (T41). The following example illustrates the use of Table A.4.

Example 2.8 Spencer (E22) reported 5 positive results among 17 subjects with a diagnosis of peptic ulcer who underwent the OEsophageal Perfusion Test described by Bernstein and Baker (E23). If we assume that this sample meets the conditions for the binomial test, we can construct the 0.95% confidence interval for the proportion of positives in the population of subjects from which the 17 may be presumed to have been drawn.

Entering Table A.4 with $n = 17$, $r = 5$, and a confidence coefficient of 0.95, we find that the lower and upper limits of the confidence interval are 0.124 and 0.544, respectively. The point estimate of p is $5/17 = 0.294$.

FURTHER READING

Clopper and Pearson (T42) have also prepared tables for determining confidence intervals for p. In addition, they present a graphical method for finding upper and lower confidence limits. Anderson and Burstein (T43, T44) have proposed methods for obtaining approximate one-sided intervals for p. Quesenberry and Hurst (T45) and Goodman (T46) discuss techniques for the construction of simultaneous confidence intervals for multinomial proportions.

Large-Sample Approximation   When np and $n(1 - p)$ are both greater than 5, we can construct confidence intervals for p by applying the large-sample approximation. The interval is

$$\hat{p} \pm z\sqrt{\hat{p}(1 - \hat{p})/n} \tag{2.9}$$

where $\hat{p}$ is the proportion of the sample with the characteristic of interest. That is, if r is the number with the characteristic of interest and n is the total number of subjects in a sample, then

$$\hat{p} = r/n \tag{2.10}$$

z is the value from Table A.2 that has $\alpha/2$ of the area under the standard normal curve to its right.

Example 2.9

In a cross-sectional sample of 974 Tunisian male workers, Sack (E24) found that 38.7% had received some vocational training beyond the primary level. If the sample meets the assumptions underlying the binomial test, we may obtain a 95% confidence interval for p by using Expression 2.9, as follows.

$$0.387 \pm 1.96\sqrt{(0.387)(0.613)/974}$$

$$0.387 \pm 0.031$$

The lower and upper limits are 0.356 and 0.418, respectively.

EXERCISES

2.16 In a sample of 216 children who were officially designated illegitimate at the time their births were registered, May (E25) found 52 delinquents. Assume that this sample meets the assumptions underlying the binomial test. Construct a 95% confidence interval for the true proportion in the sampled population who are delinquent.

2.17 Westermeyer (E26), in a study of *amok* (a Malaysian word meaning "to engage furiously in battle") in 18 Laotian males, found that eight subjects were single. Assume that this sample satisfies the assumptions underlying the binomial test, and construct the 95% confidence interval for the true proportion in the sampled population who are single.

2.18 In a survey of 14,744 male college students, Strimbu et al. (E27) reported that 44.9% claimed they had tried marijuana. Suppose that in a simple random sample of 25 males drawn from another college population, 11 claimed they had tried marijuana. Construct the 95% confidence interval for the true proportion in this population who would claim to have tried marijuana.

2.3

ONE-SAMPLE RUNS TEST FOR RANDOMNESS

In many situations we want to know whether we can conclude that a series of items or events is random. An important example is the sample of data available for some statistical analysis. A basic assumption underlying procedures for statistical inference is that the inference is based on a random sample. If the randomness of a sample is suspect, we want to have some way of deciding whether the sample is random before we proceed with the analysis. There are many other situations in which we may want to investigate the assumption of randomness. Let us look at two of them.

First, in certain quality-control procedures, control charts are constructed to study and control the fraction of defective items in the output of a manufacturing operation. Samples of output are drawn periodically, and the fraction of defectives in the sample is calculated. The investigator notes whether the fraction of defectives in the sample is greater than or less than the fraction of defectives in the entire process. Frequently the investigator wants to know whether the pattern of fractions of defectives that occurs in a series of samples can be considered random. A lack of randomness may indicate a lack of control in the manufacturing process.

Second, in regression analysis, the difference between an observed value of the dependent variable and the corresponding fitted value is called a *residual*. Residuals may be either positive or negative. When we compute residuals for a sample of data, we frequently test the pattern of occurrence of positive and negative residuals for randomness, since a lack of randomness may mean that one of the assumptions underlying the regression analysis has been violated.

Procedures for investigating randomness are based on the number and nature of the *runs* present in the data of interest. A run is defined as a sequence of like events, items, or symbols that is preceded and followed by an event, item, or symbol of a different type, or by none at all. The number of events, items, or symbols in a run is referred to as its *length*. We doubt the randomness of a series when there appear to be either too many or too few runs.

Consider, for example, a sample of 10 subjects in a psychological experiment. If the sexes of the subjects in the order of their selection were

$$M\ F\ M\ F\ M\ F\ M\ F\ M\ F$$

we would suspect that they were not selected at random, but rather by some systematic procedure. In this case we doubt the procedure's randomness because there are too many runs — 10 in this instance. If the order of occurrence were

$$M\ M\ M\ M\ M\ F\ F\ F\ F\ F$$

we would doubt the procedure's randomness, since here there are only two runs.

Let us now consider the *one-sample runs test*. This procedure helps us decide whether a sequence of events, items, or symbols is the result of a random process.

Assumptions

The data available for analysis consist of a sequence of observations, recorded in the order of their occurrence, which we can categorize into two mutually exclusive types. We let n = the total sample size, n_1 = the number of observations of one type, and n_2 = the number of observations of the other type.

Hypotheses

    A. (Two-sided)
        H_0: The pattern of occurrence of the two types of observation is determined by a random process
        H_1: The pattern of occurrence is not random
    B. (One-sided)
        H_0: The pattern of occurrence of the two types of observation is determined by a random process
        H_1: The pattern is not random (because there are too few runs to be attributed to chance)
    C. (One-sided)
        H_0: The pattern of occurrence of the two types of observation is determined by a random process
        H_1: The pattern is not random (because there are too many runs to be attributed to chance)

Test Statistic

The test statistic is r, the total number of runs.

Decision Rule

    A. Since the null hypothesis does not specify direction, a two-sided test is appropriate. We therefore obtain a lower and an upper critical value of the test statistic. If r is either less than or equal to the lower critical value, or greater than or equal to the upper critical value, we reject the null hypothesis of randomness.

        Tables A.5 and A.6 give lower and upper critical values, respectively, of the test statistic for the 0.05 level of significance and values of n_1 and n_2 through 20. [These tables have been adapted from those given by Swed and Eisenhart (T47).] To determine the lower critical value, we enter Table A.5 with n_1 and n_2. To determine the upper critical value, we enter Table A.6 with n_1 and n_2.

B. Enter Table A.5 with n_1 and n_2. If r is less than or equal to the tabulated value of the test statistic, reject H_0 at the 0.025 level of significance.

C. Enter Table A.6 with n_1 and n_2. If r is greater than or equal to the tabulated value of the test statistic, reject H_0 at the 0.025 level of significance.

Example 2.10   Table 2.18 shows the departures from normal of daily temperatures recorded at Atlanta, Georgia, during November 1974 (E28). We wish to know whether we may conclude that the pattern of departures above and below normal is the result of a nonrandom process.

TABLE 2.18   **Departures from normal of daily temperatures recorded at Atlanta, Georgia, during November, 1974**

Day	1	2	3	4	5	6	7	8	9	10	11	12	13	14	15
Departure from normal	12	13	12	11	5	2	-1	2	-1	3	2	-6	-7	-7	-12
Day	16	17	18	19	20	21	22	23	24	25	26	27	28	29	30
Departure from normal	-9	6	7	10	6	1	1	3	7	-2	-6	-6	-5	-2	-1

Source: *Local Climatological Data*, U.S. Department of Commerce, National Oceanic and Atmospheric Administration, Environmental Data Service, National Climatic Center, Federal Building, Asheville, N.C., November 1974.

These data satisfy the assumptions for using the one-sample runs test, since they are recorded in the order of their occurrence and since they may be dichotomized according to whether the departure from normal is positive or negative. Let $n_1 =$ the number of departures above normal (positive) $= 17$, $n_2 =$ the number of departures below normal (negative) $= 13$, and let $n_1 + n_2 = n = 30$.

Hypotheses

H_0: The pattern of occurrences of negative and positive departures from normal is determined by a random process

H_1: The pattern of occurrences of negative and positive departures from normal is not random

Test Statistic

Table 2.18 shows that for days 1 through 6, the departures from normal are all positive. This sequence of positive departures constitutes a run. Three runs, each of

length one, follow this run. Days 10 and 11 contribute another run. Other runs occur as follows: days 12 through 16, days 17 through 24, and days 25 through 30. Thus the data have a total of eight runs.

Decision

We enter Tables A.5 and A.6 with $n_1 = 17$ and $n_2 = 13$ and find that the critical values for this test are 10 and 22. Since 8 is less than 10, we reject H_0 and conclude that the pattern of occurrence of temperatures above and below normal is not random (P value < 0.05).

Large-Sample Approximation   When either n_1 or n_2 is greater than 20, we cannot use Tables A.5 and A.6 to test our hypothesis. However, for large samples,

$$z = \frac{r - \{[(2n_1 n_2)/(n_1 + n_2)] + 1\}}{\sqrt{\dfrac{2n_1 n_2(2n_1 n_2 - n_1 - n_2)}{(n_1 + n_2)^2(n_1 + n_2 - 1)}}} \tag{2.11}$$

is distributed approximately as the standard normal distribution when H_0 is true. We compare the computed z with the tabulated z for significance.

FURTHER READING

Mosteller (T48) and Prairie et al. (T49) discuss the use of the theory of runs in quality control. Goodman and Grunfield (T50), Moore and Wallis (T51), Sen (T52), and Wallis and Moore (T53) discuss the theory of runs in time-series applications. Mood (T54) gives a history of the use of runs, and Barton (T55) gives another note of interest.

Koppen and Verhelst (T56) present an algorithm for computing the exact distribution of the number of runs in a sequence. Grafton (T57) has written a FORTRAN algorithm for computing the runs up and runs down test found in the book by Knuth (T58).

EXERCISES

2.19   Table 2.19 shows the actual daily occurrence of sunshine in Atlanta during November 1974, as a percentage of the possible time the sun could have shone if it had not been for cloudy skies. The data are from the U.S. Department of Commerce (E28). Dichotomize the observations according to whether the amount of sunshine was more than 50% of possible or 50% or less, and test the null hypothesis that the pattern of occurrence of the two types of day is random.

TABLE 2.19   **Percentage of day during which sunshine occurred in Atlanta, November, 1974**

Day	Percentage	Day	Percentage	Day	Percentage
1	85	11	31	21	87
2	85	12	86	22	100
3	99	13	100	23	100
4	70	14	0	24	88
5	17	15	100	25	50
6	74	16	100	26	100
7	100	17	46	27	100
8	28	18	7	28	100
9	100	19	12	29	48
10	100	20	54	30	0

Source: Local Climatological Data, U.S. Department of Commerce, National Oceanic and Atmospheric Administration, Environmental Data Service, National Climatic Center, Federal Building, Asheville, N.C., November 1974.

2.20 Columns 1 and 2 of Table 2.20 show, for 15 normal fetuses, the gestational age and mean $Q - Ao$ (a measurement of the cardiac cycle) values, as reported by Murata and Martin (E29). If we perform a regression analysis on the data using gestational age as the independent variable X and mean $Q - Ao$ as the dependent variable Y, we obtain the residuals by subtracting the fitted from the observed value of Y (shown in column 3). Dichotomize the residuals according to whether they are negative or positive, and test the null hypothesis that their pattern of occurrence is random.

TABLE 2.20   **Observed age, mean $Q - Ao$ values, and residuals obtained by fitting a regression line to the data**

Gestational age	40	39	40	38	40	40	39	37
Mean $Q - Ao$	71.5	71.5	72.5	64.4	69.3	72.7	67.7	61.1
Residual	−1.4	+2.6	−0.4	−0.5	−3.6	−0.2	−1.2	+0.2
Gestational age	38	39	40	38	36	39	36	
Mean $Q - Ao$	69.5	69.5	71.8	68.3	57.5	70.7	51.6	
Residual	+4.6	+0.6	−1.1	+3.4	+0.6	+5.9	−6.5	

Source: Yuji Murata and Chester B. Martin, Jr., "Systolic Time Intervals of the Fetal Cardiac Cycle," *Obstet. Gynecol.*, 44 (1974), 224–232.

2.21 In an article on quality control, Purcell (E30) gives the set of typical data shown in Table 2.21. Categorize each observation according to whether it falls above or below 1435, and test the pattern for randomness.

2.22 Littler et al. (E31) studied the blood flow in lung capillaries in 16 patients with scoliosis or neuromuscular weakness. They reported the sex of the patients in the following order:

F F F M F F M M M F F F F F F M

Test the null hypothesis that this sequence is random.

TABLE 2.21   **Typical data for life of incandescent lamps in hours, before establishment of control**

Sample*	Median	Sample*	Median
1	1100	17	1210
2	1280	18	1620
3	1460	19	1560
4	1350	20	730
5	1060	21	1260
6	1250	22	1560
7	1440	23	1770
8	1230	24	1160
9	1630	25	1300
10	2100	26	1500
11	1210	27	1270
12	1760	28	1560
13	2410	29	1150
14	2080	30	1940
15	1500	31	840
16	1550	32	1140
		Average	1435

Source: Warren B. Purcell, "Saving Time in Testing Life," *Indust. Quality Control*, 3 (March 1947), 15–18. Copyright 1947, American Society for Quality Control, Inc.; reprinted by permission.

* Each sample contains five lamps.

2.4

COX–STUART TEST FOR TREND

In many scientific endeavors, researchers want to determine whether a sequence of observations taken over time exhibits some type of *trend*. The most familiar example is economics. Economic trends are the subject of extensive study and evaluation. Elementary trend analysis in economics is discussed in most business statistics texts, such as the one by Daniel and Terrell (T59). There are two types of trend: *upward trend* and *downward trend*. A series of observations is said to exhibit an upward trend if the magnitudes of the later observations tend to be greater than those of the earlier observations. The data exhibit a downward trend if the earlier observations tend to be larger than the later observations.

Cox and Stuart (T60) have proposed an easily applied test for detecting a trend. This test, the *Cox–Stuart test for trend*, is a modification of the sign test. To use this test, we pair one of the early observations with one of the later observations. When the later observation exceeds the earlier observation, we replace the pair by a minus sign. When the earlier observation is greater than the later observation, we replace the pair by a plus sign. A preponderance of plus signs suggests a downward trend, and a preponderance of minus signs suggests an upward trend. If positive and negative signs occur in equal number, no trend is present. We can define this

procedure more formally as follows:

Assumptions

A. The data available for analysis consist of n' independent observations $X_1, X_2, \ldots, X_{n'}$, arranged in a certain order.

B. The measurement scale is at least ordinal.

Hypotheses

A. (Two-sided)
H_0: There is no trend present in the data
H_1: There is either an upward trend or a downward trend

B. (One-sided)
H_0: There is no upward trend
H_1: There is an upward trend

C. (One-sided)
H_0: There is no downward trend
H_1: There is a downward trend

Test Statistic

We first form the pairs

$$(X_1, X_{1+C}), (X_2, X_{2+C}), \ldots, (X_{n'-C}, X_{n'})$$

where $C = n'/2$ when n' is an even number, and $C = (n' + 1)/2$ when n' is an odd number. For example, if $n' = 6$, $X_1 = 2$, $X_2 = 4$, $X_3 = 6$, $X_4 = 8$, $X_5 = 10$, and $X_6 = 12$, then $C = 6/2 = 3$. The pairs in this case are $(2, 8)$, $(4, 10)$, and $(6, 12)$. If $n' = 7$, and we add the observation $X_7 = 14$ to the other six observations, then $C = (7 + 1)/2 = 4$. The pairs in this case are $(2, 10)$, $(4, 12)$, $(6, 14)$, and we note that the middle term is not used. This is always the case when n' is an odd number.

A plus sign replaces each pair (X_i, X_{i+C}) for which X_i is greater than X_{i+C}, and a minus sign replaces each pair for which X_{i+C} is greater than X_i. We omit pairs leading to zero differences from the analysis. The number of pairs yielding nonzero differences is equal to n.

The test statistic depends on the hypothesis being tested. The test statistic for hypothesis A is the number of plus signs or the number of minus signs, whichever is smaller. The test statistic for hypothesis B is the number of plus signs, and the test statistic for hypothesis C is the number of minus signs.

Decision Rule

The decision rules for the possible hypotheses are as follows.

A. For a given n, reject H_0 at the α level of significance if the probability, when H_0 is true, of observing as few (or fewer) of the less frequently occurring sign as were actually observed is less than or equal to $\alpha/2$.

B. For a given n, reject H_0 at the α level of significance if the probability, when H_0 is true, of observing as few (or fewer) plus signs as were actually observed is less than or equal to α.

C. For a given n, reject H_0 at the α level of significance if the probability, when H_0 is true, of observing as few (or fewer) minus signs as were actually observed is less than or equal to α.

You can obtain these probabilities from Table A.1 in the same manner as for the sign test.

Example 2.11   Table 2.22 shows the length of the growing season at Atlanta, Georgia, for the years 1899 to 1938, as reported by the U.S. Department of Agriculture (E32). We wish to know if these data provide sufficient evidence to indicate the presence of a trend in length of growing season. We presume that there is no reason to suspect one type of trend (either upward or downward) rather than another. Consequently we perform a two-sided test using the Cox–Stuart test for trend.

The data consist of $n' = 40$ observations, arranged in chronological order. Thus they meet the necessary measurement scale requirement for the test. We let $\alpha = 0.05$.

TABLE 2.22   **Length of growing season, in days, in Atlanta, 1899–1938**

Year	Length of growing season	Year	Length of growing season
1899	207	1919	227
1900	223	1920	213
1901	235	1921	213
1902	254	1922	261
1903	237	1923	222
1904	217	1924	237
1905	188	1925	239
1906	204	1926	216
1907	182	1927	260
1908	230	1928	246
1909	223	1929	256
1910	227	1930	242
1911	242	1931	266
1912	238	1932	242
1913	207	1933	249
1914	201	1934	228
1915	226	1935	255
1916	243	1936	226
1917	215	1937	209
1918	259	1938	247

Source: Climate and Man, Yearbook of Agriculture, 1941, U.S. Department of Agriculture, Washington, D.C.

Hypotheses

H_0: There is no indication of a trend in the length of the growing season
H_1: There is either an upward trend or a downward trend

Test Statistic

Since $n' = 40$ is even, we have 20 pairs of observations.

(207, 227)	(223, 213)	(235, 213)	(254, 261)	(237, 222)
(217, 237)	(188, 239)	(204, 216)	(182, 260)	(230, 246)
(223, 256)	(227, 242)	(242, 266)	(238, 242)	(207, 249)
(201, 228)	(226, 255)	(243, 226)	(215, 209)	(259, 247)

When we take the differences, we have 6 plus signs and 14 minus signs. The value of the test statistic then is 6, the number of plus signs.

Decision

We will reject H_0 if the probability of observing 6 or fewer plus signs when $n = 20$ and H_0 is true (i.e., there is no trend) is less than or equal to 0.025. Table A.1 shows that this probability is $P(K \leq 6 \mid 20, 0.50) = 0.0577$. Since 0.0577 is greater than 0.025, we cannot reject H_0. We conclude that this test does not indicate the presence of a trend. Since we have a two-sided test, the P value for this example is $2(0.0577) = 0.1154$.

Large-Sample Approximation   The large-sample approximation is the same as for the sign test.

Power-Efficiency   Cox and Stuart (T60) discuss the asymptotic relative efficiency of this test. They report an ARE of 0.79 when this test is compared to rank correlation tests, and an ARE of 0.78 when compared to the best parametric test.

FURTHER READING

More detail about nonparametric trend analysis is available in Bartholomew (T61), Cox (T62), Ferguson (T63), Mann (T64), Mansfield (T65), Olshen (T66), Rao (T67), Sen (T52), Stuart (T68), and Ury (T69). Hirsch et al. (T70) and van Belle and Hughes (T71) discuss the use of nonparametric techniques for trend in the analysis of water quality data.

EXERCISES

2.23 The Federal Crop Insurance Corporation's *Annual Report to Congress* for 1973 (E33) contains the information on cotton crop insurance shown in Table 2.23. Do these data

indicate a downward trend in the number of cotton crops insured? Let $\alpha = 0.05$. What is the P value?

TABLE 2.23   **Number of U.S. crops insured each year, 1948–1972**

Year	Crops insured	Year	Crops insured
1948	19,479	1961	15,375
1949	26,667	1962	21,312
1950	63,969	1963	26,526
1951	57,715	1964	24,865
1952	38,086	1965	21,152
1953	38,434	1966	23,458
1954	24,196	1967	25,774
1955	19,319	1968	32,646
1956	29,975	1969	31,786
1957	25,451	1970	24,821
1958	20,410	1971	19,593
1959	19,910	1972	14,960
1960	15,628		

Source: Annual Report to Congress, Federal Crop Insurance Corp., U.S. Department of Agriculture, 1973.

2.24 The 1972 edition of the *FAA Statistical Handbook of Aviation* (E34) gives the information on annual United States exports of aircraft, aircraft parts, and accessories shown in Table 2.24. Do these data reflect an upward trend in exports? Let $\alpha = 0.05$. What is the P value?

TABLE 2.24   **Number of U.S. aircraft exports, 1947–1971**

Year	Aircraft exports*	Year	Aircraft exports*
1947	3125	1960	2336
1948	2259	1961	2459
1949	881	1962	2131
1950	756	1963	2251
1951	894	1964	2577
1952	1180	1965	3129
1953	1377	1966	3611
1954	1053	1967	3881
1955	1714	1968	3682
1956	1711	1969	3322
1957	2025	1970	3383
1958	1689	1971	2904
1959	1628		

Source: FAA Statistical Handbook of Aviation, Department of Transportation, Federal Aviation Administration, for sale by the Superintendent of Documents, U.S. Government Printing Office, Washington, D.C., 1972.

* 1949–1954, civil only

2.5

COMPUTER PROGRAMS

The computations required by the procedures discussed in this chapter are easily and quickly performed when the sample sizes are small. In such situations the need for a computer is not acute. If sample sizes are large and for some reason a normal approximation is not employed, the speed, accuracy, and convenience of a computer are greatly appreciated when some of the procedures presented in this chapter must be used.

Krieg (T72) has written a FORTRAN 77 program for the runs test. The required input values are the number of items of one type, the number of items of the other type, and the number of runs. Output includes the exact and normal approximations of the runs probabilities and the runs distribution parameters.

Numerous microcomputer statistical software packages will accommodate one or more of the procedures presented in this chapter. The packages that will perform the sign test include *BMDPC, SCA, STATA,* and *STATISTIX.* The Wilcoxon one-sample test is available on *EXEC*U*STAT, MICROSTAT, MINITAB, NUMBER CRUNCHER STATISTICAL SYSTEM, SPSS/PC, STATGRAPHICS,* and *STATISTIX,* among others. The binomial test is available on *SCA.* Among the packages that perform the runs test are *M/STAT-2000, MIPSAQ, SCA, SPSS/PC,* and *STATPRO.*

REVIEW EXERCISES

2.25 Fifteen heroin addicts were asked to state the age at which they first started using the drug. The results are shown in Table 2.25. Can one conclude from these data that the median age of the sampled population is not 20? Use the sign test, and determine the P value.

TABLE 2.25 | **Age at which 15 heroin addicts first used the drug**

Subject	1	2	3	4	5	6	7	8	9	10	11	12	13	14	15
Age	22	24	37	28	15	14	22	16	18	17	23	16	20	18	15

2.26 Construct a 95% confidence interval for the median of the population from which the data in Exercise 2.25 were drawn.

2.27 In a sample of 20 working wives, 6 stated that their primary reason for working was to earn money for luxuries. Do these data provide sufficient evidence to indicate that the proportion of subjects in the sampled population who work for that reason is greater than 0.25? What is the P value for this test?

2.28 Use the data in Exercise 2.27 to construct a 95% confidence interval for the population proportion.

2.29 A test to measure knowledge of current events was given to a sample of 25 elementary school children from an inner-city neighborhood. The scores are shown in Table 2.26. Can one conclude from these data that the population median is less than 70? Use the Wilcoxon test and determine the P value.

TABLE 2.26   **Scores made on a current-events test by 25 elementary school students from an inner-city neighborhood**

80	68	30	67	70	62	69	65	53	29	65	68	62	56	46	48	39	72	36	69	40	61	54	53	25

2.30 Use the data of Exercise 2.29 to construct a 95% confidence interval for the population median.

2.31 In a psychology experiment, 24 previously unacquainted adults were asked to line up against a wall. The arrangement according to sex was as follows:

M F MMMM F MMM FFFF M FF MMM FFFF

Do these data provide sufficient evidence to indicate a lack of randomness?

2.32 In a survey of 200 residents of a certain metropolitan area, 110 stated that they favored capital punishment. Do these data provide sufficient evidence to indicate that the proportion of the population holding this opinion is greater than 0.50?

2.33 In an extrasensory perception (ESP) test, 25 subjects in one room attempted to identify a target picture in another room. There were 11 "hits." Do these data provide sufficient evidence to support ESP? What is the P value for this test?

2.34 The quality-control department of a shampoo manufacturer requires the mean weight of bottles of its product to be 12 fluid ounces. A sample of 20 consecutive bottles filled by the same machine is taken from the assembly line and measured. The results (in fluid ounces) were as follows: 12.9, 12.5, 12.2, 12.3, 11.5, 11.8, 11.7, 12.2, 12.4, 12.6, 12.5, 12.8, 11.8, 11.5, 11.6, 12.7, 12.6, 12.7, 12.8, 12.2. Do these data provide sufficient evidence to indicate a lack of randomness in the pattern of overfilling and underfilling? Let $\alpha = 0.05$.

2.35 Lapp (E35) reports data on estimated aggregate retirement flow for males between the ages of 55 and 64, inclusive, for the years 1948 through 1977, shown in Table 2.27. Test for the presence of an upward trend in these data. Let $\alpha = 0.05$.

2.36 A market analyst wants to know whether he can conclude, at the 0.05 level of significance, that the median annual family income in a certain geographic area is less than $35,000. Interviews with heads of household in a random sample of 20 families from the area gave the results shown in Table 2.28. The distribution of incomes in the population is believed to be asymmetric. What can the analyst conclude from the sample data?

2.37 In a simple random sample of 15 employees, 9 stated that they had been victims of sexual harassment. Can we conclude that in the population from which the sample was drawn, more than 50% of the employees have been sexually harassed? Let $\alpha = 0.05$.

TABLE 2.27   **Aggregate retirement flows, 1948–1977, males 55–64 years of age**

Year	Number (thousands)	Year	Number (thousands)
1948	38.7	1963	171.0
1949	179.4	1964	223.4
1950	124.4	1965	272.3
1951	86.9	1966	213.8
1952	89.3	1967	207.3
1953	136.2	1968	215.1
1954	34.9	1969	254.2
1955	171.2	1970	218.1
1956	91.1	1971	246.5
1957	163.6	1972	273.0
1958	103.1	1973	398.8
1959	165.5	1974	272.3
1960	167.5	1975	383.2
1961	111.1	1976	352.3
1962	238.6	1977	290.1

Source: John S. Lapp, "The Secular Behavior of Aggregate Retirement Flows," *Atlantic Econ. J.*, 14 (March 1986), 30–38. Reprinted by permission.

TABLE 2.28   **Annual family income of a random sample of 20 families (in dollars)**

28,900	29,300	30,100	38,000	30,300	32,200	27,500
29,900	31,200	35,300	37,200	43,000	32,500	35,100
34,900	34,300	36,200	33,900	35,000		

2.38 Refer to Exercise 2.37. Construct a 95% confidence interval for the population proportion.

2.39 An industrial psychologist believes that the median test score for manual dexterity of a population of assembly-line employees with a certain handicap is greater than 70. A simple random sample of employees drawn from the population yielded the following test scores: 72, 94, 91, 84, 80, 58, 46, 47, 49, 76, 86, 64, 86, 87, 93, 65, 48, 71, 85, 59. Choose the test that uses the most information contained in the data to determine whether the sample results provide sufficient evidence to support the psychologist's belief. Let $\alpha = 0.01$.

2.40 Refer to Exercise 2.39. Construct a 99% confidence interval for the population median.

2.41 Refer to Exercise 2.36. Construct a 95% confidence interval for the population median.

2.42 A large corporation conducted a sample survey of employed adults to determine the nature of these people's attitudes toward the corporation. In a sample of 5,000 people, 2,900 reported negative attitudes. Can one conclude from these data that more than 50% of the people in the population have a negative attitude toward the corporation? Let $\alpha = 0.05$. What is the P value for the test?

2.43 Refer to Exercise 2.42. Construct a 95% confidence interval for the proportion in the population who have a negative attitude toward the corporation.

2.44 Consult a recent issue of *Forbes* magazine containing data on "Ranking the Forbes 500s." (These annual special issues usually carry a date near the end of April.) Select a variable from those listed, such as assets, sales, or the like. Then select a simple random sample of 30 or more firms from those in the list. Formulate an appropriate set of hypotheses and use the large-sample approximation of the sign test to test the null hypothesis.

2.45 Use the data collected in Exercise 2.44 to construct a confidence interval for the population median using the large-sample approximation based on the sign test.

2.46 See Exercise 2.44. Use the large-sample approximation to the Wilcoxon test to test the null hypothesis.

2.47 Repeat Exercise 2.45 using the large-sample method based on the Wilcoxon test.

2.48 Use the *Forbes* data described in Exercise 2.44. For some variable such as assets, formulate a set of hypotheses regarding the proportion of firms with values of the chosen variable less than, greater than, or equal to some chosen value. Select a simple random sample of 30 or more firms and use the large-sample approximation for proportions to test the null hypothesis.

2.49 Use the data collected for Exercise 2.48 and the large-sample approximation for proportions to construct a confidence interval for the population proportion.

2.50 Consult a large table of random digits such as the Rand Corporation's *A Million Random Digits with 100,000 Normal Deviates* (Glencoe, Illinois: The Free Press, 1955). Select a random starting point in the table and record the next 50 or more successive digits. Let the odd digits be the items of one type and the even digits be the items of the other type. Count the number of runs by observing the sequences of odd and even digits. Formulate an appropriate set of hypotheses and use the large-sample approximation of the runs test to test the null hypothesis. Compare the results with those of classmates. Based on the sample, what can you conclude about the randomness of the digits in the table consulted?

2.51 Consult a publication that contains large sets of time-series data. Select a sample of 30 or more consecutive values of some variable and use the large-sample approximation to the Cox–Stuart test for trend to test an appropriate null hypothesis. (Suggestion: try the *Statistical Abstract of the United States*. Possible variables include annual deaths from motor vehicle accidents and annual volume of passenger traffic by intercity bus. It may be necessary to consult more than one volume to obtain the necessary sample size.)

REFERENCES

T1 Arbuthnott, J., "An Argument for Divine Providence Taken from the Constant Regularity Observed in the Births of Both Sexes," *Philosophical Transactions*, 27 (1710), 186–190.

T2 Dixon, W. J., and A. M. Mood, "The Statistical Sign Test," *J. Amer. Statist. Assoc.*, 41 (1946), 557–566.

T3 Noether, Gottfried E., "Sample Size Determination for Some Common Nonparametric Tests," *J. Amer. Statist. Assoc.*, 82 (1987), 645–647.

T4 Nelson, Lloyd S., "Simplified Critical Values for the Sign Test," *J. Quality Technology*, 16 (April 1984), 125–126.

T5   Mantel, Nathan, and Alton J. Rahe, "Differentiated Sign Tests," *Int. Statist. Rev.*, 48 (1980), 19–28.

T6   Nelson, Lloyd S., "A Sign Test for Correlation," *J. Quality Technology*, 15 (October 1983), 199–200.

T7   Mackinnon, William J., "Table for Both the Sign Test and Distribution-Free Confidence Intervals of the Median for Sample Sizes to 1,000," *J. Amer. Statist. Assoc.*, 59 (1964), 935–956.

T8   Walsh, J. E., "On the Power Function of the Sign Test for Slippage of Means," *Ann. Math. Statist.*, 17 (1946), 358–362.

T9   Dixon, W. J., "Power Functions of the Sign Test and Power Efficiency for Normal Alternatives," *Ann. Math. Statist.*, 24 (1953), 467–473.

T10   Cochran, W. G., "The Efficiencies of the Binomial Series Tests of Significance of a Mean and of a Correlation Coefficient," *J. Roy. Statist. Soc.*, 100 (1937), 69–73.

T11   Gibbons, Jean Dickinson, *Nonparametric Statistical Inference*, New York: McGraw-Hill, 1971.

T12   Hodges, J. L., Jr., and E. L. Lehman, "The Efficiency of Some Nonparametric Competitors of the *t*-test," *Ann. Math. Statist.*, 27 (1956), 324–335.

T13   MacStewart, W., "A Note on the Power of the Sign Test," *Ann. Math. Statist.*, 12 (1941), 236–239.

T14   Wilcoxon, Frank, "Individual Comparisons by Ranking Methods," *Biometrics*, 1 (1945), 80–83.

T15   McCornack, R. L., "Extended Tables of the Wilcoxon Matched Pairs Signed Rank Statistics," *J. Amer. Statist. Assoc.*, 60 (1965), 864–871.

T16   Wilcoxon, Frank, "Probability Tables for Individual Comparisons by Ranking Methods," *Biometrics*, 3 (1947), 119–122.

T17   Wilcoxon, Frank, *Some Rapid Approximate Statistical Procedures*, New York: American Cyanamid, 1949.

T18   Wilcoxon, F., S. Katti, and R. A. Wilcox, *Critical Values and Probability Levels for the Wilcoxon Rank Sum Test and the Wilcoxon Signed Rank Test*, in H. L. Harter and D. B. Owen (eds.), *Selected Tables in Mathematical Statistics*, Volume 1, Chicago: Markham, 1970.

T19   Pratt, J. W., "Remarks on Zeroes and Ties in the Wilcoxon Signed Rank Procedures," *J. Amer. Statist. Assoc.*, 54 (1959), 655–667.

T20   Cureton, E. E., "The Normal Approximation to the Signed-Rank Sampling Distribution When Zero Differences Are Present," *J. Amer. Statist. Assoc.*, 62 (1967), 1068–1069.

T21   Buck, W., "Signed-Rank Tests in the Presence of Ties," *Biometrical J.*, 21 (1979), 501–526.

T22   Conover, W. J., "Methods of Handling Ties in the Wilcoxon Signed-Rank Test," *J. Amer. Statist. Assoc.*, 68 (1973), 985–988.

T23   Klotz, J., "The Wilcoxon, Ties, and the Computer," *J. Amer. Statist. Assoc.*, 61 (1966), 772–787.

T24   Putter, Joseph, "The Treatment of Ties in Some Nonparametric Tests," *Ann. Math. Statist.*, 26 (1955), 368–386.

T25   Larsen, S. Olesen, "APL Programs for *p*-Values in Wilcoxon's One-Sample and Two-Sample Tests," *Computer Prog. in Biomed.*, 12 (1980), 42–44.

T26   Arnold, H. J., "Small Sample Power for the One-Sample Wilcoxon Test for Non-Normal Shift Alternatives," *Ann. Math. Statist.*, 36 (1965), 1767–1778.

T27   Klotz, J., "Small Sample Power and Efficiency for the One-Sample Wilcoxon and Normal Scores Test," *Ann. Math. Statist.*, 34 (1963), 624–632.

T28 Klotz, J., "Alternative Efficiencies for Signed Rank Tests," *Ann. Math. Statist.*, 36 (1965), 1759–1766.

T29 Mood, A. M., "On the Asymptotic Efficiency of Certain Nonparametric Two-Sample Tests," *Ann. Math. Statist.*, 25 (1954), 514–522.

T30 Noether, Gottfried E., *Elements of Nonparametric Statistics*, New York: Wiley, 1967.

T31 Thompson, W. R., "On Confidence Ranges for the Median and Other Expectation Distributions for Populations of Unknown Distribution Form," *Ann. Math. Statist.*, 7 (1936), 122–128.

T32 Savur, S. R., "The Use of the Median in Tests of Significance," *Proc. Indian Acad. Sci.*, A5 (1937), 564–576.

T33 David, H. A., *Order Statistics*, New York: Wiley, 1970.

T34 Hollander, M., and D. A. Wolfe, *Nonparametric Statistical Methods*, New York: Wiley, 1973.

T35 Efron, Bradley, "Better Bootstrap Confidence Intervals," *J. Amer. Statist. Assoc.*, 82 (1987), 171–200.

T36 Hettmansperger, Thomas P., and Simon J. Sheather, "Confidence Intervals Based on Interpolated Order Statistics," *Statist. & Probabil. Letters*, 4 (1986), 75–79.

T37 Tukey, J. W., "The Simplest Signed-Rank Test," *Mimeographed Report Number 17*, Statistical Research Group, Princeton University, 1949. Cited in Conover, W. J., *Practical Nonparametric Statistics*, New York: Wiley, 1971.

T38 Moses, L. E., "Query: Confidence Limits from Rank Tests," *Technometrics*, 7 (1965), 257–260.

T39 Noether, G. E., "Wilcoxon Confidence Intervals for Location Parameters in the Discrete Case," *J. Amer. Statist. Assoc.*, 62 (1967), 184–188.

T40 Crow, E. L., "Confidence Intervals for a Proportion," *Biometrika*, 43 (1956), 423–435.

T41 Sterne, T. E., "Some Remarks on Confidence or Fiducial Limits," *Biometrika*, 41 (1954), 275–278.

T42 Clopper, C. J., and E. S. Pearson, "The Use of Confidence or Fiducial Limits Illustrated in the Case of the Binomial," *Biometrika*, 26 (1934), 404–413.

T43 Anderson, T. W., and H. Burstein, "Approximating the Upper Binomial Confidence Limit," *J. Amer. Statist. Assoc.*, 62 (1967), 857–861.

T44 Anderson, T. W., and H. Burstein, "Approximating the Lower Binomial Confidence Limit," *J. Amer. Statist. Assoc.*, 63 (1968), 1413–1415. Correction, 64 (1969), 669.

T45 Quesenberry, C. P., and D. C. Hurst, "Large Sample Simultaneous Confidence Intervals for Multinomial Proportions," *Technometrics*, 6 (1964), 191–195.

T46 Goodman, Leo A., "On Simultaneous Confidence Intervals for Multinomial Proportions," *Technometrics*, 7 (1965), 247–254.

T47 Swed, F. S., and C. Eisenhart, "Tables for Testing Randomness of Grouping in a Sequence of Alternatives," *Ann. Math. Statist.*, 14 (1943), 66–87.

T48 Mosteller, F., "Note on an Application of Runs to Quality Control Charts," *Ann. Math. Statist.*, 12 (1941), 228–232.

T49 Prairie, R. R., W. J. Zimmer, and J. K. Brookhouse, "Some Acceptance Sampling Plans Based on the Theory of Runs," *Technometrics*, 4 (1962), 177–185.

T50 Goodman, L. A., and Y. Grunfield, "Some Nonparametric Tests for Comovements between Time Series," *J. Amer. Statist. Assoc.*, 56 (1961), 11–26.

T51 Moore, G. H., and W. A. Wallis, "Time Series Significance Test Based on Signs of Differences," *J. Amer. Statist. Assoc.*, 38 (1943), 153–164.

T52 Sen, P. K., "Some Nonparametric Tests for m-Dependent Time Series," *J. Amer. Statist. Assoc.*, 60 (1965), 134–147.

T53 Wallis, W. A., and G. H. Moore, "A Significance Test for Time Series," *J. Amer. Statist. Assoc.*, 36 (1941), 401–409.

T54 Mood, A. M., "The Distribution Theory of Runs," *Ann. Math. Statist.*, 11 (1940), 367–392.

T55 Barton, D. E., "Query: Completed Runs of Length *k* above and below Median," *Technometrics*, 9 (1967), 682–694.

T56 Koppen, M. G. M., and N. D. Verhelst, "The Exact Runs Test and Some Large Sample Approximations," *Br. J. Math. & Statist. Psychol.*, 39 (1986), 168–182.

T57 Grafton, R. G. T., "The Runs-Up and Runs-Down Tests," *Applied Statist.*, 30 (1981), 81–85.

T58 Knuth, Donald E., *The Art of Computer Programming*, volume 2, Reading, Mass.: Addison-Wesley, 1969.

T59 Daniel, Wayne W., and James C. Terrell, *Business Statistics For Management and Economics*, fifth edition, Boston: Houghton Mifflin, 1989.

T60 Cox, D. R., and A. Stuart, "Some Quick Tests for Trend in Location and Dispersion," *Biometrika*, 42 (1955), 80–95.

T61 Bartholomew, D., "Tests for Randomness in a Series of Events Where the Alternative Is a Trend," *J. Roy. Statist. Soc. Ser. B*, 18 (1956), 234–239.

T62 Cox, D., "Some Statistical Methods Connected with Series of Events," *J. Roy. Statist. Soc. Ser. B*, 17 (1955), 129–164.

T63 Ferguson, George Andrew, *Nonparametric Trend Analysis*, Montreal: McGill University Press, 1965.

T64 Mann, H. B., "Nonparametric Tests against Trend," *Econometrika*, 13 (1945), 245–259.

T65 Mansfield, E., "Power Functions for Cox's Test of Randomness against Trend," *Technometrics*, 4 (1962), 430–432.

T66 Olshen, R. A., "Sign and Wilcoxon Tests for Linearity," *Ann. Math. Statist.*, 38 (1967), 1759–1769.

T67 Rao, T. S., "A Note on the Asymptotic Relative Efficiencies of Cox and Stuart's Tests for Testing Trend in Dispersion of a *p*-Dependent Time Series," *Biometrika*, 55 (1968), 381–386.

T68 Stuart, A., "The Efficiencies of Tests of Randomness against Normal Regression," *J. Amer. Statist. Assoc.*, 51 (1956), 285–287.

T69 Ury, H. K., "Large-Sample Sign Tests for Trend in Dispersion," *Biometrika*, 53 (1966), 289–291.

T70 Hirsch, Robert M., James R. Slack, and Richard A. Smith, "Techniques of Trend Analysis for Monthly Water Quality Data," *Water Resources Res.*, 18 (February 1982), 107–121.

T71 van Belle, Gerald, and James P. Hughes, "Nonparametric Tests for Trend in Water Quality," *Water Resources Res.*, 20 (January 1984), 127–136.

T72 Krieg, Edward F., Jr., "A FORTRAN 77 Program for the Runs Test," *Behav. Res. Methods, Instruments, & Computers*, 20 (1988), 361.

E1 Liedtke, A. James, Harvey G. Kemp, David M. Borkenhagen, and Richard Gorlin, "Myocardial Transit Times from Intra-coronary Dye—Dilution Curves in Normal Subjects and Patients with Coronary Artery Disease," *Amer. J. Cardiol.*, 32 (1973), 831–839.

E2 Lenzer, Irmingard I., and Carol A. White, "Satiation Effects in Continuous Reinforcement and Successive Sensory Discrimination Situations," *Physiolog. Psychol.*, 1 (1973), 77–82.

E3 Iwamoto, Mitsuo, "Morphological Studies of *Macaca Fuscata*: VI, Somatometry," *Primates*, 12 (1971), 151–174.

E4 *Drug Abuse in Suburbia*, Mineola, New York: Nassau County Probation Department, 1970.

E5   Malina, Robert M., "Comparison of the Increase in Body Size between 1899 and 1970 in a Specially Selected Group with That in the General Population," *Amer. J. Phys. Anthropol.*, 37 (1972), 135–142.

E6   Moore, Ruth C., and Earl J. Ogletree, "A Comparison of the Readiness and Intelligence of First Grade Children with and without a Full Year of Head Start Training," *Education*, 93 (1973), 266–270.

E7   Palmer, Alden, "The Alden Palmer Letter," *The Fraternal Monitor*, 70 (February 1960), 18–19.

E8   Agnew, Harman W., Jr., Wilse W. Webb, and Robert L. Williams, "Sleep Patterns in Late Middle Age Males: An EEG Study," *Electroencephalog. Clin. Neurophysiol.*, 23 (1967), 168–171.

E9   Armstrong, R. W., "Tracing Exposure to Specific Environments in Medical Geography," *Geog. Anal.*, 5 (1973), 122–132.

E10 Abu-Ayyash, A. Y., "The Mobile Home: A Neglected Phenomenon in Geographic Research," *Geog. Bull.*, 5 (November 1972), 28–30.

E11 *Changes in Alabama Agriculture and Related Data 1950 to 1960*, Agricultural Experiment Station of Auburn University, Auburn, Alabama, 1961.

E12 Cartright, Lillian Kaufman, "Personality Differences in Male and Female Medical Students," *Psychiatry in Med.*, 3 (1972), 213–218.

E13 Hall, F. F., C. R. Ratliff, T. Hayakawa, T. W. Culp, and N. C. Hightower, "Substrate Differentiation of Human Pancreatic and Salivary Alpha-Amylases," *Amer. J. Dig. Dis.*, 15 (1970), 1031–1038.

E14 Golde, David W., Belina Rothman, and Martin J. Cline, "Production of Colony-Stimulating Factor by Malignant Leukocytes," *Blood*, 43 (1974), 749–756.

E15 Silverman, M., E. Zeidifard, J. W. Paterson, and S. Godfrey, "The Effect of Isoprenaline on the Cardiac and Respiratory Responses to Exercise," *Quart. J. Exper. Physiol.*, 58 (1973), 7–17.

E16 Cinotti, Alfonse A., and Joseph C. Patti, "Lens Abnormalities in an Aging Population of Nonglaucomatous Patients," *Amer. J. Ophthal.*, 65 (1968), 25–32.

E17 Georgia Board of Corrections, *Annual Report July 1, 1969–June 30, 1970*, Atlanta.

E18 Bowman, P., J. C. Wilt, and H. Sayed, "Chronicity and Recurrence of Psittacosis," *Canad. J. Public Health*, 64 (1973), 167–173.

E19 Moles, Oliver C., "The Relationship of Family Circumstances and Personal History to Use of Public Assistance," *Social Work*, 16 (April 1971), 37–46.

E20 Chin, William, David M. Bear, Edward J. Colwell, and Sanong Kosakal, "A Comparative Evaluation of Sulfalene–Trimethoprim and Sulphormethoxine–Pyrimethamine against Falciparum Malaria in Thailand," *Amer. J. Trop. Med. Hyg.*, 22 (1973), 308–312.

E21 Zackler, Zack, Olga Brolnitsky, and Hyman Orbach, "Preliminary Report on a Mass Program for Detection of Gonorrhea," *Public Health Reports*, 85 (1970), 681–684.

E22 Spencer, J., "Hiatus Hernia and OEsophageal Reflux," *Proc. Roy. Soc. Med.*, 65 (1972), 30–32.

E23 Bernstein, L. M., and L. A. Baker, "A Clinical Test for Esophagitis," *Gastroenterology*, 34 (1958), 760–781.

E24 Sack, Richard, "The Impact of Education of Individual Modernity in Tunisia," *Int. J. Compar. Sociol.*, 14 (1973), 245–272.

E25 May, David, "Illegitimacy and Juvenile Court Involvement," *Int. J. Criminol. Penol.*, 1 (1973), 227–252.

E26 Westermeyer, Joseph, "Grenade-Amok in Laos: A Psycho-Social Perspective," *Int. J. Soc. Psychiatry*, 19 (1973), 251–260.

E27 Strimbu, Jerry L., Lyle F. Schoenfeldt, and O. Suthern Sims, Jr., "Sex Differences in College Student Drug Use," *J. College Student Personnel*, 14 (1973), 507–510.

E28 *Local Climatological Data*, U.S. Department of Commerce, National Oceanic and Atmospheric Administration, Environmental Data Service, National Climatic Center, Federal Building, Asheville, North Carolina, November 1974.

E29 Murata, Yuji, and Chester B. Martin, Jr., "Systolic Time Intervals of the Fetal Cardiac Cycle," *Obstet. Gynecol.*, 44 (1974), 224–232.

E30 Purcell, Warren B., "Saving Time in Testing Life," *Indust. Quality Control*, 3 (March 1947), 15–18.

E31 Littler, W. A., S. R. Reuben, and D. J. Lane, "Lung Blood Flow Studies in Patients with Scoliosis and Neuromuscular Weakness," *Thorax*, 28 (1973), 209–213.

E32 *Climate and Man, Yearbook of Agriculture*, 1941, United States Department of Agriculture, Washington, D. C.

E33 *Annual Report to Congress*, Federal Crop Insurance Corp., U.S. Department of Agriculture, 1973.

E34 *FAA Statistical Handbook of Aviation*, Department of Transportation, Federal Aviation Administration, for sale by the Superintendent of Documents, U.S. Government Printing Office, Washington, D.C., 1972.

E35 Lapp, John S., "The Secular Behavior of Aggregate Retirement Flows," *Atlantic Econ. J.*, 14 (March 1986), 30–38.

PROCEDURES THAT UTILIZE DATA FROM TWO INDEPENDENT SAMPLES

Chapter 2 covered several procedures that are appropriate for making inferences about a population based on the data of a single sample drawn from that population. This chapter focuses on two populations. The inferential procedures examined here are based on the data generated by two independent samples, one from each of the two populations of interest. The samples must be independent in two respects. First, the elements we select for the first sample must in no way depend on which elements we select for the second sample. Second, within each sample, each element must be independent of every other element in that sample. In other words, there is independence within samples as well as between samples.

The inferential objective is either to estimate the difference between certain parameters of the two populations or to test hypotheses about the two populations. For example, we may wish to test the null hypothesis that the numerical value of some parameter is the same in both populations. Readers who have had a course in elementary statistics will be familiar with parametric procedures for making inferences on the basis of results obtained from two independent samples. For example, we use the t test to test the null hypothesis that two population means are equal and the F test to test the null hypothesis that two population variances are equal under certain assumptions about the data. When the data do not meet some basic assumptions, an alternative is to employ a nonparametric procedure. The

purpose of this chapter is to introduce some nonparametric procedures that are appropriate for two independent samples.

The populations of interest may be either of two types: real or hypothetical. There may, in fact, exist two well-defined (real) populations from which we select independent random samples so that we can draw an inference about them. Population 1, for example, may consist of households in a certain community of some urban area, while population 2 may consist of households in a nearby suburban area. The variable of interest may be annual family income, and a researcher may wish to test the null hypothesis that the median annual family income is the same in both populations.

The two populations, on the other hand, may be hypothetical. For example, subjects available for study may be randomly assigned to receive one of two treatments. (The term treatment is used in its broadest sense, to designate some procedure whose effect we measure.) A treatment may even be no treatment or the application of a placebo when we want one group to serve as a control. The hypothetical populations then would be all subjects similar to those observed who could at some time receive one of the treatments.

This chapter covers procedures for estimating and testing hypotheses about the difference between two location parameters, and procedures for testing hypotheses about the equality of two dispersion parameters. In addition three other procedures are considered: (a) a general two-sample test, (b) the Hollander test for extreme reactions, and (c) the Fisher exact probability test.

Later chapters present other procedures appropriate for analyzing the data of two independent samples. Chapter 5 presents the chi-square tests for independence and homogeneity, and Chapter 8 includes discussion of some goodness-of-fit procedures.

3.1

MAKING INFERENCES ABOUT THE DIFFERENCE BETWEEN TWO LOCATION PARAMETERS

This section covers some procedures that are appropriate for making inferences about the difference between two location parameters: procedures for testing the null hypothesis that two location parameters are equal and a procedure for constructing a confidence interval for the difference between two location parameters.

The procedures are the nonparametric analogues of the familiar parametric t test. Recall that for the t test in the two-sample case to be valid, the two sampled populations must be at least approximately normally distributed and have equal variances, and that measurement must be on at least the interval scale. When the sampled populations fail to meet one or more of these assumptions, the researcher may select one of the procedures discussed in this section as an alternative method of analysis.

MEDIAN TEST

One of the simplest and most widely used procedures for testing the null hypothesis that two independent samples have been drawn from populations with equal medians is the *median test*, usually attributed to Mood (T1) and Westenberg (T2).

Assumptions

A. The data consist of two independent random samples: $X_1, X_2,$ $X_3, \ldots, X_{n_1}$, and $Y_1, Y_2, Y_3, \ldots, Y_{n_2}$. The first sample is from a population with unknown median, M_X, and the second is from a population with unknown median, M_Y.

B. The measurement scale employed is at least ordinal.

C. The variable of interest is continuous.

D. The two populations have the same shape.

E. If the two populations have the same median, then for each population the probability p is the same that an observed value will exceed the grand median.

Hypotheses

$$H_0: \ M_X = M_Y$$
$$H_1: \ M_X \neq M_Y$$

The median test may also be used with one-sided alternatives, but since the procedure is somewhat complicated, this use of the median test will not be discussed here.

Test Statistic

Before looking at the formula for the test statistic, let us first consider the rationale underlying the procedure. If the two populations have the same median, we would expect about half the observations in each of the two samples to be above the common median and about half to be below. Under the hypothesis that the two population medians are equal, we may estimate this common parameter by computing the median of the sample values in the two samples combined. In other words, we combine the observations from the two samples and compute the median of the $n_1 + n_2$ observations.

Then we may classify each sample observation according to two criteria: (a) whether it belongs to sample 1 or sample 2 and (b) whether it is above or below the computed sample median. The number of observations falling into each of the four resulting categories may be displayed in a *contingency table*, such as Table 3.1. In Table 3.1, A is the number of observations from sample 1 falling above the median, B is the number of observations from sample 2 falling above the sample median, and so on.

If H_0 is true, we expect about half the observations in each sample to fall above the

TABLE 3.1

Data display for median test

Relationship to sample median	Sample		
	1	2	Total
Above	*A*	*B*	*A* + *B*
Below	*C*	*D*	*C* + *D*
Total	*A* + *C* = n_1	*B* + *D* = n_2	*N* = n_1 + n_2

combined sample median and about half to fall below. In other words, if H_0 is true, we expect A and C each to be approximately equal to $n_1/2$, and B and D each to be approximately equal to $n_2/2$. The median test allows us to conclude, on the basis of sample data, whether it is likely that the null hypothesis is false. If the observed proportions above and below the sample median differ sufficiently from what we expect under H_0, we reject H_0. The median test enables us to decide whether the magnitude of this discrepancy is great enough for us to reject H_0.

To carry out this test, we need the sampling distribution of A and B under the null hypothesis. Mood (T1) has shown that the desired distribution may be expressed as the *hypergeometric distribution*:

$$P(A, B) = \frac{\binom{n_1}{A}\binom{n_2}{B}}{\binom{N}{A + B}} \qquad (3.1)$$

To test H_0, we may either evaluate Equation 3.1 or use tables of the hypergeometric distribution such as the one given by Owen (T3). Both these methods are rather tedious, however, unless the sample sizes are quite small. If the sample sizes are even moderately large, we prefer to use an alternative testing procedure.

The reader may recall from a previous course in statistics that under certain conditions we can approximate the hypergeometric distribution by the binomial distribution, which in turn we may approximate by the normal distribution. [See Ostle and Mensing (T4), for example.] But we have to consider the conditions under which these approximations are valid. Burr (T5) states that for the binomial approximation to the hypergeometric distribution to be useful, the population should be eight to ten times as large as the sample. The normal approximation to the binomial is best when the sample size is large and when the proportion of items in the population that have the characteristic of interest is near 0.5. A rule of thumb frequently given in elementary statistics books [see Dunn (T6), for example] is that the normal approximation to the binomial is satisfactory if np and $n(1 - p)$ are both larger than 5, where n is the sample size and p is the population proportion with the characteristic of interest.

When the sample satisfies the conditions for the normal approximation, the test statistic is

$$T = \frac{(A/n_1) - (B/n_2)}{\sqrt{\hat{p}(1 - \hat{p})(1/n_1 + 1/n_2)}} \tag{3.2}$$

$$\text{where} \quad \hat{p} = (A + B)/N \tag{3.3}$$

Decision Rule

If H_0 is true and if the sample is the required size, then T is distributed approximately as the standard normal distribution. For a given significance level α, critical values of T correspond to values of z obtained from Table A.2 in such a way that $\alpha/2$ is to the right of z and $\alpha/2$ is to the left of $-z$. If T is equal to or exceeds z or is smaller than or equal to $-z$, we reject H_0. Otherwise we do not.

Example 3.1   Russell et al. (E1) reported the stroke-index values shown in Table 3.2 for patients admitted to the myocardial-infarction research unit of a university hospital. We wish to know whether these data provide sufficient evidence to indicate that the medians of the two populations represented by the sample data are different. Let $\alpha = 0.05$.

Hypotheses

$$H_0: M_X = M_Y$$
$$H_1: M_X \neq M_Y$$

TABLE 3.2   **Stroke-index values, milliliters, for patients admitted to the myocardial-infarction research unit of a university hospital**

Diagnosis

Anterior transmural infarction and anterior necrosis (X)				Inferior transmural infarction and inferior necrosis (Y)	
25	13	9	46	31	43
25	30	17	20	21	42
17	20	37	25	38	30
26	23	20	17	19	20
18	26	11	36	38	29
30	12	32	54	41	13
24	20	16	8	68	32
21	37	31	26	28	30

Source: Richard O. Russell, Jr., David Hunt, and Charles E. Rackley, "Left Ventricular Hemodynamics in Anterior and Inferior Myocardial Infarction," *Amer. J. Cardiol.*, 32 (1973), 8–16.

Test Statistic

The median computed from the 48 sample values is $(25 + 26)/2 = 25.5$. Table 3.3 shows the number of observations in each sample falling above and below 25.5.

We use Equation 3.3 and the data in Table 3.3 to compute $\hat{p} = (12 + 12)/48 = 0.50$. By Equation 3.2 we compute

$$T = \frac{(12/32) - (12/16)}{\sqrt{(0.50)(1 - 0.50)(1/32 + 1/16)}} = -2.45$$

TABLE 3.3

Contingency table for Example 3.1

Relationship to 25.5	Anterior transmural infarction and anterior necrosis	Inferior transmural infarction and inferior necrosis	Total
Above	12	12	24
Below	20	4	24
Total	32	16	48

Decision

For $\alpha = 0.05$, the critical values of the test statistic from the standard normal distribution are ± 1.96. Since -2.45 is less than -1.96, we reject H_0 and conclude that the two population medians are not equal. The P value is $2(0.5 - 0.4929) = 0.0142$.

When the sample sizes are large, we can also test H_0 by using the chi-square test of homogeneity (see Chapter 5).

Ties Although the assumption of continuity underlies the median test, ties do occur in practice; that is, one or more sample observations may be exactly equal to the computed sample median. In fact, if the total number of sample observations $(n_1 + n_2 = N)$ is odd, at least one sample observation will be equal to the combined sample median. We can handle ties in one of three ways: (a) If $n_1 + n_2 = N$ is large, and if only a few observations are equal to the median, we may discard them before computing the test statistic. (b) We may categorize the observations in each sample as either falling above the sample median or not falling above it, so that tied observations fall in the latter category. (c) We may count observations that are equal to the sample median in the above and below categories in all possible ways. We select as the true result the contingency table that is least likely to cause us to reject H_0.

Power-Efficiency Mood (T7) has shown that the asymptotic efficiency of the median test for independent samples drawn from normally distributed populations is $2/\pi \approx 64\%$.

FURTHER READING

Bedall and Zimmerman (T8) have proposed a bivariate median test that makes use of the geometric median or median center. Fligner and Rust (T9) describe a modification of the median test that allows one to test for differences between two medians without making any assumptions about the shapes of the underlying population distributions. Brown and Mood (T10) and Westenberg (T11) also discuss the median test.

EXERCISES

3.1   Taylor (E2) collected the data shown in Table 3.4 on 116 subjects who dropped out of the Job Corps. Can we conclude from these data that the reading-test scores of subjects remaining in the Job Corps less than three months differ from the scores of those remaining three months or longer? Let $\alpha = 0.05$. Compute the P value.

TABLE 3.4   **Data on 116 subjects who dropped out of the Job Corps**

Subject	Length of stay, months	Reading score	Subject	Length of stay, months	Reading score
1	0	19	29	1	15
2	0	8	30	1	20
3	1	5	31	0	17
4	2	20	32	1	22
5	1	20	33	0	10
6	0	10	34	1	23
7	0	10	35	1	4
8	1	13	36	2	5
9	1	10	37	1	20
10	2	12	38	0	17
11	1	13	39	1	21
12	2	23	40	1	16
13	1	4	41	1	24
14	0	16	42	2	12
15	0	21	43	0	15
16	1	11	44	2	11
17	1	17	45	2	2
18	0	5	46	1	22
19	1	6	47	1	24
20	1	6	48	1	9
21	0	21	49	1	1
22	0	18	50	2	5
23	1	8	51	0	5
24	0	26	52	1	20
25	0	9	53	0	22
26	1	14	54	2	0
27	2	6	55	1	20
28	1	12	56	1	0

TABLE 3.4   (*Continued*)

Subject	Length of stay, months	Reading score	Subject	Length of stay, months	Reading score
57	1	9	87	5	4
58	2	8	88	6	10
59	0	6	89	10	17
60	1	20	90	7	10
61	0	9	91	3	16
62	2	25	92	6	17
63	1	15	93	26	19
64	0	17	94	3	16
65	0	0	95	3	24
66	0	19	96	4	20
67	1	12	97	6	11
68	0	19	98	3	6
69	2	7	99	3	7
70	1	14	100	6	6
71	0	15	101	11	0
72	0	11	102	11	13
73	1	2	103	14	19
74	7	22	104	8	14
75	5	6	105	4	16
76	4	23	106	12	10
77	10	19	107	24	8
78	7	18	108	19	23
79	7	18	109	5	20
80	3	16	110	6	18
81	3	8	111	13	6
82	7	14	112	11	13
83	7	11	113	17	15
84	3	17	114	22	10
85	6	22	115	12	19
86	4	11	116	5	20

Source: William H. Taylor, ''Correlations between Length of Stay in the Job Corps and Reading Ability of the Corpsmen,'' *J. Employment Counseling*, 9 (1972), 78–85. Copyright 1972, American Association of Counseling and Development; reprinted with permission.

3.2 The quality control manager with a drug manufacturer wishes to know whether two methods of producing a particular tablet result in a difference between the median thicknesses. A random sample of tablets is drawn from batches produced by the two methods. Table 3.5 shows the results, which have been coded for computational convenience. Do these data provide sufficient evidence to indicate that the two population medians are different? Let $\alpha = 0.05$.

TABLE 3.5   **Data for Exercise 3.2**

Method	Thickness									
A	51	42	45	48	52	44	58	41	52	44
	45	52	61	60	41					
B	40	47	36	39	37	46	43	55	53	56

MANN–WHITNEY TEST

Another procedure for testing the null hypothesis of equal population location parameters was proposed by Mann and Whitney (T12). Although Festinger (T13), White (T14), and Wilcoxon (T15) have proposed equivalent procedures, the test is usually referred to as the *Mann–Whitney test*. The test is sometimes also referred to as the Mann–Whitney–Wilcoxon test. Wilcoxon (T15) considered only the case of equal sample sizes and used a rank sum as the test statistic. Mann and Whitney (T12), who seem to have been the first to treat the case of unequal sample sizes, point out the relationship between their test statistic, as given below, and that of Wilcoxon.

Assumptions

    A. The data consist of a random sample of observations $X_1, X_2, \ldots, X_{n_1}$ from population 1 with unknown median M_X, and another random sample of observations $Y_1, Y_2, \ldots, Y_{n_2}$ from population 2 with unknown median M_Y.

    B. The two samples are independent.

    C. The variable observed is a continuous random variable.

    D. The measurement scale employed is at least ordinal.

    E. The distribution functions of the two populations differ only with respect to location, if they differ at all.

Hypotheses

These hypotheses are appropriate only when assumption E is met. Either one of the following null hypotheses may be tested against the corresponding alternative.

    A. (Two-sided)
$$H_0: M_X = M_Y$$
$$H_1: M_X \neq M_Y$$

    B. (One-sided)
$$H_0: M_X \geq M_Y$$
$$H_1: M_X < M_Y$$

    C. (One-sided)
$$H_0: M_X \leq M_Y$$
$$H_1: M_X > M_Y$$

Test Statistic

To compute the observed value of the test statistic, we combine the two samples and rank all sample observations from smallest to largest. We assign tied observations the mean of the rank positions they would have occupied had there been no ties. We then sum the ranks of the observations from population 1 (that is, the X's). If the location parameter of population 1 is smaller than the location parameter of

population 2, we expect (for equal sample sizes) the sum of the ranks of the sample observations from population 1 to be smaller than the sum of the ranks of the sample observations from population 2. Similarly, if the location parameter of population 1 is larger than the location parameter of population 2, we expect just the reverse to be true. The test statistic is based on this rationale in such a way that, depending on the null hypothesis, either a sufficiently small or a sufficiently large sum of ranks assigned to sample observations from population 1 causes us to reject the null hypothesis.

The test statistic is

$$T = S - \frac{n_1(n_1 + 1)}{2} \qquad (3.4)$$

where S is the sum of the ranks assigned to the sample observations from population 1.

Decision Rule

The choice of a decision rule depends on the null hypothesis. The possible choices for an α level of significance are as follows.

A. When we test H_0 of A, we reject H_0 for either a sufficiently small or a sufficiently large value of T. Therefore we reject H_0 if the computed value of T is less than $w_{\alpha/2}$ or greater than $w_{1-\alpha/2}$, where $w_{\alpha/2}$ is the critical value of T given in Table A.7, and $w_{1-\alpha/2}$ is given by

$$w_{1-\alpha/2} = n_1 n_2 - w_{\alpha/2} \qquad (3.5)$$

B. When we test H_0 of B, we reject it for sufficiently small values of T. Reject H_0 if the computed T is less than w_α, the critical value of T obtained by entering Table A.7 with n_1, n_2, and α.

C. When we test H_0 of C, we reject H_0 for sufficiently large values of T. Therefore we reject H_0 if the computed T is greater than $w_{1-\alpha}$, where

$$w_{1-\alpha} = n_1 n_2 - w_\alpha \qquad (3.6)$$

Example 3.2   Newmark et al. (E3) have reported the results of an attempt to assess the predictive validity of Klopfer's Prognostic Rating Scale (PRS) with subjects who received behavior modification psychotherapy. Following psychotherapy, the subjects were separated into two groups: improved and unimproved. Table 3.6 shows the PRS score for each subject before therapy.

We wish to see whether we can conclude on the basis of these data that the two represented populations are different with respect to their medians. We let $X_1 = 11.9$, $X_2 = 11.7,\ldots,X_{n_1} = X_{17} = 2.2$, and $Y_1 = 6.6$, $Y_2 = 5.8,\ldots,Y_{n_2} = Y_{10} = 1.7$. Let $\alpha = 0.05$.

TABLE 3.6   **PRS score before therapy for improved and un-improved subjects**

Improved subjects		Unimproved subjects	
Subject	Score (X)	Subject	Score (Y)
1	11.9	1	6.6
2	11.7	2	5.8
3	9.5	3	5.4
4	9.4	4	5.1
5	8.7	5	5.0
6	8.2	6	4.3
7	7.7	7	3.9
8	7.4	8	3.3
9	7.4	9	2.4
10	7.1	10	1.7
11	6.9		
12	6.8		
13	6.3		
14	5.0		
15	4.2		
16	4.1		
17	2.2		

Source: Charles S. Newmark, William Hetzel, Lilly Walker, Steven Holstein, and Martin Finklestein, "Predictive Validity of the Rorschach Prognostic Rating Scale with Behavior Modification Techniques," *J. Clin. Psychol.*, 29 (1973), 246–248.

Hypotheses

$$H_0: M_X = M_Y$$
$$H_1: M_X \neq M_Y$$

Test Statistic

Table 3.7 shows the scores of Table 3.6 in rank order with the ranks attached. By Equation 3.4, we have

$$T = 296.5 - 17(17 + 1)/2 = 143.5$$

Decision

Table A.7 shows that $w_{\alpha/2} = 46$ for $n_1 = 17$, $n_2 = 10$, and $\alpha/2 = 0.025$. Thus by Equation 3.5, $w_{1-\alpha/2} = (17)(10) - 46 = 124$. Since 143.5 is greater than 124, we reject H_0, and we conclude that the two population location parameters are different.

Finding the P Value

When the null hypothesis is true, the sampling distribution of the Mann–Whitney test statistic is symmetric. Since this is the case, we can find the two-sided P value

TABLE 3.7   **Scores and corresponding ranks, Example 3.3**

X Score	Rank	Y Score	Rank
		1.7	1
2.2	2		
		2.4	3
		3.3	4
		3.9	5
4.1	6		
4.2	7		
		4.3	8
		5.0	9.5
5.0	9.5		
		5.1	11
		5.4	12
		5.8	13
6.3	14		
		6.6	15
6.8	16		
6.9	17		
7.1	18		
7.4	19.5		
7.4	19.5		
7.7	21		
8.2	22		
8.7	23		
9.4	24		
9.5	25		
11.7	26		
11.9	27		
Total	296.5		

by doubling the P value we would have if the test had been one-sided. For the present example we consult Table A.7 for $n_1 = 17$, $n_2 = 10$. We find that the computed value of our test statistic, 143.5, is between $(17)(10) - 26 = 144$ and $(17)(10) - 35 = 135$. Consequently, for this test, $2(0.005) > P > 2(0.001)$ or $0.010 > P > 0.002$.

Ties   Noether (T16) gives an adjustment factor to use when ties occur, but he points out that the adjustment has a negligible effect unless a large proportion of observations are tied or there are ties of considerable extent.

Large-Sample Approximation   When either n_1 or n_2 is greater than 20, we cannot use Table A.7. When n_1 and n_2 are both large, however, the central limit theorem applies. Then

$$z = \frac{T - n_1 n_2/2}{\sqrt{n_1 n_2 (n_1 + n_2 + 1)/12}}$$

(3.7)

has approximately the standard normal distribution when H_0 is true. In Equation 3.7 the expected value of T is $n_1 n_2/2$, and its variance is $n_1 n_2(n_1 + n_2 + 1)/12$.

Ties may occur within groups or across them. Ties within groups have no effect on the test statistic, but those across groups do. When we use the large-sample approximation, we may adjust the formula for the test statistic. Let t be the number of ties for a given rank. Then the correction for ties is

$$\frac{n_1 n_2(\Sigma t^3 - \Sigma t)}{12(n_1 + n_2)(n_1 + n_2 - 1)}$$

We subtract this correction factor from the term under the radical of Equation 3.7, so that the denominator of that equation then becomes

$$\sqrt{\frac{n_1 n_2(n_1 + n_2 + 1)}{12} - \frac{n_1 n_2(\Sigma t^3 - \Sigma t)}{12(n_1 + n_2)(n_1 + n_2 - 1)}}$$

Power-Efficiency   Mood (T17) has shown that for very large samples, the power-efficiency of the Mann–Whitney test approaches $3/\pi \approx 95.5\%$ when the underlying populations are normally distributed. Haynam and Govindarajulu (T17) and Chanda (T18) also discuss the power-efficiency aspect of the test. Chanda gives examples for the Poisson, binomial, and geometric distributions. Rasmussen (T19) and Blair and Higgins (T20, T21) have compared the power of the test with that of Student's t statistic. Bradstreet and Lemeshow (T22) compared Monte Carlo simulated type I error rates of Student's t test and several other two-sample parametric and nonparametric tests, including the Mann–Whitney–Wilcoxon test, for the situation in which the assumption of equal variances is violated. Auble (T23), Jacobson (T24), Milton (T25), and Verdooren (T26) have constructed tables of critical values for the test. Noether (T27) gives a sample-size formula for use when employment of the Mann–Whitney test is anticipated. The Mann–Whitney test is also discussed by Buckle et al. (T28), Gelberg (T29), Nemenyi (T30), Serfling (T31), Zaremba (T32), and Zeoli and Fong (T33). Jacobson (T24) gives a bibliography of 57 references on the test.

Sampling Distribution of Mann–Whitney Statistic   We obtain the sampling distribution of the Mann–Whitney test statistic T under the assumption that the X's and Y's are identically distributed. If this is true and if the X's and Y's are also independent, each of the possible arrangements of X's and Y's that may be formed when the two samples are combined and the measurements ordered is just as likely as each of the others—that is, the possible arrangements are all equally likely.

If the assumption of independent and identically distributed X's and Y's is true, we may consider the ranks assigned to the X's in the combined sample as a random selection of n of the integers between 1 and $n_1 + n_2$. This seems intuitively reasonable, because there is no reason to think that a given X measurement should be more likely to receive one rank than another. Any given X measurement, then, is assigned a rank between 1 and $n_1 + n_2$ with equal likelihood; each X measurement

is just as likely to receive one of the ranks from 1 through $n_1 + n_2$ as another. As a result of the ranking process, n of the digits between 1 and $n_1 + n_2$ are chosen as ranks for the X measurements. Consequently the probability distribution of S, the sum of the ranks assigned to the X measurements, is the probability distribution of the sum of n_1 integers chosen randomly and without replacement from among the integers between 1 and $n_1 + n_2$.

The number of ways in which we may choose n_1 integers from a total of $n_1 + n_2$ integers is given by the number of combinations of $n_1 + n_2$ things taken n_1 at a time, which may be written

$$\binom{n_1 + n_2}{n} = (n_1 + n_2)!/n_1!n_2!$$

In accordance with the preceding line of reasoning, each of these ways of choosing n_1 integers from the $n_1 + n_2$ integers has the same probability of being realized. Thus to find the probability that S is equal to a given sum (say, s), we count the number of different sets of n_1 integers from 1 through $n_1 + n_2$ that yield the sum s and divide the result by the total number of ways of choosing n_1 integers from $n_1 + n_2$ integers.

To illustrate the procedure, let us consider the case in which $n_1 = 2$ and $n_2 = 3$. The combined sample size is $n_1 + n_2 = 2 + 3 = 5$. The number of ways of selecting 2 ranks out of 5 is

$$\binom{5}{2} = 5!/2!3! = 10$$

The ten possible sets of 2 ranks and their sums S are

Ranks	1,2	1,3	1,4	1,5	2,3	2,4	2,5	3,4	3,5	4,5
S	3	4	5	6	5	6	7	7	8	9

Since $T = S - n_1(n_1 + 1)/2 = S - 2(2 + 1)/2 = 3$, we may display the probability distribution of S and T at the same time as follows:

S	T	f(S) = f(T)	
3	0	1	$P(S = 3) = P(T = 0) =$ 1/10
4	1	1	$P(S = 4) = P(T = 1) =$ 1/10
5	2	2	$P(S = 5) = P(T = 2) =$ 2/10
6	3	2	$P(S = 6) = P(T = 3) =$ 2/10
7	4	2	$P(S = 7) = P(T = 4) =$ 2/10
8	5	1	$P(S = 8) = P(T = 5) =$ 1/10
9	6	1	$P(S = 9) = P(T = 6) =$ 1/10
			Total 10/10

OTHER TWO-SAMPLE LOCATION TESTS

When the data for analysis consist of two small independent samples, we can quickly and easily test the null hypothesis of no difference between populations by using the *randomization test* suggested by Fisher (T34) and elaborated on by Pitman (T35). When sample sizes are even moderately large, however, using the randomization test can be very tedious. Although the test does not appear to be widely used, it has received considerable attention in the statistical literature. Those interested should refer to the articles by Hoeffding (T36), Lehmann and Stein (T37), Welch (T38), Moses (T39), Gocka (T40), Ray (T41), and Ray and Schabert (T42).

Perhaps the simplest and most quickly performed nonparametric two-sample test is *Tukey's quick test* (T43). This test is based on the fact that when we compare two groups, the less overlap in the observations, the more likely we are to reach the conclusion that the groups differ. Tukey's objective in devising the test was to provide an easily performed and easily remembered procedure. To realize this objective, he devised decision rules that are easy to remember.

Although Tukey's quick test is likely to be less powerful (in the usual sense of the word) than some alternative procedures, Tukey points out that the practical power of the test may negate this apparent disadvantage. *Practical power*, a concept that Tukey attributes to Churchill Eisenhart, is the product of a procedure's mathematical power and the probability that the procedure will be used. Tukey argues that a procedure such as his quick test may be sufficiently simple in concept and execution that it will be much more frequently used than more difficult procedures. Thus its simplicity compensates for its lack of power. The power and other properties of Tukey's quick test have been studied by Neave and Granger (T44, T45, T46), Rosenbaum (T47), and Biebler and Jager (T48).

EXERCISES

3.3 West (E4) conducted an experiment with adult aphasic subjects, in which each was required to respond to one of 62 commands. Five subjects received an experimental treatment program, and five controls received conventional speech therapy. Table 3.8 shows the percentage of correct responses of each subject in the two groups following treatment. Do these data provide sufficient evidence to indicate that the experimental treatment improves the proportion of correct responses? Let $\alpha = 0.05$. What is the P value?

TABLE 3.8   **Percentage of correct responses to 62 commands by aphasic subjects in two treatment groups**

Experimental (X)	73	42	90	58	62
Control (Y)	50	23	68	40	45

Source: Joyce A. West, "Auditory Comprehension in Aphasic Adults: Improvement through Training," *Arch. Phys. Med. Rehabil.*, 54 (1973), 78–86.

3.4 Table 3.9 shows the tidal volume of 37 adults suffering from atrial septal defect. In 26 of these, pulmonary hypertension was absent, and in 11 it was present. The data were reported by Ressl et al. (E5). Do these data provide sufficient evidence to indicate a lower tidal volume in subjects without pulmonary hypertension? Let $\alpha = 0.05$. What is the P value?

TABLE 3.9 Tidal volume, in milliliters, in two groups of subjects

Pulmonary hypertension absent		Pulmonary hypertension absent	
Case	(X)	Case	(Y)
1	652	1	876
2	556	2	556
3	618	3	493
4	500	4	348
5	500	5	530
6	526	6	780
7	511	7	569
8	538	8	546
9	440	9	766
10	547	10	819
11	605	11	710
12	500		
13	437		
14	481		
15	572		
16	589		
17	605		
18	436		
19	724		
20	515		
21	552		
22	722		
23	778		
24	677		
25	680		
26	428		

Source: J. Ressl, M. Kubis, P. Lukl, J. Vykydal, and J. Weinberg, "Resting Hyperventilation in Adults with Atrial Septal Defect," *Br. Heart. J.*, 31 (1969), 118–121; used by permission of the authors and the editor.

CONFIDENCE INTERVAL FOR DIFFERENCE BETWEEN TWO LOCATION PARAMETERS

Just as we can construct a confidence interval for a single population location parameter, we can also construct one for the difference between two population location parameters. Several procedures are available. The general method consists of specifying the possible values of the parameter of interest that would not lead us to reject an appropriate null hypothesis at an α level of significance. The "acceptable" possible values of the parameter constitute the confidence interval, and the confidence coefficient is $1 - \alpha$.

Specifically, in the case of two independent samples, let us consider a test of the null hypothesis that the difference between the medians of the two relevant populations is equal to zero. The $100(1 - \alpha)\%$ confidence interval for the difference between the two population medians consists of the possible values of the difference for which we will "accept" the null hypothesis at a significance level of α. The procedure presented here is based on the Mann–Whitney test.

Assumptions The assumptions underlying the construction of a confidence interval by this procedure are as follows:

1. The data consist of two random samples: $X_1, X_2, \ldots, X_{n_1}$ from population 1 and $Y_1, Y_2, \ldots, Y_{n_2}$ from population 2.
2. The two samples are independent.
3. The distribution functions of the two populations are identical except for a possible difference in location parameters.

An assumption of continuity is not necessary. As noted in Chapter 2, Noether (T49, T50) has shown that when we sample from discontinuous distributions, confidence intervals constructed in this manner have confidence coefficients at least equal to the confidence coefficient for the continuous case if we consider the confidence intervals closed.

We may use either a graphical or an arithmetic method to construct confidence intervals based on the Mann–Whitney test.

Graphical Method To use the graphical method, we begin with an X (horizontal) axis and a Y (vertical) axis intersecting at the origin $(0, 0)$. Both axes use the same scale. We locate the n_2 Y values on the vertical axis and draw lines parallel to the horizontal axis through their corresponding points. We plot points corresponding to the X values on the horizontal axis and draw lines parallel to the vertical axis through these points. The same constant may be subtracted from each value of X and Y to bring the lines closer to the origin. Such a transformation will not affect the values of the endpoints of the interval, since they are in terms of differences rather than original measurements. We draw dots at each intersection of a horizontal line with a vertical line. To represent any duplicate values of either variable (resulting in superimposed lines), we can circle the first dot at an intersection once for each additional value. There will be a total of $n_1 n_2$ dots and circles.

For a $100(1 - \alpha)\%$ confidence interval, we locate $w_{\alpha/2}$ in Table A.7. We draw a $45°$-angle (to the horizontal axis) line through the dots in such a way that $w_{\alpha/2}$ of the dots, counting from the left, are either on the line or to its left. We draw another $45°$-angle (to the horizontal axis) line in such a way that $w_{\alpha/2}$ dots are either on the line or to its right when counting from the right. The lower endpoint L of the confidence interval is the point at which the left-hand $45°$-angle line crosses the horizontal axis. The upper endpoint U of the confidence interval is the point at which the right-hand $45°$-angle line crosses the horizontal axis. The practical interpretation of this

interval is that we are at least $100(1 - \alpha)\%$ confident that the difference in population medians is somewhere between L and U inclusive.

If, because of tied differences, several points fall on the same 45°-angle line, or if the application of the large-sample approximation yields a noninteger value of $w_{\alpha/2}$, we select the 45°-angle line in such a way that the number of dots outside the line is less than $w_{\alpha/2}$, but nearest to $w_{\alpha/2}$.

The following example illustrates the graphical method.

Example 3.3	Davidson et al. (E6) studied the responses to oral glucose in patients with Huntington's disease and in a group of control subjects. The five-hour responses are shown in Table 3.10. We wish to construct a 95% confidence interval for the difference between population medians. In Table A.7 we find that $w_{\alpha/2} = 27$ for $n_1 = 11$, $n_2 = 10$, and $\alpha/2 = 0.025$. Before drawing the graph, we subtract 60 from each observed value to give the following transformed values.

$$X: \quad 25, 29, 26, 31, 17, 33, 40, 22, 32, 26, 26$$

$$Y: \quad 23, 13, 5, 5, 30, 17, 18, 37, 25, 15$$

TABLE 3.10   **Five-hour glucose (milligrams percent) responses to oral carbohydrate in 11 patients with Huntington's disease (HD) and 10 control subjects**

HD patients (X)	85	89	86	91	77	93	100	82	92	86	86
Control (Y)	83	73	65	65	90	77	78	97	85	75	

Source: Mayer B. Davidson, Stuart Green, and John H. Menkes, "Normal Glucose, Insulin, and Growth Hormone Responses to Oral Glucose in Huntington's Disease," *J. Lab. Clin. Med.*, 84 (1974), 807–812.

Figure 3.1, which illustrates the appropriate graph for this example, shows that $L = 1$ and $U = 17$, so we can be at least 95% confident that the difference between population medians is somewhere between 1 and 17 inclusive.

Arithmetic Method   The arithmetic method is easier to use if we first place the values in each sample in numerical order. We then compute the $n_1 n_2$ differences by subtracting each Y value from each X value, and arrange the differences in order of magnitude from smallest to largest. We find $w_{\alpha/2}$ in Table A.7 corresponding to n_1, n_2, and $\alpha/2$, where $1 - \alpha$ is the desired confidence coefficient.

The lower limit L of the confidence interval is the $w_{\alpha/2}$th smallest difference, and the upper limit U is the $w_{\alpha/2}$th largest difference. We can then say that we are at least $100(1 - \alpha)\%$ confident that the difference in population medians is somewhere between L and U inclusive.

FIGURE 3.1   95% confidence interval for the difference between population medians in
Example 3.3.

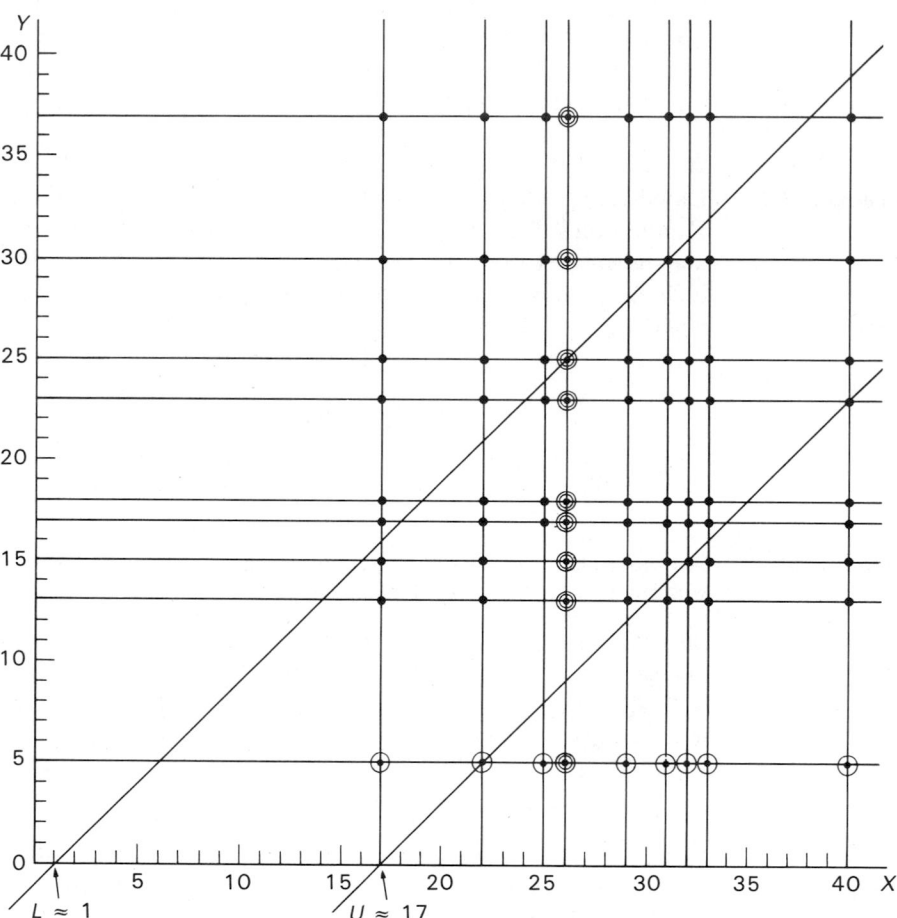

Example 3.4   Let us use the data of Example 3.3 to illustrate the arithmetic method of constructing a 95% confidence interval for the difference between population medians.
The ordered samples are as follows.

$$X: \quad 77, 82, 85, 86, 86, 86, 89, 91, 92, 93, 100$$

$$Y: \quad 65, 65, 73, 75, 77, 78, 83, 85, 90, 97$$

From Table A.7 we find that $w_{\alpha/2}$ is 27. We must therefore compute the 27 smallest
and the 27 largest differences.

TABLE 3.11   **Array of differences between X and Y values in Example 3.4**

| | X | | | | | | | | | | |
Y	77	82	85	86	86	86	89	91	92	93	100
65	12	17	20	21	21	21	24	26	27	28	35
65	12	17	20	21	21	21	24	26	27	28	35
73	4	9	12	13	13	13	16	18	19	20	27
75	2	7	10	11	11	11	14	16	17	18	25
77	0	5	8	9	9	9	12	14	15	16	23
78	−1	4	7	8	8	8	11	13	14	15	22
83	−6	−1	2	3	3	3	6	8	9	10	17
85	−8	−3	0	1	1	1	4	6	7	8	15
90	−13	−8	−5	−4	−4	−4	−1	1	2	3	10
97	−20	−15	−12	−11	−11	−11	−8	−6	−5	−4	3

First we prepare a matrix of differences, as shown in Table 3.11. Beginning with the largest difference, we can identify the 27 largest differences in the table. The largest differences appear where small values of Y are subtracted from large values of X—in this case, in the upper right-hand corner of the table. Then we can identify the 27 smallest differences in the table. These occur in the lower left-hand corner of the table, where small values of X are subtracted from large values of Y. Table 3.11 reveals that the 27 smallest differences are $-20, -15, -13, -12, -11,$ $-11, -11, -8, -8, -8, -6, -6, -5, -5, -4, -4, -4, -4, -3, -1, -1, -1,$ $0, 0, 1, 1, 1,$ so that $L = 1$. The 27 largest differences are 35, 35, 28, 28, 27, 27, 27, 26, 26, 25, 24, 24, 23, 22, 21, 21, 21, 21, 21, 21, 20, 20, 20, 19, 18, 18, 17. Therefore $U = 17$. Thus the lower and upper limits of our 95% confidence interval are 1 and 17, as before.

FURTHER READING

Moses (T51) and Walker and Lev (T52) discuss the graphical method of constructing confidence intervals based on the Mann–Whitney test, and Noether (T16) discusses the arithmetic method. Bauer (T53) discusses systematic procedures for constructing confidence bounds and point estimates based on rank statistics for the two-sample location parameter, two-sample scale, and one-sample location parameter problems.

Gibbons (T54) discusses the construction of a confidence interval for the two-sample case based on the median test. Sandelius (T55) presents a graphical version of the confidence interval based on Tukey's quick test. Confidence intervals for the two-sample case are also discussed by Hodges and Lehmann (T56), Høyland (T57),

Laan (T58), and Lehmann (T59). Hettmansperger and McKean (T60) present a graphical representation of the relationship between nonparametric estimation and hypothesis testing.

EXERCISE

3.5 In a study of the effects of vitamin D in epilepsia, Christiansen et al. (E7) reported the number of seizures in a 28-day period for 9 patients who received 4000 and 16,000 IUs of vitamin D_2 daily (group A) and 14 patients who received a placebo (group B). The results are shown in Table 3.12. Use the graphical and arithmetic methods to construct a 95% confidence interval for the difference between population medians.

TABLE 3.12 **Number of seizures experienced by two groups of subjects**

Group A (X)	4	1	1	3	4	12	19	23	7					
Group B (Y)	2	6	21	2	3	17	3	34	2	6	30	53	4	16

Source: Claus Christiansen, Paul Rødbro, and Ole Sjö, " 'Anticonvulsant Action' of Vitamin D in Epileptic Patients? A Controlled Pilot Study," *Br. Med. J.*, 5913 (1974), 258–259; reprinted by permission of the authors and the editor.

3.2

MAKING INFERENCES ABOUT THE EQUALITY OF TWO DISPERSION PARAMETERS

A question that the researcher frequently encounters has to do with the equality of two population parameters that measure *dispersion*. Synonyms for the term dispersion include *spread, scatter, variability*, and *scale*. In parametric statistical inference the F test is used to test the null hypothesis that two population dispersion parameters are equal. In the parametric case the measures of dispersion are the two population variances, usually designated σ_1^2 and σ_2^2. However, the F test is not very reliable when the populations of interest are not normally distributed, as Pearson (T61) first pointed out and Gayen (T62), Geary (T63), and Finch (T64) later corroborated. More recently, Miller (T65) used Monte Carlo sampling to show the undesirability of the F test for $H_0: \sigma_1^2 = \sigma_2^2$ when the populations are not normally distributed.

Several alternative dispersion tests have been proposed over the years, but many of them have certain drawbacks. At least one requires that the population medians be known, while others permit unknown population medians but require that they be equal. We may modify these tests by centering each sample at its sample median or mean, but as Miller (T65) points out, the tests are then not generally distribution-free.

This section presents two distribution-free alternatives to the parametric F test for testing equality of dispersion parameters. The first test assumes that the two unknown population medians are equal, whereas the second does not depend on this assumption.

ANSARI–BRADLEY TEST

The first test for dispersion considered here was proposed by Ansari and Bradley (T66), David and Barton (T67), and Freund and Ansari (T68). The test is usually referred to as the Ansari–Bradley test.

Assumptions

A. The data consist of two random samples $X_1, X_2, \ldots, X_{n_1}$ and $Y_1, Y_2, \ldots, Y_{n_2}$ from populations 1 and 2, respectively.
B. The population distributions are continuous.
C. The two samples are independent.
D. The data are measured on at least an ordinal scale.
E. The two populations are identical (including equal medians) except for a possible difference in dispersion.

Hypotheses

If we denote the dispersion parameters of populations 1 and 2 by σ_1 and σ_2, respectively, we may test the following null hypotheses against their appropriate alternatives.

A. (Two-sided)
   $H_0: \sigma_1 = \sigma_2, \qquad H_1: \sigma_1 \neq \sigma_2$
B. (One-sided)
   $H_0: \sigma_1 \leq \sigma_2, \qquad H_1: \sigma_1 > \sigma_2$
C. (One-sided)
   $H_0: \sigma_1 \geq \sigma_2, \qquad H_1: \sigma_1 < \sigma_2$

The symbol σ should not necessarily be interpreted as the population standard deviation, but rather as a general measure of dispersion.

Test Statistic

To obtain the test statistic, we arrange the combined set of $n_1 + n_2 = n'$ measurements in order from smallest to largest, while retaining the identity of each measurement with respect to the sample of which it is a member. We then assign ranks to the ordered measurements as follows: The smallest measurement and the largest measurement are each given a rank of 1; the second smallest measurement and the second largest measurement are each given a rank of 2; and we continue

in this manner until all measurements have been assigned a rank. If $n_1 + n_2$ is an even number, the array of ranks will be $1, 2, 3, \ldots, n'/2, n'/2, \ldots, 3, 2, 1$. If $n_1 + n_2$ is an odd number, the array of ranks will be $1, 2, 3, \ldots, (n' - 1)/2, (n' + 1)/2, (n' - 1)/2, \ldots, 3, 2, 1$. Let R_i be the rank of the *i*th X measurement in the set of ranks. The test statistic, then, is

$$T = \sum R_i \tag{3.8}$$

In other words, T is the sum of the ranks assigned to the X values.

The rationale underlying the test statistic is that if the two sampled populations have equal medians, we would expect the population with the greater amount of dispersion to yield a sample with dispersion greater than that of the sample from the other population. The sample with the greater amount of dispersion will receive the smaller ranks. If the sample with the greater dispersion is the sample of X measurements, the test statistic T will tend to be small. If, on the other hand, the sample of Y measurements has the greater amount of dispersion, T will tend to be large. This rationale provides the basis for the decision rules.

Decision Rule

A. Since the alternative hypothesis in A is two-sided, either a sufficiently large or a sufficiently small value of T will cause rejection of H_0. Consequently, for our chosen level of significance α, we reject H_0 if T is either greater than or equal to the larger critical value of x in Table A.8 or less than the lower critical value of x in the table. To find the upper critical value of x, we enter Table A.8 with $\alpha/2$, n_1, and n_2. The x in the leftmost column of the table corresponding to these values is the upper critical value.

To find the lower critical value of x, enter Table A.8 with $1 - \alpha/2$, n_1, and n_2. The x in the leftmost column of the table corresponding to these values is the lower critical value of x. If exact values of $\alpha/2$ and $1 - \alpha/2$ for given n_1 and n_2 are not in the table, choose appropriate neighboring values.

B. A sufficiently small value of T will cause rejection of H_0 appearing in B, since the alternative hypothesis specifies that there is more variability in the X's. We reject H_0 if T is less than the critical value of x in Table A.8 for $1 - \alpha$, n_1, and n_2.

C. The alternative hypothesis in C specifies less variability in the X's. Consequently we reject H_0 if T is greater than or equal to the critical value of x in Table A.8 for α, n_1, and n_2.

Example 3.5   Gordon et al. (E8) reported cardiac-index values for two groups of patients, as shown in Table 3.13. All patients were seen initially for severe aortic valvular disease requiring prosthetic valve replacement. Data on the cardiac index were obtained

TABLE 3.13	**Cardiac index, liters/minute/M², in two groups of patients following prosthetic valve replacement**				
Group 1 (X)	3.84	2.60	1.19	2.00	
Group 2 (Y)	3.97	2.50	2.70	3.36	2.30

Source: Richard F. Gordon, Moosa Najmi, Benedict Kingsley, Bernard L. Segal, and Joseph W. Linhart, ''Spectroanalytic Evaluation of Aortic Prosthetic Valves,'' *Chest*, 66 (1974), 44–49.

after operation. Group 1 consisted of patients with normal prosthetic valve function. Group 2 consisted of patients with abnormal prosthetic valve function. We wish to know whether the dispersion with respect to the variable of interest is different in the two populations represented by these samples. Let $\alpha = 0.05$.

Hypotheses

$$H_0: \sigma_1 = \sigma_2, \qquad H_1: \sigma_1 \neq \sigma_2$$

Test Statistic

On combining the two samples and ranking, we have the results shown in Table 3.14.

TABLE 3.14	**Data of Table 3.13 combined and ranked**								
Observation	1.19	2.00	2.30	2.50	2.60	2.70	3.36	3.84	3.97
Group	X	X	Y	Y	X	Y	Y	X	Y
Rank	1	2	3	4	5	4	3	2	1

Decision

By Equation 3.8 we have

$$T = 1 + 2 + 5 + 2 = 10$$

To find the upper critical values of T, we enter Table A.8 with $n_1 = 4$, $n_2 = 5$, and $\alpha/2 = 0.025$. Since the exact value, 0.025, is not in the table, we choose the closest value, 0.0159. The corresponding value of T is 16. To find the lower critical value of T, we enter Table A.8 with $n_1 = 4$, $n_2 = 5$, and $1 - \alpha/2 = 1 - 0.025 = 0.975$. Again, the exact probability value is not in the table, and we choose the closest value, 0.9603. The corresponding value of T is 8. The decision rule, then, is "Reject H_0 if T is greater than or equal to 16 or less than 8." Since our computed T of 10 is neither greater than or equal to 16 nor less than 8, we are unable to reject H_0. We conclude that the two population dispersion parameters may be equal.

Large-Sample Approximation   When the sample sizes exceed those found in Table A.8, we may compute

$$T^* = \frac{T - [n_1(n_1 + n_2 + 2)/4]}{\sqrt{n_1 n_2 (n_1 + n_2 + 2)(n_1 + n_2 - 2)/[48(n_1 + n_2 - 1)]}} \tag{3.9}$$

if $n_1 + n_2$ is even, and

$$T^* = \frac{T - [n_1(n_1 + n_2 + 1)^2/4(n_1 + n_2)]}{\sqrt{n_1 n_2 (n_1 + n_2 + 1)[3 + (n_1 + n_2)^2]/48(n_1 + n_2)^2}} \tag{3.10}$$

if $n_1 + n_2$ is odd. We compare T^* for significance with appropriate values of the standard normal distribution.

Ties   When the samples are small and contain ties, we use average ranks when computing T and follow the previously described procedure for small samples. When the samples are large, we modify Equation 3.9 or Equation 3.10, whichever is needed, as follows. Replace the denominator of Equation 3.9 with

$$\frac{n_1 n_2 \left[16 \sum_{j=1}^{g} t_j r_j^2 - (n_1 + n_2)(n_1 + n_2 + 2)^2 \right]}{16(n_1 + n_2)(n_1 + n_2 - 1)} \tag{3.11}$$

and replace the denominator of Equation 3.10 with

$$\frac{n_1 n_2 \left[16(n_1 + n_2) \sum_{j=1}^{g} t_j r_j^2 - (n_1 + n_2 + 1)^4 \right]}{16(n_1 + n_2)^2 (n_1 + n_2 - 1)} \tag{3.12}$$

In Equations 3.11 and 3.12, g is the number of tied groups, t_j is the size of the jth tied group, and r_j is the average rank of the measurements in the jth tied group. We treat an untied measurement as a tied group of size 1.

Power-Efficiency   Ansari and Bradley (T66) state that the relative efficiency of their statistic when compared with the parametric F test is $6/\pi^2$ when sampling is from normally distributed populations. They also note that the statistic is less efficient asymptotically than some other dispersion tests, but easier to apply.

EXERCISES

3.6   Reimer et al. (E9) studied the effect of propranolol on the severity of myocardial necrosis following 40 minutes of temporary coronary-artery occlusion in dogs. One group of dogs was untreated, and a second group received propranolol 10 minutes before the occlusion. Following this procedure Reimer et al. recorded the relative area of necrosis (percentage of fibers involved) in the posterior papillary muscle of each dog's heart. Table 3.15 gives

partial results. Can we conclude on the basis of these data that dispersion with regard to relative necrosis differs in the two populations represented? Let $\alpha = 0.05$.

TABLE 3.15	**Necrosis of projecting posterior papillary muscle (%) in two groups of subjects**									
Untreated (X)	44.4	81.0	23.6	62.1	39.1	25.5	44.2	43.3	39.8	61.3
Propranolol treated, 5.0 mg/kg, iv (Y)	0	4.5	5.6	6.1	22.6	30.8	13.4	1.3	45.0	30.3

Source: Keith A. Reimer, Margaret M. Rasmussen, and Robert B. Jennings, "Reduction by Propranolol of Myocardial Necrosis Following Temporary Coronary Artery Occlusion in Dogs," *Circ. Res.*, 33 (1973), 353–363; reprinted by permission of the American Heart Association.

3.7 Boullin and O'Brien (E10) studied uptake and loss of ^{14}C-dopamine by platelets in five autistic children and five normal controls. Part of their results are shown in Table 3.16. Can we conclude from these data that the two populations represented differ with respect to dispersion of uptake values? Let $\alpha = 0.05$.

TABLE 3.16	**Uptake of ^{14}C-dopamine in platelets of autistic children and controls**				
Autistic children (X)	433	347	328	607	478
Controls (Y)	428	372	434	425	336

Source: David J. Boullin and Robert A. O'Brien, "Uptake and Loss of ^{14}C-dopamine by Platelets from Children with Infantile Autism," *J. Aut. Child. Schizophrenia*, 2 (1972), 67–74; published by Plenum Publishing Corporation, New York.

MOSES TEST

Another test for equality of dispersion parameters was proposed by Moses (T69). Unlike the Ansari–Bradley test, the Moses test does not assume equality of location parameters, and this fact gives the test wider applicability.

Assumptions

A. The data consist of two random samples $X_1, X_2, \ldots, X_{n_1}$ and $Y_1, Y_2, \ldots, Y_{n_2}$ from populations 1 and 2, respectively.

B. The population distributions are continuous, are measured on at least an interval scale, and have the same shape.

C. The two samples are independent.

Hypotheses

A. (Two-sided)
$$H_0: \sigma_1 = \sigma_2, \qquad H_1: \sigma_1 \neq \sigma_2$$
B. (One-sided)
$$H_0: \sigma_1 \geq \sigma_2, \qquad H_1: \sigma_1 < \sigma_2$$
C. (One-sided)
$$H_0: \sigma_1 \leq \sigma_2, \qquad H_1: \sigma_1 > \sigma_2$$

Test Statistic

Briefly, we obtain the test statistic by (1) subdividing the sample of X's at random into subsamples of equal size, (2) subdividing the sample of Y's at random into subsamples of equal size, (3) computing for each subsample the sum of squared deviations of the observations from their mean, and (4) applying the Mann–Whitney location test to the results. The details of the procedure are as follows:

1. Divide the X observations at random into m_1 subsamples of size k. Discard leftover observations.

2. Divide the Y observations at random into m_2 subsamples of size k. Discard any leftover observations. Shorack (T70) recommends that k be as large as possible, but not greater than 10, and that m_1 and m_2 be large enough to permit meaningful results from the application of the location test.

3. For each subsample, obtain the sum of the squared deviations of observations from their mean. That is, compute the numerator of the familiar sample variance; the numerator has the form $\Sigma(X - \bar{X})^2$ or $\Sigma(Y - \bar{Y})^2$. Designate the m_1 sums of squares computed from the subsamples of X's by $C_1, C_2, \ldots, C_{m_1}$. Designate the m_2 sums of squares computed from the subsamples of Y's by $D_1, D_2, \ldots, D_{m_2}$.

4. Apply the Mann–Whitney test by letting the C's and D's play the roles of the X's and Y's, respectively, and letting m_1 and m_2 replace n_1 and n_2.

The test statistic is then

$$T = S - m_1(m_1 + 1)/2 \tag{3.13}$$

where S is equal to the sum of the ranks assigned to the sums of squares computed from the subsamples of X's.

Decision Rule

For a given level of significance α, the three decision rules are as follows:

A. Reject $H_0: \sigma_1 = \sigma_2$ if the computed value of T is less than $w_{\alpha/2}$ or greater than $w_{1-\alpha/2}$, where $w_{\alpha/2}$ is the critical value of T given in Table A.7 and $w_{1-\alpha/2}$ is given by Equation 3.5.

B. Reject $H_0: \sigma_1 \geq \sigma_2$ if the computed T is less than w_α, the critical value of T obtained by entering Table A.7 with m_1, m_2, and α.

C. Reject $H_0: \sigma_1 \leq \sigma_2$ if the computed T is greater than $w_{1-\alpha}$ given by Equation 3.6.

Example 3.6   In a study of the mean life span and the production of platelets in patients with idiopathic thrombocytopenic purpura (ITP) and in healthy subjects, Branehög et al. (E11) collected the data on platelet recovery shown in Table 3.17. They

TABLE 3.17   **Percentage of platelet recovery in two groups of subjects**

Patients with ITP (untreated)(X)											
26	30	32	17	21	27	26	44	35	14	18	18
17	23	29	16	13	36	28	23	24	34	52	35

Healthy subjects (Y)										
47	66	51	44	80	65	58	65	61	64	51	56
76	58	61	48	55	68	59	60	58			

Source: Ingemar Branehog, Jack Kutti, and Aleksander Weinfeld, "Platelet Survival and Platelet Production in Idiopathic Thrombocytopenic Purpura (ITP)," *Br. J. Haematol.,* 27 (1974), 127–143; published by Blackwell Scientific Publications, Oxford.

defined platelet recovery as the percentage of infused platelet-bound radioactivity remaining in the peripheral blood 15 minutes after infusion. We wish to know whether these data provide sufficient evidence to indicate a difference in dispersion between the two populations represented by the observed samples. We choose a level of significance of 0.05.

Hypotheses

$$H_0: \sigma_1 = \sigma_2 \qquad H_1: \sigma_1 \neq \sigma_2$$

Test Statistic

If we let the number k of observations in each subsample equal 4, then the 24 X observations yield $m_1 = 6$ subsamples, and the 21 Y observations yield $m_2 = 5$ subsamples. We must discard one Y observation. When we randomly divide the X observations into six subsamples, one possible set of subsamples and the corresponding sums of squares are as follows:

Subsample	Observations				Sums of squares
1	26,	32,	35,	24	78.75
2	26,	36,	18,	23	172.75
3	18,	16,	30,	13	166.75
4	35,	27,	29,	28	38.75
5	52,	17,	14,	17	978.00
6	21,	44,	23,	34	341.00

Random subdivision of the Y observations leads to the following possible set of subsamples and corresponding sums of squares.

Subsample	Observations				Sums of squares
1	60,	58,	48,	61	106.75
2	80,	58,	58,	61	336.75
3	64,	56,	51,	51	113.00
4	55,	44,	66,	65	317.00
5	59,	76,	68,	47	465.00

Table 3.18 shows the sums of squares in rank order with the rank attached. By Equation 3.13 we have $T = 34 - 6(7)/2 = 13$.

TABLE 3.18

Sums of squares and corresponding ranks, Example 3.6

Sums of squares, X group	Rank	Sums of squares, Y group	Rank
38.75	1		
78.75	2		
		106.75	3
		113.00	4
166.75	5		
172.75	6		
		317.00	7
		336.75	8
341.00	9		
		465.00	10
978.00	11		
Total	34		

Decision

Table A.7 reveals that the critical values for this test are 4 and 26. Since $4 < 13 < 26$, we cannot reject H_0, and thus we cannot conclude that the two populations differ with respect to dispersion.

Ties   We handle ties among the C's and D's as described in the discussion of the Mann–Whitney test.

Large-Sample Approximation   When we cannot use Table A.7 to obtain critical values because either the number of C's or the number of D's exceeds 20, we can use the large-sample approximation of the Mann–Whitney test.

Power-Efficiency   Moses (T69) found the asymptotic efficiency of his test to be 0.50 when $k = 3$ and the observations are drawn from normally distributed populations. Shorack (T70) refers to tests of the Moses type as "useful inefficient statistics."

Advantages and Disadvantages of Moses Test   One disadvantage of the Moses test for equality of dispersion parameters is its relative inefficiency. Another is that, since the subpopulations are obtained by a random procedure, it is highly probable that different people applying the test will obtain different values of the test statistic. Thus one subdivision may lead to significant results where another does not. It is inappropriate to try different randomization procedures until one is found that leads to a desired conclusion. We should determine the size of k according to the

criterion mentioned earlier and use an appropriate procedure for random subdivision. We should then be prepared to live with the results.

An advantage of the Moses test is that it does not depend on assumptions of known or equal location parameters. In addition, the Moses test is fairly easy to compute when compared to a more efficient alternative.

Frequently a researcher's primary objective is to decide whether two location parameters are equal. When the appropriate statistical test assumes that the dispersion parameters are equal, the researcher may wish to decide this by analyzing the same data obtained to test equality of location parameters. When samples are drawn from symmetric distributions, and when one wants to test simultaneously for equality of location parameters and equality of dispersion parameters, Hollander (T71) recommends the use of the Mann–Whitney test for testing equality of location parameters, and either the Moses test or a test due to Lehmann (T72) to test for equality of dispersion parameters. He shows that the Mann–Whitney test and the Moses and Lehmann tests are uncorrelated and asymptotically independent when H_0 is true (dispersion parameters are equal) and the sampled populations are symmetric. Under these conditions, when the Mann–Whitney test is carried out at an α_1 level of significance and the Moses test, for example, at an α_2 significance level, the probability of rejecting at least one of the two hypotheses tested is approximately $\alpha_1 + \alpha_2 - \alpha_1\alpha_2$.

OTHER DISPERSION TESTS

Other tests for equality of dispersion parameters suggested as substitutes for the F test include those by Box (T73), Box and Andersen (T74), Capon (T75), Crouse (T76), Deshpandé and Kusum (T77), Kamat (T78), Klotz (T79), Lehmann (T72), Levene (T80), Miller (T65), Mood (T7), Nemenyi (T81), Raghavachari (T82), Rosenbaum (T83), Savage (T84), Shorack (T85, T70), Siegel and Tukey (T86), Sukhatme (T87), Taha (T88), Tiku (T89), and Ulrich and Einsporn (T90). Not all these tests are distribution-free.

FURTHER READING

The literature on dispersion tests is extensive. Laubscher et al. (T91) give selected critical values for Mood's dispersion test, and Mielke (T92) proposes an adjustment for use when ties are present. The power and efficiency of the Mood test have been discussed by, among others, Basu and Woodworth (T93), Klotz (T79), Puri (T94), Rosenbaum (T47), Duran (T95), and Sukhatme (T96). Mood's test is also discussed by Crouse (T76).

Additional papers that may prove useful are those by David (T97), Duran and Mielke (T98), Gibbons (T99), Goria (T100, T101), Goria and Vorlíčková (T102), Lepage (T103), Noether (T104), Penfield (T105, T106), Puri (T107), Randles and Hogg (T108), Sen (T109), Van Eeden (T110), Wilks (T111), Kochar and Gupta (T112), Tiku and Balakrishnan (T113), and Vegelius (T114).

EXERCISES

3.8 In a study designed to determine the long-term effect of halofenate on patients with hyper-triglyceridemia, Aronow et al. (E12) collected data on triglyceride levels of patients during a control period and at the end of an experimental period. The patients were placed on an appropriate diet for hypertriglyceridemia, and their hypolipidemic medication was discontinued. After one month of this regimen, the patients went through a two-month control period. Then they were designated at random to receive either halofenate or a placebo. Table 3.19 shows the change in triglyceride levels of the two groups of patients from the control period to the end of the experimental period. Can we conclude on the basis of these data that there is a difference in dispersion between the two populations represented by the samples? Let $\alpha = 0.05$. Determine the P value.

TABLE 3.19   **Change in triglyceride levels in two groups of patients**

Patients on	1170	−16	36	371	68	−169	151	97
halofenate (X)	169	238	253	93	201	−75	201	
	−1	−377	2	658	166	66	23	
Patients on	−640	−1368	−28	−33	9	−1703	−63	−59
placebo (Y)	−68	−88	−58	−458	−117	42	4	
	652	124	2	−1720	15	60		

Source: Wilbert S. Aronow, Phillip R. Harding, Mohammed Khursheed, Jack S. Vangrow, Nicholas P. Papageorge's, and James Mays, "Effect of Halofenate on Serum Lipids," *Clin. Pharmacol. Ther.*, 14 (1973), 358–365.

Note: Positive numbers indicate a decrease, negative numbers an increase.

3.9 In a study designed to determine whether middle-aged and old subjects with maturity-onset diabetes respond to exercise by producing high levels of fasting serum growth hormone, Hansen (E13) collected the data shown in Table 3.20. Test for a significant difference in dispersion between the two groups. Let $\alpha = 0.05$. Determine the P value.

TABLE 3.20   **Levels of fasting serum growth hormone (nanograms per milliliter) in two groups of subjects**

Diabetics (X)	1.2	0.2	0.3	0.9	4.2	0.9	0.3	0.7	0.9	1.1	3.0	0.9	2.3
	1.3	0.2	1.2	1.5	2.1	7.7	20.0	1.2	3.4	2.2	0.1	4.3	0.7
	0.7	1.3	1.3	9.8	0.9	4.7	0.0	0.4	21.0	12.0	4.2	2.7	1.7
	0.5	1.0	0.9	2.1	0.1	1.7	1.0	3.9	1.0	0.5	0.7	0.2	0.9
	0.9	0.8	0.5	1.5	1.1	1.1	1.6	1.5	4.0	4.7	0.9		
Controls (Y)	1.4	1.6	1.4	4.1	2.6	1.1	0.4	1.8	2.2	0.3	1.3	1.7	1.0
	1.2	1.4	0.5	1.1	1.5	1.1	3.3	2.6	0.7	0.1	1.6	2.5	0.7
	1.7	0.3	1.9	0.0	0.5								

Source: Aage Prange Hansen, "Abnormal Serum Growth Hormone Response to Exercise in Maturity-Onset Diabetes," *Diabetes*, 24 (1973), 619–628.

3.3

SOME MISCELLANEOUS TWO-SAMPLE TESTS

In this section, we discuss three tests that are useful in certain situations for analyzing the data of two samples.

WALD–WOLFOWITZ RUNS TEST

In Chapter 2 the one-sample runs test for randomness was discussed. In that chapter a run was defined as a sequence of like events, items, or symbols that is preceded and followed by an event, item, or symbol of a different type or by none at all. The test statistic was r, the number of runs present in the set of data.

The Wald–Wolfowitz runs test (T115) uses the number of runs present in the data from two samples to test the null hypothesis that the samples come from identical populations against the alternative that the populations differ in any respect whatsoever—dispersion, or location, or skewness, for example.

Assumptions

    A. The data consist of observations $X_1, X_2, \ldots, X_{n_1}$ and $Y_1, Y_2, \ldots, Y_{n_2}$, comprising random samples drawn from population 1 and population 2, respectively.

    B. The two samples are independent.

    C. The variable of interest is continuous.

Hypotheses

    H_0: The X's and Y's come from identically distributed populations

    H_1: The population of X's and the population of Y's are not identically distributed

Test Statistic

The definition of a run in this test is the same as that given in Chapter 2. The test statistic is r, the number of runs present in the complete set of data. To obtain the value of the test statistic in a given situation, we combine the two samples and arrange the observations in order of magnitude. In the ordering, we must keep track of the sample to which each observation belongs.

If there are ties across samples—that is, if one or more X values are equal to one or more Y values—we must apply some method of dealing with the problem. One approach is to prepare two ordered arrangements, one resulting in the fewest number of runs r', and one resulting in the largest number of runs r'', and then let r equal the mean of the two. Ties among just the X's or among just the Y's do not present a problem.

The rationale underlying the Wald–Wolfowitz runs test is as follows. If the samples of X's and Y's do in fact come from identically distributed populations, we expect them to be well mixed. When this is the case, the number of runs r is relatively large. Large values of r, then, tend to support the null hypothesis.

Sufficiently few runs cause us to reject H_0 and to conclude that the two populations represented by the two samples are not identically distributed. The populations may have different medians, different dispersion parameters, or different types or degrees of skewness; or they may differ in some other respect. When we reject H_0, we cannot be more specific in hypothesizing an alternative.

Decision Rule

Reject H_0 at the 0.025 level of significance if the computed value of r is less than or equal to the tabulated value of r for n_1 and n_2 given in Table A.5.

Example 3.7

As part of a larger study conducted by Cabasso et al. (E14), subjects who had received rabies vaccine earlier, but not during the previous six months, were assigned randomly to two groups. Subjects in one group received a booster dose of a certain type of rabies vaccine, and subjects in the other group were given a dose of another type. Table 3.21 shows the antibody responses of the subjects on the fourteenth day following receipt of the vaccine. We wish to know whether we may conclude that the population distributions represented by these two samples are different. Let $\alpha = 0.05$.

TABLE 3.21

Antibody responses of subjects receiving one booster dose (international units of rabies antibody per milliliter of serum) rabies vaccine of specified type

Type I (X)	0.26	1.00	1.28	5.00	8.00	16.00	<2.56*	1.60	8.00	10.20
Type II (Y)	0.16	0.08	<0.08**	0.10	1.28	2.02	0.80	2.56	1.28	5.10

Source: V. J. Cabasso, M. B. Dobkin, R. E. Roby, and A. H. Hammar, "Antibody Response to a Human Diploid Cell Rabies Vaccine," *Appl. Microbiol.*, 27 (1974), 553–561.

* Omitted from calculations
** Treated as zero in calculations

Hypotheses

H_0: The two samples come from identically distributed populations
H_1: The two populations are not identically distributed

Test Statistic

We note in Table 3.21 that there is one tie across samples. (We have both an X value and a Y value equal to 1.28.) The ordered arrangements of the two samples yielding the smallest and largest values of r are given in Table 3.22. Since $r' = 10$ and $r'' = 12$, we have $r = (10 + 12)/2 = 11$.

TABLE 3.22	**Ordered arrangements of the data in Table 3.21 yielding smallest and largest values of r**

Ordered arrangement yielding smallest value of r

0	0.08	0.10	0.16	0.26	0.80	1.00	1.28	1.28	1.28	1.60
Y	Y	Y	Y	X	Y	X	X	Y	Y	X

2.02	2.56	5.00	5.10	8.00	8.00	10.20	16.00			
Y	Y	X	Y	X	X	X	X	$r' = 10$		

Ordered arrangement yielding largest value of r

0	0.08	0.10	0.16	0.26	0.80	1.00	1.28	1.28	1.28	1.60
Y	Y	Y	Y	X	Y	X	Y	X	Y	X

2.02	2.56	5.00	5.10	8.00	8.00	10.20	16.00			
Y	Y	X	Y	X	X	X	X	$r'' = 12$		

Decision

For $n_1 = 9$ and $n_2 = 10$, Table A.5 reveals that the critical value of the test statistic is 5. Since 11 is not less than or equal to 5, we cannot reject H_0. We conclude that the two populations represented by the two samples may have identical distributions (P value > 0.025).

Large-Sample Approximation   When either n_1 or n_2 is greater than 20, we cannot use Table A.5 to obtain critical values. When sample sizes are large, the test statistic

$$z = \frac{r - \left(\dfrac{2n_1 n_2}{n_1 + n_2} + 1\right)}{\sqrt{\dfrac{2n_1 n_2 (2n_1 n_2 - n_1 - n_2)}{(n_1 + n_2)^2 (n_1 + n_2 - 1)}}} \tag{3.14}$$

is distributed approximately as the standard normal, and we may compare it for significance with appropriate values from Table A.2.

Power-Efficiency   The Wald–Wolfowitz runs test is not a very powerful test for specific alternatives such as inequality of location parameters or dispersion parameters. One of the previously described tests is a better choice when specific alternatives have been formulated. The Wald–Wolfowitz test is most useful as a quick and easy test for analyzing data when no particular alternative is of primary interest.

Smith (T116) found that the Wald–Wolfowitz test has power-efficiency in the neighborhood of 75% when the distributions differ only with respect to means and when sample sizes are about 20. The test has also been discussed by Blumenthal (T117) and Noether (T118).

EXERCISES

3.10   Von Burg and Rustam (E15) reported values of the median nerve motor conduction veloc-ity shown in Table 3.23. Experimental subjects admitted to a hospital had been diagnosed as having methylmercury poisoning and as having eaten contaminated bread. The controls were members of the hospital personnel who had presumably escaped exposure to the contaminated bread. Do these data provide sufficient evidence to indicate that the popu-lations represented by the two samples have different distributions? Let $\alpha = 0.025$.

TABLE 3.23   **Median nerve motor conduction velocity (meters per second) in two groups of subjects**

Controls (X)	68	67	58	62	55	60	67							
Experimental subjects (Y)	60	59	72	73	56	53	43	50	65	56	56	56	57	36

Source: R. Van Burg and Hussain Rustam, ''Electrophysiological Investigations of Methylmercury Intoxication in Humans. Evaluation of Peripheral Nerve by Conduction Velocity and Electromyography,'' *Electro-encephalogr. Clin. Neurophysiol.*, 37 (1974), 381–392.

3.11   Skerfving et al. (E16) reported data on blood mercury levels shown in Table 3.24. All subjects in the exposed group had had more than three meals a week of contaminated fish (0.5–7 mg mercury as methylmercury per kilogram of fish) for more than three years. None of the control subjects had a history indicating regular consumption of con-taminated fish. Can we conclude on the basis of these data that the two populations represented have different distributions? Let $\alpha = 0.025$.

TABLE 3.24   **Mercury levels, nanograms per gram, in blood cells of two groups of subjects**

Controls (X)	5.3 4	15 3	11 12.2	5.8 6.1	17 10.2	7	8.5	9.4	7.8	12.0	8.7
Exposed (Y)	100 40 120	70 100 300	50 70 161	196 150 62	69 200 12.8	370 68	270 304	150 236	60 75	330 178	1100 41

Source: S. Skerfving, K. Hansson, C. Mangs, J. Lindsten, and N. Ryman, ''Methylmercury-Induced Chromosome Damage in Man,'' *Environ. Res.*, 7 (1974), 83–98.

HOLLANDER TEST OF EXTREME REACTIONS

In certain situations exposure to experimental conditions may cause some subjects to react in one manner, while causing others to have an opposite reaction. A drug may depress some subjects and elevate the mood of others. A certain teaching method may be effective with some students but ineffective with others. Certain economic conditions may cause some people to be more conservative in their spending habits, while creating a propensity to.spend in others. In certain psychological experiments, some subjects may exhibit a defensive reaction through a delayed response, yet other subjects may indicate defensive behavior through a rapid reaction.

If experimenters employ a test of location in such situations to determine whether there is a significant difference between the average responses of controls and experimental subjects, they are likely to find that they cannot reject the hypothesis of no difference. Reactions in one direction by some of the experimental subjects and reactions in the opposite direction by the others may cause the average response of experimental subjects to be about the same as the average response of the controls. Then they cannot reject the null hypothesis of no treatment effect when in fact they should. Hollander (T119) proposed the following test to detect differences between control and experimental subjects when some of the latter are expected to react in one way and the others in the opposite way.

Assumptions

A. The data consist of two independent random samples $X_1, X_2, \ldots, X_{n_1}$ and $Y_1, Y_2, \ldots, Y_{n_2}$ of control and experimental subjects, respectively.
B. The variable of interest is continuous.
C. The strength of the measurement is at least ordinal.

Hypotheses

H_0: The two samples may be considered as having been drawn from the same population

H_1: One population consists of observations resulting from extreme reactions in both directions

Test Statistic

To obtain the numerical value of the test statistic, we first combine the observations from the two samples and arrange them in order from smallest to largest, keeping track of which observations are X's and which are Y's. The test statistic is

$$G = \sum_{i=1}^{n_1} (r_i - \bar{r})^2 \qquad (3.15)$$

where r_i is the rank of the ith largest X value and $\bar{r}$ is the mean of the ranks assigned to the n_1 X values. That is,

$$\bar{r} = \frac{\sum_{i=1}^{n_1} r_i}{n_1} \qquad (3.16)$$

If the reactions of the experimental subjects are extreme, the responses of the control subjects tend to be compressed with respect to their ranks, and G is relatively small.

Decision Rule

Reject the null hypothesis if the computed value of G is less than or equal to C_α given in Table A.9, where C_α is the critical value of G for significance level α. Enter Table A.9 with α, n_1 and $N = n_1 + n_2$.

Example 3.8

Twenty children enrolled in a swimming class for the handicapped were randomly assigned to one of two groups. Children in group A, the control group, were taught by traditional methods, while those in group B, the experimental group, were taught by a new method. The instructor suspected that the new method might be more effective than the traditional method with some children and less effective with others. At the end of the course, a judge who was unaware of the method by which a particular child was taught graded each child's swimming ability. Table 3.25 shows the results. We wish to test for extreme reactions.

TABLE 3.25

Swimming scores of 20 handicapped children

(X) Group A, control group	66	86	80	78	77	63	62	87	75	84
(Y) Group B, experimental group	95	85	56	46	91	79	94	45	41	54

Hypotheses

H_0: The two samples of scores may be considered as having been drawn from the same population

H_1: The new method produces extreme reactions

Test Statistic

Table 3.26 shows the ordered combined sample and the corresponding ranks. By Equation 3.16, we have

$$\bar{r} = \frac{6 + 7 + 8 + 9 + 10 + 11 + 13 + 14 + 16 + 17}{10} = 11.1$$

and by Equation 3.15, the value of the test statistic is

$$G = (6 - 11.1)^2 + (7 - 11.1)^2 + \cdots + (17 - 11.1)^2 = 128.9$$

TABLE 3.26

The data of Table 3.25 and assigned ranks

Observation	41	45	46	54	56	62	63	66	75	77
Group	Y	Y	Y	Y	Y	X	X	X	X	X
Rank	1	2	3	4	5	6	7	8	9	10

Observation	78	79	80	84	85	86	87	91	94	95
Group	X	Y	X	X	Y	X	X	Y	Y	Y
Rank	11	12	13	14	15	16	17	18	19	20

Decision

Entering Table A.9 with $n_1 = 10$ and $N = 20$, we find that the critical values for $\alpha = 0.01$ and 0.05 are 153.6 and 196.4, respectively. Since 128.9 is less than 153.6, we can reject H_0 at $\alpha = 0.01$ and conclude that the new method does produce extreme reactions.

Ties Tied observations pose a problem only when they occur between an X and a Y. In such situations, we use an appropriate tie-breaking technique. For example, we may break the ties in all possible ways, compute a value of the test statistic for each arrangement, and reject H_0 if the maximum G is less than or equal to C_α.

Large-Sample Approximation For sample sizes that exceed those for which critical values are given in Table A.9, we may use the large-sample approximation.

Hollander (T119) has shown that for large samples and when H_0 is true, G is approximately normally distributed, with mean

$$E(G) = \frac{(n_1 - 1)(N^2 + N)}{12} \tag{3.17}$$

and variance

$$\sigma_G^2 = E(G^2) - [E(G)]^2 \tag{3.18}$$

where

$$E(G^2) = \frac{(n_1 - 1)^2}{720}\left[\frac{-6}{n_1}(N^4 + 2N^3 + N^2)\right.$$

$$\left. + \left(\frac{n_1 + 1}{n_1 - 1}\right)(5N^4 + 6N^3 - 5N^2 - 6N)\right] \tag{3.19}$$

Power-Efficiency The power-efficiency of the Hollander test has not been investigated.

FURTHER READING

Moses (T120) first proposed a test of extreme reactions. His test, however, has a number of shortcomings that the Hollander test overcomes. Hollander (T119) points out that his test should not be thought of as a two-sample test of dispersion.

Arnold and Briley (T121) present a test of extreme reactions based on rank sums that is equivalent to the dispersion tests of Ansari and Bradley (T66), Siegel and Tukey (T86), and others.

EXERCISES

3.12 In an experiment conducted to evaluate the effects of a proposed antidepressant drug, 19 mildly depressed subjects were randomly assigned to receive either the experimental drug or a placebo. The experimenters suspected that the drug might cause depression in some subjects. After subjects were administered the drug and placebo, they were given a test to measure their levels of depression. Table 3.27 shows the results. Do these data provide sufficient evidence to suggest that the drug produces extreme reactions? Let $\alpha = 0.05$.

TABLE 3.27   **Levels of depression in two groups of subjects**

Placebo group (X)	83	80	73	86	82	79	70	81	76	
Experimental group (Y)	85	96	97	58	84	67	72	74	75	54

3.13 To evaluate the effectiveness of a method of teaching reading, researchers randomly assigned 20 first-grade pupils either to a control group or to an experimental group. Following the experiment, each pupil was given a test to measure the extent of reading improvement. Table 3.28 shows the results. Test the null hypothesis of no difference in distribution of scores between the two populations represented against the alternative that the experimental group represents extreme reactions. Let $\alpha = 0.05$.

TABLE 3.28   .   **Reading improvement scores in two groups of pupils**

Control group (X)	50	48	41	58	51	56	57	40	42	53
Experimental group (Y)	69	88	65	38	54	85	35	55	75	43

FISHER EXACT TEST

Frequently the data from two independent samples consist of measurements that are dichotomous—that is, each observation is one or the other of two mutually exclusive types. We get such data when we compare two treatments and classify subjects as either responding or not responding. We may draw independent random samples from each of two populations and classify the items or subjects as either possessing or not possessing some characteristic.

For example, we may classify adolescents from two independent samples according to whether they have experimented with drugs. In a factory, we may categorize items produced on two different shifts as defective or not defective. Or we may randomly assign students with reading problems to one of two different remedial reading programs and later classify their reading abilities as improved or not improved. We may cast the results in a 2×2 contingency table such as Table 3.29. When, in anticipation of using the test about to be described, we put the data in the form of a table such as Table 3.29, we must arrange them in such a way that $A \geq B$ and choose the characteristic of interest so that $a/A \geq b/B$.

TABLE 3.29

A 2 × 2 contingency table

Sample	With characteristic	Without characteristic	Total
1	a	$A - a$	A
2	b	$B - b$	B
Total	$a + b$	$A + B - a - b$	$A + B$

The research objective in studies of this type is to determine whether the two populations differ with respect to the proportion of subjects (or items) that fall into the two classifications. In another words, we wish to test the null hypothesis that $p_1 = p_2$, where p_1 is the proportion with some characteristic of interest in population 1 and p_2 is the proportion with the characteristic in population 2.

A test, appropriate for use with experiments and studies yielding data with these characteristics, was proposed in the mid-1930s almost simultaneously by Fisher (T122, T123), Irwin (T124), and Yates (T125). The test is most useful when sample sizes are small. It has come to be known as the *Fisher exact test* because, if desired, it permits us to calculate the exact probability of obtaining the observed results or results that are more extreme.

Some theorists hold that Fisher's exact test is appropriate only when both marginal totals of Table 3.29 are fixed by the experiment. This specific model does not appear to arise very frequently in practice. Experimenters, however, have used the test when both marginal totals are not fixed. They usually resort to the test when samples are so small that the chi-square test (see Chapter 5) should not be used.

If we regard the marginal totals as fixed, we can obtain the probability of any given arrangement of the cell frequencies by evaluating the hypergeometric formula (see Equation 3.1). For very small samples, we may evaluate the appropriate hypergeometric formula to determine the probability of observing results as extreme as or more extreme than those actually observed, given that the null hypothesis is true. Tables are available, however, so in practice we do not have to perform these calculations.

Assumptions

A. The data consist of A sample observations from population 1 and B sample observations from population 2.

B. The samples are random and independent.

C. Each observation can be categorized as one of two mutually exclusive types.

Hypotheses

A. (Two-sided)

H_0: The proportion with the characteristic of interest is the same in both populations; that is, $p_1 = p_2$

H_1: The proportion with the characteristic of interest is not the same in both populations; $p_1 \neq p_2$

B. (One-sided)

H_0: The proportion with the characteristic of interest in population 1 is less than or the same as the proportion in population 2; $p_1 \leq p_2$

H_1: The proportion with the characteristic of interest is greater in population 1 than in population 2; $p_1 > p_2$

Test Statistic

The test statistic is b, the number in sample 2 with the characteristic of interest.

Decision Rule

Finney (T126) has prepared critical values of b for $A \leq 15$. Latscha (T127) has extended Finney's tables to accommodate values of A up to 20. Table A.10 gives these critical values of b for A between 3 and 20, inclusive. Significance levels of 0.05, 0.025, 0.01, and 0.005 are included. The specific decision rules are as follows:

A. *Two-sided test:* Enter Table A.10 with A, B, and a. If the observed value of b is equal to or less than the integer in a given column, reject H_0 at a level of significance equal to twice the significance level shown at the top of that column. For example, suppose $A = 8$, $B = 7$, $a = 7$, and the observed value of b is 1. We can reject the null hypothesis at the $2(0.05) = 0.10$, the $2(0.025) = 0.05$, and the $2(0.01) = 0.02$ levels of significance, but not at the $2(0.005) = 0.01$ level.

B. *One-sided test:* Enter Table A.10 with A, B, and a. If the observed value of b is less than or equal to the integer in a given column, reject H_0 at the level of significance shown at the top of that column. For example, suppose that $A = 16$, $B = 8$, $a = 4$, and the observed value of b is 3. We can reject the null hypothesis at the 0.05 and 0.025 levels of significance, but not at the 0.01 or 0.005 levels.

Example 3.9   Almy (E17) studied the relationship between residential location of social-class groups in American cities and the electoral cohesion displayed by such population groups. He also studied the consequences of this intragroup cohesion on intergroup conflict as manifested in electoral competition. Table 3.30 shows 14 cities classified on the basis of residential location of social-class groups and electoral cohesion within class groups at the time of an educational referendum.

To make the data conform to the arrangement specified in Table 3.29, we take the spatially segregated group as sample 1, and high electoral cohesion within class groups as the characteristic of interest. Table 3.31 shows the rearrangement.

We assume that we have a spatially segregated sample of size 10 and an independent spatially integrated sample of size 4. We wish to know whether we can conclude that the proportion of cities with high electoral cohesion within class

TABLE 3.30		**Residential location of social-class groups and electoral cohesiveness at the time of an educational referendum**

Residential location of social-class groups	*Electoral cohesion within class groups*		
	Low	High	Total
Spatially segregated	1*	9	10
Spatially integrated	3	1	4
Total	4	10	14

Source: Timothy A. Almy, "Residential Location and Electoral Cohesion: The Pattern of Urban Political Conflict," *Am. Polit. Sci. Rev.*, 67 (1973), 914–923.

* The author reported percentages in the four cells of his table. These have been applied to his row totals to obtain the cell frequencies shown here.

TABLE 3.31		**The data of Table 3.31 rearranged to conform to Table 3.29**

Residential location of social-class groups	*Electoral cohesion within class groups*		
	High	Low	Total
Spatially segregated	$9 = a$	$1 = A - a$	$10 = A$
Spatially integrated	$1 = b$	$3 = B - b$	$4 = B$
Total	$10 = a + b$	$4 = A + B - a - b$	$14 = A + B$

groups is higher in the population of cities with spatially segregated social-class groups than in the population of cities with spatially integrated social-class groups.

Hypotheses

$$H_0: p_1 \leq p_2$$
$$H_1: p_1 > p_2$$

Test Statistic

The test statistic is $b = 1$.

Decision

When we enter Table A.10 with $A = 10$, $B = 4$, and $a = 9$, we find that the critical value of b is 1 for $\alpha = 0.05$ (one-sided test). Since the observed value of $b = 1$ is

equal to the critical value of b for $\alpha = 0.05$, we can reject H_0 at the 0.05 level of significance. We cannot reject the null hypothesis at any other level of significance given in Table A.10, since $1 > 0$.

Large-Sample Approximation   For sufficiently large samples we can test the null hypothesis of the equality of two population proportions by using the normal approximation discussed earlier in connection with the median test. Compute

$$z = \frac{(a/A) - (b/B)}{\sqrt{\hat{p}(1 - \hat{p})(1/A + 1/B)}} \tag{3.20}$$

where

$$\hat{p} = (a + b)/(A + B) \tag{3.21}$$

and compare it for significance with appropriate critical values of the standard normal distribution. The use of the normal approximation is generally considered satisfactory if $a, b, A - a$, and $B - b$ are all greater than or equal to 5. Alternatively, when sample sizes are sufficiently large, we may test the null hypothesis by means of the chi-square test discussed in Chapter 5.

Power-Efficiency   The power of the Fisher exact test has been considered by Mainland and Sutcliffe (T128), Bennett and Hsu (T129), Sillitto (T130), and Gail and Gart (T131). Gail and Gart (T131) give minimal sample sizes n required to obtain a power of at least 0.50, 0.80, and 0.90, respectively, at the 0.05 and 0.01 nominal significance levels for the case of a one-sided test with equal margins ($A = B$) and selected alternative hypotheses.

FURTHER READING

The Fisher exact test has been the subject of some controversy among statisticians. Some feel that the assumption of fixed marginal totals is unrealistic in most practical applications. The controversy then centers around whether the test is appropriate when both marginal totals are not fixed. For further discussion of this and other points, see the articles by Barnard (T132, T133, T134), Fisher (T135), and Pearson (T136).

Sweetland (T137) compared the results of using the chi-square test described in Chapter 5 with those obtained using the Fisher exact test for samples of size $A + B = 3$ to $A + B = 69$. He found close agreement when A and B were close in size and the test was one-sided. Overall (T138) states that the Fisher exact test is severely conservative with reference to conventional levels of significance due to the discontinuity of the sampling distribution for 2×2 tables. He proposes an adjustment of the cell frequencies, which he says results in a correction for continuity with appropriate significance level continuity and increased power.

Carr (T139) presents an extension of the Fisher exact test to more than two samples of equal size and gives an industrial example to demonstrate the calculations. In another paper Carr (T140) offers an alternative to the test, proposing the use of a selection procedure of the type discussed by Gibbons et al. (T141). He claims that the selection procedure is simpler and requires a much smaller sample size.

Neave (T142) presents the Fisher exact test in a new format; the test is treated as one of independence rather than of homogeneity (see Chapter 5). He has prepared extensive tables for use with his approach.

The sensitivity of Fisher's exact test to minor perturbations in 2×2 contingency tables is discussed by Dupont (T143). For a good elementary discussion of the Fisher exact test, see Sokal and Rohlf (T144).

3.4

COMPUTER PROGRAMS

Dinneen and Blakesley (T145), Jung et al. (T146), Odeh (T147), Larsen (T148), Kummer (T149), and Harding (T150) discuss the use of the computer with the Mann–Whitney–Wilcoxon test. Schuster (T151) compares algorithms for the computation of the Mann–Whitney statistic. An algorithm that generates the null frequency distribution of the Ansari–Bradley statistic has been written by Dinneen and Blakesley (T152). Youngman and Daniel (T153) present a computer program for calculating the Mood dispersion test statistic. A number of computer programs have been written for the Fisher exact test. They include those by Gregory (T154), Robertson (T155), Berry and Mielke (T156, T157, T158) and Mehta and Patel (T159, T160).

Edgington et al. (T161) and Harlow and Lowry (T162) have written computer programs for the randomization test for the case of two independent samples (mentioned earlier). Edgington and Strain (T163) consider computer-time requirements for the computations involved in the randomization test counterparts of the *t* test and one-way analysis of variance. They find that using a high-speed computer to carry out these randomization tests can be relatively inexpensive.

Among the microcomputer software packages that include capabilities for the procedures covered in this chapter are *FILESTAT* (median test), *HP STATISTICS LIBRARY/2000* (median test), *SCA* (median test, Mann–Whitney test), and *STATISTIX* (Fisher exact test, Mann–Whitney test, median test). Among the other packages that will accommodate the Mann–Whitney test are *BMDPC*, *EXEC∗U∗STAT*, *MICROSTAT*, *MINITAB*, *NUMBER CRUNCHER*, *SPSS*, *STATISTIX*, *STATGRAPHICS*, and *STATPRO*.

EXERCISES

3.14 In a study of the effects of different interviewing techniques on the diastolic blood pressure of interviewees, Williams et al. (E18) obtained the results shown in Table 3.32. In one type of interview (CARD), the interviewee responded verbally to questions presented one at a time on 3 × 5 index cards. The interviewer assumed a passive role. During a second type of interview (INT), the interviewer tried to interact warmly and appropriately with the interviewee by asking questions and making appropriate comments while the interviewee answered the questions presented. Diastolic blood pressure was measured at one-minute intervals during the interviews. Table 3.32 shows the classification of subjects with regard to mean diastolic blood pressure during interview and type of interview.

Can we conclude on the basis of these data that the proportion of "significantly largers" is higher in the population represented by the INT sample than that represented by the CARD sample? Let $\alpha = 0.05$. Determine the P value.

TABLE 3.32   **Comparisons of mean diastolic pressure during interviews**

Interview	Significantly larger	Smaller and no change	Total
INT	6	0	6
CARD	1	5	6
Total	7	5	12

Source: Redford B. Williams, Jr., Chase P. Kimball, and Harold N. Williard, "The Influence of Interpersonal Interaction on Diastolic Blood Pressure," *Psychosom. Med.*, 34 (1972), 194–198.

3.15 Gill and Murray (E19) conducted an experiment designed to test discrimination of song ability among blue-winged and golden-winged warblers (*Vermivora pinus* and *V. chrysopetra*) of southeastern Michigan. Within the hearing range of territorial males, the experimenters played tape recordings of songs of the listening bird and songs of the other species. Discriminatory behavior of the birds was based on whether they responded to the recordings. Table 3.33 shows 22 birds classified according to species and discriminatory behavior. Can we conclude on the basis of these data that the proportion of nondiscriminators is higher among golden-wings than among blue-wings? Let $\alpha = 0.05$. Determine the P value.

TABLE 3.33   **Discriminatory behavior of territorial male *Vermivora***

Species	Discriminators	Nondiscriminators	Total
Michigan blue-wings	4	6	10
Michigan golden-wings	3	9	12
Total	7	15	22

Source: Frank B. Gill and Bertram G. Murray, Jr., "Discrimination Behavior and Hybridization of the Blue-Winged and Golden-Winged Warblers," *Evolution*, 26 (1972), 282–293; by permission of the Society for the Study of Evolution.

REVIEW EXERCISES

3.16 In a study of cardiovascular findings in patients with acromegaly, McGuffin et al. (E20) obtained the data shown in Table 3.34 on 13 of these patients who were hypertensive and 10 who were normotensive. We wish to know whether we can conclude on the basis of these data that there is greater dispersion in systolic blood pressure among hypertensive patients than among those who are normotensive. Use the Moses dispersion test and determine the P value.

TABLE 3.34 **Systolic blood pressure, in millimeters of mercury, for two groups of patients**

Normotensive (X)	122	110	140	130	140	110	120
	105	98	140				
Hypertensive (Y)	160	140	150	140	150	160	220
	155	150	170	180	210	150	

Source: William L. McGuffin, Barry M. Sherman, Jesse Roth, Phillip Gorden, C. Ronald Kahn, William C. Roberts, and Peter L. Frommer, "Acromegaly and Cardiovascular Disorders," *Ann. Intern. Med.*, 81 (1974), 11–18.

3.17 In a study designed to assess the performance of subjects with Parkinsonism, Cassell et al. (E21) recorded the average time delay of experimental and normal subjects performing fast tasks. Table 3.35 shows the results. The patients with Parkinsonism were categorized as severely disabled. On the basis of these data, can one conclude that there is less dispersion in the performance of normal subjects? Determine the P value, using the Ansari–Bradley dispersion test.

TABLE 3.35 **Average time delay, milliseconds, performing fast tasks**

Normal subjects (X)	321	213	211	258	267	281	317	229	206	281
Patients with Parkinson's disease (Y)	290	660	400	403	290	360	460	420		

Source: Kenneth Cassell, Kenneth Shaw, and Gerald Stern, "A Computerized Tracking Technique for the Assessment of Parkinsonian Motor Disabilities," *Brain*, 96 (1973), 815–826.

3.18 In a behavior-modification experiment, Tanner (E22) obtained the results shown in Table 3.36 on 12 subjects; half received 5 mA and half received less than 5 mA stimulation during the course of the experiment. Can we conclude on the basis of these data that subjects receiving 5 mA stimulation tend to have a higher median outcome score than subjects receiving less? Use the median test.

3.19 Keyvan-Larijarni and Tannenberg (E23) studied the effects of two different methods of treating patients who had ingested an unknown amount of a solution containing 60% methanol by volume. Three patients (group 1) were treated with a recycling single-pair twin-coil hemodialyzer and three (group 2) were treated with peritoneal dialysis. Levels of methanol for each patient were determined in the same hospital by identical procedures.

TABLE 3.36   **Outcome scores for 12 subjects participating in a behavior-modification experiment**

5 mA (X)	1.299	0.059	0.610	0.247	1.745	0.236
<5 mA (Y)	0.097	0.066	0.133	0.123	0.000	0.375

Source: Barry A. Tanner, "Shock Intensity and Fear of Shock in the Modification of Homosexual Behavior in Males by Avoidance Learning," *Behav. Res. Therapy*, 11 (1973), 213–218; reprinted with permission of Pergamon Press.

The serum methyl alcohol levels (SMAL), measured at admission and eight hours later, are shown in Table 3.37, along with the amount of reduction. We wish to know whether we can conclude on the basis of these data that the two methods differ in their ability to reduce SMAL values. Use the Mann–Whitney test and determine the P value.

TABLE 3.37   **Serum methyl alcohol levels, milligrams per 100 ml, in six patients before and after treatment by two methods**

Group 1

At admission	Eight hours later	Amount of reduction (X)
185	22	163
178	54	124
96	78	18

Group 2

At admission	Eight hours later	Amount of reduction (Y)
186	158	28
198	177	21
171	147	24

Source: Hossein Keyvan-Larijarni and Alf M. Tannenberg, "Methanol Intoxication, Comparison of Peritoneal Dialysis and Hemodialysis," *Arch. Inter. Med.*, 134 (1974), 293–296; copyright 1974, American Medical Association.

3.20   Seppälä et al. (E24) reported the data shown in Table 3.38. Use the Mann–Whitney procedure to test the null hypothesis of no difference between the location parameters of the populations represented by these samples.

TABLE 3.38   **Oxytocin concentrations, picograms per milliliter, in amniotic fluid in two groups of women**

Parturient patients with labor contractions (X)	1450	510	720	510	820	680	400	710	110
	800	380	1100	1600	650	800	670	350	1350
Women with no labor contractions (Y)	240	300	150	330	360	180	250	170	200
	430	510	800						

Source: Markku Seppälä, Ilkka Aho, Anja Tissari, and Erkki Ruoslahti, "Radio-Immunoassay of Oxytocin in Amniotic Fluid, Fetal Urine, and Meconium During Late Pregnancy and Delivery," *Am. J. Obstet. Gynecol.*, 114 (1972), 788–795.

3.21 Griffiths (E25) reported data on the extent of coffee-berry disease on farms not sprayed and on those sprayed with a fungicide. The results are shown in Table 3.39, in terms of percentage infections in test berries. Do these data provide sufficient evidence to indicate a difference in population location parameters? Let $\alpha = 0.01$. Use the Mann–Whitney test and determine the P value.

TABLE 3.39

Extent of coffee-berry disease (percentage infections in test berries) on farms sprayed and not sprayed with a fungicide

Unsprayed	6.01	2.48	1.76	5.10	0.75	7.13	4.88
Sprayed (at least 14 months prior to sampling)	5.68	5.68	16.30	21.46	11.63	44.20	33.30

Source: Ellis Griffiths, " 'Negative' Effects of Fungicides in Coffee," *Trop. Sci.*, 14 (1972), 79–89.

3.22 Garrod et al. (E26) measured the nicotine metabolites, cotinine, and nicotine-1′-N-oxide, in 24-hour urine collections from normal healthy male cigarette smokers and cigarette smokers with cancer of the urinary bladder. Table 3.40 shows the ratio of cotinine to nicotine-1′-N-oxide in the two groups of subjects. Test for a difference between population medians at the 0.01 level. Use the median test and determine the P value.

TABLE 3.40

Ratio of cotinine to nicotine-1′-N-oxide in two groups of subjects

Patients with cancer of the urinary bladder	5.0	8.3	6.7	3.0	2.5	12.5	2.4	5.5	5.2	21.3	5.1	1.6
	2.1	4.6	3.2	2.2	7.0	3.3	6.7	11.1	3.4	5.9	27.4	
Control subjects	2.3	1.9	3.6	2.5	0.75	2.5	2.1	1.1	2.3	2.2	3.5	1.8
	2.3	1.4	2.1	2.0	2.3	2.4	3.6	2.6	1.5			

Source: J. W. Garrod, P. Jenner, G. R. Keysell, and B. R. Mikhael, "Oxidative Metabolism of Nicotine by Cigarette Smokers with Cancer of the Urinary Bladder," *J. Nat. Cancer Inst.*, 52 (1974), 1421–1924.

3.23 O'Neill et al. (E27) drew activated partial thromboplastin time (APTT) specimens from 10 patients (group A) immediately after the placement of heparin locks. APTT specimens were also drawn from 10 patients (group B) prior to placement of the heparin lock. The APTT values for the two groups are shown in Table 3.41. Construct the approximate 95% confidence interval for the difference between population medians.

TABLE 3.41

APTT values for two groups of subjects in Exercise 3.23

Group A	52.0	74.0	34.0	32.5	51.5	65.0	52.0	49.0	32.0	40.0
Group B	31.0	30.0	32.5	28.0	31.5	33.0	31.0	27.0	30.5	34.5

Source: Thomas J. O'Neill, Lawrence M. Tierney, Jr., Ronald J. Proulx, "Heparin Lock-Induced Alterations in the Activated Partial Thromboplastin Time," *J. Am. Med. Assoc.*, 227 (1974), 1297–1298; copyright 1974, American Medical Association.

3.24 A self-concept scale was administered to a sample of 15 apparently normal subjects and to an independent sample of 15 patients undergoing psychiatric treatment. The scores are

shown in Table 3.42. Can one conclude from these data that the two population medians are different? Let $\alpha = 0.05$. Use the median test.

TABLE 3.42   **Scores made on a self-concept scale by 15 normal subjects and 15 psychiatric patients**

Normal subjects	63	69	70	81	90	91	90	88	82	87	83	85	85	87	86
Psychiatric patients	62	64	68	88	75	79	71	70	82	70	69	75	79	78	75

3.25 Table 3.43 shows the scores made on a verbal comprehension test by a sample of 25 educationally handicapped children and a sample of 20 educable mentally retarded children. Do these data provide sufficient evidence to indicate that the two population medians differ? Let $\alpha = 0.05$ and determine the P value. Use the median test.

TABLE 3.43   **Verbal comprehension test scores made by educationally handicapped (EH) and educable mentally retarded (EMR) children**

EH	77	78	70	72	74	68	71	70	72	71	75	78	79
	87	88	70	72	74	88	71	80	82	72	72	73	
EMR	60	62	65	71	62	70	68	65	76	72			
	68	72	78	71	70	76	79	68	66	70			

3.26 An experiment using seventh-grade pupils was conducted to compare the performance of normal readers and poor readers on a maze task. The scores are shown in Table 3.44. Do these data provide sufficient evidence to indicate that poor readers score lower on the average than normal readers on the maze task? Use the Mann–Whitney test and a significance level of 0.05. Determine the P value.

TABLE 3.44   **Maze task performance scores of normal and poor readers**

Poor readers	67	55	51	40	25	18	34	44	52	59	54	53			
Normal readers	95	87	77	73	44	64	68	70	55	59	67	88	89	90	52

3.27 To study the effects of prolonged inhalation of cadmium, Princi and Geever (E28) exposed 10 dogs to cadmium oxide, while 10 dogs serving as controls were not exposed to this substance. At the end of the experiment, they determined the levels of hemoglobin of the 20 dogs, shown in Table 3.45. Let $\alpha = 0.05$ and use the Mann–Whitney test to determine if one may conclude that, on the average, inhalation of cadmium causes a reduction in hemoglobin levels in dogs.

TABLE 3.45   **Hemoglobin determinations, grams, in twenty dogs**

Exposed to cadmium oxide (X)	14.6	15.8	16.4	14.6	14.9	14.3	14.7	17.2	16.8	16.1
Controls (Y)	15.5	17.9	15.5	16.7	17.6	16.8	16.7	16.8	17.2	18.0

Source: Frank Princi and Erving F. Geever, "Prolonged Inhalation of Cadmium," *Arch. Indust. Hyg. Occup. Med.*, 1 (1950), 651–661. Copyright 1950, American Medical Association.

3.28 Singh and Prasad (E29) studied social conformity and self-esteem in postgraduate students. They found that the mean score on self-esteem in a sample of 100 conformists and the mean self-esteem score of a sample of 100 nonconformists were significantly different at the 0.01 level of significance by the *t* test. Suppose that the study was repeated with other subjects, and that the results were those shown in Table 3.46. Can we conclude on the basis of these data that the population medians are different? Use the median test, and let $\alpha = 0.01$.

TABLE 3.46	Self-esteem scores of two groups									
Conformists	48	55	56	49	41	55	44	53	42	50
Nonconformists	59	58	48	57	59	45	59	68	61	

3.29 Jamuar and Singh (E30) used an adaptation in Hindi of Maslow's security–insecurity inventory with a sample of 50 after-care home girls and a sample of 50 college girls. A *t* test revealed a significant difference at the 0.01 level in sample means. Suppose that the study was repeated with two other samples of subjects; the results are shown in Table 3.47. Do these data provide sufficient evidence to indicate that the population medians are different? Use the Mann–Whitney test and a 0.01 level of significance.

TABLE 3.47	Scores on security–insecurity inventory													
After-care home girls	34	39	32	32	31	34	35	40	45	43	30	37	39	43
College girls	22	27	22	22	29	26	20	30	23	34	31	33		

3.30 Hoffman and Jackson (E31) administered Jackson's (E32) Differential Personality Inventory to male and female alcoholics. The authors computed *t* tests to evaluate differences of means on individual scales between males and females. Among their findings was a significant difference at the 0.01 level on the cynicism scale. Suppose that this test was administered to another group of male and female alcoholics, with the results shown in Table 3.48. Use the Mann–Whitney test to see whether these data provide sufficient evidence to indicate a difference in population medians. Let $\alpha = 0.05$.

TABLE 3.48	Scores on cynicism scale							
Males	5	6	5	4	3	2	3	4
Females	1	2	2	3	1	2	3	

3.31 A researcher gave a random sample of 15 college men and an independent random sample of college women a test to measure their knowledge of the stock market. Table 3.49 shows the scores. We wish to know if we may conclude on the basis of these samples that the two populations differ with respect to their medians. Let $\alpha = 0.05$. Use the test that makes use of the most information.

| *TABLE 3.49* | **Data for Exercise 3.31** |

Men's scores, X			Women's scores, Y		
21.50	20.00	15.40	15.00	9.10	8.00
17.00	19.00	18.20	13.00	8.75	11.00
23.00	12.50	12.50	11.20	16.10	13.50
21.00	15.00	22.25	18.00	16.50	17.25
22.50	10.00	11.00	20.00	17.75	16.30
			13.00	13.25	19.00
			20.30	21.20	

3.32 A firm wished to compare two methods of advertising a new product. Two samples of subjects were chosen to participate in the experiment. Subjects in the first sample were exposed to advertising method A, and subjects in the second sample were exposed to advertising method B. At the end of the experiment, the subjects were given a test to measure their knowledge of the new product. The results are shown in Table 3.50. Use the Mann–Whitney test to determine if these data provide sufficient evidence to indicate a difference in median scores between the two populations. Let $\alpha = 0.05$.

| *TABLE 3.50* | **Data for Exercise 3.32** |

Method A scores	55	64	65	76	85	86	85	83	77	82	78	80	80	82	81
Method B scores	57	59	63	83	70	74	66	65	77	65	64	70	74	73	70

3.33 Table 3.51 shows the tensile strengths (coded to facilitate calculations) of samples of plastic trash bags drawn from the stock of two manufacturers, A and B. Use the Mann–Whitney test to determine whether one may conclude that bags made by manufacturer B are, on the average, stronger than those made by manufacturer A. Let $\alpha = 0.05$.

| *TABLE 3.51* | **Data for Exercise 3.33** |

A	62	71	62	68	66	60	68	70	60	65	70	65	72	62	71	
B	69	63	68	72	75	72	71	72	70	71	68	74	72	70	68	67

3.34 A team of industrial psychologists studied the emotional maturity of high school graduates and high school dropouts employed in the same type of work. A standardized test was given to samples of subjects selected from the two populations. The test scores are shown in Table 3.52. Use the Mann–Whitney test to determine whether we may conclude that the two sampled populations have different medians. Let $\alpha = 0.05$ and find the P value.

| *TABLE 3.52* | **Data for Exercise 3.34** |

High school graduates	57	69	62	94	96	63	62	89	69	67	86	97	71
	64	85	68	57	78	80	57	83	73	80	79	84	
High school dropouts	49	54	47	72	78	49	74	67	58	41	63	57	51
	65	63	64	65	57	85	59	85					

3.35 A simple random sample of female applicants for an assembly-line job and an independent simple random sample of male applicants for the same job were given an aptitude test for the job. The scores are shown in Table 3.53. Can one conclude from these data that the two sampled populations of scores are not identically distributed? Let $\alpha = 0.05$.

TABLE 3.53   **Data for Exercise 3.35**

Females' scores	18.1	9.0	9.0	25.9	11.5	9.9	18.1	10.7	13.6	18.2
	10.4	28.8	32.4	28.0	24.3	17.1	21.9	9.0	9.5	
Males' scores	21.3	23.1	28.9	32.6	29.3	15.0	39.9	16.8	11.8	32.3
	20.0	13.0	14.0	12.5	13.3	23.7				

3.36 A manufacturer compared the viscosity of two brands of motor oil. A simple random sample of specimens of brand A and an independent simple random sample of specimens of brand B yielded the results (coded for ease of computation) shown in Table 3.54. Can we conclude on the basis of these data that the viscosity of brand A is more variable than the viscosity of brand B? Use the Ansari–Bradley test. Let $\alpha = 0.05$.

TABLE 3.54   **Data for Exercise 3.36**

Brand A	25	39	35	18	50	11	42	47	19	16	45	41	25	35	44	35
	24	55	36	11	11	10	38	25	52	35	25	60	51	21	17	13
Brand B	59	37	36	65	42	33	56	36	64	55	61	63	57	33	37	60
	38	41	48	62	32	55	56	35	59	66	50	38	70	46	35	54

3.37 A soap manufacturer wished to know if two different formulas would cause a difference in the variability of the specific gravity of its product. Sample bars of soap were made by each of the two formulas; the results (coded) are shown in Table 3.55. What should the manufacturer conclude on the basis of these data if the Moses test is used?

TABLE 3.55   **Data for Exercise 3.37**

Formula A	9	11	9	13	10	8	7	12	11	9					
Formula B	12	10	13	11	11	15	15	14	15	11	14	14	13	13	9

3.38 A productivity specialist conducted an experiment to investigate the effect of music on the efficiency of assembly-line employees. Sixteen employees were randomly assigned to a music group or a no-music group. Table 3.56 shows the efficiency levels of the subjects at the end of one month. Can we conclude on the basis of these data that music causes extreme reactions in employees' efficiency levels? At what tabulated α level are the sample results significant?

TABLE 3.56   **Data for Exercise 3.38**

Music group	12	35	18	57	67	40	39	58
No-music group	21	35	40	38	23	27	28	39

3.39 An advertising agency used the following procedure to test the effectiveness of two radio commercials for the same product. A format A commercial was used during the Monday morning programming of a local radio station. The next day the agency telephoned 15 listeners; nine of the listeners remembered the commercial. The following Monday morning the format B commercial was aired on the station. A call to 12 listeners found four that recalled the format B commercial. Do these data provide sufficient evidence to indicate that the format A commercial is more easily remembered? Let $\alpha = 0.01$ and find the P value.

3.40 Consult *Newspaper Rates and Data*, published by Standard Rate & Data Service, Inc., at Wilmette, Illinois. Select a simple random sample of 30 or more daily newspapers published east of the Mississippi River and an independent simple random sample of 30 or more published west of the Mississippi. Use the large-sample approximation to the Mann–Whitney test to determine whether we can conclude that daily newspapers from the two areas of the country differ with respect to their average classified display rates.

3.41 From each of the populations defined in Exercise 3.40, select a simple random sample of 30 or more daily newspapers. Use the large-sample approximation to the Ansari–Bradley test to test an appropriate null hypothesis about the dispersion of classified display rates in the two populations.

3.42 Consider the populations defined in Exercise 3.40. Let classified display rate be the variable of interest. Formulate a hypothesis about the variable in these populations that can be tested by the Wald–Wolfowitz runs test. Select a simple random sample of 30 or more newspapers from the populations and perform the test, using the large-sample approximation.

REFERENCES

T1 Mood, Alexander M., *Introduction to the Theory of Statistics*, New York: McGraw-Hill, 1950.

T2 Westenberg, J., "Significance Test for Median and Interquartile Range in Samples from Continuous Populations of Any Form," *Akad. Wetensch. Afdeeling Voor de Wis.*, 51 (1948), 252–261.

T3 Owen, D. B., *Handbook of Statistical Tables*, Reading, Mass.: Addison-Wesley, 1962.

T4 Ostle, Bernard, and Richard W. Mensing, *Statistics in Research*, third edition, Ames, Iowa: The Iowa State University Press, 1975.

T5 Burr, Irving W., *Applied Statistical Methods*, New York: Academic Press, 1974.

T6 Dunn, Olive Jean, *Basic Statistics: A Primer for the Biomedical Sciences*, second edition, New York: Wiley, 1977.

T7 Mood, Alexander M., "On the Asymptotic Efficiency of Certain, Non-Parametric Two-Sample Tests," *Ann. Math. Statist.*, 25 (1954), 514–522.

T8 Bedall, F. K., and H. Zimmermann, "A Bivariate Median Test," *Biometrical J.*, 22 (1980), 281–282.

T9 Fligner, Michael A., and Steven W. Rust, "A Modification of Mood's Median Test for the Generalized Behrens-Fisher Problem," *Biometrika*, 69 (1982), 221–226.

T10 Brown, G. W., and A. M. Mood, "On Median Tests for Linear Hypotheses," in *Proceedings of*

the Second Berkeley Symposium on Mathematical Statistics and Probability, Jerzy Neyman (ed.), Berkeley and Los Angeles: University of California Press, 1951.

T11   Westenberg, J., "A Tabulation of the Median Test with Comments and Corrections to Previous Papers," *Neder. Akad. Wetensch. Proc. Ser. A*, 55 (*Indag. Math.*) 14 (1952), 10–15.

T12   Mann, H. B., and D. R. Whitney, "On a Test of Whether One of Two Random Variables Is Stochastically Larger than the Other," *Ann. Math. Statist.*, 18 (1947), 50–60.

T13   Festinger, L., "The Significance of Differences between Means without Reference to the Frequency Distribution Function," *Psychometrika*, 11 (1946), 97–105.

T14   White, C., "The Use of Ranks in a Test of Significance for Comparing Two Treatments," *Biometrics*, 8 (1952), 33–41.

T15   Wilcoxon, F., "Individual Comparisons by Ranking Methods," *Biometrics*, 1 (1945), 80–83.

T16   Noether, Gottfried E., *Elements of Nonparametric Statistics*, New York: Wiley, 1967.

T17   Haynam, G. E., and Z. Govindarajulu, "Exact Power of the Mann–Whitney Test for Exponential and Rectangular Alternatives," *Ann. Math. Statist.*, 37 (1966), 945–953.

T18   Chanda, K. C., "On the Efficiency of Two-Sample Mann–Whitney Test for Discrete Populations," *Ann. Math. Statist.*, 34 (1963), 612–617.

T19   Rasmussen, Jeffrey Lee, "The Power of Student's *t* and Wilcoxon *W* Statistics: A Comparison," *Eval. Rev.*, 9 (1985), 505–510.

T20   Blair, R. Clifford, and James J. Higgins, "A Comparison of the Power of Wilcoxon's Rank-Sum Statistic to that of Student's *t* Statistic Under Various Nonnormal Distributions," *J. Educ. Statist.*, 5 (1980), 309–335.

T21   Blair, R. Clifford, and J. J. Higgins, "The Power of *t* and Wilcoxon Statistics: A Comparison," *Eval. Rev.*, 5 (1980), 645–656.

T22   Bradstreet, T. E., and S. A. Lemeshow, "A Comparison of Type I Error Rates of Some Two-Sample Parametric and Nonparametric Tests of Location When the Assumption of Equal Variances is Violated" (Abstract), *Biometrics*, 40 (1984), 1185.

T23   Auble, D., "Extended Tables for the Mann–Whitney Statistic," *Bull. Inst. Educ. Res. Indiana Univ.*, 1 (1953), 1–39.

T24   Jacobson, J. E., "The Wilcoxon Two-Sample Statistic: Tables and Bibliography," *J. Amer. Statist. Assoc.*, 58 (1963), 1086–1103.

T25   Milton, R. C., "An Extended Table of Critical Values for the Mann–Whitney (Wilcoxon) Two-Sample Statistic," *J. Amer. Statist. Assoc.*, 59 (1964), 925–934.

T26   Verdooren, L. R., "Extended Tables of Critical Values for Wilcoxon's Test Statistic," *Biometrika*, 50 (1963), 177–186.

T27   Noether, Gottfried E., "Sample Size Determination for Some Common Nonparametric Tests," *J. Amer. Statist. Assoc.*, 82 (1987), 645–647.

T28   Buckle, N., C. H. Kraft, and C. Van Eeden, "An Approximation to the Wilcoxon–Mann–Whitney Distribution," *J. Amer. Statist. Assoc.*, 64 (1969), 591–599.

T29   Gelberg, R. S., "Relation between Mann–Whitney's Statistics and Kendall's Correlation Coefficient Tau," *Teor. Veroyatnost. I Yeye Primenen.*, 19 (1974), 211.

T30   Nemenyi, P., "Agility Ratings and Ground Meat—Introducing Mann–Whitney Differences," *Int. Statist. Rev.*, 41 (1973), 240–244.

T31   Serfling, R. J., "The Wilcoxon Two-Sample Statistic on Strongly Mixing Processes," *Ann. Math. Statist.*, 39 (1968), 1202–1209.

T32   Zaremba, S. K., "Note on the Wilcoxon–Mann–Whitney Statistic," *Ann. Math. Statist.*, 36 (1965), 1058–1060.

T33 Zeoli, G. W., and T. S. Fong, "Performance of a 2-Sample Mann–Whitney Nonparametric Detector in a Radar Application," *IEEE Trans. Aerospace and Electron. Systems*, AES 7 (1971), 951.

T34 Fisher, R. A., *Design of Experiments*, London: Oliver and Boyd, 1936.

T35 Pitman, E. J. G., "Significance Tests Which May Be Applied to Samples from Any Populations," *Supplement to J. Roy. Statist. Soc.*, 4 (1937), 119–130.

T36 Hoeffding, W., "Large-Sample Power of Tests Based on Permutations of Observations," *Ann. Math. Statist.*, 23 (1952), 169–192.

T37 Lehmann, E. L., and C. Stein, "On Some Theory of Nonparametric Hypotheses," *Ann. Math. Statist.*, 20 (1949), 28–46.

T38 Welch, B. L., "On the *z*-Test in Randomized Blocks and Latin Squares," *Biometrika*, 29 (1937), 21–52.

T39 Moses, L. E., "Non-Parametric Statistics for Psychological Research," *Psychol. Bull.*, 49 (1952), 122–143.

T40 Gocka, E. F., "Comments on Randomization Tests," *Psychol. Rep.*, 32 (1973), 293–294.

T41 Ray, W., "Logic for a Randomization Test," *Behav. Sci.*, 11 (1966), 405–406.

T42 Ray, W. S., and S. A. Schabert, "An Algorithm for a Randomization Test," *Educ. Psychol. Measurement*, 32 (1972), 823–829.

T43 Tukey, J. W., "A Quick, Compact, Two-Sample Test to Duckworth's Specifications," *Technometrics*, 1 (1959), 31–48.

T44 Neave, H. R., and W. J. Granger, "A Monte Carlo Study Comparing Various Two-Sample Tests for Differences in Mean," *Technometrics*, 10 (1968), 509–522.

T45 Neave, H. R., "A Development of Tukey's Quick Test of Location," *J. Amer. Statist. Assoc.*, 61 (1966), 949–964. Corrigenda, *Ibid.*, 62 (1967), 1522.

T46 Neave, H. R., and C. W. J. Granger, "Two-Sample Tests for Differences in Mean—Results of a Simulation Study," Interval Report of the Department of Mathematics, University of Nottingham, 1966. Cited in Neave (T45).

T47 Rosenbaum, S., "On Some Two-Sample Nonparametric Tests," *J. Amer. Statist. Assoc.*, 60 (1965), 1118–1126.

T48 Biebler, K. E., and B. Jager, "On a Nonparametric Test of Location," *Int. J. Clin. Pharmacol., Therapy and Toxicol.*, 22 (1984), 479–480.

T49 Noether, G. E., "Wilcoxon Confidence Intervals for Location Parameters in the Discrete Case," *J. Amer. Statist. Assoc.*, 62 (1967), 184–188.

T50 Noether, G. E., "Distribution-Free Confidence Intervals," *Amer. Statist.*, 26 (February 1972), 39–41.

T51 Moses, L. E., "Query: Confidence Limits from Rank Tests," *Technometrics*, 7 (1965), 257–260.

T52 Walker, H. M., and J. Lev, *Statistical Inference*, New York: Holt, Rinehart, and Winston, 1953.

T53 Bauer, D. F., "Constructing Confidence Sets Using Rank Statistics," *J. Amer. Statist. Assoc.*, 67 (1972), 687–690.

T54 Gibbons, Jean Dickinson, *Nonparametric Statistical Inference*, New York: McGraw-Hill, 1971.

T55 Sandelius, M., "A Graphical Version of Tukey's Confidence Interval for Slippage," *Technometrics*, 10 (1968), 193–194.

T56 Hodges, J. L., Jr., and E. L. Lehmann, "Estimates of Location Based on Rank Tests," *Ann. Math. Statist.*, 34 (1963), 598–611.

T57 Høyland, A., "Robustness of the Hodges–Lehmann Estimates for Shift," *Ann. Math. Statist.*, 36 (1965), 174–197.

T58 Laan, P. Van Der, *Simple Distribution-Free Confidence Intervals for a Difference in Location*, Eindhoven, The Netherlands: Philips Research Reports Supplements Number 5, 1970.

T59 Lehmann, E. L., "Nonparametric Confidence Intervals for a Shift Parameter," *Ann. Math. Statist.*, 34 (1963), 1507–1512.

T60 Hettmansperger, Thomas P., and Joseph W. McKean, "A Graphical Representation for Non-Parametric Inference," *Amer. Statist.*, 28 (1974), 100–102.

T61 Pearson, E. S., "The Analysis of Variance in Cases of Non-Normal Variation," *Biometrika*, 23 (1931), 114–133.

T62 Gayen, A. K., "The Distribution of the Variance Ratio in Random Samples of Any Size Drawn from Non-Normal Universes," *Biometrika*, 37 (1950), 236–255.

T63 Geary, R. C., "Testing for Normality," *Biometrika*, 34 (1947), 209–242.

T64 Finch, D. J., "The Effect of Non-Normality on the *z*-test When Used to Compare the Variances in Two Populations," *Biometrika*, 37 (1950), 186–189.

T65 Miller, R. G., Jr., "Jackknifing Variances," *Ann. Math. Statist.*, 39 (1968), 567–582.

T66 Ansari, A. R., and R. A. Bradley, "Rank-Sum Tests for Dispersion," *Ann. Math. Statist.*, 31 (1960), 1174–1189.

T67 David, F. N., and D. E. Barton, "A Test for Birth-Order Effects," *Ann. Hum. Gen.*, 22 (1958), 250–257.

T68 Freund, J. E., and A. R. Ansari, *Two-Way Rank Sum Test for Variances*, Technical Report Number 34, Blacksburg, Va.: Virginia Polytechnic Institute, 1957.

T69 Moses, L. E., "Rank Tests of Dispersion," *Ann. Math. Statist.*, 34 (1963), 973–983.

T70 Shorack, G. R., "Testing and Estimating Ratios of Scale Parameters," *J. Amer. Statist. Assoc.*, 64 (1969), 999–1013.

T71 Hollander, M., "Certain Un-Correlated Nonparametric Test Statistics," *J. Amer. Statist. Assoc.*, 63 (1968), 707–714.

T72 Lehmann, E. L., "Consistency and Unbiasedness of Certain Nonparametric Tests," *Ann. Math. Statist.*, 22 (1951), 165–179.

T73 Box, G. E. P., "Non-Normality and Tests on Variances," *Biometrika*, 40 (1953), 318–335.

T74 Box, G. E. P., and S. L. Andersen, "Permutation Theory in the Derivation of Robust Criteria and the Study of Departures from Assumption," *J. Roy. Statist. Soc. Ser. B*, 17 (1955), 1–26.

T75 Capon, J., "Asymptotic Efficiency of Certain Locally Most Powerful Rank Tests," *Ann. Math. Statist.*, 32 (1961), 88–100.

T76 Crouse, C. F., "Note on Mood's Test," *Ann. Math. Statist.*, 35 (1964), 1825–1826.

T77 Deshpandé, Jayant V., and Kalpana Kusum, "A Test for the Nonparametric Two-Sample Scale Problem," *Austral. J. Statist.*, 26 (1984), 16–24.

T78 Kamat, A. R., "A Two-Sample Distribution-Free Test," *Biometrika*, 43 (1956), 377–387.

T79 Klotz, J. H., "Nonparametric Tests for Scale," *Ann. Math. Statist.*, 33 (1962), 498–512.

T80 Levene, H., "Robust Tests for Equality of Variances," in *Contributions to Probability and Statistics*, Olkin et al. (eds.), Palo Alto, Calif.: Stanford University Press, 1960, pp. 278–292.

T81 Nemenyi, P., "Variances: An Elementary Proof and a Nearly Distribution-Free Test," *Amer. Statist.*, 23 (December 1969), 35–37.

T82 Raghavachari, M., "The Two-Sample Scale Problem When Locations Are Unknown," *Ann. Math. Statist.*, 36 (1965), 1236–1242.

T83 Rosenbaum, S., "Tables for a Nonparametric Test of Dispersion," *Ann. Math. Statist.*, 24 (1953), 663–668.

138   CHAPTER 3   PROCEDURES THAT UTILIZE DATA FROM TWO INDEPENDENT SAMPLES

T84 Savage, I. R., "Contributions to the Theory of Rank Order Statistics—the Two-Sample Case," *Ann. Math. Statist.*, 27 (1956), 590–615.

T85 Shorack, G. R., "Nonparametric Tests and Estimation of Scale in the Two-Sample Problem," Technical Report Number 10 (USPHS-5TIGM 25–07), Statistics Department, Stanford University, 1965.

T86 Siegel, S., and J. W. Tukey, "A Nonparametric Sum of Ranks Procedure for Relative Spread in Unpaired Samples," *J. Amer. Statist. Assoc.*, 55 (1960), 429–445, Errata, *Ibid.*, 56 (1961), 1005.

T87 Sukhatme, B. V., "A Two-Sample Distribution-Free Test for Comparing Variances," *Biometrika*, 45 (1958), 544–548.

T88 Taha, M. A. H., "Rank Test for Scale Parameter for Asymmetrical One-Sided Distributions," *Publ. Inst. Statist. Univ. Paris*, 13 (1964), 169–180.

T89 Tiku, M. L., "Robust Statistics for Testing Equality of Means or Variances," *Communic. in Statist.—Theory and Methods*, 11 (1982), 2543–2558.

T90 Ulrich, Gary, and Richard Einsporn, "A Comparison of Nonparametric Tests for Shift in Scale," *Proc. Amer. Statist. Assoc., Statist. Computing Sec.* (1985), 242–245.

T91 Laubscher, Nico F., F. E. Steffens, and Elsie M. Delange, "Exact Critical Values for Mood's Distribution-Free Test Statistic for Dispersion and Its Normal Approximation," *Technometrics*, 10 (1968), 497–507.

T92 Mielke, P. W., Jr., "Note on Some Squared Rank Tests With Existing Ties," *Technometrics*, 9 (1967), 312–314.

T93 Basu, A. P., and G. Woodworth, "A Note on Nonparametric Tests for Scale," *Ann. Math. Statist.*, 38 (1967), 274–277.

T94 Puri, M. L., "On Some Tests of Homogeneity of Variances," *Ann. Inst. Statist. Math.*, 17 (1965), 323–330.

T95 Duran, B. S., "A Survey of Nonparametric Tests for Scale," *Communic. in Statist.—Theory and Methods*, 5 (1976), 1287–1312.

T96 Sukhatme, B. V., "On Certain Two Sample Nonparametric Tests for Variances," *Ann. Math. Statist.*, 28 (1957), 188–194.

T97 David, F. N., "A Note on Wilcoxon's and Allied Tests," *Biometrika*, 43 (1956), 485–488.

T98 Duran, B. S., and P. W. Mielke, Jr., "Robustness of the Sum of Squared Ranks Test," *J. Amer. Statist. Assoc.*, 63 (1968), 338–344.

T99 Gibbons, J. D., "Correlation Coefficients between Nonparametric Tests for Location and Scale," *Ann. Inst. Statist. Math.*, 19 (1967), 519–526.

T100 Goria, M. N., "Some Locally Most Powerful Generalized Rank Tests," *Biometrika*, 67 (1980), 497–500.

T101 Goria, M. N., "A Survey of Two-Sample Location-Scale Problem, Asymptotic Relative Efficiencies of Some Rank Tests," *Statist. Neerland.*, 36 (1982), 3–13.

T102 Goria, M. N., and Dana Vorlíčková, "On the Asymptotic Properties of Rank Statistics for the Two-Sample Location and Scale Problem," *Aplikace Matematiky*, 30 (1985), 425–434.

T103 Lepage, Y., "A Combination of Wilcoxon's and Ansari–Bradley's Statistics," *Biometrika*, 58 (1971), 213–217.

T104 Noether, Gottfried E., "Use of the Range Instead of the Standard Deviation," *J. Amer. Statist. Assoc.*, 59 (1955), 1040–1055.

T105 Penfield, Douglas A., "A Comparison of Some Nonparametric Tests for Scale," paper presented at the annual meeting of the American Educational Research Association, Chicago, April 1972.

T106 Penfield, Douglas A., and Stephen L. Koffler, "A Comparison of Some *K*-Sample Nonparametric Tests for Scale," paper presented at the annual meeting of the American Educational Research Association, Chicago, April 1974.

T107 Puri, M. L., "Multi-Sample Scale Problem: Unknown Location Parameters," *Ann. Inst. Statist. Math.*, 20 (1968), 99–106.

T108 Randles, R. H., and R. V. Hogg, "Certain Uncorrelated and Independent Rank Statistics," *J. Amer. Statist. Assoc.*, 66 (1971), 569–574.

T109 Sen, P. K., "On Weighted Rank-Sum Tests for Dispersion," *Ann. Inst. Statist. Math.*, 15 (1963), 117–135.

T110 Van Eeden, C., "Note on the Consistency of Some Distribution-Free Tests for Dispersion," *J. Amer. Statist. Assoc.*, 59 (1964), 105–119.

T111 Wilks, S. S., "Statistical Prediction with Special Reference to the Problem of Tolerance Limits," *Ann. Math. Statist.*, 13 (1942), 400–409.

T112 Kochar, Subhash C., and R. P. Gupta, "Some Competitors of the Mood Test for the Two-Sample Scale Problem," *Communic. in Statist. —Theory and Methods*, 15 (1986), 231–239.

T113 Tiku, M. L., and N. Balakrishnan, "Testing Equality of Population Variances the Robust Way," *Communic. in Statist.— Theory and Methods*, 13 (1984), 2143–2159.

T114 Vegelius, Jan, "A Note on the Tie Problem for the Siegel–Tukey Non-Parametric Dispersion Test," *Bull. Appl. Statist.*, 9 (1982), 19–23.

T115 Wald, A., and J. Wolfowitz, "On a Test Whether Two Samples Are from the Same Population," *Ann. Math. Statist.*, 11 (1940), 147–162.

T116 Smith, K., "Distribution-Free Statistical Methods and the Concept of Power Efficiency," in *Research Methods in the Behavioral Sciences*, L. Festinger and D. Katz (eds.), New York: Dryden, pp. 536–577.

T117 Blumenthal, Saul, "The Asymptotic Normality of Two Test Statistics Associated with the Two-Sample Problem," *Ann. Math. Statist.*, 34 (1963), 1513–1523.

T118 Noether, G. E., "Asymptotic Properties of the Wald–Wolfowitz Test of Randomness," *Ann. Math. Statist.*, 21 (1950), 231–246.

T119 Hollander, M., "A Nonparametric Test for the Two-Sample Problem," *Psychometrika*, 28 (1963), 395–403.

T120 Moses, L. E., "A Two-Sample Test," *Psychometrika*, 17 (1952), 239–247.

T121 Arnold, J. C., and T. S. Briley, "A Distribution-Free Test for Extreme Reactions," *Educ. Psychol. Measurement*, 33 (1973), 301–309.

T122 Fisher, R. A., *Statistical Methods for Research Workers*, fifth edition, Edinburgh: Oliver and Boyd, 1934.

T123 Fisher, R. A., "The Logic of Inductive Inference," *J. Roy. Statist. Soc. Ser. A*, 98 (1935), 39–54.

T124 Irwin, J. O., "Tests of Significance for Differences between Percentages Based on Small Numbers," *Metron*, 12 (1935), 83–94.

T125 Yates, F., "Contingency Tables Involving Small Numbers and the χ^2 Test," *J. Roy. Statist. Soc. Suppl.*, 1 (1934), 217–235.

T126 Finney, D. J., "The Fisher–Yates Test of Significance in 2×2 Contingency Tables," *Biometrika*, 35 (1948), 145–156.

T127 Latscha, R., "Tests of Significance in a 2×2 Contingency Table: Extension of Finney's Table," *Biometrika*, 40 (1955), 74–86.

T128 Mainland, D., and M. I. Sutcliffe, "Statistical Methods in Medical Research. II, Sample

Sizes in Experiments Involving All-or-None Responses," *Can. J. Med. Sci.*, 31 (1953), 406–416.

T129 Bennett, B. M., and P. Hsu, "On the Power Function of the Exact Test for the 2×2 Contingency Table," *Biometrika*, 47 (1960), 393–398. Correction, *Ibid.*, 48 (1961), 475.

T130 Sillitto, S. P., "Note on Approximations to the Power Function of the 2×2 Comparative Trial," *Biometrika*, 36 (1949), 347–352.

T131 Gail, Mitchell, and John J. Gart, "The Determination of Sample Sizes for Use with the Exact Conditional Test in 2×2 Comparative Trials," *Biometrics*, 29 (1973), 441–448.

T132 Barnard, G. A., "A New Test for 2×2 Tables," *Nature*, 156 (1945), 177.

T133 Barnard, G. A., "A New Test for 2×2 Tables," *Nature*, 156 (1945), 783–784.

T134 Barnard, G. A., "Significance Tests for 2×2 Tables," *Biometrika*, 34 (1947), 123–138.

T135 Fisher, R. A., "A New Test for 2×2 Tables," *Nature*, 156 (1945), 388.

T136 Pearson, E. S., "The Choice of Statistical Tests Illustrated on the Interpretation of Data Classed in a 2×2 Table," *Biometrika*, 34 (1947), 139–167.

T137 Sweetland, A., "A Comparison of the Chi-Square Test for 1 df and the Fisher Exact Test," Santa Monica, California: Rand Corporation, 1972.

T138 Overall, John, "Continuity Correction for Fisher's Exact Probability Test," *J. Educ. Statist.*, 5 (1980), 179–190.

T139 Carr, Wendell E., "Fisher's Exact Test Extended to More Than Two Samples of Equal Size," *Technometrics*, 22 (1980), 269–270.

T140 Carr, Wendell E., "An Exciting Alternative to Fisher's Exact Test for Two Proportions," *J. Quality Technol.*, 17 (July 1985), 128–133.

T141 Gibbons, Jean Dickinson, Ingram Olkin, and Milton Sobel, *Selecting and Ordering Populations: A New Statistical Methodology*, New York: Wiley, 1977.

T142 Neave, Henry R., "A New Look at an Old Test," *Bull. Appl. Statist.*, 9 (1982), 165–178.

T143 Dupont, William D., "Sensitivity of Fisher's Exact Test to Minor Perturbations in 2×2 Contingency Tables," *Statist. Med.*, 5 (1986), 629–635.

T144 Sokal, Robert R., and F. James Rohlf, *Biometry*, second edition, San Francisco: W. H. Freeman, 1981.

T145 Dinneen, L. C., and B. C. Blakesley, "A Generator for the Sampling Distribution of the Mann–Whitney U Statistic," *Appl. Statist.*, 22 (1973), 269–273.

T146 Jung, Steven M., Dewey Lipe, and Thomas J. Quirk, "An Alteration of Program U Test to Determine the Direction of Group Differences for the Mann–Whitney U Test," *Educ. Psychol. Measurement*, 31 (1971), 269–273.

T147 Odeh, R. E., "Generalized Mann–Whitney U Statistic," *Appl. Statist.*, 21 (1972), 348–351.

T148 Larsen, S. Olesen, "APL Programs for *p*-values in Wilcoxon's One-Sample and Two-Sample Tests," *Computer Prog. in Biomed.*, 12 (1980), 42–44.

T149 Kummer, G., "Formulas for the Computation of the Wilcoxon Test and Other Rank Statistics," *Biometrical, J.*, 23 (1981), 237–243.

T150 Harding, E. F., "An Efficient Minimal-Storage Procedure for Calculating the Mann–Whitney U, Generalized U and Similar Distributions," *Appl. Statist.*, 33 (1984), 1–6.

T151 Schuster, E., "A Comparison of Algorithms for the Computation of the Mann–Whitney Test," *Biometrical J.*, 27 (1985), 405–410.

T152 Dinneen, L. C., and B. C. Blakesley, "A Generator for the Null Distribution of the Ansari–Bradley W Statistic," *Appl. Statist.*, 25 (1976), 75–81.

T153 Youngman, Grant H., and Wayne W. Daniel, "Calculation of Mood's Dispersion Test Statistic Using Mielke's Adjustment for Ties," *Behav. Res. Methods Instrum.*, 8 (1976), 469.

T154 Gregory, R. J., "Fortran Computer Program for Fisher Exact Probability Test," *Educ. Psychol. Measurement*, 33 (1973), 697–700.

T155 Robertson, W. H., "Programming Fisher's Exact Method of Comparing Two Percentages," *Technometrics*, 2 (1960), 103–107.

T156 Berry, Kenneth J., and Paul W. Mielke, Jr., "A Rapid FORTRAN Subroutine for the Fisher Exact Probability Test," *Educ. and Psychol. Measurement*, 43 (1983), 167–171.

T157 Berry, Kenneth J., and Paul W. Mielke, Jr., "Subroutines for Computing Exact Chi-Square and Fisher's Exact Probability Tests," *Educ. and Psychol. Measurement*, 45 (1985), 153–159.

T158 Berry, Kenneth J., and Paul W. Mielke, Jr., "Exact Chi-Square and Fisher's Exact Probability Test for 3 by 2 Cross-Classification Tables," *Educ. and Psychol. Measurement*, 47 (1987), 631–636.

T159 Mehta, Cyrus R., and Nitin R. Patel, "A Network Algorithm for Performing Fisher's Exact Test in $r \times c$ Contingency Tables," *J. Amer. Statist. Assoc.*, 78 (1983), 427–434.

T160 Mehta, Cyrus R., and Nitin R. Patel, "FEXACT: A FORTRAN Subroutine for Fisher's Exact Test on Unordered $r \times c$ Contingency Tables," *ACM Transac. Math. Software*, 12 (1986), 154–161.

T161 Edgington, E. S., T. Taerum, F. Pysh, A. R. Strain, "Computer-Program for a Randomization Test of a Difference between Independent Groups," *Behav. Res. Methods Instrum.*, 6 (1974), 352–353.

T162 Harlow, D. N., and J. Lowry, "Program for Pitman's Randomization Test for Two Independent Samples," *Behav. Sci.*, 15 (1970), 206.

T163 Edgington, E. S., and A. R. Strain, "Randomization Tests—Computer Time Requirements," *J. Psychol.*, 85 (1973), 89–95.

E1 Russell, Richard O., Jr., David Hunt, and Charles E. Rackley, "Left Ventricular Hemodynamics in Anterior and Inferior Myocardial Infarction," *Amer. J. Cardiol.*, 32 (1973), 8–16.

E2 Taylor, William H., "Correlations between Length of Stay in the Job Corps and Reading Ability of the Corpsmen," *J. Employment Counseling*, 9 (1972), 78–85.

E3 Newmark, Charles S., William Hetzel, Lilly Walker, Steven Holstein, and Martin Finklestein, "Predictive Validity of the Rorschach Prognostic Rating Scale with Behavior Modification Techniques," *J. Clin. Psychol.*, 29 (1973), 246–248.

E4 West, Joyce A., "Auditory Comprehension in Aphasic Adults: Improvement through Training," *Arch. Phys. Med. Rehabil.*, 54 (1973), 78–86.

E5 Ressl, J., M. Kubis, P. Lukl, J. Vykydal, and J. Weinberg, "Resting Hyperventilation in Adults with Atrial Septal Defect," *Br. Heart. J.*, 31 (1969), 118–121.

E6 Davidson, Mayer B., Stuart Green, and John H. Menkes, "Normal Glucose, Insulin, and Growth Hormone Responses to Oral Glucose in Huntington's Disease," *J. Lab. Clin. Med.*, 84 (1974), 807–812.

E7 Christiansen, Claus, Paul Rødbro, and Ole Sjö, "'Anticonvulsant Action' of Vitamin D in Epileptic Patients? A Controlled Pilot Study," *Br. Med. J.*, 5913 (1974), 258–259.

E8 Gordon, Richard F., Moosa Najmi, Benedict Kingsley, Bernard L. Segal, and Joseph W. Linhart, "Spectroanalytic Evaluation of Aortic Prosthetic Valves," *Chest*, 66 (1974), 44–49.

E9   Reimer, Keith A., Margaret M. Rasmussen, and Robert B. Jennings, "Reduction by Propranolol of Myocardial Necrosis Following Temporary Coronary Artery Occlusion in Dogs," *Circ. Res.*, 33 (1973), 353–363.

E10   Boullin, David J., and Robert A. O'Brien, "Uptake and Loss of ^{14}C-dopamine by Platelets from Children with Infantile Autism," *J. Autism Child. Schizophrenia*, 2 (1972), 67–74.

E11   Branehög, Ingemar, Jack Kutti, and Aleksander Weinfeld, "Platelet Survival and Platelet Production in Idiopathic Thrombocytopenic Purpura (ITP)," *Br. J. Haematol.*, 27 (1974), 127–143.

E12   Aronow, Wilbert S., Phillip R. Harding, Mohammed Khursheed, Jack S. Vangrow, Nicholas P. Papageorge's, and James Mays, "Effect of Halofenate on Serum Lipids," *Clin. Pharmacol. Ther.*, 14 (1973), 358–365.

E13   Hansen, Aage Prange, "Abnormal Serum Growth Hormone Response to Exercise in Maturity-Onset Diabetes," *Diabetes*, 22 (1973), 619–628.

E14   Cabasso, V. J., M. B. Dobkin, R. E. Roby, and A. H. Hammar, "Antibody Response to a Human Diploid Cell Rabies Vaccine," *Appl. Microbiol.*, 27 (1974), 553–561.

E15   Burg, R. Von, and Hussain Rustam, "Electrophysiological Investigations of Methylmercury Intoxication in Humans. Evaluation of Peripheral Nerve by Conduction Velocity and Electromyography," *Electroencephalog. Clin. Neurophysiol.*, 37 (1974), 381–392.

E16   Skerfving, S., K. Hansson, C. Mangs, J. Lindsten, and N. Ryman, "Methylmercury-Induced Chromosome Damage in Man," *Environ. Res.*, 7 (1974), 83–98.

E17   Almy, Timothy A., "Residential Location and Electoral Cohesion: The Pattern of Urban Political Conflict," *Amer. Polit. Sci. Rev.*, 67 (1973), 914–923.

E18   Williams, Redford B., Jr., Chase P. Kimball, and Harold N. Williard, "The Influence of Interpersonal Interaction on Diastolic Blood Pressure," *Psychosom. Med.*, 34 (1972), 194–198.

E19   Gill, Frank B., and Bertram G. Murray, Jr., "Discrimination Behavior and Hybridization of the Blue-Winged and Golden-Winged Warblers," *Evolution*, 26 (1972), 282–293.

E20   McGuffin, William L., Barry M. Sherman, Jesse Roth, Phillip Gorden, C. Ronald Kahn, William C. Roberts, and Peter L. Frommer, "Acromegaly and Cardiovascular Disorders," *Ann. Intern. Med.*, 81 (1974), 11–18.

E21   Cassell, Kenneth, Kenneth Shaw, and Gerald Stern, "A Computerized Tracking Technique for the Assessment of Parkinsonian Motor Disabilities," *Brain*, 96 (1973), 815–826.

E22   Tanner, Barry A., "Shock Intensity and Fear of Shock in the Modification of Homosexual Behavior in Males by Avoidance Learning," *Behav. Res. Ther.*, 11 (1973), 213–218.

E23   Keyvan-Larijarni, Hossein, and Alf M. Tannenberg, "Methanol Intoxication, Comparison of Peritoneal Dialysis and Hemodialysis," *Arch. Intern. Med.*, 134 (1974), 293–296.

E24   Seppälä, Markku, Ilkka Aho, Anja Tissari, and Erkki Ruoslahti, "Radio-Immunoassay of Oxytocin in Amniotic Fluid, Fetal Urine, and Meconium During Late Pregnancy and Delivery," *Amer. J. Obstet. Gynecol.*, 114 (1972), 788–795.

E25   Griffiths, Ellis, " 'Negative' Effects of Fungicides in Coffee," *Trop. Sci.*, 14 (1972), 79–89.

E26   Garrod, J. W., P. Jenner, G. R. Keysell, and B. R. Mikhael, "Oxidative Metabolism of Nicotine by Cigarette Smokers With Cancer of the Urinary Bladder," *J. Natl. Cancer Inst.*, 52 (1974), 1421–1924.

E27   O'Neill, Thomas J., Lawrence M. Tierney, Jr., Ronald J. Proulx, "Heparin Lock-Induced Alterations in the Activated Partial Thromboplastin Time," *J. Amer. Med. Assoc.*, 227 (1974), 1297–1298.

E28   Princi, Frank, and Erving F. Geever, "Prolonged Inhalation of Cadmium," *Arch. Ind. Hyg. Occup. Med.*, 1 (1950), 651–661.

E29 Singh, Udai P., and Tapeshwar Prasad, "Self-Esteem, Social-Esteem and Conformity Behaviour," *Psychologia*, 16 (1973), 61–68.

E30 Jamuar, K. K., and S. Singh, "Adaptation in Hindi of Maslow's Security–Insecurity Inventory," *Psychologia*, 16 (1973), 214–217.

E31 Hoffman, Helmut, and Douglas N. Jackson, "Differential Personality Inventory for Male and Female Alcoholics," *Psychol. Rep.*, 34 (1974), 21–22.

E32 Jackson, D. N., *Differential Personality Inventory*, London, Ontario: Author, 1972.

PROCEDURES THAT UTILIZE DATA FROM TWO RELATED SAMPLES

Chapter 3 introduced several procedures for making inferences about populations based on the data from two independent samples. This chapter is devoted to hypothesis-testing procedures and the construction of confidence intervals when the data are from two related samples. Data of this type can arise from a variety of situations. One of the most common examples is the "before and after" or "pre- and post-test" experiment in which measurements are taken on the same subjects both before and after they have been exposed to some intervening treatment or experimental manipulation. For example, obese subjects weighed before and after a period of dieting would yield two related samples of observations. The two samples are related in the sense that they are composed of measurements taken on the same subjects.

We may generate related samples by pairing subjects on the basis of certain characteristics. One member of each pair receives one treatment, while the other member receives a different treatment. One of the "treatments" may be no treatment at all, or a placebo—that is, one member of each pair may serve as a control. The objective of matching is to make each pair of subjects as much alike as possible with respect to extraneous variables that may affect the results of the experiment. Suppose, for example, that a researcher wishes to assess the effect of some drug on the reaction time to a certain stimulus. As a minimum, the researcher will want to

match subjects on the basis of age, since that variable is likely to have an effect on reaction time.

Natural pairs, such as human twins or siblings and animal litter mates, may be available to the researcher. If the use of such natural pairs is not feasible, unrelated subjects must be carefully matched on as many relevant variables as practical. Industries frequently obtain paired samples by dividing specimens of material in half and randomly assigning one half to one treatment and the other half to another treatment. For example, a detergent manufacturer may uniformly soil specimens of fabric, cut them in half, and randomly assign one half of each specimen for laundering with one detergent and the other half for laundering with the other detergent.

The variable of interest in the analysis of two related samples is the difference between the two measurements within pairs. If the measurements within sample pairs are sufficiently disparate, the researcher can conclude that the two treatments have different effects or that one treatment is more effective than another or no treatment, as the case may be. This chapter covers procedures for testing hypotheses of no treatment effect or no difference in treatments, as well as procedures for constructing confidence intervals for the median difference between pairs of measurements. Also presented is a hypothesis-testing procedure that is appropriate when the data consist of frequencies rather than measured variables.

4.1

PROCEDURES FOR TESTING HYPOTHESES ABOUT LOCATION PARAMETERS

This section presents some procedures for testing the null hypothesis of no difference in treatment effects when we are comparing two treatments. Also presented are procedures for testing the hypothesis of no treatment effect when we compare a single treatment with a placebo, or with no treatment at all. These tests are sometimes called *paired comparisons tests* when the data consist of two related samples of measurements.

We usually use the *t* test for paired data when a parametric inferential procedure is appropriate. We obtain the difference between each pair of measurements and analyze the resulting sample of differences. We are interested, then, in knowing whether we can conclude that the mean difference is significantly different from zero. When the *t* test is not appropriate, we can use one of the nonparametric analogues of this section. We seek an alternative to the *t* test if the assumptions underlying its use are not met, if the measurement scale employed is not at least interval, or if we need results in a hurry and a more rapidly executed procedure is available.

The first test considered is the *sign test*, which was presented in Chapter 2 as an inferential procedure for use with data from a single sample. We can easily adapt the test to the case in which the data consist of measurements from two related samples.

SIGN TEST FOR TWO RELATED SAMPLES

The sign test for two related samples is particularly useful if the measurement scale is only ordinal, so that within each pair of measurements we can determine only whether one is larger than the other and, if so, which is larger. If we measure the data on a stronger scale, we may be able to use a more powerful test.

Since the hypothesis-testing procedure and rationale for the case of two related samples are similar to those for the single-sample case discussed in Chapter 2, the discussion here will be somewhat abbreviated. It may be helpful to reread the discussion of the sign test in Chapter 2 before reading this section.

Assumptions

A. The data consist of a random sample of n pairs of measurements $(X_1, Y_1), (X_2, Y_2), \ldots, (X_n, Y_n)$, where each pair of measurements is taken on the same subject or subjects that have been paired with respect to one or more variables. The variable of interest is $X_i - Y_i = D_i$, the difference between pairs of measurements. The parameter about which we make inferences is M_D, the median of a population of differences between pairs of measurements.

B. The n pairs of measurements are independent.

C. The measurement scale is at least ordinal within each pair, so that one can determine which of the two members is larger (unless they are equal).

D. The variable under study is continuous.

Hypotheses

A. (Two-sided)
   $H_0: M_D = 0$
   $H_1: M_D \neq 0$

B. (One-sided)
   $H_0: M_D \leq 0$
   $H_1: M_D > 0$

C. (One-sided)
   $H_0: M_D \geq 0$
   $H_1: M_D < 0$

Test Statistic

For each pair (X_i, Y_i) record the sign of the difference obtained by subtracting Y_i from X_i. In other words, record a plus if $X_i - Y_i > 0$, and record a minus if $X_i - Y_i < 0$. If ties occur—that is, if $X_i = Y_i$ for any pair—eliminate those pairs from the analysis and reduce n accordingly.

A. If H_0 is true, we expect a sample of differences to include about as many plus signs as minus signs. Either a sufficiently small number of

plus signs or a sufficiently small number of minus signs causes us to reject H_0. The test statistic for this hypothesis, then, is the number of plus signs or the number of minus signs, whichever is smaller.
B. Since a sufficiently small number of minus signs causes us to reject H_0, the test statistic for this hypothesis is the number of minus signs.
C. A sufficiently small number of plus signs causes us to reject H_0; consequently the test statistic is the number of plus signs.

Decision Rule

We can state the hypotheses in terms of probabilities of plus and minus signs.

A. $H_0: P(+) = P(-) = 0.50, \qquad H_1: P(+) \neq P(-) \neq 0.50$
B. $H_0: P(+) \leq P(-), \qquad H_1: P(+) > P(-)$
C. $H_0: P(+) \geq P(-), \qquad H_1: P(+) < P(-)$

We see again that the sign test is a special case of the binomial test (discussed in Chapter 2) where we are testing $H_0: p = 0.50$ against an appropriate alternative. Whether we reject H_0 or not depends on the magnitude of

$$P(K \leq k \mid n, 0.50) \qquad \text{(4.1)}$$

where K is the random variable (the number of signs of interest under H_0), k is the observed value of the test statistic, and n is the effective sample size. The reader may recognize this as Equation 2.1. We evaluate Equation 4.1 by referring to Table A.1. The specific decision rules for the three hypotheses are as follows:

A. (Two-sided): Reject H_0 at the α level of significance if

$$P(K \leq k \mid n, 0.50) \leq \alpha/2.$$

B, C. (One-sided): Reject H_0 at the α level of significance if

$$P(K \leq k \mid n, 0.50) < \alpha.$$

Example 4.1   Latané and Cappell (E1) studied the effects of togetherness on heart rate in rats. They recorded the heart rates of 10 rats while they were alone and while in the presence of another rat. The results are shown in Table 4.1. The authors, who used a t test with the data, were able to conclude at the 0.05 level of significance that togetherness in rats increases heart rate. Let us see whether we can reach the same conclusion by using the sign test. Let $\alpha = 0.05$.

TABLE 4.1   **Heart rate, beats per minute, of 10 rats alone and in the presence of another rat**

Rat	1	2	3	4	5	6	7	8	9	10
Alone rate (X)	463	462	462	456	450	426	418	415	409	402
Together rate (Y)	523	494	461	535	476	454	448	408	470	437

Source: Bibb Latané and Howard Cappell, "The Effects of Togetherness on Heart Rate in Rats," *Psychon. Sci.*, 29 (1972), 177–179.

Hypotheses

$$H_0: \; M_D \geq 0 \text{ or } P(+) \geq P(-)$$
$$H_1: \; M_D < 0 \text{ or } P(+) < P(-)$$

Test Statistic

The differences between alone and together heart rates and their signs are shown in Table 4.2. The value of the test statistic is 2, the number of plus signs.

TABLE 4.2   **Alone and together heart rates, differences, and signs of differences for the data of Table 4.1**

Alone rate (X)	463	462	462	456	450	426	418	415	409	402
Together rate (Y)	523	494	461	535	476	454	448	408	470	437
Difference ($X_i - Y_i$)	−60	−32	+1	−79	−26	−28	−30	+7	−61	−35
Sign of difference	−	−	+	−	−	−	−	+	−	−

Decision

Table A.1 reveals that $P(K \leq 2 \,|\, 10, 0.50) = 0.0547$. The sign test does not quite allow us to reject H_0 at the 0.05 level, as did the t test. The P value for this example is 0.0547.

Power-Efficiency   For a discussion of the power-efficiency of the sign test, see Chapter 2.

FURTHER READING

Bennett (T1) proposed an asymptotically nonparametric extension of the sign test for testing the median of a multivariate distribution. The power properties of Bennett's test have been studied by Bhattacharyya (T2). Krauth (T3) has shown that the one-sided sign test proposed by Putter (T4) is an asymptotic, uniformly most powerful test. Chatterjee (T5) has proposed a distribution-free sign test for testing that several independent pairs of random variables have specified locations. Sign tests for symmetry and a sequential sign test are discussed by Gastwirt (T6) and Groeneve (T7), respectively.

EXERCISES

4.1 Van Duijn (E2) studied the effect of clonazepam on cobalt-induced focal seizures in alert cats. Before and after administering clonazepam, he recorded focal paroxysmal activity as

mean seconds with spikes/10 seconds recording. Table 4.3 shows the results of this part of the experiment. Can we conclude on the basis of these data that clonazepam decreases focal spiking? Let $\alpha = 0.05$. What is the P value?

TABLE 4.3

Effect of clonazepam on focal spikes produced by cobalt

Cat	1	2	3	4	5	6	7	8	9	10
Before (X)	2.7	4.2	3.3	5.3	4.2	3.5	6.5	4.8	3.7	7.1
After (Y)	4.5	2.6	1.4	2.5	2.5	2.3	3.0	2.6	1.9	0.4

Source: H. Van Duijn, "Superiority of Clonazepam over Diazepam in Experimental Epilepsy," *Epilepsia*, 14 (1973), 195–202.

4.2 Shani et al. (E3) studied the effect of phenobarbital on liver functions in patients with Dubin–Johnson syndrome. Table 4.4 shows the total bilirubin in the sera of these patients before and after treatment with phenobarbital. Can we conclude on the basis of these data that phenobarbital reduces total bilirubin level? Let $\alpha = 0.05$. What is the P value?

TABLE 4.4

Total bilirubin, milligrams per 100 ml, in sera of patients with Dubin–Johnson syndrome, before and after treatment with phenobarbital

Patient	1	2	3	4	5	6	7	8	9	10	11	12	13
Before (X)	4.0	3.2	3.8	1.8	3.0	5.3	5.7	3.0	2.7	2.9	2.8	1.8	2.6
After (Y)	3.1	3.0	3.5	1.0	1.8	3.9	2.2	2.1	1.4	2.9	2.6	1.4	2.5

Source: Mordechai Shani, Uri Seligsohn, and Judith Ben-Ezzer, "Effect of Phenobarbital on Liver Functions in Patients with Dubin–Johnson Syndrome," *Gastroenterology*, 67 (1974), 303–308; copyright 1974, Williams & Wilkins, Baltimore.

4.3 Smith and Di Girolamo (E4) examined the morphological changes in epididymal fat cells of mature, moderately obese rats during reduction of body weight and adipose tissue mass. Table 4.5 shows part of their results. Do these data provide sufficient evidence at the 0.05 level of significance to indicate that reduction of body weight reduces the diameter of fat cells?

TABLE 4.5

Mean* fat cell diameters, in micrometers, of mature, moderately obese rats before and after 20% reduction in body weight by underfeeding

Rat	1	2	3	4	5	6
Before (X)	84.4	86.0	87.9	93.9	95.2	96.6
After (Y)	62.9	75.4	78.2	83.6	57.6	58.0
Rat	7	8	9	10	11	12
Before (X)	97.5	101.4	103.8	115.2	116.4	134.5
After (Y)	69.6	76.5	73.9	88.0	73.8	94.1

Source: Jerry E. Smith, and Mario Di Girolamo, "Effect of Weight Reduction in the Rat on Epididymal Fat-Cell Size and Relative Dispersion," *Amer. J. Physiol.*, 227 (1974), 420–424.

* Derived from measurements of diameters in 300 fat cells.

WILCOXON MATCHED-PAIRS SIGNED-RANKS TEST

The only information that the sign test for analyzing paired observations uses is whether X is larger than, smaller than, or equal to Y. If the measurement scale is so weak that the raw data do not provide more information, the sign test may be the best test for making inferences based on the data. If, however, the data contain more information, the sign test may not be the best choice, because it sacrifices additional information. Usually when we use a test that ignores available information, we lose statistical power. What we need is a test that uses more of the available information.

The *Wilcoxon matched-pairs signed-ranks test* (T8) fulfills this need for the case of two related samples when the measurement scale allows us to determine not only whether the members of a pair of observations differ, but also the magnitude of any difference. In other words, the Wilcoxon matched-pairs signed-ranks test is appropriate when we can determine the amount of any difference between pairs of observations X_i and Y_i, as well as the direction of the difference. When we can determine the magnitudes of differences, we can rank them. It is through the rankings of the differences that the Wilcoxon test utilizes the additional information.

Recall that in Chapter 2 the Wilcoxon signed-ranks test was used as a median test. In this application differences between paired observations are analyzed, rather than differences between a single observation and a hypothesized median.

Assumptions

   A. The data for analysis consist of n values of the difference $D_i = Y_i - X_i$. Each pair of measurements (X_i, Y_i) is taken on the same subject or on subjects that have been paired with respect to one or more variables. The sample of (X_i, Y_i) pairs is random.
   B. The differences represent observations on a continuous random variable.
   C. The distribution of the population of differences is symmetric about their median, M_D.
   D. The differences are independent.
   E. The differences are measured on at least an interval scale.

Hypotheses

   A. (Two-sided)
   $H_0: M_D = 0$
   $H_1: M_D \neq 0$
   B. (One-sided)
   $H_0: M_D \leq 0$
   $H_1: M_D > 0$
   C. (One-sided)
   $H_0: M_D \geq 0$
   $H_1: M_D < 0$

Test Statistic

The procedure for obtaining the numerical value of the test statistic is as follows:

1. Obtain each of the signed differences

$$D_i = Y_i - X_i \qquad\qquad (4.2)$$

2. Rank the absolute values of these differences from smallest to largest; that is, rank

$$|D_i| = |Y_i - X_i| \qquad\qquad (4.3)$$

3. Assign to each of the resulting ranks the sign of the difference whose absolute value yielded that rank.
4. Compute

$$T_+ = \text{the sum of the ranks with positive signs} \qquad (4.4)$$

and

$$T_- = \text{the sum of the ranks with negative signs} \qquad (4.5)$$

T_+ or T_- is the test statistic, depending on the alternative hypothesis.

Ties   There are two types of ties; one or both may occur in a given situation. The first type occurs when $X_i = Y_i$ for a given pair. We eliminate from the analysis all pairs of observations having $D_i = Y_i - X_i = 0$ and reduce n accordingly. The other type of tie occurs when two or more values of $|D_i|$ are equal. For ties of this type, the $|D_i|$ receive the average of the ranks that otherwise would be assigned to them.

Decision Rule

If H_0 is true—that is, if the median of the population of differences is zero—we expect to find among the large ranks about as many positive signs as negative signs. Similarly, among the small ranks, we expect to find about equal representation of positive and negative signs. Thus if H_0 is true, we expect the sum of the ranks with positive signs to be about equal to the sum of the ranks with negative signs. A sufficiently large observed departure from this expectation casts doubt on the null hypotheses. The specific decision rules for the three possible sets of hypotheses are as follows.

A. Either a sufficiently small value of T_+ or a sufficiently small value of T_- will cause rejection of H_0 that the median of the population of differences is equal to zero. Therefore the test statistic in this case is T_+ or T_-, whichever is smaller. As in Chapter 2, the test statistic will be referred to as T, for notational simplicity when the test is two-sided. Reject H_0 at the α level of significance if calculated T is smaller than or equal to tabulated T for n and preselected $\alpha/2$ given in Table A.3. If we reject H_0, we conclude that the median of the population of

differences is not zero. Note the practical implications of being able to reject or not to reject the null hypothesis.

Suppose, for example, that the X observations are taken on subjects before receiving some treatment, and the Y observations are made after they receive the treatment. Rejecting H_0 enables us to conclude that the treatment has an effect. If we do not reject H_0, we conclude that the treatment may not have an effect. If the data result from an experiment designed to compare two treatments, rejecting H_0 enables us to conclude that the treatments have different effects.

B. Sufficiently small values of T_- will cause rejection of H_0 that the median of the population of differences is equal to or less than zero. Then we should reject H_0 at the α level of significance if T_- is less than or equal to tabulated T for n and preselected α (one-sided) given in Table A.3. If the data are generated by an experiment in which two treatments are being compared, rejecting H_0 leads to the conclusion that one treatment is more effective than the other.

C. Sufficiently small values of T_+ will cause rejection of H_0 that the median of the population of differences is greater than or equal to zero. Reject H_0 at the α level of significance if the computed value of T_+ is less than or equal to tabulated T for n and preselected α (one-sided) given in Table A.3. Again, if the data are generated by an experiment in which two treatments are being compared, rejecting H_0 enables us to conclude that one treatment is more effective than the other.

Example 4.2	Dickie et al. (E5) studied hemodynamic changes in patients with acute pulmonary thromboembolism. Table 4.6 shows the mean pulmonary artery pressure of nine of these patients before and 24 hours after urokinase therapy. We wish to know whether these data provide sufficient evidence to indicate that urokinase therapy lowers pulmonary artery pressure. Let $\alpha = 0.05$.

TABLE 4.6   **Mean pulmonary artery pressure, millimeters of mercury**

Patient	1	2	3	4	5	6	7	8	9
0 hours (X)	33	17	30	25	36	25	31	20	18
24 hours (Y)	21	17	22	13	33	20	19	13	9

Source: Kenneth J. Dickie, William J. de Groot, Robert N. Cooley, Ted P. Bond, and M. Mason Guest, "Hemodynamic Effects of Bolus Infusion of Urokinase in Pulmonary Thromboembolism," *Amer. Rev. Respir. Dis.*, 109 (1974), 48–56.

Hypotheses

$$H_0: \ M_D \geq 0$$
$$H_1: \ M_D < 0$$

Test Statistic

The calculation of the test statistic is shown in Table 4.7.

TABLE 4.7   **Calculation of test statistic for Example 4.2**

Before therapy (X)	After 24 hours (Y)	$D_i = Y_i - X_i$	Signed rank of $|D_i|$
33	21	−12	−7
17	17	0	omit
30	22	−8	−4
25	13	−12	−7
36	33	−3	−1
25	20	−5	−2
31	19	−12	−7
20	13	−7	−3
18	9	−9	−5
			$T_+ = 0$

Decision

Table A.3 with $n = 8$ reveals that the probability of observing a value of $T_+ = 0$, when H_0 is true, is 0.0039. Since 0.0039 is less than 0.05, we reject H_0. Thus we conclude that the median of the population of differences is less than zero and hence that urokinase therapy lowers pulmonary artery pressure. Since $T_+ = 0$, the P value is 0.0039.

Large-Sample Approximation   When n is greater than 30, we cannot use Table A.3 to obtain critical values of the test statistics. For samples greater than 20, we may compute

$$z = \frac{T - [n(n + 1)]/4}{\sqrt{n(n + 1)(2n + 1)/24}} \qquad (4.6)$$

z is distributed approximately as the standard normal. Table A.2 gives critical values of z.

Power-Efficiency   The power-efficiency aspects of the Wilcoxon test were discussed in Chapter 2. Basu (T9), who also discusses the Wilcoxon test, considers the large-sample properties of a generalized Wilcoxon–Mann–Whitney statistic.

Buckle et. al. (T10) and Claypool and Holbert (T11) discuss approximations to the Wilcoxon distribution, and Hodges and Lehmann (T12) discuss the Wilcoxon and t test for matched pairs when the data are of a certain type. Hollander et al. (T13) consider the robustness of the Wilcoxon test under certain special conditions. Jurečková (T14) discusses a central-limit theorem for the Wilcoxon rank statistic process, and Noether (T15) considers the efficiency of the two-sample statistic for randomized blocks.

FURTHER READING

Lepage (T16) has proposed a two-sample test that combines the Wilcoxon test for location and the Ansari–Bradley test for dispersion. In a later paper (T17), he presents tables of critical points and significance levels for his test. Hollander (T18) considers the asymptotic efficiency of two nonparametric competitors of the Wilcoxon two-sample test. Gehan (T19, T20) discusses generalized Wilcoxon tests for use with censored data. Alling (T21) discusses the use of the Wilcoxon two-sample statistic in a sequential testing procedure, and Stone (T22) considers the asymptotic behavior of the extreme tail probabilities of the null distribution of the two-sample Wilcoxon statistic. The Wilcoxon test is also discussed in Wilcoxon (T23) and by Bühler (T24), Claypool (T25), and Verdooren (T26).

The *randomization test for matched pairs* is also available, but the computations for this test are very tedious and time-consuming, except with very small samples or with the aid of a computer.

Kempthorne and Doerfler (T27) compare the randomization test with the sign test and the Wilcoxon test; they conclude that the randomization test is always preferable to the other two. David and Kim (T28) suggest a randomization technique for analyzing matched-pair data in s dimensions. Edgington (T29) considers the special problem of applying randomization tests as substitutes for analysis of covariance or partial correlation (T30). For other treatments of randomization tests, see Edgington (T31, T32), Cleroux (T33), Klauber (T34), and Alf and Abraham (T35). Walsh (T36, T37) has proposed another test appropriate for paired data.

EXERCISES

4.4 Piggott et al. (E6) paired 10 psychotic and 10 normal children on age and gender. They then compared subjects for differences in respiratory sinus arrythmia under conditions of spontaneous and 5-, 10-, and 15-second interval breathing. They recorded cardiac rate and respiratory changes simultaneously. Table 4.8 shows the differences in duration of the cardiac acceleratory phase following the beginning of inspiration (psychotics compared to the controls for the third respiration). Do these data provide sufficient evidence to indicate a difference between psychotic and normal children? Let $\alpha = 0.05$. What is the P value for this test?

TABLE 4.8 **Duration of cardiac acceleration, seconds, timed respiration means for 15-second-interval breathing**

Pair	1	2	3	4	5	6	7	8	9	10
Psychotic (X)	1.74	1.44	2.12	1.80	2.00	2.70	1.96	1.46	1.82	1.40
Control (Y)	2.46	1.88	2.38	1.94	2.14	1.60	1.96	1.82	1.80	1.84

Source: Leonard R. Piggott, Albert F. Ax, Jacqueline L. Bamford, and Joanne M. Fetzner, "Respiration Sinus Arrhythmia in Psychotic Children," *Psychophysiology*, 10 (1973), 401–414; copyright 1973, The Society for Psychophysiological Research; reprinted with permission of the publisher.

4.5 Bhatia et al. (E7) reported the data shown in Table 4.9. Can we conclude on the basis of these data that treatment lowers the stroke index in patients of this type? Let $\alpha = 0.05$.

TABLE 4.9 **Stroke index (ml/beat/m²) in pre- and post-treatment studies of coronary circulation in chronic severe anaemia**

Case	1	2	3	4	5	6	7	8
Before treatment (X)	109	57	53	57	68	72	51	65
After treatment (Y)	56	44	55	40	62	46	49	41

Source: M. L. Bhatia, S. C. Manchanda, and Sujoy B. Roy, "Coronary Haemodynamic Studies in Chronic Severe Anaemia," *Br. Heart. J.*, 31 (1969), 365–374. Reprinted by permission of the authors and the editor.

4.6 Hall et al. (E8) reported the Lintner Soluble Starch (LSS) to Amylase Azure (AA) ratios shown in Table 4.10. Do these data provide sufficient evidence to indicate that saliva and duodenal fluid LSS/AA ratios are, on the average, different? Let $\alpha = 0.05$. Find the P value for this test.

TABLE 4.10 **LSS/AA ratios for saliva and duodenal fluid in 11 patients**

Subject	A	B	C	D	E	F	G	H	I	J	K
Saliva (X)	0.78	0.86	0.82	1.10	0.71	1.00	0.65	0.85	0.64	0.86	0.70
Duodenal fluid (Y)	0.46	0.37	0.41	0.46	0.45	0.67	0.47	0.37	0.25	0.33	0.42

Source: F. F. Hall, C. R. Ratliff, T. Hayakawa, T. W. Culp, and N. C. Hightower, "Substrate Differentiation of Human Pancreatic and Salivary Alpha-Amylases," *Amer. J. Dig. Dis.*, 15 (1970), 1031–1038.

4.2

CONFIDENCE INTERVAL PROCEDURES FOR THE MEDIAN DIFFERENCE

When the data available for analysis are paired observations (X_i, Y_i), we may want to determine a confidence interval for the median of the population of differences, $D_i = Y_i - X_i$ (alternatively, we may define D_i as $D_i = X_i - Y_i$). This section deals

with two methods of constructing confidence intervals for the median difference. The first method discussed is based on the sign test, and the second is based on the Wilcoxon test.

CONFIDENCE INTERVAL FOR MEDIAN DIFFERENCE BASED ON THE SIGN TEST

The procedure for constructing a symmetric, two-sided $100(1 - \alpha)\%$ confidence interval for the median of a population of differences based on matched pairs is identical to the method described in Chapter 2 for constructing a confidence interval (based on the sign test) for a population median. In this procedure the random sample of differences D_i in the case of paired data replaces the random sample of observations in the single-sample case. We assume that the data satisfy the assumptions underlying the sign test for paired data.

We may summarize the procedure in the following steps:

1. For each of n pairs of observations (X_i, Y_i), determine the differences $D_i = (Y_i - X_i)$.
2. Refer to Table A.1 to find the largest K' such that $P(K \leq K' \,|\, n, 0.50) \leq \alpha/2$. (In other words, select K' as in the two-sided sign test.)
3. Arrange the sample differences in order of magnitude. Then the $(K' + 1)$th difference is M_L, the lower limit of the $100(1 - \alpha)\%$ confidence interval.
4. If we count backward from the largest toward the smallest, the $(K' + 1)$th difference is M_U, the upper limit of the confidence interval.

As shown in Chapter 2, we cannot obtain confidence intervals with the usual exact confidence coefficients of 0.90, 0.95, 0.99. We must be content with intervals with approximately these confidence coefficients and select the K' corresponding to an approximate value of $\alpha/2$.

Example 4.3   Vagenakis et al. (E9) reported the data in Table 4.11 showing the effect of iodide administration on serum concentrations of serum thyroxine (T_4) in nine male and three female normal volunteers. Subjects were given 190 mg iodide daily for 10 days. We wish to construct a 95% confidence interval for the median of the population of differences from which the sample of differences is presumed to have been drawn.

1. The observed differences $D_i = (Y_i - X_i)$ in order of magnitude are $-0.1, 0.2, 0.8, 1.1, 1.4, 1.5, 1.6, 2.3, 2.3, 2.3, 2.3, 2.4$.
2. Referring to Table A.1, we find that $P(K \leq 2 \,|\, 12, 0.50) = 0.0192$.
3. $K' + 1 = 2 + 1 = 3$, so the third sample difference from the low end of the ordered array of differences is 0.8. Thus the lower limit of the confidence interval is $M_L = 0.8$.

4. The third sample difference from the high end of the ordered array of differences is 2.3, and the upper limit of the confidence interval is $M_U = 2.3$.

TABLE 4.11   **Serum concentrations of T_4 before (control) and after administration of iodide (190 mg daily for 10 days) in 12 subjects**

Subject	1	2	3	4	5	6	7	8	9	10	11	12
Iodide* (X)	7.9	9.1	9.2	8.1	4.2	7.2	5.4	4.9	6.6	4.7	5.2	7.3
Control[†] (Y)	10.2	10.2	11.5	8.0	6.6	7.4	7.7	7.2	8.2	6.2	6.0	8.7

Source: Apostolos G. Vagenakis, Basil Rapoport, Fereidoun Azizi, Gary I. Portnay, Lewis E. Braverman, and Sidney H. Ingbar, "Hyperresponse to Thyrotropin-Releasing Hormone Accompanying Small Decreases in Serum Thyroid Concentrations," *J. Clin. Invest.*, 54 (1974), 913–918.

* Mean value of last three days of iodide administration

[†] Mean value of three successive days

Thus the $100[1 - 2(0.0192)] = 96.16\%$ confidence interval for the median difference in serum concentrations of T_4 between the control and iodide conditions is 0.8 to 2.3.

The method for handling large samples is discussed in Chapter 2.

EXERCISES

4.7   Angseesing (E10) conducted an experiment in which 18 slugs were offered both cyanogenic (Ac) and acyanogenic (ac) plants. Table 4.12 shows the quantities of leaflets eaten from each type of plant by each slug. Construct an approximate 95% confidence interval for the median difference.

TABLE 4.12   **Quantities of cyanogenic (Ac) and acyanogenic (ac) leaflets eaten by slugs**

Slug	1	2	3	4	5	6	7	8	9
Ac total (X)	5.00	9.25	7.50	3.00	1.50	3.50	6.25	5.75	5.00
ac total (Y)	17.50	8.50	14.75	13.50	5.00	17.25	17.25	8.75	11.25
Slug	10	11	12	13	14	15	16	17	18
Ac total (X)	6.75	3.50	3.75	5.00	2.25	4.25	7.50	1.25	4.50
ac total (Y)	10.25	9.25	16.00	22.00	18.25	13.50	8.50	9.00	13.25

Source: J. P. A. Angseesing, "Selective Eating of the Acyanogenic Form of Trifolium Repens," *Heredity*, 32 (1974), 73–83.

4.8   Weis and Peak (E11) studied the effects of oxytocin on blood pressure during anesthesia. The subjects were 11 women, 19 to 31 years of age, who weighed 103 to 251 pounds and were in the first trimester of pregnancy. They had been anesthetized for dilation and

curettage, and were injected with 0.1 unit/kg of oxytocin. The mean arterial blood pressures before and after oxytocin are shown in Table 4.13. Construct an approximate 90% confidence interval for the median difference.

TABLE 4.13   **Mean arterial blood pressure in 11 anesthetized subjects before and after receiving oxytocin**

Subject	1	2	3	4	5	6	7	8	9	10	11
Before	95	173	94	97	81	100	97	104	72	101	83
After	55	90	36	59	46	46	49	92	23	55	49

Source: F. Robert Weis, Jr., and Jerome Peak, "Effects of Oxytocin on Blood Pressure during Anesthesia," *Anesthesiology*, 40 (1974), 189–190.

CONFIDENCE INTERVAL FOR MEDIAN DIFFERENCE BASED ON THE WILCOXON TEST

We may construct a confidence interval for the median difference based on the Wilcoxon matched-pairs signed-ranks test if the assumptions underlying that test are met. We may construct the interval graphically or arithmetically. The procedures are the same as described in Chapter 2 for constructing a confidence interval for a population median, except that in the matched-pairs case, differences play the role of original observations used in the single-sample case.

    Arithmetic Procedure   The arithmetic procedure consists of the following steps:

1. For each of the n matched pairs (X_i, Y_i), obtain the difference $D_i = Y_i - X_i$.
2. Form all possible averages

$$u_{ij} = \frac{D_i + D_j}{2}, \qquad 1 \le i \le j \le n \qquad (4.7)$$

   There will be a total of $[n(n-1)/2] + n$ such averages.
3. Arrange the u_{ij}'s in order of magnitude from smallest to largest.
4. The median of the u_{ij}'s provides a point estimate of the population median difference.
5. Locate in Table A.3 the sample size and the appropriate value of P as determined by the desired level of confidence. When $(1 - \alpha)$ is the confidence coefficient, $P = \alpha/2$.

   When the exact value of $\alpha/2$ cannot be found in Table A.3, choose a neighboring value—either the closest value or one that is larger or smaller than $\alpha/2$, depending on whether a slightly wider or slightly narrower interval than desired is more acceptable.
6. The endpoints of the confidence interval are the Kth smallest and the Kth largest values of u_{ij}. $K = T + 1$, where T is the value in the column labeled T corresponding to the value of P selected in Step 5.

Example 4.4

Adamson et al. (E12) studied the effects of strenuous daytime physical exercise on healthy male volunteers. For eight of the subjects Table 4.14 shows the nocturnal plasma corticosteroid levels on control nights and on nights following exercise. We wish to construct an approximate 95% confidence interval for the median difference.

TABLE 4.14

Nocturnal plasma corticosteroid levels in eight subjects on control nights and on nights following exercise

Subject	1	2	3	4	5	6	7	8
Control nights (Y)	69.9	46.0	63.7	55.9	53.9	72.9	53.9	36.5
Post-exercise nights (X)	49.9	45.9	47.5	57.9	47.1	50.3	36.7	31.4

Source: Liisi Adamson, W. M. Hunter, O. O. Ogunremi, I. Oswald, and I. W. Percy-Robb, "Growth Hormone Increase during Sleep after Daytime Exercise," *J. Endocrinol.*, 62 (1974), 473–478

1. The differences are 20.0, 0.1, 16.2, −2.0, 6.8, 22.6, 17.2, 5.1.
2. The $[8(7)/2] + 8 = 36$ averages arranged in order of magnitude are shown in Table 4.15.
3. The median of the u_{ij}'s, 10.90, is our point estimate of the median difference.
4. In Table A.3 we find that $T = 4$ for $n = 8$ and $P = 0.0273 \approx 0.025 = 1 − 0.05/2$. The endpoints of our interval, then, are the ($T + 1 = 4 + 1 = 5$)th smallest and largest of the ordered averages shown in Table 4.15.
5. The lower limit of our confidence interval, then, is 2.4, and the upper limit is 19.4. Thus we are $100[1 − 2(0.0273)] = 94.54\%$ confident that the population median difference is somewhere between 2.4 and 19.4.

TABLE 4.15

Ordered array of averages, $u_{ij} = (D_i + D_j)/2$, for Example 4.4

1 −2.0	**7** 3.45	**13** 8.15	**19** 11.15	**25** 13.85	**31** 18.6
2 −0.95	**8** 5.1	**14** 8.65	**20** 11.35	**26** 14.7	**32** 19.4
3 0.1	**9** 5.95	**15** 9.0	**21** 11.5	**27** 16.2	**33** 19.9
4 1.55	**10** 6.8	**16** 10.05	**22** 12.0	**28** 16.7	**34** 20.0
5 2.4	**11** 7.1	**17** 10.3	**23** 12.55	**29** 17.2	**35** 21.3
6 2.6	**12** 7.6	**18** 10.65	**24** 13.4	**30** 18.1	**36** 22.6

We may save some time in constructing the confidence interval if we order the D_i's before calculating the u_{ij}'s. Then we can compute the K smallest and K largest values fairly quickly. If we want the sample median difference, we let a equal the total number of averages and compute only the $(a + 1)/2$ smallest (or largest) averages if a is odd and the $(a/2) + 1$ smallest (or largest) averages if a is even.

FIGURE 4.1   Graphical procedure for finding 95% confidence interval in Example 4.4

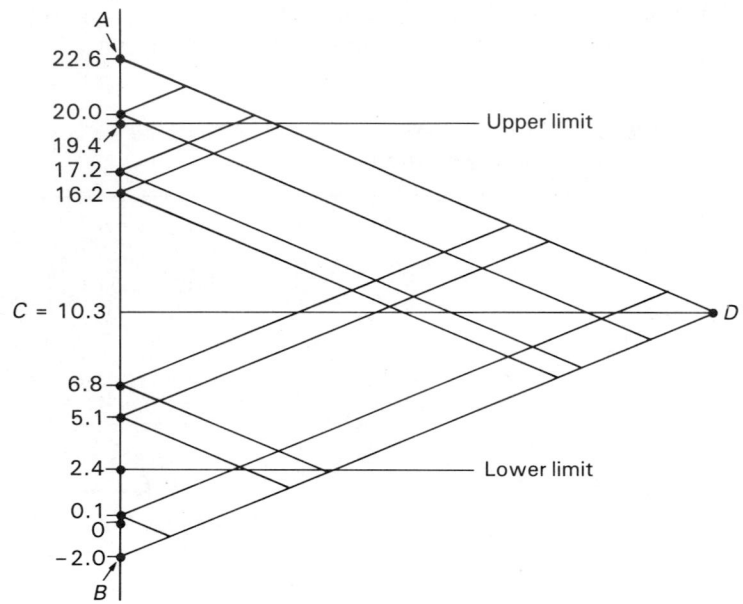

Graphical Procedure   In some instances the graphical method described in Chapter 2 may be more desirable for finding the confidence limits for a median difference. The following steps in the procedure, as applied to Example 4.4, are illustrated in Figure 4.1.

1. For each of the n matched pairs (X_i, Y_i), obtain the differences $D_i = Y_i - X_i$.
2. Plot the differences on the vertical axis of the graph.
3. Label the largest difference A and the smallest difference B. In Example 4.4, $A = 22.6$ and $B = -2.0$.
4. Find the point halfway between A and B and label it C. In the example, C appears on the graph at 10.3.
5. Draw a line of convenient length through C, perpendicular to the vertical axis. Label the endpoint of this line D.
6. Connect AD and DB with straight lines to form the triangle ADB.
7. Draw a line parallel to BD from each data point on the vertical axis to line AD.
8. Draw a line parallel to AD from each data point on the vertical axis to line BD.

9. To locate the upper confidence limit, consult Table A.3 to determine K. Beginning with point A, count down to the Kth intersection. In counting, include intersections with the vertical axis. Draw a horizontal line from the Kth intersection to the vertical axis. The point of intersection of this line and the vertical axis is the upper limit of the confidence interval. For Example 4.4 K is 5. A horizontal line from the fifth intersection from the top intersects the vertical axis of the graph at about 19.4.

10. To locate the lower limit, begin with point B and count up to the Kth intersection. Draw a horizontal line from this intersection to the vertical axis. The point at which this line intersects the vertical axis is the lower limit of the confidence interval. For Example 4.4 the lower limit is about 2.4. Thus the upper and lower limits of 19.4 and 2.4 obtained graphically agree with those obtained by the arithmetic procedure.

As pointed out in Chapter 2, if a value in the original plotted data occurs more than once, the horizontal lines through the repeated points on the vertical axis will coincide. Then we have to be careful to count the intersections made by all such lines, even though they appear as a single line on the graph.

Large-Sample Approximation For a discussion of the large-sample approximation, see Chapter 2.

Høyland (T38) has investigated the robustness of the Wilcoxon estimate of location when only a few observations can be collected per day and when several days are required to obtain the necessary number of observations. In such a situation some of the assumptions underlying the Wilcoxon procedure may be violated. Srivastava and Sen (T39) discuss sequential confidence intervals based on Wilcoxon-type estimators.

EXERCISES

4.9 Scott and Patterson (E13) measured cardiac output in 15 patients before and three days after direct-current conversion of atrial fibrillation to sinus rhythm. The results are shown in Table 4.16. Construct a 99% confidence interval for the median difference.

TABLE 4.16 **Effect of dc conversion on cardiac output, liters per minute**

Patient	1	2	3	4	5	6	7	8	9	10	11	12	13	14	15
Before conversion	6.3	3.6	4.5	6.3	5.7	4.7	4.8	5.6	6.6	5.5	5.8	4.0	5.4	3.1	4.2
3 hours after conversion	6.4	3.4	4.8	6.6	5.4	5.8	5.1	6.2	6.1	6.8	6.3	4.7	6.8	4.4	4.8

Source: M. E. Scott and G. C. Patterson, "Cardiac Output after Direct Current Conversion of Atrial Fibrillation," *Br. Heart J.*, 31 (1969), 87–90; reprinted by permission of the editor and publisher.

4.10 Yau et al. (E14) reported the details of surgical correction of tuberculosis kyphosis of the spine in 30 patients. Table 4.17 shows the pre- and postoperative lung capacities of 20 of their patients. Construct an approximate 99% confidence interval for the median difference.

TABLE 4.17   **Total lung capacity, milliliters, of 20 patients before and after surgical correction of tuberculosis kyphosis**

Patient	1	2	3	4	5	6	7	8	9	10
Preoperative	1160	1870	1980	1520	3155	1485	1150	1740	3260	4950
Postoperative	1500	2220	2080	2160	3040	2030	1370	2370	4060	5070
Patient	11	12	13	14	15	16	17	18	19	20
Preoperative	1440	1770	2850	2860	1530	3770	2260	3370	2570	2810
Postoperative	1680	1750	3730	3430	1570	3750	2840	3500	2640	3260

Source: A. C. M. C. Yau, L. C. S. Hsu, J. P. O'Brien, and A. R. Hodgson, "Tuberculosis Kyphosis, Correction with Spinal Osteotomy, Halo-Pelvic Distraction, and Anterior and Posterior Fusion," *J. Bone Joint Surg.*, 56-A (1974), 1419–1434.

4.3

TEST FOR TWO RELATED SAMPLES WHEN DATA CONSIST OF FREQUENCIES

The responses of subjects consisting of paired samples frequently fall into one of two categories: yes or no, respond or do not respond, survived or did not survive, and so on. In such cases, we are interested in the frequencies of occurrence (or proportions) of subjects falling into the two categories. For example, we may match subjects on relevant variables and then teach them to perform some task, using the standard method for one member of each pair (control subject), and a new or experimental method for the other member (experimental subject). The response of interest might be whether an individual subject, after training, can complete the task correctly within a specified length of time. The results might be displayed in a 2×2 contingency table such as Table 4.18, where the letters have the following meanings:

N = the total number of matched pairs

A = the number of pairs in which both members are able to perform the task within the specified length of time

D = the number of pairs in which neither of the members is able to perform the task within the specified length of time

B = the number of pairs in which the control subject is able to perform the task in the specified length of time, but the experimental subject is not

TABLE 4.18   **Results of two methods of teaching experimental subjects**

		Experimental subjects		
		Yes	No	Total
Control subjects	Yes	A	B	A + B
	No	C	D	C + D
	Total	A + C	B + D	N

C = the number of pairs in which the experimental subject is able to perform the task in the specified length of time, but the control subject is not

$A + B$ = the total number of pairs in which the control subject is able to perform the task in the specified length of time

$C + D$ = the total number of pairs in which the control subject is not able to perform the task in the specified length of time

$A + C$ = the total number of pairs in which the experimental subject is able to perform the task in the specified length of time

$B + D$ = the total number of pairs in which the experimental subject is not able to perform the task in the specified length of time

Paired observations may arise in other situations. The same subjects may be given a placebo at one time and a real treatment at another. The numbers of subjects responding and not responding under the two conditions may be displayed in a contingency table similar to Table 4.18, where N = the total number of subjects, A = the number responding to both the placebo and the treatment, D = the number responding to neither the placebo nor the treatment, and so on. In an industrial experiment, N specimens of a certain material may be divided in half, one half treated by protective agent A and the other half by protective agent B. To evaluate the relative merits of the two agents, the experimenter may expose each half to the same potentially destructive conditions. Again the results may be portrayed in a table such as Table 4.18. In this case, N = the number of specimens, A = the number of specimens in which both halves withstand the experimental conditions, and D = the number of specimens in which both halves fail to withstand the experimental conditions.

MCNEMAR TEST FOR TWO RELATED SAMPLES

The proper test for analyzing frequency data from two related samples is the *McNemar test for related samples*, introduced by McNemar (T40) in 1947.

Assumptions

The data consist of N randomly selected subjects (or items) or pairs of subjects, depending on whether subjects act as their own controls or whether experimental subjects are paired with a matched control. The data available for analysis may be displayed in a table similar to Table 4.18.

The measurement scale is nominal, with four categories. Using the notation of Table 4.18, the four categories are Yes–Yes, Yes–No, No–Yes, and No–No.

When subjects are their own controls, they are independent of each other. Of course, the two observations made on the same subject are related, since they are made on the same individual. When matched pairs are used, the pairs are independent, but observations within a given pair are related.

Hypotheses

A. We may wish to test the null hypothesis that the proportion of items or subjects with the characteristic of interest is the same under two conditions (or treatments). We let p_1 be the proportion with the characteristic of interest under one condition, and p_2 the proportion with the characteristic of interest under the other condition. We may state the null and alternative hypotheses symbolically as follows:

$$H_0: p_1 = p_2 \quad \text{or} \quad p_1 - p_2 = 0,$$
$$H_1: p_1 \neq p_2 \quad \text{or} \quad p_1 - p_2 \neq 0$$

B. If we want to see whether we can conclude that the proportion with the characteristic of interest under condition 1 is higher than under condition 2, we have a one-sided test, and the hypotheses are

$$H_0: p_1 \leq p_2, \qquad H_1: p_1 > p_2$$

C. The hypotheses for the other one-sided test are

$$H_0: p_1 \geq p_2, \qquad H_1: p_1 < p_2$$

Test Statistic

The sample proportions with the characteristic of interest may be designated $\hat{p}_1$ and $\hat{p}_2$. Using the notation of Table 4.18, we designate the control subjects sample 1 and the experimental subjects sample 2. Then the sample proportions are

$$\hat{p}_1 = \frac{A + B}{N} \tag{4.8}$$

$$\hat{p}_2 = \frac{A + C}{N} \tag{4.9}$$

where the characteristic of interest is a Yes response.

The difference between sample proportions is

$$\hat{p}_1 - \hat{p}_2 = \frac{A + B}{N} - \frac{A + C}{N} = \frac{B - C}{N} \qquad \text{(4.10)}$$

The null hypothesis is that the expectation of $(B - C)/N$ is zero. McNemar (T40) shows that an appropriate test statistic is

$$z = \frac{B - C}{\sqrt{B + C}} \qquad \text{(4.11)}$$

When $H_0: p_1 = p_2$ is true, z is distributed approximately as the standard normal variate, provided that $B + C$ is at least 10.

Decision Rule

A. Reject H_0 at the α level of significance if the computed z is equal to or greater than the z (Table A.2) to the right of which lies $\alpha/2$ of the area under the standard normal curve (two-sided test).

B. Reject H_0 at the α level of significance if the computed z is greater than or equal to the tabulated value that has α of the area to its right.

C. Reject H_0 at the α level of significance if the computed z is less than or equal to the tabulated z that has α of the area to its left.

The McNemar test has been recommended (T40, T41) for use in certain before-and-after experiments when the experimenter is interested in the number of subjects who respond differently after they are exposed to some intervening condition or treatment. For example, a random sample of subjects may be asked to indicate whether they are for or against a certain candidate for public office. The subjects may then be exposed to a speech or debate by the candidate. After that, they are again asked whether they are for or against the candidate. We are interested in knowing whether the intervening speech or debate caused a significant change of opinion. Because of this use, the test is frequently referred to as the *McNemar test for the significance of changes.*

Example 4.5   Pike and Smith (E15) use the data of Johnson and Johnson (E16) to illustrate the use of the McNemar test. They matched each of 85 patients treated for Hodgkin's disease with a nonpatient sibling of the same sex and within five years of age of the patient. The question to be answered is whether there is a differential rate of tonsillectomies in the two groups. The results are shown in Table 4.19.

Hypotheses

The characteristic of interest is a history of a tonsillectomy in a given subject. The appropriate null hypothesis is that the proportion of tonsillectomies in the

TABLE 4.19   **History of tonsillectomy in patients with Hodgkin's disease and in matched controls**

		Tonsillectomy in matched controls		
		Yes	No	Total
	Yes	26	15	41
Tonsillectomy in patients	No	7	37	44
	Total	33	52	85

Source: Sandra K. and Ralph E. Johnson, "Tonsillectomy History and Hodgkin's Disease," *N. Eng. J. Med.*, 287 (1972), 1122–1125: reprinted by permission from *The New England Journal of Medicine.*

population of Hodgkin's-disease patients is the same as in the population of matched siblings. We let $\alpha = 0.05$, and we state the hypotheses as follows.

$$H_0: p_1 = p_2, \qquad H_1: p_1 \neq p_2$$

Test Statistic

Using the data in Table 4.19 and Equation 4.11, we find that the computed value of the test statistic is

$$z = \frac{15 - 7}{\sqrt{15 + 7}} = 1.71$$

Decision

Since 1.71 is less than the critical z of 1.96, we cannot reject H_0. Since we have a two-sided test, the P value (determined from Table A.2) is $2(0.0436) = 0.0872$.

Power-Efficiency   Bennett and Underwood (T42) have investigated the power function of the McNemar test. Nam (T43) derived the power functions of the McNemar test and a test due to Gart (T44) on the basis of a comparison of the two tests. Nam (T43) suggests that Gart's test is preferable if the sample size is small and an order effect exists. The results of Nam's study indicate that Gart's test should be used, because under these conditions the McNemar test provided a distorted P value and was found to be biased.

FURTHER READING

Bennett (T45) has extended the McNemar test to situations in which there are more than two categories of matching. An extension of the test by Cochran (T46) is discussed in Chapter 7. Fink (T47) considers a use of McNemar's test when one wishes to have an additional classification for doubtful cases.

McNemar (T41) recommends the use of a continuity-correction factor, but Bennett and Underwood (T42) found that in general the continuity correction

recommended by McNemar should be avoided. They give an alternative correction, which they say is better.

Schemper (T48) presents a brief survey of nonparametric procedures and computer programs for use with paired data. The power of the component randomization test for the paired-sample location test is compared to the power of the parametric test for paired comparisons in a paper by Deutsch and Schmeiser (T49). The analysis is presented in the form of operating characteristic curves for normal, exponential, uniform, and absolute lambda observations over a range of small sample sizes.

A computer-based algorithm for Fisher's component randomization test for paired comparisons is developed in a paper by Schmeiser and Deutsch (T50). They describe the program operation and present an example along with the program listing. The program provides a graphical comparison of the randomization distribution to the corresponding Student's *t* distribution.

In a paper that explores the problem of ties in the sign test when the comparisons within pairs are based on ordered classes, Hay and Peck (T51) suggest an alternative to the sign test, which they say may be more powerful.

Lam and Longnecker (T52) propose a function of the Wilcoxon rank sum statistic for testing the equality of the marginal distributions when sampling is from a bivariate population. The test statistic is asymptotically nonparametric. The authors compare their proposed statistic with several parametric and nonparametric tests with respect to Pitman relative efficiency and small-sample performance. They state that their test statistic has power comparable to the paired *t* test when sampling from the bivariate normal distribution and greater power in several important cases.

Lachenbruch and Woolson (T53) compare three tests for analyzing paired survival data. The McNemar test is discussed in some detail in the book on contingency table analysis by Everitt (T54).

EXERCISES

4.11 Waters (E17) conducted a controlled clinical trial for treating migraine headache in women. The subjects received tablets of ergotamine (1 mg) and placebo (lactose) in random order for periods of eight weeks each. Table 4.20 shows the results of the experi-

TABLE 4.20   **Response in 79 women treated with ergotamine and placebo**

		Ergotamine benefit?		
		Yes	No	Total
	Yes	22	24	46
Placebo benefit?	**No**	18	15	33
	Total	40	39	79

Source: W. E. Waters, "Controlled Clinical Trial of Ergotamine Tartrate," *Br. Med. J.*, 2 (1970), 325–327; used by permission of the editor and publisher.

ment. Can we conclude on the basis of these data that the experimental treatment is effective in the treatment of headache of the type studied? Let $\alpha = 0.05$. What is the P value?

4.12 Sokal and Rohlf (E18) cite a study by Nelson (E19) in which the response of the rabbit tick (*Haemaphysalis leporis-palustris*) to light was measured. Individual ticks were placed first in an arena that was 1 inch in diameter, then in an arena 2 inches in diameter. The response of interest was whether the tick left the arena on the side toward the light or away from the light. The results are shown in Table 4.21. Can we conclude from these data that the size of the arena affects the response of rabbit ticks to light? Let $\alpha = 0.05$. What is the P value?

TABLE 4.21 **Response of rabbit ticks to light**

		Left toward light from 1-inch arena		
		Yes	No	Total
Left toward light from 2-inch arena	Yes	8	9	17
	No	5	8	13
	Total	13	17	30

Source: V. E. Nelson, unpublished laboratory data; used by permission of Professor Nelson.

4.4

COMPUTER PROGRAMS

Among the microcomputer software packages on which the procedures covered in this chapter will run are *BETASTAT* (McNemar test), *SCA* (sign test and Wilcoxon test), *SPSS/PC* (McNemar test), and *STATA* (Wilcoxon test).

REVIEW EXERCISES

4.13 August et al. (E20) studied collagen metabolism in children deficient in growth hormone before and after growth-hormone therapy. They reported the data on hydroxyproline shown in Table 4.22. Can we conclude on the basis of these data that growth-hormone therapy increases heat-insoluble hydroxyproline in the skin? Let $\alpha = 0.05$. Use the sign test, and determine the P value.

TABLE 4.22

Total quantity of heat-insoluble hydroxyproline, micromoles per gram of dry weight, in skin of seven children before and three months after growth-hormone therapy

Child	1	2	3	4	5	6	7
Before (X)	349	400	520	490	574	427	435
After (Y)	425	533	362	628	463	427	449

Source: Gilbert P. August, Wellington Hung, and John C. Houck, "The Effects of Growth Hormone Therapy on Collagen Metabolism in Children," *J. Clin. Endocrinol. Metabol.*, 39 (1974), 1103–1109.

4.14 Goldzimer et al. (E21) studied the role of vasoconstriction in zones of nonnecrotizing pneumonia. In an experimental pneumonia model, they measured the extent to which infusion of potent pulmonary vasodilators corrected the perfusion deficit. The subjects were otherwise healthy mongrel dogs in which lobar pneumococcal pneumonia was induced. Lung perfusion was studied both immediately before and during the infusion of isoproterenol and/or prostaglandin E_1. Change in relative blood flow to the infected lung zone was measured by serial perfusion scintiphotographs and serial shunt fractions.

Table 4.23 shows the shunt-fraction values (Qs/Qt %) for five dogs prior to any pharmacologic manipulation and during prostaglandin E_1 infusion at 10 μg $\cdot$ min^{-1} (high-dose PGE_1 infusion). Use the Wilcoxon test to determine whether these data provide sufficient evidence to indicate that high-dose PGE_1 infusion increases shunt fractions. Let $\alpha = 0.05$. Find the P value.

TABLE 4.23

Shunt fractions, Qs/Qt %, for dogs with pneumococcal pneumonia before and during high-dose (10 μg $\cdot$ min^{-1}) infusion of PGE,

Dog	1	2	3	4	5
Before	20.7	22.7	19.9	20.4	18
During	37.4	28.6	31	29.1	26.5

Source: Edward L. Goldzimer, Ronald G. Konopka, and Kenneth M. Moser, "Reversal of the Perfusion Defect in Experimental Canine Lobar Pneumococcal Pneumonia," *J. Appl. Physiol.*, 37 (1974), 85–91.

4.15 Del Greco and Burgess (E22) describe their experience with patients in terminal renal failure observed before and after nephrectomy. Table 4.24 shows the supine diastolic blood pressure values in patients with controllable hypertension before and two to six months after nephrectomy. Use the procedure based on the sign test to construct a 95% confidence interval for the median difference.

TABLE 4.24

Supine diastolic blood pressure values in terminal renal patients with controllable hypertension before and two to six months after bilateral nephrectomy

Patient	1	2	3	4	5
Before	107	102	95	106	112
After	87	97	101	113	80

Source: Francesco del Greco and Janis L. Burgess, "Hypertension in Terminal Renal Failure, Observations Pre and Post Bilateral Nephrectomy," *J. Chronic Dis.*, 26 (1973), 471–501; reprinted with permission from Pergamon Press.

4.16 Heiman et al. (E23) randomly placed one member of each of seven matched pairs of students reading below grade level in a supplementary reading program. Subjects were matched as nearly as possible on the discrepancy between their reading level and their current grade level. The supplementary program consisted of a point system to reward attention to and identification of letter and word combinations. Table 4.25, derived from the authors' results, shows the differences between the scores made by the subjects on a reading test after and before the program. Use the procedure based on the Wilcoxon test to construct a 95% confidence interval for the median difference.

TABLE 4.25   **Differences in reading test scores made by seven matched pairs of subjects, one member of which was assigned to an experimental program and the other to a control group (before scores subtracted from after scores)**

Pair	1	2	3	4	5	6	7
Experimental	0.5	1.0	0.6	0.1	1.3	0.1	1.0
Control	0.8	1.1	−0.1	0.2	0.2	1.5	0.8

Source: Julia R. Heiman, Mark J. Fischer, and Alan O. Ross, ''A Supplementary Behaviorial Program to Improve Deficient Reading Performance,'' *J. Abnormal Child Psychol.*, 1 (1973), 390–399; published by Plenum Publishing Corporation, New York.

4.17 A psychologist rated subjects undergoing withdrawal from narcotics on the basis of the extent of their depression before and one hour after receiving a dose of methadone. Degree of depression was rated as N (not depressed), M (mildly depressed), or S (severely depressed). The results are shown in Table 4.26. Can one conclude from these data that subjects undergoing withdrawal from narcotics tend to be more depressed after receiving a dose of methadone than before? Let $\alpha = 0.05$, and determine the P value.

TABLE 4.26   **Extent of depression observed in subjects undergoing withdrawal from narcotics before and after a dose of methadone**

Subject	1	2	3	4	5	6	7	8	9	10
Before	N	N	M	N	M	M	M	N	N	N
After	M	M	S	M	M	N	S	S	M	M

Note: N = not depressed, M = mildly depressed, S = severely depressed.

4.18 Fifteen high school dropouts were given a test to measure their attitudes toward "the Establishment" shortly after they dropped out and six months after they were readmitted to school. The results are shown in Table 4.27. Do these data provide sufficient evidence to

TABLE 4.27   **Test scores showing attitude toward "the Establishment" of high school dropouts before and after being readmitted to school**

Subject	1	2	3	4	5	6	7	8	9	10	11	12	13	14	15
Before	63	75	78	84	58	58	70	76	74	88	74	94	99	79	93
After	84	86	75	94	50	95	97	98	72	100	101	98	105	84	90

indicate that high school dropouts tend to score higher on the attitude scale after being re-admitted to school? Let $\alpha = 0.05$, and determine the P value. Use the method that uses the most information contained in the data.

4.19 Twenty healthy high school students participated in an encounter-group experience in which half the participants were handicapped persons of about the same age. Before and after the group experience, the 20 nonhandicapped students took a test designed to measure their understanding of handicapped persons. Table 4.28 shows the results of the tests. Can one conclude from the data that such an experience increases students' understanding (as indicated by a higher score) of handicapped persons? Let $\alpha = 0.05$, and determine the P value. Use the test that uses the most information contained in the data.

TABLE 4.28 **Scores made by 20 high school students on a test to measure understanding of handicapped persons before and after an encounter-group experience**

Student	1	2	3	4	5	6	7	8	9	10	11	12	13	14	15	16	17	18	19	20
Before	55	63	54	61	63	60	59	67	55	64	68	57	81	84	90	97	80	72	64	55
After	60	68	69	64	67	61	63	62	58	62	70	57	89	88	94	96	80	65	74	70

4.20 Fifteen sixth-grade students reading at or below the third-grade level participated in a remedial reading program for six months. Table 4.29 shows the oral reading test scores of these students before and after the program. Use the best test to determine whether one can conclude that the program is effective. Let $\alpha = 0.05$, and determine the P value.

TABLE 4.29 **Oral reading scores of 15 sixth-grade students before and after a remedial reading program**

Student	1	2	3	4	5	6	7	8	9	10	11	12	13	14	15
Before	22	20	19	14	17	20	24	18	23	20	22	22	22	25	30
After	50	30	29	17	17	40	41	17	38	32	42	31	30	30	38

4.21 In a study of factors related to the development of rapport between male juvenile offenders and social workers, two types of interview were conducted with each of 12 subjects. In the first interview the social worker wore a beard and long hair, dressed in a very casual manner, and used the language of the youth subculture. Each juvenile offender was then interviewed by another male social worker who was about the same age as the first, whose dress, language, and appearance identified him as a member of "the Establishment." The interviews were otherwise conducted in as similar a manner as possible. After each interview, subjects rated the interviewer on the basis of how well they would like to have him as a permanent counselor (1 = would very much want him, 2 = would want him, 3 = would not want him, and 4 = would very much not want him). The results are shown in Table 4.30. Do these data provide sufficient information to indicate that social workers who are perceived as antiestablishment are more popular with juvenile offenders? Let $\alpha = 0.05$, and determine the P value.

TABLE 4.30   **Juvenile offenders' ratings of two social workers**

Subject	1	2	3	4	5	6	7	8	9	10	11	12
Antiestablishment social worker	1	2	2	1	1	2	2	1	1	1	2	2
Establishment social worker	4	2	1	3	4	4	3	2	4	3	3	4

4.22 A sample of 150 college students indicated on a questionnaire whether they believed smoking causes lung cancer. The students then attended a lecture and exhibit conducted by a health team explaining the hazards of smoking. After the health team's presentation, the students were again asked their opinions on the cause–effect relationship of smoking and lung cancer, with the results shown in Table 4.31. Do these data provide sufficient evidence to indicate that the lecture and exhibit on smoking hazards are effective in changing people's opinions on the relationship between smoking and lung cancer? Let $\alpha = 0.05$, and determine the P value.

TABLE 4.31   **Opinions of students on cause–effect relationship of smoking and lung cancer before and after exposure to lecture and exhibit on hazards of smoking**

Does smoking cause lung cancer?

		Before		
		Yes	No	Total
	Yes	30	67	97
After	No	10	43	53
	Total	40	110	150

4.23 To compare the effectiveness of two detergents in cleaning cotton fabric, researchers uniformly soiled and then cut in half twelve pieces of the fabric. One half of each piece was randomly assigned to be washed in detergent A; the other half was washed in detergent B. After the fabric halves had been washed and dried, each was examined to determine the effectiveness of the detergent. Table 4.32 shows the results. An x or a y shows which half of each piece of fabric was considered to be cleaner than the half washed in the other detergent. There was one instance in which the two halves of a piece were judged to be equally clean. Can we conclude from these data that detergent B is better than detergent A? Let $\alpha = 0.05$, and find the P value.

TABLE 4.32   **Data for Exercise 4.23**

Fabric sample	1	2	3	4	5	6	7	8	9	10	11	12
Detergent A	x			x						x		
Detergent B		y	y		y	y	y	y	y	y	y	y

4.24 Each of a random sample of 16 shoppers who participated in a survey was given a case of soft drinks for their efforts. They could choose between two brands. Thirteen chose brand

A. Can one conclude on the basis of this information that the sampled population of shoppers prefers brand A soft drinks? Let $\alpha = 0.05$, and find the P value.

4.25 A company conducted an experiment to determine which of two training methods is most effective for their employees. Nine pairs of employees were matched on age, gender, and other relevant variables. One member of each pair was randomly assigned to a training course taught by method A. The other was assigned to the same type of training course taught by method B. At the end of the course, each employee was given an examination to test retention of the material presented. Table 4.33 shows the results. May we conclude on the basis of these data that method A is better than method B? A higher score indicates a better performance. Let $\alpha = 0.01$, and find the P value. Use the procedure that uses the most information contained in the data.

TABLE 4.33 **Test scores made by employees following a training course**

Pair	1	2	3	4	5	6	7	8	9
Method A	91	87	84	93	89	84	86	94	89
Method B	79	71	69	81	81	85	86	87	84

4.26 A researcher with an advertising agency wished to assess the effect of a particular advertising strategy on the future purchasing habits of consumers. A random sample of 25 adults were matched with respect to age, gender, socioeconomic status, and other relevant variables with 25 other adults to form a set of 25 matched pairs. One member of each pair was randomly assigned to watch, once a week for six weeks, a television commercial for a certain new brand of toothpaste. The other member of each pair, serving as controls, did not see the commercials. At the end of the experiment all 50 subjects were given enough money to buy a tube of the new brand of toothpaste or a tube of several competing brands. In three pairs of subjects both members bought the new brand; in two pairs neither member bought the new brand; and in 15 pairs the member who had seen the commercials bought the new brand, but the member who had not seen the commercials bought another brand. Can we conclude from these data that the advertising strategy was effective? Let $\alpha = 0.05$, and find the P value.

4.27 Ten restaurants located in a large city were selected at random to participate in an experiment conducted by the county health department as part of an effort to improve sanitation among the city's eating establishments. The restaurant managers were paid to attend a three-day seminar that emphasized the benefits and techniques of operating a clean restaurant. The restaurants were inspected by a health department sanitarian before and six months after the seminar. The inspection scores are shown in Table 4.34. Can we conclude from these data that the seminar was effective? Use the Wilcoxon test. Let $\alpha = 0.05$, and find the P value.

TABLE 4.34 **Inspection score data for Exercise 4.27**

Restaurant	1	2	3	4	5	6	7	8	9	10
Before	80	83	82	81	77	77	65	67	75	85
After	90	85	87	78	75	82	75	85	90	95

4.28 Researchers wished to compare two methods of teaching employees to operate an intricate machine. One member in each of 30 pairs of matched subjects was randomly assigned to method A, and the other subject was taught by method B. At the end of the training period, the subjects were given a test to determine whether they could operate the machine satisfactorily. In four pairs both subjects passed the test. In sixteen pairs the member who was taught by method B passed, and the other member failed. In one pair both subjects failed. Can we conclude from these data that method B is more effective than method A? Let $\alpha = 0.05$, and determine the P value.

4.29 Consult *Newspaper Rates and Data*, published by Standard Rate & Data Service, Inc., of Wilmette, Illinois. Select a simple random sample of 30 or more daily newspapers that are published east of the Mississippi River. Match on the basis of circulation each newspaper in this sample with a daily newspaper published west of the Mississippi. Use the large-sample approximation to the Wilcoxon matched-pairs signed-ranks test to test an appropriate null hypothesis regarding classified display rates.

REFERENCES

T1 Bennett, B. M., "On Multivariate Sign Tests," *J. Roy. Statist. Soc. Ser. B*, 24 (1962), 159–161.

T2 Bhattacharyya, G. K., "A Note on Asymptotic Efficiency of Bennett's Bivariate Sign Test," *J. Roy. Statist. Soc. Ser. B*, 28 (1966), 146–149.

T3 Krauth, J., "Asymptotic UMP Sign Test in Presence of Ties," *Ann. Statist.* 1 (1973), 166–169.

T4 Putter, J., "The Treatment of Ties in Some Nonparametric Tests," *Ann. Math. Statist.*, 26 (1955), 368–386.

T5 Chatterjee, S. K., "A Bivariate Sign Test for Location," *Ann. Math. Statist.*, 37 (1966), 1771–1782.

T6 Gastwirt, J. L., "Sign Test for Symmetry," *J. Amer. Statist. Assoc.*, 66 (1971), 821–823.

T7 Groeneve, R. A., "Note on Sequential Sign Test," *Amer. Statist.*, 25 (April 1971), 15–16.

T8 Wilcoxon, F., "Individual Comparisons by Ranking Methods," *Biometrics* 1 (1945), 80–83.

T9 Basu, A. P., "On Large Sample Properties of a Generalized Wilcoxon–Mann–Whitney Statistic," *Ann. Math. Statist.*, 38 (1967), 905–915.

T10 Buckle, N., C. Kraft, and C. Van Eeden, "An Approximation to the Wilcoxon–Mann–Whitney Distribution," *J. Amer. Statist. Assoc.*, 64 (1969), 591–599.

T11 Claypool, P. L., and D. Holbert, "Accuracy of Normal and Edgeworth Approximations to Distribution of Wilcoxon Signed-Rank Statistic," *J. Amer. Statist. Assoc.*, 69 (1974), 255–258.

T12 Hodges, J. L., and E. L. Lehmann, "Wilcoxon and *t* Test for Matched Pairs of Typed Subjects," *J. Amer. Statist. Assoc.*, 68 (1973), 151–158.

T13 Hollander, M., G. Pledger, and P. E. Lin, "Robustness of Wilcoxon Test to a Certain Dependency between Samples," *Ann. Statist.*, 2 (1974), 177–181.

T14 Jurečkova, J., "Central Limit Theorem for Wilcoxon Rank Statistics Process," *Ann. Statist.*, 1 (1973), 1046–1060.

T15 Noether, G. E., "Efficiency of the Wilcoxon Two-Sample Statistic for Randomized Blocks," *J. Amer. Statist. Assoc.*, 58 (1963), 894–898.

T16 Lepage, Y., "A Combination of Wilcoxon's and Ansari–Bradley's Statistics," *Biometrika*, 58 (1971), 213–217.

T17 Lepage, Y., "A Table for a Combined Wilcoxon Ansari–Bradley Statistic," *Biometrika* 60 (1973), 113–116.

T18 Hollander, M., "Asymptotic Efficiency of Two Nonparametric Competitors of Wilcoxon's Two-Sample Test," *J. Amer. Statist. Assoc.*, 62 (1967), 939–949.

T19 Gehan, E. A., "A Generalized Wilcoxon Test for Comparing Arbitrarily Singly Censored Samples," *Biometrika*, 52 (1965), 203–223.

T20 Gehan, E. A., "A Generalized Two-Sample Wilcoxon Test for Doubly Censored Data." *Biometrika*, 52 (1965), 650–653.

T21 Alling, D., "Early Decision in the Wilcoxon Two-Sample Test," *J. Amer. Statist. Assoc.*, 58 (1963), 713–720.

T22 Stone, M., "Extreme Tail Probabilities for the Null Distribution of the Two-Sample Wilcoxon Statistic," *Biometrika*, 54 (1967), 629–640.

T23 Wilcoxon, F., *Some Rapid Approximate Statistical Procedures*, Stamford, Conn.: American Cyanamid Company, 1949.

T24 Bühler, W. J., "The Treatment of Ties in the Wilcoxon Test," *Ann. Math. Statist.*, 38 (1967), 519–522.

T25 Claypool, P. L., "Linear Interpolation within McCornack's Table of the Wilcoxon Matched-Pair Signed-Rank Statistic," *J. Amer. Statist. Assoc.*, 65 (1970), 974–975.

T26 Verdooren, L. R., "Extended Tables of Critical Values for Wilcoxon's Test Statistic," *Biometrika*, 50 (1963), 177–186.

T27 Kempthorne, O., and T. E. Doerfler, "The Behavior of Some Significance Tests under Experimental Randomization," *Biometrika*, 56 (1969), 231–248.

T28 David, F. N., and P. J. Kim, "Matched Pairs and Randomization Sets," *Ann. Hum. Genet.*, 31 (1967), 21–27.

T29 Edgington, E. S., "Approximate Randomization Tests," *J. Psychol.*, 72 (1969), 143–149.

T30 Edgington, E. S., "Randomization Tests with Statistical Control over Concomitant Variables," *J. Psychol.*, 79 (1971), 13–19.

T31 Edgington, E. S., "Randomization Tests," *J. Psychol.*, 57 (1964), 445–449.

T32 Edgington, E. S. "The Random-Sampling Assumption in 'Comment on Component–Randomization Tests,'" *Psychol. Bull.*, 80 (1973), 84–85.

T33 Cleroux, R., "First and Second Moments of the Randomization Test in Two-Associate PBIB Designs," *J. Amer. Statist. Assoc.*, 64 (1969), 1424–1433.

T34 Klauber, M. R., "Two-Sample Randomization Tests for Space–Time Clustering," *Biometrics*, 27 (1971), 129–142.

T35 Alf, E. F., and N. M. Abraham, "Comment on Component Randomization Tests," *Psychol. Bull.*, 77 (1972), 223–224.

T36 Walsh, J. E., "Some Significance Tests for the Median Which Are Valid under Very General Conditions," *Ann. Math. Statist.*, 20 (1949), 64–81.

T37 Walsh, J. E., "Applications of Some Significance Tests for the Median Which Are Valid under Very General Conditions," *J. Amer. Statist. Assoc.*, 44 (1949), 342–355.

T38 Høyland, A., "Robustness of the Wilcoxon Estimate of Location against a Certain Dependence," *Ann. Math. Statist.*, 39 (1968), 1196–1201.

T39 Srivastava, M. S., and A. K. Sen, "Sequential Confidence Intervals Based on Wilcoxon Type Estimates," *Ann. Statist.*, 1 (1973), 1200–1202.

T40　McNemar, Quinn, "Note on the Sampling Error of the Difference between Correlated Proportions or Percentages," *Psychometrika*, 12 (1947), 153–157.

T41　McNemar, Quinn, *Psychological Statistics*, fourth edition, New York: Wiley, 1969.

T42　Bennett, B. M., and R. E. Underwood, "On McNemar's Test for the 2 × 2 Table and Its Power Function," *Biometrics*, 26 (1970), 339–343.

T43　Nam, Jun-Mo, "On Two Tests for Comparing Matched Proportions," *Biometrics*, 27 (1971), 945–959.

T44　Gart, J. J., "An Exact Test for Comparing Matched Proportions in Crossover Designs," *Biometrika*, 56 (1969), 75–80.

T45　Bennett, B. M., "Tests of Hypotheses Concerning Matched Samples," *J. Roy. Statist. Soc., Ser. B*, 29 (1967), 468–474.

T46　Cochran, W. G., "The Comparison of Percentages in Matched Samples," *Biometrika*, 37 (1950), 256–266.

T47　Fink, H., "McNemar's Test Using an Additional Classification for Doubtful Cases," *Arzneimittel–Forschung*, 22 (1972), 606–609.

T48　Schemper, M., "Statistical Methods and Programs for Nonparametric Analysis of Pairs," *Statist. Software Newsl.*, 11 (1985), 128–129.

T49　Deutsch, Stuart Jay, and Bruce Wayne Schmeiser, "The Power of Paired Sample Tests," *Naval Res. Logist. Q.*, 29 (1982), 635–649.

T50　Schmeiser, Bruce, and Stuart J. Deutsch, "Computation of the Component Randomization Test for Paired Comparisons," *J. Quality Technol.*, 15 (1983), 94–98.

T51　Hay, Alan, and Francis Peck, "An Alternative to the Sign Test in a Matched Pairs Design," *Statistician*, 33 (1984), 201–204.

T52　Lam, F. C., and M. T. Longnecker, "A Modified Wilcoxon Rank Sum Test for Paired Data," *Biometrika*, 70 (1983), 510–513.

T53　Lachenbruch, Peter A., and Robert F. Woolson, "The Generalized Signed Rank Test, the Generalized Sign Test and the Stratified Log Rank Test," in P. K. Sen (ed.), *Biostatistics: Statistics in Biomedical, Public Health and Environmental Sciences*, North-Holland: Elsevier 1985, 389–398.

T54　Everitt, B. S., *The Analysis of Contingency Tables*, London: Chapman and Hall, 1977.

E1　Latané, Bibb, and Howard Cappell, "The Effects of Togetherness on Heart Rate in Rats," *Psychon. Sci.*, 29 (1972), 177–179.

E2　Van Duijn, H., "Superiority of Clonazepam over Diazepam in Experimental Epilepsy," *Epilepsia*, 14 (1973), 195–202.

E3　Shani, Mordechai, Uri Seligsohn, and Judith Ben-Ezzer, "Effect of Phenobarbital on Liver Functions in Patients with Dubin–Johnson Syndrome," *Gastroenterology*, 67 (1974), 303–308.

E4　Smith, Jerry E., and Mario Di Girolamo, "Effect of Weight Reduction in the Rat on Epididymal Fat-Cell Size and Relative Dispersion," *Amer. J. Physiol.*, 227 (1974), 420–424.

E5　Dickie, Kenneth J., William J. de Groot, Robert N. Cooley, Ted P. Bond, and M. Mason Guest, "Hemodynamic Effects of Bolus Infusion of Urokinase in Pulmonary Thromboembolism," *Amer. Rev. Respir. Dis.*, 109 (1974), 48–56.

E6　Piggott, Leonard R., Albert F. Ax, Jacqueline L. Bamford, and Joanne M. Fetzner, "Respiration Sinus Arrhythmia in Psychotic Children," *Psychophysiology*, 10 (1973), 401–414.

E7　Bhatia, M. L., S. C. Manchanda, and Sujoy B. Roy, "Coronary Haemodynamic Studies in Chronic Severe Anaemia," *Br. Heart J.*, 31 (1969), 365–374.

E8 Hall, F. F., C. R. Ratliff, T. Hayakawa, T. W. Culp, and N. C. Hightower, "Substrate Differentiation of Human Pancreatic and Salivary Alpha-Amylases," *Amer. J. Dig. Dis.*, 15 (1970), 1031–1038.

E9 Vagenakis, Apostolos G., Basil Rapoport, Fereidoun Azizi, Gary I. Portnay, Lewis E. Braverman, and Sidney H. Ingbar, "Hyperresponse to Thyrotropin-Releasing Hormone Accompanying Small Decreases in Serum Thyroid Concentrations," *J. Clin. Invest.*, 54 (1974), 913–918.

E10 Angseesing, J. P. A., "Selective Eating of the Acyanogenic Form of Trifolium Repens," *Heredity*, 32 (1974), 73–83.

E11 Weis, F. Robert, Jr., and Jerome Peak, "Effects of Oxytocin on Blood Pressure during Anesthesia," *Anesthesiology*, 40 (1974), 189–190.

E12 Adamson, Liisi, W. M. Hunter, O. O. Ogunremi, I. Oswald, and I. W. Percy-Robb, "Growth Hormone Increase during Sleep after Daytime Exercise," *J. Endocrinol.*, 62 (1974), 473–478.

E13 Scott, M. E., and G. C. Patterson, "Cardiac Output after Direct Current Conversion of Atrial Fibrillation," *Br. Heart J.*, 31 (1969), 87–90.

E14 Yau, A. C. M. C., L. C. S. Hsu, J. P. O'Brien, and A. R. Hodgson, "Tuberculosis Kyphosis, Correction with Spinal Osteotomy, Halo-Pelvic Distraction, and Anterior and Posterior Fusion," *J. Bone Joint Surg.*, 56-A (1974), 1419–1434.

E15 Pike, M. C., and P. G. Smith, "Tonsillectomy and Hodgkin's Disease," *Lancet*, 1 (1973), 434.

E16 Johnson, Sandra K., and Ralph E. Johnson, "Tonsillectomy History and Hodgkin's Disease," *N. Engl. J. Med.*, 287 (1972), 1122–1125.

E17 Waters, W. E., "Controlled Clinical Trial of Ergotamine Tartrate," *Br. Med. J.*, 2 (1970), 325–327.

E18 Sokal, Robert R., and F. James Rohlf, *Biometry*, San Francisco: W. H. Freeman, 1969.

E19 Nelson, V. E., unpublished laboratory data.

E20 August, Gilbert P., Wellington Hung, and John C. Houck, "The Effects of Growth Hormone Therapy on Collagen Metabolism in Children," *J. Clin. Endocrinol. Metabol.*, 39 (1974), 1103–1109.

E21 Goldzimer, Edward L., Ronald G. Konopka, and Kenneth M. Moser, "Reversal of the Perfusion Defect in Experimental Canine Lobar Pneumococcal Pneumonia," *J. Appl. Physiol.*, 37 (1974), 85–91.

E22 del Greco, Francesco, and Janis L. Burgess, "Hypertension in Terminal Renal Failure, Observations Pre and Post Bilateral Nephrectomy," *J. Chronic Dis.*, 26 (1973), 471–501.

E23 Heiman, Julia R., Mark J. Fischer, and Alan O. Ross, "A Supplementary Behavioral Program to Improve Deficient Reading Performance," *J. Abnorm. Child Psychol.*, 1 (1973), 390–399.

5

CHI-SQUARE TESTS OF INDEPENDENCE AND HOMOGENEITY

Among the most widely used of all statistical procedures are the chi-square tests of independence and homogeneity. These tests are based on a technique introduced in 1900 by Karl Pearson (T1), who has been called the founder of the science of statistics (T2). Pearson was concerned with the problem of the goodness of fit of observed data to theoretical frequency curves. In his important paper (T1), he derived the chi-square test of goodness of fit that is discussed in more detail in Chapter 6.

The chi-square tests of independence and homogeneity are essentially tests of goodness of fit. In each case the test procedure consists of comparing observed frequencies with frequencies that are expected if some null hypothesis is true. We compute a measure of the goodness of fit of observed to expected frequencies. If the computed measure indicates that the fit is sufficiently poor, we reject the null hypothesis; otherwise, we do not. We judge the goodness of the fit of observed frequencies to expected frequencies by comparing the computed measure of fit to an appropriate value of a distribution known as the *chi-square distribution*.

Many statisticians do not consider the chi-square tests of independence and homogeneity to be nonparametric procedures. In the two-sample case, these tests are just another way of testing the null hypothesis that two population proportions

are equal. In the multisample case, the chi-square tests allow us to make inferences about the parameters of the multinomial distribution. However, it is customary to discuss the chi-square tests in nonparametric statistics texts, and this book will not break tradition by omitting them.

The next section examines briefly some of the mathematical properties of the chi-square distribution, and the two sections that follow cover the tests for independence and homogeneity. The final section is devoted to a series of special techniques that enable investigators to make better use of chi-square tests.

5.1

MATHEMATICAL PROPERTIES OF THE CHI-SQUARE DISTRIBUTION

As the reader may know from an elementary course in statistics, any normal distribution can be transformed to the standard normal distribution (which has a mean of 0 and a standard deviation of 1) by the formula

$$Z_i = \frac{X_i - \mu}{\sigma} \tag{5.1}$$

where Z_i is a value from the standard normal distribution, X_i is an observation from the normal distribution to be transformed, and μ and σ are the mean and standard deviation, respectively, of this distribution.

If we square the Z_i variates of Equation 5.1, computing

$$Z_i^2 = \left(\frac{X_i - \mu}{\sigma}\right)^2 \tag{5.2}$$

then Z^2 follows a chi-square distribution. Thus if we randomly and independently select values of some normally distributed random variable X, standardize the values, and square the resulting standardized values, we have a random variable that has a chi-square distribution.

If we randomly and independently select a sample of two values of some normally distributed random variable X, standardize each by computing $Z_1 = (X_1 - \mu)/\sigma$ and $Z_2 = (X_2 - \mu)/\sigma$, square each resulting Z, and obtain the sum of the squared Z's, we have the variable $Z_1^2 + Z_2^2$, which also has a chi-square distribution. In general, if we follow this procedure for a sample of size n, we obtain the variable

$$Z_1^2 + Z_2^2 + Z_3^2 + \cdots + Z_n^2$$

This, too, follows a chi-square distribution.

One chi-square distribution differs from another on the basis of the number of degrees of freedom associated with each. The term *degrees of freedom* refers to the number of independent standard normal variates that we square and add to obtain the variable that has the chi-square distribution. The chi-square variable is

FIGURE 5.1   Chi-square distributions for selected degrees of freedom

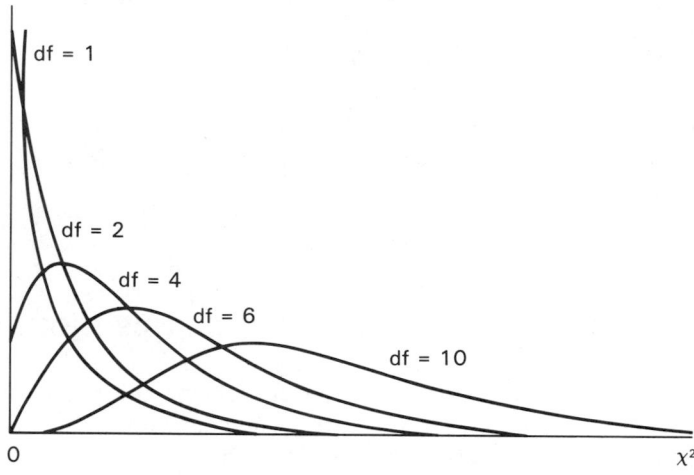

designated by the symbol χ^2, which was first used by Pearson (T3). A subscript, indicating degrees of freedom, may be added to distinguish one distribution from another. Thus χ_1^2, χ_2^2, and χ_n^2 designate variables distributed as chi-square with 1, 2, and n degrees of freedom, respectively. Chi-square distributions for selected degrees of freedom are shown in Figure 5.1. As a result of the central limit theorem, the chi-square distribution approaches the normal distribution as n increases.

Table A.11 gives chi-square values for various degrees of freedom values. The subscripts on χ^2 given as the column headings indicate the proportion of the area under the chi-square curve that is to the left of the value appearing in the body of the table. For example, $\chi_{0.95}^2$ indicates that 95% of the area under the chi-square curve for 1 degree of freedom is to the left of 3.841.

The mean of a chi-square distribution is equal to its degrees of freedom, and its variance is equal to two times its degrees of freedom. For example, the mean of the chi-square distribution with 10 degrees of freedom is 10, and its variance is 20.

A more extensive discussion of the mathematical properties of the chi-square distribution may be found in most textbooks on mathematical statistics. Lancaster's book (T4) is devoted entirely to the chi-square distribution.

The following sections show that under some null hypothesis, the measure of the goodness of fit of observed data to expected results has approximately a chi-square distribution when the null hypothesis is true. In each test, then, the chi-square distribution with the appropriate degrees of freedom provides a standard against which to compare the computed value of the test statistic so that we can decide whether to reject the null hypothesis.

5.2

CHI-SQUARE TEST OF INDEPENDENCE

A research question that frequently arises is whether two variables are associated. For example, a sociologist may wish to know whether level of formal education is associated with income. A consumer protection official may be interested in knowing whether price is associated with the quality of small household appliances. A school nutritionist may want to know whether the nutritional status of students is associated with their academic performance.

If there is no association between two variables, we say that they are *independent*. Two variables are independent if the distribution of one in no way depends on the distribution of the other. If two variables are not associated (that is, if they are independent), then knowing the value of one variable for some subject will not help us determine the value of the other variable for the same subject. For example, if the price and quality of small household appliances are independent, a person who knows the price of an appliance is in no better position to predict its quality than the person who does not know its price. On the other hand, if two variables are associated, knowledge of one is helpful in predicting the value the other is likely to assume.

We use the chi-square test of independence as follows to decide whether two variables in a population are independent.

Assumptions

    A. The data consist of a simple random sample of size *n* from some population of interest.

    B. The observations in the sample may be cross-classified according to two criteria, so that each observation belongs to one and only one category of each criterion. The criteria are the variables of interest in a given situation.

    C. The variables may be inherently categorical, or they may be quantitative variables whose measurements are capable of being classified into mutually exclusive numerical categories.

The data may be displayed in a contingency table such as Table 5.1, in which the observed number n_{ij} of subjects characterized by one category of each criterion is placed in the cell formed by the intersection of the ith row and jth column. The cell entries are referred to as *observed cell frequencies*, and they are usually designated O_{ij}—that is, $O_{ij} = n_{ij}$. The observed cell frequency O_{ij} represents the joint occurrence in the sample subjects of the ith category of the first criterion of classification with the jth category of the second.

TABLE 5.1   **Contingency table for chi-square test of independence**

First criterion of classification	Second criterion of classification						
	Category						
Category	1	2	$\cdots$	j	$\cdots$	c	Total
1	n_{11}	n_{12}	$\cdots$	n_{1j}	$\cdots$	n_{1c}	$n_{1.}$
2	n_{21}	n_{22}	$\cdots$	n_{2j}	$\cdots$	n_{2c}	$n_{2.}$
$\vdots$							$\vdots$
i	n_{i1}	n_{i2}	$\cdots$	n_{ij}	$\cdots$	n_{ic}	$n_{i.}$
$\vdots$							$\vdots$
r	n_{r1}	n_{r2}	$\cdots$	n_{rj}	$\cdots$	n_{rc}	$n_{r.}$
Total	$n_{.1}$	$n_{.2}$	$\cdots$	$n_{.j}$	$\cdots$	$n_{.c}$	n

Hypotheses

H_0: The two criteria of classification are independent

H_1: The two criteria of classification are not independent

Test Statistic

We compute the test statistic under the assumption that H_0 is true—that is, that the two criteria of classification are independent. As already pointed out, the chi-square test of independence compares observed results with results that are expected when H_0 is true. Specifically, the test involves a comparison of the observed cell frequencies with cell frequencies that are expected when H_0 is true.

To obtain the expected cell frequencies, we use the following elementary law of probability: If two events are independent, the probability of their joint occurrence is equal to the product of their individual probabilities. If H_0 is true—that is, if the two criteria of classification are independent—the probability that a subject in a sample of size n will belong in cell ij is equal to the probability that the subject will belong in the ith row times the probability that the subject will belong in the jth column. We estimate these probabilities from the sample data by $n_{i.}/n$ and $n_{.j}/n$, respectively. Then we can write the estimated probability that a subject will belong in cell ij as follows:

$$P \text{ (subject belongs in cell } ij) = \left(\frac{n_{i.}}{n}\right)\left(\frac{n_{.j}}{n}\right) \tag{5.3}$$

To obtain E_{ij}, the expected frequency for cell ij, we multiply this estimated probability by the total sample size. Thus the expected frequency for cell ij of the contingency table shown as Table 5.1 is

$$E_{ij} = n\left(\frac{n_{i.}}{n}\right)\left(\frac{n_{.j}}{n}\right) \tag{5.4}$$

Since the n in the numerator of Equation 5.4 cancels one of the n's in the denominator, the equation reduces to

$$E_{ij} = n_{i.}n_{.j}/n \qquad (5.5)$$

This form of the equation indicates that we can easily compute an expected cell frequency by multiplying together the appropriate row and column totals and dividing the product by the total sample size.

When we have the observed cell frequencies and the corresponding expected cell frequencies, we are interested in the magnitudes of the differences between them. Specifically, we wish to know whether the differences are small enough to be attributed to chance (sampling variability) when H_0 is true, or if the differences are so large that some other explanation (namely, that H_0 is false) is necessary. From the expected and observed frequencies, we may compute a test statistic that reflects the magnitudes of differences between these two quantities. When H_0 is true, this statistic has approximately a χ^2 distribution with $(r-1)(c-1)$ degrees of freedom, where r is the number of rows and c is the number of columns in the contingency table. The statistic is

$$X^2 = \sum_{i=1}^{r} \sum_{j=1}^{c} \left[\frac{(O_{ij} - E_{ij})^2}{E_{ij}} \right] \qquad (5.6)$$

When the differences between observed and expected frequencies are large, X^2 is large; when there is close agreement between them, X^2 is small.

Decision Rule

We may reject the null hypothesis of independence at the α level of significance if the computed value of the test statistic X^2 exceeds the tabulated value of $\chi^2_{1-\alpha}$ for $(r-1)(c-1)$ degrees of freedom.

Example 5.1

The data in Table 5.2, reported by Monteiro (E1), show 2764 Rhode Island residents classified according to income and time elapsed since they last consulted a physician. We wish to know whether these data provide sufficient evidence to indicate that there is an association between income and length of time since the last consultation with a physician. In other words, we wish to know whether we can conclude that the two variables are not independent.

Hypotheses

H_0: Income and time elapsed since last physician consultation are independent

H_1: The two variables are not independent

TABLE 5.2

Time elapsed since last consultation with a physician, by income (Rhode Island, 1971)

	Last consulted physician			
Income	Within previous 6 months	Within 7 months to one year	More than one year	Total
Less than $3000	186	38	35	259
$3000–$4999	227	54	45	326
$5000–$6999	219	78	78	375
$7000–$9999	355	112	140	607
$10,000 plus	653	285	259	1197
Total	1640	567	557	2764

Source: Lois A. Monteiro, "Expense is No Object...: Income and Physician Visits Reconsidered," J. Health Soc. Behav., 14 (1973), 99–115.

Test Statistic

By Equation 5.5, we find that the expected cell frequency for cell 11 is

$$E_{11} = (259)(1640)/2764 = 153.68$$

Similar calculations yield the other expected cell frequencies, as shown in parentheses in Table 5.3.

TABLE 5.3

Observed frequencies and expected frequencies for Example 5.1 (expected frequencies are in parentheses)

	Last consulted physician			
Income	Within previous 6 months	Within 7 months to one year	More than one year	Total
Less than $3000	186 (153.68)	38 (53.13)	35 (52.19)	259
$3000–$4999	227 (193.43)	54 (66.87)	45 (65.70)	326
$5000–$6999	219 (222.50)	78 (76.93)	78 (75.57)	375
$7000–$9999	355 (360.16)	112 (124.52)	140 (122.32)	607
$10,000 plus	653 (710.23)	285 (245.55)	259 (241.22)	1197
Total	1640	567	557	2764

The computed value of the test statistic by Equation 5.6 is

$$X^2 = \frac{(186 - 153.68)^2}{153.68} + \frac{(227 - 193.43)^2}{193.43} + \cdots + \frac{(259 - 241.22)^2}{241.22}$$

$$= 47.90$$

The degrees of freedom are $(5 - 1)(3 - 1) = 8$.

Decision

Since 47.90 is greater than $\chi^2_{0.995} = 21.955$, we can reject H_0 at the 0.005 level of significance. We conclude that income and the time elapsed since the last consultation with a physician are not independent. The P value is less than 0.005.

Small Expected Frequencies   The statistic

$$X^2 = \sum_{i=1}^{r} \sum_{j=1}^{c} \left[\frac{(O_{ij} - E_{ij})^2}{E_{ij}} \right]$$

is distributed approximately as χ^2 when H_0 is true only if the expected frequencies E_{ij} are large. Just how large the E_{ij} should be in order for the application of the chi-square test to be valid has not been unanimously agreed on by statisticians. Some writers have recommended minimum values as high as 10 for the E_{ij}. For a contingency table with more than one degree of freedom, however, Cochran (T5, T6) recommends that a minimum expected cell frequency as low as 1 be allowed if no more than 20% of the cells have expected frequencies less than 5. If X^2 has less than 30 degrees of freedom and the minimum expected frequency is 2 or more, Cochran (T6) states that the use of the ordinary χ^2 tables is usually adequate.

Adjacent rows and/or adjacent columns in a contingency table may be combined to achieve the minimum expected cell frequencies. The problem of small expected frequencies has been dealt with also by Katti and Sastry (T7), Nass (T8), Roscoe and Byars (T9), Tate and Hyer (T10, T11), and Yarnold (T12).

Roscoe and Byars (T9), for example, are less restrictive than Cochran in their recommendations with respect to minimum expected frequency sizes. With data drawn from a uniform population, they obtained acceptable results at the 0.05 significance level with an average expected frequency as low as 2. An average expected frequency as low as 4 gave acceptable results at the 0.01 level. They recommend that when sampling from populations exhibiting moderate departures from uniformity, the average expected frequency should be maintained at 4 or more for the 0.05 level and at 6 or more for the 0.01 level. Other researchers are even less conservative, suggesting that under fairly general conditions, expected frequencies in all cells as low as 1 may be tolerated without endangering the validity of the test.

2 × 2 Contingency Table   When there are two categories of each of two criteria of classification, the resulting contingency table has two rows and two columns forming four cells. Such a table is referred to as a 2 × 2, or *fourfold*, contingency table. A general 2 × 2 contingency table is shown as Table 5.4. We find that there is 1 degree of freedom associated with a 2 × 2 contingency table when we apply the $(r - 1)(c - 1)$ rule. We can use the following shortcut formula to compute X^2 from the data of a 2 × 2 table.

$$X^2 = \frac{n(ad - bc)^2}{(a + c)(b + d)(c + d)(a + b)} \tag{5.7}$$

TABLE 5.4   **A 2 × 2 contingency table**

First criterion of classification	Second criterion of classification		Total
	1	**2**	**Total**
1	a	b	a + b
2	c	d	c + d
Total	a + c	b + d	n

The following example illustrates the procedure for analyzing the data of a 2 × 2 table.

Example 5.2   Abse et al. (E2) reported the data shown in Table 5.5 on nighttime smoking and lung cancer in 56 subjects. We wish to know whether we may conclude from these data that nighttime smoking and lung cancer are related.

TABLE 5.5   **Nighttime smoking and lung cancer status in 56 subjects**

Lung cancer	Nighttime smoking		Total
	Yes	**No**	**Total**
Yes	20	16	36
No	6	14	20
Total	26	30	56

Source: D. Wilfred Abse, Marilyn M. Wilkins, Gordon Kirshner, Don L. Weston, Robert S. Brown, and W. D. Buxton, "Self-Frustration, Nighttime Smoking and Lung Cancer," *Psychosom. Med.*, 34 (1972), 395–404.

Hypotheses

H_0: Nighttime smoking and lung cancer are independent

H_1: The two variables are related (not independent)

Test Statistic

From the data of Table 5.5, we use Equation 5.7 to compute

$$X^2 = \frac{56[(20)(14) - (16)(6)]^2}{(26)(30)(20)(36)} = 3.376$$

Decision

Since $X^2 = 3.376$ is less than 3.841, the tabulated value of chi-square for 1 degree of freedom and $\alpha = 0.05$, we cannot reject H_0 at the 0.05 level of significance. We therefore conclude at the 0.05 level of significance that the two variables—nighttime smoking and lung cancer—may be independent ($0.10 > P$ value > 0.05).

The problem of small expected frequencies may also be encountered in situations involving 2×2 contingency tables. In these cases the recommendation of Cochran (T5, T6) is again frequently followed. He recommends not using the chi-square test when $n < 20$. For $20 < n < 40$, Cochran recommends not using the chi-square test if the smallest expected frequency is less than 5. For $n > 40$, no expected frequency in a 2×2 contingency table should be less than 1.

Yates's Correction   In 1934, Yates (T13) proposed a correction procedure for use when computing X^2 from the data of a 2×2 contingency table. The procedure, known as *Yates's correction for continuity*, consists of subtracting $0.5n$ from the absolute value of $ad - bc$ in the numerator of Equation 5.7. The purpose of this procedure is to "correct" for the use of a continuous distribution χ^2 to approximate the discrete distribution X^2. When we use the correction, Equation 5.7 becomes

$$X_c^2 = \frac{n(|ad - bc| - 0.5n)^2}{(a + c)(b + d)(c + d)(a + b)} \tag{5.8}$$

Although Yates's correction has been used extensively in the past, there have been critics of the procedure, including Conover (T14), Grizzle (T15), Lancaster (T16), Pearson (T17), and Plackett (T18). As a result of their findings, there currently appears to be a trend away from using the correction. For other comments on Yates's correction, see the articles by Liddell (T19) and Mantel and Greenhouse (T20).

If we apply Yates's correction to the data of Example 5.2 shown in Table 5.5, we have

$$X_c^2 = \frac{56[|(20)(14) - (16)(6)| - 0.5(56)]^2}{(26)(30)(20)(36)} = 2.427$$

Thus, at the 0.05 level of significance, we reach the same conclusion both with and without the correction (P value > 0.10).

EXERCISES

5.1   Highman and Davidson (E3) surveyed nonprofit, acute, general hospitals of 100 beds or more in the United States to determine their planning methods. Table 5.6 shows the

number of questionnaires returned and not returned by size of hospital surveyed. Can we conclude from these data that there is an association between a hospital's size and its willingness to return survey questionnaires? What is the P value?

TABLE 5.6   **Number of hospitals responding and not responding to a survey questionnaire, by size of hospital**

Response	Size of hospital, in beds					Total
	100–199	**200–299**	**300–399**	**400–699**	**700 and over**	
Questionnaires returned	108	94	62	67	14	345
Questionnaires not returned	334	151	112	53	6	656
Questionnaires sent	442	245	174	120	20	1001

Source: Arthur Highman and Leth Davidson, "Plans and Planning Methods of Hospital Administrators," *Hosp. Topics*, 50 (June 1972), 30–42; reprinted by permission from *Hospital Topics*.

5.2   As part of a study designed to identify possible subcultural, ethnic, and family influences on mobility after training, Lieberman (E4) obtained the data for underclass whites shown in Table 5.7. Are these data sufficient evidence for us to reject the null hypothesis that opportunity level is independent of labor force mobility? What is the P value?

TABLE 5.7   **Opportunity levels for underclass whites and labor force mobility**

Opportunities	Labor force mobility		Total
	Low	**High**	
Low	45	19	64
High	6	43	49
Total	51	62	113

Source: Leonard Lieberman, "Atomism and Mobility among Underclass Chippewas and Whites," *Hum. Org.*, 32 (1973), 337–347; reproduced by permission of the Society for Applied Anthropology.

5.3   In a study of the association between hypoglycemia and increasing mean daily dosage (MDD) of insulin, researchers with the Boston Collaborative Drug Surveillance Program (E5) obtained the data shown in Table 5.8. Can we conclude from these data that mean daily dosage of insulin is related to the presence or absence of hypoglycemia? Find the P value.

5.4   The data shown in Table 5.9 on fertility status and the presence or absence of a psychiatric diagnosis for 100 married women were reported by Mai et al. (E6). Do these data provide sufficient evidence to indicate a relationship between fertility status and the presence of a psychiatric diagnosis? Determine the P value.

TABLE 5.8   **Mean daily dosage (MDD) of insulin, units per kilogram of body weight, and presence or absence of hypoglycemia in 325 patients**

	Hypoglycemia		
MDD	**Present**	**Absent**	**Total**
< 0.25	4	40	44
0.25–0.49	21	74	95
0.50–0.74	28	59	87
0.75–0.99	15	26	41
≥ 1.0	12	46	58
Total	80	245	325

Source: "Special Communications: Relation of Body Weight and Insulin Dose to the Frequency of Hypogly-cemia, A Report from the Boston Collaborative Drug Surveillance Program," *J. Amer. Med. Assoc.*, 228 (1974), 192–194; copyright 1974, American Medical Association.

TABLE 5.9   **Fertility status and psychiatric diagnosis status in 100 married women**

	Fertility status		
Psychiatric diagnosis	**Infertile**	**Fertile**	**Total**
Yes	31	17	48
No	19	33	52
Total	50	50	100

Source: Francois M. Mai, Robert N. Munday, and Eric E. Rump, "Psychiatric Interview Comparisons between Infertile and Fertile Couples," *Psychosom. Med.*, 34 (1972), 431–440.

5.5   In a study of the attitudes of whites toward blacks, Heller and Redente (E7) focused on attitudes toward potentially integrated neighborhoods. From the responses to a questionnaire administered in 231 households, the authors derived an "Attitudinal Index" in which a score of 10 represents the most positive attitude. Table 5.10 shows the attitude index and the distance in miles from the sample point to the center of the black neighborhood. Do these data provide sufficient evidence to indicate that distance from black neighborhood and attitude are not independent? Determine the P value.

TABLE 5.10   **Attitude index and distance from black neighborhood of 231 residents**

	Attitude index				
Distance, in miles	**0–3**	**4–6**	**7–8**	**9–10**	**Total**
0.55–1.23	5	11	13	21	50
1.54–1.78	3	16	27	9	55
1.97–3.02	10	26	20	22	78
3.88–7.12	14	15	11	8	48
Total	32	68	71	60	231

Source: Charles F. Heller, Jr., and Anthony L. Redente, "Residential Location and White Attitudes toward Mixed-Race Neighborhoods in Kalamazoo, Michigan," *J. Geog.*, 72 (March 1973), 15–25.

5.6   Steadman (E8) examined data from 256 cases of incompetent, indicted felony defendants reviewed under a New York State statute that necessitated determinations of dangerousness. Subjects were cross-classified according to (a) whether they were considered dangerous on the basis of a psychiatric examination and (b) their alleged offense. Table 5.11 shows the results. Can one conclude on the basis of these data that the two variables are related? Determine the *P* value.

TABLE 5.11   **Psychiatric findings of dangerousness by alleged offense**

| Psychiatrically dangerous | Alleged offense | | | | |
	Violent crime against a person	Robbery and burglary	Crime against property	Other	Total
Yes	75	46	23	11	155
No	30	32	24	15	101
Total	105	78	47	26	256

Source: Henry J. Steadman, "Some Evidence on the Inadequacy of the Concept and Determination of Dangerousness in Law and Psychiatry," *J. Psychiatry and the Law*, 1 (1973), 409–426; copyright 1973 by Federal Legal Publications, 95 Morton Street, New York, N.Y.

5.7   In a study of the influence of childhood background factors on adult levels of participation in hunting and fishing activities, Sofranko and Nolan (E9) reported the data shown in Table 5.12 on 440 respondents. Test the null hypothesis of independence and determine the *P* value.

TABLE 5.12   **Residence in youth and source of introduction to the sport by 440 fishermen**

| Source of introduction | Residence in youth | | |
	Rural	Nonrural	Total
Parents	118	47	165
Other relatives	32	24	56
Friends	56	40	96
No one	44	24	68
Combination	34	12	46
Spouse	2	7	9
Total	286	154	440

Source: Andrew J. Sofranko and Michael F. Nolan, "Early Life Experiences and Adult Sports Participation," *J. Leisure Res.*, 4 (1972), 6–18; copyright National Recreation and Park Association, 1972.

5.8   Farrell (E10) studied 108 offenders who received legal treatment during a five-year period in a large urban jurisdiction in the United States. Table 5.13 shows the subjects cross-classified by social status and legal treatment. Test the null hypothesis that these two criteria of classification are independent. Find the *P* value.

TABLE 5.13 **Legal treatment and social status of offenders**

	Social status			
Legal treatment	**Middle class**	**Upper-lower class**	**Lower-lower class**	**Total**
Incarceration	2	11	26	39
Probation	8	17	6	31
Fine or less	16	15	7	38
Total	26	43	39	108

Source: Ronald A. Farrell, "Class Linkages of Legal Treatment of Homosexuals," *Criminology*, 9 (May 1971), 57; reprinted by permission of Sage Publications.

5.9 Hazelrigg (E11) classified a sample of respondents in Italy on the basis of social class and political interests. Social class membership was determined on the basis of occupation. Respondents holding nonmanual positions were classified as middle class, those holding manual positions were classified as working class, and small-farm owners and sharecroppers were classified as agrarian. Political interest classifications were based on the respondents' answer to the question: "In your opinion, which party in Italy best defends the interests of people like you?" The results are shown in Table 5.14; the numbers in each column are percentages of the sample size shown at the bottom of the column. Convert these percentages to frequencies and determine whether the resulting data justify the conclusion that social class and political interests are related. Determine the *P* value.

TABLE 5.14 **Sample of respondents classified by social class and political interest**

	Social class		
Political interest	**Working**	**Middle**	**Agrarian**
Leftist	68	47	37
Centrist	28	34	57
Rightist	4	19	6
	$n_{.1} = 392$	$n_{.2} = 230$	$n_{.3} = 230$

Source: Lawrence E. Hazelrigg, "Religious and Class Bases of Political Conflict in Italy," *Amer. J. Sociol.*, 75 (1970), 496–511; copyright 1971 by The University of Chicago; all rights reserved.

Note: Numbers in each column are percentages of column *n*'s

5.10 Potter and Hoeke (E12) studied the investment behavior of 756 investors in common stocks. Table 5.15 shows the subjects in the study classified by value of stock shares owned (in dollars) and level of financial sophistication. Level I investors consisted of 512 subjects who subscribed to an investors' magazine. The authors refer to these as "heterogeneous" investors. Level II investors consisted of 172 account executives in stock brokerages. Level III investors were 72 financial analysts who were members of a chapter of the Society for Financial Analysts. Convert the percentages in Table 5.15 to numbers, and determine whether the resulting data provide sufficient evidence to indicate that value of stock shares owned and level of financial sophistication are related. Find the *P* value.

TABLE 5.15 | **Percentage of investors in three financial sophistication groups who own varying values of shares of stocks**

Value of shares ($)	Level I	Level II	Level III
0–999	16.3	8.1	5.1
1,000–4,999	27.5	14.6	5.1
5,000–19,999	26.7	27.3	17.0
20,000–99,999	19.1	28.5	20.3
Over 1,000,000	10.2	21.4	52.6

Source: Roger E. Potter and Robert S. Hoeke, "Intergroup Comparison of Stock Investor Dimensions," *Oklahoma Bus. Bull.*, 41 (April 1973), 6–11.

5.11 Table 5.16 shows a sample of 178 contact-lens wearers classified by their sex and their age when they first wore contact lenses [Bailey (E13)]. Do these data indicate that sex and age when contact lenses were first worn are not independent? Determine the P value.

TABLE 5.16 | **Age when contact lenses were first worn and sex of 178 subjects**

Age when contact lenses first worn	Male	Female	Total
< 15	2	8	10
15–19	38	93	131
20 and older	22	15	37
Total	62	116	178

Source: Neal J. Bailey, "Contact Lens Design—A Survey," *Amer. J. Optom. and Arch. Amer. Acad. Optom.*, 45 (1968), 96–102; used by permission of the publishers.

5.3

CHI-SQUARE TEST OF HOMOGENEITY

In elementary courses in statistics, students learn to test the null hypothesis that two population proportions p_1 and p_2 are equal by using the normal approximation to the binomial. The null hypothesis is usually stated as $H_0: p_1 = p_2$. Alternatively, the null hypothesis can be stated as H_0: The two populations are homogeneous (with respect to the proportion of subjects possessing some characteristic of interest). Customarily, a sample is drawn from each of the populations under study, and the subjects are classified according to whether they possess the characteristic of interest. The results may be displayed in a 2×2 contingency table such as Table 5.17. The resemblance of Table 5.17 to the 2×2 contingency table (Table 5.4) of the previous section is obvious. As might be suspected, we may compute an X^2 statistic from the data of Table 5.17 by using Equation 5.7.

Although we compute X^2 for both the test of homogeneity and the test for independence by using the same formula, the tests differ in two important respects:

TABLE 5.17 **A 2 × 2 contingency table**

| Sample | Characteristic of interest present | | Total |
	Yes	No	
1	a	b	a + b
2	c	d	c + d
Total	a + c	b + d	n

the sampling procedure and the rationale underlying the calculation of expected frequencies. The discussion that follows covers differences in greater detail.

We test the null hypothesis that the two populations represented by the two samples are homogeneous by comparing the computed value of X^2 with the tabulated value of χ^2 with 1 degree of freedom. Such a test is called the *chi-square test of homogeneity*. The testing procedure described above for the 2 × 2 contingency table is a special case of a more general procedure. When the contingency table consists of r rows and c columns, we may summarize the procedure as follows:

Assumptions

A. The samples are independent.

B. The samples are random.

C. Each subject in the population may be classified into one of two mutually exclusive categories, according to whether it has or does not have the characteristic of interest.

D. The variable whose measurements yield the categories may be inherently a categorical variable, or it may be a quantitative variable whose measurements are capable of being classified into mutually exclusive numerical categories.

Hypotheses

H_0: The sampled populations are homogeneous

H_1: The sampled populations are not homogeneous

Test Statistic

The test statistic is

$$X^2 = \sum_{i=1}^{r} \sum_{j=1}^{c} \left[\frac{(O_{ij} - E_{ij})^2}{E_{ij}} \right]$$

(5.9)

For a 2 × 2 contingency table, we may compute X^2 by using the computational formula, Equation 5.7.

Decision Rule

We reject H_0 if the computed value of X^2 is greater than or equal to the tabulated value of χ^2 with $(r - 1)(c - 1)$ degrees of freedom.

The following example illustrates the procedure for the case of the 2×2 contingency table.

Example 5.3 Richardson et al. (E14) reported the presence and absence of respiratory distress syndrome (RDS) in two groups of infants. Group 1 consisted of 42 infants whose fetal membranes ruptured 24 hours or less before delivery, while group 2 was composed of 22 infants whose membranes ruptured more than 24 hours before delivery. The data are shown in Table 5.18. We may use their results to test the null hypothesis that the two populations are homogeneous. Let $\alpha = 0.05$.

TABLE 5.18 **Incidence of respiratory distress syndrome (RDS) in two groups of infants**

	RDS		
Group	Yes	No	Total
1	27	15	42
2	7	15	22
Total	34	30	64

Source: C. Joan Richardson, Jeffrey J. Pomerance, M. Douglas Cunningham, and Louis Gluck, "Acceleration of Fetal Lung Maturation Following Prolonged Rupture of the Membranes," *Amer. J. Obstet. Gynecol.*, 118 (1974), 1115–1118.

Hypotheses

H_0: The two populations represented by the two groups in the study are homogeneous with respect to the presence of RDS

H_1: The two populations are not homogeneous

Test Statistic

By Equation 5.7, we have

$$X^2 = \frac{64[(27)(15) - (15)(7)]^2}{(34)(30)(22)(42)} = 6.112$$

Decision

Since $6.112 > 3.841$, reject H_0 and conclude that the two populations are not homogeneous ($0.025 > P$ value > 0.01).

The comments regarding small expected frequencies and Yates's correction in connection with 2×2 tables and the test of independence also apply in the case of the 2×2 test of homogeneity.

The $r \times 2$ Contingency Table   We can extend the chi-square test of homogeneity to the case of three or more populations whose subjects can be classified into one of two mutually exclusive categories. The data for such a test are typically obtained by drawing independent random samples from each of r populations, and they can be displayed in an $r \times 2$ contingency table, where the r rows represent the r populations and the two columns represent the two classifications of the characteristic of interest. Alternatively, the words or symbols that serve to identify the populations may provide the row headings and the categories of the variable of interest may provide the column headings.

The following example illustrates the test of homogeneity in this situation.

Example 5.4   Mims et al. (E15) studied the characteristics of subjects attending a five-day human sexuality program. The results are shown in Table 5.19. In the following test of homogeneity, we assume that the four groups of subjects constitute independent simple random samples from the four populations identified by the row headings.

TABLE 5.19   **Marital status of four groups of subjects attending a course on human sexuality**

Group	Single	Married or divorced	Total
Medical students	50	20	70
Nursing students	12	25	37
Other students	6	8	14
Group leaders	1	21	22
Total	69	74	143

Source: Fern Mims, Rosalee Yeaworth, and Stephen Hornstein, "Effectiveness of an Interdisciplinary Course in Human Sexuality," *Nursing Res.*, 23 (1974), 248–253. Copyright 1974, The American Journal of Nursing Company.

Hypotheses

H_0: The four populations represented in the study are homogeneous with respect to marital status

H_1: The four populations are not homogeneous with respect to marital status

Test Statistic

The expected cell frequencies are shown in Table 5.20.

TABLE 5.20 | **Expected cell frequencies for Example 5.4**

Group	Single	Married or divorced	Total
Medical students	33.776	36.224	70
Nursing students	17.853	19.147	37
Other students	6.755	7.245	14
Group leaders	10.615	11.385	22
Total	69	74	143

From the expected frequencies and the observed frequencies of Table 5.19, we compute

$$X^2 = \frac{(50 - 33.776)^2}{33.776} + \frac{(20 - 36.224)^2}{36.224} + \cdots + \frac{(21 - 11.385)^2}{11.385}$$

$$= 35.761$$

Decision

Since $35.761 > \chi_3^2 = 12.838$, we can reject H_0 at the 0.005 level of significance. We conclude on the basis of the reported data that the four populations are not homogeneous with respect to marital status. The P value is less than 0.005.

The $r \times c$ Contingency Table Frequently, subjects of a population can be placed into three or more mutually exclusive categories of some characteristic. When subjects in samples from two populations are classified into one of three or more categories, the results can be displayed in a $2 \times c$ contingency table. In the more general case, there are three or more populations represented and three or more mutually exclusive categories into which subjects can be placed. Under these conditions the data may be displayed in an $r \times c$ contingency table such as Table 5.21, where the r rows correspond to the r populations and the c columns correspond to the c categories of classification. Alternatively, the rows may correspond to the categories of classification and the columns to the identified populations. We may perform a chi-square test of homogeneity on the resulting data. Mechanically this test is identical to the test of independence for an $r \times c$ table, discussed in the previous section.

One important way in which the chi-square test of homogeneity differs from the chi-square test of independence is in the rationale underlying the calculation of expected frequencies. In the test of independence described in the previous section, we calculated expected cell frequencies for contingency tables under the assumption that the two criteria of classification are independent. Then according to a law of

TABLE 5.21 $\qquad$ **An $r \times c$ contingency table for chi-square test of homogeneity**

	Categories of classification						
Population	1	2	$\cdots$	j	$\cdots$	c	Total
1	n_{11}	n_{12}	$\cdots$	n_{1j}	$\cdots$	n_{1c}	$n_{1.}$
2	n_{21}	n_{22}	$\cdots$	n_{2j}	$\cdots$	n_{2c}	$n_{2.}$
$\vdots$							$\vdots$
i	n_{i1}	n_{i2}		n_{ij}		n_{ic}	$n_{i.}$
$\vdots$							$\vdots$
r	n_{r1}	n_{r2}	$\cdots$	n_{rj}	$\cdots$	n_{rc}	$n_{r.}$
Total	$n_{.1}$	$n_{.2}$	$\cdots$	$n_{.j}$	$\cdots$	$n_{.c}$	n

probability, the probability of the joint occurrence of two levels of each is equal to the product of the two individual probabilities.

We calculate expected cell frequencies in the test of homogeneity under the assumption that the populations represented in the contingency table are homogeneous with respect to the variable of interest. If this is true, we may conbine the several samples and treat them together as one single large sample as far as the critcrion or variable of classification is concerned. Consequently, for any one population represented in the study, we obtain the best estimate of the proportion of subjects falling into a given category by dividing the total from all samples falling into that category by the total of all subjects in the study.

For example, in Table 5.21, the best estimate of the proportion of subjects in population 1 falling into category 1 of the criterion of classification is equal to $n_{.1}/n$, the sample grand total in category 1 divided by the total number of subjects in the study. We refer to $n_{.1}/n$ as the expected proportion of the $n_{1.}$ subjects falling into category 1. Then, to obtain the expected number of the $n_{1.}$ sample subjects falling into category 1, we multiply the expected proportion in that category by $n_{1.}$ to obtain

$$ E_{11} = \left(\frac{n_{.1}}{n} \right) n_{1.} $$

Thus any expected cell frequency can be obtained by dividing the product of corresponding marginal totals by the grand total. Recall that this is the same procedure used to calculate expected cell frequencies in the test of independence. Thus, although the test of independence and the test of homogeneity are computationally identical, the rationales underlying the calculation of expected frequencies differ.

The two tests also differ with regard to the manner in which the data are usually gathered. For the test of independence, the investigator typically draws a single sample of subjects from a population and then cross-classifies the subjects on the basis of two criteria of interest. For the test of homogeneity, on the other hand, the

investigator usually identifies two or more populations of interest before collecting the data and draws an independent sample from each of the identified populations. After collecting the data, the investigator places subjects in each sample, separately, in one of the two or more categories of the criterion of classification. In both cases, we may summarize the results in a contingency table. The following is an example of a chi-square test of homogeneity.

Example 5.5   To determine public awareness and concern for atmospheric pollution, Wall (E16) interviewed a sample of 40 residents of each of three areas in Great Britain. Table 5.22 shows their responses to the question, "Is there an air pollution problem in this neighborhood?" Let us use the hypothesis-testing procedure described earlier in this section. Let $\alpha = 0.05$.

TABLE 5.22   **Responses to the question: "Is there an air pollution problem in this neighborhood?"**

Area of residence	No	Yes	Doubtful	Don't know	Total
Rawmarsh	5	31	2	2	40
Treeton	10	21	4	5	40
Wath	11	20	7	2	40
Total	26	72	13	9	120

Source: Geoffrey Wall, "Public Response to Air Pollution in South Yorkshire, England," *Environ. and Behav.*, 5 (June 1973), p. 239, copyright 1973 by Sage Publications, Inc. Reprinted by permission of Sage Publications, Inc.

Hypotheses

H_0: The three populations of residents are homogeneous with respect to knowledge of air pollution problems

H_1: The three populations are not homogeneous

Test Statistic

The expected cell frequencies are shown in Table 5.23. From these expected frequencies and the observed frequencies of Table 5.22, we compute

$$X^2 = \frac{(5 - 8.6667)^2}{8.6667} + \frac{(31 - 24)^2}{24} + \cdots + \frac{(2 - 3)^2}{3} = 10.391$$

Decision

Since $10.391 < \chi_6^2 = 12.592$, we cannot reject H_0, and we conclude that the populations may be homogeneous with respect to knowledge of air pollution problems (P value > 0.10).

TABLE 5.23	**Expected cell frequencies for Example 5.5**				
Area of residence	**No**	**Yes**	**Doubtful**	**Don't know**	**Total**
Rawmarsh	8.6667	24	4.3333	3	40
Treeton	8.6667	24	4.3333	3	40
Wath	8.6667	24	4.3333	3	40
Total	26.0001	72	13.0000	9	120

Methods for handling the problem of small expected frequencies in the case of the $r \times c$ contingency table given in the previous section apply also to the case of the test of homogeneity.

EXERCISES

5.12 Rosenwald and Stonehill (E17) hypothesized that women's early postpartum breakdowns differ from later breakdowns in that each represents the pathological resolution of a different crisis of motherhood. Table 5.24 shows, for 12 women experiencing early postpartum breakdowns and 14 experiencing late breakdowns, the number who suffered high and low levels of withdrawal. Can we conclude on the basis of these data that the two groups differ with respect to the extent of withdrawal? Determine the P value.

TABLE 5.24	**Extent of withdrawal in women experiencing early and late postpartum breakdowns**		
	Extent of withdrawal		
Time of breakdown	**High**	**Low**	**Total**
Early	10	2	12
Late	2	12	14
Total	12	14	26

Source: George C. Rosenwald and Marshall W. Stonehill, "Early and Late Postpartum Illnesses," *Psychosom. Med.*, 34 (1972), 129–137.

5.13 van Vliet and Gupta (E18) report the results of a controlled trial for treatment of severe idiopathic respiratory distress syndrome (RDS). Babies considered to have severe RDS were randomly treated with either molar sodium bicarbonate ($NaHCO_3$) solution or 0.3 molar tri-hydroxy-methyl-amino-methane (THAM). The outcome of treatment in the two groups is shown in Table 5.25. Can we conclude that the two groups differ with respect to rate of recovery? Find the P value.

5.14 Kaufmann and Raaheim (E19) used 93 male psychology students in an experiment to see whether increasing the subjects' level of activity increased the number of solutions to a certain problem. A control group of 46 subjects were given basic instructions on solving

TABLE 5.25   **Outcome of treatment in two groups treated for severe respiratory distress syndrome**

Group	Recovered from respiratory distress	Died from respiratory distress	Total
THAM treated	18	7	25
NaHCO₃ treated	10	15	25
Total	28	22	50

Source: P. K. J. van Vliet and J. M. Gupta, "THAM v. Sodium Bicarbonate in Idiopathic Respiratory Distress Syndrome," *Arch. Dis. in Childhood*, 48 (1973), 249–255; reprinted by permission of the editor and publisher.

the problem, while an experimental group of 47 subjects were given additional instructions to be active—that is, to make constructive attempts at solution. Table 5.26 shows the number in each group who solved the problem within 30 minutes. Do these data provide sufficient evidence to indicate that the experimental and control groups differ with respect to the proportion of solutions? Determine the P value.

TABLE 5.26   **Number of subjects solving a problem within 30 minutes**

	Experimental group	Control group	Total
Solved	33	23	56
Not solved	14	23	37
Total	47	46	93

Source: Reproduced with permission of the author and publisher from Kaufmann, G., & Raaheim, K. "Effect of inducing activity upon performance in an unfamiliar task." *Psychological Reports*, 1973, 32, 303–306, Table 1.

5.15 Bayer (E20) reported the data, shown in Table 5.27, on the number of married and unmarried freshmen men entering college who were still enrolled in college three years later. Can we conclude from these data that males who were married and males who were unmarried when entering college differ with respect to the proportion who are still in college three years later?

TABLE 5.27   **Male students classified by marital status at time of initial college enrollment and enrollment status three years later**

	Unmarried	Married	Total
Enrolled	775	687	1462
Not enrolled	76	170	246
Total	851	857	1708

Source: Alan E. Bayer, "College Impact on Marriage," *J. Marriage Fam.*, 34 (1972), 600–609; copyright 1972 by National Council on Family Relations; reprinted by permission.

5.16 In a study of drug usage by students at a large university, Garfield and Garfield (E21) obtained the data regarding hard liquor experience of marijuana users and nonusers shown

in Table 5.28. Can we conclude on the basis of these data that marijuana users and non-users differ with respect to their use of alcohol? Find the P value.

TABLE 5.28   **Hard-liquor experience of marijuana users compared with nonusers**

	Marijuana use		
Hard-liquor use	Users	Nonusers	Total
Once or more	66	23	89
Never	3	8	11
Total	69	31	100

Source: Mark D. Garfield and Emily F. Garfield, "A Longitudinal Study of Drugs on a Campus," *Int. J. Addictions*, 8 (1973), 599–611.

5.17   In an analysis of the records of heroin abusers, Newmeyer and Gay (E22) classified each subject as either Old-Style Junkie (OSJ), Transition Junkie (TJ), New Junkie (NJ), or Very New Junkie (VNJ), on the basis of how much time had elapsed since his or her first use of the drug. Table 5.29 shows the number of middle-class, lower-middle-class, and working-class subjects falling into each of these four categories. They determined class membership on the basis of the job for which the individual was trained. Do these data provide sufficient evidence to indicate a lack of homogeneity among the three classes with respect to addict classification?

TABLE 5.29   **Job trained for the addict category**

	Middle class	Lower middle class	Working class	Total
OSJ	25 (11.8%)	35 (16.5%)	152 (71.7%)	212 (100.0%)
TJ	16 (12.7%)	23 (18.3%)	87 (69.0%)	126 (100.0%)
NJ	27 (11.3%)	50 (20.8%)	163 (67.9%)	240 (100.0%)
VNJ	56 (11.6%)	126 (26.1%)	300 (62.3%)	482 (100.0%)
All	124 (11.7%)	234 (22.1%)	702 (66.2%)	1060 (100.0%)

Source: John A. Newmeyer and George R. Gay, "The Traditional Junkie, the Aquarian Age Junkie, and the Nixon Era Junkie," *Drug Forum*, 2 (number 1, Fall 1972), 17–30; copyright 1972, Baywood Publishing Company.

5.18   In research designed to study the effects of different sociocultural environments on the child's feelings and evaluations of the other family members, Oliverio (E23) used as subjects 36 lower-class children from schools in Rome, Sardinia, and the Ivory Coast. Each child was instructed to draw a family. Table 5.30 shows the leading figure in the drawings made by each group. Test the null hypothesis that these three populations are homogeneous with respect to the relationship of the leading figure in drawings. Determine the P value.

5.19   In a study designed to explore pre-retirement attitudes of older female workers, Jacobson (E24) reported the data shown in Table 5.31. Convert the percentages to frequencies. Do these data provide sufficient evidence to indicate a lack of homogeneity with respect to

TABLE 5.30   **Leading figure in children's drawings, by area of residence**

	Rome	Ivory Coast	Sardinia	Total
Parents	33	19	34	86
Relatives	0	16	1	17
Self	3	1	1	5
Total	36	36	36	108

Source: Anna Ferraris Oliverio, "Children's Evaluations of Family Roles, A Cross Cultural Comparison," *Int. J. Psychol.*, 8 (1973), 153–158; reprinted by permission of the International Union of Psychological Science and Dunod, Editeur, Paris.

TABLE 5.31   **Women employees' willingness to retire, by location of principal social contacts**

	Location of principal social contacts			
	Total (%) ($n = 66$)	In work (%) ($n = 28$)	Out of work (%) ($n = 21$)	No "real" friends (%) ($n = 17$)
Willing to retire	42.4	14.3	61.9	64.7
Reluctant to retire	57.6	85.7	38.1	35.3

Source: Dan Jacobson, "Rejection of the Retiree Role: A Study of Female Industrial Workers in Their 50's," *Hum. Rel.*, 27 (1974), 477–492; published by Plenum Publishing Corporation, New York.

location of principal social contacts between women willing to retire and those not willing to retire? What is the *P* value?

5.20   Steinitz et al. (E25), in a study of modes of political thought among "bright" working-class adolescents, obtained the data in Table 5.32. Can we conclude on the basis of these data that adolescents from the different communities differ with respect to the occupation of their fathers? Determine the *P* value.

TABLE 5.32   **Adolescents from three communities classified according to father's occupation**

	Community		
Father's occupation	Mill town	Townline	City Ville
Unskilled and semi-skilled blue collar	8	5	5
Skilled blue collar	4	7	8
White collar and sales	2	7	8
Business and professional	6	1	0
Total	20	·20	21

Source: Victoria A. Steinitz, Prudence King, Ellen Solomon, and Ellen Shapiro, "Ideological Development in Working-Class Youth," *Harvard Educ. Rev.*, 43 (1973), 333–361; copyright 1973, President and Fellows of Harvard College.

5.21 In a study designed to evaluate the effects of social class on the academic careers and vocational choices of male students at an "elite" college, Goldstein (E26) collected the data shown in Table 5.33. Do these data provide sufficient evidence to indicate a difference in choice of major subject between the two social classes?

TABLE 5.33 **Choice of major subject among seniors, by social class**

| Social class | Major subject | | | | | |
	Humanities	Social sciences	Biological sciences	Physical sciences	Engineering	Total
Upper middle class	20	40	9	14	10	93
Working class	27	39	10	16	12	104
Total	47	79	19	30	22	197

Source: Michael S. Goldstein, "Academic Careers and Vocational Choices of Elite and Non-Elite Students at an Elite College," *Sociol. Educ.*, 47 (1974), 491–510.

5.4

MISCELLANY

This section introduces some topics of interest to users of chi-square tests of independence and homogeneity. Familiarity with these topics will greatly increase the investigator's analytical potential. Exhaustive treatment of these topics cannot be provided here, but there are sufficient references to help the interested reader pursue each of the topics in greater detail.

Partitioning Chi-Square When a chi-square test of independence or homogeneity is carried out on the data of a 2×2 contingency table, the interpretation of the results is straightforward and unambiguous. In a test of independence, we conclude either that the two variables under study may be independent or that they are related. In a test for homogeneity, we conclude either that the two sampled populations may be homogeneous or that they are not homogeneous, depending on whether or not we reject the null hypothesis. When we perform chi-square tests of independence and homogeneity on the data of contingency tables yielding more than one degree of freedom, the interpretations of the outcomes are not as clear-cut. If we reject the null hypothesis of independence, for example, we do not know whether a lack of independence occurs throughout the table or only among certain categories of the two variables under study. If we do not reject a hypothesis of independence, we do not know whether independence occurs uniformly among all categories or only among certain categories, since independence among some categories masks or dilutes dependence among others. The result may be an overall computed X^2 value that is not significant.

We can gain some insight into this problem by using a technique known as *partitioning of chi-square*. Basically, the procedure is to break down a large $r \times c$ contingency table into smaller tables. We then analyze these smaller tables separately to determine whether the larger table contains areas in which we can reject the hypothesis of independence or the hypothesis of homogeneity.

The pioneer work in this area was carried out by Irwin (T21), Kastenbaum (T22), Kimball (T23), and Lancaster (T24, T25). Their methods have been discussed and illustrated by Castellan (T26) and Maxwell (T27). More recent papers on the subject have been written by Bresnahan and Shapiro (T28), and Shaffer (T29). Brunden (T30) describes a partitioning method for a $2 \times c$ table that allows for the comparison of a placebo group with all other treatment groups through the analysis of appropriately chosen 2×2 tables.

Multidimensional Contingency Tables   The discussion so far has been concerned with the analysis of data that can be displayed in a two-way contingency table. However, it is not uncommon for an experiment or a sample survey to yield data that can be displayed meaningfully in a contingency table of higher dimensions. Hoyt et al. (T31), for example, analyze a four-way contingency table that classifies 13,968 high school graduates on the basis of (a) position by thirds in high school graduating class, (b) their status at a certain date after high school (i.e., enrolled in college, enrolled in noncollegiate school, employed full-time, or other), (c) sex, and (d) father's occupational level. The marginal headings for their table are shown in Table 5.34. The body of the table consisted of $2 \times 7 \times 4 \times 3 = 168$ cells holding frequency counts corresponding to indicated subgroups of the 13,968 subjects.

TABLE 5.34   **Marginal headings of contingency table classifying 13,968 high school graduates**

Post-high school status		High school rank											
		Lowest third				Middle third				Upper third			
		1	2	3	4	1	2	3	4	1	2	3	4
Sex (1)	*1*												
Father's	*2*												
occupational	*3*												
status	*4*												
	5												
	6												
	7												
Sex (2)	*1*												
Father's	*2*												
occupational	*3*												
status	*4*												
	5												
	6												
	7												

Lewis (T32) presents a general review of the more important methods of analyzing multidimensional tables, along with a selection of procedures that he says are computationally the simplest. Other articles of interest include those by Lancaster (T33), Plackett (T34), Shaffer (T35), Smith (T36), Sutcliffe (T37), and Goodman (T38, T39, T40, T41). Additional references may be found in Lancaster (T4). Freeman (T42) and Paik (T43) discuss graphical techniques for representing multidimensional contingency tables.

Combining Contingency Tables   Contingency table data relating to the same research may sometimes be available. A logical question then is whether and how the several pieces of evidence can be pooled to test the hypothesis of independence or homogeneity among rows and columns.

Armitage (T44) and Cochran (T6) discuss several methods for combining contingency tables that have been employed in the past. Cochran (T6), who cites the inadequacies of some of the procedures, suggests a method using weights in the pooling process. Radhakrishna (T45) elaborates on Cochran's method. A note by Nelson (T46) may also be of interest.

Testing for Trend in Contingency Tables   When the c classification categories of a $2 \times c$ contingency table fall into a natural order, as they do when the categories are age groups, a component of chi-square due to linear trend can be isolated and tested for significance. Armitage (T44, T47) and Cochran (T6) discuss this procedure in detail. Smith (T48) illustrates the procedure using data from a study of consumer preference. An overall chi-square test of the null hypothesis that preference for a certain product and age of consumer are independent could not be rejected at the 0.10 level of significance. Applying the test for trend, however, he was able to conclude, at the 0.01 level of significance, that the proportion of consumers preferring the product increases with age. Steel and Torrie (T49) also give a numerical example illustrating the technique. The choice of appropriate groupings when testing for trend in this context is discussed by Connor (T50).

Simultaneous Statistical Tests Using Categorical Data   In the typical sample survey, responses to a variety of questions are summarized in several two-way contingency tables. When we perform a chi-square test on each table and compare the computed X^2 for significance to tabulated values of the chi-square distribution in the usual way, a problem of interpretation arises because the various tests are generally not independent—the same individuals are involved in all tests. Jensen et al. (T51) have presented a procedure for performing several chi-square tests simultaneously despite the presence of correlations among the test statistics. The authors also provide special tables that facilitate the application of the proposed procedure.

Determining Sample Size   When any statistical technique is to be used, the size of the sample is an important consideration. This is also true when chi-square

analysis is to be used. Holt et al. (T52) give a formula for estimating the sample size needed for a 2×2 contingency table. In another paper Holt (T53) presents the derivation of a formula for sample size for an $r \times c$ table.

Hornick and Overall (T54) evaluated estimates of sample size requirements for 2×2 contingency tables, derived from three approximate formulas, by comparison with exact calculations. The authors describe their strategy for exact determination of the minimum sample size required to provide specified power for tests of significance of differences in proportions in two treatment groups. A simple approximation to the sample size required for the Yates-corrected chi-square test to have specified power is given by Ury and Fleiss (T55). They compare their approximation with other approximations and with the exact sample size for the equal sample case.

Power   A number of papers concerned with the power of chi-square tests appear in the literature, including those by Diamond (T56), Harkness and Katz (T57), Meng and Chapman (T58), and Mitra (T59). The power of chi-square tests for linear trend in proportions has been considered by Chapman (T60). Overall (T61) has investigated the power of chi-square tests for 2×2 contingency tables with small expected frequencies. He notes that small expected frequencies in 2×2 contingency tables often occur because one of the categorical outcomes represents a relatively rare event. The author presents formulas for estimating the number of relatively rare events required to provide adequate power for chi-square tests. He concludes that the power of chi-square tests of relationship in a 2×2 table with small expected frequencies in one row or column is so low that the tests are rendered essentially useless.

Haber (T62, T63) discusses the asymptotic power of the chi-square test for particular classes of alternatives and for multidimensional contingency tables. In another paper by Haber (T64), three approximations to the power function of the chi-square test for the hypotheses of "no three factor interaction" in a $2 \times 2 \times 2$ contingency table are introduced and compared.

Improper Use of Chi-Square Analysis   Since chi-square analysis is one of the most widely used statistical techniques, it is not surprising that it is also misused. Lewis and Burke (T65, T66) discuss several examples of what they consider misuses of chi-square. Their paper has prompted additional comments by Peters (T67), Pastore (T68), Edwards (T69), and Burke (T70). The entire series of papers is available in a book of readings edited by Steger (T71). Alleged misuses of chi-square analysis have been cited also by Rich et al. (T72) and Wright (T73).

The 1949 article by Lewis and Burke (T65) was updated in 1983 by Delucchi (T74). In addition, the author examines the research on the application of the chi-square statistic and discusses supplementary and alternative approaches that may be employed by researchers who often deal with qualitative data.

FURTHER READING

Of interest to users of the chi-square technique is a discussion by Feinstein and Ranshaw (T75) on a procedure for the rapid mental calculation of a fourfold chi-square test. Smith (T76) presents a nomogram for computing the chi-square statistic that may prove useful in the field or when results are needed in a hurry. The problem of the optimum choice of classes for contingency tables has been considered by Hamdan (T77). Chernoff (T78) discusses degrees of freedom for chi-square.

As mentioned earlier, many situations involving contingency table analysis arise in which the categories for one or more variables represent points on an ordered scale. Such situations have been the focus of research by Hirotsu (T79, T80), Takeuchi and Hirotsu (T81), and Goodman (T82).

In 1959 Mantel and Haenszel (T83) proposed a method for analyzing sets of 2×2 contingency tables that compare the outcome of two treatments in several strata. The power of the Mantel–Haenszel test is investigated by Wittes and Wallenstein (T84). Mantel and Fleiss (T85) propose a rule regarding the minimum expected cell frequency for the Mantel–Haenszel test for independence in several fourfold tables. Alternative formulas for computing the chi-square statistic have been proposed by Boynton and Poe (T86), McDonald-Schlichting (T87), and Cox (T88).

Lackritz (T89) describes a method for finding the exact P value for chi-square tests with the aid of a hand calculator and a standard normal table. The method also allows the investigator to find values of the chi-square statistic that are missing from most readily available tables.

The analysis of the data of 2×2 contingency tables by chi-square and other techniques is the subject of a paper by Brown (T90), who uses examples from medicine to illustrate the procedures. The use of chi-square analysis in physical therapy research is discussed by Witt and McGrain (T91). A multiple testing procedure using the chi-square statistic is proposed by O'Brien and Fleming (T92) for use in clinical trials.

Schouten et al. (T93) propose a continuity correction, which they claim is more appropriate and more powerful than Yates's correction. Haber (T94) reports the results of a study in which he compared the uncorrected chi-square test with four methods that adjust for continuity.

Suggestions on the graphical display of the significant components in two-way contingency tables is the subject of a paper by Cohen (T95). Of historical interest is a paper by Plackett (T96), who discusses Karl Pearson's relevant activities during the first decade of his statistical career and describes some of the work by his contemporaries and predecessors.

More extensive discussions of chi-square tests and contingency table analysis in general are to be found in the books by Aickin (T97), Everitt (T98), Fienberg (T99), Fingleton (T100), Gokhale and Kullback (T101), Goodman (T82, T102), and Reynolds (T103).

5.5

COMPUTER PROGRAMS

Russell (T104) has written a computer program that constructs frequency and contingency tables from a questionnaire and other data stored on punch cards. Hill and Pike (T105) present an algorithm for finding the probability that χ^2 for specified degrees of freedom exceeds a designated value. Another algorithm discussed by Goldstein (T106) evaluates the quantile at a specified probability level for the chi-square distribution with designated degrees of freedom.

A BASIC program for calculating the chi-square statistic with more than 1 degree of freedom has been written by Forer (T107). The author states that the program requires minimum computer sophistication and is easily adapted to spreadsheet software packages.

Tsai (T108) describes an interactive computer program that allows a researcher to calculate the sample size necessary to control for both type I and type II errors or, on a post hoc basis, to analyze the power of a test. The program is written in BASIC and requires 7K of storage.

CHITEST, described by Romesburg and Marshall (T109), is an interactive FORTRAN 77 program that uses Monte Carlo methods to test the null hypothesis that the row and column factors of an $r \times k$ contingency table are independent of each other. The program is designed to give more accurate P value estimates when expected cell frequencies are small.

Arena (T110) has written a FORTRAN program for computing the generalized Mantel–Haenszel chi-square test statistic for multiple $2 \times J$ tables. In the analysis of 2×2 tables, the program checks the expected cell frequencies to assure that the asymptotic assumptions are met. As an option, Yates's correction for continuity can be employed in the analysis of 2×2 tables.

Bergan et al. (T111) describe a computer program for constructing multidimensional contingency tables. The program can accept up to 500 variables as data and construct a frequency table of from one to eight dimensions. Each variable may have as many as eight categories. In addition, the program can combine, recode, and create formal variables at the user's discretion.

The use of computers in contingency table analysis is discussed in the books by Everitt (T98) and Goodman (T82, T102). Thakur et al. (T112) have written a FORTRAN program for testing linear trend and homogeneity in proportions. Trend is evaluated by the Cochran–Armitage method, and homogeneity is tested by an overall chi-square test as well as by multiple pairwise comparisons by the Fisher–Irwin exact method. The authors state that the program should be easy to implement on any size of computer with a FORTRAN compiler.

Berry and Mielke (T113) have written an APL function for an exact chi-square test. A program for computing chi-square probabilities as well as probabilities for other widely used distributions has been made available by von Collani (T114).

Since chi-square analysis is such a widely used statistical technique, most commercial computer software packages for general-purpose statistical analysis are equipped to perform chi-square tests.

REVIEW EXERCISES

For Exercises 5.22 through 5.27, indicate whether the null hypothesis should be one of independence or homogeneity.

5.22 A psychologist wished to know whether certain professional groups hold different opinions about the legalization of marijuana. A random sample was selected from each of the following groups: physicians, clergymen, lawyers, teachers, and nurses. Each potential respondent was asked to answer either Yes, No, or Don't care to the question: "Do you think the sale and use of marijuana should be legalized?"

5.23 A random sample of adult residents of a certain community contained 30% single persons, 20% married persons with no children, and 50% married persons with one or more children. Each person in the sample was asked to answer either Yes, No, or Don't care to the question: "Do you think abortion on demand should be made legal?"

5.24 A random sample of 300 residents of a certain metropolitan area who participated in a survey were categorized on the basis of their answers to the following questions: (1) What is your occupation? (2) Do you think people who commit suicide are mentally ill?

5.25 In a personal-interview survey of adult residents of a certain metropolitan area, respondents were asked: "What do you consider to be the most pressing problem facing this country today?" During the survey interviewers made a note of the apparent ethnic group to which the respondent belonged.

5.26 A sample of students was selected from the student body of each of three county high schools. Each student was then classified on the basis of whether his or her parents belonged to the low, middle, or upper social class.

5.27 A public school psychologist wishes to know whether to conclude that high school freshmen, sophomores, juniors, and seniors differ with respect to their attitudes toward school counselors.

5.28 Arlinghaus and Salzarulo (E27) examined the perceived need for post-baccalaureate education for tax professionals. The researchers mailed a questionnaire to all of the tax professionals employed in Ohio by the eight largest national accounting firms. Of the 605 questionnaires sent out, 384 were returned. One of the questions required respondents to indicate if they thought that their educational background adequately prepared them for an entry-level position at the time they entered the tax department. After eliminating the answers of respondents with more than one degree, 318 questionnaires were available for the construction of Table 5.35.

Can one conclude from these data that there is a relationship between educational background and perception of preparedness for the entry-level position? Let $\alpha = 0.01$, and find

the *P* value. Comment on the design of the study and the assumptions underlying your analysis of the data.

TABLE 5.35   **Opinions of respondents regarding whether they were adequately prepared for entry-level position at the time they entered the tax department**

Educational background	Response			
	Yes	No	Uncertain	Total
Bachelor's degree in accounting	110	69	17	196
Law degree	71	1	4	76
MBA degree	24	4	1	29
Master of Taxation degree	12	3	2	17
Total	217	77	24	318

Source: Barry P. Arlinghaus and W. Peter Salzarulo, "The Importance of Post-Baccalaureate Education for Tax Professionals," *Akron Bus. Econ. Rev.*, 17 (Winter 1986), 8–17. Reprinted by permission.

5.29 In a study of the activities of a group of small business firms over a six-year period to determine the manner in which their performance reflected the results of a counseling program, Rocha and Khan (E28) collected the data shown in Table 5.36.

TABLE 5.36   **Extent of implementation by functional area**

Recommendations	Area of recommendation					
	A	B	C	D	E	Total
Implemented	25	17	8	5	5	60
Rejected	2	10	3	3	0	18
Total offered	27	27	11	8	5	78

Source: Joseph R. Rocha, Jr., and M. Riaz Khan, "The Human Resource Factor in Small Business Decision Making," *Amer. J. Small Bus.*, 10 (Fall 1985), 53–62. Reprinted by permission.

A = accounting and finance, B = marketing, C = personnel, D = operations, E = other

The firms studied participated in counseling services offered by the Small Business Institute of the University of Lowell in Massachusetts. After studying the operations of the firms participating in the program, student teams made a total of 78 recommendations to the firms in the study. Can we conclude from the data in Table 5.36 that there is a lack of homogeneity among the functional areas with respect to response to recommendations? Comment on the appropriateness of using a chi-square test with these data. To what population does the inference apply? Let $\alpha = 0.05$, and find the *P* value.

5.30 According to Bellur (E29) retailers in the United States annually lose goods worth more than five billion dollars because of shoplifting. To obtain additional information on shoplifting, shoplifters, and their victims, the author designed a questionnaire that was sent to a sample of midwestern retailers. He received 106 completed questionnaires. Seven types

of stores (department, grocery, clothing, card and gift, drug, variety and discount, and specialty) were represented in the sample. One of the questions asked of the retailers was, "Do you have a shoplifting problem?" The response was either Yes or No. In a test of the null hypothesis that the type of store is independent of the presence of a shoplifting problem, Bellur computed a chi-square value of 32.60. What are the degrees of freedom for the test? Should the null hypothesis be rejected at the 0.05 level of significance? Why? What is the *P* value for the test? What conclusions can one draw from these results? What assumptions are necessary to validate any inferences that are drawn?

5.31 Denisoff and Bridges (E30) state that the recording artist is one of the least examined aspects of the music industry. In a study that examined the "who-what-where" of performers and their music, they analyzed biographical data on 667 artists, who were classified according to their musical style: (1) rock, (2) soul/rhythm and blues/disco, (3) easy listening, (4) country and Western, (5) jazz, (6) classical, and (7) "other," which included comedy artists and purely electronic artists. The researchers also categorized the musicians on the basis of several other variables, including gender. As part of their statistical analysis, Denisoff and Bridges cross-classified 580 artists on the basis of gender and the previously listed musical styles. They computed a chi-square statistic of 41.13. What null hypothesis can one test with these data? Should the null hypothesis be rejected? Why? Let $\alpha = 0.05$. Compute the *P* value for the test. What assumptions are necessary to validate any inferences that are drawn?

5.32 Researchers wished to know whether white-collar and blue-collar employees differ with respect to their opinions regarding the implementation of a no-smoking policy in their workplace. A random sample drawn from each of the two populations yielded the information shown in Table 5.37.

Can we conclude from these data that the two populations differ with respect to their opinions regarding the implementation of a no-smoking policy in their workplace? Let $\alpha = 0.01$, and find the *P* value.

TABLE 5.37 **Opinions of workers with respect to implementation of a no-smoking policy in their workplace**

| | Employee category | |
Opinion	White collar	Blue collar
For	80	45
Against	30	95

5.33 Nonresponse to mail questionnaires is one of the biggest problems with which users of sample surveys have to contend. Among the proposed solutions to the problem are monetary incentives, personalization of cover letters, shortening of questionnaires, and premailing contact with prospective respondents. In the belief that mailed questionnaires printed on colored paper would be more likely to be returned than those printed on white paper, a market analyst conducted an experiment. A random sample of 400 subscribers to a general-interest magazine was selected, and each subscriber was sent the same questionnaire. Half were printed on white paper and half on colored paper. Of the 183

questionnaires returned, 50 were printed on white paper. Do these results provide sufficient evidence to indicate that people are more likely to respond to questionnaires printed on colored paper? Let $\alpha = 0.01$, and find the P value.

5.34 An advertising researcher wished to know if there is any truth to the claim that certain kinds of advertisements tend to be more male-oriented than female-oriented. The researcher selected a sample of 200 ads from the media available in a large city. The ads were then categorized on the basis of sex orientation and content. Eighty of the ads were classified as having a male orientation and 70 a female orientation. The remainder were considered to be neutral. Seventy-five of the ads had an adventure content, of which 60 were male-oriented and 10 were female-oriented. Of the 65 that had a domestic situation content, 5 were male-oriented and 59 were female-oriented. The other content category was nostalgia. On the basis of these data, can one conclude that sex orientation and content of the advertisements from which the sample was drawn are not independent? Let $\alpha = 0.05$, and find the P value. What is the target population in this situation?

5.35 Do you believe the rate of inflation will be higher during the coming year than last year?" That question was posed to 500 business executives by a researcher in the field of economics. Five industries were equally represented in the survey. The results are shown in Table 5.38.

   Can one conclude on the basis of these results that there is a lack of homogeneity among the industries with respect to the executives' opinions regarding inflation? Let $\alpha = 0.01$, and find the P value.

TABLE 5.38   **Business executives' responses to the question, "Do you believe the rate of inflation will be higher during the coming year than last year?"**

Response	Industry				
	A	B	C	D	E
Yes	25	75	50	45	65
No	75	25	50	55	35

REFERENCES

T1 Pearson, Karl, "On the Criterion that a Given System of Deviations from the Probable in the Case of a Correlated System of Variables Is Such that It Can Be Reasonably Supposed to Have Arisen from Random Sampling," *The London, Edinburgh and Dublin Philosoph. Mag. J. Sci.* (5th series), 50 (1900), 157–175. Reprinted in Karl Pearson's *Early Statistical Papers*, London: Cambridge University Press, 1948.

T2 Wilks, Samuel S., "Karl Pearson: Founder of the Science of Statistics," *Scientific Monthly*, 53 (1941), 249–253.

T3 Pearson, Karl, "Mathematical Contributions to the Theory of Evolution. III, Regression, Heredity, and Panmixia," *Philosoph. Trans. Roy. Soc. A*, 187 (1896), 253–318.

T4 Lancaster, H. O., *The Chi-Squared Distribution*, New York: Wiley, 1969.

T5 Cochran, W. G., "The χ^2 Test of Goodness of Fit," *Ann. Math. Statis.*, 23 (1952), 315–345.

T6 Cochran, W. G., "Some Methods for Strengthening the Common χ^2 Tests," *Biometrics*, 10 (1954), 417–451.

T7 Katti, S. K., and A. N. Sastry, "Biological Examples of Small Expected Frequencies and the Chi-Square Test," *Biometrics*, 21 (1965), 49–54.

T8 Nass, C. A. G., "The χ^2 Test for Small Expectations in Contingency Tables, with Special Reference to Accidents and Absenteeism," *Biometrika*, 46 (1959), 365–385.

T9 Roscoe, J. T., and J. A. Byars, "An Investigation of the Restraints with Respect to Sample Size Commonly Imposed on the Use of the Chi-Square Statistic," *J. Amer. Statist. Assoc.*, 66 (1971), 755–759.

T10 Tate, Merle W., and Leon A. Hyer, "Significance Values for an Exact Multinomial Test and Accuracy of the Chi-Square Approximation," U.S. Department of Health, Education, and Welfare, Office of Education, Bureau of Research, August 1969.

T11 Tate, Merle W., and Leon A. Hyer, "Inaccuracy of the χ^2 Test of Goodness of Fit When Expected Frequencies Are Small," *J. Amer. Statist. Assoc.*, 68 (1973), 836–841.

T12 Yarnold, James K., "The Minimum Expectation in χ^2 Goodness-of-Fit Tests, and the Accuracy of Approximations for the Null Distribution," *J. Amer. Statist. Assoc.*, 65 (1970), 864–886.

T13 Yates, Frank, "Contingency Tables Involving Small Numbers and the χ^2 Test," *J. Roy. Statist. Soc.*, 1 (1934), 217–235.

T14 Conover, W. J., "Some Reasons for Not Using the Yates Continuity Correction on 2 × 2 Contingency Tables," with comments by E. Frank Starmer, James E. Grizzle, P. K. Sen, Nathan Mantel, Olli S. Miettinen, and a rejoinder by W. J. Conover, *J. Amer. Statist. Assoc.*, 69 (1974), 374–382.

T15 Grizzle, J. E., "Continuity Correction in the χ^2 Test for 2 × 2 Tables," *Amer. Statist.*, 21 (October 1967), 28–32.

T16 Lancaster, H. O., "The Combination of Probabilities Arising from Data in Discrete Distributions," *Biometrika*, 36 (1949), 370–382.

T17 Pearson, E. S., "The Choice of Statistical Test Illustrated on the Interpretation of Data Classed in a 2 × 2 Table," *Biometrika*, 34 (1947), 139–167.

T18 Plackett, R. L., "The Continuity Correction in 2 × 2 Tables," *Biometrika*, 51 (1964), 327–338.

T19 Liddell, F. D. K., "Correcting Correction in Chi-Squared Test in 2 × 2 Tables," *Biometrics*, 28 (1972), 268–269.

T20 Mantel, N., and S. W. Greenhouse, "What Is the Continuity Correction?" *Amer. Statist.*, 22 (December 1968), 27–30.

T21 Irwin, J. O., "A Note on the Subdivision of χ^2 into Components," *Biometrika*, 36 (1949), 130–134.

T22 Kastenbaum, M. A., "A Note on the Additive Partitioning of Chi-Square in Contingency Tables," *Biometrics*, 16 (1960), 416–422.

T23 Kimball, A. W., "Short-Cut Formulas for the Exact Partitioning of χ^2 in Contingency Tables," *Biometrics*, 10 (1954), 452–458.

T24 Lancaster, H. O., "The Derivation and Partition of χ^2 in Certain Discrete Distributions," *Biometrika*, 36 (1949), 117–129.

T25 Lancaster, H. O., "The Exact Partition of χ^2 and Its Application to the Problem of Pooling Small Expectations," *Biometrika*, 37 (1950), 267–270.

T26   Castellan, N. J., Jr., "On the Partitioning of Contingency Tables," *Psychol. Bull.*, 64 (1965), 330–338.

T27   Maxwell, A. E., *Analysing Qualitative Data*, London: Methuen, 1961.

T28   Bresnahan, J. L., and M. M. Shapiro, "A General Equation and Technique for Exact Partitioning of Chi-Square Contingency Tables," *Psychol. Bull.*, 66 (1966), 252–262.

T29   Shaffer, J. P., "Testing Specific Hypotheses in Contingency Tables—Chi-Square Partitioning and Other Methods," *Psychol. Rep.*, 33 (1973), 343–348.

T30   Brunden, Marshall N., "The Analysis of Non-Independent 2 × 2 Tables from 2 × c Tables Using Rank Sums," *Biometrics*, 28 (1972), 603–607.

T31   Hoyt, C. J., P. R. Krishnaiah, and E. P. Torrance, "Analysis of Complex Contingency Data," *J. Experim. Educ.*, 27 (1959), 187–194.

T32   Lewis, B. N., "On the Analysis of Interaction in Multi-Dimensional Contingency Tables," *J. Roy. Statist. Soc. Ser. A*, 125 (1962), 88–117.

T33   Lancaster, H. O., "Complex Contingency Tables Treated by the Partition of χ^2," *J. Roy. Statist. Soc. Ser. B*, 13 (1951), 242–249.

T34   Plackett, R. L., "A Note on Interactions in Contingency Tables," *J. Roy. Statist. Soc. Ser. B*, 24 (1962), 162–166.

T35   Shaffer, J. P., "Defining and Testing Hypotheses in Multidimensional Contingency Tables," *Psychol. Bull.*, 79 (1973), 127–141. Correction, *Ibid.*, p. 340.

T36   Smith, R. A., M. James, and W. B. Michael, "Fortran IV Program to Compute Post Hoc Comparisons for Multilevel Chi-Square Tests," *Educ. Psychol. Measurement*, 34 (1974), 159–160.

T37   Sutcliffe, J. P., "A General Method of Analysis of Frequency Data for Multiple Classification Designs," *Psychol. Bull.*, 54 (1957), 134–137.

T38   Goodman, L. A., "On Methods for Comparing Contingency Tables," *J. Roy. Statist. Soc. Ser. A*, 126 (1963), 94–108.

T39   Goodman, L. A., "On Partitioning χ^2 and Detecting Partial Association in Three-Way Contingency Tables," *J. Roy. Statist. Soc. Ser. B*, 31 (1969), 486–498.

T40   Goodman, L. A., "Partitioning of Chi-Square, Analysis of Marginal Contingency Tables, and Estimation of Expected Frequencies in Multidimensional Contingency Tables," *J. Amer. Statist. Assoc.*, 66 (1971), 339–344.

T41   Goodman, L. A., "Analysis of Multidimensional Contingency Tables—Stepwise Procedures and Direct Estimation Methods for Building Models for Multiple Classifications," *Technometrics*, 13 (1971), 33–61.

T42   Freeman, Daniel H., Jr., "Graphical Representation of Multiway Contingency Tables: Alternative Measures of Association," *Proc. Soc. Statist. Sec. of Amer. Statist. Assoc.*, Washington, D.C.: American Statistical Association, 1983, 544–548.

T43   Paik, Minja, "A Graphic Representation of a Three-Way Contingency Table: Simpson's Paradox and Correlation," *Amer. Statist.*, 39 (1985), 53–54.

T44   Armitage, P., *Statistical Methods in Research*, Oxford: Blackwell Scientific Publications, 1971.

T45   Radhakrishna, S., "Combination of Results from Several 2 × 2 Contingency Tables," *Biometrics*, 21 (1965), 86–98.

T46   Nelson, L. S., "Query: Combining Values of Observed χ^2's," *Technometrics*, 8 (1966), 709.

T47   Armitage, P., "Tests for Linear Trends in Proportions and Frequencies," *Biometrics*, 11 (1955), 375–385.

T48   Smith, Harry, Jr., "Problem 6-66," *Indust. Quality Control*, 23 (1966), 285–286.

T49 Steel, Robert G. D., and James H. Torrie, *Principles and Procedures of Statistics*, New York: McGraw-Hill, 1960.

T50 Connor, Robert J., "Grouping for Testing Trends in Categorical Data," *J. Amer. Statist. Assoc.*, 67 (1972), 601–604.

T51 Jensen, D. R., G. B. Beus, and G. Storm, "Simultaneous Statistical Tests on Categorical Data," *J. Experim. Educ.*, 36 (Summer 1968), 46–56.

T52 Holt, W. R., B. H. Kennedy, and J. W. Peacock, "Formulae for Estimating Sample Size for Chi-Square Test," *J. Econ. Entomol.*, 60 (1967), 286–288.

T53 Holt, W. R., "A Further Note on Sample Size for Chi-Square Test—$R \times C$ Table," *J. Econ. Entomol.*, 61 (1968), 853–854.

T54 Hornick, Chris W., and John E. Overall, "Evaluation of Three Sample Size Formulae for 2×2 Contingency Tables," *J. Educ. Statist.*, 5 (1980), 351–362.

T55 Ury, Hans K., and Joseph L. Fleiss, "On Approximate Sample Sizes for Comparing Two Independent Proportions with the Use of Yates' Correction," *Biometrics*, 36 (1980), 347–351.

T56 Diamond, E. L., "The Limiting Power of Categorical Data Chi-Square Tests Analogous to Normal Analysis of Variance," *Ann. Math. Statist.*, 34 (1963), 1432–1440.

T57 Harkness, W. L., and L. Katz, "Comparison of the Power Functions for the Test of Independence in 2×2 Contingency Tables," *Ann. Math. Statist.*, 35 (1964), 1115–1127.

T58 Meng, Rosa C., and Douglas G. Chapman, "The Power of Chi-Square Tests for Contingency Tables," *J. Amer. Statist. Assoc.*, 61 (1966), 965–975.

T59 Mitra, S. K., "On the Limiting Power Function of the Frequency Chi-Square Test," *Ann. Math. Statist.*, 29 (1958), 1221–1233.

T60 Chapman, D. G., "Asymptotic Power of Chi-Square Tests for Linear Trends in Proportions," *Biometrics*, 24 (1968), 315–327.

T61 Overall, John E., "Power of Chi-Square Tests for 2×2 Contingency Tables with Small Expected Frequencies," *Psychol. Bull.*, 87 (1980), 132–135.

T62 Haber, Michael, "On the Asymptotic Power and Relative Efficiency of the Frequency X^2 Test," *J. Statist. Planning and Inference*, 5 (1981), 299–308.

T63 Haber, M., "The Large-Sample Power of the χ^2 Test for Multidimensional Contingency Tables," *Metrika*, 31 (1984), 195–202.

T64 Haber, M., "The Power Function of the Test for 'No Three Factor Interaction' in $2 \times 2 \times 2$ Contingency Tables," *Biometrical J.*, 27 (1985), 231–235.

T65 Lewis, D., and C. J. Burke, "The Use and Misuse of the Chi-Square Test," *Psychol. Bull.*, 46 (1949), 433–489.

T66 Lewis, Don, and C. J. Burke, "Further Discussion of the Use and Misuse of the Chi-Square Test," *Psychol. Bull.*, 47 (1950), 347–355.

T67 Peters, Charles C., "The Misuse of Chi-Square—A Reply to Lewis and Burke," *Psychol. Bull.*, 47 (1950), 331–337.

T68 Pastore, Nicholas, "Some Comments on 'The Use and Misuse of the Chi-Square Test,'" *Psychol. Bull.*, 47 (1950), 338–340.

T69 Edwards, Allen L., "On 'The Use and Misuse of the Chi-Square Test'—The Case of the 2×2 Contingency Table," *Psychol. Bull.*, 47 (1950), 341–346.

T70 Burke, C. J., "Letter to the Editor on Peters' Reply to Lewis and Burke," *Psychol. Bull.*, 48 (1951), 81–82.

T71 Steger, Joseph A. (ed.), *Readings in Statistics for the Behavioral Scientist*, New York: Holt, Rinehart, and Winston, 1971.

T72　Rich, H., A. L. Luhby, H. M. Babikian, and M. Gordon, "Misuse of the Chi-Squared Test," *Lancet*, 1 (1974), 1294–1295.

T73　Wright, P. B., "Erroneous Use of Chi-Square Test," *Meteorol. Mag.*, 100 (1971), 301–303.

T74　Delucchi, Kevin L., "The Use and Misuse of Chi-Square: Lewis and Burke Revisited," *Psychol. Bull.*, 94 (1983), 166–176.

T75　Feinstein, A. R., and W. A. Ranshaw, "Procedure for the Rapid Mental Calculation of a Four-Fold Chi-Square Test," *J. Chronic Dis.*, 25 (1972), 551–553.

T76　Smith, D. B., "A Chi-Squared Nomogram," *Ecology*, 53 (1972), 529–530.

T77　Hamdan, M. A., "Optimum Choice of Classes for Contingency Tables," *J. Amer. Statist. Assoc.*, 63 (1968), 291–297.

T78　Chernoff, H., "Degrees of Freedom for Chi-Square," *Technometrics*, 9 (1967), 489–490.

T79　Hirotsu, C., "Use of Cumulative Efficient Scores for Testing Ordered Alternatives in Discrete Models," *Biometrika*, 69 (1982), 567–577.

T80　Hirotsu, C., "Defining the Pattern of Association in Two-Way Contingency Tables," *Biometrika*, 70 (1983), 579–589.

T81　Takeuchi, Kei, and Chihiro Hirotsu, "The Cumulative Chi-Squares Method against Ordered Alternatives in Two-Way Contingency Tables," *Rep. Statist. Appl. Res.*, 29 (September 1982), 1–13.

T82　Goodman, Leo A., *The Analysis of Cross-Classified Data Having Ordered Categories*, Cambridge, Mass.: Harvard University Press, 1984.

T83　Mantel, Nathan, and William Haenszel, "Statistical Aspects of the Analysis of Data from Retrospective Studies of Disease," *J. Nat. Cancer Institute*, 22 (1959), 719–784.

T84　Wittes, Janet, and Sylvan Wallenstein, "The Power of the Mantel–Haenszel Test," *J. Amer. Statist. Assoc.*, 82 (1987), 1104–1109.

T85　Mantel, Nathan, and Joseph L. Fleiss, "Minimum Expected Cell Size Requirements for the Mantel–Haenszel One-Degree-of-Freedom Chi-Square Test and a Related Rapid Procedure," *Amer. J. Epidemiology*, 112 (1980), 129–134.

T86　Boynton, Kathryn M., and Nancy M. Poe, "Alternate Formulas for Chi Square," *Perceptual and Motor Skills*, 48 (1979), 556–558.

T87　McDonald-Schlichting, U., "Note on Simply Calculating Chi-Square for a $r \times c$-Contingency Table," *Biometrical J.*, 21 (1979), 787–789.

T88　Cox, C. Philip, "An Alternative Way of Calculating the χ^2 Independence or Association Test Statistic for a $2 \times k$ Contingency Table," *Amer. Statist.*, 36 (1982), 133.

T89　Lackritz, James R., "Exact *P*-Values for Chi-Squared Tests," *Proc. Statist. Educ. Sec. Amer. Statist. Assoc.*, Washington, D.C.: American Statistical Association, 1983, 130–132.

T90　Brown, George W., "2 × 2 Tables," *Amer. J. Dis. of Children*, 139 (1985), 410–416.

T91　Witt, Philip L., and Peter McGrain, "Nonparametric Testing Using the Chi-Square Distribution," *Physical Therapy*, 66 (1986), 264–268.

T92　O'Brien, Peter C., and Thomas R. Fleming, "A Multiple Testing Procedure for Clinical Trials," *Biometrics*, 35 (1979), 549–556.

T93　Schouten, H. J. A., I. W. Molenaar, R. van Strik, and A. Boomsma, "Comparing Two Independent Binomial Proportions by a Modified Chi Square Test," *Biometrical J.*, 22 (1980), 241–248.

T94　Haber, Michael, "A Comparison of Some Continuity Corrections for the Chi-Squared Test on 2 × 2 Tables," *J. Amer. Statist. Assoc.*, 75 (1980), 510–515.

T95 Cohen, Ayala, "On the Graphical Display of the Significant Components in Two-Way Contingency Tables," *Communic. Statist.—Theory and Methods*, A9 (1980), 1025–1041.

T96 Plackett, R. L., "Karl Pearson and the Chi-Squared Test," *Int. Statist. Rev.*, 51 (1983), 59–72.

T97 Aickin, Mikel, *Linear Statistical Analysis of Discrete Data*, New York: Wiley, 1983.

T98 Everitt, B. S., *The Analysis of Contingency Tables*, London: Chapman and Hall, 1977.

T99 Fienberg, Stephen, *The Analysis of Cross-Classified Categorical Data*, Cambridge, Mass.: MIT Press, 1977.

T100 Fingleton, B., *Models of Category Counts*, Cambridge, England: Cambridge University Press, 1984.

T101 Gokhale, D. V., and Solomon Kullback, *The Information in Contingency Tables*, New York: Marcel Dekker, 1978.

T102 Goodman, Leo A., *Analyzing Qualitative/Categorical Data*, Cambridge, Mass.: Abt Books, 1978.

T103 Reynolds, H. T., *The Analysis of Cross-Classifications*, New York: The Free Press, 1977.

T104 Russell, P. N., "Cross Tabulating—An IBM 360/44 Fortran 4 Program for Construction of Contingency Tables and Calculation of Chi-Square and Contingency Coefficients," *Behav. Sci.*, 14 (1969), 166–167.

T105 Hill, J. D., and M. C. Pike, "Chi-Squared Integral," *Comm. ACM*, 10 (1967), 243–244.

T106 Goldstein, R. B., "Chi-Square Quantiles," *Comm. ACM*, 16 (1973), 483–485.

T107 Forer, Bertram R., "Computerized Chi-Squared: Multiple Degrees of Freedom," *Perceptual and Motor Skills*, 58 (1985), 525–526.

T108 Tsai, San-Yun W., "A Computerized Aid for Calculating the Power of Chi-Squared Tests," *Amer. Statist.*, 34 (1980), 184–185.

T109 Romesburg, H. Charles, and Kim Marshall, "CHITEST: A Monte-Carlo Computer Program for Contingency Table Tests," *Comput. & Geosci.*, 11 (1985), 69–78.

T110 Arena, Vincent C., "A Program to Compute the Generalized Mantel–Haenszel Chi-Square for Multiple 2 × *J* Tables," *Computer Programs in Biomed.*, 17 (1983), 65–72.

T111 Bergan, John R., Olga M. Towstopiat, and Joh W. Luiten, "A Computer Program for Multidimensional Contingency Table Construction," *Educ. Psychol. Measurement*, 40 (1980), 127–132.

T112 Thakur, Ajit K., Kenneth J. Berry, and Paul W. Mielke, Jr., "A FORTRAN Program for Testing Trend and Homogeneity in Proportions," *Computer Programs in Biomed.*, 19 (1985), 229–233.

T113 Berry, Kenneth J., and Paul Mielke, Jr., "An APL Function for Radlow and Alf's Exact Chi-Square Test," *Behav. Res. Methods, Instruments, & Computers*, 17 (1985), 131–132.

T114 von Collani, Gernot, "Computing Probabilities for *F*, *t*, Chi-Square, and *z* in BASIC," *Behav. Res. Methods & Instrumentation*, 15 (1983), 543–544.

E1 Monteiro, Lois A., "Expense Is No Object . . . : Income and Physician Visits Reconsidered," *J. Health Soc. Behav.*, 14 (1973), 99–115.

E2 Abse, D. Wilfred, Marilyn M. Wilkins, Gordon Kirshner, Don L. Weston, Robert S. Brown, and W. D. Buxton, "Self-Frustration, Nighttime Smoking and Lung Cancer," *Psychosom. Med.*, 34 (1972), 395–404.

E3 Highman, Arthur, and Leth Davidson, "Plans and Planning Methods of Hospital Administrators," *Hosp. Topics*, 50 (June 1972), 30–42.

E4　Lieberman, Leonard, "Atomism and Mobility among Underclass Chippewas and Whites," *Hum. Org.*, 32 (1973), 337–347.

E5　"Special Communications: Relation of Body Weight and Insulin Dose to the Frequency of Hypoglycemia, A Report from the Boston Collaborative Drug Surveillance Program," *J. Amer. Med. Assoc.*, 228 (1974), 192–194.

E6　Mai, Francois M., Robert N. Munday, and Eric E. Rump, "Psychiatric Interview Comparisons between Infertile and Fertile Couples," *Pyschosom. Med.*, 34 (1972), 431–440.

E7　Heller, Charles F., Jr., and Anthony L. Redente, "Residential Location and White Attitudes toward Mixed-Race Neighborhoods in Kalamazoo, Michigan," *J. Geog.*, 72 (March 1973), 15–25.

E8　Steadman, Henry J., "Some Evidence on the Inadequacy of the Concept and Determination of Dangerousness in Law and Psychiatry," *J. Psychiatry and the Law*, 1 (1973), 409–426.

E9　Sofranko, Andrew J., and Michael F. Nolan, "Early Life Experiences and Adult Sports Participation," *J. Leisure Res.*, 4 (1972), 6–18.

E10　Farrell, Ronald A., "Class Linkages of Legal Treatment of Homosexuals," *Criminology*, 9 (May 1971), 49–68.

E11　Hazelrigg, Lawrence E., "Religious and Class Bases of Political Conflict in Italy," *Amer. J. Sociol.*, 75 (1970), 496–511.

E12　Potter, Roger E., and Robert S. Hoeke, "Intergroup Comparison of Stock Investor Dimensions," *Okla. Bus. Bull.*, 41 (April 1973), 6–11.

E13　Bailey, Neal J., "Contact Lens Design—A Survey," *Amer. J. Optom. Arch. Amer. Acad. Optom.*, 45 (1968), 96–102.

E14　Richardson, C. Joan, Jeffrey J. Pomerance, M. Douglas Cunningham, and Louis Gluck, "Acceleration of Fetal Lung Maturation Following Prolonged Rupture of the Membranes," *Amer. J. Obstet. Gynecol.*, 118 (1974), 1115–1118.

E15　Mims, Fern, Rosalee Yeaworth, and Stephen Hornstein, "Effectiveness of an Interdisciplinary Course in Human Sexuality," *Nursing Res.*, 23 (1974), 248–253.

E16　Wall, Geoffrey, "Public Response to Air Pollution in South Yorkshire, England," *Environ. and Behav.*, 5 (1973), 219–248.

E17　Rosenwald, George C., and Marshall W. Stonehill, "Early and Late Postpartum Illnesses," *Psychosom. Med.*, 34 (1972), 129–137.

E18　van Vliet, P. K. J., and J. M. Gupta, "THAM v. Sodium Bicarbonate in Idiopathic Respiratory Distress Syndrome," *Arch. Dis. in Childhood*, 48 (1973), 249–255.

E19　Kaufmann, Geir, and Kjell Raaheim, "Effect of Inducting Activity upon Performance in an Unfamiliar Task," *Psychol. Rep.*, 32 (1973), 303–306.

E20　Bayer, Alan E., "College Impact on Marriage," *J. Marriage Fam.*, 34 (1972), 600–609.

E21　Garfield, Mark D., and Emily F. Garfield, "A Longitudinal Study of Drugs on a Campus," *Int. J. Addict.*, 8 (1973), 599–611.

E22　Newmeyer, John A., and George R. Gay, "The Traditional Junkie, the Aquarian Age Junkie, and the Nixon Era Junkie," *Drug Forum*, 2 (Fall 1972), 17–30.

E23　Oliverio, Anna Ferraris, "Children's Evaluations of Family Roles, A Cross Cultural Comparison," *Int. J. Psychol.*, 8 (1973), 153–158.

E24　Jacobson, Dan, "Rejection of the Retiree Role: A Study of Female Industrial Workers in Their 50's," *Hum. Rel.*, 27 (1974), 477–492.

E25　Steinitz, Victoria A., Prudence King, Ellen Solomon, and Ellen Shapiro, "Ideological Development in Working-Class Youth," *Harv. Educ. Rev.*, 43 (1973), 333–361.

E26 Goldstein, Michael S., "Academic Careers and Vocational Choices of Elite and Non-Elite Students at an Elite College," *Sociol. Educ.*, 47 (1974), 491–510.

E27 Arlinghaus, Barry P., and W. Peter Salzarulo, "The Importance of Post-Baccalaureate Education for Tax Professionals," *Akron Bus. Econ. Rev.*, 17 (Winter 1986), 8–17.

E28 Rocha, Joseph R., Jr., and M. Riaz Khan, "The Human Resource Factor in Small Business Decision Making," *Amer. J. Small Bus.*, 10 (Fall 1985), 53–62.

E29 Bellur, Venkatakrishna V., "Shoplifting: Can It Be Prevented?" *J. Acad. Marketing Sci.*, 9 (1981), 78–87.

E30 Denisoff, Serge, and John Bridges, "Popular Music: Who Are the Recording Artists?" *J. Communic.*, 32 (Winter 1982), 132–142.

6

PROCEDURES THAT UTILIZE DATA FROM THREE OR MORE INDEPENDENT SAMPLES

Chapter 3 presented several nonparametric techniques for analyzing the data from two independent samples. This chapter introduces some procedures that are appropriate for use with data from three or more independent samples. As in Chapter 3, the word independence is used to refer to independence both within and among the several samples under study.

In most cases in this chapter, we shall be interested in testing the null hypothesis that the several samples have been drawn from the same population or from populations with equal location parameters. The parametric counterpart to most of these procedures is the one-way analysis of variance F test. The reader may recall from a course in elementary statistics that the one-way analysis of variance assumes that the samples are randomly and independently drawn from normally distributed populations with equal variances. An advantage of the procedures discussed in this chapter is that their validity does not depend on such restrictive assumptions.

Chapter 5 covered another test that uses data from three or more independent samples: the chi-square test of homogeneity, which tests the null hypothesis that the samples have been drawn from populations with equal proportions.

6.1

EXTENSION OF THE MEDIAN TEST

Chapter 3 introduced the median test for testing the null hypothesis that two independent samples have been drawn from populations with the same median. Now this test is extended to the case in which three or more independent samples are available for analysis.

Assumptions

A. Each sample is a random sample of size n_i drawn from one of c populations of interest with unknown medians $M_1, M_2, \ldots, M_c$.
B. The observations are independent both within and among samples.
C. The measurement scale employed is at least ordinal.
D. If all populations have the same median, then for each population the probability p is the same that an observed value exceeds the grand median.

Hypotheses

H_0: $M_1 = M_2 = \cdots = M_c$
H_1: At least one population has a median different from at least one of the others.

Test Statistic

We combine the c samples, order them, and compute the combined sample median. We then classify each observation according to the sample (or population) to which it belongs and according to whether it is larger than, equal to, or less than the median. We can display the results in a two-way contingency table such as Table 6.1. In this table O_{ij} is the observed frequency of observations falling in the ith group of the jth sample, a is the total number of observations larger than the combined

	Sample					
	1	**2**	**3**	$\cdots$	**c**	**Total**
> *Sample median*	O_{11}	O_{12}	O_{13}	$\cdots$	O_{1c}	a
≤ *Sample median*	O_{21}	O_{22}	O_{23}	$\cdots$	O_{2c}	b
Total	n_1	n_2	n_3	$\cdots$	n_c	N

TABLE 6.1 **Contingency table for median test**

sample median, and b is the total number of observations that are equal to or smaller than the combined sample median. As the table shows, $\Sigma_{i=1}^{c} n_i = a + b = N$.

Testing the null hypothesis that the sampled populations have the same median is equivalent to testing the null hypothesis that the populations are homogeneous with respect to the proportion of observations falling above and below the common population median. Consequently, if the expected cell frequencies of the contingency table meet the minimum size requirements specified in Chapter 5, we may test the null hypothesis by using the chi-square test of homogeneity. We calculate expected cell frequencies as described in Chapter 5, compute X^2 by Equation 5.9, and compare it for significance with tabulated values of chi-square for $c - 1$ degrees of freedom given in Table A.11.

Decision Rule

If the calculated value of the test statistic X^2 exceeds the tabulated value of chi-square for $c - 1$ degrees of freedom and α, then we reject the null hypothesis of equal population medians at the α level of significance.

Example 6.1   In a study designed to determine the distribution of myocardial water and the cellular concentrations of cardiac electrolytes, Polimeni (E1) used the tracer method to measure the extracellular space in the ventricular muscle of two groups of nephrectomized rats and one group of intact rats. Table 6.2 shows the results. We wish to know whether we may conclude from these data that the population medians are different.

TABLE 6.2   **Extracellular spaces, tracer method (g/g) of ventricular muscle of nephrectomized and intact rats**

Nephrectomized rats		Intact rats
Group I	Group II	Group III
0.185	0.189	0.219
0.187	0.193	0.204
0.209	0.176	0.219
0.194	0.195	0.234
0.175	0.169	0.233
0.197	0.183	0.194
0.188	0.185	0.209
0.185	0.179	0.195

Source: Phillip I. Polimeni, "Extracellular Space and Ionic Distribution in Rat Ventricle," *Amer. J. Physiol.*, 227 (1974), 676–683.

Hypotheses

$$H_0: M_1 = M_2 = M_3$$

H_1: The three population medians are not all equal

Test Statistic

The median for the three samples combined is 0.1935. Table 6.3 shows the numbers in each sample falling above and below the sample median.

TABLE 6.3   **Contingency table for Example 6.1**

	Sample			
	I	II	III	Total
> 0.1935	3	1	8	12
≤ 0.1935	5	7	0	12
Total	8	8	8	24

The expected frequency for each cell is $(8)(12)/24 = 4$, so that we may compute

$$X^2 = \frac{(3 - 4)^2}{4} + \frac{(1 - 4)^2}{4} + \cdots + \frac{(0 - 4)^2}{4} = 13.00$$

Decision

Since 13.00 is larger than 10.597, the tabulated chi-square value for $\alpha = 0.005$ and two degrees of freedom, we can reject H_0 at the 0.005 level of significance, and we conclude that the three population medians are not equal (P value < 0.005).

EXERCISES

6.1 Sattler (E2) studied the effects of olfactory and auditory stress in mice. The variable of interest was the ratio of the weight of adrenal glands (micrograms) to body weight (grams). Sattler used the following three groups of nulliparous albino mice in the experiment.

Group I: The crowded group, consisting of 12 males and 12 females, which provided a source of sounds and scents for the pseudocrowded group.

Group II: The pseudocrowded group, consisting of five isolated pairs, that received olfactory and auditory stimuli from group I.

Group III: The control group, consisting of five pairs, raised under conditions identical to those of the pseudocrowded mice, but without the stimuli from group I.

The results are shown in Table 6.4. Test for a significant difference among the sample groups with respect to the median adrenal ratios. Let $\alpha = 0.05$. Determine the P value.

TABLE 6.4

Ratios, micrograms adrenal glands per gram body weight, in three groups of albino mice

Group I	86.4	128.8	138.8	140.6	140.8	154.1
	158.8	165.7	170.4	190.6	194.4	213.1
Group II	182.8	190.3	191.6	234.0	252.5	
Group III	139.8	142.8	144.8	185.5	207.7	

Source: Krista M. Sattler, "Olfactory and Auditory Stress in Mice (*mus musculus*)," *Psychon. Sci.*, 29 (1972), 294–296.

6.2 Mameesh et al. (E3) determined the availability of iron from isotopically labeled wheat, chickpea, broad bean, and okra in anemic blood donors. The results are shown in Table 6.5. Can we conclude from these data that the percentage of iron absorbed differs, on the average, according to the source? Find the P value.

TABLE 6.5

Iron absorption from specified source in anemic blood donors

Source	Percent iron absorption									
Whole wheat arabic bread	27,	16,	19,	4,	2,	16,	30,	9,	16	
Chickpea fried patties (folafel)	44,	34,	43,	47,	22,	35,	51,	22,	37,	29
Broad bean fried patties (folafel)	17,	45,	28,	13,	36,	3,	42,	41,	15	
Okra with tomato juice	51,	29,	30,	50,	47,	40,	43,	44,	54	

Source: Mostafa Mameesh, Simon Aprahamian, Joseph P. Salji, and James W. Cowan, "Availability of Iron from Labeled Wheat, Chickpea, Broad Bean, and Okra in Anemic Blood Donors," *Amer. J. Clin. Nutr.*, 23 (1970), 1027–1032.

6.3 The data of Table 6.6 were reported by Schapira et al. (E4). Do these data provide sufficient evidence to indicate that the three represented populations differ with respect to median serum aldolase levels? Determine the P value.

6.4 In a study of the relationship between dietary deficiency and myopia, Young et al. (E5) collected the data shown in Table 6.7. The subjects consisted of 20 Rhesus monkeys maintained on a low-protein diet for an average of 32 months, 17 Rhesus monkeys maintained on a high-protein diet for an average of 28 months, and 40 Rhesus monkeys raised on a commercial monkey chow. Do these data provide sufficient evidence to indicate that, on the average, diet affects vertical spectacle refraction values?

6.5 Chandra (E6) obtained the data shown in Table 6.8 on 22 children categorized as either healthy, suffering from moderate iron deficiency, or suffering from severe iron deficiency. Do these data provide sufficient evidence to indicate a difference among the medians of the three populations? Determine the P value for the test.

TABLE 6.6

Values of serum aldolase in muscle disorders expressed in Meyerhof's units, milligrams of alkali-labile phosphorus released per minute, per liter of serum

Control	0.20	0.20	0.20	0.30	0.30	0.40	0.40	0.40
children	0.40	0.50	0.50	0.50	0.50	0.60	0.60	0.60
Progressive	0.30	0.34	0.50	0.60	0.70	0.90	1	1
muscular	1.1	1.2	1.2	1.2	1.2	1.3	1.3	1.3
dystrophy	1.4	1.4	1.5	1.5	1.5	1.6	1.6	1.7
children	1.7	1.8	2	2	2	2.1	2.2	2.2
	2.4	2.4	2.4	2.4	2.8	2.8	2.8	2.8
	3	3	3.8	3.9	4	4	4.2	4.3
	4.8	5	5	5.2	5.2	5.4	6.2	7
	7	7	10	12	13			
Poliomyelitis	0.2	0.2	0.3	0.35	0.4	0.4	0.4	0.45
children	0.5	0.5	0.5	0.5	0.6	0.6	0.7	0.7
	0.75	0.75	0.75	1.4				

Source: Georges Schapira, Jean-Claude Dreyfus, Fanny Schapira, and Jacques Kruh, "Glycogenolytic Enzymes in Human Progressive Muscular Dystrophy," *Amer. J. Phys. Med.*, 34 (1955), 313–319; copyright 1955, Williams & Wilkins.

TABLE 6.7

Vertical spectacle refraction for the right eye for the low-, normal-, and high-protein groups of Rhesus monkeys

Low-protein	+1.27	−4.98	−0.50	+1.25	−0.25	+0.75	−2.75	+0.75	+1.00	+3.00
group	+2.25	+0.53	+1.25	−1.50	−5.00	+0.75	+1.50	+0.50	+1.75	+1.50
Normal-protein	−6.00	−2.00	−4.00	+0.50	0.00	+2.50	+2.75	+3.00	+1.50	−1.50
*group**	+0.75	+1.75	+2.50	−3.00	+0.25	+1.00	+0.75	+1.25	−1.75	
	−5.50	−3.50	+0.25	+1.25	+1.00	−1.00	+1.50	+1.00	−0.75	+2.00
	+1.25	+3.59	+5.00	+3.00	−1.25	−1.75	+1.75	−1.00	+4.75	−0.50
High-protein	−6.00	+0.25	+1.25	−2.00	+3.14	+2.00	+0.75	+1.75	0.00	+0.75
group	+0.75	+0.25	+1.25	+1.25	+1.00	+0.50	−2.25			

Source: Francis A. Young, George A. Leary, Robert R. Zimmerman, and David Strobel, "Diet and Refractive Characteristics," *Amer. J. Optom. and Arch. Amer. Acad. Optom.*, 50 (1973), 226–233; used by permission.

* Data available on 39 out of 40 subjects

TABLE 6.8

Results of quantitative nitro-blue tetrazolium test (Δ optical density) in 22 children

Normal controls	0.32	0.23	0.20	0.14	0.19	0.32	0.31	0.34	0.37	0.39
With moderate iron deficiency	0.06	0.06	0.19	0.13	0.11	0.03				
With severe iron deficiency	0.03	0.02	0.06	0.01	0.02	0.01				

Source: R. K. Chandra, "Reduced Bactericidal Capacity of Polymorphs in Iron Deficiency," *Arch. Dis. Child.*, 48 (1973), 864–866; used by permission of the author and the editor.

6.2

KRUSKAL–WALLIS ONE-WAY ANALYSIS OF VARIANCE BY RANKS

Perhaps the most widely used nonparametric technique for testing the null hypothesis that several samples have been drawn from the same or identical populations is the Kruskal–Wallis *one-way analysis of variance by ranks* (T1). When only two samples are being considered, the Kruskal–Wallis test is equivalent to the Mann–Whitney test discussed in Chapter 3.

The Kruskal–Wallis test uses more information than the median test. As a consequence, the Kruskal–Wallis test is usually more powerful, and is preferred when the available data are measured on at least the ordinal scale.

Assumptions

    A. The data for analysis consist of k random samples of sizes $n_1, n_2, \ldots, n_k$.

    B. The observations are independent both within and among samples.

    C. The variable of interest is continuous.

    D. The measurement scale is at least ordinal.

    E. The populations are identical except for a possible difference in location for at least one population.

Hypotheses

    H_0: The k population distribution functions are identical

    H_1: The k populations do not all have the same median

Test Statistic

We may display the data available for analysis in a table such as Table 6.9. We replace each original observation by its rank relative to all the observations in the k

TABLE 6.9   **Data display for Kruskal–Wallis one-way analysis of variance by ranks**

Sample			
1	**2**	$\cdots$	**k**
$X_{1,1}$	$X_{2,1}$	$\cdots$	$X_{k,1}$
$X_{1,2}$	$X_{2,2}$	$\cdots$	$X_{k,2}$
$\vdots$	$\vdots$		$\vdots$
X_{1,n_1}	X_{2,n_2}	$\cdots$	X_{k,n_k}

samples. If we let $N = \Sigma_{i=1}^{k} n_i$ be the total number of observations in the k samples, we assign the rank 1 to the smallest of these, the rank 2 to the next in size, and so on to the largest, which is given the rank N. In case of ties we assign the tied observations the average of the ranks that would be assigned if there were no ties.

If the null hypothesis is true, we expect the distribution of ranks over the groups to be a matter of chance, so that either the small ranks or the large ranks do not tend to be concentrated in one sample. Therefore, if the null hypothesis is true, we expect the k sums of ranks (that is, the sums of the ranks in each sample) to be about equal when adjusted for unequal sample sizes. An intuitively appealing test statistic is one that determines whether the sums of the ranks are sufficiently disparate that they are not likely to have been derived from samples from identical populations—leading to rejection of H_0—or whether they are so close in magnitude that we cannot discredit the hypothesis of identical population distributions. Just such a statistic is the Kruskal–Wallis test statistic. It is a weighted sum of squares of deviations of sums of ranks from the expected sum of ranks, using reciprocals of sample sizes as the weights. We write the Kruskal–Wallis test statistic as

$$H = \frac{12}{N(N+1)} \sum_{i=1}^{k} \frac{1}{n_i} \left[R_i - \frac{n_i(N+1)}{2} \right]^2 \tag{6.1}$$

where R_i is the sum of the ranks assigned to observations in the ith sample, and $n_i(N+1)/2$ is the expected sum of ranks for the ith treatment under H_0. A computationally more convenient form of Equation 6.1 is

$$H = \frac{12}{N(N+1)} \sum_{i=1}^{k} \frac{R_i^2}{n_i} - 3(N+1) \tag{6.2}$$

Decision Rule

When we are considering three samples, and each·sample has 5 or fewer observations, we compare the computed value of H for significance with tabulated values of the test statistic given in Table A.12. When the number of samples and/or observations per sample are such that we cannot use Table A.12, we compare the computed value of H for significance with the chi-square values for $k - 1$ degrees of freedom given in Table A.11. We do so since Kruskal (T2) shows that for large n_i and k, H is distributed approximately as chi-square with $k - 1$ degrees of freedom. The adequacy of the chi-square approximation for small samples has been investigated by Gabriel and Lachenbruch (T3).

For the number of samples and sample sizes that can be accommodated by Table A.12, we reject H_0 if the computed value of H exceeds the critical value listed in that table for the preselected value of α. When the preselected value of α is not in the table (the usual case), choose a neighboring value, perhaps the closest. When the chi-square table must be used, reject H_0 if the computed value of H exceeds the tabulated value of chi-square for preselected α and $k - 1$ degrees of freedom.

Example 6.2

Cawson et al. (E7) reported the data shown in Table 6.10 on cortisol levels in three groups of patients who were delivered between 38 and 42 weeks gestation. Group I was studied before the onset of labor at elective Caesarean section, group II was studied at emergency Caesarean section during induced labor, and group III consisted of patients in whom spontaneous labor occurred and who were delivered either vaginally or by Caesarean section. We wish to know whether these data provide sufficient evidence to indicate a difference in median cortisol levels among the three populations represented.

TABLE 6.10

Antecubital vein cortisol levels in three groups of patients studied at time of delivery

Group I	262	307	211	323	454	339	304	154	287	356
Group II	465	501	455	355	468	362				
Group III	343	772	207	1048	838	687				

Source: M. J. Cawson, Anne B. M. Anderson, A. C. Turnbull, and L. Lampe, "Cortisol, Cortisone, and 11-Deoxycortisol Levels in Human Umbilical and Maternal Plasma in Relation to the Onset of Labour," *J. Obstet. Gynaecol. Brit. Commonw.*, 81 (1974), 737–745.

Hypotheses

H_0: The three populations represented by the data are identical

H_1: The three populations do not have the same median

Test Statistic

The ranks replacing the original observations of Table 6.10 are displayed in Table 6.11, along with the three rank sums. From these data we use Equation 6.2 to compute

$$H = \frac{12}{22(22 + 1)} \left[\frac{69^2}{10} + \frac{90^2}{6} + \frac{94^2}{6} \right] - 3(22 + 1) = 9.232$$

TABLE 6.11

Ranks corresponding to data of Table 6.10

Group	Ranks										Rank sums
I	4	7	3	8	14	9	6	1	5	12	$R_1 = 69$
II					16	18	15	11	17	13	$R_2 = 90$
III					10	20	2	22	21	19	$R_3 = 94$

Decision

Since the sample sizes all exceed 5, we must use the chi-square table to determine whether the sample medians are significantly different. The critical value of chi-

square for $k - 1 = 3 - 1 = 2$ degrees of freedom is 9.210 for $\alpha = 0.01$. Thus with $H = 9.232$, we can reject H_0 at that level of significance. We conclude that the medians of the populations represented are not all equal—that is, the median cortisol levels are not the same for all three types of patient. The P value is between 0.01 and 0.005.

Correction for Ties   If there are a substantial number of ties, we may want to adjust the test statistic. The adjustment factor is

$$1 - \frac{\Sigma T}{N^3 - N} \tag{6.3}$$

where $T = t^3 - t$ and t is the number of tied observations in a group of tied scores. The adjusted test statistic becomes

$$H_C = \frac{H}{1 - \Sigma T/(N^3 - N)} \tag{6.4}$$

where H is the value computed from Equation 6.2. The effect of the adjustment is to inflate the value of the test statistic. Thus if H is significant at the desired level of α without the adjustment, there is no point in computing H_C. Furthermore, Kruskal and Wallis (T1) point out that with 10 or fewer samples, a value of H of 0.01 or more does not change more than 10% when the adjusted value H_C is computed, provided that not more than one-fourth of the observations are involved in ties.

Power-Efficiency   Andrews (T4) found the asymptotic efficiency of the Kruskal–Wallis test, relative to the usual parametric F test, to be 0.955 when sampling is from normally distributed populations. If the distribution functions have identical shapes and differ only in location, Hodges and Lehmann (T5) have shown that the asymptotic relative efficiency is never less than 0.864 and may be greater than 1 under certain conditions.

FURTHER READING

Iman, Quade, and Alexander (T6) present more extensive tables for the three-sample Kruskal–Wallis problem. Breslow (T7) discusses a generalized Kruskal–Wallis test for comparing several samples subject to unequal patterns of censorship. Rust and Fligner (T8) propose a modification of the Kruskal–Wallis test, which they contend requires fewer assumptions about the shapes of the populations. The authors state that the modified procedure is asymptotically distribution-free when the populations are assumed to be symmetric. On the basis of a small sample study, they conclude that use of the modification rather than the original Kruskal–Wallis statistic results in little or no loss of power when it is used in situations in which the Kruskal–Wallis procedure is appropriate.

   Citing the need for a satisfactory rank-sum method for factorial experimental designs, Toothaker and Chang (T9) report on their use of Monte Carlo methods to

evaluate such a procedure that was previously proposed by Scheirer et al. (T10). They found that the distributions of the suggested test statistics are a function of effects other than those being tested except under the completely null situation. They recommend against the use of the procedures described by Scheirer et al. (T10).

In a paper concerned with the use of classical and nonparametric techniques in the analysis of experiments, van der Laan and Verdooren (T11) discuss a number of problems encountered in the application of both approaches.

SAMPLING DISTRIBUTION OF THE KRUSKAL–WALLIS STATISTIC

To find the distribution of the Kruskal–Wallis test statistic, H, we assume that all measurements are drawn from the same or identical populations. To find the distribution of H, we use the method of randomization that was used to find the distribution of the Mann–Whitney statistic. The number of possible sets of ranks of sizes $n_1, n_2, \ldots n_k$, respectively, that can be formed from the ranks 1 through N is equal to $N!/n_1!n_2! \ldots n_k!$. Each set is equally likely under the assumption that measurements come from the same or identical populations. The probability of the occurrence of a given set of ranks is equal to 1 divided by $N!/n_1!n_2! \ldots n_k!$. We calculate H for each set of ranks, determine the frequency of occurrence of each distinct value of H, and compute the probability of each value of H by dividing each frequency by $N!/n_1!n_2! \ldots n_k!$. The result is the sampling distribution of H.

The construction of the sampling distribution of the Kruskal–Wallis test statistic is illustrated with an example in which we have three samples of size $n_1 = 2, n_2 = 2$, and $n_3 = 1$. Since $N = 2 + 2 + 1 = 5$, the number of possible sets of ranks is $5!/2!2!1! = 30$. In other words, the number of ways in which the ranks 1, 2, 3, 4, and 5 can be assigned to three groups containing, respectively, 2, 2, and 1 ranks is 30. These possible sets of ranks, along with the values of H that can be computed from them, are as follows:

Set	Sample 1	2	3	H	Set	Sample 1	2	3	H
1	1, 2	3, 4	5	3.6	16	2, 4	1, 3	5	2.4
2	1, 2	3, 5	4	3.0	17	2, 4	1, 5	3	0.0
3	1, 2	4, 5	3	3.6	18	2, 4	3, 5	1	2.4
4	1, 3	2, 4	5	2.4	19	2, 5	1, 3	4	1.4
5	1, 3	2, 5	4	1.4	20	2, 5	1, 4	3	0.4
6	1, 3	4, 5	2	3.0	21	2, 5	3, 4	1	2.0
7	1, 4	2, 3	5	2.0	22	3, 4	1, 2	5	3.6
8	1, 4	2, 5	3	0.4	23	3, 4	1, 5	2	0.6
9	1, 4	3, 5	2	1.4	24	3, 4	2, 5	1	2.0
10	1, 5	2, 3	4	0.6	25	3, 5	1, 2	4	3.0
11	1, 5	2, 4	3	0.0	26	3, 5	1, 4	2	1.4
12	1, 5	3, 4	2	0.6	27	3, 5	2, 4	1	2.4
13	2, 3	1, 4	5	2.0	28	4, 5	1, 2	3	3.6
14	2, 3	1, 5	4	0.6	29	4, 5	1, 3	4	3.0
15	2, 3	4, 5	1	3.6	30	4, 5	2, 3	1	3.6

The following is the sampling distribution of H for the conditions described in our example:

H	Frequency of H	$P(H)$
0.0	2	2/30
0.4	2	2/30
0.6	4	4/30
1.4	4	4/30
2.0	4	4/30
2.4	4	4/30
3.0	4	4/30
3.6	6	6/30
Total	30	30/30

SOME ALTERNATIVE PROCEDURES

A number of alternatives to the k-sample procedures presented in this chapter are available. Some of them are described in the paragraphs that follow.

1. Expected Normal-Scores Test When we use this test, we replace each of the N original observations by its rank relative to all other observations in the k samples, as with the Kruskal–Wallis test. At this point, we further transform the data by replacing each rank with the expected value of the observation having that rank in a random sample of N observations from a standard normal distribution. We need special tables for this transformation. We then construct the test statistic from these expected normal scores and compare it for significance with tabulated values of the chi-square statistic.

The expected normal-scores test is one of several types of normal-scores test that have been proposed. Fisher and Yates (T12) proposed the technique of replacing ranked data by "normal scores" before performing the statistical analysis as early as 1938 in the introduction to Table XX in their book of statistical tables. Since then, considerable attention has been focused on the use of normal scores in the two-sample location problem. The work of Hoeffding (T13) and Terry (T14) is of particular importance. A test employing an inverse normal-scores transformation sensitive to unequal locations in the two-sample case has been proposed by van der Waerden (T15).

Puri (T16) has shown that the distribution of the test statistic for the expected normal-scores test under the null hypothesis of identical distributions tends to approach the chi-square distribution with $k - 1$ degrees of freedom as the sample sizes tend to infinity. See also Hájek and Šidák (T17).

McSweeney and Penfield (T18) consider the extension of the expected normal-scores technique to the case of three or more samples. They present a rationale for and derivation of the test statistic for the $k(>2)$-sample case. These authors cite the work of Puri (T16) to substantiate their contention that the normal-scores test is at

least as efficient as the parametric F test and more efficient than the Kruskal–Wallis test when samples have been drawn from distributions that are likely to be encountered in practice. Their overall conclusions favor the normal-scores test on the basis of high efficiency and power. Feir-Walsh and Toothaker (T19), on the other hand, arrive at opposite conclusions. On the basis of their study, they do not recommend the normal-scores test.

Tables of expected normal scores have been made available by David et al. (T20), Harter (T21, T22, T23), and Owen (T24). Bell and Doksum (T25, T26) have proposed a series of tests that use random normal observations themselves, as replacements for ranks, instead of expected or inverse normal scores. These techniques, along with numerical examples, are discussed in considerable detail by Conover (T27). Bradley (T28) devotes an entire chapter of his book on nonparametric statistics to the topic of normal-scores tests.

2.   Rank Test Due to Barbour et al.   Barbour et al. (T29) suggest a generalization of the two-sample Wilcoxon rank-sum statistic to provide a k-sample test derived from an expression for the treatment sum of squares in a one-way analysis of variance. The authors report that the relevant statistic is essentially the same as one proposed by Crouse (T30, T31), and that for practical applications its empirical distribution is well approximated by a beta distribution.

3.   Pairwise Ranking Procedure by Fligner   Fligner (T32) proposes the construction of a test statistic for the k-sample location problem by appropriately combining all pairwise two-sample Wilcoxon tests. On the basis of a simulation study, he concludes that the resulting procedure performs at least as well as the Kruskal–Wallis test.

4.   Intrinsic Rank Test by Kannemann   To overcome the sensitivity of the Kruskal–Wallis test to certain perturbations, Kannemann (T33) proposes the intrinsic rank test for k independent samples. The test statistic is computed not from the ordinal ranks of the observations directly, as is the Kruskal–Wallis statistic, but from the *intrinsic* ranks. The intrinsic rank of an observation is defined as the ordinal number of the rank interval into which its ordinal rank falls. For k samples of equal size n, these rank intervals are determined by subdividing the ordinal ranks of the observations into k intervals containing n divisions of the ordinal ranks scale. A $k \times k$ matrix is then constructed, with cells containing the frequency of occurrence of intrinsic ranks within each sample. A chi-square-like test statistic computed from the table is referred to the gamma distribution for significance.

5.   Median-Test-Based Procedure by Shoemaker   Using the extension of the median test as a starting point, Shoemaker (T34) introduces a nonparametric procedure that he claims is highly resistant to outliers, computationally simple, and comprehensible to anyone with a rudimentary knowledge of classical analysis of variance. The author discusses the construction of variations on the median test to

test for interactions and combinations of effects in the two-way and higher-order models.

Also of interest is a permutation test procedure for use with the one-way analysis of variance layout for binary data, proposed by Soms (T35).

EXERCISES

6.6 Ali and Sweeney (E8) determined the protoporphyrin levels in 15 normal, healthy laboratory workers and in 26 patients admitted with acute alcoholism. Their results appear in Table 6.12. Can we conclude from these data that the normal workers and the two groups of alcoholics differ with respect to average protoporphyrin level? Determine the *P* value.

TABLE 6.12 **Protoporphyrin level, milligrams/100 ml RBC, in three groups of subjects**

Normal	22	27	47	30	38	78	28	58	72	56	30	39	53	50	36
Alcoholics with ring sideroblasts in bone marrow	78	172	286	82	453	513	174	915	84	153	780				
Alcoholics without ring sideroblasts in bone marrow	37	28	38	45	47	29	34	20	68	12	37	8	76	148	11

Source: M. A. M. Ali and G. Sweeney, "Erythrocyte Corproporphyrin and Protoporphyrin in Ethanol-Induced Sideroblastic Erythropoiesis," *Blood*, 43 (1974), 291–295; used by permission.

6.7 Torre et al. (E9) recorded the changes in rat cerebral and extracerebral (platelet) serotonin (5-HT) after intraperitoneal administration of LSD-25 and 1-methyl-d-lysergic acid butanolamide (UML). They also took measurements on 11 controls. The results are shown in Table 6.13. Do these data provide sufficient evidence to indicate a difference among the three groups? Determine the *P* value.

TABLE 6.13 **Brain serotonin (5-HT), nanograms per gram, in three groups of mice**

Controls	340	340	356	386	386	402	402	417	433	495	557
LSD, 0.5 mg/kg	294	325	325	340	356	371	385	402			
UML, 0.5 mg/kg	263	309	340	356	371	371	402	417			

Source: Michele Torre, Filippo Bogetto, and Eugenio Torre, "Effect of LSD-25 and 1-Methyl-d-Lysergic Acid Butanolamide on Rat Brain and Platelet Serotonin Levels," *Psychopharmacologia*, 36 (1974), 117–122.

6.8 Kando (E10) ranked males, females, and feminized transsexuals according to the strength of their endorsement of traditional sex ascriptions. The rankings are shown in Table 6.14. Can we conclude from these data that the three groups differ with respect to strength of endorsement?

TABLE 6.14

Sample ranked according to strength of endorsement of traditional sex ascriptions ("ought" scale)

Males	1	10	11	12	13	14	15	16	17
	20	23	25	26.5	30	32	44.5	51	
Females	5	9	22	26.5	33.5	33.5	36	38	39
	42	43	44.5	46	47	48	49	50	
Transsexuals	2	3.5	3.5	6	7	8	18	19	21
	24	28	29	31	35	37	40	41	

Source: Thomas M. Kando, "Role Strain: A Comparison of Males, Females, and Transsexuals," *J. Marriage Fam.,* 34 (1972), 459–464; copyright 1972 by National Council on Family Relations; reprinted by permission.

6.9 Stern et al. (E11) studied the uptake of tritiated thymidine by the dorsal epidermis of the fetal and newborn rat. To determine whether the method of application of ^{3}H-thymidine affected the results, the authors performed a Kruskal–Wallis test on the data shown in Table 6.15. They obtained a value of the test statistic equal to 3.28. Verify these results and determine the P value.

TABLE 6.15

Data for Exercise 6.9

Method								
1	125.6	123.8	123.3	132.4	156.6	99.9	60.4	135.1	72.5
2	140.4	50.0	101.0	70.4	149.7			
3	74.2	72.4	55.8	95.1	134.6			
4	125.5	102.4	80.6	82.9	95.4	93.7		
5	72.9	90.0	137.5	65.5				
6	88.0	96.0	106.4					

Source: I. B. Stern, L. Dayton, and J. Duecy, "The Uptake of Tritiated Thymidine by the Dorsal Epidermis of the Fetal and Newborn Rat," *Anatom. Rec.,* 170 (1971), 225–234.

6.3

JONCKHEERE–TERPSTRA TEST FOR ORDERED ALTERNATIVES

In some applications of parametric statistical procedures, it is appropriate to test the null hypothesis of equality among population means (μ_j's)

$$H_0: \mu_1 = \mu_2 = \cdots = \mu_k$$

against an alternative in which order is specified—say,

$$H_1: \mu_1 \leq \mu_2 \leq \cdots \leq \mu_k$$

(where at least one of the inequalities is strict; that is, at least one population mean is less than at least one of the other population means). This alternative is sometimes

more meaningful than the alternative

H_1: Not all μ's are equal

which merely negates at least one of the equalities in H_0. Analogous situations arise when the application of nonparametric statistical procedures is appropriate and the location parameters of interest are medians.

In a study of the efficacy of some drug, for example, the investigator may wish to know whether the sample data indicate that increased response accompanies increased dosage. An educator may wish to know whether levels of distraction varying from none to moderate to excessive during an examination result in scores in the reverse order of magnitude. A sociologist may be interested in knowing whether people in low, middle, and high socioeconomic groups possess low, middle, and high knowledge of certain current events. Alternative hypotheses of this type are referred to as *ordered alternatives*. The Kruskal–Wallis test and the extension of the median test discussed in Sections 6.1 and 6.2 are not appropriate when the alternative is ordered, since these tests are not designed to detect differences in a particular direction.

In the two-sample case involving location parameters, we achieve the objectives of an ordered alternative by using a one-sided alternative instead of a two-sided alternative. When the data available for analysis consist of three or more samples of observations, however, the distinction between one-sided and two-sided tests is not maintained. Consequently we need a procedure that specifically allows for ordered alternatives in the k-sample case.

Terpstra (T36) and Jonckheere (T37) have proposed a test that can be used when an ordered alternative is appropriate.

Assumptions

    A. The data for analysis consist of k random samples of sizes $n_1, n_2, \ldots, n_k$ from populations $1, 2, \ldots, k$, with unknown medians $M_1, M_2, \ldots, M_k$, respectively.

    B. The observations are independent, both within and among samples.

    C. The variable of interest is continuous.

    D. The measurement scale is at least ordinal.

    E. The sampled populations are identical except for a possible difference in location parameters.

Hypotheses

H_0: $M_1 = M_2 = \cdots = M_k$

H_1: $M_1 \leq M_2 \leq \cdots \leq M_k$,   with at least one strict inequality.

If the expected direction of inequality is not as specified in this alternative hypothesis, relabel and reorder the samples to achieve conformity.

Test Statistic

The test statistic is

$$J = \sum_{i<j} U_{ij} \tag{6.5}$$

where U_{ij} is the number of pairs of observations (a, b) for which X_{ia} is less than X_{jb}. In other words, we compare observations in all pairs of samples. We compare each observation in the first sample in the pair of samples with each observation in the second sample in the pair, and if the observation from the first sample is less than the observation in the second sample, we record a score of 1. We record a score of 0 if the observation from the first sample is greater than the observation from the second sample.

Decision Rule

Reject H_0 at the α level of significance if the computed J is greater than or equal to the critical value of J for α, k, and $n_1, n_2, \ldots, n_k$ given in Table A.13. The critical values for J given in Table A.13 for $k = 3$ are tabulated for sample sizes of $n_1 \leq n_2 \leq n_3$. Because the distribution of J has certain symmetry properties, however, we may obtain critical values for configurations not in that order by rearranging the three sample sizes so that they are in order of increasing size before we enter the table. For example, if we wish critical values for sample sizes $n_1 = 5$, $n_2 = 7$, $n_3 = 3$, we enter Table A.13 at $n_1 = 3$, $n_2 = 5$, $n_3 = 7$.

Ties   In computing U_{ij}, record a score of 1/2 for each case where $X_{ia} = X_{jb}$. In other words, each time a tie is encountered when comparing observations, record a score of 1/2 rather than 1.

Example 6.3   Nappi (E12) investigated the changes occurring in the haemocytes of larvae of *Drosophila algonquin* during parasitization by the hymenopterous parasite (parasitoid) *Pseudeucoila bochei*. Twenty-seven hours after parasitization of *Drosophila algonquin* larvae, differential counts (%) of plasmatocytes were made on three groups: host larvae in which reaction was successful (S), those in which the reaction was unsuccessful (U), and those in which there was no visible host reaction (N). The results are shown in Table 6.16. We wish to test the null hypothesis of no difference among the three groups against the alternative that the differential counts of plasmatocytes (%) decrease in the three groups from group N to group S.

Hypotheses

$$H_0\colon\ M_S = M_U = M_N$$

$$H_1\colon\ M_S \leq M_U \leq M_N, \qquad \text{with at least one strict inequality}$$

TABLE 6.16 **Differential plasmatocyte counts, percent from larvae of**
Drosophila algonquin 27 hours after parasitization by
Pseudeucoila bochei (host age 91 hours when parasitized)

Successful host reactions (S)	Unsuccessful host reactions (U)	No visible host reactions (N)
54.0	79.8	98.6
67.0	82.0	99.5
47.2	88.8	95.8
71.1	79.6	93.3
62.7	85.7	98.9
44.8	81.7	91.1
67.4	88.5	94.5
80.2		

Source: A. J. Nappi, "Cellular Immune Reactions of Larvae of *Drosophila algonquin*," *Parasitology,* 70 (1975), 189–194: published by Cambridge University Press.

Test Statistic

We first compare observations in group S with those in group U to obtain $U_{SU} = 54$. A comparison of observations in group S with those in group N yields $U_{SN} = 56$. Finally, when we compare observations in group U with those in group N, we obtain $U_{UN} = 49$. Thus we have

$$J = 54 + 56 + 49 = 159$$

Decision

When we compare our computed value of J with the critical values for $k = 3$ and sample sizes 7, 7, 8 in Table A.13, we find that the probability of obtaining a value of J as large as 159 when H_0 is true is less than 0.00494 (since $159 > 123$). Consequently we reject H_0 in favor of the altered alternative. The P value is less than 0.00494.

Large-Sample Approximation For large sample sizes, J is approximately normally distributed with mean 0 and variance 1. When we use the normal approximation, we compute

$$z = \frac{J - [(N^2 - \Sigma_{i=1}^k n_i^2)/4]}{\sqrt{[N^2(2N + 3) - \Sigma_{i=1}^k n_i^2(2n_i + 3)]/72}} \tag{6.6}$$

and compare it for significance with tabulated values of the standard normal distribution given in Table A.2.

To illustrate the use of the normal approximation, consider the preceding example where

$$J = 159$$

$$N = 8 + 7 + 7 = 22$$

$$\sum_{i=1}^{k} n_i^2 = 8^2 + 7^2 + 7^2 = 162$$

$$\sum_{i=1}^{k} n_i^2(2n_i + 3) = 8^2(2 \cdot 8 + 3) + 7^2(2 \cdot 7 + 3) + 7^2(2 \cdot 7 + 3) = 2882$$

Substituting these values into Equation 6.6 yields

$$z = \frac{159 - [(22^2 - 162)/4]}{\sqrt{[22^2(2 \cdot 22 + 3) - 2882]/72}} = 4.73$$

The P value for $z = 4.73$ is less than 0.001, and we reject H_0. We see that the large-sample approximation leads to the same conclusion as the exact test.

Power-Efficiency   The power of the Jonckheere–Terpstra test for ordered alternatives has been considered by Odeh (T38). Puri (T39) discusses the test's efficiency. Potter and Sturm (T40) show how the maximum attainable power of the test may be computed for some simple alternatives. They found that the power of the test is bounded significantly away from one under certain shift alternatives and sample sizes.

FURTHER READING

An alternative test procedure for ordered alternatives has been proposed by Chacko (T41) and extended by Shorack (T42). Lehmann (T43) gives a numerical example illustrating this procedure and points out that in certain situations the test proposed by Chacko seems preferable to the Jonckheere–Terpstra test.

Roth and Daniel (T44) present a table of critical values for use with Chacko's test. The tables will accommodate from 3 to 15 samples for nominal significance levels of 0.005, 0.01, 0.025, 0.05, 0.10, and 0.20. Parsons (T45) computes the exact distribution of the statistic as extended by Shorack (T42) for three small populations with small sample sizes.

Further discussion of the Jonckheere–Terpstra test can be found in the articles by Odeh (T46) and Tryon and Hettmansperger (T47). Cuzick (T48) has developed an ordered alternative test that is an extension of the Wilcoxon rank-sum test. The author illustrates its application with two examples from cancer research. Also of interest are the papers by Hirotsu (T49) and Rothe (T50). A more complete and general discussion of ordered alternatives is to be found in the book by Barlow et al. (T51).

EXERCISES

6.10 Wohl et al. (E13) studied the physiological properties of the lungs in patients who received therapeutic bilateral pulmonary irradiation in early childhood. Their subjects consisted of three groups of children. Group 1 was composed of six children who received a single course of bilateral pulmonary irradiation. Group 2 contained six children who had received additional pulmonary radiotherapy or thoracic surgery or both. Group 3 consisted of eight children who received no irradiation directed primarily to the lung. Vital capacity values, expressed as percentages of predicted values based on standing height, for subjects in the three groups are shown in Table 6.17. Test the null hypothesis of no difference among the three populations represented against the ordered alternative that the median vital capacity values are in the order: group 2 $\leq$ group 1 $\leq$ group 3. Determine the P value.

TABLE 6.17 **Vital capacity values, expressed as percentage of predicted values based on standing height, in three groups of subjects**

Group 1: Bilateral pulmonary irradiation

71	57	85	67	66	79

Group 2: Additional pulmonary irradiation or thoracic surgery

76	94	61	36	42	49

Group 3: No pulmonary irradiation

80	104	81	90	93	85	101	83

Source: Mary Ellen B. Wohl, N. Thornton Griscom, Demetrius G. Traggis, and Norman Jaffee, ''Effects of Therapeutic Irradiation Delivered in Early Childhood upon Subsequent Lung Function,'' *Pediatrics*, 55 (1975), 507–516.

6.11 Davis (E14) investigated the performance of hard-of-hearing school children on a task involving knowledge of 50 basic concepts considered necessary for satisfactory academic achievement during kindergarten, first grade, and second grade. The raw scores made by 24 hard-of-hearing school children (kindergarten to third grade) on the Boehm Test of Basic Concepts (E15) are shown in Table 6.18 by age. Do these data provide sufficient evidence to indicate that, on the average, scores tend to increase with age? Find the P value.

TABLE 6.18 **Raw scores for 24 hearing-impaired children by age groups**

Age 6 Raw scores	17	20	24	34	34	38						
Age 7 Raw scores	23	25	27	34	38	47						
Age 8 Raw scores	22	23	26	32	34	34	36	38	38	42	48	50

Source: Julia Davis, ''Performance of Young Hearing-Impaired Children on a Test of Basic Concepts,'' *J. Speech Hear. Res.*, 17 (1974), 342–351.

6.4

MULTIPLE COMPARISONS

When a hypothesis-testing procedure such as the Kruskal–Wallis test leads us to reject the null hypothesis and thus to conclude that not all sampled populations are identical, we naturally question which populations are different from which others.

To be more specific, let us again consider Example 6.2, in which the application of the Kruskal–Wallis test led us to conclude that the mean cortisol levels were not the same in three types of parturient subjects. In this, as in most similar instances, it is probably of greater interest and importance to be able to say more about the differences. In this case, for example, we would like to know whether the medians M_1, M_2, and M_3 are all different from each other or if the difference is between M_1 and M_2 only, between M_1 and M_3 only, or between M_2 and M_3 only.

The logical approach to answering this question might appear to be to use some procedure such as the Mann–Whitney test, discussed in Chapter 3, to test for a significant difference between each of all the possible pairs of samples. There is, however, a problem inherent in following such a course: Testing all possible pairs of means in the usual way affects the probability of rejecting a true null hypothesis.

If we carry out C independent comparisons between pairs of samples, each at a stated significance level of α, the probability of declaring at least one difference significant as a result of chance is equal to $1 - (1 - \alpha)^C$, which is approximately equal to $C\alpha$ for small values of α. Consequently, in the typical situation, the probability of finding at least one counterfeit significant outcome increases as the number of independent comparisons increases. Finding the corresponding probability in the case of nonindependent comparisons is more complicated.

One way to circumvent this problem is to use a *multiple-comparison* procedure that incorporates an adjustment for the problem regarding the level of significance. Several such procedures are available. The one considered here is suggested by Dunn (T52); it is appropriate for use following a Kruskal–Wallis test.

When we apply this multiple-comparison procedure, we use what is known as an *experimentwise error rate*. The experimentwise error rate, which represents a conservative approach in making multiple comparisons, holds the probability of making only correct decisions at $1 - \alpha$ when the null hypothesis of no difference among populations is true. This approach protects well against error when H_0 is true, but it makes more difficult the task of detecting differences that are significant when the null hypothesis is false. The justification for using an experimentwise error rate is discussed by Kurtz et al. (T53).

To use the multiple-comparison procedure, we first obtain the mean of the ranks for each sample, and let $\bar{R}_i$ be the mean of the ranks of the ith sample and $\bar{R}_j$ be the mean of the ranks of the jth sample. We next select an experimentwise error rate of α, which we think of as an overall level of significance. Our choice of α is determined in part by k, the number of samples involved, and is larger for larger k. If we have k samples, there will be a total of $k(k - 1)/2$ pairs of samples that can be compared a

pair at a time. If we have 5 samples, for example, there will be a total of $5(4)/2 = 10$ pairwise comparisons that we can make. When making multiple comparisons with an experimentwise error rate, we usually select a value of α larger than those customarily encountered in single-comparison inference procedures—for example, 0.15, 0.20, or perhaps 0.25, depending on the size of k.

The next step is to find in Table A.2 the value of z that has $\alpha/k(k - 1)$ area to its right. Finally, we form the inequality

$$|\bar{R}_i - \bar{R}_j| \le z_{(1 - [\alpha/k(k-1)])}\sqrt{\frac{N(N + 1)}{12}\left(\frac{1}{n_i} + \frac{1}{n_j}\right)} \tag{6.7}$$

where N is the number of observations in all samples combined. The probability that Inequality 6.7 holds for all pairs of means, when H_0 is true, is at least $1 - \alpha$.

If the k samples are all of the same size, Inequality 6.7 reduces to

$$|\bar{R}_i - \bar{R}_j| \le z_{(1 - [\alpha/k(k-1)])}\sqrt{k(N + 1)/6} \tag{6.8}$$

Any difference $|\bar{R}_i - \bar{R}_j|$ that is larger than the right-hand side of Inequality 6.7 (or Inequality 6.8 if applicable) is declared significant at the α level.

Thus the procedure allows us to forget about the direction of the differences between mean ranks when performing the calculations. The direction of the differences should, of course, be taken into account in our interpretation of the results. Also, in following the procedure, we in effect cut off the same amount of area in each tail of the standard normal distribution. It is for that reason that we divide α by $k(k - 1)$ rather than by $k(k - 1)/2$, the number of possible pairs of differences, when analyzing the data from k samples.

Let us now illustrate the procedure with an example.

Example 6.4 We refer again to the data of Example 6.2, which we analyzed by using the Kruskal–Wallis test. A computed value of the test statistic of $H = 9.232$ allowed us to reject, at the 0.01 level of significance, the null hypothesis that three populations were identical. As a result, we concluded that median cortisol levels are not the same for all three types of patients studied.

To make all possible comparisons in order to locate just where the differences occur, let us choose an error rate of $\alpha = 0.15$. There are $k = 3$ samples involved, so that there will be $3(2)/2 = 3$ comparisons to make, and $\alpha/k(k - 1) = 0.15/3(2) = 0.025$. From Table A.2, we find the z with 0.025 of the area to its right to be 1.96.

The means of the ranks for the three samples are $\bar{R}_1 = 69/10 = 6.9$, $\bar{R}_2 = 90/6 = 15$, and $\bar{R}_3 = 94/6 = 15.67$.

To compare groups I and II, we evaluate the right-hand side of Inequality 6.7 to obtain

$$1.96\sqrt{\frac{22(22 + 1)}{12}\left(\frac{1}{10} + \frac{1}{6}\right)} = 6.57$$

Since $|6.9 - 15| = 8.1 > 6.57$, this comparison is significant. From the data, we conclude that subjects for elective Caesarean section before the onset of labor tend

to have lower cortisol levels than women experiencing induced labor at emergency Caesarean section. The sign of the difference indicates its direction.

To compare groups I and III, we again compute

$$1.96 \sqrt{\frac{22(22 + 1)}{12}\left(\frac{1}{10} + \frac{1}{6}\right)} = 6.57$$

Since $|6.9 - 15.67| = 8.77 > 6.57$, we conclude that the women in population I also tend to have lower cortisol values than the women in population III.

Finally, we compare groups II and III by first computing

$$1.96 \sqrt{\frac{22(22 + 1)}{12}\left(\frac{1}{6} + \frac{1}{6}\right)} = 7.35$$

Since $|15 - 15.67| = 0.67$ is less than 7.35, we cannot conclude that the populations represented by samples II and III differ with respect to their median cortisol levels.

Ties   If there are extensive ties in the data, we can adjust Inequalities 6.7 and 6.8 to ensure a conservative result. When we adjust for ties, the appropriate inequality for unequal sample sizes is

$$|\bar{R}_i - \bar{R}_j| \le z \sqrt{\frac{[N(N^2 - 1) - (\Sigma t^3 - \Sigma t)]\left[\dfrac{1}{n_i} + \dfrac{1}{n_j}\right]}{12(N - 1)}} \qquad (6.9)$$

The appropriate inequality for equal sample sizes is

$$|\bar{R}_i - \bar{R}_j| \le z \sqrt{\frac{k[N(N^2 - 1) - (\Sigma t^3 - \Sigma t)]}{6N(N - 1)}} \qquad (6.10)$$

In these inequalities, t is the number of values in the combined sample that are tied at a given rank. The adjustment for ties usually has a negligible effect on the results.

COMPARING ALL TREATMENTS WITH A CONTROL

Sometimes the research situation is such that one of the k treatments is a control condition. When this is the case, the investigator is frequently interested in comparing each treatment with the control condition without regard to whether the overall test for a treatment effect is significant, and irrespective of any potential significant differences between other pairs of treatments. When interest focuses on comparing all treatments with a control condition, there will be $k - 1$ comparisons to be made. The procedure is the same as described for the case in which all possible pairs of treatments are compared except for the method of obtaining z for Expressions 6.7 through 6.10. In order to in effect cut off the appropriate equal amount of area in each tail of the standard normal distribution, we divide α by $2(k - 1)$. The following example illustrates the procedure when we wish to compare all treatments with a control.

Example 6.5

A fertilizer manufacturer conducted an experiment to compare the effects of four types of fertilizer on the yield of a certain grain. Homogeneous equal-size experimental plots of soil were made available for the experiment. They were randomly assigned to receive one of the five fertilizers, and plots receiving no fertilizer served as controls. At harvest time nine plots were randomly selected from those previously assigned to each of the fertilizers and the control plots. The yields (in coded form) for each plot are shown in Table 6.19.

TABLE 6.19   **Grain yield data for Example 6.5**

Fertilizer				
1 None (0)	2 A	3 B	4 C	5 D
58	68	96	101	124
29	67	90	110	114
37	69	90	90	111
40	58	92	103	113
44	62	99	100	114
37	48	86	91	102
49	62	79	100	114
49	76	96	114	112
38	66	75	94	103

We wish to know which fertilizers are superior to no fertilizer. A significance level of 0.20 will be used. The ranks, rank sums, and mean ranks of the yields for each treatment are shown in Table 6.20.

TABLE 6.20   **Ranks, rank totals, and mean ranks for data in Table 6.19**

	Fertilizer				
	1 None (0)	2 A	3 B	4 C	5 D
	10.5	16	28.5	33	45
	1	15	23	37	42.5
	2.5	17	23	23	38
	5	10.5	26	35.5	40
	6	12.5	30	31.5	42.5
	2.5	7	21	25	34
	8.5	12.5	20	31.5	42.5
	8.5	19	28.5	42.5	39
	4	14	18	27	35.5
Total (R)	48.5	123.5	218.0	286.0	359.0
Mean ($\bar{R}$)	5.39	13.72	24.22	31.78	39.89

Since the data contain several ties, and since the samples are all the same size, we will use Expression 6.10. The adjustment for ties is obtained as follows:

Rank position where tie occurred	t	t^3
2	2	8
8	2	8
10	2	8
12	2	8
22	3	27
28	2	8
31	2	8
35	2	8
41	4	64
Total	21	147

We now compute $\Sigma t^3 - \Sigma t = 147 - 21 = 126$. Since there are five treatments (four fertilizers and no fertilizer) under study, we have four comparisons to make. To find the appropriate z value for $\alpha = 0.20$, we compute $0.20/2(4) = 0.025$. From Table A.2 we obtain a z of 1.96. We now use Expression 6.10 to compute

$$1.96 \sqrt{\frac{5[45(45^2 - 1) - 126)]}{6(45)(45 - 1)}} = 12.127$$

Table 6.21 shows the absolute values of the differences between $\bar{R}_0$, the mean rank of the yields of the plots to which no fertilizer was applied, and the $\bar{R}_j$'s, the mean ranks of the plots receiving fertilizers $j = A, B, C, D$.

TABLE 6.21   **Comparison of yields of plots receiving fertilizer to yields of plots receiving no fertilizer**

Fertilizer	$|\bar{R}_0 - \bar{R}_j|$	Better than no fertilizer?
$A(2)$	$|5.39 - 13.72| = 8.33$	No
$B(3)$	$|5.39 - 24.22| = 18.83$	Yes
$C(4)$	$|5.39 - 31.78| = 26.39$	Yes
$D(5)$	$|5.39 - 39.89| = 34.50$	Yes

Since 8.33 is less than 12.127, we cannot conclude that fertilizer A is better than no fertilizer. Since 18.83, 26.39, and 34.50 are all greater than 12.127, we conclude that fertilizers B, C, and D will all result in higher yields than if no fertilizer at all were used.

CONTRAST ESTIMATION

Researchers are frequently interested not only in testing hypotheses about the differences between pairs of population parameters such as means and medians, but in estimating the differences as well. In Example 6.5, for instance, we might wish to estimate the difference between the median yield when fertilizer A is applied and the median yield when fertilizer D is applied. Such a comparison is also referred to as a *contrast*. In general, a contrast among k medians is of the form $w_1 M_1 + w_2 M_2 + \cdots + w_k M_k$, with the restriction that $w_1 + w_2 + \cdots + w_k = 0$. If, in Example 6.5 (ignoring the no fertilizer treatment), we wished to estimate the difference between the median effects of fertilizer $A = 1$ and fertilizer $D = 4$, we could write the contrast among all the medians, including the median for fertilizer $B = 2$ and fertilizer $C = 3$, as $(-1)M_1 + (0)M_2 + (0)M_3 + (1)M_4$. In this contrast, $w_1 = -1, w_2 = 0, w_3 = 0$, and $w_4 = 1$, and $-1 + 0 + 0 + 1 = 0$. The contrast $(-1)M_1 + (0)M_2 + (0)M_3 + (1)M_4$ can be written as $M_1 - M_2$. In this form it is clear that we wish to compare the median effects of fertilizers A and D. To estimate a simple contrast of this type, we use a statistic proposed by Spjøtvoll (T54) which is a modification of an estimator developed by Lehmann (T55). The procedure consists of the following steps:

1. We calculate the $k(k - 1)/2$ medians of the form

$$Z_{hj} = \text{median}\{x_{rh} - x_{sj}, r = 1, \ldots, n_h, s = 1, \ldots, n_j\}, \qquad h < j$$

(6.11)

We call Z_{hj} the unadjusted estimator of $M_h - M_j$, the difference between the median of population h and the median of population j. It is necessary to calculate the $k(k - 1)/2$ estimators only for $h < j$, rather than also those for $h > j$, because $Z_{hj} = -Z_{jh}$. For example, Z_{12} is the median of the $n_1 n_2$ differences $x_{r1} - x_{s2}$ obtained from the measurements in samples 1 and 2. Suppose sample 1 consists of measurements 4, 5, and 6, and sample 2 consists of measurements 1 and 2. The $3 \times 2 = 6$ differences $x_{r1} - x_{s2}$ are $(4 - 1) = 3, (4 - 2) = 2, (5 - 1) = 4, (5 - 2) = 3, (6 - 1) = 5$, and $(6 - 2) = 4$. The median of these differences is $Z_{12} = 3.5$. Thus $Z_{21} = -3.5$, since the differences from which it is calculated are negative.

2. We obtain the weighted averages

$$\bar{m}_h = \frac{\sum_{j=1}^{k} n_j Z_{hj}}{\sum_{j=1}^{k} n_j}, \qquad h = 1, \ldots, k$$

(6.12)

Note that application of Equation 6.12 yields $Z_{hh} = 0$.

When the sample sizes are all equal, Equation 6.12 reduces to

$$\bar{m}_h = \frac{\sum\limits_{j=1}^{k} Z_{hj}}{k} \qquad (6.13)$$

3. The weighted and adjusted estimator of the difference $M_h - M_j$ is

$$\bar{m}_h - \bar{m}_j \qquad (6.14)$$

We now illustrate the procedure by means of an example.

Example 6.6   An assembly-line operation can be performed under three different kinds of setup. In an effort to gain insight into the most efficient way of performing the operation, a time and motion specialist had an employee perform the task under each of the three setups. The times required for the performance are shown in Table 6.22.

TABLE 6.22   **Times in seconds required to perform an assembly-line task under three different setups**

Setup		
1	2	3
18.2	17.1	32.9
16.9	19.4	25.8
17.6	20.4	31.9
	17.8	

We wish to estimate $M_1 - M_2$, the difference between the long-run median times required to perform the task under setups 1 and 2. We proceed as follows:

1. We first calculate the following medians:

$$
\begin{aligned}
Z_{12} = \text{median} \{&(18.2 - 17.1) = 1.1, (18.2 - 19.4) = -1.2, \\
&(18.2 - 20.4) = -2.2, (18.2 - 17.8) = 0.4, \\
&(16.9 - 17.1) = -0.2, (16.9 - 19.4) = -2.5, \\
&(16.9 - 20.4) = -3.5, (16.9 - 17.8) = -0.9, \\
&(17.6 - 17.1) = 0.5, (17.6 - 19.4) = -1.8, \\
&(17.6 - 20.4) = -2.8, (17.6 - 17.8) = -0.2\} \\
= &-1.05
\end{aligned}
$$

$$Z_{13} = \text{median} \{(18.2 - 32.9) = -14.7, (18.2 - 25.8) = -7.6,$$
$$(18.2 - 31.9) = -13.7, (16.9 - 32.9) = -16.0,$$
$$(16.9 - 25.8) = -8.9, (16.9 - 31.9) = -15.0,$$
$$(17.6 - 32.9) = -15.3, (17.6 - 25.8) = -8.2,$$
$$(17.6 - 31.9) = -14.3\}$$

$$= -14.3$$

$$Z_{23} = \text{median} \{(17.1 - 32.9) = -15.8, (17.1 - 25.8) = -8.7,$$
$$(17.1 - 31.9) = -14.8, (19.4 - 32.9) = -13.5,$$
$$(19.4 - 25.8) = -6.4, (19.4 - 31.9) = -12.5,$$
$$(20.4 - 32.9) = -12.5, (20.4 - 25.8) = -5.4,$$
$$(20.4 - 31.9) = -11.5, (17.8 - 32.9) = -15.1,$$
$$(17.8 - 25.8) = -8.0, (17.8 - 31.9) = -14.1\}$$

$$= -12.5$$

$$Z_{21} = 1.05$$

$$Z_{31} = 14.3$$

$$Z_{32} = 12.5$$

$$Z_{11} = 0$$

$$Z_{22} = 0$$

$$Z_{33} = 0$$

2. We now calculate the weighted averages

$$\bar{m}_1 = \frac{3(0) + 4(-1.05) + 3(-14.3)}{10}$$

$$= -4.71$$

$$\bar{m}_2 = \frac{3(1.05) + 4(0) + 3(-12.5)}{10}$$

$$= -3.435$$

3. Our estimate of $M_1 - M_2$ is

$$\bar{m}_1 - \bar{m}_2 = -4.71 - (-3.435) = -1.275$$

It is possible to formulate and estimate more complex contrasts than the simple pairwise ones discussed here. For example, in some situation, we might wish to estimate the contrast

$$\frac{M_1 + M_2}{2} - M_3$$

For further information on such contrasts, the reader is referred to textbooks on analysis of variance and the relevant references cited in the subsection that follows.

FURTHER READING

Nonparametric multiple-comparison procedures are discussed by Anscombe (T56), Dwass (T57), Gabriel (T58), Marascuilo and McSweeney (T59), McDonald and Thompson (T60), Miller (T61), Sherman (T62), and Steel (T63, T64). Rhyne and Steel (T65) discuss a multiple-comparisons sign test for treatment versus control.

Campbell and Skillings (T66) discuss nonparametric stepwise multiple-comparison procedures. The authors propose procedures and compare them among each other as well as with nonstepwise procedures on the basis of type I error levels and comparisonwise power. The authors conclude that the stepwise procedures control type I error levels and that they have superior pairwise power when compared to the commonly used nonstepwise procedures.

Planned and post hoc comparisons for tests of homogeneity in cases where the dependent variable is categorical and ordered are discussed in a paper by Marascuilo and Dagenais (T67).

Skillings (T68) addresses the question of whether a joint ranking or a separate ranking multiple-comparison procedure performs better as a test and as a device for treatment separation. He concludes that the joint ranking procedure does slightly better as a test; for treatment separation, the decision depends on the situation.

Levy (T69) introduces Tukey-type multiple comparison procedures for use in both the one-way and two-way analysis of variance contexts. By means of Monte Carlo studies, Wike and Church (T70) determined the type I error rates for Levy's tests. Also using Monte Carlo techniques, Wike and Church (T71, T72) compared four nonparametric multiple-comparison tests to determine their type I error rates when overall Kruskal–Wallis tests were and were not significant.

Zwick and Marascuilo (T73) address a number of issues related to the use of nonparametric multiple-comparison procedures and make extensive recommendations to the researcher contemplating the use of these techniques.

Hurwitz (T74) presents programmable calculator programs for multiple comparisons and nonparametric tests. A bibliography by Daniel (T75) includes references to publications concerned with nonparametric multiple-comparison procedures.

For a discussion of the experimentwise and other error rates, see the articles by Balaam (T76), Federer (T77), Petrinovich and Hardyck (T78), Ryan (T79), Steel (T80), and Wilson (T81), and the book by Kirk (T82).

Treatment versus control multiple comparisons is the subject of a paper by Fligner (T83), who compares the use of the joint ranking procedure with the pairwise ranking method. He concludes that although joint ranking has the advantage of simplicity and possibly computational ease, the pairwise ranking of each treatment against the control proposed by Steel (T84) is superior in other respects.

For both the one-way and two-way classification schemes, Fligner and Wolfe (T85) propose an alternative to the multiple-comparison approach for assessing the merits of several treatments over a control condition. They introduce a single distribution-free test of the null hypothesis of no differences among the treatments and the control against the one-sided alternative that at least one treatment yields a greater or smaller response than the control.

The properties of the contrast estimator presented in this section are discussed by Lehmann (T55) and Spjøtvoll (T54). For more on contrast estimation, see the papers by Bhuchongkul and Puri (T86), Lehmann (T87), and Sen (T88).

EXERCISES

6.12 Make all possible comparisons in Exercise 6.6. Let $\alpha = 0.15$.

6.13 Make all possible comparisons in Exercise 6.7. Let $\alpha = 0.20$.

6.14 Make all possible comparisons in Exercise 6.8. Let $\alpha = 0.15$.

6.5

COMPUTER PROGRAMS

Individual computer programs have been written and reported in the journal literature for most of the procedures discussed in this chapter. Smith et al. (T89), Roberge (T90, T91, T92), Rock (T93), and Theodorsson–Norheim (T94) have written programs for the Kruskal–Wallis test. The program by Theodorsson–Norheim also includes subroutines for performing multiple comparisons. Programs written by Roberge (T91, T95) also handle multiple-comparison procedures. A FORTRAN program by Thakur (T96) performs the Jonckheere–Terpstra test that can be implemented on a microcomputer. Blaker (T97) presents a program written for the IBM-PC that allows for the application of the Kruskal–Wallis test and Dunn's procedure for comparing several treatments to a control.

A generalized nonparametric analysis-of-variance program written by Roberge and Roberge (T98) computes various nonparametric analysis-of-variance statistics, compares the value of a given statistic with the chi-square values required for significance at the 0.05 and 0.01 significance levels, performs a priori or post hoc trend analyses, and carries out post hoc multiple comparisons.

Borys and Corrigan (T99) present a BASIC program that computes post hoc multiple comparisons for a variety of multivariate nonparametric tests, including the chi-square test of homogeneity, the Kruskal–Wallis analysis-of-variance test, and two procedures covered in Chapter 7, Cochran's Q test and the Friedman test.

The Kruskal–Wallis test is available in most microcomputer software packages, including *BMDPC, MINITAB, SCA, SPSS/PC, STATA, STATISTIX*, and *STATPRO*.

REVIEW EXERCISES

6.15 A study of fecal excretion of rose bengal I^{131} (RB I^{131}) in the diagnosis of obstructive jaundice in infancy was conducted by Maksoud et al. (E16). Subjects consisted of 5 normal controls, 6 subjects with nonatresic jaundice, and 16 with atresic jaundice. Table 6.23 shows the 48-hour fecal excretion of RB I^{131} expressed as the percentage of the administered dose. We wish to know whether these data provide sufficient evidence to indicate a difference among the three groups. Use the Kruskal–Wallis test to determine the P value, and make all possible comparisons at the 0.15 level of significance.

TABLE 6.23 **Forty-eight-hour percentage fecal excretion of rose bengal I^{131} in 27 newborns and infants**

Controls	37	22.1	44.4	71.5	70			
Neonatal hepatitis	9.5	22	10.5	11.3	16.9	17.3*		
Biliary	2.6	4	4.7	4.7	0.9	6.6	1.6	3.3
atresia	3.2	4.1	0.6	2.9	3.6	2.3	3.7	3*

Source: João Gilberto Maksoud, Anneliese Fischer Thom, Julio Kieffer, and Virgilio A. Carvalho Pinto, "Fecal Excretion of Rose Bengal I^{131} in the Diagnosis of Obstructive Jaundice in Infancy with Special Reference to Biliary Atresia," *Pediatrics*, 48 (1971), 966–969.

* Second determinations on these subjects reported by the authors are not included here.

6.16 To investigate the influence of light stimulation on the multiple-unit pineal activity (MUA) of the quail, Herbuté and Baylé (E17) studied spontaneous multiple-unit discharges in the pineal glands of three groups of quail. Group I consisted of intact birds in which the tip of the recording electrodes was 30 μm in diameter, group II consisted of intact birds in which the tip of the recording electrodes was 15 μm in diameter, and group III consisted of birds with a bilateral section of the optic nerves performed three weeks prior to pineal electrophysiological exploration. The tip of the recording electrodes in group III was 15 μm in diameter. Table 6.24 shows the basal MUA (spikes/10 seconds) reported for the three

TABLE 6.24 **Basal MUA (spikes/10 sec) in the pineal body in three groups of quail**

Group I	82	127	53	89	81		
Group II	32	24	16	22	30	27	30
Group III	117	63	72	96	117	45	

Source: S. Herbuté and J. D. Baylé, "Multiple-Unit Activity in the Pineal Gland of the Japanese Quail: Spontaneous Firing and Responses to Photic Stimulations," *Neuroendocrinology*, 16 (1974), 52–64; used by permission of S. Karger AG, Basel.

groups. Do these data provide sufficient evidence to indicate a difference among the three groups? Use the Kruskal–Wallis test to determine the P value, and make all possible comparisons at the 0.20 level of significance.

6.17 The data in Table 6.25 were reported by Kaklamanis et al. (E18), who conducted a study to clarify the role of T lymphocytes in regulating the response of the host to HBAg (hepatitis B antigen). They studied three groups of subjects. Group A was composed of five HBAg carriers with no clinical evidence of liver disease, group B was made up of eight subjects who had HBAb (hepatitis B antibody) and no history of hepatitis, and group C consisted of six healthy controls with neither HBAg nor HBAb. Table 6.25 shows the response to phytohaemagglutinin (PHA) of T lymphocytes from the 19 subjects. Do these data provide sufficient evidence to indicate a difference among the three populations represented in the study? Use the Kruskal–Wallis test to determine the P value, and make all possible comparisons at the 0.15 level of significance.

TABLE 6.25   **Incorporation of ^{14}C-thymidine by PHA-stimulated lymphocytes of peripheral blood, according to presence of HB antigen (Ag +, Ab −) and antibodies (Ag −, Ab +)**

Carriers (Ag+, Ab−)	4163	9420	6428	6322	5919			
HB-antibody positive (Ag−, Ab+)	7124	10,698	13,722	7435	8600	10,443	10,094	8720
Controls (Ag−, Ab−)	16,864	16,767	17,427	16,733	17,972	15,720		

Source: E. Kaklamanis, D. Trichopoulos, G. Papaevangelou, M. Drouga, and D. Karalis, "T Lymphocytes in HBAg Carriers and Responders," *Lancet*, Vol. I, No. 7908 (March 22, 1975), 689.

6.18 Svenningsen (E19) reported the results of renal acid-base titration studies performed in 24 infants randomly selected from a population of 516 newborn infants in whom metabolic acidosis was being studied. The infants in the study under discussion were divided into three groups as follows:

> *Group I* consisted of 6 infants with normal acid-base balance in the neonatal and postneonatal period.
> *Group II* (called group IIa in the study) contained 10 preterm infants with normal acid-base values.
> *Group III* (called group IIb in the study) was made up of 8 preterm infants whose blood acid-base values indicated a condition of so-called late metabolic acidosis.

A certain chemical analysis of the urine of these infants yielded the values shown in Table 6.26. Test the null hypothesis of no difference in the populations represented against the ordered alternative that the chemical values tend to decrease from group III to group I.

6.19 A psychology research team administered a test designed to measure neuroticism in four groups of subjects who differed on the basis of their smoking habits. The results are shown in Table 6.27. Do these data suggest a difference in neuroticism level among the four groups? Use the Kruskal–Wallis test.

TABLE 6.26   **Results of a certain chemical analysis of the urine of 24 infants**

Group I (term infants)	4.5	3.9	5.0	4.8	4.1	4.6			
Group II (preterm infants)	4.1	3.9	3.2	4.6	5.1	4.9	5.0	4.3	5.2	5.3
Group III (preterm infants with acidosis at 1–3 weeks of age)	7.3	8.4	6.9	7.3	8.2	6.2	8.2	7.9		

Source: N. W. Svenningsen, "Renal Acid-Base Titration Studies in Infants with and without Metabolic Acidosis in the Postneonatal Period," *Pediatric Res.*, 8 (1974), 659–672.

TABLE 6.27   **Neuroticism scores of subjects classified according to smoking habits**

Nonsmokers	7.6	7.7	7.5	7.8	7.6	7.3	7.1	8.0	7.5	8.0
Light smokers	8.9	8.2	8.1	8.0	8.6	8.6	8.6	8.4		
Medium smokers	8.0	8.8	8.7	8.6	9.0	8.8	8.5			
Heavy smokers	9.9	9.1	9.8	9.8	9.9	9.6	9.2	9.8		

6.20   A test designed to measure a person's level of mental health was administered to three groups of subjects. The results are shown in Table 6.28. Can one conclude from these data that the status of mental health is different among these three groups? Use the extension of the median test.

TABLE 6.28   **Scores made on a level of mental health test by three groups of subjects**

Unwed pregnant girls	30	40	70	75	15	20	20	30	80	25	25	25
	75	30	90	30	60	85	100	90	75	70	90	15
	15											
Married pregnant women	15	17	25	30	35	45	50	50	55	60	60	60
	100	95	80	75	70	65	60	80	30	30	25	50
	60	65	65	65	70	70	75	75	90	35	75	30
	25	45	65	60	20	75	100	80	90	90		
Unmarried girls, not pregnant	20	15	60	45	15	35	55	20	70	65	30	100
	90	85	10	55	25	90	80	75	25	30	65	80
	55	100	15	100	50	90	30	40	15	40	50	65
	70											

6.21   Table 6.29 shows the verbal IQ scores of samples of first-grade children residing in four types of community. Use the extension of the median test to determine whether one should conclude that the populations differ with respect to median verbal IQ.

6.22   Each of three groups of 12-year-old children—10 normal, 6 moderately retarded, and 6 severely retarded—were given 100 tasks to perform. Table 6.30 shows the proportion of tasks correctly completed by each child. Do these data provide sufficient evidence to indicate a difference among the three groups with respect to the variable of interest? Determine the *P* value.

TABLE 6.29 **Verbal IQ scores of first-grade children residing in four types of community**

Very	21	22	22	20	25	28	22	23	20	44	27	30
isolated	30	22	21	25	21	23	26	23	23	28	37	
Moderately	33	27	29	29	26	29	26	35	33	34	36	40
isolated	45	34	26	34	33	33	32	42	28	34	44	
Rural	36	35	33	35	39	37	30	33	35	35	27	42
nonisolated	26	36	30	39	30	37	39	33	40	36	41	
Urban	41	43	44	37	36	36	42	32	43	25	25	38
ghetto	34	42	40	45	37	28	24	40	42	41	45	

TABLE 6.30 **Proportion of tasks correctly completed by three groups of children**

Normal	Moderately retarded	Severely retarded
0.64	0.15	0.08
0.25	0.10	0.01
0.14	0.08	0.03
0.20	0.11	0.11
0.75	0.13	0.10
0.62	0.12	0.09
0.25		
0.85		
0.90		
0.55		
0.82		

6.23 Refer to Exercise 6.22. Is there a significant trend from normal to severely retarded?

6.24 In a study of the effects on children of films portraying aggression, a group of psychologists randomly assigned a group of third-grade students to view one of three films with varying degrees of aggressive content. Group I saw a film with no aggressive content, the content of group II's film was moderately aggressive, and group III saw a film with highly aggressive content. After the children had viewed the films, investigators observed each group separately for a period of one hour and kept a record of the number of aggressive acts engaged in by each child. The results are shown in Table 6.31. We wish to know if these data provide sufficient evidence to indicate a difference in medians among the three populations represented.

TABLE 6.31 **Number of aggressive acts performed by three groups of children**

Group I	15, 18, 13, 19, 25, 20, 17, 10, 16, 23
Group II	28, 32, 26, 22, 30, 24
Group III	21, 40, 12, 42, 39, 36

6.25 Refer to Exercise 6.24. Is there a significant trend in these data?

6.26 Refer to Exercise 6.24. Treat group I as a control condition, compare the other two "treatments" with it, and test for significance.

6.27 Four tour guides are employed at a state-operated tourist attraction of historical interest. Each guide conducts tours of exactly 20 persons each during peak season. Each tour is timed. A researcher with the state department of tourism conducted a study to compare the tour guides with respect to efficiency. For each guide the researcher selected a simple random sample of 6 tours conducted during the month of June, and the time required for each was recorded. The results are shown in Table 6.32.

TABLE 6.32   **Time required to conduct tours by four tour guides**

Tour guide

A	B	C	D
37	36	38	30
31	33	38	34
33	38	39	31
30	36	36	30
39	37	38	35
31	39	37	34

Can the researcher conclude on the basis of these data that the four tour guides differ with respect to the average time it takes them to conduct a tour? Let $\alpha = 0.05$, and find the P value.

6.28 Refer to Exercise 6.27. Make all possible multiple comparisons. What additional information does this exercise provide?

6.29 A researcher is interested in how well three types of training programs prepare their graduates for entrance into a certain profession. The criterion measure was performance on a standardized examination required for certification of persons entering the profession. The three types of training programs are state-supported college programs, private college programs, and private noncollegiate programs. Samples of scores made by graduates of the three programs on the certification examination are shown in Table 6.33.

TABLE 6.33   **Scores made on a certifying examination by graduates of three types of training programs**

Training program	Score			
A	572	664	600	564
B	795	715	609	
C	755	823	920	

Treat program A as the "control" and perform a test to see if you can conclude that, on the average, graduates of each of the other two programs make higher scores than graduates of program A. Use an overall significance level of 0.15.

6.30 Of interest to a clothing manufacturer is the tensile strength of synthetic fibers used to make cloth. Researchers suspect that the strength of the fiber is affected by the percentage of a natural fiber that is present. Four percentage levels of natural fiber are of interest:

20%, 30%, 40%, and 50%. Three specimens of fiber are manufactured at each level of natural-fiber percentage, and the tensile strength of each is recorded. To serve as a control, three specimens of fiber are manufactured without the addition of the natural fiber. The results are shown in Table 6.34.

TABLE 6.34

Tensile strength of synthetic fiber (lb/sq in.)

Percentage of natural fiber	Tensile strength
None	7, 7, 9
20	16, 13, 17
30	15, 18, 19
40	24, 25, 23
50	25, 24, 24

On the basis of these data, can one conclude that the addition of the natural fiber in any tested percentages increases tensile strength? Use an overall significance level of 0.20.

REFERENCES

T1 Kruskal, W. H., and W. A. Wallis, "Use of Ranks in One-Criterion Variance Analysis," *J. Amer. Statist. Assoc.*, 47 (1952), 583–621. Addendum, *Ibid.*, 48 (1953), 907–911.

T2 Kruskal, W. H., "A Nonparametric Test for the Several Sample Problem," *Ann. Math. Statist.*, 23 (1952), 525–540.

T3 Gabriel, K. R., and P. A. Lachenbruch, "Non-Parametric ANOVA in Small Samples: A Monte Carlo Study of the Adequacy of the Asymptotic Approximation," *Biometrics*, 25 (1969), 593–596.

T4 Andrews, F. C., "Asymptotic Behavior of Some Rank Tests for Analysis of Variance," *Ann. Math. Statist.*, 25 (1954), 724–736.

T5 Hodges, J. L., Jr., and E. Lehmann, "The Efficiency of Some Nonparametric Competitors of the *t*-Test," *Ann. Math. Statist.*, 27 (1956), 324–335.

T6 Iman, Ronald L., Dana Quade, and Douglas A. Alexander, "Exact Probability Levels for the Kruskal–Wallis Test," in H. L. Harter, D. B. Owen, and J. M. Davenport (eds.), *Selected Tables in Mathematical Statistics*, Vol. III, Providence, R. I.: American Mathematical Society, 1975, pp. 329–384.

T7 Breslow, Norman, "A Generalized Kruskal–Wallis Test for Comparing *K* Samples Subject to Unequal Patterns of Censorship," *Biometrika*, 57 (1970), 579–594.

T8 Rust, Steven W., and Michael A. Fligner, "A Modification of the Kruskal–Wallis Statistic for the Generalized Behrens–Fisher Problem," *Communic. in Statist.—Theory and Methods*, 13 (1984), 2013–2027.

T9 Toothaker, Larry E., and Horng-shing Chang, "On 'The Analysis of Ranked Data Derived from Completely Randomized Factorial Designs,'" *J. Educ. Statist.*, 5 (1980), 169–176.

T10 Scheirer, C. J., W. S. Ray, and N. Hare, "The Analysis of Ranked Data Derived from Completely Randomized Factorial Designs," *Biometrics*, 32 (1976), 429–434.

T11 van der Laan, Paul, and L. Rob Verdooren, "Classical Analysis of Variance Methods and Nonparametric Counterparts," *Biometrical J.*, 29 (1987), 635–665.

T12 Fisher, R. A., and F. Yates, *Statistical Tables for Biological, Agricultural and Medical Research*, Edinburgh: Oliver and Boyd, 1938.

T13 Hoeffding, W., " 'Optimum' Nonparametric Tests," in Jerzy Neyman (ed.), *Proceedings of the Second Berkeley Symposium on Mathematical Statistics and Probability*, Berkeley and Los Angeles: University of California Press, 1951, pp. 83–92.

T14 Terry, M. E., "Some Rank Order Tests Which Are Most Powerful against Specific Parametric Alternatives," *Ann. Math. Statist.*, 23 (1952), 346–366.

T15 Van der Waerden, B. L., "Order Tests for the Two-Sample Problem and Their Power," *Nederl. Akad. Wetensch. Proc. Ser. A*, 55 (1952). (*Indag. Math.* 14), 453–458, and *Indag. Math.*, 150 (1953), 303–316. Errata, *Ibid.* (1953), 80.

T16 Puri, Madan Lal, "Asymptotic Efficiency of a Class of *c*-Sample Tests," *Ann. Math. Statist.*, 35 (1964), 102–121.

T17 Hájek J., and Z. Šedák, *Theory of Rank Tests*, Prague: Academic Press and Academia, 1967.

T18 McSweeney, Maryellen, and Douglas Penfield, "The Normal Scores Test for the *c*-Sample Problem," *Br. J. Math. Statist. Psychol.*, 22 (1969), 177–192.

T19 Feir-Walsh, Betty J., and Larry E. Toothaker, "An Empirical Comparison of the ANOVA *F*-test, Normal Scores Test, and Kruskal–Wallis Test under Violations of Assumptions," *Educ. and Psychol. Measurement*, 34 (1974), 789–799.

T20 David, F. N., D. E. Barton, S. Ganeshalingam, H. L. Harter, P. J. Kim, and M. Merrington, *Normal Centroids, Medians and Scores for Ordinal Data*, London: Cambridge University Press, 1968.

T21 Harter, Harman Leon, *Expected Values of Normal Order Statistics*, Wright-Patterson Air Force Base, Ohio: Aeronautical Research Laboratory, Office of Aerospace Research, U.S. Air Force, 1960 (ARL TR 60–292).

T22 Harter, H. Leon, "Expected Values of Normal Order Statistics," *Biometrika*, 48 (1961), 151–165.

T23 Harter, H. Leon, *Order Statistics and Their Use in Testing and Estimation*, Vol. 2, Washington, D.C.: U.S. Government Printing Office, 1969.

T24 Owen, D. B., *Handbook of Statistical Tables*, Reading, Mass.: Addison-Wesley, 1962.

T25 Bell, C. B., and K. A. Doksum, "Some New Distribution-Free Statistics," *Ann. Math. Statist.*, 36 (1965), 203–214.

T26 Bell, C. B., and K. A. Doksum, "Distribution-Free Tests of Independence," *Ann. Math. Statist.*, 38 (1967), 429–446.

T27 Conover, W. J., *Practical Nonparametric Statistics*, New York: Wiley, 1971.

T28 Bradley, James V., *Distribution-Free Statistical Tests*, Englewood Cliffs, N.J.: Prentice-Hall, 1968.

T29 Barbour, A. D., D. I. Cartwright, J. B. Donnelly, and G. K. Eagleson, "A New Rank Test for the *k*-Sample Problem," *Communic. in Statist.—Theory and Methods*, 14 (1985), 1471–1484.

T30 Crouse, C. F., "A Non-Null Ranking Model for a Sequence of *m* Alternatives," *Biometrika*, 48 (1961), 441–444.

T31 Crouse, C. F., "Distribution Free Tests Based on the Sample Distribution Function," *Biometrika*, 53 (1966), 99–108.

T32 Fligner, Michael, "Pairwise versus Joint Ranking: Another Look at the Kruskal–Wallis Statistic," *Biometrika*, 72 (1985), 705–709.

T33 Kannemann, K., "An Intrinsic Rank Test for k Independent Samples," *Biometrical J.*, 22 (1980), 229–239.

T34 Shoemaker, Lewis H., "A Nonparametric Method for Analysis of Variance," *Communic. in Statist.—Simulation*, 15 (1986), 609–632.

T35 Soms, Andrew P., "Permutation Tests for k-Sample Binomial Data with Comparisons of Exact and Approximate P-Levels," *Communic. in Statist.—Theory and Methods*, 14 (1985), 217–233.

T36 Terpstra, T. J., "The Asymptotic Normality and Consistency of Kendall's Test against Trend, When Ties Are Present in One Ranking," *Indag. Math.*, 14 (1952), 327–333.

T37 Jonckheere, A. R., "A Distribution-Free k-Sample Test against Ordered Alternatives," *Biometrika*, 41 (1954), 133–145.

T38 Odeh, R. E., "On the Power of Jonckheere's k-Sample Test against Ordered Alternatives," *Biometrika*, 59 (1972), 467–471.

T39 Puri, M. L., "Some Distribution-Free k-Sample Rank Tests of Homogeneity against Ordered Alternatives," *Comm. Pure Appl. Math.*, 18 (1965), 51–63.

T40 Potter, R. W., and G. W. Sturm, "The Power of Jonckheere's Test," *Amer. Statist.*, 35 (1981), 249–250.

T41 Chacko, V. J., "Testing Homogeneity against Ordered Alternatives," *Ann. Math. Statist.*, 34 (1963), 945–956.

T42 Shorack, Galen R., "Testing against Ordered Alternatives in Model I Analysis of Variance; Normal Theory and Nonparametric," *Ann. Math. Statist.*, 38 (1967), 1740–1752.

T43 Lehmann, E. L., *Nonparametrics: Statistical Methods Based on Ranks*, San Francisco: Holden-Day, 1975.

T44 Roth, Gary L., and Wayne W. Daniel, "Critical Values for Chacko's Homogeneity Test against Ordered Alternatives," *Educ. Psychol. Measurement*, 38 (1978), 889–891.

T45 Parsons, Van L., "Small Sample Distribution for a Nonparametric Test for Trend," *Communic. in Statist.: Simulation and Computation*, 10 (1981), 289–302.

T46 Odeh, R. E., "On Jonckheere's k-Sample Test against Ordered Alternatives," *Technometrics*, 13 (1971), 912–918.

T47 Tryon, P. V., and T. P. Hettmansperger, "A Class of Nonparametric Tests for Homogeneity against Ordered Alternatives," *Ann. Statist.*, 1 (1973), 1061–1070.

T48 Cuzick, Jack, "A Wilcoxon-Type Test for Trend," *Statist. in Med.*, 4 (1985), 87–90.

T49 Hirotsu, C., "The Cumulative Chi-squares Method and a Studentized Maximal Contrast Method for Testing an Ordered Alternative in a One-way Analysis of Variance Model," *Rep. Statist. Appl. Res.*, 26 (December 1979), 12–21.

T50 Rothe, G., "Linear Trend Test versus Global Test: A Comparison," *Statistica Neerlandica*, 40 (1986), 1–7.

T51 Barlow, R. E., D. J. Bartholomew, J. M. Bremner, and H. D. Brunk, *Statistical Inference under Order Restrictions*, New York: Wiley, 1972.

T52 Dunn, O. J., "Multiple Comparisons Using Rank Sums," *Technometrics*, 6 (1964), 241–252.

T53 Kurtz, T. E., R. F. Link, J. W. Tukey, and D. L. Wallace, "Short-Cut Multiple Comparisons for Balanced Single and Double Classification, Part 1, Results," *Technometrics*, 7 (1965), 95–161.

T54 Spjøtvoll, E., "A Note on Robust Estimation in Analysis of Variance," *Ann. Math. Statist.*, 39 (1968), 1486–1492.

T55 Lehmann, E. L., "Robust Estimation in Analysis of Variance," *Ann. Math. Statist.*, 34 (1963), 957–966.

T56 Anscombe, J. J., "Comments on Kurtz–Link–Tukey–Wallace Paper," *Technometrics*, 7 (1965), 167–168.

T57 Dwass, M., "Some *k*-Sample Rank-Order Tests," in I. Olkin et al. (eds.), *Contributions to Probability and Statistics*, Palo Alto, Calif.: Stanford University Press, 1960, pp. 198–202.

T58 Gabriel, K. R., "Simultaneous Test Procedures—Some Theory of Multiple Comparisons," *Ann. Math. Statist.*, 40 (1969), 224–250.

T59 Marascuilo, L. A., and M. McSweeney, "Nonparametric Post Hoc Comparisons for Trend," *Psychol. Bull.*, 67 (1967), 401–412.

T60 McDonald, B. J., and W. A. Thompson, "Rank Sum Multiple Comparisons in One- and Two-Way Classifications," *Biometrika*, 54 (1967), 487–498.

T61 Miller, R. G., Jr., *Simultaneous Statistical Inference*, second edition, New York: Springer-Verlag, 1981.

T62 Sherman, Ellen, "A Note on Multiple Comparisons Using Rank Sums," *Technometrics*, 7 (1965), 255–256.

T63 Steel, R. G. D., "Some Rank Sum Multiple Comparisons Tests," *Biometrics*, 17 (1961), 539–552.

T64 Steel, Robert G. D., "A Rank Sum Test for Comparing All Pairs of Treatments," *Technometrics*, 2 (1960), 197–207.

T65 Rhyne, A. L., and R. G. D. Steel, "Tables for a Treatment versus Control Multiple Comparisons Sign Test," *Technometrics*, (1965), 293–306.

T66 Campbell, Gregory, and John H. Skillings, "Nonparametric Stepwise Multiple Comparison Procedures," *J. Amer. Statist. Assoc.*, 80 (1985), 998–1003.

T67 Marascuilo, Leonard A., and Fred Dagenais, "Planned and Post-hoc Comparisons for Tests of Homogeneity Where the Dependent Variable Is Categorical and Ordered," *Educ. Psychol. Measurement*, 42 (1982), 777–782.

T68 Skillings, John H., "Nonparametric Approaches to Testing and Multiple Comparisons in a One-Way ANOVA," *Communic. in Statist.—Simulation and Computation*, 12 (1983), 373–387.

T69 Levy, Kenneth J., "Nonparametric Large-Sample Pairwise Comparisons," *Psychol. Bull.*, 86 (1979), 371–375.

T70 Wike, Edward L., and James D. Church, "Monte Carlo Studies of Levy's 'Nonparametric Large-Sample Pairwise Comparisons,'" *Psychol. Bull.*, 88 (1980), 607–613.

T71 Wike, Edward L., and James D. Church, "Further Comments on Nonparametric Multiple-Comparison Tests," *Perceptual and Motor Skills*, 45 (1977), 917–918.

T72 Wike, Edward L., and James D. Church, "A Monte Carlo Investigation of Four Nonparametric Multiple-Comparison Tests for *k* Independent Groups," *Bull. Psychonomic Soc.*, 11 (1978), 25–28.

T73 Zwick, Rebecca, and Leonard A. Marascuilo, "Selection of Pairwise Multiple Comparison Procedures for Parametric and Nonparametric Analysis of Variance Models," *Psychol. Bull.*, 95 (1984), 148–155.

T74 Hurwitz, Aryeh, "Multiple Comparisons and Nonparametric Statistical Tests on a Programmable Calculator," *J. Pharmacol. Methods*, 17 (1987), 17–38.

T75 Daniel, Wayne W., *Multiple Comparison Procedures*, Monticello, Ill.: Vance Bibliographies, June 1980.

T76 Balaam, L. N., and W. T. Federer, "Error Rate Bases," *Technometrics*, 7 (1965), 260–262.

T77 Federer, W. T., "Experimental Error Rates," *Proc. Am. Soc. Hortic. Sci.*, 78 (1961), 605–615.

T78 Petrinovich, L. F., and C. D. Hardyck, "Error Rates for Multiple Comparison Methods: Some

Evidence Concerning the Frequency of Erroneous Conclusions," *Psychol. Bull.*, 71 (1969), 43–54.

T79 Ryan, T. A., "The Experiment as the Unit for Computing Rate of Error," *Psychol. Bull.*, 59 (1962), 301–305.

T80 Steel, R. D. G., "Query 163: Error Rates in Multiple Comparisons," *Biometrics*, 17 (1961), 326–328.

T81 Wilson, W. A., "A Note on the Inconsistency Inherent in the Necessity to Perform Multiple Comparisons," *Psychol. Bull.*, 59 (1962), 296–300.

T82 Kirk, Roger E., *Experimental Design: Procedures for the Behavioral Sciences*, Belmont, Calif.: Brooks/Cole, 1968.

T83 Fligner, Michael A., "A Note on Two-Sided Distribution-Free Treatment versus Control Multiple Comparisons," *J. Amer. Statist. Assoc.*, 79 (1984), 208–211.

T84 Steel, R. G. D., "A Multiple Comparison Rank Sum Test: Treatment versus Control," *Biometrics*, 15 (1959), 560–572.

T85 Fligner, M. A., and D. A. Wolfe, "Distribution-Free Tests for Comparing Several Treatments with a Control," *Statistica Neerlandica*, 36 (1982), 119–127.

T86 Bhuchongkul, S., and M. L. Puri, "On the Estimation of Contrasts in Linear Models," *Ann. Math. Statist.*, 36 (1965), 847–858.

T87 Lehmann, E. L., "Asymptotically Nonparametric Inference: An Alternative Approach to Linear Models," *Ann. Math. Statist.*, 34 (1963), 1494–1506.

T88 Sen, P. K., "On Nonparametric Simultaneous Confidence Regions and Tests for the One Criterion Analysis of Variance Problem," *Ann. Inst. Statist. Math.*, 18 (1966), 319–336.

T89 Smith, Robert A., Young B. Lee, and William B. Michael, "Fortran IV Program to Compute the Kruskal–Wallis Statistic," *Educ. Psychol. Measurement*, 30 (1970), 735–736.

T90 Roberge, James J., "A Computer Program for Nonparametric Analysis of Variance," *Educ. Psychol. Measurement*, 30 (1970), 731, 733.

T91 Roberge, James J., "A Generalized Nonparametric Analysis of Variance Program," *Educ. Psychol. Measurement*, 32 (1972), 805–809.

T92 Roberge, James J., "A Generalized Non-Parametric Analysis of Variance Program," *Br. J. Math. Statist. Psychol.*, 25 (1972), 128.

T93 Rock, N. M. S., "*NPSTAT*: A FORTRAN-77 Program to Perform Nonparametric Variable-by-Variable Comparisons on Two or More Independent Groups of Data," *Comput. & Geosci.*, 12 (1986), 757–777.

T94 Theodorsson–Norheim, Elvar, "Kruskal–Wallis Test: BASIC Computer Program to Perform Nonparametric One-Way Analysis of Variance and Multiple Comparisons on Ranks of Several Independent Samples," *Comput. Methods & Programs in Biomed.* 23 (1986), 57–62.

T95 Roberge, James J., "A Computer Program for Nonparametric Post Hoc Comparisons for Trend," *Educ. Psychol. Measurement*, 31 (1971), 275–278.

T96 Thakur, Ajit K., "A FORTRAN Program to Perform the Nonparametric Terpstra–Jonckheere Test," *Comput. Programs in Biomed.*, 18 (1984), 235–240.

T97 Blaker, William D., "Computer Program for the Parametric and Nonparametric Comparison of Several Groups to a Control," *Comput. in Biol. & Med.*, 17 (1987), 37–44.

T98 Roberge, James J., and James Roberge, "A Generalized Nonparametric ANOVA Program (Version 2)," *Behav. Res. Methods & Instrumentation*, 9 (1977), 28.

T99 Borys, Suzanne V., and James G. Corrigan, "A BASIC Program for Nonparametric Post Hoc Comparisons," *Behav. Res. Methods & Instrumentation*, 12 (1980), 635.

E1 Polimeni, Phillip I., "Extracellular Space and Ionic Distribution in Rat Ventricle," *Amer. J. Physiol.*, 227 (1974), 676–683.

E2 Sattler, Krista M., "Olfactory and Auditory Stress in Mice (*mus musculus*)," *Psychon. Sci.*, 29 (1972), 294–296.

E3 Mameesh, Mostafa, Simon Aprahamian, Joseph P. Salji, and James W. Cowan, "Availability of Iron from Labeled Wheat, Chickpea, Broad Bean, and Okra in Anemic Blood Donors," *Amer. J. Clin. Nutr.*, 23 (1970), 1027–1032.

E4 Schapira, Georges, Jean-Claude Dreyfus, Fanny Schapira, and Jacques Kruh, "Glycogenolytic Enzymes in Human Progressive Muscular Dystrophy," *Amer. J. Phys. Med.*, 34 (1955), 313–319.

E5 Young, Francis A., George A. Leary, Robert R. Zimmerman, and David Strobel, "Diet and Refractive Characteristics," *Amer. J. Optom. Arch. Amer. Acad. Optom.*, 50 (1973), 226–233.

E6 Chandra, R. K., "Reduced Bactericidal Capacity of Polymorphs in Iron Deficiency," *Arch. Dis. Child.*, 48 (1973), 864–866.

E7 Cawson, M. J., Anne B. M. Anderson, A. C. Turnbull, and L. Lampe, "Cortisol, Cortisone, and 11-Deoxycortisol Levels in Human Umbilical and Maternal Plasma in Relation to the Onset of Labour," *J. Obstet. Gynaecol. Br. Commonw.*, 81 (1974), 737–745.

E8 Ali, M. A. M., and G. Sweney, "Erythrocyte Corproporphyrin and Photoporphyrin in Ethanol-Induced Sideroblastic Erythropoiesis," *Blood*, 43 (1974), 291–295.

E9 Torre, Michele, Filippo Bogetto, and Eugenio Torre, "Effect of LSD-25 and 1-Methyl-d-Lysergic Acid Butanolamide on Rat Brain and Platelet Serotonin Levels," *Psychopharmacologia*, 36 (1974), 117–122.

E10 Kando, Thomas M., "Role Strain: A Comparison of Males, Females, and Transsexuals," *J. Marriage Fam.*, 34 (1972), 459–464.

E11 Stern, I. B., L. Dayton, and J. Duecy, "The Uptake of Tritiated Thymidine by the Dorsal Epidermis of the Fetal and Newborn Rat," *Anatom. Rec.*, 170 (1971), 225–234.

E12 Nappi, A. J., "Cellular Immune Reactions of Larvae of *Drosophila algonquin*," *Parasitology*, 70 (1975), 189–194.

E13 Wohl, Mary Ellen B., N. Thornton Griscom, Demetrius G. Traggis, and Norman Jaffee, "Effects of Therapeutic Irradiation Delivered in Early Childhood upon Subsequent Lung Function," *Pediatrics*, 55 (1975), 507–516.

E14 Davis, Julia, "Performance of Young Hearing-Impaired Children on a Test of Basic Concepts," *J. Speech Hear. Res.*, 17 (1974), 342–351.

E15 Boehm, A. E., *Boehm Test of Basic Concepts Manual*, New York: Psychological Corporation, 1971.

E16 Maksoud, João Gilberto, Anneliese Fischer Thom, Julio Kieffer, and Virgilio A. Carvalho Pinto, "Fecal Excretion of Rose Bengal I^{131} in the Diagnosis of Obstructive Jaundice in Infancy with Special Reference to Biliary Atresia," *Pediatrics*, 48 (1971), 966–969.

E17 Herbuté, S., and J. D. Baylé, "Multiple-Unit Activity in the Pineal Gland of the Japanese Quail: Spontaneous Firing and Responses to Photic Stimulations," *Neuroendocrinology*, 16 (1974), 52–64.

E18 Kaklamanis, E., D. Trichopoulos, G. Papaevangelou, M. Drouga, and D. Karalis, "T Lymphocytes in HBAg Carriers and Responders," *Lancet*, Vol. I, Number 7908 (March 22, 1975), 689.

E19 Svenningsen, N. W., "Renal Acid-Base Titration Studies in Infants with and without Metabolic Acidosis in the Postneonatal Period," *Pediatric Res.*, 8 (1974), 659–672.

PROCEDURES THAT UTILIZE DATA FROM THREE OR MORE RELATED SAMPLES

Chapter 6 introduced procedures that are applicable when the data come from three or more independent samples. If the subjects exhibit a great deal of variability, it may be difficult to detect differences in the variable of interest among groups by the methods of Chapter 6. The variability among subjects in the same group may mask any differences in the variable of interest that may exist among the groups.

Suppose, for example, that a drug manufacturer wishes to compare the effects of four drugs on reaction time of adult subjects to some stimulus. Let us assume that the purpose of the experiment is to determine which drugs are best at creating a calming effect on subjects. If the completely randomized experimental design discussed in Chapter 6 is employed, it is possible that the subjects assigned at random to one of the drugs (perhaps the one having the poorest calming effect) are all older. Assuming a natural tendency for older subjects to have slower reaction times, such a fortuitous assignment of subjects might make this drug appear to have an acceptable calming effect, when in reality the favorable showing was due to the age of the subjects rather than the effect of the drug. Such a circumstance can be avoided by employing a better experimental design.

Frequently, we can greatly improve the ability to detect group differences in the variable of interest by dividing subjects into homogeneous subgroups, called *blocks*,

and then making comparisons among subjects within the subgroups. We can do this by using an experimental design known as the *randomized complete block design*. This technique extends the two-sample paired comparison model discussed in Chapter 4 to the case in which several samples are available for analysis. Thus, for three or more samples, a block is composed of three or more subjects, more generally referred to as *experimental units*, who are more homogeneous with respect to each other than with respect to subjects in another block. Referring again to the drug/reaction time example, we could form blocks on the basis of age, thus assuring that each age group (block) is equally represented in each treatment (drug) group.

A block may consist of animals drawn from the same litter, a batch of material produced according to the same formula, or subjects who have been carefully matched on the basis of certain relevant variables such as age, education, and physical condition. In certain situations a single subject may be a block. For example, in a study of the effects of different dosages of a drug, each of several subjects (blocks) may be given varying amounts of the drug at different times (sufficiently spaced to avoid possible "carryover" effects).

The reader may recall that the parametric technique used to analyze data resulting from the randomized complete block design is two-way analysis of variance. This technique utilizes the actual measurements (or an appropriate transformation) resulting from the experiment.

Section 7.1 presents a nonparametric analogue of the parametric two-way analysis of variance, called the Friedman *two-way analysis of variance by ranks*. As the name implies, it is based on ranks. Section 7.2 introduces a technique for making multiple comparisons using the same ranking procedure. Section 7.3 is devoted to a procedure that is appropriate when the alternative hypothesis is ordered. Section 7.4 presents a test for use with a design called the *incomplete block design*, and Section 7.5 presents a test that is appropriate when the observations are dichotomous.

7.1

FRIEDMAN TWO-WAY ANALYSIS OF VARIANCE BY RANKS

The test presented in this section is a nonparametric analogue of the parametric two-way analysis of variance. We perform calculations on ranks, which may be derived from observations measured on a higher scale or may be the original observations themselves. The procedure, which was introduced by Friedman (T1, T2), may be used when for one reason or another it is undesirable to use the parametric two-way analysis of variance. For example, the investigator may be unwilling to assume that the sampled populations are normally distributed, a requirement for the valid use of the parametric test. Also, in some cases only ranks may be available for analysis..

The objective is to determine if we may conclude from sample evidence that there is a difference among treatment effects. We reason that if the treatments do not differ

in their effects, the median response of a population of subjects receiving a given treatment will be the same as the median response of a population of subjects receiving any one of the other treatments under study, after the effect of the blocking variable has been removed. Thus, if we are comparing k treatments that have identical effects, $M_1 = M_2 = \cdots = M_k$, where M_j is the median of the population receiving the jth treatment, and $1 \leq j \leq k$.

Assumptions

A. The data consist of b mutually independent samples (blocks) of size k. The typical observation X_{ij} is the jth observation in the ith sample (block). The data may be displayed as in Table 7.1, where the rows represent the blocks and the columns are called treatments. The term treatment has a very general meaning; it may refer to a treatment in the usual sense of the word, or it may refer to some other condition such as socioeconomic status or educational level.

B. The variable of interest is continuous.

C. There is no interaction between blocks and treatments.

D. The observations within each block may be ranked in order of magnitude.

Hypotheses

H_0: $M_1 = M_2 = \cdots = M_k$

H_1: At least one equality is violated

Test Statistic

The first step in calculating the test statistic is to convert the original observations to ranks. (This step, or course, is unnecessary if the original observations *are* ranks.) The ranking procedure for the Friedman test differs from that of the Kruskal–

TABLE 7.1   **Data display for the Friedman two-way analysis of variance by ranks**

Block	Treatment						
	1	2	3	...	j	...	k
1	X_{11}	X_{12}	X_{13}	...	X_{1j}		X_{1k}
2	X_{21}	X_{22}	X_{23}		X_{2j}		X_{2k}
3	X_{31}	X_{32}	X_{33}		X_{3j}		X_{3k}
$\vdots$							$\vdots$
i	X_{i1}	X_{i2}	X_{i3}		X_{ij}		X_{ik}
$\vdots$							$\vdots$
b	X_{b1}	X_{b2}	X_{b3}	...	X_{bj}	...	X_{bk}

Wallis test, in which the observations in all samples combined are ranked relative to each other. In the Friedman test the observations within each block are ranked separately from smallest to largest, so each block contains a separate set of k ranks.

If H_0 is true and all treatments have identical effects, the rank that appears in a particular column when the data are displayed as in Table 7.1 is merely a matter of chance. Consequently, when H_0 is true, neither small nor large ranks should tend to show a "preference" for a particular column; that is, the ranks in each block should be randomly distributed over the columns (treatments) in each block. We expect the true state of the null hypothesis, whether true or false, to be reflected in the way in which the ranks within blocks are distributed over the columns. If H_0 is false, we expect a lack of randomness in this distribution. If one treatment is better than the others, for example, we expect large or small ranks to "favor" a particular column. A useful test of the null hypothesis, then, is one that is sensitive to such a tendency. The Friedman test is such a test because it detects departures from expectation under H_0 on the basis of the magnitudes of the sums of the ranks by column.

The second step in calculating the test statistic is to obtain the sums of the ranks R_j in each column. If H_0 is true, we expect the sums to be fairly close in size—so close that we can attribute differences to chance. When H_0 is false, however, we expect at least one sum to be sufficiently different in size from at least one other sum that we are reluctant to attribute the difference to chance alone. In other words, if H_0 is false, we expect to see at least one difference between pairs of rank sums so large that we cannot reasonably attribute it to sampling variability. We have to attribute it to some other cause—namely, a false null hypothesis. Differences among rank sums of sufficient magnitude give rise to a value of the test statistic sufficiently large to cause us to reject H_0.

The Friedman test statistic is defined as

$$\chi_r^2 = \frac{12}{bk(k+1)} \sum_{j=1}^{k} \left[R_j - \frac{b(k+1)}{2} \right]^2$$

in which $b(k+1)/2$ is the mean of the R_j's under H_0. Inspection of the formula shows that large discrepancies between the R_j's and the mean have the effect of inflating χ_r^2. A sufficiently large value of χ_r^2 will cause rejection of H_0.

The usual computational formula for the test statistic is

$$\chi_r^2 = \frac{12}{bk(k+1)} \sum_{j=1}^{k} R_j^2 - 3b(k+1)$$

Alternatively, we may use as our test statistic

$$W = \frac{12 \sum_{j=1}^{k} R_j^2 - 3b^2 k(k+1)^2}{b^2 k(k^2 - 1)} \tag{7.1}$$

where $W = \chi_r^2 / b(k-1)$.

Decision Rule

When b and k are small, we compare W for significance with appropriate critical values in Table A.14. If the computed W is greater than or equal to the tabulated W for b, k, and $\alpha = P$, we can reject H_0 at the α level of significance. For values of b and/or k not included in Table A.14, we compare χ_r^2 for significance with tabulated values of chi-square (Table A.11) with $k - 1$ degrees of freedom. Reject H_0 at the α level of significance if the $\chi_r^2 = b(k - 1)W$ computed from the data is greater than or equal to the tabulated value of $\chi_{(1-\alpha)}^2$ for $k - 1$ degrees of freedom.

Ties   Theoretically, no ties should occur, since the variable whose values are ranked is assumed to be continuous. In practice, however, ties do occur, and we give tied observations the mean of the rank positions for which they are tied. Note that only ties within a given block are of concern.

When ties occur, we may adjust the test statistic to take them into account by replacing the denominator of W in Equation 7.1 by

$$b^2 k(k^2 - 1) - b(\Sigma t^3 - \Sigma t) \tag{7.2}$$

where t is the number of observations tied for a given rank in any block.

Lehmann (T3) gives a limit theorem that supports the approximation of χ_r^2 adjusted for ties by the chi-square distribution with $k - 1$ degrees of freedom.

Example 7.1   Hall et al. (E1) compared three methods of determining serum amylase values in patients with pancreatitis. The results are shown in Table 7.2. We wish to know whether these data indicate a difference among the three methods.

TABLE 7.2   **Serum amylase values (enzyme units per 100 ml of serum) in patients with pancreatitis**

	Method of determination		
Specimen	A	B	C
1	4000	3210	6120
2	1600	1040	2410
3	1600	647	2210
4	1200	570	2060
5	840	445	1400
6	352	156	249
7	224	155	224
8	200	99	208
9	184	70	227

Source: F. F. Hall, T. W. Culp, T. Hayakawa, C. R. Ratliff, and N. C. Hightower, "An Improved Amylase Assay Using a New Starch Derivative," *Amer. J. Clin. Pathol.*, 53 (1970), 627–634: reproduced with permission.

Hypotheses

H_0: $M_A = M_B = M_C$

H_1: At least one equality is violated

Test Statistic

The specimens in this example are the blocks, so $b = 9$. Since we analyzed each specimen by each of three methods, we have $k = 3$. When we replace the original measurements shown in Table 7.2 by their ranks, we obtain the data displayed in Table 7.3, which also shows the sums of the ranks by treatment.

TABLE 7.3   **The data of Table 7.2 replaced by ranks before calculation of the Friedman test statistic**

Specimen	Method of determination		
	A	B	C
1	2	1	3
2	2	1	3
3	2	1	3
4	2	1	3
5	2	1	3
6	3	1	2
7	2.5	1	2.5
8	2	1	3
9	2	1	3
	$R_A = 19.5$	$R_B = 9$	$R_C = 25.5$

By Equation 7.1, we have

$$W = \frac{12(19.5^2 + 9^2 + 25.5^2) - 3(9)^2 3(3 + 1)^2}{(9)^2(3)(3^2 - 1)}$$

$$= 1674/1944 = 0.8611$$

Since we have one tie, we adjust W.

$$W \text{ (adjusted for ties)} = \frac{1674}{1944 - 9(2^3 - 2)} = 0.8857$$

Decision

Table A.14 with $k = 3$ and $b = 9$ shows that the probability of obtaining a value of W as large as or larger than 0.8857 when H_0 is true is less than 0.001. Consequently we reject H_0, and we conclude that the three methods do not all yield identical results. The P value is less than 0.001.

At this point, the reader may want to know which methods are different from which others. Section 7.2 presents a procedure to assist in determining these differences.

Since adjusting for ties inflates W, we do not have to adjust for ties when W unadjusted is of sufficient magnitude to cause us to reject H_0.

USE OF ALIGNED RANKS

The Friedman test is based on b sets of ranks, and the treatments are ranked separately in each set. Such a ranking scheme allows for intrablock comparisons only, since interblock comparisons are not meaningful. When the number of treatments is small, this may pose a disadvantage. When situations arise in which comparability among blocks is desirable, the method of *aligned ranks* may be employed. The technique involves subtracting from each observation within a block some measure of location such as the block mean or median. The resulting differences, called aligned observations, which keep their identities with respect to the block and treatment combination to which they belong, are then ranked from 1 to kb relative to each other. In other words, the ranking scheme is the same as that employed with the Kruskal–Wallis test. The ranks assigned to the aligned observations are called aligned ranks.

If there is no treatment effect, we would expect each of the blocks to receive approximately the same sequence of aligned ranks. Consequently we would expect the treatment rank totals to be about equal. The test statistic is derived in such a way that sufficient disparity among the treatment rank sums will cause rejection of the null hypothesis of no treatment effects.

In the absence of ties, the aligned-ranks test statistic for the randomized complete block design portrayed in Table 7.1 may be written as

$$T = \frac{(k-1)\left[\sum\limits_{j=1}^{k} \hat{R}_{.j}^2 - (kb^2/4)(kb+1)^2\right]}{\{[kb(kb+1)(2kb+1)]/6\} - (1/k)\sum\limits_{i=1}^{b} \hat{R}_{i.}^2} \tag{7.3}$$

where $\hat{R}_{i.}$ = rank total of the ith block, and $\hat{R}_{.j}$ = rank total of the jth treatment. If ties (which are broken in the usual way) are present, replace the denominator of T with

$$\sum_{i=1}^{b} \sum_{j=1}^{k} \hat{R}_{ij}^2 - (1/k)\sum_{i=1}^{b} \hat{R}_{i.}^2$$

where $\hat{R}_{ij}$ = the aligned rank of the jth measurement in the ith block. The test statistic T is compared for significance with tabulated chi-square for $k - 1$ degrees of freedom.

We illustrate the aligned ranks technique by means of the following example.

Example 7.2   In order to assess the effect of different amounts of cobalt (Co) on the tensile strength of steel, researchers conducted an experiment employing a completely randomized experimental design. The treatments consisted of four different levels (expressed as percentages) of Co, and the eight crucibles in which the alloying process took place served as the blocks. The tensile strengths in thousands of psi of the resulting 32 specimens of steel are shown in Table 7.4.

TABLE 7.4   **Tensile strength of steel for four levels of Co and eight crucibles**

Block (crucible)	Treatment (% Co)			
	A	**B**	**C**	**D**
1	43.3	45.8	45.5	44.7
2	48.3	48.7	46.9	48.8
3	49.8	48.7	56.0	48.6
4	49.8	51.3	55.3	58.6
5	56.6	56.1	58.6	54.6
6	57.6	57.5	58.1	57.7
7	72.0	74.2	89.6	82.1
8	88.1	88.7	92.6	88.2

For the purpose of comparison, let us first compute the Friedman test statistic by Equation 7.1. The necessary ranks and rank totals are displayed in Table 7.5.

TABLE 7.5   **Ranks and rank totals for Example 7.2**

Block	Treatment			
	A	**B**	**C**	**D**
1	1	4	3	2
2	2	3	1	4
3	3	2	4	1
4	1	2	3	4
5	3	2	4	1
6	2	1	4	3
7	1	2	4	3
8	1	3	4	2
Total	14	19	27	20

By Equation 7.1 we have

$$W = \frac{12(14^2 + 19^2 + 27^2 + 20^2) - 3(8^2)(4)(4 + 1)^2}{(8)^2(4)(4^2 - 1)}$$

$$= 0.26875$$

Entering Table A.14 with $k = 4$ and $b = 8$, we find that $P = 0.091$. We cannot reject the null hypothesis of no treatment differences at the 0.05 level of significance.

Now let us use the aligned-ranks method to test for a significant difference among treatment effects. For the original measurements in Table 7.4, the block means are, respectively, 44.825, 48.175, 50.775, 53.75, 56.475, 57.725, 79.475, and 89.4. When we subtract each block mean from the measurements from which it was computed, we obtain the aligned observations shown in Table 7.6.

TABLE 7.6   **Aligned observations for Example 7.2**

	Treatment			
Block	A	B	C	D
1	−1.525	0.975	0.675	−0.125
2	0.125	0.525	−1.275	0.625
3	−0.975	−2.075	5.225	−2.175
4	−3.95	−2.45	1.55	4.85
5	0.125	−0.375	2.125	−1.875
6	−0.125	−0.225	0.375	−0.025
7	−7.475	−5.275	10.125	2.625
8	−1.3	−0.7	3.2	−1.2

The aligned ranks, along with the treatment and block rank totals, are displayed in Table 7.7.

TABLE 7.7   **Aligned ranks and totals for Example 7.2**

	Treatment				
Block	A	B	C	D	Total
1	8	25	24	16.5	73.5
2	19.5	22	10	23	74.5
3	12	6	31	5	54
4	3	4	26	30	63
5	19.5	14	27	7	67.5
6	16.5	15	21	18	70.5
7	1	2	32	28	63
8	9	13	29	11	62
Total	88.5	101	200	138.5	528

In preparation for using Equation 7.3, adjusted for the ties that occurred in the ranking of the aligned ranks, we compute the following preliminary results to

substitute into the equation:

$$\sum_{j=1}^{k} \hat{R}_{.j}^2 = 88.5^2 + 101^2 + 200^2 + 138.5^2 = 77215.5$$

$$\sum_{i=1}^{b} \hat{R}_{i.}^2 = 73.5^2 + 74.5^2 + \cdots + 62^2 = 35177$$

$$\sum_{i=1}^{b} \sum_{j=1}^{k} \hat{R}_{ij}^2 = 8^2 + 19.5^2 + \cdots + 11^2 = 11439$$

Now, by Equation 7.3, adjusted for ties, we compute

$$T = \frac{(4-1)[77215.5 - (4 \cdot 8^2/4)(4 \cdot 8 + 1)^2]}{11439 - \frac{1}{4}(35177)}$$

$$= 8.53$$

Reference to Table A.11 with $k - 1 = 3$ degrees reveals that $0.05 > P > 0.025$. Whereas we were unable to reject the null hypothesis of no treatment differences using the Friedman test, the aligned-ranks procedure allows us to do so.

Power-Efficiency   The asymptotic relative efficiency of the Friedman test is discussed by Noether (T4), who points out that the asymptotic efficiency of the test depends on the number of observations per block k. He shows that relative to the parametric F test, it is $0.955k/(k + 1)$ when the populations are normally distributed, $k/(k + 1)$ when they are uniformly distributed, and $3k/2(k + 1)$ when the populations follow the double exponential distribution.

SAMPLING DISTRIBUTION OF χ_r^2

We find the sampling distribution of χ_r^2 by assuming that each set of ranks within a given block is just as likely to occur as any other set of ranks within the block. When there are k treatments, the number of possible sets of ranks within a block is equal to $k!$. When we have b blocks, there are $(k!)^b$ possible sets of ranks among the b blocks. When the null hypothesis is true, each of these $(k!)^b$ sets of ranks is equally likely. To find the sampling distribution of χ_r^2 for a given number of samples k and blocks b, we proceed as follows:

1. List all the possible sets of ranks.
2. Compute χ_r^2 for each set of ranks.
3. Determine the frequency of each distinct value of χ_r^2.
4. Divide each frequency by $(k!)^b$.

Let us illustrate by constructing the sampling distribution of χ_r^2 when there are $k = 2$ samples and $b = 3$ blocks, so that the number of possible sets of ranks

is $(2!)^3 = 8$. These 8 sets of ranks, along with the values of χ_r^2 that can be computed from them, are as follows:

Block	Sets of ranks							
	1	**2**	**3**	**4**	**5**	**6**	**7**	**8**
1	1, 2	1, 2	1, 2	2, 1	2, 1	2, 1	2, 1	1, 2
2	1, 2	1, 2	2, 1	1, 2	2, 1	2, 1	1, 2	2, 1
3	1, 2	2, 1	1, 2	1, 2	2, 1	1, 2	2, 1	2, 1
χ_r^2	3	1/3	1/3	1/3	3	1/3	1/3	1/3

The sampling distribution of χ_r^2, then, is

χ_r^2	Frequency of χ_r^2	$P(\chi_r^2)$
3	2	2/8
1/3	6	6/8
Total	8	8/8

Jensen and Hui (T5) have studied the efficiencies of Friedman's test when observations within blocks are independent and when they are exchangeably dependent. In a paper by Rothe (T6) an expression for the sharp lower bound for the efficiency of any Friedman-type test with respect to the Friedman-type test with optimal scores under weak model restrictions is given and calculated explicitly for the classical Friedman test.

FURTHER READING

Competitors of the Friedman test abound. An equivalent statistic has been proposed by Kendall and Babbington-Smith (T7) and Wallis (T8). Wilcoxon (T9) extended the Friedman test to test for interaction, as did de Kroon and van der Laan (T10). Additional competitors and extensions include those by Cooley and Cooley (T11), Mack and Skillings (T12), Mack (T13), de Kroon and van der Laan (T14), Rinaman (T15), Haux et al. (T16), and Shoemaker (T17). Comparisons among various rank tests for the two-way layout have been reported on by Iman and Conover (T18), Iman et al. (T19), Lemmer (T20), and Hora and Iman (T21). Three different ranking methods used with the Friedman test were investigated by van der Laan and de Kroon (T22).

Iman and Davenport (T23), who present two alternative approximations to the distribution of the Friedman statistic, recommend the use of the F distribution as opposed to the usual chi-square distribution for determining the critical region in hypothesis testing. Multivariate extensions have been considered by Gerig (T24, T25) and Jensen (T26). Other articles of interest are those by Mehra and Sarangi

(T27), Sen (T28), Likes and Laga (T29), Meddis (T30), Wei (T31), Brits and Lemmer (T32), and van der Laan (T33).

The aligned-ranks procedure was introduced by Hodges and Lehmann (T34). For the derivation of the formulas used in this chapter, see Lehmann (T3). Tardif (T35, T36, T37) has studied the asymptotic efficiency and other aspects of aligned-ranks tests in randomized block designs.

EXERCISES

7.1   A study of the effects of three drugs on reaction time of human subjects resulted in the data in Table 7.8. Do these data provide sufficient evidence to indicate that the three drugs differ in their effects? Determine the P value.

TABLE 7.8   **Change in response time (milliseconds) of 10 subjects after receiving one of three drugs**

Drug	Subject									
	1	**2**	**3**	**4**	**5**	**6**	**7**	**8**	**9**	**10**
A	10	10	11	8	7	15	14	10	9	10
B	10	15	15	12	12	10	12	14	9	14
C	15	20	12	10	9	15	18	17	12	16

7.2   Perry et al. (E2) determined plasma epinephrine concentrations during isoflurane, halothane, and cyclopropane anesthesia in 10 dogs. The results are shown in Table 7.9. Do these data suggest a difference in treatment effects? Find the P value.

TABLE 7.9   **Concentrations, nanograms per milliliter, of free catecholamines in arterial plasma in response to isoflurane, halothane, and cyclopropane**

Dog	**1**	**2**	**3**	**4**	**5**	**6**	**7**	**8**	**9**	**10**
Isoflurane	0.28	0.51	1.00	0.39	0.29	0.36	0.32	0.69	0.17	0.33
Halothane	0.30	0.39	0.63	0.38	0.21	0.88	0.39	0.51	0.32	0.42
Cyclopropane	1.07	1.35	0.69	0.28	1.24	1.53	0.49	0.56	1.02	0.30

Source: Lawrence B. Perry, Russell A. Van Dyke, and Richard A. Theye, "Sympathoadrenal and Hemodynamic Effects of Isoflurane, Halothane, and Cyclopropane in Dogs," *Anesthesiology*, 40 (1974), 465–470.

7.3   Syme and Pollard (E3) conducted an experiment to investigate the effect of different motivation levels on measures of food-getting dominance in the laboratory rat. The data shown in Table 7.10 are the amounts of food in grams eaten by eight male hooded rats following 0, 24, and 72 hours of food deprivation. Do these data provide sufficient evidence to indicate a difference in the effects of the three levels of food deprivation? Find the P value.

TABLE 7.10 **Amount of food, grams, eaten by eight rats under three levels of food deprivation**

	Hours of food deprivation		
Subject	0	24	72
1	3.5	5.9	13.9
2	3.7	8.1	12.6
3	1.6	8.1	8.1
4	2.5	8.6	6.8
5	2.8	8.1	14.3
6	2.0	5.9	4.2
7	5.9	9.5	14.5
8	2.5	7.9	7.9

Source: G. J. Syme and J. S. Pollard, "The Relation between Differences in Level of Food Deprivation and Dominance in Food-Getting in the Rat," *Psychon. Sci.*, 29 (1972), 297–298.

7.4 A systematic study of certain variables of the rubella hemagglutination-inhibition test system and their effect on antigen and antibody titers was conducted by Schmidt and Lennette (E4). To compare the effectiveness of various methods of removing nonspecific inhibitors, they treated sera in parallel by four different methods and then tested for rubella hemagglutination-inhibition antibody. The results are shown in Table 7.11. Can we conclude on the basis of these data that the four methods produce different results? Find the *P* value.

TABLE 7.11 **Effect of four methods of removing nonspecific inhibitors on rubella hemagglutination-inhibition antibody titers**

	HI titer after treatment by methods A, B, C, and D											
Serum*	9	10	11	12	13	14	15	16	17	18	19	20
A	32	32	32	64	64	64	32	128	128	128	128	256
B	16	32	32	64	64	32	32	128	64	128	128	128
C	16	32	32	32	64	32	32	64	64	128	128	128
D	<8	8	16	8	32	32	16	32	64	32	32	64

Serum*	21	22	23	24	25	26	27
A	256	256	128	512	512	1024	1024
B	128	128	128	256	128	512	1024
C	128	64	64	256	128	256	1024
D	128	32	32	128	64	256	1024

Source: Nathalie J. Schmidt and Edwin H. Lennette, "Variables of the Rubella Hemagglutination-Inhibition Test System and Their Effect on Antigen and Antibody Titers," *Appl. Microbiol.*, 19 (1970), 491–504.

* Sera 1–8 included in the authors' original table not included here.

7.5 Basmajian and Super (E5) studied the effects of a certain muscle relaxant when muscle spasticity was caused by upper motor neuron disease. They took measurements on patients when they were selected for study and one week later (control periods 1 and 2). Each subject subsequently received at different times a placebo and the test drug. The results

under the four conditions are shown in Table 7.12. From these data, can we conclude that there is a difference in response under the four conditions? Find the P value.

TABLE 7.12   **Peak ft/lb force readings in four periods in patients with muscle spasticity**

Patient	Controls 1	2	Placebo	Test drug
1	5.140	9.927	6.592	3.814
2	8.112	12.381	10.779	8.112
3	4.608	3.954	5.019	1.859
5	11.418	12.061	10.992	3.495
6	15.154	15.367	13.765	9.607
7	12.371	13.368	11.931	11.156
9	8.262	7.962	6.403	5.276
10	6.578	5.450	2.193	3.069
12	5.629	6.166	5.963	4.705
13	8.272	9.491	7.521	4.705
14	13.325	12.177	12.081	8.436
15	16.582	11.539	14.593	14.462
18	3.601	4.990	6.079	8.146
19	5.629	3.533	5.310	1.781
22	19.955	17.076	19.316	16.650

Source: J. V. Basmajian, and Gail A. Super, "Dantrolene Sodium in the Treatment of Spasticity," *Arch. Phys. Med. Rehabil.*, 54 (1973), 60–64.

7.2

MULTIPLE-COMPARISON PROCEDURE FOR USE WITH FRIEDMAN TEST

Investigators are usually not satisfied to know simply that their data allow them to conclude that not all sampled populations or all treatment effects are identical. For example, when the application of the Friedman test leads us to reject H_0, we are usually interested in exactly where the differences are located. What we need, then, is a multiple-comparison procedure to use after the Friedman test.

When we compare all possible differences between pairs of samples, when the experimentwise error rate is α, and when the number of blocks is large, then we declare R_j and $R_{j'}$ significantly different if

$$|R_j - R_{j'}| \geq z \sqrt{\frac{bk(k+1)}{6}} \tag{7.4}$$

where R_j and $R_{j'}$ are the jth and j'th treatment rank totals, and z is a value from Table A.2 corresponding to $\alpha/k(k-1)$. For an alternative multiple-comparisons formula, see Hollander and Wolfe (T38).

Example 7.3   To illustrate the use of this procedure, let us consider again the data of Example 7.1. Since we rejected H_0, we wish to know specifically which methods are different from which others. Suppose we choose an experimentwise error rate of $\alpha = 0.10$. With $k = 3$ and $\alpha = 0.10$ ($0.10/6 = 0.0167 \approx 0.02$), we find in Table A.2 that $z = 2.05$. When we make appropriate substitutions in the right-hand side of the inequality in 7.4, we have

$$2.05 \sqrt{\frac{9(3)(3 + 1)}{6}} = 8.697$$

The rank totals were $R_A = 19.5$, $R_B = 9$, and $R_C = 25.5$. The three pairs of differences $|R_j - R_{j'}|$ are

$$|19.5 - 9| = 10.5, \qquad |19.5 - 25.5| = 6, \qquad |9 - 25.5| = 16.5$$

Thus we conclude that methods A and B yield different results, methods B and C yield different results, but methods A and C do not.

COMPARING ALL TREATMENTS WITH A CONTROL

As discussed in Chapter 6, many research situations involve the comparison of two or more treatments, of which one is a control condition. Once an appropriate statistical procedure allows the researcher to conclude that there is a difference among the treatment effects, interest usually focuses on determining which of the other treatments exhibit an effect that is different from the control effect.

The following procedure, proposed by Rhyne and Steel (T39), allows us to compare all of the other treatments in a randomized block design with a control condition that is labeled treatment 0. The procedure may be used with both a one-sided and a two-sided alternative with an experimentwise error rate of α. The technique, an extension of the familiar sign test, was developed along the lines of the one for the one-way layout presented by Steel (T40). The steps employed in carrying out the procedure are as follows:

1. Represent by x_{i0} and x_{ij} ($i = 1, \ldots, b$, and $j = 1, \ldots, k$) the responses to the control and the jth treatment in the ith block of a randomized complete block design. Here k is the number of treatments, excluding the control condition.

2. Compute the signed differences $d_{ij} = x_{ij} - x_{i0}$. In other words, pair each treatment with the control condition, and in each block of this pairing, subtract the control measurement from the treatment measurement. There will be k pairings, each containing b differences.

3. Let r_j be the number of differences, d_{ij}, that have the less frequently occurring sign (either positive or negative) within a pairing of a treatment with the control.

4. Let M_0 be the median response of a population of subjects or objects experiencing the control condition and M_j be the median response of a population of objects or subjects receiving the jth treatment. Apply one of the following decision rules:

   a. For testing $H_0: M_j \geq M_0$ against $H_1: M_j < M_0$, reject H_0 if the number of plus signs is less than or equal to the critical value of r_j appearing in Table A.15 for k (the number of treatments excluding the control), b, and the chosen experimentwise error rate.

   b. For testing $H_0: M_j \leq M_0$ against $H_1: M_j > M_0$, reject H_0 if the number of minus signs is equal to or less than the critical value of r_j appearing in Table A.15 for k, b, and the chosen experimentwise error rate.

   c. For testing $H_0: M_j = M_0$ against $H_1: M_j \neq M_0$, reject H_0 if the number of minus signs or the number of plus signs (whichever is fewer) is equal to or less than the critical value of r_j in Table A.16 for k, b, and the chosen experimentwise error rate.

The technique is illustrated by means of the following example.

Example 7.4   A paint manufacturer wished to compare three newly proposed formulas with the standard formula currently in use for manufacturing a particular type of house paint. The response variable was a composite score that incorporated several properties that measure paint quality. Since the effective quality of paint is affected by the surface to which it is applied, the company researchers applied paint manufactured by the standard and the three new formulas to four different surfaces, which served as the blocks for the experiment. The resulting quality scores are shown in Table 7.13. A high score indicates a higher quality. For each treatment the differences, d_{ij}, are in parentheses.

TABLE 7.13   **Quality scores of paint manufactured by a standard and three experimental formulas**

Surface	Formula 0 (standard)	1	2	3
A	13	25(+12)	17(+6)	25(+12)
B	12	27(+15)	15(+3)	25(+13)
C	15	29(+14)	19(+4)	23(+8)
D	14	21(+7)	9(−5)	13(−1)
E	13	31(+18)	27(+14)	21(+8)
Number of minuses, r_j		0	1	1
Number of pluses, $b - r_j$		5	4	4

Suppose we choose an experimentwise error rate of 0.10 and let our hypotheses be $H_0: M_j \leq M_0$ and $H_1: M_j > M_0$. Reference to Table A.15 for $k = 3$ and $b = 5$ reveals that the critical value of r_j is 0. Since formula 1, when compared to the standard, yielded zero minus signs, we conclude that it is superior to the standard. Since the other two treatments yielded more than zero minus signs, we conclude that neither formula 2 nor formula 3 is better than the standard.

CONTRAST ESTIMATION

Following the analysis of the data resulting from the use of a randomized complete block design in an experiment, the researcher frequently wishes to estimate the difference between two treatment effects. A procedure for this purpose has been proposed by Doksum (T41). We assume that treatment effects are reflected by the magnitudes of the medians of the populations represented by our samples. Consequently we are interested in estimating the contrast between population medians as discussed in Chapter 6. To estimate the difference between two treatment effects, we proceed as follows:

1. For every pair of the k treatments in the experiment, we compute the difference between the responses to the two treatments in each of the b blocks. In other words, we compute the differences

   $$D_{i(uv)} = x_{iu} - x_{iv}$$

   where $i = 1, \ldots, b; u = 1, \ldots, k;$ and $v = 1, \ldots, k$. We form treatment pairs only for those in which $u < v$. For example, $D_{5(12)}$ is the difference between the observations for treatments 1 and 2 in block 5.

2. We find the median of each set of differences and call it Z_{uv}. For example, Z_{12} is the median of the $D_{i(12)}$ values. We call Z_{uv} the *unadjusted estimator* of $M_u - M_v$. Since $Z_{vu} = -Z_{uv}$, we have only to calculate Z_{uv} for the case where $u < v$. There are $k(k - 1)/2$ of these medians. Also note that $Z_{uu} = 0$.

3. We compute the mean of each set of unadjusted medians having the same first subscript and call the result m_u; that is, we compute

   $$m_{u.} = \frac{\sum\limits_{j=1}^{k} Z_{uj}}{k}, \qquad u = 1, \ldots, k$$

4. The estimator of $M_u - M_v$ is $m_{u.} - m_{v.}$, where u and v range from 1 through k. For example, the estimator of $M_1 - M_2$, the difference between M_1 and M_2, is $m_{1.} - m_{2.}$.

The procedure is illustrated by means of the following example:

Example 7.5

Let us refer to Example 7.1, in which the "treatments" were three methods of determining serum amylase values in patients with pancreatitis, and serum specimens from the patients were the blocks. Let $A = 1, B = 2$, and $C = 3$. We wish to estimate $M_1 - M_2$.

1. From the measurements in Table 7.2, we compute for each pair of treatments (1 and 2, 1 and 3, and 2 and 3) the differences between pairs of measurements in each block. They are shown in Table 7.14.

2. From the differences in Table 7.14, we find that the three medians are $Z_{12} = 395$, $Z_{13} = -560$, and $Z_{23} = -955$.

3. We now compute the following averages:

$$m_{1.} = \frac{0 + 395 + (-560)}{3} = -55$$

$$m_{2.} = \frac{-395 + 0 + (-955)}{3} = -450$$

$$m_{3.} = \frac{560 + 955 + 0}{3} = 505$$

4. Our estimate of $M_1 - M_2$ is $m_{1.} - m_{2.} = -55 - (-450) = 395$.

TABLE 7.14

Differences between pairs of measurements in each block for different pairs of treatments (data from Table 7.2)

Block i	Differences, $D_{i(uv)}$		
	$D_{i(12)}$	$D_{i(13)}$	$D_{i(23)}$
1	790	−2120	−2910
2	560	−810	−1370
3	953	−610	−1563
4	630	−860	−1490
5	395	−560	−955
6	196	103	−93
7	69	0	−69
8	101	−8	−109
9	114	−43	−157

Thus we estimate that method A yields values that are, on the average, 395 units larger than those resulting from the use of method B.

FURTHER READING

Multiple-comparison procedures for the two-way classification are discussed by McDonald and Thompson (T42), Dunn-Rankin and Wilcoxon (T43), and Rosenthal and Ferguson (T44). Nemenyi (T45) discusses an alternative procedure employing intrablock ranks that allows us to compare all treatments with a control in the two-way layout. Wilcoxon and Wilcox (T46) discuss the comparison of all possible pairs of treatments, as well as the comparison of several treatments with a control, in both the one-way and the two-way layouts.

EXERCISES

7.6 Apply the multiple-comparison procedure of this section to Exercise 7.2. Let $\alpha = 0.10$.

7.7 Apply the multiple-comparison procedure of this section to Exercise 7.3. Let $\alpha = 0.05$.

7.8 Apply the multiple-comparison procedure of this section to Exercise 7.4. Let $\alpha = 0.10$.

7.3

PAGE'S TEST FOR ORDERED ALTERNATIVES

As mentioned in Chapter 6, there are certain multisample situations in which an ordered alternative hypothesis is more meaningful than one in which order is ignored. Chapter 6 presented a procedure for testing H_0 against ordered alternatives when the data conform to the format for one-way analysis of variance. This section introduces a procedure that is appropriate in two-way analysis of variance hypothesis-testing situations in which an ordered alternative is meaningful. This procedure, *Page's test for ordered alternatives*, was introduced by Page (T47).

Assumptions

The assumptions are the same as those for the Friedman test discussed in Section 7.1.

Hypotheses

If we let τ_j designate the effect of the jth treatment, then we may state the hypotheses as follows:

$$H_0: \tilde{\tau}_1 = \tilde{\tau}_2 = \cdots = \tilde{\tau}_k$$
$$H_1: \text{The treatment effects } \tau_1, \tau_2, \ldots, \tau_k \text{ are ordered in the following way:}$$
$$\tau_1 \leq \tau_2 \leq \cdots \leq \tau_k$$

Test Statistic

The test statistic is

$$L = \sum_{j=1}^{k} jR_j = R_1 + 2R_2 + \cdots + kR_k \qquad (7.5)$$

where $R_1, \ldots, R_k$ are the treatment rank sums obtained in the manner explained in the discussion of the Friedman test.

If the treatment effects are ordered as specified in H_1, then R_j tends to be larger than $R_{j'}$ for $j' < j$. In other words, if there are three treatments and their effects are ordered according to H_1, then R_1 tends to be smaller than R_2, and R_2 in turn tends to be smaller than R_3. Since the treatment rank sums are weighted by the index of their position in the ordering specified by H_1, L tends to be large when H_1 is true.

Decision Rule

Reject H_0 at the α level of significance if the computed value of L is greater than or equal to the critical value of L for k, b, and α given in Table A.17.

Example 7.6

Cromer (E6) reported the scores made by 36 children who performed a certain task as part of an experiment. The children, matched by chronological age and sex, were divided into three groups. Children in group 1 were congenitally blind, those in group 2 were sighted children who performed the task blindfolded, and those in group 3 consisted of sighted children who performed the task without visual obstruction. The results are shown in Table 7.15. We wish to test the null hypothesis of identical results against the alternative that children in group 1 tend to score lower than those in group 2, and that those in group 2 tend to score lower than those in group 3.

TABLE 7.15

Conservation scores for 12 children in three groups matched for chronological age

Age	Sex	Blind	Blindfolded	Seeing
5:7	F	0	0	0
6:0	M	0	8	1
6:4	F	0	0	8
6:6	M	0	0	8
6:11	F	1	2	0
7:9	F	8	8	8
7:11	F	8	5	8
8:0	F	8	6	8
8:5	F	0	8	8
8:6	F	8	8	8
8:10	F	8	3	8
9:6	M	8	8	8

Source: Richard F. Cromer, "Conservation by the Congenitally Blind," *Br. J. Psychol.*, 64 (1973), 241–250; published by Cambridge University Press.

Hypotheses

H_0: $\tilde{\tau}_1 = \tilde{\tau}_2 = \tilde{\tau}_3$

H_1: The "treatment" effects are ordered, that is, $\tau_1 \leq \tau_2 \leq \tau_3$

Test Statistic

When we replace the scores of Table 7.15 with ranks, we obtain Table 7.16, which also shows the treatment rank sums. From these data we compute $L = 22.5 + 2(22.5) + 3(27) = 148.5$.

TABLE 7.16   **The scores of Table 7.15 ranked within blocks**

Block	1 (blind)	2 (blindfolded)	3 (seeing)
1	2	2	2
2	1	3	2
3	1.5	1.5	3
4	1.5	1.5	3
5	2	3	1
6	2	2	2
7	2.5	1	2.5
8	2.5	1	2.5
9	1	2.5	2.5
10	2	2	2
11	2.5	1	2.5
12	2	2	2
	$R_1 = 22.5$	$R_2 = 22.5$	$R_3 = 27$

Decision

Table A.17 reveals that 148.5 is less than 153, the critical value of L for $k = 3$, $b = 12$, and $\alpha = 0.05$. Therefore we cannot reject H_0 at the 0.05 level of significance, and we are unable (at that level) to conclude that the experimental results are ordered as specified by the alternative hypothesis. The P value is greater than 0.05.

Large-Sample Approximation   For large samples, the statistic

$$z = \frac{L - [bk(k + 1)^2/4]}{\sqrt{b(k^3 - k)^2/144(k - 1)}} \tag{7.6}$$

is distributed approximately as the standard normal. When we use the large-sample approximation, we reject H_0 at the α level of significance if the computed z is greater than or equal to the z value which has α area to its right (Table A.2).

Power-Efficiency   The power and efficiency of Page's test for ordered alternatives have been studied by Hollander (T48). Pirie (T49) has compared two classes

of rank tests for ordered alternatives: those in which the ranking is carried out within blocks (W tests) and those based on rankings among blocks (A tests). Page's test is an example of a W test, as is a normal-scores test proposed by Pirie and Hollander (T50). A tests include those proposed by Hollander (T48) and Puri and Sen (T51). The asymptotic relative efficiency results obtained by Pirie (T49) did not provide sufficient evidence on which to base a definitive choice between W tests and A tests. However, he recommends the W tests in most cases, for the following reasons:

1. The W tests are distribution free, while the A tests are not.
2. The W tests are easier to calculate.
3. The W tests are amenable to adjustment for certain situations, whereas the A tests are not.

FURTHER READING

Other distribution-free tests for ordered alternatives have been proposed by Jonckheere (T52) and Shorack (T53). Hettmansperger (T54) has proposed a rank test for ordered alternatives in the randomized block design with more than one observation per cell. The author also discusses relevant multiple-comparison and estimation procedures. Berenson (T55) proposes two A-type tests for ordered alternatives, which he compares with other competitors.

Kepner and Robinson (T56) suggest a distribution-free rank test for detecting ordered alternatives in randomized complete block designs and repeated-measures designs for four or fewer treatments. The null distribution of their test statistic is the same as that of the Wilcoxon signed-rank statistic. The authors report good power properties in both small- and large-sample situations.

For the case of unequally spaced ordered alternatives in two-way layouts, Ghiassi and Govindarajulu (T57) propose an asymptotically distribution-free test statistic that is a linear function of the ranks of residuals when the nuisance parameters are estimated. Gore et al. (T58) propose some median statistics to test the usual null hypothesis of homogeneity versus ordered location alternatives in the one-way, two-way classified data, and incomplete block designs.

A large-sample test for comparing increasing doses of a substance with a zero-dose control response is given for observations from a randomized block experimental design by House (T59). The method, which is a nonparametric version of a test due to Williams (T60, T61), is based on Friedman-type ranks. The author investigates the validity of the test and presents a numerical example.

Hirotsu (T62) defines several types of ordered alternatives for the case of the two-factor interaction, with one factor indicating the treatments and the other expressing the stages or degrees of the effects of the first factor. For these ordered alternatives, the author proposes methods for testing the null hypothesis and examines the power of the procedures for seven different patterns of the alternatives. A real drug experiment is presented as an illustration.

For both the one-way and the two-way layout, Berenson (T63, T64) investigates the small-sample power of several tests designed for tests with ordered location alternatives. Also of interest to researchers concerned with ordered alternatives is a paper by Rao and Gore (T65).

EXERCISES

7.9 Henry et al. (E7) conducted an investigation to determine whether experimentally induced pulmonary embolism in animals without underlying heart disease consistently leads to alterations in serum creatine phosphokinase (CPK) activity. They produced experimental pulmonary embolism in conscious mongrel dogs by injecting whole blood or plasma clots directly into the left main pulmonary artery. They measured serum CPK during the control period and serially after induction of pulmonary embolism. Their results are shown in Table 7.17. Test the null hypothesis of no change in CPK activity against the alternative that it increases with time up to 120 minutes. What is the P value?

TABLE 7.17   **Serum creatine phosphokinase activity after pulmonary embolization in mongrel dogs**

		After embolization, minutes		
Dog	Before embolization	15	60	120
A*	28	97	126	158
B	23	45	48	48
C	26	22	87	97
D	24	32	33	52
E	25	68	60	80

Source: Philip D. Henry, Colin M. Bloor, and Burton E. Sobel, "Increased Serum Creatine Phosphokinase Activity in Experimental Pulmonary Embolism," *Amer. J. Cardiol.*, 26 (1970), 151–155.

* Four experiments were run on dog A. Only the data for experiment 4 are reported here.

7.10 In a chemistry survey described by Gilbert (E8), eight vials of lyophilized human serum were analyzed by a large number of laboratories for 17 constituents, one of which was glucose. Seven methods of analysis—including "other"—were employed to analyze the eight vials of serum for glucose, with a total of 1602 laboratories performing the analysis. The mean values, by method of analysis, over all hospitals employing a given method were computed. These means were ranked in order for each vial, with a rank of 1 assigned to the smallest mean. The results are shown in Table 7.18. Test the null hypothesis of no difference among methods against the alternative that the methods yield results ordered as follows: A < B < C < D < E < F < G. Determine the P value.

7.11 Stern et al. (E9) reported the serum indirect bilirubin values in 10 normal infants shown in Table 7.19. Do these data provide sufficient evidence to indicate that levels decrease over time between 4 and 10 days of age? What is the P value?

TABLE 7.18   **Ranks of means reported for the analysis of eight vials of human serum for glucose by seven methods and 1602 laboratories**

Vial	Method						
	A	**B**	**C**	**D**	**E**	**F**	**G**
1	1	3	2	4	5	7	6
2	1	3	2	5	4	6	7
3	1	2	4	3	5	7	6
4	1	3	2	4	5	6	7
5	1	2	3	4	5	6	7
6	1	3	2	5	6	4	7
7	1	2	4	3	6	5	7
8	1	2	3	4	6	5	7

Source: Roger K. Gilbert, "Analysis of Results of the 1969 Comprehensive Chemistry Survey of the College of American Pathologists," *Amer. J. Clin. Pathol.*, 54 (1970), 463–482; reproduced with permission.

TABLE 7.19   **Serum indirect bilirubin levels, milligrams per 100 cc, in 10 normal infants**

Case	Age, days						
	4	**5**	**6**	**7**	**8**	**9**	**10**
1	10.80	6.15	4.10	5.00	5.00	3.40	2.60
2	12.50	11.80	13.20	11.00	8.20	6.80	6.00
3	13.70	16.80	16.80	15.60	11.70	12.50	10.55
4	11.50	6.80	4.00	3.50	1.66	1.60	1.60
5	10.20	6.40	3.10	3.00	2.60	2.20	1.98
6	8.00	7.85	7.45	7.00	3.60	4.00	3.00
7	10.80	11.10	6.15	7.00	3.80	4.30	5.60
8	14.90	10.80	9.90	9.40	10.50	7.70	7.60
9	16.20	16.40	15.40	10.20	8.30	10.70	7.40
10	10.80	10.00	6.80	4.60	4.20	3.80	3.50

Source: L. Stern, N. N. Khanna, G. Levey, S. J. Yaffe, "Effect of Phenobarbital on Hyperbilirubinemia and Glucuronide Formation in Newborns," *Amer. J. Dis. Child.*, 120 (1970), 26–31; copyright 1970, American Medical Association.

7.4

DURBIN TEST FOR INCOMPLETE BLOCK DESIGNS

In designing an experiment, the investigator may find that it is impossible or impractical to construct a randomized complete block design of the type discussed so far. It may be impossible or impractical to apply all treatments to each block. This becomes an important problem when the number of treatments is large and the size of the blocks is limited. Suppose, for example, that we are going to compare the effects of seven treatments by administering the treatments to laboratory animals, with litters serving as blocks. Because the subjects must meet certain criteria, we can use only three animals from each litter. These conditions suggest that we use an

incomplete block design, since we cannot administer each treatment to an animal from each litter.

The particular type of incomplete block design with which we are concerned is the *balanced incomplete block design*. In this design every possible pair of treatments appears the same number of times. Further, the balanced incomplete block design requires that each block contain the same number of subjects and that each treatment occur the same number of times.

Let us consider again our experiment comparing seven treatments and using three litter mates as subjects. Table 7.20 shows a possible layout of the data that meets the requirements of a balanced incomplete block design. In Table 7.20 note that every possible pair of treatments appears once, each block contains three subjects, and each treatment appears three times.

TABLE 7.20   **Data layout for a balanced incomplete block design**

	Treatment						
Block	A	B	C	D	E	F	G
1	×	×		×			
2		×	×		×		
3			×	×		×	
4				×	×		×
5	×				×	×	
6		×				×	×
7	×		×				×

Note: × = response of a subject in a given block to indicated treatment

A valid test of the null hypothesis of no differences among treatments in the incomplete block design using parametric techniques requires that the population distributions meet certain assumptions. When the distributions do not meet these assumptions, the following test, proposed by Durbin (T66), allows the researcher to test the null hypothesis of no differences among treatments.

Assumptions

    A. The blocks are mutually independent of each other.
    B. The observations within each block may be ranked in order of magnitude.

Hypotheses

    H_0: The treatments have equal effects
    H_1: The responses to at least one treatment tend to be larger than the responses to at least one other treatment

Test Statistic

It is convenient to display the data in a table similar to Table 7.20. We then rank observations within each block from smallest to largest. We assign tied observations the mean of the rank positions for which they are tied. A moderate number of ties does not greatly affect the results.

The test statistic for the Durbin test is

$$T = \frac{12(t - 1)}{rt(k - 1)(k + 1)} \sum_{j=1}^{t} R_j^2 - \frac{3r(t - 1)(k + 1)}{k - 1} \qquad \text{(7.7)}$$

where

$t = $ the number of treatments under investigation
$k = $ the number of subjects per block $(k < t)$
$r = $ the number of times each treatment occurs
$R_j = $ the sum of the ranks appearing under the jth treatment

If the treatments have the same effect—that is, if H_0 is true—the ranks tend to be randomly distributed over the treatments, and the R_j values tend to be of about the same magnitude. If, however, one or more treatments has a different effect, relatively large ranks or relatively small ranks tend to appear under one treatment. This has the effect of increasing the differences among the R_j's, which in turn causes T to be large. A sufficiently large T, then, causes us to reject the null hypothesis.

Decision Rule

Reject H_0 at the approximate level of significance α if the computed value of the test statistic exceeds the tabulated value of chi-square for α and $t - 1$ degrees of freedom.

The chi-square approximation is good only when r is large, and it should be realized that the results are probably very crude when r is small.

Example 7.7 Moore and Bliss (E10) compared the toxicity of each of seven chemicals applied to *Aphis rumicis*, a black aphid found on nasturtiums. The logarithm of the dose ($+3.806$) required to kill 95% of the insects exposed to a chemical was the measurement reported. Since the experimenters could test only three chemicals a day, they used a balanced incomplete block design requiring seven days for completion of the experiment. The toxicities are shown in Table 7.21. We wish to know whether we may conclude from these data that the effectiveness of the seven chemicals differs.

Hypotheses

H_0: The chemicals are all equally effective
H_1: At least one chemical is more effective than at least one of the others

TABLE 7.21 **Toxicities of seven chemicals applied to *Aphis rumicis* in a balanced incomplete block design**

Chemical	Day 1	2	3	4	5	6	7
A	0.465	0.602		0.423			
B	0.343				0.652	0.536	
C		0.873	0.875		1.142		
D	0.396		0.325				0.609
E		0.634				0.409	0.417
F				0.987	0.989		0.931
G			0.330	0.426		0.309	

Source: W. Moore and C. I. Bliss, "A Method for Determining Insecticidal Effectiveness Using *Aphis rumicis* and Certain Organic Compounds," *J. Econ. Entomol.*, 35 (1942), 544–553: used by permission of the Entomological Society of America.

Test Statistic

We have 7 blocks (days), $t = 7$ treatments (chemicals), $k = 3$ observations per block, and $r = 3$, the number of times that each treatment occurs. The assignment of ranks within blocks and the sum of the ranks by treatment are shown in Table 7.22. When we substitute these data into Equation 7.7, we have

$$T = \frac{12(7-1)}{3(7)(3-1)(3+1)}(5^2 + 5^2 + \cdots + 5^2) - \frac{3(3)(7-1)(3+1)}{3-1}$$

$$= 7.71$$

TABLE 7.22 **The data of Table 7.21 ranked in order of magnitude within blocks**

Chemical	Day 1	2	3	4	5	6	7	R_j
A	3	1		1				5
B	1				1	3		5
C		3	3		3			9
D	2		1				2	5
E		2				2	1	5
F				3	2		3	8
G			2	2		1		5

Decision

When we enter Table A.11 with $t - 1 = 7 - 1 = 6$ degrees of freedom, we find that the probability of observing a value of T as large as 7.71 when H_0 is true is greater than 0.10. Consequently these data do not provide convincing evidence of a difference in effectiveness among chemicals (P value > 0.10).

Power-Efficiency   Noether (T4) has shown that the asymptotic relative efficiency of the Durbin test relative to the corresponding parametric analysis of variance test is the same as that of the Friedman test relative to its corresponding parametric analysis of variance test.

FURTHER READING

Benard and van Elteren (T67) have generalized the Durbin test to the case in which several observations are taken on some experimental units. Puri and Sen (T68) discuss the case of $k = 2$ (the paired comparison test). Skillings and Mack (T69) present a distribution-free test that may be used in lieu of the Durbin test under certain circumstances and that is also appropriate for unbalanced block designs. In certain cases the calculation of the test statistic is cumbersome. In the same paper the authors present distribution-free multiple-comparison procedures for balanced incomplete block designs and for designs that have missing observations in only one treatment.

Since their introduction by Yates (T70) in 1936, incomplete block designs have been used extensively in agricultural research. Industrial researchers also have utilized these designs, but to a lesser extent. A general discussion of incomplete block designs may be found in most experimental design textbooks, such as those by Cochran and Cox (T71), Hicks (T72), Federer (T73), and Kirk (T74).

EXERCISES

7.12 Davies (E11) describes a road test conducted to assess the effect of four different compounding ingredients on the life of automobile tires. Researchers used a balanced incomplete block design to compare four compounds using a single test car. They obtained the relative wear values shown in Table 7.23. Can we conclude from these data that there is a difference among the compounds? Determine the P value.

7.13 Anderson and Bancroft (E12) cite an experiment conducted by Paul (E13) to compare the effects of cold storage on the tenderness and flavor of beef roasts. Six periods of storage (0, 1, 2, 4, 9, and 18 days) were tested. The scores for tenderness are shown in Table 7.24. On the basis of these data, can one conclude that storage periods affect flavor?

TABLE 7.23

Road tests on tires, relative wear

Compound (treatment)	Tire (block)			
	1	2	3	4
A	238	196	254	
B	238	213		312
C	279		334	421
D		308	367	412

Source: Owen L. Davies (ed.), *The Design and Analysis of Industrial Experiments*, second edition, New York: Hafner, 1956.

TABLE 7.24

Tenderness scores of beef roasts following different periods of storage

Blocks	Periods of storage (treatments)					
	0	1	2	4	9	18
1	7	17				
2			26	25		
3					33	29
4	17		27			
5		23		27		
6				29		30
7	10			25		
8		26				37
9			24		26	
10	25				40	
11		25		34		
12			34			32
13	11					27
14		24	21			
15				26	32	

Source: Pauline Paul, "Changes in Palatability, Microscopic Appearance and Electrical Resistance in Beef during the Onset and Passing of Rigor and during Subsequent Storage," unpublished thesis, Ames, Iowa: Iowa State College, 1943.

TABLE 7.25

Measurements, percentage elongation −300, of specimens of rubber stressed at 400 psi

Blocks	Treatments				
	1	2	3	4	5
1	35	16			
2	20		10		
3	13			26	
4	25				21
5		16	5		
6		21		24	
7		27			16
8			20	37	
9			15		20
10				31	17

Source: Carl A. Bennett and Norman L. Franklin, *Statistical Analysis in Chemistry and the Chemical Industry*, New York: Wiley, 1954.

7.14 The data shown in Table 7.25 (see p. 289) are given by Bennett and Franklin (E14). The measurements recorded are (percentage elongation $- 300$) of specimens of rubber stressed at 400 psi. The blocks are 10 bales of rubber; from each, two specimens were taken. Each specimen was subjected to one of a series of five tests (treatments). Test the null hypothesis of no difference among treatments. What is the P value?

7.5

COCHRAN'S TEST FOR RELATED OBSERVATIONS

In some investigations that utilize the randomized complete block design, the response to a treatment may take on only one of two values. We may arbitrarily designate these two possible outcomes "success," or 1, and "failure," or 0. For example, we may assess the effectiveness of four pain-relieving drugs by giving each of several patients each of the drugs. The patients are the blocks in the design. If a given patient obtains relief from pain after receiving a drug, the response is assigned a score of 1. If the patient does not obtain relief from pain, the response is assigned a score of 0.

Cochran (T75) proposed a procedure for testing the null hypothesis of equal treatment effectiveness in this situation, which is a problem of correlated proportions. Cochran's procedure is a generalization of McNemar's (T76) technique, discussed in Chapter 4, to three or more treatments. The test, known as *Cochran's Q test*, consists of the following steps.

Assumptions

A. The data for analysis consist of the responses of r blocks to c independently applied treatments.

B. The responses are 1 for "success" or 0 for "failure." The results may be displayed in a contingency table such as Table 7.26, where the X_{ij}'s are either 0's or 1's.

C. The blocks are a random selection of blocks from a population of all possible blocks.

TABLE 7.26 Contingency table for data layout for Cochran's Q test

Block	Treatment					Block totals
	1	2	3	$\cdots$	c	
1	X_{11}	X_{12}	X_{13}	$\cdots$	X_{1c}	R_1
2	X_{21}	X_{22}	X_{23}		X_{2c}	R_2
3	X_{31}	X_{32}	X_{33}		X_{3c}	R_3
$\vdots$						$\vdots$
r	X_{r1}	X_{r2}	X_{r3}	$\cdots$	X_{rc}	R_r
Treatment totals	C_1	C_2	C_3	$\cdots$	C_c	N = grand total

Hypotheses

H_0: The treatments are equally effective

H_1: The treatments do not all have the same effect

Test Statistic

Cochran (T75) points out that the total number of successes (1's) in a given block is considered fixed. If the null hypothesis is true, every one of the treatments is considered equally likely to contain one of these successes. The objective of the test is to determine whether the treatment totals differ significantly among the treatments.

The test statistic is

$$Q = \frac{c(c-1)\sum_{j=1}^{c} C_j^2 - (c-1)N^2}{cN - \sum_{i=1}^{r} R_i^2} \tag{7.8}$$

Decision Rule

Cochran (T75) showed that the limiting distribution of Q as r increases is the chi-square distribution with $c - 1$ degrees of freedom. Since, as Cochran (T75) points out, blocks containing either all 0's or all 1's do not affect the value of Q, Tate and Brown (T77) recommend the following procedure for testing a computed value of Q for significance.

When the data have been displayed as in Table 7.26, delete all blocks containing only 0's or 1's. If the product of the remaining blocks by the number of treatments is 24 or more, and the number of blocks is at least 4, compare the computed Q for significance with tabulated values of chi-square with $c - 1$ degrees of freedom. If the product is less than 24, construct the exact distribution or use special tables. Tate and Brown (T78) and Patil (T79) have prepared tables for certain values of r and c.

Example 7.8   Gustafson et al. (E15) compared the abilities of three computer-aided diagnostic systems (called models) and physician opinions (majority opinion) in diagnosing on the basis of symptoms, physical signs, and laboratory information. Table 7.27 shows the results obtained with 11 hypothyroid patients. To test the null hypothesis that the four diagnostic methods give the same results, we use Cochran's Q test as follows.

Hypotheses

H_0: The four methods of diagnosing yield identical results

H_1: The four methods differ in their ability to diagnose correctly

Test Statistic

When we delete rows containing all 0's and all 1's from Table 7.27, we have Table 7.28, which also contains the row and column totals needed to compute Q.

TABLE 7.27

Individual mistakes made by physicians and models

Hypothyroid patients	Majority opinion	Actuarial PIP	Subjective PIP	Semi-PIP
1	1	0	0	0
2	1	1	1	1
3	0	0	0	0
4	0	1	1	1
5	1	1	1	1
6	1	0	0	1
7	1	0	1	1
8	1	0	0	1
9	1	0	0	0
10	1	0	0	0
11	1	1	1	1

Source: David H. Gustafson, John J. Kestly, Robert L. Ludke, and Frank Larson, "Probabilistic Information Processing: Implementation and Evaluation of a Semi-PIP Diagnostic System," *Comput. Biomed. Res.*, 6 (1973), 355–370.

Note: 1 indicates correct diagnosis, 0 indicates mistake

TABLE 7.28

Table 7.27 with rows containing all 0's and all 1's deleted

Hypothyroid patients	Majority opinion	Actuarial PIP	Subjective PIP	Semi-PIP	Total
1	1	0	0	0	1
4	0	1	1	1	3
6	1	0	0	1	2
7	1	0	1	1	3
8	1	0	0	1	2
9	1	0	0	0	1
10	1	0	0	0	1
Total	6	1	2	4	13

When we substitute data from Table 7.28 into Equation 7.8, we have

$$Q = \frac{4(4-1)(6^2 + 1^2 + 2^2 + 4^2) - (4-1)(13)^2}{4(13) - (1^2 + 3^2 + 2^2 + 3^2 + 2^2 + 1^2 + 1^2)} = 7.70$$

Since the product of rows by columns in the reduced table is $7 \times 4 = 28$, we compare our computed value of $Q = 7.70$ with tabulated chi-square values with $4 - 1 = 3$ degrees of freedom.

Decision

Table A.11 shows that when H_0 is true, the probability of a Q value as large as 7.70 is between 0.05 and 0.10.

Power-Efficiency The power of Cochran's Q test has been investigated by Wallenstein and Berger (T80).

FURTHER READING

Marascuilo and McSweeney (T81) have proposed a multiple-comparison procedure for use with Cochran's Q test. Shah and Claypool (T82) suggest an alternative statistic to Cochran's Q, as well as an alternative derivation of the latter statistic. Other articles of interest on matched samples and Cochran's Q test include those by Bennett (T83), Berger and Gold (T84), Blomqvist (T85), Fleiss (T86), Grizzle et al. (T87), Madansky (T88), and Ramsey and Ramsey (T89).

EXERCISES

7.15 A manufacturer wished to compare the effectiveness of four methods of treating raw fabric to render it water-repellent. Six types of fabric were used in the operation. A specimen of each type was divided into fourths, and each fourth was randomly assigned to one of the four methods. After treatment, each item was tested for water-repellency and scored 0 if unsatisfactory and 1 if satisfactory. The results shown in Table 7.29 were obtained. Do these data provide sufficient evidence to indicate a difference among the methods?

TABLE 7.29   **Data for Exercise 7.15**

| Fabric | Method | | | |
	A	B	C	D
I	1	1	0	0
II	1	1	0	1
III	1	0	0	0
IV	1	1	1	0
V	1	1	0	1
VI	1	1	0	1

7.16 A horticulturist designed the following experiment to determine whether five liquid fertilizers differ in effectiveness. A fertilizer is defined as effective if the plant to which it is applied achieves certain minimum standards of health, growth, and so forth. Fifteen greenhouse benches were set up as blocks. On each bench were placed five plants of uniform size, age, health, etc., and the fertilizers were applied at random. At the end of a specified period of time, a panel of experts graded each plant. If a plant had achieved the minimum standards of health, growth, etc., it was given a score of 1; if not, it was given a score of 0. The results of Table 7.30 were obtained. Do these data provide sufficient evidence to indicate a difference among fertilizers? What is the P value?

TABLE 7.30   **Data for Exercise 7.16**

| Fertilizers | Blocks | | | | | | | | | | | | | | |
	1	2	3	4	5	6	7	8	9	10	11	12	13	14	15
A	0	0	0	0	0	0	1	0	0	1	0	0	1	0	0
B	0	0	1	0	1	0	0	0	0	1	0	0	0	1	0
C	1	0	1	1	0	0	1	1	1	1	1	1	1	0	0
D	0	1	1	0	0	0	0	1	0	1	0	1	0	1	1
E	0	0	1	1	0	0	0	1	0	1	0	1	0	1	1

7.17   A manufacturer is considering the purchase of four machines for the assembly of a certain product. An experiment is conducted to determine acceptability of the machines to employees. A random sample of 10 employees is selected, and each is assigned to operate each machine (in random order) during a complete assembly cycle. Employees give each machine a score of 0 if they don't like the machine, or a 1 if they do like it. The results shown in Table 7.31 are obtained. Do these data provide sufficient evidence to indicate that the four machines are not equally accepted? Determine the P value.

TABLE 7.31   **Data for Exercise 7.17**

| | Employee | | | | | | | | | |
	1	2	3	4	5	6	7	8	9	10
Machine A	1	1	0	0	1	1	1	0	1	0
Machine B	1	1	1	0	1	1	1	0	1	1
Machine C	1	1	0	1	0	1	1	0	0	0
Machine D	0	0	1	1	0	0	0	1	0	0

7.6

COMPUTER PROGRAMS

Theodorsson–Norheim (T90) describes a computer program written in an elementary subset of BASIC which performs the Friedman test and a two-way analysis of variance on ranks test [the Quade (T91) test], followed by multiple comparisons. The paper includes a program listing as well as sample output.

Roberge (T92, T93) has programmed Cochran's Q test for the computer.

Among the software packages for microcomputers, Friedman's test is well represented. Cochran's Q test can be found in *DAISY PROFESSIONAL*, *M/STAT-2000*, and *SPSS/PC*. Programs for the other procedures discussed in this chapter are less likely to be found.

REVIEW EXERCISES

7.18 Eighteen 7th-grade students participated in an experiment to evaluate the effects of different levels of teacher support during the completion of four units of programmed instruction in mathematics. The units were comparable in difficulty and required background. Each student completed each unit with one of four degrees of teacher support. After completing each unit, the student's mastery of the skills and concepts covered was measured. The results in the form of ranks are shown in Table 7.32. Do these data indicate a difference in the effects of the four different levels of teacher support?

TABLE 7.32   **Ranked achievement scores of students completing programmed instruction units in mathematics with different levels of teacher support**

| Student | Level of teacher support | | | |
	None	Minimal	Moderate	Strong
1	1	2	4	3
2	1.5	1.5	3	4
3	1	3	2	4
4	1	3	4	2
5	1	3	4	2
6	1	4	3	2
7	1	4	3	2
8	2	4	3	1
9	1	2	4	3
10	1	3	4	2
11	1	4	3	2
12	2	4	1	3
13	4	1	2	3
14	1	4	2	3
15	1	4	2	3
16	3	2	4	1
17	2	1	4	3
18	2	4	3	1

7.19 Make all possible comparisons using the data of Exercise 7.18. Let $\alpha = 0.18$.

7.20 In a study of the effects of crowding on the timidity of mice, a psychologist placed one young mouse (all mice were the same sex) from each of 15 litters at random into one of three living situations: no crowding, moderate crowding, and extreme crowding. At the end of the experimental period, a psychologist who was unaware of the condition to which the individual mice had been exposed measured the degree of timidity exhibited by each. The results (ranked) are shown in Table 7.33. Do these data provide sufficient evidence to indicate that different levels of crowding affect timidity in mice?

7.21 Make all possible comparisons using the data of Exercise 7.20 and a significance level of 0.15.

TABLE 7.33   **Level of timidity of mice reared in different conditions of crowding**

	Litter														
	1	2	3	4	5	6	7	8	9	10	11	12	13	14	15
No crowding	3	3	3	2.5	1	3	3	2	2.5	3	2	2	2	3	3
Moderate crowding	2	1	2	2.5	3	2	2	3	2.5	1	3	3	3	1	1
Extreme crowding	1	2	1	1	2	1	1	1	1	2	1	1	1	2	2

7.22   A research psychologist wished to compare the effects of six methods of teaching male rats to be aggressive. The experimenter wished to eliminate hereditary factors as an extraneous source of variation by using rats from each of several litters. The sex composition of available litters necessitated the use of a balanced incomplete block design, with three male rats from each of 10 litters. After the training period the researcher measured the level of aggression exhibited by each rat. The results are shown in Table 7.34. Can one conclude from these data that the training methods have different effects?

TABLE 7.34   **Aggression scores of rats under different treatment conditions**

Litter (block)	Training methods (treatments)					
	A	B	C	D	E	F
1	8	14			11	
2	15	23				29
3	16		26	24		
4	16		19			21
5	5			10	2	
6		24	26	17		
7		29	18		18	
8		20		14		31
9			24		14	32
10				6	5	13

7.23   A psychology research team at a large university wished to determine the extent of frustration exhibited by early adolescent boys attempting to perform a complicated task after receiving one of four different types of instruction. In an attempt to eliminate the effects of as many hereditary and environmental variables as possible, the researchers decided to use siblings. Because of the scarcity of subjects fulfilling the criterion, they used a balanced incomplete block design. A level of frustration score was assigned each boy under each treatment condition. The results are shown in Table 7.35. Can one conclude from these data that the treatments have different effects on level of frustration?

7.24   A research team in a residential facility for the mentally retarded conducted an experiment to assess the differential effectiveness of four different methods of reducing patients' destructive behavior. Ten sets of four patients each were formed, and one patient from each set was randomly assigned to each of the treatments. Within a set, patients were carefully matched on the basis of age, sex, degree of retardation, and other relevant factors. At the end of one month, each patient was assigned a score of 1 if there had been an appreciable improvement in behavior and a score of 0 if there had been no appreciable improvement.

TABLE 7.35
Level of frustration exhibited by early adolescent boys during performance of a complicated task under different methods of instruction

| Family (block) | Methods of instruction (treatments) | | | |
	General verbal instructions	Specific verbal instruction	Direct assistance	Demonstration of task performance
1	50		20	30
2		20	30	35
3	50	40	15	
4	60	50		40

The results are shown in Table 7.36. Do these data provide sufficient evidence to indicate a difference in treatments?

TABLE 7.36
Results of experiments to study effectiveness of different methods of reducing destructive behavior in mentally retarded subjects

| Matched set | Method | | | |
	Overcorrection	Isolation	Electric shock	Token rewards and forfeits
1	1	0	0	1
2	1	1	0	0
3	1	0	1	1
4	0	0	1	1
5	0	0	1	1
6	0	1	1	1
7	0	1	1	0
8	1	1	1	1
9	0	0	1	0
10	0	0	1	1

7.25 A research team compared four methods of promoting relaxation in tense subjects. The subjects were 12 college seniors majoring in psychology. The four treatments were solitary meditation (SM), massage (M), group meditation (GM), and warm bath (WB). Each subject received each treatment with a 48-hour period between treatments. Each subject received a score of 1 when an appreciable degree of relaxation was achieved, and a score of 0 when the treatment failed to appreciably reduce tension. The results are shown in Table 7.37. On the basis of these data, could one conclude that the treatments are not equally effective?

7.26 Refer to Exercise 7.1. Perform all possible multiple comparisons, using an experimentwise error rate of 0.10.

7.27 Refer to Exercise 7.1. Use the method of aligned ranks to test for a difference among treatment effects. Let $\alpha = 0.05$, and find the P value.

7.28 Refer to Exercise 7.2. Use the method of aligned ranks to test for a difference among treatment effects. Let $\alpha = 0.05$, and find the P value.

TABLE 7.37   **Results of experiments to compare four methods of promoting relaxation in tense subjects**

	Subject											
	1	2	3	4	5	6	7	8	9	10	11	12
Treatment SM	1	0	0	0	1	0	0	1	1	1	0	0
Treatment M	1	1	1	1	1	0	1	1	1	1	0	0
Treatment GM	1	1	0	0	1	1	1	0	1	0	1	1
Treatment WB	0	0	1	1	0	0	1	1	1	0	0	0

7.29 Refer to Exercise 7.3. Use the method of aligned ranks to test for a difference among treatment effects. Let $\alpha = 0.05$, and find the P value.

7.30 Refer to Exercise 7.1. Compute an estimate of $M_A - M_C$.

7.31 Refer to Exercise 7.2. Compute an estimate of the difference between the median effect of cyclopropane and the median effect of isoflurane.

7.32 Refer to Exercise 7.3. Compute an estimate of the difference between M_{72} and M_0.

7.33 Refer to Exercise 7.3. Use zero hours of deprivation as the control condition, and perform a test to determine whether one can conclude that the other two conditions yield higher levels of the response variable. What is the smallest level of significance at which the null hypotheses can be rejected?

7.34 Refer to Exercise 7.5. Let control period 1 serve as the control condition and perform a test to see whether the responses for the other conditions are significantly different from the control.

7.35 Refer to Exercise 7.2. Test for order in treatment effects according to the pattern isoflurane $\leq$ halothane $\leq$ cyclopropane.

7.36 Refer to Exercise 7.3. Test for order in treatment effects according to the pattern suggested by the data.

7.37 A researcher wished to compare the effect on current flow of different nonconductive materials used in the manufacture of resistors. The treatments consisted of six different materials, and the blocks were 10 kinds of electronic device. The logistics of the experiment made it necessary to limit the number of treatments to be used per device. To solve the problem, a balanced incomplete block design was employed. The results were as shown in Table 7.38. Can one conclude from these data that the materials have a differential effect on current flow?

7.38 A director of marketing with a large firm conducted an experiment to compare four different sales strategies. Fifteen salespersons participated in the experiment. Each salesperson tried each of the four strategies for a week, determining at random which strategy would be used during a given week. A record was kept of whether during each week the salesperson exceeded (1) or failed to exceed (0) his or her weekly quota. The results are shown in Table 7.39. Can one conclude from these data that the strategies differ in their effectiveness? Let $\alpha = 0.05$, and find the P value.

TABLE 7.38 **Data for Exercise 7.37**

Devices	Materials					
	A	B	C	D	E	F
1			12	15		20
2	7			10		30
3		4		18	20	
4	8	7		16		
5			18	20	13	
6		9			17	26
7		13	16			24
8	12				20	28
9	10	11	12			
10	9		15		13	

TABLE 7.39 **Data for Exercise 7.38**

Salesperson	Strategy			
	A	B	C	D
1	1	0	0	1
2	0	1	0	0
3	1	1	0	1
4	0	1	1	1
5	1	1	0	1
6	0	1	0	0
7	1	1	1	0
8	1	0	0	1
9	0	1	0	0
10	0	1	1	1
11	0	0	0	1
12	0	1	0	0
13	1	1	1	0
14	1	1	0	1
15	0	1	0	0

REFERENCES

T1 Friedman, M., "The Use of Ranks to Avoid the Assumption of Normality Implicit in the Analysis of Variance," *J. Amer. Statist. Assoc.*, 32 (1937), 675–701.

T2 Friedman, M., "A Comparison of Alternative Tests of Significance for the Problem of *m* Rankings," *Ann. Math. Statist.*, 11 (1940), 86–92.

T3 Lehmann, E. L., *Nonparametrics: Statistical Methods Based on Ranks*, San Francisco: Holden-Day, 1975.

T4 Noether, Gottfried E., *Elements of Nonparametric Statistics*, New York: Wiley, 1967.

T5 Jensen, D. R., and Y. V. Hui, "Efficiency of Friedman's χ_r^2 Test under Dependence," *J. Amer. Statist. Assoc.*, 77 (1982), 468–474.

T6   Rothe, Gunter, "A Lower Bound for the Pitman Efficiency of Friedman Type Tests," *Statist. & Probabil. Letters*, 1 (1983), 239–242.

T7   Kendall, M. G., and B. Babington-Smith, "The Problem of *m* Rankings," *Ann. Math. Statist.*, 10 (1939), 275–287.

T8   Wallis, W. A., "The Correlation Ratio for Ranked Data," *J. Amer. Statist. Assoc.*, 34 (1939), 533–538.

T9   Wilcoxon, Frank, *Some Rapid Approximate Statistical Procedures*, New York: American Cyanamid Company, 1949.

T10 de Kroon, J., and P. van der Laan, "Distribution-Free Test Procedures in Two-Way Layouts; A Concept of Rank-Interaction," *Statistica Neerlandica*, 35 (1981), 189–213.

T11 Cooley, Belva J., and John W. Cooley, "Data Analysis for Simulation Experiments: Application of a Distribution-Free Multiple Comparisons Procedure," *Decision Sci.*, 11 (1980), 483–492.

T12 Mack, Gregory A., and John H. Skillings, "A Friedman-Type Rank Test for Main Effects in a Two-Factor ANOVA," *J. Amer. Statist. Assoc.*, 75 (1980), 947–951.

T13 Mack, Gregory A., "A Quick and Easy Distribution-Free Test for Main Effects in a Two-Factor ANOVA," *Communic. in Statist.—Simulation and Computation*, 10 (1981), 571–591.

T14 de Kroon, J., and P. van der Laan, "A Generalization of Friedman's Rank Statistic," *Statistica Neerlandica*, 37 (1983), 1–14.

T15 Rinaman, William C., Jr., "On Distribution-Free Rank Tests for Two-Way Layouts," *J. Amer. Statist. Assoc.*, 78 (1983), 655–659.

T16 Haux, R., M. Schumacher, and G. Weckesser, "Rank Tests for Complete Block Designs," *Biometrical J.*, 26 (1984), 567–582.

T17 Shoemaker, Lewis H., "A Nonparametric Method for Analysis of Variance," *Communic. in Statist.—Simulation and Computation*, 15 (1986), 609–632.

T18 Iman, Ronald L., and William J. Conover, *A Comparison of Several Rank Tests for the Two-Way Layout*, Springfield, Va.: National Technical Information Service, U.S. Department of Commerce, 1976.

T19 Iman, Ronald L., Stephen C. Hora, and W. J. Conover, "Comparison of Asymptotically Distribution-Free Procedures for the Analysis of Complete Blocks," *J. Amer. Statist. Assoc.*, 79 (1984), 674–685.

T20 Lemmer, H. H., "Some Empirical Results on the Two-Way Analysis of Variance by Ranks," *Communic. in Statist.—Theory and Methods*, 9 (1980), 1427–1438.

T21 Hora, Stephen C., and Ronald L. Iman, "Asymptotic Relative Efficiencies of the Rank-Transformation Procedure in Randomized Complete Block Designs," *J. Amer. Statist. Assoc.*, 83 (1988), 462–470.

T22 van der Laan, Paul, and Jos de Kroon, "Ranks, Standardized Ranks and Aligned Ranks in the Analysis of Friedman's Block Design," in Dieter Rasch and Moti Lal Tiku (eds.), *Robustness of Statistical Methods and Nonparametric Statistics*, Dordrecht, Netherlands: D. Reidel, 1984, pp. 66–69.

T23 Iman, Ronald L., and James M. Davenport, "Approximations of the Critical Region of the Friedman Statistic," *Communic. in Statist.—Theory and Methods*, 9 (1980), 571–595.

T24 Gerig, T. M., "Multivariate Extension of Friedman's χ_r^2 Test with Random Covariates," *J. Amer. Statist. Assoc.*, 70 (1975), 443–447.

T25 Gerig, T. M., "A Multivariate Extension of Friedman's χ_r^2 Test," *J. Amer. Statist. Assoc.*, 64 (1969), 1595–1608.

T26 Jensen, D. R., "On the Joint Distribution of Friedman's χ_r^2 Statistic," *Ann. Statist.*, 2 (1974), 311–322.

T27 Mehra, K. L., and J. Sarangi, "Asymptotic Efficiency of Certain Rank Tests for Comparative Experiments," *Ann. Math. Statist.*, 38 (1967), 90–107.

T28 Sen, P. K., "A Note on the Asymptotic Efficiency of Friedman's χ_r^2 Test," *Biometrika*, 54 (1967), 677–679.

T29 Likes, J., and J. Laga, "Probabilities $P(S \geq s)$ for the Friedman Statistic," *Biometrical J.*, 22 (1980), 433–440.

T30 Meddis, Ray, "Unified Analysis of Variance by Ranks," *Br. J. Math. Statist. Psychol.*, 33 (1980), 84–98.

T31 Wei, L. J., "Asymptotically Distribution-Free Simultaneous Confidence Region of Treatment Differences in a Randomized Complete Block Design," *J. Roy. Statist. Soc., Ser. B*, 44 (1982), 201–208.

T32 Brits, Susannah J. M., and H. H. Lemmer, "Nonparametric Tests for Treatment Effects after a Preliminary Test on Block Effects in a Randomized Block Design," *S. African Statist. J.*, 20 (1986), 45–65.

T33 van der Laan, Paul, "Extensive Tables with Exact Critical Values of a Distribution-Free Test for Rank-Interaction in a Two-Way Layout," *Biuletyn Oceny Odmian*, 12 (1987), 196–202.

T34 Hodges, J. L., and Lehmann, E. L., "Rank Methods for Combination of Independent Experiments in Analysis of Variance," *Ann. Math. Statist.*, 33 (1962), 482–497.

T35 Tardif, Serge, "On the Asymptotic Distribution of a Class of Aligned Rank Order Test Statistics in Randomized Block Designs," *Canad. J. Statist.*, 8 (1980), 7–25.

T36 Tardif, Serge, "On the Almost Sure Convergence of the Permutation Distribution for Aligned Rank Test Statistics in Randomized Block Designs," *Ann. Statist.*, 9 (1981), 190–193.

T37 Tardif, Serge, "On the Asymptotic Efficiency of Aligned-Rank Tests in Randomized Block Designs," *Canad. J. Statist.*, 13 (1985), 217–232.

T38 Hollander, Myles, and Douglas A. Wolfe, *Nonparametric Statistical Methods*, New York: Wiley, 1973.

T39 Rhyne, A. L., Jr., and R. G. D. Steel, "Tables for a Treatments versus Control Multiple Comparisons Sign Test," *Technometrics*, 7 (1965), 293–306.

T40 Steel, Robert G. D., "A Multiple Comparison Sign Test: Treatments versus Control," *J. Amer. Statist. Assoc.*, 54 (1959), 767–775.

T41 Doksum, Kjell, "Robust Procedures for Some Linear Models with One Observation Per Cell," *Ann. Math. Statist.*, 38 (1967), 878–883.

T42 McDonald, B. J., and W. A. Thompson, "Rank Sum Multiple Comparisons in One- and Two-Way Classifications," *Biometrika*, 54 (1967), 487–497.

T43 Dunn–Rankin, P., and F. Wilcoxon, "The True Distribution of the Range of Rank Totals in the Two-Way Classification," *Psychometrika*, 31 (1966), 573–580.

T44 Rosenthal, Irene, and Thomas S. Ferguson, "An Asymptotically Distribution-Free Multiple Comparison Method with Application to the Problem of n Rankings of m Objects," *Br. J. Math. Statist. Psychol.*, 18 (1965), 243–254.

T45 Nemenyi, P., *Distribution-Free Multiple Comparisons*, Ph.D. thesis, Princeton University, 1963.

T46 Wilcoxon, Frank, and Roberta A. Wilcox, *Some Rapid Approximate Statistical Procedures* (revised), Pearl River, N.Y.: Lederle Laboratories, 1964.

T47 Page, E. B., "Ordered Hypotheses for Multiple Treatments: A Significance Test for Linear Ranks," *J. Amer. Statist. Assoc.*, 58 (1963), 216–230.

T48   Hollander, Myles, "Rank Tests for Randomized Blocks When the Alternatives Have an *a Priori* Ordering," *Ann. Math. Statist.*, 38 (1967), 867–877.

T49   Pirie, W. R., "Comparing Rank Tests for Ordered Alternatives in Randomized Blocks," *Ann. Statist.*, 2 (1974), 374–382.

T50   Pirie, W. R., and M. Hollander, "A Distribution-Free Normal Scores Test for Ordered Alternatives in Randomized Block Design," *J. Amer. Statist. Assoc.*, 67 (1972), 855–857.

T51   Puri, M. L., and P. K. Sen, "On Chernoff–Savage Tests for Ordered Alternatives in Randomized Blocks," *Ann. Math. Statist.*, 39 (1968), 967–972.

T52   Jonckheere, A. R., "A Test of Significance for the Relation between *m* Rankings and *k* Ranked Categories," *Br. J. Statist. Psychol.*, 7 (1954), 93–100.

T53   Shorack, G. R., "Testing against Ordered Alternatives in Model I Analysis of Variance; Normal Theory and Nonparametric," *Ann. Math. Statist.*, 38 (1967), 1740–1752.

T54   Hettmansperger, T. P., "Nonparametric Inference for Ordered Alternatives in a Randomized Block Design," *Psychometrika*, 40 (1975), 53–62.

T55   Berenson, Mark L., "Some Useful Nonparametric Tests for Ordered Alternatives in Randomized Block Experiments," *Communic. in Statist.—Theory and Methods*, 11 (1982), 1681–1693.

T56   Kepner, James L., and David H. Robinson, "A Distribution-Free Rank Test for Ordered Alternatives in Randomized Complete Block Designs," *J. Amer. Statist. Assoc.*, 79 (1984), 212–217.

T57   Ghiassi, S. H. Mansouri, and Z. Govindarajulu, "An Asymptotically Distribution-Free Test for Ordered Alternatives in Two-Way Layouts," *J. Statist. Planning and Inference*, 13 (1986), 239–249.

T58   Gore, A. P., K. S. Madhav Rao, and M. N. Sahasrabudhe, "Median Tests for Ordered Alternatives," *Gujarat Statist. Rev.*, 13 (April 1986), 55–63.

T59   House, Dennis E., "A Nonparametric Version of Williams' Test for a Randomized Block Design," *Biometrics*, 42 (1986), 187–190.

T60   Williams, D. A., "A Test for Differences between Treatment Means When Several Dose Levels Are Compared with a Zero Dose Control," *Biometrics*, 27 (1971), 103–117.

T61   Williams, D. A., "The Comparison of Several Dose Levels with a Zero Dose Control," *Biometrics*, 28 (1972), 519–531.

T62   Hirotsu, C., "Ordered Alternatives for Interaction Effects," *Biometrika*, 65 (1978), 561–570.

T63   Berenson, Mark L., "A Study of Several Useful Tests for Ordered Alternatives in the Randomized Block Design," *Communic. in Statist.—Simulation and Computation*, 11 (1982), 563–581.

T64   Berenson, Mark L., "A Comparison of Several *k* Sample Tests for Ordered Alternatives in Completely Randomized Designs," *Psychometrika*, 47 (1982), 265–280. Errata, *ibid.*, 535–539.

T65   Rao, K. S. Madhav, and A. P. Gore, "A Note on Optimality of Certain Rank Tests for Ordered Alternatives in Randomized Blocks," *Biometrical J.*, 28 (1986), 267–271.

T66   Durbin, J., "Incomplete Blocks in Ranking Experiments," *Br. J. Psychol.* (Statistical Section), 4 (1951), 85–90.

T67   Benard, A., and P. van Elteren, "A Generalization of the Method of *m* Rankings," *Nederl. Akad. Wetensch. Proc. Ser. A,* 56 (*Indag. Math.* 15) (1953), 358–369.

T68   Puri, M. L., and P. K. Sen, "On the Asymptotic Theory of Rank Order Tests for Experiments Involving Paired Comparisons," *Ann. Inst. Statist. Math.*, 21 (1969), 163–173.

T69   Skillings, John H., and Gregory A. Mack, "On the Use of a Friedman-Type Statistic in Balanced and Unbalanced Block Designs," *Technometrics*, 23 (1981), 171–177.

T70   Yates, F., "A New Method of Arranging Variety Trials Involving a Large Number of Varieties," *J. Agric. Sci.*, 26 (1936), 424–455.

T71 Cochran, William G., and Gertrude M. Cox, *Experimental Designs*, New York: Wiley, 1957.

T72 Hicks, Charles R., *Fundamental Concepts in the Design of Experiments*, New York: Holt, Rinehart, and Winston, 1964.

T73 Federer, Walter T., *Experimental Design*, New York: Macmillan, 1955.

T74 Kirk, Roger E., *Experimental Design Procedures for the Behavioral Sciences*, Belmont, Calif.: Brooks/Cole, 1968.

T75 Cochran, W. G., "The Comparison of Percentages in Matched Samples," *Biometrika*, 37 (1950), 256–266.

T76 McNemar, W., "Note on the Sampling Error of the Difference between Correlated Proportions or Percentages," *Psychometrika*, 12 (1947), 153–157.

T77 Tate, M. W., and S. M. Brown, "Note on the Cochran's Q Test," *J. Amer. Statist. Assoc.*, 65 (1970), 155–160.

T78 Tate, M. W., and S. M. Brown, "Table for Comparing Related Sample Percentages and for the Median Test," Philadelphia: Graduate School of Education, University of Pennsylvania, 1964 (monograph).

T79 Patil, K. D., "Cochran's Q Test: Exact Distribution," *J. Amer. Statist. Assoc.*, 70 (1975), 186–189.

T80 Wallenstein, Sylvan, and Agnes Berger, "On the Asymptotic Power of Tests for Comparing k Correlated Proportions," *J. Amer. Statist. Assoc.*, 76 (1981), 114–118.

T81 Marascuilo, L. A., and M. McSweeney, "Nonparametric Post Hoc Comparisons for Trend," *Psychol. Bull.*, 67 (1967), 401–412.

T82 Shah, Arvind K., and P. L. Claypool, "Analysis of Binary Data in the Randomized Complete Block Design," *Communic. in Statist.—Theory and Methods*, 14 (1985), 1175–1179.

T83 Bennett, B. M., "Tests of Hypotheses Concerning Matched Samples," *J. Roy. Statist. Soc., Ser. B*, 29 (1967), 468–474.

T84 Berger, A., and R. Z. Gold, "Cochran's Q-Test for the Comparison of Correlated Proportions," *J. Amer. Statist. Assoc.*, 68 (1973), 989–993.

T85 Blomqvist, N., "Some Tests Based on Dichotomization," *Ann. Math. Statist.*, 22 (1951), 362–371.

T86 Fleiss, J. L., "A Note on Cochran's Q-Test," *Biometrics*, 21 (1965), 1008–1010.

T87 Grizzle, J. E., C. F. Starmer, and G. G. Koch, "Analysis of Categorical Data by Linear Models," *Biometrics*, 25 (1969), 489–504.

T88 Madansky, A., "Tests of Homogeneity for Correlated Samples," *J. Amer. Statist. Assoc.*, 58 (1963), 97–119.

T89 Ramsey, Patricia P., and Philip H. Ramsey, "Minimum Sample Sizes for Cochran's Test," *Proc. Amer. Statist. Assoc., Sec. on Survey Res. Methods*, Washington, D.C.: American Statistical Association, 1981.

T90 Theodorsson–Norheim, Elvar, "Friedman and Quade Tests: BASIC Computer Program to Perform Nonparametric Two-Way Analysis of Variance and Multiple Comparisons on Ranks of Several Related Samples," *Comput. Biol. Med.*, 17 (1987), 85–99.

T91 Quade, D., "Using Weighted Rankings in the Analysis of Complete Blocks with Additive Block Effects," *J. Amer. Statist. Assoc.*, 74 (1979), 680–683.

T92 Roberge, James J., "A Generalized Non-Parametric Analysis of Variance Program," *Br. J. Math. Statist. Psychol.*, 25 (1972), 128.

T93 Roberge, James J., "A Generalized Non-Parametric Analysis of Variance Program," *Educ. and Psychol. Measurement*, 32 (1972), 805–809.

E1   Hall, F. F., T. W. Culp, T. Hayakawa, C. R. Ratliff, and N. C. Hightower, "An Improved Amylase Assay Using a New Starch Derivative," *Amer. J. Clin. Pathol.*, 53 (1970), 627–634.

E2   Perry, Lawrence B., Russell A. Van Dyke, and Richard A. Theye, "Sympathoadrenal and Hemodynamic Effects of Isoflurane, Halothane, and Cyclopropane in Dogs," *Anesthesiology*, 40 (1974), 465–470.

E3   Syme, G. J., and J. S. Pollard, "The Relation between Differences in Level of Food Deprivation and Dominance in Food Getting in the Rat," *Psychon. Sci.*, 29 (1972), 297–298.

E4   Schmidt, Nathalie J., and Edwin H. Lennette, "Variables of the Rubella Hemagglutination-Inhibition Test System and Their Effect on Antigen and Antibody Titers," *Appl. Microbiol.*, 19 (1970), 491–504.

E5   Basmajian, J. V., and Gail A. Super, "Dantrolene Sodium in the Treatment of Spasticity," *Arch. Phys. Med. Rehabil.*, 54 (1973), 60–64.

E6   Cromer, Richard F., "Conservation by the Congenitally Blind," *Br. J. Psychol.*, 64 (1973), 241–250.

E7   Henry, Philip D., Colin M. Bloor, and Burton E. Sobel, "Increased Serum Creatine Phosphokinase Activity in Experimental Pulmonary Embolism," *Amer. J. Cardiol.*, 26 (1970), 151–155.

E8   Gilbert, Roger K., "Analysis of Results of the 1969 Comprehensive Chemistry Survey of the College of American Pathologists," *Amer. J. Clin. Pathol.*, 54 (1970), 463–482.

E9   Stern, L., N. N. Khanna, G. Levey, S. J. Yaffe, "Effect of Phenobarbital on Hyperbilirubinemia and Glucuronide Formation in Newborns," *Amer. J. Dis. Child.*, 120 (1970), 26–31.

E10 Moore, W., and C. I. Bliss, "A Method for Determining Insecticidal Effectiveness Using *Aphis rumicis* and Certain Organic Compounds," *J. Econ. Entomol.*, 35 (1942), 544–553.

E11 Davies, Owen L. (ed.), *The Design and Analysis of Industrial Experiments*, second edition, New York: Hafner, 1956.

E12 Anderson, R. L., and T. A. Bancroft, *Statistical Theory in Research*, New York: McGraw-Hill, 1952.

E13 Paul, Pauline, "Changes in Palatability, Microscopic Appearance and Electrical Resistance in Beef during the Onset and Passing of Rigor and during Subsequent Storage," unpublished thesis, Ames, Iowa: Iowa State College, 1943.

E14 Bennett, Carl A., and Norman L. Franklin, *Statistical Analysis in Chemistry and the Chemical Industry*, New York: Wiley, 1954.

E15 Gustafson, David H., John J. Kestly, Robert L. Ludke, and Frank Larson, "Probabilistic Information Processing: Implementation and Evaluation of a Semi-PIP Diagnostic System," *Comput. Biomed. Res.*, 6 (1973), 355–370.

GOODNESS-OF-FIT TESTS

In many situations a researcher's interest is focused on the nature of one or more population distributions. The validity of parametric statistical inference procedures for example, depends on the shape of the populations from which the samples have been drawn. When we do not know the functional forms of these populations, we first want to test whether the population of interest is likely to be distributed according to the assumptions underlying the proposed parametric procedure.

In genetic research, probability models have been constructed to account for the structure of the populations resulting from certain plant and animal matings. These models presume that certain conditions or assumptions prevail. Samples are drawn from unknown populations, and methods of goodness of fit are employed to determine how well the observed sample data "fit" some proposed model. In many areas of quantitative analysis, model-building is an essential activity. Goodness-of-fit tests can be useful tools for evaluating how well a model approximates the real-world situation it is designed to describe.

In this text three types of goodness-of-fit problems are considered. The first type considered is the problem in which the investigator wishes to know whether sample data support the hypothesis that the sampled population follows some specified distribution. The second type is the problem that arises when an investigator wishes

to know whether two independent samples come from identically distributed populations. Finally, a procedure is presented for constructing a confidence band for a population distribution.

Over the years many procedures for goodness-of-fit problems have been proposed. This chapter covers in detail the two best known of these: the *chi-square goodness-of-fit test* and the *Kolmogorov–Smirnov test*. Section 8.6 mentions briefly some additional goodness-of-fit procedures and refers the reader to appropriate references containing more detailed information about them.

8.1

CHI-SQUARE GOODNESS-OF-FIT TEST

The chi-square tests of independence and homogeneity were discussed in Chapter 5. The chi-square test of goodness of fit of this chapter is similar to these other chi-square tests in that the test statistic results from a comparison of expected and observed frequencies.

Assumptions

A. The data available for analysis consist of a random sample of n independent observations.

B. The measurement scale may be nominal.

C. The observations can be classified into r nonoverlapping categories that exhaust all classification possibilities; that is, the categories are mutually exclusive and exhaustive. The number of observations falling into a given category is called the *observed frequency* of that category.

The data for analysis may be displayed in tabular form, as shown in Table 8.1. The categories may be either nominal or numerical. For example, a sample of individuals may be placed into one of the two nominal categories, male and female. If age is the variable of interest, these same individuals may be placed in one or the other of several age (numerical) categories, such as less than 15 years, 15–24, 25–34, 35–44, 45–54, and 55 and older.

For each category there is a probability that an observation randomly selected from the hypothesized population will fall in that category. We may designate the

---

TABLE 8.1   **Data display for chi-square goodness-of-fit test**

Categories	1	2	3	$\cdots$	i	$\cdots$	r	Total
Observed frequencies	O_1	O_2	O_3	$\cdots$	O_i	$\cdots$	O_r	n

probabilities for categories $1, 2, \ldots, i, \ldots, r$, respectively, as $p_1, p_2, \ldots, p_i, \ldots, p_r$. When the null hypothesis is true, we can obtain the *expected frequency* for each category by computing the product of n and the corresponding category probability. For example, the products np_1, np_i, and np_r yield the *expected frequencies*, under H_0, for categories 1, i, and r.

Hypotheses

H_0: The sample has been drawn from a population that follows a specified distribution

H_1: The sample has not been drawn from a population that follows the specified distribution

Note that the alternative hypothesis does not indicate how the true distribution differs from the hypothesized distribution.

Test Statistic

We expect random samples drawn from populations to reflect the characteristics of those populations. Thus, if we have drawn a sample from a specified (hypothesized) population, we expect a close (in some appropriate sense) "fit" between observed and expected frequencies falling into the various categories. If the null hypothesis is true, then, we expect close agreement between the observed and expected frequencies falling into the categories.

As seen in the discussion of chi-square tests of independence and homogeneity, an appropriate measure of the agreement (or disagreement) between observed and expected frequencies is the test statistic computed by dividing the squared difference between observed and expected frequencies in each set of such frequencies by the expected frequency and summing over all sets. Using the notation of this chapter, we find that this test statistic is

$$X^2 = \sum_{i=1}^{r} \frac{(O_i - E_i)^2}{E_i} \qquad \text{(8.1)}$$

Decision Rule

For large samples X^2 is distributed approximately as chi-square with $r - 1$ degrees of freedom. Thus, if the computed value of X^2 is equal to or greater than the tabulated value of chi-square for $r - 1$ degrees of freedom and significance level α, we can reject the null hypothesis at the α level of significance.

Sample Size   As we already noted, the statistic X^2 is distributed approximately as chi-square only if n is large. A sample size of at least 30 is adequate in most practical applications, provided that none of the expected frequencies is too small. A frequently followed rule of thumb regarding the minimum size of expected frequencies was proposed by Cochran (T1, T2). He recommends that in goodness-of-fit tests of the type described here, no expected frequency should be less than 1.

When expected frequencies less than 1 occur, we usually combine the categories in which they occur with adjacent categories until the minimum frequency requirement has been met. When we follow this procedure, we have to recompute the degrees of freedom on the basis of the new number of categories.

Estimation of Parameters   When we apply the chi-square goodness-of-fit test, we frequently have to compute estimates of certain parameters of the hypothesized population before we can compute the p_i's, the expected relative frequencies. In rarer instances, these parameters are specified as part of the null hypothesis. When we have to estimate parameters from the sample data, we compute X^2 in the way previously described, but we compute the degrees of freedom differently. In addition to subtracting 1 from r, we also subtract an additional 1 for each parameter that we have to estimate. If g is the number of parameters that we have to estimate from sample data, the degrees of freedom are $r - g - 1$.

The following examples illustrate the chi-square goodness-of-fit test.

Example 8.1 (Uniform Distribution)   Stranges and Riccio (E1) report the results of a study to determine whether counselees with different racial and ethnic backgrounds preferred being counseled by persons with similar backgrounds. Thirty-six black trainees in a Manpower Development and Training Program were allowed to choose freely a counselor from among the following: a black male, a northern white male, an Appalachian white female, a northern white female, a black female, and an Appalachian white male. The frequencies with which the counselors were chosen are shown in column 2 of Table 8.2. The expected frequencies under the null hypothesis of no racial preference on the part of the counselees are shown in column 1 of the table. We wish to know whether we can conclude that there is a lack of uniformity in counselees' choices.

TABLE 8.2   **Black trainees' choices of counselors**

Counselor choice	Expected frequencies	Observed frequencies
Black male	6	13
Northern white male	6	6
Appalachian white female	6	0
Northern white female	6	3
Black female	6	11
Appalachian white male	6	3
Total	36	36

Source: Richard J. Stranges and Anthony C. Riccio, "Counselee Preferences for Counselors: Some Implications for Counselor Education," *Counselor Education and Supervision*, 10 (1970), 39–45, Copyright 1970. American Personnel and Guidance Association; reprinted with permission.

Hypotheses

H_0: The available counselors are equally preferred by counselees (that is, the distribution of the population is uniform)

H_1: At least one available counselor is preferred over at least one other (the distribution is not uniform)

Test Statistic

Under the null hypothesis that all counselors are equally preferred, the p_i are the same for all categories (counselors) and each is equal to $p_i = 1/6 = 0.1666667$. When we multiply 36 by 0.1666667, we get 6, the expected frequency for each category, as shown in Table 8.2. By Equation 8.1, we compute

$$X^2 = \frac{(13 - 6)^2}{6} + \frac{(6 - 6)^2}{6} + \cdots + \frac{(3 - 6)^2}{6} = 21.33$$

Decision

Since we did not have to estimate any parameters from the sample data, we have $r - 1 = 6 - 1 = 5$ degrees of freedom. Table A.11 reveals that the probability of obtaining a value of X^2 as large as 21.33 when H_0 is true is less than 0.005. Since this is such a small probability, we reject H_0 and conclude that the available counselors are not equally preferred.

Example 8.2 (Normal Distribution) An important characteristic to be considered in the selection and breeding of alfalfa is the level of Fraction 1 Protein (FR1P) in the plant. Miltimore et al. (E2) investigated the distribution of FR1P levels in alfalfa clones. Table 8.3 shows the results of their study. We wish to see whether these data are compatible with the hypothesis that the variable under consideration is normally distributed.

TABLE 8.3 Frequency distribution of FR1P levels in 1372 alfalfa seedlings from diverse genetic sources

FR1P class interval, % oven-dry weight	Number in class	FR1P class interval, % oven-dry weight	Number in class
0.61–1.20	1	4.81–5.40	284
1.21–1.80	3	5.41–6.00	83
1.81–2.40	4	6.01–6.60	13
2.41–3.00	65	6.61–7.20	1
3.01–3.60	180	7.21–7.80	1
3.61–4.20	328	7.81–8.40	0
4.21–4.80	408	8.41–9.00	1

Source: J. E. Miltimore, J. M. McArthur, B. P. Goplen, W. Majak, and R. E. Horwath, "Variability of Fraction 1 Protein and Total Phenolic Constituents in Alfalfa," *Agron. J.*, 66 (1974), 384–386; reproduced by permission of the American Society of Agronomy.

Hypotheses

H_0: The distribution of FR1P, by percentage of oven-dry weight, in alfalfa seedlings follows a normal distribution

H_1: The population is not normally distributed

Test Statistic

Before computing the test statistic, we must first estimate the mean, μ, and the variance, σ^2, of the population represented by our sample. When we compute estimates of parameters for hypothesized distributions from sample data grouped into class intervals, it is appropriate to compute the estimates from the grouped data rather than from the ungrouped data.

The following formulas for computing the sample mean and variance from grouped data are found in most elementary statistics texts:

$$\text{Mean: } \hat{\mu} = \frac{\sum_{i=1}^{r} f_i x_i}{n} \tag{8.2}$$

$$\text{Variance: } \hat{\sigma}^2 = \frac{n \sum_{i=1}^{r} f_i x_i^2 - (\sum_{i=1}^{r} f_i x_i)^2}{n(n-1)} \tag{8.3}$$

where f_i is the frequency in the ith class interval, x_i is the midpoint of the ith class interval, and $n = \sum_{i=1}^{r} f_i$. When we use these formulas with the data of Table 8.3, we find $\hat{\mu} = 4.32$ and $\hat{\sigma} = 0.81$.

The next step is to find for each class interval the expected frequency under the assumption of a normally distributed population. We first find the expected relative frequency or expected proportion for each class interval. Then we multiply these expected relative frequencies by 1372 to obtain expected frequencies for each class interval.

To find the expected relative frequencies, we convert the lower limit of each class interval to the corresponding value of the standard normal distribution by using the formula

$$z_{Li} = (x_{Li} - \hat{\mu})/\hat{\sigma} \tag{8.4}$$

in which x_{Li} is the lower limit of the ith class interval. We use $\hat{\mu}$ and $\hat{\sigma}$ to estimate μ and σ. We then obtain expected relative frequencies from the standard normal distribution (Table A.2). We use the computed values of z_{Li} to establish the boundaries of the class intervals expressed in terms of the standard normal variable. Table 8.4 shows the steps involved in finding the expected frequencies. (Because of the limitations of Table A.2, we must combine some of the class intervals at each end of the observed distribution in order to reach values of z that are given in the table.)

TABLE 8.4   **Calculations necessary to find expected frequencies for Example 8.2**

Class interval	z_{Li}	Expected relative frequency	Expected frequency	Observed frequency
< 0.61	∞			0
0.61 − 1.20	−4.58	0.0091	12.49	1
1.21 − 1.80	−3.84			3
1.81 − 2.40	−3.10			4 } 8
2.41 − 3.00	−2.36	0.0435	59.68	65
3.01 − 3.60	−1.62	0.1368	187.69	180
3.61 − 4.20	−0.88	0.2549	349.72	328
4.21 − 4.80	−0.14	0.2814	386.08	408
4.81 − 5.40	0.60	0.1858	254.92	284
5.41 − 6.00	1.35	0.0702	96.31	83
6.01 − 6.60	2.09	0.0160	21.95	13
6.61 − 7.20	2.83			1
7.21 − 7.80	3.57			1
7.81 − 8.40	4.31	0.0023	3.16	0 } 3
8.41 − 9.00	5.05			1
> 9.00	5.78			0
		1.0000	1372.00	1372

Table 8.4 shows that $z_{L5} = (2.41 - 4.32)/0.81 = -2.36$ is the smallest value of z_{Li} contained in Table A.2. From Table A.2, the area to the left of -2.36 is equal to 0.0091. Multiplying 0.0091 by 1372 gives 12.49 as the expected frequency for the class interval resulting from the combining of the first four class intervals.

We can now use Equation 8.1 to compute our test statistic.

$$X^2 = \frac{(8 - 12.49)^2}{12.49} + \frac{(65 - 59.68)^2}{59.68} + \cdots + \frac{(3.00 - 3.16)^2}{3.16} = 13.81$$

Decision

Since we have nine categories (class intervals) after combining those at the ends, and since we estimated two parameters from the data, our degrees of freedom are $9 - 3 = 6$. We refer to Table A.11, and since 13.81 is greater than 12.592, we can reject H_0 at the 0.05 level of significance. Then we can conclude at the 0.05 level of significance that the data did not come from a normally distributed population ($0.025 < P$ value < 0.05).

Example 8.3 (Binomial Distribution)

Gore (E3) describes the mandrel bend test, used to determine brittleness in nylon. The test consists of bending a nylon test bar 1/2 inch wide, 5 inches long, and 1/8 inch thick around a mandrel (spindle) 1/2 inch in diameter. Breaks occur at brittle points. Each of 280 bars was bent in five places, and the number of breaks

(0, 1, 2, 3, 4, 5) for a given bar was recorded. The author reports that two important results of the investigation were (a) a characterization of the brittleness phenomenon, and (b) a reevaluation of the mandrel bend test. The results are shown in Table 8.5. We wish to test the null hypothesis that the data come from a binomial distribution.

TABLE 8.5

Breaks per bar and observed number of bars breaking that number of times for nylon bars subjected to mandrel bend test

Breaks per bar	0	1	2	3	4	5
Number of bars	157	69	35	17	1	1

Source: W. L. Gore, "Quality Control in the Chemical Industry IV: Statistical Methods in Plastics Research and Development," *Indust. Quality Control*, 4 (September 1947), 5–8; copyright 1947, American Society for Quality Control; reprinted by permission.

Hypotheses

H_0: The data come from a binomial distribution

H_1: The data do not come from a binomial distribution

Test Statistic

The test statistic is X^2, as computed by Equation 8.1, but first we must compute the expected frequencies. A total of 280(5) = 1400 bends were made. The number of bends producing breaks is given by

$$(0)(157) + (1)(69) + (2)(35) + (3)(17) + (4)(1) + (5)(1) = 199$$

Under the hypothesis that each bend is equally likely to produce a break, an estimate $\hat{p}$ of the probability p that a bend will produce a break is given by $\hat{p} = 199/1400 = 0.14$. We obtain the expected relative frequencies $f(x)$ by evaluating the following binomial function for $x = 0, 1, 2, 3, 4$, and 5:

$$f(x) = \binom{5}{x}(0.14)^x(0.86)^{5-x}$$

Table A.1 reveals that these expected relative frequencies are, respectively, 0.4704, 0.3829, 0.1247, 0.0203, 0.0017, and 0.0001. Multiplying each of these expected relative frequencies by 280 gives the following expected frequencies for 0, 1, 2, 3, 4, and 5 breaks, respectively: 131.71, 107.21, 34.92, 5.68, 0.48, and 0.03. Since the last two expected frequencies are less than 1, we combine them with the fourth expected frequency to get 6.19.

The computed value of the test statistic, then, is

$$X^2 = \frac{(157 - 131.71)^2}{131.71} + \frac{(69 - 107.21)^2}{107.21}$$

$$+ \frac{(35 - 34.92)^2}{34.92} + \frac{(19 - 6.19)^2}{6.19} = 44.99$$

Decision

We lost one degree of freedom when we estimated the binomial parameter p from the sample data, and after combining we have only 4 categories. Thus we compare the test statistic for significance with the chi-square distribution with $4 - 1 - 1 = 2$ degrees of freedom. We find from Table A.11 that the probability of observing a value of X^2 as large as 44.99 when H_0 is true is less than 0.005. Consequently we reject H_0 and conclude that the data did not come from a binomial distribution.

Example 8.4 (Poisson Distribution)

In an innovative physics course, Lafleur et al. (E4) challenged their students to investigate any phenomenon the students thought might follow a Poisson distribution. In one such investigation, a student counted the number of people in the entrance of a dormitory during the time it took the student to walk through it (30 seconds). The results are shown in Table 8.6. To test the goodness of fit of these data to a Poisson distribution, we proceed as follows.

Hypotheses

H_0: The data from which the sample was drawn follow a Poisson distribution

H_1: The data do not follow a Poisson distribution

TABLE 8.6

Number of students in lobby of a dormitory during 30-second intervals and number of intervals in which each number was observed

Observed number of students	Number of intervals
0	20
1	54
2	74
3	67
4	45
5	25
6	11
7	4
	300

Source: M. S. Lafleur, P. F. Hinrichsen, P. C. Landry, and R. B. Moore, "The Poisson Distribution: An Experimental Approach to Teaching Statistics," *Physics Teacher*, 10 (1972), 314–321. Copyright 1972, American Association of Physics Teachers.

Test Statistic

The test statistic is computed from Equation 8.1. Since the Poisson parameter λ is not given, we must compute it from the sample data before we can determine the expected frequencies. We compute the estimate $\hat{\lambda}$ of the parameter λ as follows:

$$\hat{\lambda} = \frac{(0)(20) + (1)(54) + (2)(74) + (3)(67)}{300}$$

$$+ \frac{(4)(45) + (5)(25) + (6)(11) + (7)(4)}{300}$$

$$= 2.67$$

We find the expected relative frequencies for $0, 1, \ldots, 7$ intervals by evaluating the function

$$f(x) = \frac{e^{-\hat{\lambda}}\hat{\lambda}^x}{x!}$$

for $x = 0, 1, 2, \ldots, 7$, respectively. Values for $f(x)$ for given values of x and certain $\hat{\lambda}$ may be found in most sets of mathematical and/or statistical tables and in the appendixes of statistics textbooks. You can evaluate the Poisson function quite easily by using a hand-held calculator that has functions of the form e^x, y^x, and $x!$. The expected relative frequencies for this example were computed on such a calculator; they are 0.069, 0.185, 0.247, 0.220, 0.147, 0.078, 0.035, and 0.013. We obtain expected frequencies corresponding to the eight categories of observed frequencies by multiplying the expected relative frequencies by 300. This procedure yields the following expected frequencies: 20.7, 55.5, 74.1, 66.0, 44.1, 23.4, 10.5, and 3.9.

The observed frequencies and their corresponding expected frequencies, then, are as follows:

Frequencies							
Observed	20	54	74	67	45	25	11	4
Expected	20.7	55.5	74.1	66.0	44.1	23.4	10.5	3.9

We are now able to compute the value of the test statistic.

$$X^2 = \frac{(20 - 20.7)^2}{20.7} + \frac{(54 - 55.5)^2}{55.5} + \frac{(74 - 74.1)^2}{74.1} + \frac{(67 - 66.0)^2}{66.0}$$

$$+ \frac{(45 - 44.1)^2}{44.1} + \frac{(25 - 23.4)^2}{23.4} + \frac{(11 - 10.5)^2}{10.5} + \frac{(4 - 3.9)^2}{3.9}$$

$$= 0.234$$

Decision

Since there are eight categories and we estimated one parameter from the sample data, we have $8 - 1 - 1 = 6$ degrees of freedom. When we consult Table A.11, we find that the probability of obtaining a value of X^2 as large as or larger than 0.234 is greater than 0.995. We cannot reject H_0 at any reasonable level of significance. The data fit a Poisson distribution almost perfectly.

Power-Efficiency The power of the chi-square goodness-of-fit test has been considered by Broffitt and Randles (T3) and Schorr (T4, T5). Slakter (T6) presents the results of a Monte Carlo study of the accuracy of an approximation to the power of the chi-square goodness-of-fit test with small but equal expected frequencies.

FURTHER READING

As might be suspected, the chi-square goodness-of-fit test does not yield unique results, since the value of the test statistics depends on the number of class intervals and the choice of the endpoints of these intervals. This problem has been the subject of considerable research. Viollaz (T7) studied the variability of the chi-square statistic due to the freedom in the choice of the position of the endpoints in the case of a fixed number of class intervals, each of equal expected frequency under the null hypothesis.

Dahiya and Gurland (T8) suggest the number of class intervals to use in testing for normality against certain alternatives; in another article (T9), they discuss the use of the chi-square test with random intervals. Moore (T10) discusses the case of random cell boundaries.

The power of chi-square goodness-of-fit tests based on different partitions has been investigated by Kallenberg et al. (T11). The problem of choice of number of class intervals is also discussed by Quine and Robinson (T12) and Schorr (T4, T5).

Chase (T13) points out that if the parameters are estimated independently of the sample, the chi-square test statistic for a goodness-of-fit test has a limiting distribution that is stochastically larger than that of the test of fit for a completely specified distribution. The author notes that if the critical values for the test of fit for a completely specified distribution are incorrectly used, the probability that the null hypothesis will be rejected when it is true is greater than the desired level of significance.

Lawal and Upton (T14) propose an approximation to the distribution of the chi-square goodness-of-fit statistic for use with small expected frequencies. They report that the approximation is valid provided that the smallest expected frequency is greater than $r/d^{3/2}$, where r is the number of expected frequencies that are less than 5 and d is the number of degrees of freedom.

Brock and Kshirsagar (T15) discuss a chi-square goodness-of-fit test for Markov renewal processes.

Hewett and Tsutakawa (T16) construct a two-stage test to use when sampling

occurs in two stages. They give tables of critical values for 1% and 5% level tests for 1 to 10 degrees of freedom. Moore and Spruill (T17) present a unified large-sample theory of general chi-square tests of goodness of fit under composite null hypotheses and Pitman alternatives.

Moore (T18) has written a wide-ranging paper on chi-square-type goodness-of-fit tests.

Ritchey (T19) illustrates the application of the chi-square goodness-of-fit test to discrete common-stock returns and at the same time presents a method for adjusting cell boundaries.

Other articles dealing with the chi-square goodness-of-fit test include those by O'Reilly and Quesenberry (T20), Slakter (T21, T22), Albrecht (T23), and Jammalamadaka and Tiwari (T24).

EXERCISES

8.1 Yousuf (E5) administered two inventories designed to measure risk orientation in 200 male undergraduates. The first, the Choice Dilemmas Questionnaire [Kogan and Wallach (E6) and Wallach and Kogan (E7, E8)], assesses risk-taking behavior under uncertain outcomes representing socially approved acts. The second, the Behavior Prediction Scale [Rettig and Rawson (E9)], consists of items measuring unethical risk-taking behavior. The scores for the two inventories are shown in Table 8.7. Test the goodness of fit of each of these two distributions to the normal distribution. Determine the P value.

TABLE 8.7   **Frequency distributions of scores on Choice Dilemma Questionnaire (CDQ) and Behavior Prediction Scale (BPS) by 200 male undergraduates**

CDQ		*BPS*	
Scores	**Frequency**	**Scores**	**Frequency**
37–39	3	84–90	3
34–36	15	77–83	5
31–33	7	70–76	6
28–30	17	63–69	20
25–27	6	56–62	20
22–24	36	49–55	18
19–21	23	42–48	29
16–18	34	35–41	17
13–15	16	28–34	14
10–12	25	21–27	13
7–9	5	14–20	12
4–6	13	7–13	13
	200	0–6	30
			200

$\hat{\mu} = 19.91,$* $\hat{\sigma} = 8.31$ $\hat{\mu} = 38.28,$ $\hat{\sigma} = 23.24$

Source: S. M. Anwar Yousuf, "Two Measures of Risk-Taking in India," *Psychologia,* 16 (1973), 46–48.

* Means and standard deviations reported by author of article.

8.2 Ibrahim et al. (E10) studied genetic blood markers of the inhabitants of Siwa Oasis, which lies in the Western Desert of Egypt and has about 6000 inhabitants. The sample studied consisted of school boys and girls between the ages of 6 and 18 years. The observed and expected frequencies of four blood groups in 191 subjects are shown in Table 8.8. Do a chi-square test of goodness of fit on these data. Determine the *P* value.

TABLE 8.8 **Observed and expected frequencies of four blood groups in 191 inhabitants of Siwa Oasis**

Blood group	Numbers observed	Numbers expected
O	72	72.52
A	63	62.40
B	43	42.40
AB	13	13.68
Total	191	191.00

Source: W. N. Ibrahim, K. Kamel, O. Salim, A. Azim, M. F. Gaballah, F. Sabry, A. El-Naggar, and K. Hoerman, "Hereditary Blood Factors and Anthropometry of the Inhabitants of the Egyptian Siwa Oasis," *Hum. Biol.*, 46 (1974), 57–68.

8.3 From an epidemiologic study of cleft lip and related conditions in Hungary, Czeizel and Tusnadi (E11) reported the data shown in Table 8.9. Can we conclude from these data that cleft lip occurs nonuniformly over time? Determine the *P* value.

TABLE 8.9 **Frequency of cleft lip in births occurring in Budapest, 1962–1967**

Year	Number
1962	7
1963	12
1964	6
1965	10
1966	12
1967	6
Total	53

Source: A. Czeizel and G. Tusnadi, "An Epidemiologic Study of Cleft Lip with or without Cleft Palate and Posterior Cleft Palate in Hungary," *Hum. Hered.*, 21 (1971), 17–38; used by permission of S. Karger AG, Basel.

8.4 As part of a study of Alzheimer's disease, Dusheiko (E12) reported the data shown in Table 8.10. Are the data compatible with the hypothesis that the brain weights of victims of Alzheimer's disease are normally distributed? What is the *P* value?

8.5 Table 8.11 shows the number of males in the first seven children born to 1334 Swedish ministers of religion. The data were reported by Edwards and Fraccaro (E13). Test the null hypothesis that the number of males is binomially distributed with $p = 0.51$. Find the *P* value.

8.6 An illustration of the Poisson distribution considered to be the description of a random process is contained in a discussion of a paper by Kendall (E14). The reported data, shown

TABLE 8.10   **Weight of the brain in Alzheimer's disease**

Weight, grams	800–900	900–1000	1000–1100	1100–1200	1200–1300	Above 1300
Number of cases	9	23	59	42	20	7*

Source: S. D. Dusheiko, "Some Questions Concerning the Pathological Anatomy of Alzheimer's Disease," *Sov. Neurol. Psychiat.*, 7 (Summer 1974), 56–64; published by International Arts and Sciences Press, White Plains, N.Y.; reprinted by permission.

* Delete from analysis.

TABLE 8.11   **Number of males in first seven births of children born to a sample of 1334 Swedish ministers of religion**

Number of males	0	1	2	3	4	5	6	7
Number of families in which that number occurs	6	57	206	362	365	256	69	13

Source: A. W. Edwards and M. Fraccaro, "Distribution and Sequences of Sexes in a Selected Sample of Swedish Families," *Ann. Hum. Genet.*, 24 (1960), 245–252; used by permission of Cambridge University Press.

in Table 8.12, are the numbers of taxis arriving at Euston Station, London, in one-minute intervals between 9:45 and 10:00. Test the null hypothesis that the number of arrivals follows a Poisson distribution. Determine the *P* value.

TABLE 8.12   **Data on taxicab arrivals at Euston Station, London, between 9:45 and 10:00**

Observed number of arrivals in 1-minute interval	0	1	2	3	4	5	6	7
Number of intervals in which that number arrived	18	18	14	7	3	0	0	0

Source: Dr. J. H. Howlett in David G. Kendall, "Some Problems in the Theory of Queues," *J. Roy. Statist. Soc., Ser. B*, 13 (1951), 151–185.

8.7 The data shown in Table 8.13 were reported by Neyman (E15). Test the goodness of fit of these data to a Poisson distribution. Determine the *P* value.

TABLE 8.13   **Distribution of European corn borers in 120 groups of eight hills each**

Number of borers	0	1	2	3	4	5	6	7	8	9	10	11	12
Observed frequency	24	16	16	18	15	9	6	5	3	4	3	0	1

Source: J. Neyman, "On a New Class of 'Contagious' Distributions, Applicable in Entomology and Bacteriology," *Ann. Math. Statist.*, 10 (1939), 35–57; data provided by Dr. Beall.

8.2

KOLMOGOROV–SMIRNOV ONE-SAMPLE TEST

The chi-square test of goodness of fit discussed in Section 8.1 was designed for use with nominal data. This section introduces a procedure that was designed for testing the goodness of fit for continuous data. Hence it can be used with data measured on at least an ordinal scale.

This goodness-of-fit procedure was introduced in 1933 by the Russian mathematician A. N. Kolmogorov (T25). In 1939 another Russian mathematician, N. V. Smirnov (T26), introduced a procedure for use with the data of two samples. This procedure allows the investigator to test the null hypothesis that two samples have come from the same or identical populations; it is similar in nature to the Kolmogorov test. Because of the similarities between the Kolmogorov test and the Smirnov test, the former has come to be known as the *Kolmogorov–Smirnov one-sample test* and the latter as the *Kolmogorov–Smirnov two-sample test*. The Kolmogorov–Smirnov one-sample test is discussed in this section, and the Kolmogorov–Smirnov two-sample test in Section 8.3.

When we apply the Kolmogorov–Smirnov one-sample goodness-of-fit test, we focus on two *cumulative distribution functions*: a hypothesized cumulative distribution and the observed cumulative distribution. The reader may recall that we use a capital letter such as $F(x)$ to designate a cumulative distribution function. For a given x, $F(x)$ is the probability that the value of the random variable X is less than or equal to x; that is, $F(x) = P(X \leq x)$.

Suppose that we draw a random sample from some unknown distribution function $F(x)$. We are interested in determining whether we can conclude that $F(x) \neq F_0(x)$ for all x. If $F(x) = F_0(x)$, we expect close agreement (except for sampling variability) between $F_0(x)$ and $S(x)$, the sample (observed) or empirical distribution function. The objective of the Kolmogorov–Smirnov one-sample goodness-of-fit test is to determine whether the lack of agreement between $F_0(x)$ and $S(x)$ is sufficient to cast doubt on the hypothesis that $F(x) = F_0(x)$.

The Kolmogorov–Smirnov one-sample test may be summarized in the following steps:

Assumptions

The data consist of the independent observations $X_1, X_2, \ldots, X_n$, constituting a random sample of size n from some unknown distribution function designated by $F(x)$.

Hypotheses

If we let $F_0(x)$ be the hypothesized distribution function (cumulative probability function), then we can state the null hypotheses and the corresponding alternatives as follows:

A. (Two-sided)
   H_0: $F(x) = F_0(x)$ for all values of x
   H_1: $F(x) \neq F_0(x)$ for at least one value of x

B. (One-sided)
   H_0: $F(x) \geq F_0(x)$ for all values of x
   H_1: $F(x) < F_0(x)$ for at least one value of x

C. (One-sided)
   H_0: $F(x) \leq F_0(x)$ for all values of x
   H_1: $F(x) > F_0(x)$ for at least one value of x

Test Statistic

We let $S(x)$ designate the sample (or empirical) distribution function; that is, $S(x)$ is the cumulative probability function computed from the sample data. Specifically,

$S(x)$ = the proportion of sample observations less than or equal to x

$$= \frac{\text{the number of sample observations less than or equal to } x}{n} \tag{8.5}$$

The test statistic depends on the hypothesis under consideration.

A. For the two-sided test the test statistic is

$$D = \sup_x |S(x) - F_0(x)| \tag{8.6}$$

which we read "D equals the supremum, over all x, of the absolute value of the difference $S(x) - F_0(x)$." When the two functions are represented graphically, D is the greatest vertical distance between $S(x)$ and $F_0(x)$.

B. For the one-sided test where the alternative specifies that $F(x) < F_0(x)$, the test statistic is

$$D^+ = \sup_x [F_0(x) - S(x)] \tag{8.7}$$

Graphically, this statistic denotes the greatest vertical distance between $F_0(x)$ and $S(x)$, where the hypothesized function $F_0(x)$ is above the sample function $S(x)$.

C. For the one-sided test where the alternative hypothesis specifies that $F(x) > F_0(x)$, the test statistic is

$$D^- = \sup_x [S(x) - F_0(x)] \tag{8.8}$$

When graphed, this statistic is the greatest vertical distance between $S(x)$ and $F_0(x)$ when $S(x)$ is above $F_0(x)$.

Decision Rule

Reject H_0 at the α level of significance if the test statistic under consideration, D, D^+, or D^-, exceeds the $1 - \alpha$ quantile shown in Table A.18.

If the sample data have been drawn from the hypothesized distribution, the discrepancies between $S(x)$ and $F_0(x)$ for the observed values of x should not be too large. In other words, the agreement between $S(x)$ and $F_0(x)$ for all observed values of x should be fairly close if H_0 is true. On the other hand, if H_0 is false—that is, if the sample did not come from the hypothesized distribution—we expect to observe larger discrepancies between $S(x)$ and $F_0(x)$. If D, the maximum of these differences, is too large, we reject H_0. To determine whether D is sufficiently large in a given situation to cause us to reject H_0, we compare the computed D with tabulated values given in Table A.18.

Estimation of Parameters When we must estimate parameters of the hypothesized distribution from sample data, the Kolmogorov–Smirnov one-sample test no longer applies in the strict sense. Massey (T27) suggests that when parameters are estimated from sample data, the test is conservative in the sense that the probability of a type I error will be smaller than that given in most available tables of the test statistic. Lilliefors (T28, T29) verifies this assumption when the hypothesized distributions are normal and exponential.

Example 8.5 Grundmann et al. (E16) reported the weights of the kidneys of 36 mongrel dogs before they were used in an experiment. The data are shown in Table 8.14. We wish to test the null hypothesis that these data are from a normally distributed population with a mean of 85 grams and a standard deviation of 15 grams.

TABLE 8.14 **Kidney weights, in grams, of 36 mongrel dogs**

58	78	84	90	97	70	90	86	82
59	90	70	74	83	90	76	88	84
68	93	70	94	70	110	67	68	75
80	68	82	104	92	112	84	98	80

Source: R. Grundmann, M. Raab, E. Meusel, R. Kirchhoff, and H. Pichlmaier, "Analysis of the Optimal Perfusion Pressure and Flow Rate of the Renal Vascular Resistance and Oxygen Consumption in Hypothermic Perfused Kidney," *Surgery,* 77 (1975), 451–461.

Hypotheses

H_0: $F(x) = F_0(x)$, where $F(x)$ is the distribution function of the population represented by the sample and $F_0(x)$ is the distribution function of a normally distributed population with $\mu = 85$ and $\sigma = 15$.

H_1: $F(x) \neq F_0(x)$

Test Statistic

Since the test is two-sided, we can obtain the test statistic from Equation 8.6.

$$D = \sup_{x} |S(x) - F_0(x)|$$

We first obtain values of $S(x)$ from Equation 8.5. The procedure is summarized in Table 8.15. To obtain values of $F_0(x)$, we convert each observed value of x to a value of the standard normal variable z. Then we use Table A.2 to find the area between 0 and z. From these areas, we compute values of $F_0(x)$. The procedure, which is summarized in Table 8.16, is very similar to that employed in testing goodness of fit to a normal distribution by means of the chi-square test.

TABLE 8.15   **Calculation of $S(x)$ for Example 8.5**

x	58	59	67	68	70	74
Frequency	1	1	1	3	4	1
Cumulative frequency	1	2	3	6	10	11
S(x)	0.0278	0.0556	0.0833	0.1667	0.2778	0.3056
x	75	76	78	80	82	83
Frequency	1	1	1	2	2	1
Cumulative frequency	12	13	14	16	18	19
S(x)	0.3333	0.3611	0.3889	0.4444	0.5000	0.5278
x	84	86	88	90	92	93
Frequency	3	1	1	4	1	1
Cumulative frequency	22	23	24	28	29	30
S(x)	0.6111	0.6389	0.6667	0.7778	0.8056	0.8333
x	94	97	98	104	110	112
Frequency	1	1	1	1	1	1
Cumulative frequency	31	32	33	34	35	36
S(x)	0.8611	0.8889	0.9167	0.9444	0.9722	1.0000

Before computing the test statistic arithmetically, let us graph the two functions $S(x)$ and $F_0(x)$ and determine D by actually measuring the largest vertical distance between the two curves. The graphs are shown in Figure 8.1.

From Figure 8.1 we can determine that D is approximately equal to 0.15.

Decision

Entering Table A.18 with $n = 36$, and keeping in mind that the test is two-sided, we find that the probability of obtaining a value of D as extreme as or more extreme than 0.15 is greater than 0.20. Hence these data do not provide sufficient evidence to warrant the conclusion that the weights of mongrel dog kidneys are not normally

TABLE 8.16 **Calculation of $F_0(x)$ for Example 8.5**

x	z = (x − 85)/15	P(0 ≤ Z ≤ z)	$F_0(x)$
58	−1.80	0.4641	0.0359 = 0.5 − 0.4641
59	−1.73	0.4582	0.0418 = 0.5 − 0.4582
67	−1.20	0.3849	0.1151 = 0.5 − 0.3849
68	−1.13	0.3708	0.1292 = 0.5 − 0.3708
70	−1.00	0.3413	0.1587 = 0.5 − 0.3413
74	−0.73	0.2673	0.2327 = 0.5 − 0.2673
75	−0.67	0.2486	0.2514 = 0.5 − 0.2486
76	−0.60	0.2257	0.2743 = 0.5 − 0.2257
78	−0.47	0.1808	0.3192 = 0.5 − 0.1808
80	−0.33	0.1293	0.3707 = 0.5 − 0.1293
82	−0.20	0.0793	0.4207 = 0.5 − 0.0793
83	−0.13	0.0517	0.4483 = 0.5 − 0.0517
84	−0.07	0.0279	0.4721 = 0.5 − 0.0279
86	0.07	0.0279	0.5279 = 0.5 + 0.0279
88	0.20	0.0793	0.5793 = 0.5 + 0.0793
90	0.33	0.1293	0.6293 = 0.5 + 0.1293
92	0.47	0.1808	0.6808 = 0.5 + 0.1808
93	0.53	0.2019	0.7019 = 0.5 + 0.2019
94	0.60	0.2257	0.7257 = 0.5 + 0.2257
97	0.80	0.2881	0.7881 = 0.5 + 0.2881
98	0.87	0.3078	0.8078 = 0.5 + 0.3078
104	1.27	0.3980	0.8980 = 0.5 + 0.3980
110	1.67	0.4525	0.9525 = 0.5 + 0.4525
112	1.80	0.4641	0.9641 = 0.5 + 0.4641

FIGURE 8.1 $S(x)$ and $F_0(x)$ for Example 8.5

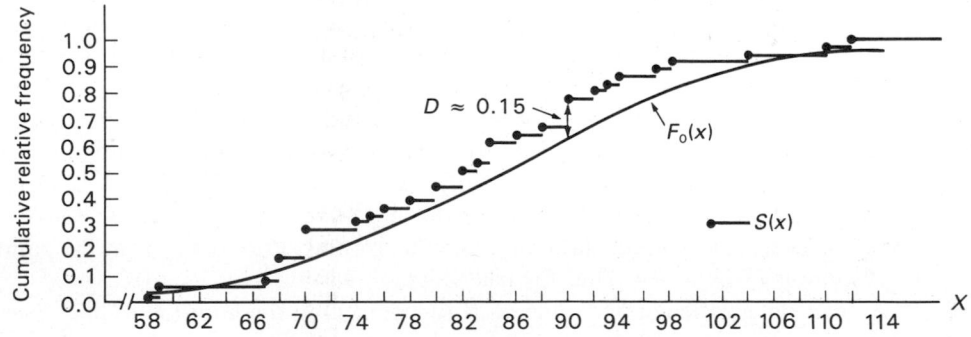

distributed. In other words, we conclude that the sampled population may be normally distributed with mean $\mu = 85$ and standard deviation $\sigma = 15$.

Now let us determine D arithmetically. All values of $|S(x) - F_0(x)|$ that can be computed from the sample data are shown in Table 8.17. This table reveals that the maximum of these occurs at $x = 90$ and is equal to 0.1485. This is the same but a more accurate value than was indicated by the graph.

TABLE 8.17   **Calculation of $|S(x) - F_0(x)|$ for Example 8.5**

x	S(x)	F_0(x)	$|S(x) - F_0(x)|$
58	0.0278	0.0359	0.0081
59	0.0556	0.0418	0.0138
67	0.0833	0.1151	0.0318
68	0.1667	0.1292	0.0375
70	0.2778	0.1587	0.1191
74	0.3056	0.2327	0.0729
75	0.3333	0.2514	0.0819
76	0.3611	0.2743	0.0868
78	0.3889	0.3192	0.0697
80	0.4444	0.3707	0.0737
82	0.5000	0.4207	0.0793
83	0.5278	0.4483	0.0795
84	0.6111	0.4721	0.1390
86	0.6389	0.5279	0.1110
88	0.6667	0.5793	0.0874
90	0.7778	0.6293	$D = 0.1485$
92	0.8056	0.6808	0.1248
93	0.8333	0.7019	0.1314
94	0.8611	0.7257	0.1354
97	0.8889	0.7881	0.1008
98	0.9167	0.8078	0.1089
104	0.9444	0.8980	0.0464
110	0.9722	0.9525	0.0197
112	1.0000	0.9641	0.0359

A Word of Caution When we determine D arithmetically, *it is not always sufficient to compute and choose from the possible values of $|S(x) - F_0(x)|$. The largest vertical distance between $S(x)$ and $F_0(x)$ may not occur at an observed value x, but at some other value of X.* As an illustration of what we mean, consider the graph of fictitious data shown in Figure 8.2. If we consider only values of $|S(x) - F_0(x)|$ at the left endpoints of the horizontal bars as candidates for D, we erroneously conclude that $D = |0.2 - 0.4| = 0.2$. In other words, if we determine values of $|S(x) - F_0(x)|$ at observed values of X only, we may not correctly identify D. Careful examination of Figure 8.2 reveals that the largest vertical distance between $S(x)$ and $F_0(x)$ occurs at the right endpoint of the horizontal bar originating at the point corresponding to $x = 0.4$. Thus the true value of D is $|0.5 - 0.2| = 0.3$.

We can obtain the correct value of D by computing the additional differences $|S(x_{i-1}) - F_0(x_i)|$ for all values of $i = 1, 2, \ldots, r + 1$, where $r = $ the number of different values of x and $S(x_0) = 0$. The correct value of the test statistic is then

$$D = \underset{1 \le i \le r}{\text{maximum}} \{ \text{maximum} \, [|S(x_i) - F_0(x_i)|, |S(x_{i-1}) - F_0(x_i)|] \} \qquad (8.9)$$

To illustrate this procedure, let us compute the additional differences $|S(x_{i-1}) - F_0(x)|$ for Example 8.5, and display them in Table 8.18 along with the

FIGURE 8.2
Graph of fictitious data showing correct calculation of D

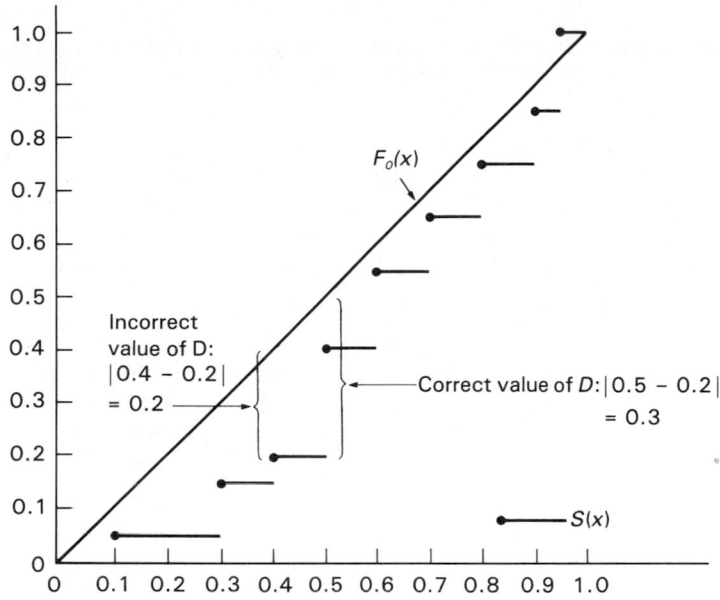

TABLE 8.18
Calculation of $|S(x_{i-1}) - F_0(x)|$ for Example 8.5

x_i	$S(x_i)$	$F_0(x_i)$	$|S(x_i) - F_0(x_i)|$	$|S(x_{i-1}) - F_0(x_i)|$
58	0.0278	0.0359	0.0081	0.0359
59	0.0556	0.0418	0.0138	0.0140
67	0.0833	0.1151	0.0318	0.0595
68	0.1667	0.1292	0.0375	0.0459
70	0.2778	0.1587	0.1191	0.0080
74	0.3056	0.2327	0.0729	0.0451
75	0.3333	0.2514	0.0819	0.0542
76	0.3611	0.2743	0.0868	0.0590
78	0.3889	0.3192	0.0697	0.0419
80	0.4444	0.3707	0.0737	0.0182
82	0.5000	0.4207	0.0793	0.0237
83	0.5278	0.4483	0.0795	0.0517
84	0.6111	0.4721	0.1390	0.0557
86	0.6389	0.5279	0.1110	0.0832
88	0.6667	0.5793	0.0874	0.0596
90	0.7778	0.6293	0.1485	0.0374
92	0.8056	0.6808	0.1248	0.0970
93	0.8333	0.7019	0.1314	0.1037
94	0.8611	0.7257	0.1354	0.1076
97	0.8889	0.7881	0.1008	0.0730
98	0.9167	0.8078	0.1089	0.0811
104	0.9444	0.8980	0.0464	0.0187
110	0.9722	0.9525	0.0197	0.0081
112	1.0000	0.9641	0.0359	0.0081

corresponding values of $|S(x) - F_0(x)|$ that have already been computed. Thus $|S(x_0) - F_0(x)| = |0 - 0.0359| = 0.0359, |S(x_1) - F_0(x_2)| = |0.0278 - 0.0418| = 0.0140$, and so on.

Since none of the values of $|S(x_{i-1}) - F_0(x)|$ is greater than 0.1485, these last calculations were not necessary in this example. However, this will not always be the case; when testing the goodness of fit of sample data to a continuous hypothesized distribution, it is wise to calculate both $|S(x_i) - F_0(x)|$ and $|S(x_{i-1}) - F_0(x_i)|$. When testing the goodness of fit of sample data to a discrete distribution, only values of $|S(x_i) - F_0(x_i)|$ need be computed.

Power-Efficiency   The power of the Kolmogorov–Smirnov one-sample test has been discussed by Massey (T27), Lee (T30), and Quade (T31). The efficiency of the test is discussed by Capon (T32) and Ramachandramurty (T33). Anděl (T34) discusses local asymptotic power and efficiency of the test.

Gleser (T35) shows that the algorithms for calculating the exact powers of Kolmogorov-type tests for true cumulative distribution functions $F(x)$ that are continuous can also be used to calculate the exact power and level of significance of Kolmogorov-type goodness-of-fit tests when $F(x)$ is discontinuous. A paper by Nikitin (T36) is devoted in part to the computation of the Hodges–Lehmann asymptotic relative efficiency (ARE) of the Kolmogorov–Smirnov goodness-of-fit tests.

Kolmogorov–Smirnov One-Sample Test and Discrete Distributions   As noted earlier, the Kolmogorov–Smirnov test was designed for use with continuous distributions. In the tests of the type discussed in this section, we can determine exact probabilities of obtaining a value of the test statistic as extreme as that observed when H_0 is true if the hypothesized distribution is continuous. When used with discrete distributions, the Kolmogorov–Smirnov one-sample test is conservative. See, for example, Noether (T37, T38), and Slakter (T39).

Conover (T40) cites the works of Schmid (T41), Carnal (T42), and Taha (T43) in obtaining asymptotic distributions of the test statistic for some discontinuous cases, but he points out that these results have some undesirable properties. Conover (T40) derives a method for finding the exact critical level (approximate in the two-sided case) of the Kolmogorov–Smirnov one-sample test statistic when the hypothesized distribution function is not continuous. The procedure is appropriate when the sample size is 30 or less.

Others who have investigated the use of the Kolmogorov–Smirnov statistics with discrete data include Azzalini and Diana (T44), Zelterman (T45), and Guilbaud (T46).

Lilliefors Test for Normality (μ and/or σ^2 unknown)   As noted previously, when we use the Kolmogorov–Smirnov one-sample procedure to test the null hypothesis that a sample was drawn from a population with unspecified parameters, the test is conservative. For several important situations, however, tables of critical

values of the Kolmogorov–Smirnov one-sample test have been constructed to allow the option of a better test when there is a need to estimate the parameters of a hypothesized population distribution. Lilliefors (T28, T29), for example, is the author of procedures that allow us to compute estimates of unknown population parameters when performing hypothesis tests in which the population specified in the null hypothesis is either normally or exponentially distributed.

The Lilliefors test for normality consists of the following steps:

Assumptions

The data consist of the independent observations $X_1, X_2, \ldots, X_n$, constituting a random sample of size n from some unknown distribution function, $F(x)$, with unknown mean μ and/or unknown variance σ^2. We use Equations 8.2 and 8.3, as needed, to compute estimates of μ and σ^2, respectively.

Hypotheses

H_0: The sampled population is normally distributed

H_1: The sampled population is not normally distributed

Test Statistic

We use the notation adopted in connection with the Kolmogorov–Smirnov one-sample test, described earlier. The test statistic in the present case is

$$D = \sup_x |S(x) - F_0(x)| \tag{8.10}$$

In other words, the test statistic is identical to the test statistic for the two-sided Kolmogorov–Smirnov one-sample test.

Decision Rule

Reject H_0 if the computed value of D is greater than the critical value for n and preselected α shown in Table A.19(a), A.19(b), or A.19(c), depending on whether the unknown parameter is μ only, σ^2 only, or both μ and σ^2.

The application of the Lilliefors test for normality is illustrated by the following example.

Example 8.6 The following table shows the times in seconds that it took a sample of 16 assembly-line employees to perform a certain operation:

5.8	7.3	8.9	7.1	8.8	6.4	7.2	5.2
10.1	8.6	9.0	9.3	6.4	7.0	9.9	6.8

Can we conclude that the times required for the population of assembly-line employees to perform the operation are not normally distributed? Let $\alpha = 0.05$.

To test the null hypothesis that the sampled population is normally distributed, without specifying μ and σ^2, we first compute the sample mean and standard deviation. By Equation 8.2 we have $\mu = 7.7375$, and by Equation 8.3 we find that $\sigma^2 = 1.4966$. From these parameter estimates and the sample data, we construct Table 8.19.

TABLE 8.19   **Calculations for Example 8.6**

x_i	$z = (x_i - 7.7375)/1.4966$	$F_0(x_i)$	$S(x_i)$	$S(x_i) - F_0(x_i)$	$S(x_{i-1}) - F_0(x_i)$
5.2	−1.70	0.0446	0.0625	0.0179	−0.0446
5.8	−1.29	0.0985	0.1250	0.0265	−0.0360
6.4	−0.89	0.1867	0.2500	0.0633	−0.0617
6.4	−0.89	0.1867	0.2500	0.0633	0.0633
6.8	−0.63	0.2643	0.3125	0.0428	−0.0143
7.0	−0.49	0.3121	0.3750	0.0629	0.0004
7.1	−0.43	0.3336	0.4375	0.1039	0.0414
7.2	−0.36	0.3594	0.5000	0.1406	0.0781
7.3	−0.29	0.3859	0.5625	0.1766	0.1141
8.6	0.58	0.7190	0.6250	−0.0940	−0.1565
8.8	0.71	0.7611	0.6875	−0.0736	−0.1361
8.9	0.78	0.7823	0.7500	−0.0323	−0.0948
9.0	0.84	0.7995	0.8125	0.0130	−0.0495
9.3	1.04	0.8508	0.8750	0.0242	−0.0383
9.9	1.44	0.9251	0.9375	0.0124	−0.0501
10.1	1.58	0.9429	1.0000	0.0571	−0.0054

Reference to Table 8.19 shows that the value of the computed test statistic is $D = 0.1766$, the largest difference in absolute value in the table. We find in Table A.19(c) that the critical value of D for $n = 16$ and $\alpha = 0.05$ is 0.212. Since 0.1766 is not greater than 0.212, we are unable to reject the null hypothesis. We conclude, therefore, that the sample observations may have been drawn from a population that is normally distributed.

FURTHER READING

Over the years goodness of fit has proved to be a fertile field for research. Schuster (T47) considers the test of $H_0: F(x) = F_0(x)$ when $F_0(x)$ is completely specified, continuous, and symmetric about some point θ against alternatives that are also continuous and symmetric about θ.

Schafer et al. (T48) discuss and give corrections for a Kolmogorov–Smirnov-type test proposed by Srinivasan (T49) for testing the goodness of fit in the case of exponential and normal distributions with parameters unspecified. Finkelstein and

Schafer (T50) and Lohrding (T51) present some goodness-of-fit tests that are more powerful than the Kolmogorov–Smirnov one-sample test. They have demonstrated the greater power of these tests by simulation.

Riedwyl (T52) defines a class of distribution-free measures of goodness of fit for which exact distributions for small samples can be calculated on a computer. Two of the measures have the same asymptotic distribution as the Kolmogorov–Smirnov statistic.

Following Riedwyl (T52), Maag et al. (T53) generalize some one-sample statistics so that they can be used to test grouped data for goodness of fit. Other articles of interest on one-sample goodness-of-fit tests are those by D'Agostino (T54), Heathcote (T55), Maag and Dicaire (T56), Suzuki (T57), Harter et al. (T58), Mantel (T59), Guilbaud (T60), and Inglot and Ledwina (T61).

Iman (T62) presents graphs on which the empirical distribution function can be plotted to test the assumption of normality by the Lilliefors test. The graphs allow for tests at the 10%, 5%, and 1% levels of significance. The paper also contains graphs for use with the Lilliefors test when the hypothesized distribution is the exponential.

EXERCISES

8.8   Table 8.20 shows the age at menarche of 324 women as reported by Treloar (E17). Use the Kolmogorov–Smirnov one-sample procedure to test the null hypothesis that these data came from a normally distributed population. What is the P value?

TABLE 8.20   **Age at menarche of 324 women**

Age, years	10	11	12	13	14	15	16	17	18	19
Number of women	4	30	79	104	60	32	11	2	1	1

Source: Alan E. Treloar, "Menarche, Menopause, and Intervening Fecundity," *Hum. Biol.*, 46 (1974), 89–107; copyright 1974, Wayne State University Press; reprinted by permission.

8.9   Moreno et al. (E18) reported the data shown in Table 8.21 on free portal pressure in patients with cirrhosis of the liver. Use the Kolmogorov–Smirnov one-sample procedure to determine whether we can conclude that these data did not come from a normal distribution.

8.10   Refer to Exercise 8.5, and use the Kolmogorov–Smirnov one-sample test to test the null hypothesis that the number of males is binomially distributed with $p = 0.51$. Find the P value.

8.11   Refer to Exercise 8.6, and use the Kolmogorov–Smirnov one-sample test to test the null hypothesis that the data are from a Poisson distribution. What is the P value?

TABLE 8.21   **Free portal pressure, millimeters water, in patients with cirrhosis of the liver**

390	410	400	420	385	380	420	440	470	400	420	360	410
435	430	410	365	360	200	285	380	450	460	365	390	480
430	320	350	325	355	450	350	410	430	420	270	285	380
430	400	460	430	400	340	430	430	320	360	380	465	370
300	440	300	430	490	370	410	500	435	365	420	460	310
350	460	450	380	300	440	400	300	370	460	405	375	320
300	410	460	370	420	400	450	310	360	390	380	335	400
330	410	350	420	430	330	480	350	300	410	480	395	
410	390	320	405	430	350	450	400	450	420	310	375	
390	490	405	340	410	410	435	340	440	345	445	330	
365	485	485	300	485	380	390	245	300	380	365	270	
340	375	320	413	390	410	365	435	340	360	400	420	

Source: A. H. Moreno, A. R. Burchell, R. V. Reddy, W. F. Panke, and T. F. Nealon, Jr., "The Hemodynamics of Portal Hypertension Revisited: Determinants and Significance of Occluded Portal Pressures," *Surgery,* 77 (1975), 167–179.

8.12 The following are the number of miles driven while on vacation as reported by a random sample of 15 families who registered at a state welcome center.

1112	1435	1789	1489	1805	1738	1932	750	2513	3201
1205	935	2085	988	2450					

Use the Lilliefors test to see if you can conclude that the sampled population is normally distributed. Let $\alpha = 0.05$.

8.3

KOLMOGOROV–SMIRNOV TWO-SAMPLE TEST

In some of the tests previously discussed, we were concerned with testing the null hypothesis that two independent samples come from populations that are identical with respect to location and dispersion. This section covers the Kolmogorov–Smirnov two-sample test, which may be referred to as a general or omnibus test, since it is sensitive to differences of all types that may exist between two distributions. The chi-square test of homogeneity discussed in Chapter 5 is another general test for independent samples.

The Kolmogorov–Smirnov two-sample test was developed by Smirnov (T26). The test also carries the name of Kolmogorov because of its similarity to the one-sample test developed by Kolmogorov (T25).

Assumptions

    A. The data for analysis consist of two independent random samples of sizes m and n. The observations may be designated $X_1, X_2, \ldots, X_m$ and $Y_1, Y_2, \ldots, Y_n$, respectively.

    B. The data are measured on at least an ordinal scale.

Hypotheses

We let $F_1(x)$ and $F_2(x)$ designate the unknown distribution functions of the X's and the Y's, respectively. The following two-sided and one-sided tests may be performed.

    A. (Two-sided)
        $H_0: F_1(x) = F_2(x)$ for all x
        $H_1: F_1(x) \neq F_2(x)$ for at least one value of x

    B. (One-sided)
        $H_0: F_1(x) \leq F_2(x)$ for all x
        $H_1: F_1(x) > F_2(x)$ for at least one value of x

    C. (One-sided)
        $H_0: F_1(x) \geq F_2(x)$ for all x
        $H_1: F_1(x) < F_2(x)$ for at least one value of x

Test Statistic

We let $S_1(x)$ and $S_2(x)$, respectively, designate the sample or empirical distribution functions of the observed X's and the observed Y's.

$$S_1(x) = (\text{number of observed } X\text{'s} \leq x)/m$$

$$S_2(x) = (\text{number of observed } Y\text{'s} \leq x)/n$$

    The test statistic for our three sets of hypotheses are as follows:

    A. (Two-sided): $D = \text{maximum}\, |S_1(x) - S_2(x)|$ **(8.11)**

    B. (One-sided): $D^+ = \text{maximum}\, [S_1(x) - S_2(x)]$ **(8.12)**

    C. (One-sided): $D^- = \text{maximum}\, [S_2(x) - S_1(x)]$ **(8.13)**

Decision Rule

If the two samples have been drawn from identical populations, $S_1(x)$ and $S_2(x)$ should be fairly close for all values of x. The test statistics D, D^+, and D^- are measures of the extent to which $S_1(x)$ and $S_2(x)$ fail to agree. If the test statistic, which is equal to the maximum difference at some x between $S_1(x)$ and $S_2(x)$, is small, differences at all other values of x are also small, and H_0 is supported (if D is sufficiently small). On the other hand, if D is sufficiently large (that is, too large to be

a reasonable occurrence when H_0 is true), we reject H_0. To determine whether we should reject H_0, we observe the following decision rule.

Reject H_0 at the α level of significance if the appropriate test statistic D, D^+, or D^- exceeds the $1 - \alpha$ quantile given in Table A.20(a) if $m = n$, and in Table A.20(b) if $m \neq n$.

If the two population distributions represented by the two samples are continuous, the test as outlined is exact. If the distributions are discrete, the test is conservative, as Noether (T38) has shown.

Example 8.7   Burrus et al. (E19) reported the basal metabolic rates shown in Table 8.22. The subjects were five nonathletic males and six champion runners. We wish to know whether we may conclude that the two populations represented by the two samples have different distribution functions.

TABLE 8.22   **Basal metabolic rates, milliliters oxygen per minute, in athletic and nonathletic males**

Athletic males (Y)	Nonathletic males (X)
236	206
209	238
278	224
276	257
252	230
251	

Source: Reprinted from "Observations at Sea Level and Altitude on Basal Metabolic Rate and Related Cardio-Pulmonary Functions," *Human Biology*, 46 (1974), 677–692 by S. Kay Burrus, et al. by permission of the Wayne State University Press. Copyright 1974 by Wayne State University Press.

Hypotheses

$$H_0: F_1(x) = F_2(x), \qquad H_1: F_1(x) \neq F_2(x)$$

We use a two-sided test, since a difference in any direction is of interest.

Test Statistic

We refer to the five observations on nonathletes as $X_1, X_2, \ldots, X_5$, and to the six observations on athletes as $Y_1, Y_2, \ldots, Y_6$. Table 8.23 shows the observations of the two samples in order of magnitude, as well as $|S_1(x) - S_2(x)|$. The test statistic as shown in the table is $D = 14/30 = 0.47$.

TABLE 8.23	**Ordered sample observations and $\|S_1(x) - S_2(x)\|$ for Example 8.7**		

X_i	Y_i	$S_1(x) - S_2(x)$
206		$1/5 - 0 = 6/30$
	209	$1/5 - 1/6 = 1/30$
224		$2/5 - 1/6 = 7/30$
230		$3/5 - 1/6 = 13/30$
	236	$3/5 - 2/6 = 8/30$
238		$4/5 - 2/6 = 14/30$
	251	$4/5 - 3/6 = 9/30$
	252	$4/5 - 4/6 = 4/30$
257		$5/5 - 4/6 = 10/30$
	276	$5/5 - 5/6 = 5/30$
	278	$5/5 - 6/6 = 0$

Decision

When we enter Table A.20(b) with $m = 5$ and $n = 6$, we find that the probability of observing a value of D as large as 0.47 when H_0 is true is greater than 0.20. Therefore we conclude that the observed data do not provide convincing evidence in support of the alternative hypothesis. In other words, the observed data are more supportive of the null hypothesis.

Graphical Analysis An alternate method of calculating the Kolmogorov–Smirnov two-sample statistic is described by Hodges (T63) and Quade (T64). The method employs a graphing technique that results in a graph known as a *pair chart*. Quade (T64) credits Drion (T65) with the invention of the pair chart.

To construct a pair chart from the data contained in two independent samples consisting of m values of X and n values of Y, we proceed as follows.

1. Draw a rectangle m units wide by n units high.
2. If the smallest of the $m + n$ observations is an X, draw a line from the lower left-hand corner of the rectangle one unit to the right. If the smallest value is a Y, draw a line from the lower left-hand corner of the rectangle one unit up instead.
3. From the end of the line drawn in Step 2, draw another line one unit to the right if the second smallest of the $m + n$ values is an X, or one unit up if the second smallest value is a Y.
4. Continue drawing lines in this fashion until a line segment has been drawn for each of the $m + n$ observations. The order of drawing line segments corresponds to the order of magnitude of the observations, so that the lines for X values are horizontal and those for Y values are vertical. The resulting line traces a path from the lower left-hand corner

FIGURE 8.3   Pair chart for data of Example 8.7

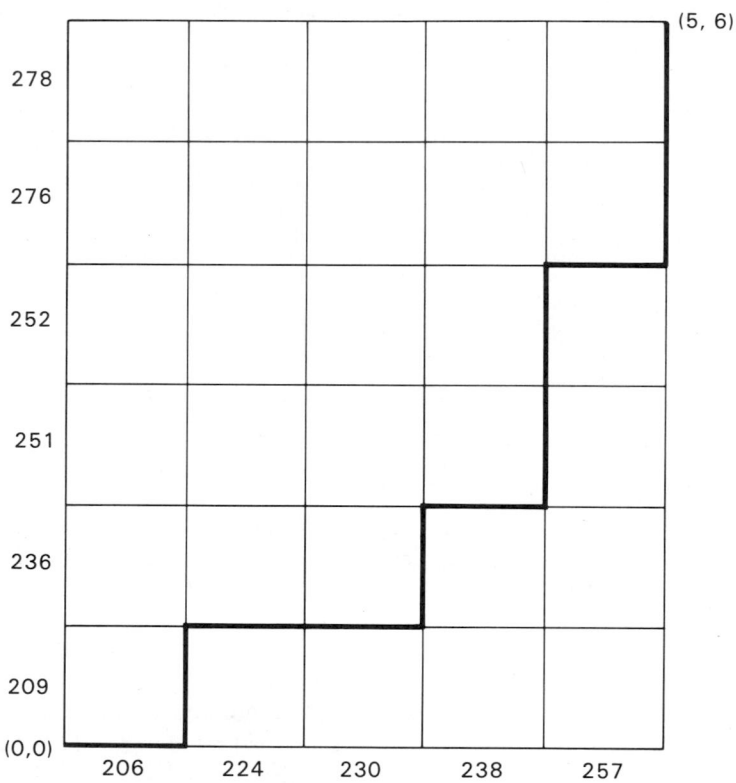

of the rectangle, which we designate as the origin $(0, 0)$, to the upper right-hand corner of the rectangle (m, n). The pair chart for the data of Example 8.7 given in Table 8.22 is shown as Figure 8.3.

Ties Ties occurring within one or both samples do not present a problem, but ties between samples introduce complications, since in their presence the path cannot be determined uniquely. For each between-sample tie that occurs, the path may follow one of several routes, depending on how we resolve the tie. The set of all possible routes describes a box. For some purposes the path may be made to follow the diagonals of the boxes.

To use the pair chart to calculate the Kolmogorov–Smirnov test statistic, we proceed as follows:

1. Draw a diagonal line on the chart from the lower left-hand corner to the upper right-hand corner, as shown in Figure 8.4, for the data of Example 8.7.

FIGURE 8.4   Pair chart for Example 8.7 showing diagonal line needed to calculate
                        Kolmogorov–Smirnov test statistic

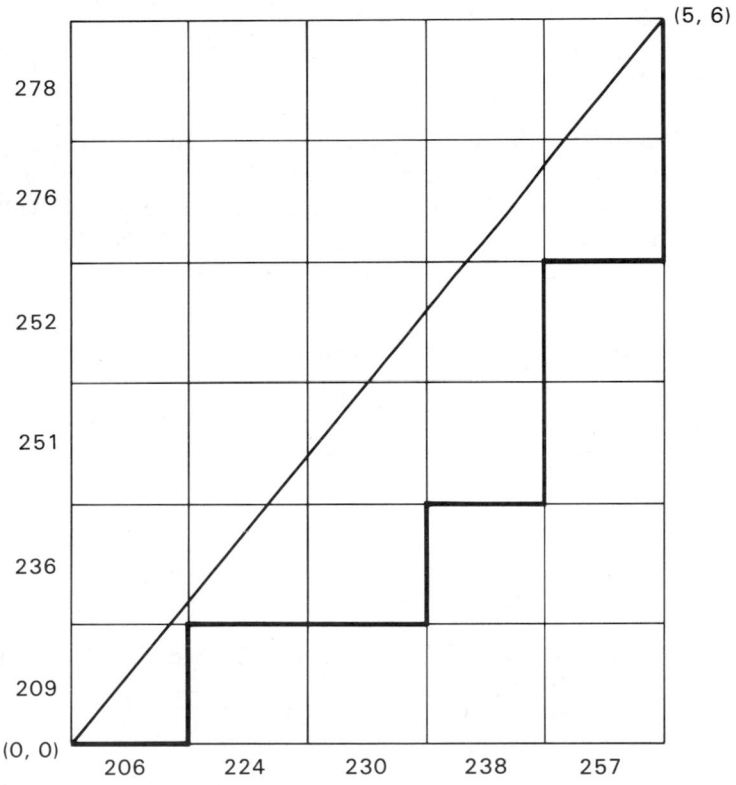

2. Let (X_x, Y_x) be the point on the path farthest below the diagonal. If two
    or more points are equally far below the diagonal, designate any one of
    them (X_x, Y_x). If the path is nowhere below the diagonal, let (X_x, Y_x)
    $= (0, 0)$.

3. Let (X_y, Y_y) be the point on the path farthest above the diagonal of the
    rectangle. If the path is nowhere above the diagonal, let (X_y, Y_y)
    $= (0, 0)$.

4. The test statistics as defined earlier are

$$D^+ = |X_x/m - Y_x/n| \qquad\qquad\qquad\text{(8.14)}$$

$$D^- = |X_y/m - Y_y/n| \qquad\qquad\qquad\text{(8.15)}$$

$$D = \text{maximum}\,(D^+, D^-) \qquad\qquad\text{(8.16)}$$

Now the data of Example 8.7 will be used to illustrate the use of the pair chart to compute D. The point on the path farthest below the diagonal in Figure 8.4 is $(X_x, Y_x) = (4, 2)$. Thus $D^+ = |4/5 - 2/6| = 14/30 = 0.47$. Since the path is nowhere above the diagonal, we let $(X_y, Y_y) = (0, 0)$ and $D^- = |0/m - 0/n| = 0$. We now compute $D = \max(0.47, 0) = 0.47$, the value found by the arithmetic method. We determine the significance of D from Table A.20(a) or Table A.20(b).

Ties   When the data contain ties between samples, we compute the appropriate test statistic by letting the path follow the diagonals of the boxes formed by drawing the paths corresponding to all possible ways of resolving the ties. In the following example, we show how to use a pair chart to calculate the Kolmogorov–Smirnov two-sample test statistic when ties occur between samples.

Example 8.8

Glovsky and Rigrodsky (E20) analyzed and compared the developmental histories of mentally deficient children, some of whom had been diagnosed as aphasics. Subjects consisted of 42 children enrolled in a speech therapy program at the Training School at Vineland, New Jersey. Twenty-one of the children had been diagnosed as aphasic sometime during their early developmental years. The remaining subjects were a random sample of 21 mentally retarded children who were also enrolled in the speech therapy program. The social quotient scores made by these children on the Vineland Social Maturity Scale are presented in Table 8.24.

TABLE 8.24   **Social quotient scores made by two groups of children on the Vineland social maturity scale**

Aphasic* (X)	90	53	32	44	47	42	58	16	49	54	81
	59	35	81	41	24	41	61	31	20		
Mentally	56	43	30	97	67	24	76	49	46	29	46
retarded (Y)	83	93	38	25	44	66	71	54	20	25	

Source: Leon Glovsky and Seymour Rigrodsky, "A Developmental Analysis of Mentally Deficient Children with Early Histories of Aphasia," *Training School Bull.*, 61 (1964), 76–96.

* One score not reported.

We wish to know whether we may conclude that the distributions of the populations represented by these samples are different; that is, we wish to test

$$H_0: F_1(x) = F_2(x) \quad \text{for all } x$$

against the alternative

$$H_1: F_1(x) \neq F_2(x) \quad \text{for at least one value of } x$$

We use the pair-chart method to compute the Kolmogorov–Smirnov two-sample test statistic.

TABLE 8.25   **Observations of Table 8.24 combined and ordered from smallest to largest**

16	(20	20)	(24	24)	25	25	29	30				
X	(X	Y)	(X	Y)	Y	Y	Y	Y				
31	32	35	38	41	41	42	43	(44	44)	46	46
X	X	X	Y	X	X	X	Y	(X	Y)	Y	Y
47	(49	49)	53	(54	54)	56	58	59				
X	(X	Y)	X	(X	Y)	Y	X	X				
61	66	67	71	76	81	81	83	90	93	97	
X	Y	Y	Y	Y	X	X	Y	X	Y	Y	

Note: Between-sample ties are given in parentheses.

To facilitate the construction of the pair chart, we combine and order the observations of the two samples, from smallest to largest, as shown in Table 8.25. Figure 8.5, the pair chart for the data reveals that the point on the path farthest below the diagonal is $(X_x, Y_x) = (17, 14)$, so that

$$D^+ = |17/20 - 14/21| = 0.18$$

From the figure, we also determine that the point on the path farthest above the diagonal is $(X_y, Y_y) = (3, 6)$, allowing us to compute

$$D^- = |3/20 - 6/21| = 0.14$$

Consequently $D = \text{maximum}(0.18, 0.14) = 0.18$.

Since we cannot use Table A.20(b) when one of the samples is as large as 21, we must compute critical values from the large-sample approximation formulas given at the bottom of the table. For a two-sided test, the critical values and the corresponding quantiles are as follows:

Quantile   (two-sided test)   **Critical value**

$p = 0.80$ $\qquad 1.07\sqrt{\dfrac{20 + 21}{(20)(21)}} = 0.33$

$p = 0.90$ $\qquad 1.22\sqrt{\dfrac{20 + 21}{(20)(21)}} = 0.38$

$p = 0.95$ $\qquad 1.36\sqrt{\dfrac{20 + 21}{(20)(21)}} = 0.42$

$p = 0.98$ $\qquad 1.52\sqrt{\dfrac{20 + 21}{(20)(21)}} = 0.47$

$p = 0.99$ $\qquad 1.63\sqrt{\dfrac{20 + 21}{(20)(21)}} = 0.51$

FIGURE 8.5   Pair chart for data of Example 8.8

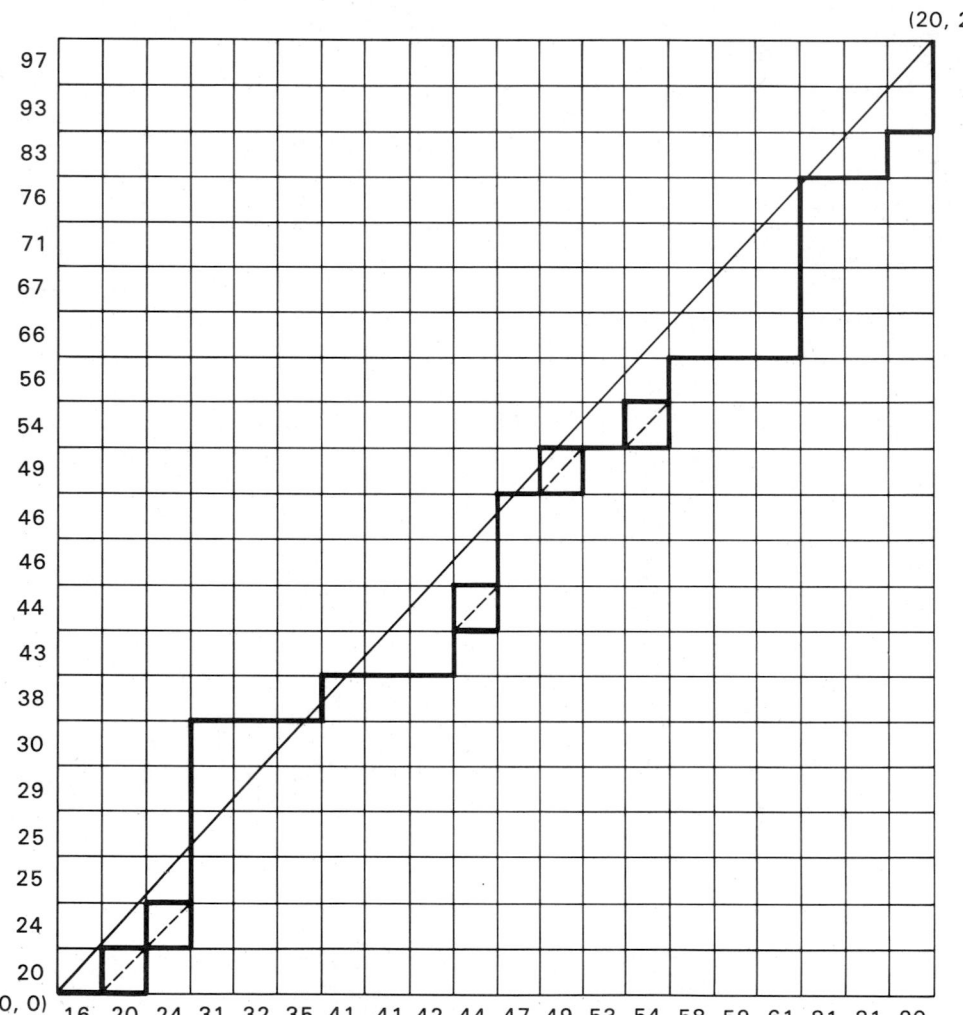

Thus the probability of obtaining a value of D as large as 0.18, when H_0 is true, is greater than 0.20. We cannot reject H_0 in favor of the alternative, and consequently we conclude that the two population distributions may not be different.

Power-Efficiency   For a discussion of the efficiency of the Kolmogorov–Smirnov two-sample test, see Capon (T32), Ramachandramurty (T33), and Yu (T66). Klotz (T67) uses a different limiting efficiency to obtain asymptotic efficiency

results for the two-sample case. Bickel (T68) discusses a distribution-free version of the Kolmogorov–Smirnov two-sample test in the *p*-variate case. The distributions of the test statistics are considered by Steck (T69).

FURTHER READING

Angus (T70) demonstrates the equivalence of the two-sample test to another goodness-of-fit test introduced by Barnard (T71) and Birnbaum (T72). An application of a generalization of the Kolmogorov–Smirnov two-sample statistic is contained in a paper by Wieand (T73).

EXERCISES

8.13 Edwards (E21) investigated the role of systemic hypertension in 10 emphysematous subjects with left ventricular hypertrophy. Table 8.26 shows the mean medial thickness of internal mammary arteries (expressed as a percentage of the diameter) in the 10 study subjects and in the 10 normal subjects. Can we conclude from these data that the two population distributions represented by these samples differ? Determine the *P* value.

TABLE 8.26   **Mean medial thicknesses of internal mammary arteries expressed as a percentage of the diameter in two groups of subjects**

Normal subjects	11.1	12.2	10.9	11.2	11.9	11.4	10.2	11.3	11.3	11.8
Subjects with emphysema and left ventricular hypertrophy	14.5	11.2	14.2	11.4	11.9	11.6	9.3	13.8	10.4	13.2

Source: C. W. Edwards, "Left Ventricular Hypertrophy in Emphysema," *Thorax*, 29 (1973), 75–80; reprinted by permission of the editor and publisher.

8.14 Zinc concentrations, micrograms per gram of dry tissue, in human skin biopsies were reported by Gooden (E22) for 17 normal subjects and 3 subjects with skin diseases (basal cell carcinoma, scleroderma, and psoriasis erythroderma). The results are shown in Table 8.27.

TABLE 8.27   **Zinc concentrations, micrograms per gram of dry tissue, in skin biopsies from two groups of subjects**

Normal	437	358	72	43	107	223	60	72	54	35	70	20	34	24	24	51	23
Diseased	18	16	18														

Source: D. S. Gooden, "Non-Destructive Neutron Activation Analysis for the Determination of Manganese and Zinc in Human Skin Biopsies," *Phys. Med. Biol.*, 17 (1972), 26–31; copyright 1972, the Institute of Physics.

Do these data provide sufficient evidence to indicate that values tend to be lower in subjects with skin diseases? Find the *P* value.

8.15 Wallenberg (E23) conducted an investigation to study whether changes in plasma oncotic pressure would influence the cold-induced natriuresis in man. Subjects included 17 healthy males; each underwent water diuresis. Subjects were exposed to cold by means of a closed hypothermic operating table in which the temperature was lowered to $+15°C$. Eight of the subjects received an infusion of human albumin. The control group of nine subjects did not receive albumin. Table 8.28 shows the urinary sodium excretion, $\mu Eq/min$, under 30 to 60 minutes in cold. Do these data provide sufficient evidence to indicate that the population distributions represented by these samples differ? What is the *P* value?

TABLE 8.28 **Urinary sodium excretion, $\mu Eq/min$, in two groups of subjects exposed to cold**

Control group	315.7	314.0	336.9	797.5	118.5	261.7	352.7	135.0	565.5
Albumin-infused group	324.3	182.7	304.0	351.2	347.9	297.7	273.0	204.0	

Source: L. R. Wallenberg, "Reduction in Cold-Induced Natriuresis Following Hyperoncotic Albumin Infusion in Man Undergoing Water Diuresis," *Scand. J. Clin. Lab. Invest.*, 34 (1974), 233–239; used by permission of the publisher, Universitetsforlaget, Oslo.

8.16 Higby et al. (E24) describe a study in which patients with thrombocytopenia and acute leukemia were randomized in a double blind study to receive either platelets or platelet-poor plasma as prophylaxis against bleeding. Table 8.29 shows the average platelet count during the study of the two groups. Can we conclude from these data that the treatments differ with respect to their effects on average platelet counts?

TABLE 8.29 **Average platelet counts, $\times 10^3$, for patients with acute myelocytic leukemia treated either with platelets or with platelet-poor plasma**

Treated with platelets	20.30	22.53	25.70	13.23	29.67	24.46
	26.07	19.35	17.813	16.00	13.50	32.90
Treated with platelet-poor plasma	10.56	28.13	19.94	11.03	8.093	12.95
	21.14	32.50	10.90			

Source: D. J. Higby, E. Cohen, J. F. Holland, and L. Sinks, "The Prophylactic Treatment of Thrombocytopenic Leukemic Patients with Platelets: A Double Blind Study," *Transfusion*, 14 (1974), 440–446.

8.17 Akerfeldt (E25) found that serum from schizophrenic patients oxidized the N,N-dimethyl form of *p*-phenylenediamine (PPD) at a faster rate than did serum from normal controls. Friedhoff et al. (E26), who attempted to repeat Akerfeldt's results, reported the data on oxidation of PPD shown in Table 8.30. The investigators measured oxidative activity in terms of the increase in optical density after the PPD had oxidized for six minutes. Do these data support the contention that optical density values are different in schizophrenics and in normal subjects? Determine the *P* value.

<table>
<thead>
<tr><td>TABLE 8.30</td><td colspan="3">Distribution of results of oxidation of PPD* by serum</td></tr>
</thead>
</table>

Optical density, × 10	Normal group	Schizophrenic group
Below 100	5	6
100–149	5	7
150–199	9	10
200–249	8	13
250–299	12	9
300–349	6	12
350–399	2	6
400–449	0	7
450–499	0	4
500–549	0	3
Over 550	0	4
	47	81

Source: Arnold J. Friedhoff, Myra Palmer, and Christine Simmons, "An Effect of Exercise, Skin Shock, and Ascorbic Acid on Serum Oxidase Activity," *Arch. Neurol. Psychiatry*, 81 (1959), 620–626. Copyright 1959, American Medical Association.

* N,N-dimethyl-*p*-phenylenediamine

8.4

CONFIDENCE BAND FOR THE POPULATION DISTRIBUTION FUNCTION

We may use the Kolmogorov–Smirnov one-sample two-sided statistic to construct a $100(1 - \alpha)\%$ confidence band for an unknown population distribution function. Given a random sample of n observations $X_1, X_2, \ldots, X_n$ from some population of unknown functional form, we may use $S(x)$ as a point estimate of $F(x)$, the unknown distribution function. As noted previously, $S(x)$ portrayed graphically is a step function. The desired confidence band consists of an upper boundary located some distance above $S(x)$ and a lower boundary located some distance below $S(x)$. The locations of the boundaries for a two-sided confidence band are determined by adding to and subtracting from $S(x)$ a quantity, $w_{1-\alpha}$, which we obtain from the two-sided test entry of Table A.18. The value of $w_{1-\alpha}$, and hence the distance of the boundaries from $S(x)$, depends on the desired level of confidence, $1 - \alpha$, and the sample size, n.

To obtain the upper bound for a $100(1 - \alpha)\%$ two-sided confidence band, we add $w_{1-\alpha}$ from Table A.18 to $S(x)$ at each x and call the bound $U(x)$. Thus, if the sum is less than or equal to 1,

$$U(x) = S(x) + w_{1-\alpha}$$

If $S(x) + w_{1-\alpha}$ is greater than 1, then $U(x) = 1$, since a distribution function cannot exceed 1.

To obtain the lower bound, $L(x)$, of the confidence band, we find for all x

$$L(x) = S(x) - w_{1-\alpha}$$

If the result is negative, we let $L(x) = 0$, since a distribution function cannot be less than zero. When graphed, $U(x)$ and $L(x)$ are also step functions.

If the variable under consideration is continuous, the confidence coefficient associated with the confidence band is exact. For discrete variables, the stated confidence coefficient is smaller than the true, but unknown, confidence coefficient so that the obtained band is conservative.

Example 8.9   The data shown in Table 8.31 are part of a larger set of data reported by Reed and Lanphere (E27). The entries in the table represent the SiO_2 content of plutonic rock samples taken from the Alaska–Aleutian Range batholith. Let us construct a 95% confidence band for the distribution function of the population from which the sample was drawn.

TABLE 8.31   **SiO_2 content, percent by weight, of plutonic rocks of the Alaska–Aleutian Range batholith (Merrill pass sequence)**

77.7	71.4	74.9	73.4	74.2	76.1	72.6
73.6	75.6	76.3	65.2	73.4	72.8	69.8
77.7	75.4	74.8	74.6	75.0	70.8	75.4

Source: Bruce L. Reed and Marvin A. Lanphere, "Chemical Variations across the Alaska–Aleutian Range Batholith," *J. Res. U.S. Geol. Survey*, 2 (1974), 343–352.

TABLE 8.32   **Calculation of 95% confidence band for Example 8.9**

x	S(x)	S(x) − 0.287	S(x) + 0.287	L(x)	U(x)
65.2	0.048	−0.239	0.335	0	0.335
69.8	0.095	−0.192	0.382	0	0.382
70.8	0.143	−0.144	0.430	0	0.430
71.4	0.190	−0.097	0.477	0	0.477
72.6	0.238	−0.049	0.525	0	0.525
72.8	0.286	−0.001	0.573	0	0.573
73.4	0.381	0.094	0.668	0.094	0.668
73.4	0.381	0.094	0.668	0.094	0.668
73.6	0.429	0.142	0.716	0.142	0.716
74.2	0.476	0.189	0.763	0.189	0.763
74.6	0.524	0.237	0.811	0.237	0.811
74.8	0.571	0.284	0.858	0.284	0.858
74.9	0.619	0.332	0.906	0.332	0.906
75.0	0.667	0.380	0.954	0.380	0.954
75.4	0.762	0.475	1.049	0.475	1
75.4	0.762	0.475	1.049	0.475	1
75.6	0.810	0.523	1.097	0.523	1
76.1	0.857	0.570	1.144	0.570	1
76.3	0.905	0.618	1.192	0.618	1
77.7	1.000	0.713	1.287	0.713	1
77.7	1.000	0.713	1.287	0.713	1

FIGURE 8.6   Confidence band for $F(x)$, Example 8.9

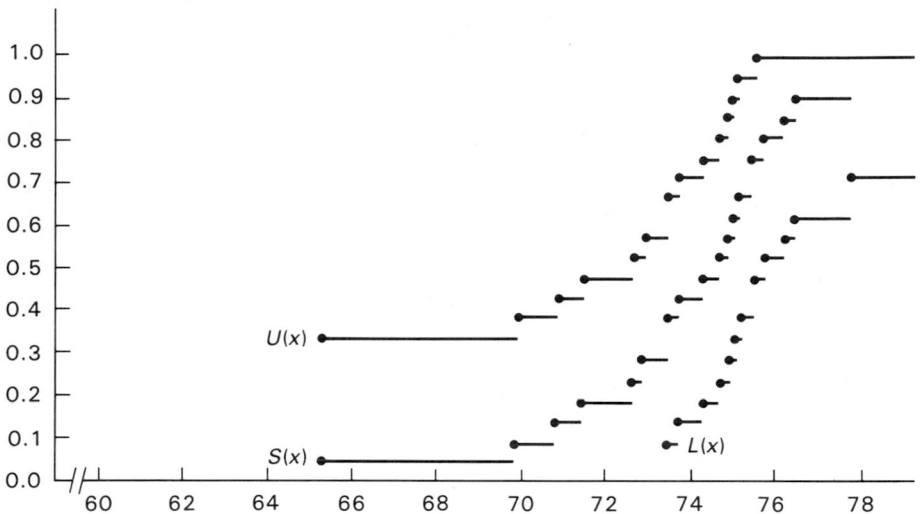

The 0.95 two-sided quantile from Table A.18 for $n = 21$ is $w_{0.95} = 0.287$. The necessary calculations are shown in Table 8.32. The graph of the confidence band is shown in Figure 8.6.

EXERCISES

8.18 Construct the 95% confidence band for $F(x)$ in Exercise 8.8.

8.19 Construct the 0.90% confidence band for $F(x)$ in Exercise 8.9.

8.5

CHI-SQUARE AND KOLMOGOROV–SMIRNOV GOODNESS-OF-FIT TESTS: A COMPARISON

The investigator faced with a problem requiring analysis by some goodness-of-fit technique may not know which procedure to use. In many instances, the choice is between the chi-square test and the Kolmogorov–Smirnov test. Birnbaum (T74), Goodman (T75), Massey (T27), and Slakter (T39) have addressed this problem. The following are some important points of comparison.

1. The chi-square test was designed for use with frequency data, while the Kolmogorov–Smirnov test was designed for use with continuous data. The chi-square test is appropriate for use with nominal data and when the hypothesized distribution is discrete. Such data occur frequently in practice. When used with discrete data, the Kolmogorov–Smirnov test is not exact, but conservative.

2. The Kolmogorov–Smirnov test allows for one-sided tests as well as for two-sided tests. The chi-square test does not distinguish the direction of discrepancies between observation and expectation.

3. The Kolmogorov–Smirnov test can be used to construct confidence bands for $F(x)$, as described in Section 8.3.

4. The exact sampling distribution of the Kolmogorov–Smirnov test statistic is known and tabulated for the case of continuous population distributions that are completely specified in the null hypothesis. The chi-square test statistic X^2 is only approximately distributed as chi-square for finite samples.

5. The chi-square test requires that data be grouped into categories, whereas the Kolmogorov–Smirnov test does not. Thus the Kolmogorov–Smirnov test makes more complete use of data that are available in ungrouped form.

6. A clear-cut correction procedure is available for use with the chi-square test when parameters have to be estimated from sample data. No similar adjustment is available for use with the Kolmogorov–Smirnov test.

8.6

OTHER GOODNESS-OF-FIT TESTS

Although the chi-square and Kolmogorov–Smirnov tests are by far the most frequently used tests for goodness of fit, others are available. One alternative, the Cramér–von Mises test, was suggested by Cramér (T76) in 1928 and by von Mises (T77) in 1931. This test considers not only the largest vertical difference between $F_0(x)$ and $S(x)$, but n differences for samples of size n. The asymptotic distribution of the Cramér–von Mises one-sample test statistic under an alternative has been studied by Angus (T78).

An analogue of the Kolmogorov–Smirnov two-sided two-sample test appropriate for several independent samples has been proposed by Birnbaum and Hall (T79). Unfortunately, the distribution of the test statistic for this procedure has been obtained only for the case of three equal-size samples.

Gibbons (T80) has proposed a distribution-free two-sample goodness-of-fit test

for general alternatives. The test is a group randomization test for which the test statistic is a function of the sum of squared deviations between group relative frequencies in the samples.

Hora (T81) presents a method for developing distribution-free goodness-of-fit tests for a completely specified continuous distribution using regression. The technique involves the transformation of the n order statistics of a random sample into a set of n statistics that, under the null hypothesis, are identically and independently distributed as standard normal random variables. The author argues that if an alternative distribution is true, the transformed values tend to exhibit systematic behavior the absence of which may be tested by a regression of the transformed values on their orders. The test procedures presented have either a chi-square or variance ratio distribution. Efficiencies for some of the tests are developed. Monte Carlo results are given, comparing the power of the proposed statistics to several well-known goodness-of-fit tests.

Some alternative procedures for testing goodness of fit for discrete distributions are discussed by Kocherlakota and Kocherlakota (T82). The procedures, based on the probability generating functions, are reported to be quite general and applicable in multidimensional situations. Zelterman (T83) introduces a goodness-of-fit statistic for use in multinomial distributions.

A paper by Kumazawa (T84) considers goodness-of-fit statistics for the case in which the underlying distribution is known to be symmetric about zero or some unknown location parameter. The author derives the asymptotic distributions of the statistics and introduces a test for normality. By Monte Carlo methods, power comparisons among the statistics against several alternatives are carried out.

Numerous goodness-of-fit tests for normality, of which some have already been mentioned, are available to the researcher. Oja (T85) presents a family of statistics for testing for normality against skewness, kurtosis, and bimodality alternatives. The tests are characterized as being simple to use and having good power properties.

Best and Rayner (T86) investigate the power properties of a test for normality previously proposed by Lancaster (T87). Three test statistics for the composite goodness-of-fit hypothesis of normality are presented by LaRiccia (T88). The proposed test statistics are designed to be especially sensitive to skewness and/or kurtic departures from normality. On the basis of a simulation study, the author reports that in the cases considered the tests compare favorably to other widely used procedures. The tests are recommended for cases in which prior information suggests the likelihood of skewness and/or kurtic alternatives to the normal.

The book by Shapiro (T89) contains extensive discussion and illustration of several goodness-of-fit tests for normality. Best and Rayner (T90) discuss a test for bivariate normality. Three tests for exponentiality are proposed by Angus (T91). Studies in which three or more goodness-of-fit tests were compared with each other have been reported by Haber (T92), Jammalamadaka and Tiwari (T93), and McKinley and Mills (T94).

For other goodness-of-fit tests, see the articles by Conover (T95, T96), Oja (T97),

Chmielewski (T98), and Hirotsu (T99). Also of interest are the book on goodness-of-fit techniques edited by D'Agostino and Stephens (T100) and the bibliography by Daniel (T101).

8.7

COMPUTER PROGRAMS

Romesburg et al. (T102) have written a FORTRAN IV computer program that performs the exact chi-square test that was proposed by Radlow and Alf (T103). The authors present a modification of the Radlow–Alf algorithm, describe the program's input and output, and provide a complete program listing.

Theodorsson (T104) has made available a program, written in Microsoft BASIC for microcomputers, that performs the Anderson–Darling (T105) test for normality.

The chi-square goodness-of-fit test is widely available in software packages for microcomputers. Those that contain the Kolmogorov–Smirnov tests include the following: *HP STATISTICS LIBRARY/2000, SCA, SPSS/PC, STATA, STATGRAPHICS, STATISTIX, STAT-PAC* (Science Software), and *STATPRO*.

REVIEW EXERCISES

8.20 A social worker believes that the age distribution of regular users of marijuana in a certain population is as follows: 20 and under, 30%; 21–30, 60%; 31–40, 8%; and over 40, 2%. A random sample of 300 drawn from the population yielded the age breakdown shown in Table 8.33. Do these data provide sufficient evidence to cast doubt on the social worker's belief?

TABLE 8.33   **Age distribution of 300 regular users of marijuana**

Age, years	Number
20 and under	96
21–30	171
31–40	22
Over 40	11

8.21 The distribution of IQ scores of a sample of 300 elementary school children is shown in Table 8.34. Can one conclude from these data that in the sampled population IQ scores are not normally distributed?

8.22 A graduate student in psychology recorded the number of people contributing to a solicitor for a charity organization stationed in a shopping mall during the Christmas season. The numbers of people contributing during five-minute time intervals were counted. The results are shown in Table 8.35. Test the goodness of fit of these data to a Poisson distribution.

TABLE 8.34	IQ scores of 300 elementary school children					
Score	50–59	60–69	70–79	80–89	90–99	100–109
Number	2	6	19	40	49	62
Score	110–119	120–129	130–139	140–149	150–159	
Number	50	42	21	7	2	

TABLE 8.35	Number of people in a shopping mall contributing to a solicitor for a charitable organization during five-minute intervals									
Number of contributors (X_i)	0	1	2	3	4	5	6	7	8	9
Number of intervals in which X_i was observed	10	27	36	33	22	12	6	4	2	1

8.23 A psychology research team set up a temporary research facility in a busy shopping mall and conducted experiments using volunteers over 12 years of age in groups of 10. They paid each volunteer two dollars for participating. Table 8.36 shows the frequency distribution of number of males in each group of 10 volunteers. Test the null hypothesis that number of males per group of 10 is distributed as the binomial with $p = 0.50$.

TABLE 8.36	Distribution of males in groups of 10 volunteers for a psychology experiment										
Number of males	0	1	2	3	4	5	6	7	8	9	10
Observed frequency	0	1	5	7	21	27	20	12	4	2	1

8.24 Table 8.37 shows the verbal reasoning test scores made by a sample of 36 high school seniors. Test the null hypothesis that the population of scores is normally distributed with a mean of 80 and a standard deviation of 6.

TABLE 8.37	Verbal reasoning scores of 36 high school seniors																	
	87	76	80	87	77	86	77	86	77	92	80	78	84	77	81	77	75	81
	75	92	80	80	84	72	80	92	72	77	78	76	68	78	92	68	80	81

8.25 A random sample of 150 male and 150 female white-collar workers gave the results shown in Table 8.38. Can one conclude from these data that women tend to receive lower salaries than men?

8.26 Table 8.39 shows the distributions of post-treatment adjustment scores of 150 alcoholics treated in a residential facility and 100 alcoholics treated on an outpatient basis. Can we conclude from these data that alcoholics treated on an outpatient basis tend to make a better post-treatment adjustment (as indicated by higher scores) than those treated in a residential facility?

TABLE 8.38   **Income distributions of 150 male and 150 female white-collar workers**

Monthly salary, $	450–499	500–549	550–599	600–649	650–699	700–749
Number of men	3	5	10	22	12	33
Number of women	14	29	40	18	21	8

Monthly salary, $	750–799	800–849	850–899	900–949	950–999
Number of men	22	15	16	9	3
Number of women	6	5	5	2	2

TABLE 8.39   **Distribution of post-treatment adjustment scores of two groups of alcoholics**

Score	Treated in facility	Treated as outpatient
45–49	2	3
50–54	3	4
55–59	2	5
60–64	10	15
65–69	17	20
70–74	25	13
75–79	30	8
80–84	15	9
85–89	29	5
90–94	5	16
95–99	12	2
	150	100

8.27   The following are the melting points in degrees Celsius of 21 specimens of a substance used in the manufacture of plastic. Are these sample measurements compatible with the hypothesis that they came from a population of normally distributed measurements? Let $\alpha = 0.05$, and use the Lilliefors test to arrive at a conclusion.

111.5	109.2	108.8	109.1	108.3	107.6	105.5	107.1	108.0
109.4	109.1	108.4	108.3	108.2	108.0	107.7	107.3	106.2
107.0	106.6	107.1						

8.28   The following are the shelf lives in days of 12 specimens of a perishable food product. Construct the 95% confidence band for $F(x)$.

5.6	4.3	6.0	4.9	5.1	5.0	5.6	4.1	6.4	4.6	5.1	5.6

8.29   Table 8.40 shows the ages of samples of 10 men and 10 women who are employed by a state government. Can one conclude on the basis of these data that the age compositions of the two populations are different? Use a 0.05 level of significance.

8.30   Two methods of performing an assembly-line task were compared by having a sample of 10 workers perform the task by method A and an independent sample of workers perform the task by method B. The times in seconds required to complete the task are shown in

TABLE 8.40	**Ages of 10 men and 10 women employed by a state government**									
Women	21	37	35	54	31	33	42	27	47	26
Men	56	25	43	42	54	33	34	35	47	39

Table 8.41. Do these data provide sufficient evidence to indicate a difference in population distributions? Use a 0.05 level of significance.

TABLE 8.41	**Times required to perform a task by two methods**									
Method A	25	22	17	16	19	24	26	21	19	27
Method B	21	22	14	13	22	22	15	21	23	21

REFERENCES

T1 Cochran, William G., "The χ^2 Test of Goodness of Fit," *Ann. Math. Statist.*, 23 (1952), 315–345.

T2 Cochran, William G., "Some Methods for Strengthening the Common χ^2 Tests," *Biometrics*, 10 (1954), 417–451.

T3 Broffitt, J. D., and R. H. Randles, "A Power Approximation for the Chi-Square Goodness-of-Fit Test; Simple Hypothesis Case," Report No. 23, Iowa City, Iowa: University of Iowa, Department of Statistics, 1973.

T4 Schorr, B., "On the Choice of the Class Intervals in the Application of the Chi-Square Test," *Math. Operationsforsch. Statist.*, 5 (1974), 357–377.

T5 Schorr, B., "On the Choice of Class Intervals for Chi-Square Test of Goodness of Fit," *Z. Angew. Math. Mech.*, 54 (1974), T249–T251.

T6 Slakter, M. J., "Accuracy of an Approximation to Power of Chi-Square Goodness of Fit Test with Small but Equal Expected Frequencies," *J. Amer. Statist. Assoc.*, 63 (1968), 912–918.

T7 Viollaz, A. J., "On the Reliability of the Chi-Square Test," *Metrika*, 33 (1986), 135–142.

T8 Dahiya, R. C., and J. Gurland, "How Many Classes in the Pearson Chi-Square Test?" *J. Amer. Statist. Assoc.*, 68 (1973), 707–712.

T9 Dahiya, R. C., and J. Gurland, "Pearson Chi-Squared Test of Fit with Random Intervals," *Biometrika*, 59 (1972), 147–153.

T10 Moore, D. S., "Chi-Square Statistic with Random Cell Boundaries," *Ann. Math. Statist.*, 42 (1971), 147–156.

T11 Kallenberg, W. C. M., J. Oosterhoff, and B. F. Schriever, "The Number of Classes in Chi-Squared Goodness-of-Fit Tests," *J. Amer. Statist. Assoc.*, 80 (1985), 959–968.

T12 Quine, M. P., and J. Robinson, "Efficiencies of Chi-Square and Likelihood Ratio Goodness-of-Fit Tests," *Ann. Statist.*, 13 (1985), 727–742.

T13 Chase, G. R., "The Chi-Square Test When Parameters Are Estimated Independently of the Sample," *J. Amer. Statist. Assoc.*, 67 (1972), 609–611.

T14 Lawal, H. Bayo, and Graham J. G. Upton, "An Approximation to the Distribution of the χ^2 Goodness-of-Fit Statistic for Use with Small Expectations," *Biometrika*, 67 (1980), 447–453.

T15 Brock, D. B., and A. M. Kshirsagar, "A χ^2 Goodness-of-Fit Test for Markov Renewal Process," *Ann. Inst. Statist. Math.*, 25 (1973), 643–654.

T16 Hewett, J. E., and R. K. Tsutakawa, "Two-Stage Chi-Square Goodness-of-Fit Test," *J. Amer. Statist. Assoc.*, 67 (1972), 395–401.

T17 Moore, David S., and M. C. Spruill, "Unified Large-Sample Theory of General Chi-Squared Statistics for Tests of Fit," *Ann. Statist.*, 3 (1975), 599–616.

T18 Moore, David S., "Tests of Chi-Squared Type," in Ralph B. D'Agostino and Michael A. Stephens (eds.), *Goodness-of-Fit Techniques*, New York: Marcel Dekker, 1986, pp. 63–95.

T19 Ritchey, Robert J., "An Application of the Chi-Squared Goodness-of-Fit Test to Discrete Common Stock Returns," *J. Bus. & Econ. Statist.*, 4 (1986), 243–254.

T20 O'Reilly, F. J., and C. P. Quesenberry, "Conditional Probability Integral Transformation and Applications to Obtain Composite Chi-Square Goodness-of-Fit Tests," *Ann. Statist.*, 1 (1973), 74–83.

T21 Slakter, M. J., "Comparative Validity of Chi-Square Goodness-of-Fit Tests for Small but Equal Expected Frequencies," *Biometrika*, 53 (1966), 619–622.

T22 Slakter, M. J., "Large Values for the Number of Groups with the Pearson Chi-Squared Goodness-of-Fit Tests," *Biometrika*, 60 (1973), 420–421.

T23 Albrecht, Peter, "On the Correct Use of the Chi-Square Goodness-of-Fit Test," *Scand. Actuarial J.* (1980), 149–160.

T24 Jammalamadaka, S. Rao, and Ram C. Tiwari, "Efficiencies of Some Disjoint Spacings Tests Relative to a χ^2 Test," in Madan L. Puri, Jose Perez Vilaplana, and Wolfgang Wertz (eds.), *New Perspectives in Theoretical and Applied Statistics*, New York: Wiley, 1987, pp. 311–317.

T25 Kolmogorov, A. N., "Sulla Determinazione Empirica di una Legge di Distribuizione," *Giorn. Ist. Ital. Attuari*, 4 (1933), 83–91.

T26 Smirnov, N. V., "Estimate of Deviation between Empirical Distribution Functions in Two Independent Samples" (Russian), *Bull. Moscow Univ.*, 2 (1939), 3–16.

T27 Massey, F. J., "The Kolmogorov–Smirnov Test for Goodness-of-Fit," *J. Amer. Statist. Assoc.*, 46 (1951), 68–78.

T28 Lilliefors, W. H., "On the Kolmogorov–Smirnov Test for Normality with Mean and Variance Unknown," *J. Amer. Statist. Assoc.*, 62 (1967), 399–402.

T29 Lilliefors, W. H., "On the Kolmogorov–Smirnov Test for the Exponential Distribution with Mean Unknown," *J. Amer. Statist. Assoc.*, 64 (1969), 387–389.

T30 Lee, S. W., "The Power of the One-Sided and One-Sample Kolmogorov–Smirnov Test," unpublished Masters' Report, Kansas State University, 1966.

T31 Quade, D., "On the Asymptotic Power of the One-Sample Kolmogorov–Smirnov Tests," *Ann. Math. Statist.*, 36 (1965), 1000–1018.

T32 Capon, J., "On the Asymptotic Efficiency of the Kolmogorov–Smirnov Test," *J. Amer. Statist. Assoc.*, 60 (1965), 843–853.

T33 Ramachandramurty, P. V., "On the Pitman Efficiency of One-sided Kolmogorov and Smirnov Tests for Normal Alternatives," *Ann. Math. Statist.*, 37 (1966), 940–944.

T34 Anděl, J., "Local Asymptotic Power and Efficiency of Tests of Kolmogorov–Smirnov Type," *Ann. Math. Statist.*, 38 (1967), 1705–1725.

T35 Gleser, Leon Jay, "Exact Power of Goodness-of-Fit Tests of Kolmogorov Type for Discontinuous Distributions," *J. Amer. Statist. Assoc.*, 80 (1985), 954–958.

T36 Nikitin, Ya. Yu., "On the Hodges–Lehmann Asymptotic Efficiency of Nonparametric Tests of Goodness of Fit and Homogeneity," *Theory Probabil. & Appl.*, 32 (1987), 77–85.

T37 Noether, G. E., "Note on the Kolmogorov Statistic in the Discrete Case," *Metrika*, 7 (1963), 115–116.

T38 Noether, G. E., *Elements of Nonparametric Statistics*, New York: Wiley, 1967.

T39 Slakter, M. J., "A Comparison of Pearson Chi-Square and Kolmogorov Goodness-of-Fit Tests with Respect to Validity," *J. Amer. Statist. Assoc.*, 60 (1965), 854–858. Corrections, *Ibid.*, 61 (1966), 1249.

T40 Conover, W. J., "A Kolmogorov Goodness-of-Fit Test for Discontinuous Distributions," *J. Amer. Statist. Assoc.*, 67 (1972), 591–596.

T41 Schmid, P., "On the Kolmogorov and Smirnov Limit Theorems for Discontinuous Distribution Functions," *Ann. Math. Statist.*, 29 (1958), 1011–1027.

T42 Carnal, H., "Sur les Théorèmes de Kolmogorov et Smirnov dans le Cas d'une Distribution Discontinue," *Comment. Math. Helv.*, 37 (1962), 19–35.

T43 Taha, M. A. H., "Ueber die verallgemeinerten Teste von Kolmogorov und Smirnov für Unstetige Verteilungen," *Mitt. Verein. Schweiz. Versicherungsmath.*, 64 (1964), 145–174.

T44 Azzalini, A., and G. Diana, "The Distribution of Kolmogorov's Statistic for Discrete Variates," *Metron*, 39 (1981), 133–140.

T45 Zelterman, Daniel, "Approximating the Distribution of Goodness of Fit Tests for Discrete Data," *Comput. Statist. & Data Analysis*, 2 (1984), 207–214.

T46 Guilbaud, Olivier, "Stochastic Inequalities for Kolmogorov and Similar Statistics, with Confidence Region Applications," *Scand. J. Statist.*, 13 (1986), 301–305.

T47 Schuster, E. F., "Goodness-of-Fit Problem for Continuous Symmetric Distributions," *J. Amer. Statist. Assoc.*, 68 (1973), 713–715. Corrigenda, *Ibid.*, 69 (1974), 288.

T48 Schafer, R. E., J. M. Finkelstein, and J. Collins, "Goodness-of-Fit Test for Exponential Distribution with Mean Unknown," *Biometrika*, 59 (1972), 222–224.

T49 Srinivasan, R., "An Approach to Testing the Goodness-of-Fit of Incompletely Specified Distributions," *Biometrika*, 57 (1970), 605–611.

T50 Finkelstein, J. M., and R. W. Schafer, "Improved Goodness-of-Fit Tests," *Biometrika*, 58 (1971), 641–645.

T51 Lohrding, R. K., "Three Kolmogorov–Smirnov-Type One-Sample Tests with Improved Power Properties," *J. Statist. Comp. and Simulation*, 2 (1973), 139–148.

T52 Riedwyl, H., "Goodness of Fit," *J. Amer. Statist. Assoc.*, 62 (1967), 390–398.

T53 Maag, U. R., F. Streit, and P. A. Drouilly, "Goodness-of-Fit Tests for Grouped Data," *J. Amer. Statist. Assoc.*, 68 (1973), 462–465.

T54 D'Agostino, R. B., "Omnibus Test of Normality for Moderate and Large Size Samples," *Biometrika*, 58 (1971), 341–348.

T55 Heathcote, C. R., "A Test of Goodness-of-Fit for Symmetric Random Variables," *Austral. J. Statist.*, 14 (1972), 172–181.

T56 Maag, U. R., and G. Dicaire, "On Kolmogorov–Smirnov Type One-Sample Statistics," *Biometrika*, 58 (1971), 653–656.

T57 Suzuki, G., "On Exact Probabilities of Some Generalized Kolmogorov's *D*-Statistics," *Ann. Inst. Statist. Math.*, 19 (1967), 373–388.

T58 Harter, H. Leon, Harry J. Khamis, and Richard E. Lamb, "Modified Kolmogorov–Smirnov Tests of Goodness of Fit," *Communic. in Statist.—Simulation and Computation*, 13 (1984), 293–323.

T59 Mantel, Nathan, "Kolmogorov–Smirnov Statistics—Weightings, Modifications, and

Variations," in P. K. Sen (ed.), *Biostatistics: Statistics in Biomedical, Public Health and Environmental Sciences*, Amsterdam: Elsevier, 1985, pp. 399–412.

T60 Guilbaud, Olivier, "Exact Kolmogorov-Type Tests for Left-Truncated and/or Right-Censored Data," *J. Amer. Statist. Assoc.*, 83 (1988), 213–221.

T61 Inglot, Tadeusz and Teresa Ledwina, "On Probabilities of Excessive Deviations for Kolmogorov–Smirnov, Cramér–Von Mises and Chi-Square Statistics," Report Series No. 8, Institute of Mathematics, Technical University of Wroclaw, Wroclaw, Poland.

T62 Iman, Ronald L., "Graphs for Use with the Lilliefors Test for Normal and Exponential Distributions," *Amer. Statist.*, 36 (1982), 109–112.

T63 Hodges, J. L., Jr., "The Significance Probability of the Smirnov Two-Sample Test," *Ark. Mat.*, 3 (1958), 469–486.

T64 Quade, Dana, "The Pair Chart," *Statistica Neerlandica*, 27 (1973), 29–45.

T65 Drion, E. F., "Some Distribution-Free Tests for the Difference between Two Empirical Cumulative Distribution Functions," *Ann. Math. Statist.*, 23 (1952), 563–574.

T66 Yu, C. S., "Pitman Efficiencies of Kolmogorov–Smirnov Tests," *Ann. Math. Statist.*, 42 (1971), 1595–1605.

T67 Klotz, J., "Asymptotic Efficiency of the Two Sample Kolmogorov–Smirnov Test," *J. Amer. Statist. Assoc.*, 62 (1967), 932–938.

T68 Bickel, P. J., "A Distribution-Free Version of the Smirnov Two-Sample Test in the p-Variate Case," *Ann. Math. Statist.*, 40 (1969), 1–23.

T69 Steck, G. P., "The Smirnov Two-Sample Tests as Rank Tests," *Ann. Math. Statist.*, 40 (1969), 1449–1466.

T70 Angus, J. E., "The Connection between the Barnard–Birnbaum Monte Carlo Test and the Two-Sample Kolmogorov–Smirnov Test," *Math. & Comput. Simulation*, 26 (1984), 20–22.

T71 Barnard, G. A. (In discussion), *J. Roy. Statist. Assoc., Ser. B*, 25 (1963), 294.

T72 Birnbaum, Z. W., "Computers and Unconventional Test Statistics," in Frank Proschan and R. J. Serfling (eds.), *Reliability and Biometry*, Philadelphia: Society for Industrial and Applied Mathematics, 1974, pp. 441–458.

T73 Wieand, H. S., "Application of Nonparametric Statistics to Cancer Data," in P. R. Krishnaiah and P. K. Sen (eds.), *Handbook of Statistics*, Vol. IV, Amsterdam: Elsevier, 1984, pp. 771–790.

T74 Birnbaum, Z. W., "Numerical Tabulation of the Distribution of Kolmogorov's Statistic for Finite Sample Size," *J. Amer. Statist. Assoc.*, 47 (1952), 425–441.

T75 Goodman, L. A., "Kolmogorov–Smirnov Tests for Psychological Research," *Psychol. Bull.*, 51 (1954), 160–168.

T76 Cramér, H., "On the Composition of Elementary Errors," *Skandinavisk Aktuarietidskrift*, 11 (1928), 13–74 and 141–180.

T77 von Mises, R., *Wahrscheinlichkeitsrechung und ihre Anwendung in der Statistik und theoretischen Physik*, Leipzig: F. Deuticke, 1931.

T78 Angus, John E., "On the Asymptotic Distribution of Cramér–von Mises One-Sample Test Statistics under an Alternative," *Communic. in Statist.—Theory and Methods*, 12 (1983), 2477–2482.

T79 Birnbaum, Z. W., and R. A. Hall, "Small Sample Distribution for Multi-Sample Statistics of the Smirnov Type," *Ann. Math. Statist.*, 31 (1960), 710–720.

T80 Gibbons, Jean D., "A Distribution-Free Two-Sample Goodness-of-Fit Test for General Alternatives," *Br. J. Math. Statist. Psychol.*, 25 (1972), 95–106.

T81 Hora, Stephen C., "Goodness of Fit Tests Using Regression," *Communic. in Statist.—Theory and Methods*, 14 (1985), 307–332.

T82 Kocherlakota, S., and K. Kocherlakota, "Goodness of Fit Tests for Discrete Distributions," *Communic. in Statist.—Theory and Methods*, 15 (1986), 815–829.

T83 Zelterman, Daniel, "Goodness-of-Fit Tests for Large Sparse Multinomial Distributions," *J. Amer. Statist. Assoc.*, 82 (1987), 624–629.

T84 Kumazawa, Yoshiki, "Goodness-of-Fit Statistics for Symmetric Distributions," *J. Japan Statist. Soc.*, 12 (1982), 25–38.

T85 Oja, Hannu, "New Tests for Normality," *Biometrika*, 70 (1983), 297–299.

T86 Best, D. J., and J. C. W. Rayner, "Lancaster's Test of Normality," *J. Statist. Planning and Inference*, 12 (1985), 395–400.

T87 Lancaster, H. O., *The Chi-Squared Distribution*, New York: Wiley, 1969.

T88 LaRiccia, Vincent N., "Optimal Goodness-of-Fit Tests for Normality against Skewness and Kurtosis Alternatives," *J. Statist. Planning and Inference*, 13 (1986), 67–79.

T89 Shapiro, Samuel S., *How to Test Normality and Other Distributional Assumptions*, Milwaukee: American Society for Quality Control, 1980.

T90 Best, D. J., and J. C. W. Rayner, "A Test for Bivariate Normality," *Statist. & Probabil. Letters*, 6 (1988), 407–412.

T91 Angus, John E., "Goodness-of-Fit Tests for Exponentiality Based on a Loss-of-Memory Type Functional Equation," *J. Statist. Planning and Inference*, 6 (1982), 241–251.

T92 Haber, Michael, "A Comparative Simulation Study of the Small Sample Powers of Several Goodness of Fit Tests," *J. Statist. Comput. and Simulation*, 11 (1980), 241–250.

T93 Jammalamadaka, S. Rao, and Ram C. Tiwari, "Asymptotic Comparison of Three Tests for Goodness of Fit," *J. Statist. Planning and Inference*, 12 (1985), 295–304.

T94 McKinley, Robert L., and Craig N. Mills, "A Comparison of Several Goodness-of-Fit Statistics," *Appl. Psychol. Measurement*, 9 (1985), 49–57.

T95 Conover, W. J., "Several k-Sample Kolmogorov–Smirnov Tests," *Ann. Math. Statist.*, 36 (1965), 1019–1026.

T96 Conover, W. J., "A k-Sample Extension of the One-Sided Two-Sample Smirnov Test Statistic," *Ann. Math. Statist.*, 38 (1967), 1726–1730.

T97 Oja, Hannu, "Two Location and Scale-Free Goodness-of-Fit Tests," *Biometrika*, 68 (1981), 637–640.

T98 Chmielewski, Margaret A., "A Re-Appraisal of Tests for Normality," *Communic. in Statist.—Theory and Methods*, 10 (1981), 2005–2014.

T99 Hirotsu, C., "Cumulative Chi-Squared Statistic as a Tool for Testing Goodness of Fit," *Biometrika*, 73 (1986), 165–173.

T100 D'Agostino, Ralph B., and Michael A. Stephens (eds.), *Goodness-of-Fit Techniques*, New York: Marcel Dekker, 1986.

T101 Daniel, Wayne W., *Goodness of Fit: A Selected Bibliography for the Statistician and the Researcher*, Monticello, Ill.: Vance Bibliographies, 1980.

T102 Romesburg, H. Charles, "FITEST: A Computer Program for 'Exact Chi-Square' Goodness-of-Fit Significance Tests," *Computers & Geosciences*, 7 (1981), 47–58.

T103 Radlow, Robert, and Edward F. Alf, Jr., "An Alternate Multinomial Assessment of the Accuracy of the χ^2 Test of Goodness of Fit," *J. Amer. Statist. Assoc.*, 70 (1975), 811–813.

T104 Theodorsson, Elvar, "BASIC Computer Program to Summarize Data Using Nonparametric

and Parametric Statistics Including Anderson–Darling Test for Normality," *Comput. Methods and Programs in Biomed.*, 26 (1988), 207–214.

T105   Anderson, T. W., and D. A. Darling, "Asymptotic Theory of Certain 'Goodness of Fit' Criteria Based on Stochastic Processes," *Ann. Math. Statist.*, 23 (1952), 193–212.

E1   Stranges, Richard J., and Anthony C. Riccio, "Counselee Preferences for Counselors: Some Implications for Counselor Education," *Counselor Educ. and Supervision*, 10 (1970), 39–45.

E2   Miltimore, J. E., J. M. McArthur, B. P. Goplen, W. Majak, and R. E. Horwath, "Variability of Fraction 1 Protein and Total Phenolic Constituents in Alfalfa," *Agron. J.*, 66 (1974), 384–386.

E3   Gore, W. L., "Quality Control in the Chemical Industry IV: Statistical Methods in Plastics Research and Development," *Indust. Quality Control*, 4 (September 1947), 5–8.

E4   Lafleur, M. S., P. F. Hinrichsen, P. C. Landry, and R. B. Moore, "The Poisson Distribution: An Experimental Approach to Teaching Statistics," *Physics Teacher*, 10 (1972), 314–321.

E5   Yousuf, S. M. Anwar, "Two Measures of Risk-Taking in India," *Psychologia*, 16 (1973), 46–48.

E6   Kogan, N., and M. A. Wallach, "The Effect of Anxiety on Relations between Subjective Age and Caution in an Older Sample," in P. H. Hoch and J. Zubin (eds.), *Psychopathology of Aging*, New York: Grune and Stratton, 1961.

E7   Wallach, M. A., and N. Kogan, "Sex Differences and Judgement Processes," *J. Personality*, 27 (1959), 555–564.

E8   Wallach, M. A., and N. Kogan, "Aspects of Judgement and Decision Making: Interrelationships and Change with Age," *Behav. Sci.*, 6 (1961), 23–36.

E9   Rettig, S., and H. W. Rawson, "The Risk Hypothesis in Predictive Judgements of Unethical Behaviour," *J. Abnorm. Soc. Psychol.*, 66 (1963), 243–248.

E10   Ibrahim, W. N., K. Kamel, O. Salim, A. Azim, M. F. Gaballah, F. Sabry, A. El-Naggar, and K. Hoerman, "Hereditary Blood Factors and Anthropometry of the Inhabitants of the Egyptian Siwa Oasis," *Hum. Biol.*, 46 (1974), 57–68.

E11   Czeizel, A., and G. Tusnadi, "An Epidemiologic Study of Cleft Lip with or without Cleft Palate and Posterior Cleft Palate in Hungary," *Hum. Hered.*, 21 (1971), 17–38.

E12   Dusheiko, S. D., "Some Questions Concerning the Pathological Anatomy of Alzheimer's Disease," *Sov. Neurol. Psych.* 7 (Summer 1974), 56–64.

E13   Edwards, A. W., and M. Fraccaro, "Distribution and Sequences of Sexes in a Selected Sample of Swedish Families," *Ann. Hum. Genet.*, 24 (1960), 245–252.

E14   Kendall, David G., "Some Problems in the Theory of Queues," *J. Roy. Statist. Soc., Ser. B*, 13 (1951), 151–185.

E15   Neyman, J., "On a New Class of 'Contagious' Distributions, Applicable in Entomology and Bacteriology," *Ann. Math. Statist.*, 10 (1939), 35–57.

E16   Grundmann, R., M. Raab, E. Meusel, R. Kirchhoff, and H. Pichlmaier, "Analysis of the Optimal Perfusion Pressure and Flow Rate of the Renal Vascular Resistance and Oxygen Consumption in Hypothermic Perfused Kidney," *Surgery*, 77 (1975), 451–461.

E17   Treloar, Alan E., "Menarche, Menopause, and Intervening Fecundity," *Hum. Biol.*, 46 (1974), 89–107.

E18   Moreno, A. H., A. R. Burchell, R. V. Reddy, W. F. Panke, and T. F. Nealon, Jr., "The Hemodynamics of Portal Hypertension Revisited: Determinants and Significance of Occluded Portal Pressures," *Surgery*, 77 (1975), 167–179.

E19   Burrus, S. K., D. B. Dill, Dianna L. Burk, David V. Freeland, and William C. Adams, "Obser-

vations at Sea Level and Altitude on Basal Metabolic Rate and Related Cardio-Pulmonary Functions," *Hum. Biol.*, 46 (1974), 677–692.

E20 Glovsky, Leon, and Seymour Rigrodsky, "A Developmental Analysis of Mentally Deficient Children with Early Histories of Aphasia," *Training School Bull.*, 61 (1964), 76–96.

E21 Edwards, C. W., "Left Ventricular Hypertrophy in Emphysema," *Thorax*, 29 (1974), 75–80.

E22 Gooden, D. S., "Non-Destructive Neutron Activation Analysis for the Determination of Manganese and Zinc in Human Skin Biopsies," *Phys. Med. Biol.*, 17 (1972), 26–31.

E23 Wallenberg, L. R., "Reduction in Cold-Induced Natriuresis Following Hyperoncotic Albumin Infusion in Man Undergoing Water Diuresis," *Scand. J. Clin. Lab. Invest.*, 34 (1974), 233–239.

E24 Higby, D. J., E. Cohen, J. F. Holland, and L. Sinks, "The Prophylactic Treatment of Thrombocytopenic Leukemic Patients with Platelets: A Double Blind Study," *Transfusion*, 14 (1974), 440–446.

E25 Akerfeldt, A., "Oxidation of N, N-Dimethyl-*p*-Phenylenediamine by Serum from Patients with Mental Disease," *Science*, 125 (1957), 117–118.

E26 Friedhoff, Arnold J., Myra Palmer, and Christine Simmons, "An Effect of Exercise, Skin Shock, and Ascorbic Acid on Serum Oxidase Activity," *Arch. Neurol. Psychiatry*, 81 (1959), 620–626.

E27 Reed, Bruce L., and Marvin A. Lanphere, "Chemical Variations across the Alaska–Aleutian Range Batholith," *J. Res. U. S. Geol. Survey*, 2 (1974), 343–352.

RANK CORRELATION
AND OTHER MEASURES
OF ASSOCIATION

This chapter presents some of the better-known procedures for investigating the presence (or absence) of association between variables. One frequently encounters research hypotheses asserting the presence of a relationship between two variables or among several variables. Here are typical examples of research hypotheses from several fields.

Agriculture: The yield of a certain crop is related to the amount of some particular chemical in the soil.

Biology: The wing length and tail length in a certain species of bird are directly correlated.

Business: There is an inverse relationship between an employee's job satisfaction and the noise level in the working area.

Engineering: There is a relationship between the strength of a certain product and the amount of impurities in the raw material.

Health and medicine: Air pollution and lung disease are associated.

Psychology: Self-concept and behavior in certain social settings are related.

Sociology: There is a relationship between social status and attitude toward certain deviant behaviors.

Additional illustrations from these and other areas are discussed in the examples and exercises that follow.

Chapter 5 covered the association between variables that are categorical in nature (that is, are measured on a nominal scale) through the use of the chi-square test of independence, a procedure that would fit logically in the present chapter. Because of the widespread use of the chi-square test of independence and other chi-square tests, however, the technique is discussed in a separate chapter.

This chapter considers two aspects of association analysis. First, we are interested in deciding whether observed sample data provide sufficient evidence to conclude that, in the sampled population, the variables of interest are associated. We want to be able to reach a decision regarding the *existence* of an association between variables; that is, we want to compute a measure that in some sense expresses the *degree* or *strength* of the relationship between variables. Such measures, computed from sample data, usually serve four purposes.

1. They measure the strength of the relationship among the sample observations.
2. They provide a point estimate of the measure of the strength of the relationship between the variables in the population.
3. They provide the basis for constructing a confidence interval for the measure of the strength of the relationship between the variables in the population.
4. They allow the investigator to reach a conclusion about the presence of a relationship in the population from which the sample was drawn.

From previous courses in classical statistics, the reader may recall that the sample correlation coefficient measures the strength of the relationship among observations drawn from bivariate populations. This statistic, known as the *Pearson product–moment correlation coefficient*, is usually designated r, and defined as

$$r = \frac{\sum_{i=1}^{n}(X_i - \bar{X})(Y_i - \bar{Y})}{\sqrt{\sum_{i=1}^{n}(X_i - \bar{X})^2 \sum_{i=1}^{n}(Y_i - \bar{Y})^2}}$$

where X and Y are the variables of interest. It is used to estimate ρ, the population correlation coefficient. Unfortunately, for making inferences, r is appropriate only if we assume that the distribution of the sampled population is a bivariate normal distribution. The measures of association discussed in this chapter were developed, in part, to provide valid statistical techniques for use with data that do not meet the assumptions necessary for parametric correlation analysis.

The Pearson product–moment correlation coefficient has the following characteristics that are considered desirable for any measure of correlation between two variables X and Y.

1. If larger values of X tend to be paired with larger values of Y (and consequently, smaller values of X and Y tend to be paired), the measure of correlation should be positive and should approach 1 as this tendency

becomes more pronounced. In such situations the relationship between X and Y is called a *direct* relationship.

2. If small values of X tend to be paired with large values of Y (and vice versa), the measure of correlation should be negative and should approach -1 as the tendency becomes more pronounced. This is an *inverse* relationship.

3. If large values of X seem just as likely to be paired with small values of Y as with large values of Y, the measure of correlation should be close to zero. When X and Y are independent, the measure of correlation should be zero. In that case, X and Y are not *related*, and we say that they are independent. If X and Y are independent, their correlation is zero, but a zero correlation does not necessarily imply independence. [See Mood et al. (T1, p. 161).]

FURTHER READING

For further general discussions of measures of association, see the article by Kruskall (T2) and the books by Kendall (T3) and Liebetrau (T4).

9.1

SPEARMAN RANK CORRELATION COEFFICIENT

The first measure of association we discuss is the well-known and widely used *Spearman rank correlation coefficient*, introduced by Spearman (T5) in 1904.

Assumptions

A. The data consist of a random sample of n pairs of numeric or non-numeric observations.

B. Each pair of observations represents two measurements taken on the same object or individual, called the *unit of association*.

In preparation for computing the Spearman rank correlation coefficient, we subject our data to the following procedures:

1. If the data consist of observations from a bivariate population, we designate the n pairs of observations $(X_1, Y_1), (X_2, Y_2), \ldots, (X_n, Y_n)$.

2. Each X is ranked relative to all other observed values of X, from smallest to largest in order of magnitude. The rank of the ith value of X is denoted by $R(X_i)$, and $R(X_i) = 1$ if X_i is the smallest observed value of X.

3. Each Y is ranked relative to all other observed values of Y, from smallest to largest in order of magnitude. The rank of the ith value of Y is

denoted by $R(Y_i)$, and $R(Y_i) = 1$ if Y_i is the smallest observed value of Y.

4. If ties occur among the X's or among the Y's, each tied value is assigned the mean of the rank positions for which it is tied.

5. If the data consist of nonnumeric observations, they must be capable of being ranked as described.

Hypotheses

A. (Two-sided)
H_0: X and Y are independent
H_1: X and Y are either directly or inversely related

B. (One-sided)
H_0: X and Y are independent
H_1: There is a direct relationship between X and Y

C. (One-sided)
H_0: X and Y are independent
H_1: There is an inverse relationship between X and Y

Test Statistic

The test statistic is

$$r_S = 1 - \frac{6 \Sigma d_i^2}{n(n^2 - 1)} \tag{9.1}$$

where

$$\Sigma d_i^2 = \sum_{i=1}^{n} [R(X_i) - R(Y_i)]^2 \tag{9.2}$$

The test statistic is also the measure of association. As such, it is strictly speaking a measure of the degree of correspondence between the ranks of the sample observations rather than between the observations themselves. It is, however, thought of as a measure of the strength of the relationship between the sample X and Y values and as an estimate of the strength of the relationship between X and Y in the sampled population. Unfortunately, the exact measure of strength that r_S is estimating is difficult to interpret. The matter will not be pursued further here, since it is primarily of theoretical interest.

When the rank of X is the same as the rank of Y for every pair of observations (*perfect direct relationship*), all the differences d_i will be equal to zero, and r_S will be equal to $+1$. Kendall (T3) has shown that in general $r_S = -1$ when the rank of one variable within each pair of observations (X_i, Y_i) is the reverse of the other (*perfect inverse relationship*). Thus if

$$[R(X) = 1, R(Y) = n]$$

$$[R(X) = 2, R(Y) = n - 1], \ldots, [R(X) = n, R(Y) = 1]$$

for n pairs of observations, $r_S = -1$. This may be illustrated by means of a simple example. Suppose we have the following pairs of observations of (X_i, Y_i): $(0, 10)$, $(8, 3)$, $(2, 9)$, $(5, 6)$. The ranks are

$R(X_i)$	1	4	2	3
$R(Y_i)$	4	1	3	2

The sum of the d_i^2 values is $(-3)^2 + (3)^2 + (-1)^2 + (1)^2 = 20$, and when we substitute into Equation 9.1, we have

$$r_S = 1 - [6(20)/4(16 - 1)] = 1 - (120/60) = 1 - 2 = -1$$

Kendall (T3) also shows that r_S can never be greater than $+1$ or less than -1.

Decision Rule

Table A.21 gives critical values of r_S for sample sizes 4 through 100, prepared by Zar (T6). The following are the decision rules for the three sets of hypotheses.

    A. (Two-sided): Reject H_0 at the α level if the computed value of r_S is greater than the tabulated value for n and $\alpha(2)$ given in Table A.21 or less than the negative of this value

    B. (One-sided): Reject H_0 at the α level if the computed value of r_S is greater than the tabulated value for n and $\alpha(1)$ in Table A.21

    C. (One-sided): Reject H_0 at the α level if the computed value of r_S is less than the negative of the tabulated value for n and $\alpha(1)$ in Table A.21

Example 9.1   Pincherle and Robinson (E1) note a marked interobserver variation in blood pressure readings. They found that doctors who read high on systolic tended to read high on diastolic. Table 9.1 shows mean systolic and diastolic blood pressure readings by 14 doctors. We wish to compute a measure of the strength of the

TABLE 9.1   **Mean blood pressure readings, millimeters mercury, by doctor**

Doctor	1	2	3	4	5	6	7
Systolic	141.8	140.2	131.8	132.5	135.7	141.2	143.9
Diastolic	89.7	74.4	83.5	77.8	85.8	86.5	89.4
Doctor	8	9	10	11	12	13	14
Systolic	140.2	140.8	131.7	130.8	135.6	143.6	133.2
Diastolic	89.3	88.0	82.2	84.6	84.4	86.3	85.9

Source: G. Pincherle and D. Robinson, "Mean Blood Pressure and Its Relation to Other Factors Determined at a Routine Executive Health Examination," J. Chronic Dis., 27 (1974), 245–260; used with permission of Pergamon Press.

relationship between the two variables. Under the assumption that these 14 doctors constitute a random sample from a population of doctors, we wish to know whether we may conclude from the data that there is a direct relationship between systolic and diastolic readings. Suppose we let $\alpha = 0.05$.

Hypotheses

H_0: Systolic and diastolic blood pressure readings by doctors are independent

H_1: There is a direct relationship between systolic and diastolic blood pressure readings by doctors

The alternative hypothesis, in other words, states that doctors who read high on systolic tend to read high on diastolic.

Test Statistic

The intermediate calculations necessary for computing r_S are given in Table 9.2. When we substitute $\Sigma d_i^2 = 132.50$ from Table 9.2 into Equation 9.1, we have

$$r_S = 1 - \frac{6(132.50)}{14(14^2 - 1)} = 0.71$$

TABLE 9.2   **Intermediate calculations for computing r_s in Example 9.1**

Systolic (X_i)	Diastolic (Y_i)	$R(X_i)$	$R(Y_i)$	$d_i = R(X_i) - R(Y_i)$	d_i^2
141.8	89.7	12	14	−2	4
140.2	74.4	8.5	1	7.5	56.25
131.8	83.5	3	4	−1	1
132.5	77.8	4	2	2	4
135.7	85.8	7	7	0	0
141.2	86.5	11	10	1	1
143.9	89.4	14	13	1	1
140.2	89.3	8.5	12	−3.5	12.25
140.8	88.0	10	11	−1	1
131.7	82.2	2	3	−1	1
130.8	84.6	1	6	−5	25
135.6	84.4	6	5	1	1
143.6	86.3	13	9	4	16
133.2	85.9	5	8	−3	9
					$\Sigma d_i^2 = 132.50$

Decision

Table A.21 reveals that, for $n = 14$ and $\alpha(1) = 0.05$, the critical value of r_S is 0.464. Since our computed value of 0.71 is greater than 0.464, we reject H_0 and conclude

that doctors who read high on systolic tend to read high on diastolic blood pressures. Since our computed value of 0.71 is between the tabulated values of 0.679 and 0.723, the P value for this test is between 0.005 and 0.0025.

Ties Earlier, it was recommended that ties occurring in the X's or in the Y's be broken by assigning to each tied observation the mean rank of the rank positions for which it is tied.

Ties affect the magnitude of the computed value of r_S very little unless their number is excessive. We may use the following correction for ties if desired. Let

$$T_x = \frac{t_x^3 - t_x}{12} \qquad (9.3)$$

$$T_y = \frac{t_y^3 - t_y}{12} \qquad (9.4)$$

$$\Sigma x^2 = \frac{n^3 - n}{12} - \Sigma T_x \qquad (9.5)$$

$$\Sigma y^2 = \frac{n^3 - n}{12} - \Sigma T_y \qquad (9.6)$$

where t_x and t_y are, respectively, the number of X observations and the number of Y observations that are tied for a particular rank. When the correction for ties is used, the test statistic is

$$r_S = \frac{\Sigma x^2 + \Sigma y^2 - \Sigma d_i^2}{2\sqrt{\Sigma x^2 \Sigma y^2}} \qquad (9.7)$$

Glasser and Winter (T7) have prepared other tables of critical values of the Spearman rank correlation coefficient. Litchfield and Wilcoxon (T8) present a nomograph that permits direct reading of the rank correlation coefficient.

Large-Sample Approximation When the sample size is greater than 100, we cannot use Table A.21 to test the significance of r_S. Then we may compute

$$z = r_S \sqrt{n - 1}$$

which is distributed approximately as the standard normal.

Power-Efficiency The asymptotic relative efficiency of the test based on r_S relative to the parametric t test based on the Pearson product–moment correlation coefficient is $9/\pi^2 = 0.912$, when both tests are applied under the same conditions in which the assumptions underlying the parametric test are met. See Stuart (T9) and Bhattacharyya et al. (T10). Woodworth (T11) discusses the Bahadur (T12) efficiency of the test based on r_S.

FURTHER READING

Daniels (T13) has suggested the use of r_S as a test for trend, and Kraemer (T14) discusses the nonnull distribution of r_S.

A paper by Zar (T15) describes the method by which the critical values in Table A.21 were calculated. Iman and Conover (T16) have also prepared tables of critical values for use with the Spearman statistic. A paper by Griffiths (T17) is also of interest.

EXERCISES

9.1 Bakos (E2) reported observations made on comet Bennett, which are recorded in Table 9.3. The author plotted a histogram of reduced visual magnitude (H) values against the log of the heliocentric distances r. Compute r_S for the pairs of observations (r, H), and test for significance.

TABLE 9.3   **Observational data on comet Bennett**

Heliocentric distance (r)	.685	.720	.735	.750	.798	.810	.828	.950
Reduced visual magnitude (H)	1.72	1.80	2.53	2.15	2.81	2.92	2.80	3.82
Heliocentric distance (r)	.988	1.210	1.228	1.244	1.267	1.295	1.312	1.330
Reduced visual magnitude (H)	3.77	5.06	5.21	5.19	5.39	5.61	5.57	5.89

Source: Gustav A. Bakos, "Photoelectric Observations of Comet Bennett," *J. Roy. Astron. Soc. Can.*, 67 (1973), 183–189.

9.2 Daniel (E3) investigated the nature of the relationship between intelligence and social dominance in albino mice. Table 9.4 shows the results of the experiment. Do these data provide sufficient evidence to indicate a relationship between intelligence and social dominance in albino mice?

What is the P value for this test?

9.3 Table 9.5 shows the serum and bone magnesium levels of 14 patients as reported by Alfrey et al. (E4). Can we conclude from these data that a relationship exists between serum magnesium and bone magnesium in the sampled population?

Find the P value.

9.4 Gentry and Pike (E5) report the data on mean rate of return and value of common stock portfolios for 32 life insurance companies for 1956 through 1969 shown in Table 9.6. Compute r_S and test for significance. Determine the P value.

TABLE 9.4

Intelligence and social dominance in albino mice

Mouse	Intelligence score	Social dominance score
1	45	63.7
2	26	0.1
3	20	15.6
4	40	101.2
5	36	25.4
6	23	1.8

Source: Jean Daniel, ''A Study of the Relationship between Social Dominance and Intelligence in Albino Mice,'' unpublished research report, 1975.

TABLE 9.5

Serum and bone magnesium levels in 14 patients

Serum Mg (m Eq./L.)	3.60	2.85	2.80	2.70	2.60	2.55	2.55
Bone Mg (m Eq./kg ash)	672	610	621	567	570	638	612
Serum Mg (m Eq./L.)	2.45	2.25	1.80	1.45	1.35	1.40	0.90
Bone Mg (m Eq./kg ash)	552	524	400	277	294	338	230

Source: Allen C. Alfrey, Nancy L. Miller, and Donald Butkus, ''Evaluation of Body Magnesium Stores,'' *J. Lab. Clin. Med.*, 84 (1974), 153–162.

TABLE 9.6

Mean rate of return of common stock portfolios for 32 life insurance companies, 1956–1969, and 1969 value of each equity portfolio

Mean rate of return	Value of common stock portfolio, December 31, 1969, million dollars	Mean rate of return	Value of common stock portfolio, December 31, 1969, million dollars
18.83	96.0	10.44	111.9
16.98	54.6	10.44	179.8
15.36	84.4	10.33	29.2
14.65	251.5	10.30	279.5
14.21	131.8	10.22	166.6
13.68	37.3	10.05	194.3
13.65	109.9	10.04	40.8
13.07	13.5	9.57	428.4
12.99	76.3	9.50	7.0
12.81	72.6	9.48	485.6
11.60	42.1	9.29	165.3
11.51	41.5	9.21	343.8
11.50	56.2	9.04	35.4
11.41	59.3	8.82	24.7
11.26	1184.0	8.78	2.7
10.67	144.0	7.26	8.9

Source: James Gentry and John Pike, ''An Empirical Study of the Risk–Return Hypothesis Using Common Stock Portfolios of Life Insurance Companies,'' *J. Finan. Quant. Anal.*, 5 (1970), 179–185.

9.2

KENDALL'S TAU

Let us now discuss a measure of correlation called *Kendall's tau* and its accompanying test. Kendall (T18) independently proposed the measure in 1938, although Kruskal (T2) traces the idea back as far as 1899. In the literature, Kendall's tau is represented by various symbols, including τ, T, and t. This book uses the symbol τ for this measure of association when referring to the population (that is, to designate the population parameter). The symbol $\hat{\tau}$ is used to designate the corresponding sample statistic.

Like the Spearman rank correlation coefficient, Kendall's $\hat{\tau}$ is based on the ranks of observations, and it can assume values between -1 and $+1$. Despite these similarities, $\hat{\tau}$ and r_S, when computed from the same data, usually have different numerical values. As will become apparent, this occurs because the two statistics measure association in different ways.

One of the important differences between $\hat{\tau}$ and r_S is that $\hat{\tau}$ provides an unbiased estimator of a population parameter, while the sample statistic r_S does not provide an estimate of a population coefficient of rank correlation. For a discussion of a parameter for which r_S can be considered an estimate, see Gibbons (T19), pages 235–240.

The parameter estimated by $\hat{\tau}$ may be defined as the probability of concordance minus the probability of discordance. Let us designate this parameter by the symbol τ. The observation pairs (X_i, Y_i) and (X_j, Y_j) are said to be *concordant* if the difference between X_i and X_j is in the same direction as the difference between Y_i and Y_j. In other words, if either $X_i > X_j$ *and* $Y_i > Y_j$ or $X_i < X_j$ *and* $Y_i < Y_j$, we have concordance. The observation pairs (X_i, Y_i) and (X_j, Y_j) are said to be *discordant* if the directions of the differences are not the same. If $X_i = X_j$ and/or $Y_i = Y_j$, the observation pairs are neither concordant nor discordant.

The objective when we use Kendall's $\hat{\tau}$ for inferential purposes is to test the null hypothesis that X and Y are independent (which implies $\tau = 0$) against one of the following alternatives: $\tau \neq 0$, $\tau > 0$, or $\tau < 0$. We may interpret the alternative $\tau \neq 0$ to mean that there is an association between X and Y. We interpret $\tau > 0$ to indicate a direct association between X and Y, and $\tau < 0$ to mean that X and Y are inversely associated.

Assumptions

A. The data consist of a random sample of n observation pairs (X_i, Y_i) of numeric or nonnumeric observations. Each pair of observations represents two measurements taken on the same unit of association.

B. The data are measured on at least an ordinal scale, so that we can rank each X observation in relation to all other observed X's and each Y observation in relation to all other observed Y's.

Hypotheses

A. (Two-sided)
   H_0: X and Y are independent
   H_1: $\tau \neq 0$
B. (One-sided)
   H_0: X and Y are independent
   H_1: $\tau > 0$
C. (One-sided)
   H_0: X and Y are independent
   H_1: $\tau < 0$

Test Statistic

The test statistic, which is also the measure of association in the sample, is given by

$$\hat{\tau} = \frac{S}{n(n-1)/2} \tag{9.8}$$

where n is the number of (X, Y) observations (or ranks). To obtain S, and consequently $\hat{\tau}$, we proceed as follows:

1. Arrange the observations (X_i, Y_i) in a column according to the magnitude of the X's, with the smallest X first, the second smallest second, and so on. Then we say that the X's are in *natural order*.

2. Compare each Y value, one at a time, with each Y value appearing below it. In making these comparisons, we say that a pair of Y values (a Y being compared and the Y below it) is in natural order if the Y below is larger than the Y above. We say that a pair of Y values is in *reverse natural order* if the Y below is smaller than the Y above.

3. Let P be the number of pairs in natural order and Q the number of pairs in reverse natural order.

4. $S = P - Q$; that is, S in Equation 9.8 is equal to the difference between P and Q.

A total of $\binom{n}{2} = n(n-1)/2$ possible comparisons of Y values can be made in this manner. If all the Y pairs are in natural order, then $P = n(n-1)/2$, $Q = 0$, $S = [n(n-1)/2] - 0 = n(n-1)/2$, and we have

$$\hat{\tau} = \frac{n(n-1)/2}{n(n-1)/2} = 1$$

indicating perfect direct correlation between the rankings of X and Y. On the other hand, if all the Y pairs are in reverse natural order, we have $P = 0$, $Q = n(n-1)/2$,

$S = 0 - [n(n - 1)/2] = -n(n - 1)/2$, and

$$\hat{\tau} = \frac{-n(n - 1)/2}{n(n - 1)/2} = -1$$

indicating a perfect inverse correlation between the X and Y rankings.

Thus $\hat{\tau}$ cannot be greater than $+1$ or smaller than -1. We can think of $\hat{\tau}$ as a relative measure of the extent of the disagreement between the observed order of the Y observations and the two orderings that represent a perfect correlation between the X and Y rankings. If the number of Y pairs that are in natural order exceeds the number in reverse natural order, we have a direct correlation between the X and Y rankings, and $\hat{\tau}$ is positive. If the number of Y pairs that are in reverse natural order exceeds the number in natural order, we have an inverse correlation between the X and Y rankings, and $\hat{\tau}$ is negative. The strength of the correlation is indicated by the magnitude of the absolute value of $\hat{\tau}$.

Decision Rule

The decision rules for the three sets of hypotheses are as follows:

A. (Two-sided): Refer to Table A.22. Reject H_0 at the α level of significance if the computed value of $\hat{\tau}$ is either positive and larger than the τ^* entry for n and $\alpha/2$, or negative and smaller than the negative of the τ^* entry for n and $\alpha/2$.

B. (One-sided): Refer to Table A.22. Reject H_0 at the α level of significance if the computed value of $\hat{\tau}$ is positive and larger than the τ^* entry for n and α.

C. (One-sided): Refer to Table A.22. Reject H_0 at the α level of significance if the computed value of $\hat{\tau}$ is smaller than the negative of the τ^* entry for n and α.

Example 9.2

Cravens and Woodruff (E6) conducted a study to design and test a methodology for analytically determining standards of sales performance. They reported the data on benchmark achievement and management rating for 25 sales territories shown in Table 9.7. They computed benchmark achievement as being sales volume divided by benchmark sales, and based management ratings on salesperson motivation and effort.

We wish to compute $\hat{\tau}$ for these data to see whether there is sufficient evidence to conclude that benchmark achievement and management rating are directly related. Although the data are reported as ranks, we follow the same procedure in computing $\hat{\tau}$ as we would if the data were reported in absolute quantities.

TABLE 9.7   **Territory rankings based on benchmark achievement and performance ratings**

Territory	Benchmark achievement (X)	Management rating (Y)	Territory	Benchmark achievement (X)	Management rating (Y)
1	2	4	14	11	10
2	9	2	15	1	1
3	7	20	16	21	14
4	23	17	17	14	15
5	5	5	18	3	11
6	17	7	19	13	13
7	16	6	20	18	19
8	25	24	21	22	25
9	4	3	22	19	16
10	10	21	23	24	23
11	20	18	24	6	22
12	15	9	25	12	12
13	8	8			

Source: David W. Cravens and Robert B. Woodruff, "An Approach for Determining Criteria of Sales Performance," *J. Appl. Psychol.*, 57 (1973), 242–247; copyright 1973. American Psychological Association; reprinted by permission.

Hypotheses

H_0: Benchmark achievement and management rating are independent

H_1: Benchmark achievement and management rating are directly related $(\tau > 0)$

Test Statistic

We first arrange the data as in the first column of Table 9.8, so that the X ranks are in natural order. The numbers of Y pairs in natural and reverse natural order with respect to each Y are shown in the second and third columns, respectively.

From the data in Table 9.8, we compute $S = P - Q = 218 - 82 = 136$, so that, by Equation 9.8, we have

$$\hat{\tau} = \frac{136}{25(24)/2} = \frac{136}{300} = 0.45$$

Decision

With $n = 25$, Table A.22 reveals that we can reject H_0 at the 0.005 level, since $\hat{\tau} = 0.45$ is larger than $\tau^* = 0.367$. We can conclude that there is a direct relation-

TABLE 9.8   **Arrangement of data for computing $\hat{\tau}$ in Example 9.2**

(X, Y) rankings	Y pairs in natural order	Y pairs in reverse natural order
(1, 1)	24	0
(2, 4)	21	2
(3, 11)	14	8
(4, 3)	20	1
(5, 5)	19	1
(6, 22)	3	16
(7, 20)	4	14
(8, 8)	14	3
(9, 2)	16	0
(10, 21)	3	12
(11, 10)	11	3
(12, 12)	10	3
(13, 13)	9	3
(14, 15)	7	4
(15, 9)	8	2
(16, 6)	9	0
(17, 7)	8	0
(18, 19)	3	4
(19, 16)	5	1
(20, 18)	3	2
(21, 14)	4	0
(22, 25)	0	3
(23, 17)	2	0
(24, 23)	1	0
(25, 24)	0	0
	$P = 218$	$Q = 82$

ship between benchmark achievement and management ranking in the sampled population. [As a matter of interest, we note that the authors compute a value of $r_S = 0.61$ for these data, which, when compared with critical values in Table A.21, is found to be significant at the 0.001 level.]

Ties   The hypothesis tests for Kendall's τ assume that the variables under study are continuous. However, ties do occur in practice, either within the X observations, within the Y observations, or both. Ties of X observations with Y observations are of no concern. When there are ties, the simplest procedure is to assign tied observations the mean of the rank positions for which they are tied. Although we

need not assign ranks explicitly in computing $\hat{\tau}$ by the method described earlier, the assignment of ranks is implicit in the procedure. When ties occur, the following procedure for calculating $\hat{\tau}$ is a convenient one. It does not require explicit assignment of ranks, but rank assignment is implicit in the procedure, as is the recommended technique for handling ties.

1. List the observations in ascending (natural) order, according to the magnitude of the X's.
2. Within the tied observations of the X's, arrange the Y values in ascending order of magnitude.
3. Count the number of Y pairs in natural order and the number of Y pairs in reverse natural order, as described before, but do not compare a Y value accompanying a tied X value (say, X_a) with any other Y value accompanying another X value that is tied with X_a.

If very many ties are present, we may compute $\hat{\tau}$ by using the following special formula, which adjusts for ties:

$$\hat{\tau} = \frac{S}{\sqrt{\frac{1}{2}n(n-1) - T_x}\sqrt{\frac{1}{2}n(n-1) - T_y}} \tag{9.9}$$

where
$T_x = \frac{1}{2}\Sigma t_x(t_x - 1), \qquad T_y = \frac{1}{2}\Sigma t_y(t_y - 1)$
$t_x = $ the number of X observations that are tied at a given rank
$t_y = $ the number of Y observations that are tied at a given rank

Sillitto (T20) has tabulated the distribution of the $\hat{\tau}$ statistic for any number of tied pairs or tied triplets up to and including $n = 10$. These tables may be used where applicable; otherwise use Table A.22. Burr (T21) and Smid (T22) have also considered the problem of ties.

The following example illustrates the calculation of $\hat{\tau}$ and the accompanying hypothesis test when ties are present.

Example 9.3   Krippner (E7) reported the data shown in Table 9.9 on 30 children (26 boys and 4 girls) who attended a summer reading clinic sponsored by a university child-study center. The data were generated as part of an investigation to determine which of several variables appear to be related to reading improvement manifested in a remedial program. We wish to compute $\hat{\tau}$ from these data and test the null hypothesis that there is no association between IQ and reading improvement.

Hypotheses

H_0: Reading improvement and IQ are independent
H_1: There is either a direct or inverse relationship between reading improvement and IQ ($\tau \neq 0$)

TABLE 9.9   **Data on 30 subjects enrolled in a five-week summer reading clinic**

Client	Improvement (X)	WISC IQ full scale (Y)
Alvin	0.6	86
Barry	0.2	107
Chester	1.6	102
Dick	0.5	104
Earl	0.9	104
Floyd	0.5	89
Gregg	0.8	109
Harry	0.8	109
Ivan	0.8	101
Jacob	0.4	96
Karl	1.8	113
Lewis	0.1	85
Marvin	0.9	100
Ned	0.2	94
Oscar	1.6	104
Peter	1.6	104
Quincy	0.0	98
Ralph	1.6	115
Rita	0.2	109
Simon	0.3	94
Tony	0.0	112
Uriah	1.0	96
Victor	1.3	113
Waldo	0.6	110
Walter	0.6	97
Wanda	0.5	107
Xavier	1.7	113
York	1.6	109
Yvonne	2.2	98
Zohra	1.5	106

Source: Stanley Krippner, "Correlates of Reading Improvement," *J. Devel. Reading*, 7 (1963), 29–39. Copyright 1963, Purdue Research Foundation; reprinted by permission.

Test Statistic

Before computing $\hat{\tau}$, we arrange the data as shown in the first two columns of Table 9.10. We next compute

$$T_x = \frac{2(1) + 3(2) + 3(2) + 3(2) + 3(2) + 2(1) + 5(4)}{2} = 24$$

and

$$T_y = \frac{2(1) + 2(1) + 2(1) + 4(3) + 2(1) + 4(3) + 3(2)}{2} = 19$$

By Equation 9.9, we have

$$\hat{\tau} = \frac{250 - 144}{\sqrt{\frac{30(29)}{2} - 24}\ \sqrt{\frac{30(29)}{2} - 19}} = 0.2564$$

TABLE 9.10   **Arrangement of data for computing $\hat{\tau}$ in Example 9.3**

Improvement (X)		IQ (Y)	Y pairs in natural order	Y pairs in reverse natural order
Tie	0.0	98	19	8
	0.0	112	4	24
	0.1	85	27	0
	0.2	94	21	2
Tie	0.2	107	8	15
	0.2	109	5	16
	0.3	94	21	2
	0.4	96	19	2
	0.5	89	18	1
Tie	0.5	104	9	7
	0.5	107	8	11
	0.6	86	16	0
Tie	0.6	97	15	1
	0.6	110	4	12
	0.8	101	10	3
Tie	0.8	109	4	8
	0.8	109	4	8
	0.9	100	9	2
Tie	0.9	104	6	3
	1.0	96	10	0
	1.3	113	1	6
	1.5	106	4	4
	1.6	102	2	1
	1.6	104	2	1
Tie	1.6	104	2	1
	1.6	109	2	1
	1.6	115	0	3
	1.7	113	0	1
	1.8	113	0	1
	2.2	98	0	0
			$P = 250$	$Q = 144$

Decision

Since our computed value of $\hat{\tau}$ (0.256) is greater than 0.218, the tabulated value of τ^* for $n = 30$ given in Table A.22, we can reject H_0 at the 0.10 level of significance (two-sided test).

Large-Sample Approximation For large samples, the statistic

$$z = \frac{3\hat{\tau}\sqrt{n(n-1)}}{\sqrt{2(2n+5)}} \tag{9.10}$$

is approximately normally distributed with mean 0 and variance 1. This normal approximation can be used for sample sizes not shown in Table A.22. Kendall (T3) gives a modification that may be used when there are ties. Results obtained by Robillard (T23) and Best (T24) suggest that the normal approximation is good even when there are several tied observations.

Power-Efficiency The hypothesis test utilizing Kendall's $\hat{\tau}$ has an asymptotic relative efficiency of $9/\pi^2 = 0.912$ when compared with the t test utilizing the Pearson product–moment correlation coefficient under conditions for which the Pearson test is valid. For further discussion of the power and efficiency of Kendall's tau test, see Bhattacharyya (T10), Farlie (T25), and Konijn (T26).

Choosing between r_S and $\hat{\tau}$ Spearman's rank correlation coefficient r_S and Kendall's $\hat{\tau}$ are the two most frequently encountered measures of association between variables measured on an ordinal scale. Many researchers probably wonder which of the two measures they should use in a given situation. A careful study of the presently known properties of the two statistics would probably lead most investigators to conclude that there is usually very little basis for choosing one over the other. The following points of comparison are worth considering.

1. When hand methods must be used, Kendall's $\hat{\tau}$ is considered more tedious to compute than the Spearman rank correlation coefficient.
2. The distribution of $\hat{\tau}$ approaches the normal more rapidly than does that of r_S. Thus when the normal approximation is used with samples of intermediate size, $\hat{\tau}$ may provide a more reliable test statistic.
3. As already noted, hypothesis tests associated with the two statistics have the same asymptotic relative efficiency when compared to the test utilizing the Pearson product–moment correlation coefficient under conditions for which the Pearson test is valid.
4. In general, when computed from the same data, r_S and $\hat{\tau}$ have different numerical values, but in a hypothesis-testing situation they usually lead to the same decision.
5. As noted previously, $\hat{\tau}$ may be interpreted as an estimator of a population parameter, while r_S does not have a corresponding population

parameter that is a coefficient of rank correlation. Thus using $\hat{\tau}$ may be more attractive to many investigators.

6. Strahan (T27) shows that when sampling is from a normally distributed population, r_S is very close in numerical size to r, the Pearson product–moment correlation coefficient. Consequently, he argues, r_S squared is a good indicator of r^2, the parametric coefficient of determination. In simple linear regression analysis, r^2 measures the proportion of variability in the dependent variable that is explained by the independent variable. Therefore, the argument continues, when studying the relationship between two variables on the basis of a sample drawn from a bivariate normal distribution, r_S^2 may be used to indicate the proportion of variability in the dependent variable that is explained by the independent variable.

Fieller et al. (T28) discuss the relative merits of the two statistics in certain special situations.

r_S and $\hat{\tau}$ as Measures of Trend   The Kendall $\hat{\tau}$ statistic and the Spearman coefficient of rank correlation are frequently used as tests for trend. Used for this purpose, these statistics provide alternatives to the Cox–Stuart test for trend described in Chapter 2. For example, to test for trend in time series data, we may denote the time variable by X and the time-dependent variable (measured on at least an ordinal scale) by Y. We can then think of the (X_i, Y_i) pairs as a sample of bivariate observations in which each variable can be ordered. If the sample data exhibit a positive trend, the Y values tend to increase over time. If the data exhibit a negative trend, the Y values tend to decrease over time. Consequently accepting the alternative hypothesis that X and Y are directly associated indicates the presence of an upward trend, while accepting the alternative of an inverse association between X and Y indicates the presence of a downward trend.

FURTHER READING

Griffin (T29) presents a graphical method for computing $\hat{\tau}$ when it is interpreted as a coefficient of disarray. Shah (T30) comments on Griffin's proposal. Knight (T31) describes a computer method for calculating $\hat{\tau}$. Noether (T32) gives a formula for determining sample size when use of Kendall's statistic is anticipated.

Best et al. (T33) consider the situation in which the researcher who has measurements on k characteristics for each of n individuals wishes to know whether any of the characteristics are associated and, if they are, to decide which ones. The authors provide a test procedure for the hypothesis that all the k characteristics are independent when one wants to use Kendall's $\hat{\tau}$ as a measure of association. Their arguments are analogous to those of Eagleson (T34), whose concern was Spearman's rank correlation coefficient.

Kendall's rank correlation $\hat{\tau}$ between two variables, in the presence of a third

blocking variable, is defined by Korn (T35). The author gives an estimator of the common $\hat{\tau}$ within the blocks and uses it to test conditional independence of the two variables, given the blocking variable. The procedure is illustrated by means of an example. Following the work of Korn (T35), Taylor (T36) compares the use of the weighted sum of Kendall's $\hat{\tau}$ with the weighted sum of Spearman's rank correlation coefficient for testing association in the presence of a blocking variable. On the basis of a Monte Carlo study, the author concludes that the two have essentially the same power with the optimal choice of weights. In the presence of ties, the weighted sum of Spearman's statistics is preferred because of the much simpler form of its variance. An example using the Spearman statistic is given.

Other papers on Kendall's rank correlation coefficient that may be of interest include those by Schumacher (T37), who discusses the use of the statistic as a coefficient of disarray between permutations with unoccupied places; Wilkie (T38), who gives a pictorial representation of the statistic; and Silverstone (T39), who discusses the cumulants of the Kendall distribution. Also of interest is an article by Noether (T40).

EXERCISES

9.5 Johnson (E8) conducted a study to determine whether, in collegiate schools of nursing, relationships between certain variables could be identified. Two variables of interest for which indexes were constructed were "extent of agreement (between the dean and the faculty) on the responsibilities for decision making" and "faculty satisfaction." The ranks on the two variables of the 12 institutions that participated in the study are shown in Table 9.11. The author computed a value of $r_s = -0.336$ from the data, which she declared not significant. Compute $\hat{\tau}$ from the data and test for significance against the alternative that $\tau < 0$. What is the P value?

TABLE 9.11

Decision-making agreement and faculty satisfaction ranks of 12 schools of nursing

School	A	B	C	D	E	F	G	H	I	J	K	L
Rank on faculty satisfaction	1	7	6	2	8	4	10	12	11	5	9	3
Rank on decision-making agreement	12	11	10	9	8	7	6	5	4	3	2	1

Source: Betty M. Johnson, "Decision Making, Faculty Satisfaction, and the Place of the School of Nursing in the University," *Nursing Res.*, 22 (1973), 100–107; copyright 1973, American Journal of Nursing.

9.6 Chaiklin and Frank (E9) studied 22 girls between 12 and 15 from a poverty-level population. Through record reviews and home interviews, the investigators were able to obtain a measure of each girl's self-perception, the mother's evaluation of her daughter, and a

measure of family functioning in terms of family adequacy. In addition, they computed measures of congruency between the daughter's self-ranking and the mother's ranking of her daughter. Table 9.12 shows the rankings of the 22 girls on the congruency measure and the measure of family functioning. From the data the authors computed a value of $r_S = 0.42$, which they indicate has a P value less than 0.05. From the data, compute $\hat{\tau}$ and find the P value for $H_1: \tau > 0$.

TABLE 9.12   **Family functioning and congruency between daughter's self-ranking and mother's ranking of daughter**

Congruency rankings	1	2	3	4	5	6	7	8	9	10	11
Family functioning rankings	5.5	5.5	5.5	1.0	16.0	5.5	5.5	5.5	16.0	16.0	11.0
Congruency rankings	12	13	14	15	16	17	18	19	20	21	22
Family functioning rankings	16.0	16.0	20.5	5.5	11.0	16.0	20.5	11.0	16.0	5.5	22.0

Source: Harris Chaiklin and Carol Landau Frank, "Separation Service Delivery and Family Functioning," *Public Welfare*, 31 (Winter 1973), p. 4, Table 2, columns 2 and 3.

9.7   Pierce (E10) points out that in most investigations of lightning discharges to earth, the estimated quantity of electricity passing from the cloud to the ground is around 20 to 30 coulombs. However, Pierce cites the data of Meese and Evans (E11), who reported much larger values. Their data as reported by Pierce (E10) are shown in Table 9.13, along with the distance of the observing site from the discharge. Pierce computes a Pearson product–moment correlation coefficient of $r = 0.877$ and a P value of 0.01. Compute $\hat{\tau}$ and the corresponding P value for $H_1: \tau > 0$.

TABLE 9.13   **Distance of lightning flash versus charge transferred to earth**

Distance, kilometers	6	6	6	6	6	7	9	10	10	10	11	12	15	15	18	23
Charge, coulombs	23	46	46	47	94	80	133	81	114	274	260	378	197	234	1035	1065

Source: A. D. Meese and W. H. Evans, "Charge Transfer in the Lightning Stroke as Determined by the Magnetograph," *J. Franklin Inst.*, 273 (1962), 375–382.

9.8   Murgatroyd (E12) investigated the traffic capacity of roundabouts (rotary intersections) in England in an attempt to establish the authenticity of the design formula in current use. He compared actual traffic flow, in passenger car units per hour, with the theoretical flow derived from the formula used in designing the roundabouts. Part of his data are shown in Table 9.14. Compute $\hat{\tau}$ and determine whether the data provide sufficient evidence to indicate a relationship between the two variables.

TABLE 9.14   **Measured and theoretical traffic flow, passenger car units per hour, on roundabouts**

Measured flow (*Y*)	Theoretical flow (*X*)
2290	3060
2100	2520
1830	2260
3290	3350
3130	3440
3400	3460

Source: B. Murgatroyd, "An Investigation into the Practical Capacity of Roundabout Weaving Sections," *Highway Engineer*, 20 (March 1973), 6–13.

9.9 From a study of the correlation between measures of the rapid eye movements of wakefulness and sleep, de la Peña et al. (E13) reported the data on eye-track lengths and fixation rates shown in Table 9.15. They defined eye-track length as the sum of the lengths, in inches, of all distinct eye movements that traveled from one sector of a viewed picture space to another. They obtained the fixation rate by counting the number of fixation changes from one given sector to another sector, and dividing by the number of seconds of viewing time. Subjects were males between the ages of 17 and 24. The authors computed a correlation coefficient of 0.85 (P value < 0.01). Compute $\hat{\tau}$ and the associated P value for $H_1: \tau > 0$.

TABLE 9.15   **Mean waking REM values for 10 seconds of inspecting pictures**

Eye track, inches	980.8	926.4	892.9	870.2	854.6
Fixation rate, fixations per second	4.85	4.41	3.80	4.53	4.33
Eye track, inches	777.2	772.6	702.4	561.7	
Fixation rate, fixations per second	3.81	3.97	3.68	3.43	

Source: A. de la Peña, V. Zarcone, and W. C. Dement," Correlation between Measures of the Rapid Eye Movements of Wakefulness and Sleep," *Psychophysiology*, 10 (1973), 488–500. Copyright 1973, the Society for Psychophysiological Research; reprinted with permission of the publisher.

9.3

CONFIDENCE INTERVAL FOR τ

Noether (T41) describes a method for constructing an approximate two-sided confidence interval for Kendall's τ, the population counterpart of $\hat{\tau}$. This section considers the procedure that is appropriate when there are no ties in either the X's or the Y's. When ties are present, the formulas given by Noether (T41) are slightly more complicated.

To compute the standard error needed in the confidence-interval formula, we must determine for each observation pair (X_i, Y_i) the number of other observation

pairs with which it is concordant. As noted, two observation pairs, (X_i, Y_i) and (X_j, Y_j), are concordant if either of the following relationships between the X's and Y's is true:

$$X_i > X_j \text{ and } Y_i > Y_j, \qquad X_i < X_j \text{ and } Y_i < Y_j$$

If the direction of any one of the inequalities in these expressions is reversed, the two observation pairs are said to be discordant. For example, the two pairs $(25, 15)$ and $(20, 9)$ are *con*cordant, since $25 > 20$ *and* $15 > 9$. The two pairs $(30, 18)$ and $(25, 20)$ are *dis*cordant, since $30 > 25$ *but* $18 < 20$.

Suppose that we let C_i be the number of pairs (X_j, Y_j) that are concordant with the pair (X_i, Y_i), where both i and j take on values between 1 and n, the number of observation pairs available for analysis. When there are no ties, the estimate of the appropriate variance for the confidence interval formula is

$$\hat{\sigma}^2 = 4\Sigma C_i^2 - 2\Sigma C_i - \frac{2(2n - 3)(\Sigma C_i)^2}{n(n - 1)} \tag{9.11}$$

For large samples an approximate $100(1 - \alpha)\%$ confidence interval for the parameter $\hat{\tau}$ is given by

$$\hat{\tau} \pm \frac{2}{n(n - 1)}\, \hat{\sigma} z \tag{9.12}$$

where z is the upper $(\alpha/2)$th percentile of the standard normal distribution (Table A.2).

The following example illustrates the construction of a confidence interval for the parameter τ.

Example 9.4

Clarke (E14) reported the results of a study conducted to investigate the relationship between whole-blood riboflavin levels in the mother and in the prenate. As part of the investigation, 11 mothers in active labor were given 1 mg of riboflavin per 30 pounds of body weight at varying intervals before delivery. Samples of maternal venous blood and blood from the umbilical cord were collected at the end of the second stage of labor. The results are shown in Table 9.16. Let us compute $\hat{\tau}$ and construct an approximate 95% confidence interval for τ.

By methods described in Section 9.2, we find that $\hat{\tau} = 0.67$. The next step is to find $C_1, C_2, \ldots, C_{11}$. To illustrate the method of finding the C_i, examine the details for finding C_6 in Table 9.17. Note that we first compare (X_6, Y_6) with each observation pair below and then with each pair above, as listed in Table 9.16. Similarly, we find

that the remaining values of C_i are

$$C_1 = 9, \quad C_2 = 8, \quad C_3 = 8, \quad C_4 = 9, \quad C_5 = 9,$$

$$C_7 = 8, \quad C_8 = 9, \quad C_9 = 9, \quad C_{10} = 8, \quad C_{11} = 9$$

TABLE 9.16

Blood concentration of riboflavin at end of second stage of labor in mother and prenate following oral treatment with riboflavin

Mother	1	2	3	4	5	6
Maternal blood MCRC*	38.6	44.7	54.2	35.3	28.0	43.0
Cord blood MCRC	44.4	44.5	56.8	40.3	27.8	64.0
Mother	**7**	**8**	**9**	**10**	**11**	
Maternal blood MCRC*	46.0	41.5	27.7	43.9	37.1	
Cord blood MCRC	57.6	40.6	32.7	48.3	33.0	

Source: H. Courtney Clarke, "Relationship between Whole-Blood Riboflavin Levels in the Mother and in the Prenate," *Amer. J. Obstet. Gynecol.*, 111 (1971), 43–46.

* MCRC = mean corpuscular riboflavin concentration or micrograms percent of hematocrit.

TABLE 9.17

Details for finding C_6 in Example 9.4

Observation pair (X_6, Y_6) compound to observation pair (X_j, Y_j)	Contribution to C_6
(43.0, 64.0), (46.0, 57.6) 43.0 < 46.0 *but* 64.0 > 57.6	0
(43.0, 64.0), (41.5, 40.6) 43.0 > 41.5 *and* 64.0 > 40.6	1
(43.0, 64.0), (27.7, 32.7) 43.0 > 27.7 *and* 64.0 > 32.7	1
(43.0, 64.0), (43.9, 48.3) 43.0 < 43.9 *but* 64.0 > 48.3	0
(43.0, 64.0), (37.1, 33.0) 43.0 > 37.1 *and* 64.0 > 33.0	1
(43.0, 64.0), (28.0, 27.8) 43.0 > 28.0 *and* 64.0 > 27.8	1
(43.0, 64.0), (35.3, 40.3) 43.0 > 35.3 *and* 64.0 > 40.3	1
(43.0, 64.0), (54.2, 56.8) 43.0 < 54.2 *but* 64.0 > 56.8	0
(43.0, 64.0), (44.7, 44.5) 43.0 < 44.7 *but* 64.0 > 44.5	0
(43.0, 64.0), (38.6, 44.4) 43.0 > 38.6 *and* 64.0 > 44.4	1
	$C_6 = 6$

We next compute

$$\Sigma C_i = 9 + 8 + 8 + 9 + 9 + 6 + 8 + 9 + 9 + 8 + 9 = 92$$

$$\Sigma C_i^2 = 9^2 + 8^2 + 8^2 + 9^2 + 9^2 + 6^2$$
$$+ 8^2 + 9^2 + 9^2 + 8^2 + 9^2 = 778$$

By Equation 9.11, we compute

$$\hat{\sigma}^2 = 4(778) - 2(92) - \frac{2(22 - 3)(92)^2}{11(10)} = 4.0727$$

The z value from Table A.2 for a 95% confidence interval is 1.96, so our approximate 95% confidence interval is

$$0.67 \pm \frac{2}{11(10)} \sqrt{4.0727}\,(1.96), \qquad 0.67 \pm 0.07$$

This gives a lower limit of 0.60 and an upper limit of 0.74. Thus we are about 95% confident that τ is somewhere between 0.60 and 0.74.

Alternative procedures for constructing confidence intervals for τ are proposed by Fligner and Rust (T42) and Samara and Randles (T43).

EXERCISES

9.10 Physicians in medical practice and research want to know the total body fat composition of individuals. If we assume that the percentage of water in the lean body is constant, we can calculate total body fat from the body's water content. However, the usual methods of determining total body water are not always practical, nor are they appropriate for use with children. Brook (E15) measured total body water in 23 children and estimated lean body mass by measuring skinfold thickness, with results as shown in Table 9.18. From the data, the author computed a correlation coefficient of 0.985. This result led to the conclusion that skinfold thickness can be used to estimate lean body mass and, consequently, total body fat in children. Compute $\hat{\tau}$ and construct a 95% confidence interval for τ.

TABLE 9.18   **Total body water and estimated lean body mass in 23 children**

Subject	1	2	3	4	5	6	7	8
Total body water, liters	7.56	11.73	10.12	10.59	12.33	18.96	7.65	9.91
Estimated lean body mass, kilograms	11.5	18.0	14.4	17.4	16.6	31.4	10.7	12.3

Table 9.18	*(Continued)*							
Subject	**9**	**10**	**11**	**12**	**13**	**14**	**15**	**16**
Total body water, liters	22.75	14.43	15.83	9.03	10.52	7.35	15.97	10.22
Estimated lean body mass, kilograms	29.0	21.6	22.9	12.9	15.1	11.0	21.7	14.0
Subject	**17**	**18**	**19**	**20**	**21**	**22**	**23**	
Total body water, liters	17.49	12.62	11.86	30.24	27.18	27.20	13.52	
Estimated lean body mass, kilograms	25.7	19.7	15.3	44.3	40.0	43.4	19.2	

Source: C. G. D. Brook, "Determination of Body Composition of Children from Skinfold Measurements," *Arch. Dis. Child.*, 46 (1971), 182–184; reprinted by permission of the editor.

9.11 Robbins et al. (E16) carried out an experiment to determine the effect of digital massage on vitreous weight. The authors' results have implications in the area of cataract surgery. As part of the experiment, they measured the vitreous body weight for each eye of 10 New Zealand albino rabbits. The results are shown in Table 9.19. Compute $\hat{\tau}$ and construct a 90% confidence interval for τ.

TABLE 9.19	**Weight, in grams, of vitreous humor in 10 rabbits**									
Rabbit	**1**	**2**	**3**	**4**	**5**	**6**	**7**	**8**	**9**	**10**
Left eye	1.915	1.374	1.735	1.635	2.040	1.540	1.850	1.910	1.340	1.315
Right eye	1.905	1.379	1.750	1.625	2.032	1.540	1.840	1.915	1.345	1.310

Source: Richard Robbins, Michael Blumenthal, and Miles A. Galin, "Reduction of Vitreous Weight by Ocular Massage," *Amer. J. Ophthalmol.*, 69 (1970), 603–607.

9.4

OLMSTEAD–TUKEY CORNER TEST OF ASSOCIATION

Olmstead and Tukey (T44) developed a test of association that is both quickly and easily applied. Sometimes referred to as the *quadrant sum test*, this test is perhaps best known as the *corner test of association*.

This test is designed to detect the presence of a correlation between two variables *X* and *Y*. It places heavy emphasis on the extreme values of the variables. Since extreme values are often the most sensitive indicators of a relationship between two variables (provided they are not merely outliers), the corner test of association provides a useful and desirable alternative to other nonparametric tests of association. In addition, the computations involved are easy to perform.

Assumptions

A. The n pairs of observations $(X_1, Y_1), (X_2, Y_2), \ldots, (X_n, Y_n)$ constitute a random sample.
B. Measurement is at least ordinal.
C. The variables are continuous.

Hypotheses

H_0: X and Y are independent
H_1: X and Y are correlated

Test Statistic

To apply the corner test of association, we use the following steps:

1. Plot the data points as a scatter diagram.
2. Draw a horizontal line through the median Y_m of the Y values and a vertical line through the median X_m of the X values.
3. Label the upper right and lower left quadrants with a plus sign and the upper left and lower right quadrants with a minus sign.
4. Beginning at the top of the scatter diagram, proceed downward. Count the number of points encountered before (in order to count the next point) it is necessary to cross the vertical median. Record the number of points counted, and affix the sign of the quadrant in which they occur.
5. Begin at the right of the scatter diagram, move to the left, and count points until it is necessary to cross the horizontal median. Record the number of points counted, and affix the sign of the quadrant in which they occur.
6. Repeat Steps 4 and 5, beginning at the bottom and left of the scatter diagram. Ignore points lying exactly on one of the median lines in the counting procedure, and proceed with the counting as if those points were not present.
7. Add the four numbers, observing signs. Take the absolute value of the sum and call it S. This is the test statistic; that is, the test statistic is

$$S = |\text{quadrant sum}| \tag{9.13}$$

Decision Rule

Reject H_0 at the α level of significance if the entry in the body of Table A.23 corresponding to S and n is equal to or less than α. For values of n greater than 14, use the entry for ∞.

Ties   Ties occur in the counting procedure when we reach a point on one side of a median line that aligns perfectly with one or more points on the other side of the

same median line. We say that such a point is *favorable* for inclusion in the sum, and we say that the aligned points on the other side are *unfavorable* for inclusion in the sum. One or more points on the counting side of a median may align with a tied group. These are also favorable for inclusion. Olmstead and Tukey (T44) recommend treating tied groups as if the number of their points preceding the crossing of the median were equal to

$$\frac{\text{Number in tied group favorable for inclusion}}{1 + \text{number unfavorable}}$$

The following example illustrates the application of the corner test of association.

Example 9.5

To estimate the broad heritability of characteristics, geneticists have compared monozygotic twins who are reared in separate environments. For example, Jensen (E17) cites the work of Burt (E18), who compared the IQs of 53 pairs of twins who had been separated at birth or during their first six months of life. The data are shown in Table 9.20. Jensen reported a Pearson product–moment correlation coefficient of 0.88. We wish to apply the Olmstead–Tukey test to the data.

TABLE 9.20

IQs for monozygotic twins reared apart

X	Y	X	Y	X	Y	X	Y	X	Y
68	63	94	86	93	99	115	101	104	114
71	76	87	93	94	94	102	104	125	114
77	73	97	87	96	95	106	103	108	115
72	75	89	102	96	93	105	109	116	116
78	71	90	80	96	109	107	106	116	118
75	79	91	82	97	92	106	108	121	118
86	81	91	88	95	97	108	107	128	125
82	82	91	92	112	97	101	107	117	129
82	93	96	92	113	108	108	95	132	131
86	83	87	93	105	99	98	111		
83	85	99	93	88	100	116	112		

Source: C. Burt, "The Genetic Determination of Differences in Intelligence: A Study of Monozygotic Twins Reared Together and Apart," *Br. J. Psychol.*, 57 (1966), 137–153. Cited in Arthur R. Jensen, "IQ's of Identical Twins Reared Apart," *Behav. Genet.*, 1 (1970), 133–148. Published by Plenum Publishing Corporation, New York.

Hypotheses

H_0: The IQ score of one twin (X) is independent of the IQ score of the other (Y)

H_1: X and Y are correlated

Test Statistic

The median of the X values is 96, and the median of the Y values is 97. The original observations and the median lines are plotted in Figure 9.1. Solid dots indicate

FIGURE 9.1   Scatter diagram for Example 9.5

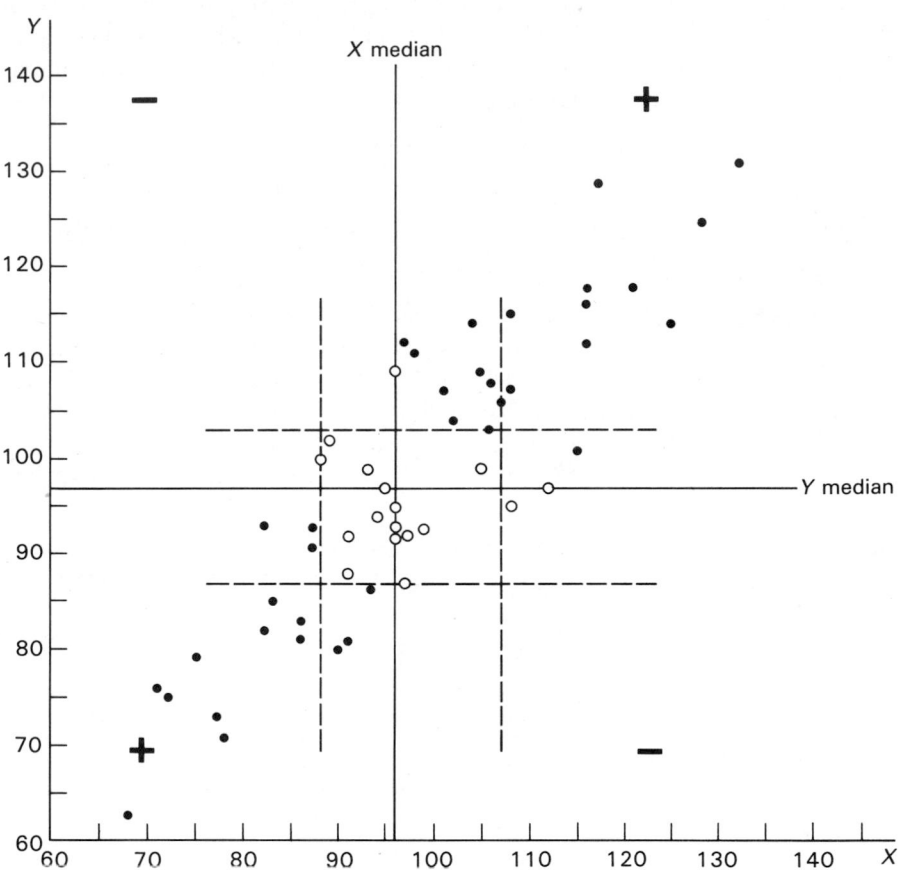

observations that enter into the sum, while hollow dots indicate all other observations. The dashed lines indicate where counting stops. Note that a point may be counted twice, once in counting from the top (say) and again when counting from the right.

In Figure 9.1, note that four dots fall on the X median and two fall on the Y median. We eliminate these six dots from the analysis, so that we have an effective n of $53 - 6 = 47$. In counting from the top, we count 19 dots before having to cross the vertical median. From the right, we count 9 dots plus a tied group of 3 to give $9 + 2/(1 + 1) = 10$. From the bottom, we count 13 dots before crossing the vertical median. Counting from the left, we also count 13 dots. All counted dots occur in the positive quadrants, so that $S = |19 + 10 + 13 + 13| = 55$.

Decision

Entering Table A.23 with $n = 47$ and $S = 55$, we find that the probability of observing a value of S as large as or larger than 55 when X and Y are independent is equal to 0.000000. We reject H_0 and conclude that X and Y are correlated.

EXERCISES

9.12 Bhatia et al. (E19) studied the coronary hemodynamics of 14 patients with chronic severe anemia. The determinations of hemoglobin and coronary blood flow of these patients are shown in Table 9.21. Are these data sufficient evidence to indicate that these variables are correlated?

TABLE 9.21 **Hemoglobin, grams per 100 ml, and coronary blood flow, ml/100 g Lv/min, determinations for 14 patients with chronic severe anemia**

Hemoglobin	1.6	2.4	3.0	3.5	3.5	3.7	4.7
Coronary blood flow	222	198	160	193	208	151	155
Hemoglobin	4.8	5.0	5.1	6.0	6.1	6.1	6.4
Coronary blood flow	139	136	177	121	109	140	122

Source: M. L. Bhatia, S. C. Manchanda, and Sujoy B. Roy, "Coronary Haemodynamic Studies in Chronic Severe Anaemia," *Br. Heart. J.,* 31 (1969), 365–374; reprinted by permission of the authors and the editor.

9.13 Poland et al. (E20) studied the effect of intensive occupational exposure to DDT on drug and steroid metabolism. Subjects participating in the study had been employed in a DDT plant for at least five years; a group of control subjects were mostly policemen and firemen. Two variables of interest were (a) the urinary excretion of 6 β-hydroxycortisol and (b) total DDT in the serum of the study subjects. Table 9.22 shows the observed values of these variables in the DDT-plant employees. We wish to know whether the two are correlated.

TABLE 9.22 **Total serum DDT and the 24-hour urinary excretion of 6 β-hydroxycortisol in DDT factory workers**

6β-hydroxycortisol excretion ($\mu g/24$ hr)	106	134	171	173	190	192	198	231	248
Total DDT in serum ($m\mu g/ml$)	751	781	579	1001	1172	826	920	816	2049
6β-hydroxycortisol excretion ($\mu g/24$ hr)	254	300	335	345	351	380	447	449	741
Total DDT in serum ($m\mu g/ml$)	2725	2914	1013	835	1986	1382	1809	1335	1565

Source: Alan Poland, Donald Smith, R. Kuntzman, M. Jacobson, and A. H. Conney, "Effect of Intensive Occupational Exposure to DDT on Phenylbutazone and Cortisol Metabolism in Human Subjects," *Clin. Pharmacol. Ther.,* 11 (1970), 724–732.

9.5

KENDALL'S COEFFICIENT OF CONCORDANCE *W*

In the previous discussion of the Spearman rank correlation coefficient r_S and Kendall's τ, the main concern was the extent to which two sets of rankings of k objects or individuals agree or disagree. In many practical situations, we may be interested in the degree of agreement among several—say, b (where $b > 2$)—sets of rankings of k objects or individuals. We may obtain a set of b rankings in one of two ways.

1. We may rank a group of k objects or individuals on the basis of each of b characteristics. For example, we might rank a group of $k = 10$ students according to their aptitude scores in each of the following $b = 6$ areas: mechanical, artistic, literary, musical, mathematical, and clerical. The display of the results of this example might look like Table 9.23. For now, disregard the row and column totals.

2. A panel of b judges or observers may rank a group of k objects or individuals on the basis of the same characteristic. For example, a group of $b = 3$ supervisors might rank $k = 5$ employees on the basis of leadership ability. The results of such a procedure might be displayed as in Table 9.24.

For situations such as these, we would like to have a measure of the strength of the agreement among the b sets of rankings. We would also like to be able to test the null hypothesis of no association among the rankings. We can achieve these objectives by using Kendall's *coefficient of concordance*. This statistic, which is designated W, was introduced independently in 1939 by Kendall and Babington–Smith (T45) and by Wallis (T46).

TABLE 9.23   **Ten students ranked according to aptitude scores in each of six areas**

Aptitude	Student										Total
	A	B	C	D	E	F	G	H	I	J	
Mechanical	4	6	1	2	8	10	9	3	5	7	55
Artistic	5	2	8	6	1	3	7	4	9	10	55
Literary	7	1	9	5	2	4	6	3	8	10	55
Musical	6	5	2	10	8	3	4	1	7	9	55
Mathematical	5	7	2	1	9	8	10	4	6	3	55
Clerical	1	4	9	7	5	3	2	8	10	6	55
Total	28	25	31	31	33	31	38	23	45	45	330

TABLE 9.24   **Five employees ranked on the basis of leadership ability by each of three supervisors**

Supervisor	Employee				
	A	**B**	**C**	**D**	**E**
I	3	2	4	1	5
II	2	1	4	3	5
III	5	1	3	2	4

Assumptions

A. The data consist of b complete sets of observations or measurements on k objects or individuals.

B. The measurement scale is at least ordinal.

C. The observations as collected or recorded may consist of ranks. If the orginal data are not ranks, they must be capable of being converted to ranks.

Hypotheses

H_0: The b sets of rankings are not associated

H_1: The b sets of rankings are associated

Note that with more than two sets of rankings, it is impossible to have an inverse relationship of the type that is possible with only two sets. Suppose, for example, that we have three judges, A, B, and C. If A disagrees with B and C on a comparison, then B and C must agree. Complete disagreement is impossible.

Alternatively, we may state the null hypothesis in terms of independence. We may think of the testing procedure, when judges or observers are involved, as testing the null hypothesis that the b judges are assigning ranks to the subjects independently and at random.

Test Statistic

The computationally most convenient form of the test statistic is

$$W = \frac{12 \sum_{j=1}^{k} R_j^2 - 3b^2 k(k+1)^2}{b^2 k(k^2 - 1)} \tag{9.14}$$

where b is the number of sets of rankings, k is the number of individuals or objects that are ranked, and R_j is the sum of the ranks assigned to the jth object or individual.

When we express W as in Equation 9.14, the true nature of the statistic is not

immediately obvious. An examination of another way of writing W reveals its important characteristics and, at the same time, helps explain the rationale behind this statistic.

Refer to Table 9.23 and assume that a researcher is interested in these hypotheses.

H_0: There is no association among the characteristics (aptitudes)

H_1: There is an association among the characteristics (aptitudes)

It seems natural that our first step should be to devise some measure of the relationship or association among the characteristics in our sample of subjects. If the characteristics are unrelated, we expect the occurrence of rank values within a given column to be a random phenomenon. Consequently we expect the column totals to be approximately equal. If, on the other hand, there is a relationship among the characteristics, we expect some columns to contain predominantly large ranks and others predominantly small ranks, so that some column totals would be relatively large and others relatively small. If the column totals were exactly equal, each would be equal to $330/10 = 33$. In other words, when H_0 is true—that is, when there is no association among the six sets of rankings—the expected value of each column total is 33. A measure of the extent to which the column totals deviate from expectation in a sense measures the extent of association among the six sets of rankings.

We may obtain a measure of the departure of the column totals from expectation by computing the sum of the squared deviations of observed totals from their expected value. For the ranks in Table 9.23, this quantity is

$$S = (28 - 33)^2 + (25 - 33)^2 + \cdots + (45 - 33)^2 = 514$$

This procedure is reminiscent of the manner in which we compute the numerator of the sample variance when we calculate the sum of squared deviations of sample observations from their mean.

In this case a desirable measure of association is one that takes on values between 0 and 1 (-1 would be meaningless in this case, since complete disagreement among ranks is impossible). A 0 occurs when there is a total absence of association, and a 1 results from perfect association or agreement among the sets of rankings. Then the desired measure of association is the ratio of the degree of agreement between observed and expected column totals, as measured by S, to the value of S that results from perfect agreement among the sets of rankings.

What we need, then, is the denominator of this ratio—that is, the value of S that we obtain if there is perfect agreement among the sets of rankings. In this example, if each student has equal aptitude (as measured by equal scores) in each of the six areas, each receives the same rank for all aptitude scores. For example, the student who ranked first in mechanical aptitude would also rank first in artistic aptitude, first in literary aptitude, and so on. Consequently the sum of the column of ranks for that student would equal $6(1) = 6$. The sum of the column of ranks for the student who ranked second in all aptitude areas would be $6(2) = 12$. In other words, if there is complete agreement among sets of ranks (that is, if there is a relationship among

aptitudes), the column totals are

$$6(1) = 6, \qquad 6(2) = 12, \qquad 6(3) = 18, \qquad \ldots, \qquad 6(10) = 60$$

though not necessarily in that order. In this example, then, if there is perfect agreement among the six sets of ranks, we compute

$$S = (6 - 33)^2 + (12 - 33)^2 + \cdots + (60 - 33)^2 = 2970$$

The ratio of the observed value of S to the value of S that would result if there were perfect agreement among the rankings is

$$\frac{514}{2970} = 0.173$$

Thus we see that a ratio of this type equals 1 when there is perfect agreement among the sets of ranks, and equals 0 (since the numerator is 0) if there is a total lack of agreement among the sets of ranks.

In general, the formula for S is given by

$$S = \sum_{j=1}^{k} \left[R_j - \frac{b(k + 1)}{2} \right]^2 \tag{9.15}$$

where R_j, b, and k are as defined for Equation 9.14.

When there is perfect agreement among the sets of ranks, the column totals are $1b, 2b, \ldots, jb, \ldots, kb$, but not necessarily in that order. The sum of the squared deviations of these column totals from their expected value is

$$\sum_{j=1}^{k} \left[jb - \frac{b(k + 1)}{2} \right]^2 = b^2 \sum_{j=1}^{k} \left[j - \frac{(k + 1)}{2} \right]^2 = \frac{b^2 k(k^2 - 1)}{12} \tag{9.16}$$

The ratio of the quantity computed by Equation 9.15 to the quantity computed by Equation 9.16 is our test statistic W. We write W in the form

$$W = \frac{\sum_{j=1}^{k} \left\{ R_j - \left[\frac{b(k + 1)}{2} \right] \right\}^2}{b^2 k(k^2 - 1)/12} \tag{9.17}$$

After appropriate algebraic manipulation, we can rewrite the conceptual or definitional formula for W given by Equation 9.17 as Equation 9.14. Usually this is the more convenient computational formula.

Decision Rule

When the observed sets of rankings are in close agreement, the computed value of S tends to be large. When S is large, W is large (close to 1). When there is poor agreement among the sets of ranks, S is relatively small, and consequently so is W (it will be close to 0). Sufficiently large values of W, therefore, lead us to reject the null hypothesis of no association.

For small values of b and k, we may use Table A.14 to decide whether to reject H_0.

We may reject the null hypothesis at the α level of significance if the value of P in Table A.14 associated with the appropriate values of W, b, and k is less than or equal to α.

For values of b and k not covered by Table A.14, compute

$$X^2 = b(k - 1)W \tag{9.18}$$

and compare it for significance with tabulated values of chi-square (Table A.11) with $k - 1$ degrees of freedom. Kendall (T3) recommends Equation 9.18 for $k > 7$, and gives another approximation that is generally applicable.

Example 9.6   Goby et al. (E21) conducted a study at an alcoholic-rehabilitation center. Their purpose was to determine how patients and staff perceived the relative importance of each component of a certain treatment program. Each of 60 patients who participated in the study was asked to rank the parts of the treatment program in the order of helpfulness, from most to least helpful. The following components of the program were evaluated: Alcoholics Anonymous (AA), the Fifth Step of AA, consultation with individual counselor, information on alcohol and drug dependence, involvement with significant others, other patients, rehabilitation community, small-group experience, small-group leader, and leaders and staff. The authors report a coefficient of concordance of $W = 0.103$ with an associated P value < 0.001. They conclude that within the patient group there was a consensus on the relative effectiveness of the structural components of the treatment program.

Suppose that the study were conducted in another alcoholic-rehabilitation center with a sample of 15 patients, yielding the results shown in Table 9.25. We wish to compute W and determine whether the data are compatible with the hypothesis that there is a consensus among the patients of this center.

Hypotheses

H_0: There is no agreement among the patients regarding the relative effectiveness of the different components of the treatment program

H_1: There is agreement among the patients regarding this matter

Test Statistic

By Equation 9.14, we compute

$$W = \frac{12(118^2 + 124^2 + \cdots + 59^2) - 3(15)^2(10)(10 + 1)^2}{(15)^2(10)(10^2 - 1)}$$

$$= 0.4036$$

Decision

Since $b = 15$ and $k = 10$ are not given in Table A.14, we use the large-sample approximation to decide whether to reject H_0. By Equation 9.18, we compute

$$X^2 = 15(10 - 1)(0.4036) = 54.486$$

TABLE 9.25   **Patients' ranking in order of effectiveness of 10 components of an alcoholic rehabilitation center**

Patient	Component									
	A	**B**	**C**	**D**	**E**	**F**	**G**	**H**	**I**	**J**
1	9	10	6	1	7	8	2	4	3	5
2	10	9	7	2	6	8	3	5	4	1
3	9	8	6	1	5	10	4	3	2	7
4	4	9	6	8	10	3	7	2	5	1
5	10	9	6	1	7	8	3	5	2	4
6	9	10	6	1	7	8	3	4	2	5
7	6	7	1	4	9	2	10	5	8	3
8	10	9	7	2	5	8	3	6	4	1
9	9	10	6	1	8	7	3	5	2	4
10	8	6	2	5	10	9	7	4	3	1
11	9	8	6	1	5	10	4	3	2	7
12	9	10	6	1	7	8	4	5	3	2
13	1	6	7	10	9	8	2	4	3	5
14	9	10	6	1	8	7	3	5	2	4
15	6	3	10	8	1	2	5	4	7	9
Total	118	124	88	47	104	106	63	64	52	59

For $k - 1 = 9$ degrees of freedom, Table A.11 reveals that the probability of obtaining a value of $X^2 = 54.486$ by chance alone when H_0 is true is less than 0.005. Consequently we reject H_0 and conclude that there is a consensus among the patients.

Ties   In a set of observations to be ranked, if two or more observations are equal, we assign each the mean of the rank positions for which it is tied. We adjust the test statistic for ties by replacing the denominator of W in Equation 9.14 by

$$b^2 k(k^2 - 1) - b \sum (t^3 - t) \tag{9.19}$$

where t is the number of observations in any set of rankings tied for a given rank.

FURTHER READING

Schucany (T47) and Schucany and Frawley (T48) introduced a statistic that allows one to test the hypothesis that several judges agree on the ranking of items within each of two groups and between the two groups. Li and Schucany (T49) discuss the properties of this test statistic.

An alternative approach to intergroup concordance was presented by Hollander and Sethuraman (T50). Intergroup concordance is also the subject of papers by

Kraemer (T51) and Snell (T52), who discuss the papers by Schucany and Frawley (T48) and Hollander and Sethuraman (T50). Alternative approaches to the two-group concordance problem have also been proposed by Baldessari and Gallo (T53), Shirahata (T54), Palachek and Kerin (T55), and Costello and Wolfe (T56).

The use of average Kendall tau as a way to measure concordance was suggested by Ehrenberg (T57). A computationally convenient way to find average tau is pointed out by Hays (T58) who makes other contributions to the idea. Alvo and Cabilio (T59) provide further justification for the use of average Kendall tau.

Palachek and Schucany (T60) propose an approximate confidence interval for the strength of concordance.

The extension of the idea of intergroup concordance to more than two groups is discussed by Serlin and Marascuilo (T61) and Feigin and Alvo (T62). Lewis (T63) discusses the use of Kendall's coefficient of concordance for rankings with self excluded. Critical values are given for S for small sample sizes.

Other articles of interest on this and related topics include those by Linhart (T64), Lyerly (T65), Willerman (T66), Wood (T67), Friedman (T68), Stewart et al. (T69), and Wakimoto and Shirahata (T70).

EXERCISES

9.14 Vine (E22) describes an experiment in which 63 judges (first-year psychology under-graduates) ranked handwriting samples according to the judged relative positions of the writers on dimensions of introversion–extraversion or stability–neuroticism. The authors computed Kendall's coefficient of concordance for each case. From these coefficients the authors were able to conclude that there was strong agreement among judges (P value < 0.001 in each case). Suppose that part of the experiment were repeated, with the results shown in Table 9.26. Compute W and test the null hypothesis of no agreement among judges.

TABLE 9.26   **Judges' rankings of sets of handwriting samples according to the judged relative positions of the writers on the introversion–extraversion dimension**

	Handwriting specimen				
Judge	**1**	**2**	**3**	**4**	**5**
A	4	3	2	5	1
B	3	4	5	2	1
C	4	3	1	5	2
D	4	3	5	2	1

9.15 As part of a larger study reported by Viney et al. (E23), 37 knowledgeable informants from Oklahoma City were asked to rank the relative influence of 10 civic groups and aggregates on the outcomes of four community issues. The rankings of the groups on the four issues,

based on mean ranks, are given in Table 9.27. Verify the values of $W = 0.57$, $X^2 = 20.45$, and P value < 0.05 reported by the authors. State appropriate hypotheses for the test.

TABLE 9.27 Rank order of influence in Oklahoma City of 10 civic groups on each of four common community issues

Civic groups	City college	Urban renewal	Race relations	Sports facility
Mayor and council	3	2	4	3
News media	2	3	3	2
Chamber of commerce	1	1	5	1
Democrats	8	8	6	6.5
Labor groups	7	5	7	5
Service groups	4	9	8	4
Republicans	9	10	10	6.5
Churches	6	6	2	8
League of voters	5	7	9	9
Civil rights	10	4	1	10

Source: Wayne Viney, Ross Loomis, Jacob Hautaluoma, and Stanley Wagner, "A Comparison of Perceived Organizational Influence in Two Metropolitan Communities," *Rocky Mountain Soc. Sci. J.*, 11 (January 1974), 81–86.

9.16 The data on university faculty salaries shown in Table 9.28 were reported by Robinson (E24). The ranks are based on total faculty compensation for nine months. Compute W and test for significance.

TABLE 9.28 Faculty compensation rankings in Big Eight Institutions, academic year 1971–1972

Institution	*Rank of average compensation, by academic rank*			
	Professor	Associate professor	Assistant professor	Instructor
Iowa State University	1	1	1	2
University of Colorado	2	2	2	1
University of Missouri	3	3	3	3
University of Nebraska	4	5	5	5.5
University of Kansas	5.5	7	6	5.5
Oklahoma State University	5.5	4	4	7
University of Oklahoma	7	8	7.5	8
Kansas State University	8	6	7.5	4

Source: Jack L. Robinson, "Faculty Compensation in the Big Eight, 1971–1972," *Oklahoma Bus. Bull.*, 40 (September 1972), 10–12.

9.17 Gibson and Reeves (E25) examined the functional bases of a sample of Arizona towns to determine whether functional complexity of settlements is closely related to a number of

TABLE 9.29 **1970 population figures and survey data for 37 sample towns in Arizona**

Town	1970 population	Number of establishments	Number of functions	Number of primary functions	Number of functional units
Tolleson	3881	110	72	68	121
Miami	3394	145	80	68	176
El Mirage	3258	83	49	40	100
Benson	2839	160	85	69	200
Kearney	2829	56	52	45	66
Cottonwood	2815	216	126	112	251
Wickenburg	2698	207	98	84	253
Buckeye	2499	189	98	80	233
Willcox	2568	213	106	91	257
Surprise	2427	35	25	20	41
Williams	2386	171	106	84	228
Thatcher	2320	51	37	35	58
Show Low	2285	174	96	87	220
Somerton	2225	75	38	34	89
Florence	2173	103	66	60	128
Goodyear	2140	54	41	37	68
Gilbert	1971	71	51	45	83
Mammoth	1953	68	44	40	71
Parker	1948	194	112	91	249
Youngtown	1885	133	77	73	152
Snowflake	1833	77	60	47	100
Gila Bend	1795	145	80	60	189
St. Johns	1320	74	62	53	99
Hayden	1283	49	44	33	66
Eager	1279	29	32	24	44
Tombstone	1241	107	57	52	121
Huachuca City	1233	37	26	23	40
Pima	1184	40	42	33	55
Springerville	1038	94	64	55	130
Winkleman	974	45	33	28	50
Welton	967	66	54	43	86
Clarkdale	892	41	38	33	47
Taylor	888	24	31	23	34
Fredonia	798	34	28	27	35
Duncan	773	65	45	35	83
Patagonia	630	38	31	29	44
Jerome	290	36	26	23	39

Source: Lay James Gibson and Richard W. Reeves, "The Roles of Hinterland Composition, Externalities, and Variable Spacing as Determinants of Economic Structure in Small Towns," *Professional Geographer*, 26 (1974), 152–158; reproduced by permission of the Association of American Geographers.

simple variables that conceivably affect aggregate demand. The results are shown in Table 9.29. Compute W and test the null hypothesis of no association among the five characteristics.

9.6

PARTIAL RANK CORRELATION

The reader who has had a course in elementary statistics may be familiar with the concept of partial correlation within the parametric multiple regression context. The reader who is unfamiliar with the concept may now wish to consult a basic applied statistics text such as the one by Blalock (T71), in which a good discussion of the subject is to be found. The need for a measure of partial correlation arises when the researcher is investigating the relationships among three or more random variables. In such a situation the multiple correlation coefficient provides an overall measure of the correlation among all the variables considered together. We may obtain a measure of the correlation between any combination of two of the three or more variables by computing the simple correlation coefficient from the data of the two variables selected for study. A disadvantage of the simple correlation coefficient in this context is the fact that it does not take into account the influences on the two selected variables of any of the other variables in the study. The partial correlation coefficient provides a solution to this problem, since it measures the correlation between two variables while holding constant one or more of the other variables. Some writers use the term *control* and describe the partial correlation coefficient as a measure of the correlation between two variables while one or more other variables are controlled. The concept is similar to that on which blocking in two-way analysis of variance is based. As previously shown, blocking allows the researcher to remove the effect of an extraneous variable.

The simplest situation in which the partial correlation coefficient may be of interest is the one in which there are only three variables under study. Suppose, for example, that we are studying the relationships among three random variables, X, Y, and Z. If the joint distribution of (X, Y, Z) is multivariate normal, the partial correlation between X and Y while holding Z constant may be written as

$$r_{xy.z} = \frac{r_{xy} - r_{xz}r_{yz}}{\sqrt{(1 - r_{xz}^2)(1 - r_{yz}^2)}} \tag{9.20}$$

where r_{xy}, r_{xz}, and r_{yz} are the Pearson product–moment correlation coefficients computed from the sample measurements on variables X and Y, the variables X and Z, and the variables Y and Z, respectively. The other two possible partial correlation coefficients, $r_{xz.y}$ and $r_{yz.x}$, may be computed by formulas analogous to Equation 9.20.

If the joint distribution of (X, Y, Z) is unknown or if, for some other reason, Equation 9.20 cannot be used, an alternative method of computing the partial correlation coefficient is needed. Both the Spearman r_S and the Kendall $\hat{\tau}$ are

available to meet this need. Of the two, the simplest to use is r_S. To compute the partial correlation coefficients based on r_S, one can merely substitute appropriate ranks for actual measurements and perform the usual parametric multiple correlation analysis, preferably through the use of a computer software package. The partial correlation coefficients ordinarily computed as part of this procedure automatically become the partial rank correlation coefficients.

When the partial rank correlation coefficients are based on $\hat{\tau}$ and there are no ties, $\hat{\tau}_{xy.z}$, the partial rank correlation coefficient measuring the correlation between X and Y while controlling for Z, may be computed by the following formula:

$$\hat{\tau}_{xy.z} = \frac{\hat{\tau}_{xy} - \hat{\tau}_{xz}\hat{\tau}_{yz}}{\sqrt{(1 - \hat{\tau}_{xz}^2)(1 - \hat{\tau}_{yz}^2)}} \tag{9.21}$$

The partial correlation coefficient defined by Equation 9.21 can be generalized to the case in which more than three variables are under consideration.

The following example illustrates the calculation of a partial rank correlation coefficient by means of Equation 9.21.

Example 9.7　For 15 market areas a marketing researcher collected data on sales (X), advertising expenditures (Y), and market share (Z) for a certain consumer product. The data are shown in Table 9.30.

TABLE 9.30　**Data for Example 9.7**

Area	1	2	3	4	5
X	149.0	152.0	155.7	159.0	163.3
Y	21.00	21.79	22.40	23.00	23.70
Z	42.50	43.70	44.75	46.00	47.00

Area	6	7	8	9	10
X	166.0	169.0	172.0	174.5	176.1
Y	24.30	24.92	25.50	25.80	26.01
Z	47.90	48.95	49.90	50.30	50.90

Area	11	12	13	14	15
X	176.5	179.0	170.4	145.0	158.8
Y	26.15	26.30	25.75	20.50	23.10
Z	50.85	51.10	49.50	43.50	46.25

Suppose we compute the partial correlation coefficient between sales (X) and advertising expenditures (Y) while controlling for market share (Z). In other words, we wish to compute $r_{xy.z}$. We first compute the $\hat{\tau}$ values for each pair of variables. To compute $\hat{\tau}_{xy}$, we count the number of Y pairs, when Y is paired with X, that are in

natural order and the number that are in reverse natural order. The procedure, as described in Section 9.2, is shown in Table 9.31.

TABLE 9.31   **Arrangement of data for computing $\hat{\tau}_{xy}$ in Example 9.7**

(X, Y) pairs	Y pairs in natural order	Y pairs in reverse natural order
(145.0, 20.50)	14	0
(149.0, 21.00)	13	0
(152.0, 21.79)	12	0
(155.7, 22.40)	11	0
(158.8, 23.10)	9	1
(159.0, 23.00)	9	0
(163.3, 23.70)	8	0
(166.0, 24.30)	7	0
(169.0, 24.92)	6	0
(170.4, 25.75)	4	1
(172.0, 25.50)	4	0
(174.5, 25.80)	3	0
(176.1, 26.01)	2	0
(176.5, 26.15)	1	0
(179.0, 26.30)	0	0
	$P = 103$	$Q = 2$

From the results of the procedure shown in Table 9.31, we compute $S = 103 - 2 = 101$. By Equation 9.8, we compute

$$\hat{\tau}_{xy} = \frac{101}{(15)(14)/2} = 0.9619$$

The intermediate calculations for obtaining $\hat{\tau}_{xz}$ are shown in Table 9.32.

TABLE 9.32   **Arrangement of data for computing $\hat{\tau}_{xz}$ for Example 9.7**

(X, Z) pairs	Z pairs in natural order	Z pairs in reverse natural order
(145.0, 43.50)	13	1
(149.0, 42.50)	13	0
(152.0, 43.70)	12	0
(155.7, 44.75)	11	0
(158.8, 46.25)	9	1
(159.0, 46.00)	9	0
(163.3, 47.00)	8	0
(166.0, 47.90)	7	0
(169.0, 48.95)	6	0
(170.4, 49.50)	5	0
(172.0, 49.90)	4	0
(174.5, 50.30)	3	0
(176.1, 50.90)	1	1
(176.5, 50.85)	1	0
(179.0, 51.10)	0	0
	$P = 102$	$Q = 3$

From the results shown in Table 9.32, we compute $S = 102 - 3 = 99$. By Equation 9.8, we compute

$$\hat{\tau}_{xz} = \frac{99}{(15)(14)/2} = 0.9428$$

Finally, the intermediate calculations for computing $\hat{\tau}_{yz}$ are shown in Table 9.33.

TABLE 9.33   **Arrangement of data for computing $\hat{\tau}_{yz}$ for Example 9.7**

(Y, Z) pairs	Z pairs in natural order	Z pairs in reverse natural order
(20.50, 43.50)	13	1
(21.00, 42.50)	13	0
(21.79, 43.70)	12	0
(22.40, 44.75)	11	0
(23.00, 46.00)	10	0
(23.10, 46.25)	9	0
(23.70, 47.00)	8	0
(24.30, 47.90)	7	0
(24.92, 48.95)	6	0
(25.50, 49.90)	4	1
(25.75, 49.50)	4	0
(25.80, 50.30)	3	0
(26.01, 50.90)	1	1
(26.15, 50.85)	1	0
(26.30, 51.10)	0	0
	$P = 102$	$Q = 3$

From the results shown in Table 9.33, we compute $S = 102 - 3 = 99$. By Equation 9.8, we compute

$$\hat{\tau}_{yz} = \frac{99}{(15)(14)/2} = 0.9428$$

By Equation 9.21, we compute

$$\hat{\tau}_{xy.z} = \frac{0.9810 - (0.9428)(0.9428)}{\sqrt{[1 - (0.9428)^2][1 - (0.9428)^2]}} = 0.6572$$

Thus we have a measure of the strength of the relationship between sales and advertising expenditures when market share is held constant.

The measure of partial correlation for the sample may be used as an estimate of the population partial correlation coefficient counterpart, $\tau_{xy.z}$. We may also use $\hat{\tau}_{xy.z}$ for hypothesis-testing purposes. The null hypotheses that may be tested and

their corresponding alternatives are as follows:

1. $H_0: \tau_{xy.z} = 0$
   $H_1: \tau_{xy.z} \neq 0$
2. $H_0: \tau_{xy.z} \leq 0$
   $H_1: \tau_{xy.z} > 0$
3. $H_0: \tau_{xy.z} \geq 0$
   $H_1: \tau_{xy.z} < 0$

Critical values for certain sample sizes and preselected levels of significance are given in Table A.24. The decision rules for testing hypotheses about $\tau_{xy.z}$ are as follows:

1. For $H_1: \tau_{xy.z} \neq 0$, reject H_0 if the computed value of $\hat{\tau}_{xy.z}$ is greater than the value of $\hat{\tau}_{xy.z}$ for n and $1 - \alpha/2$ given in Table A.24.
2. For $H_1: \tau_{xy.z} > 0$, reject H_0 if the computed value of $\hat{\tau}_{xy.z}$ is greater than the value of $\hat{\tau}_{xy.z}$ for n and $1 - \alpha$ given in Table A.24.
3. For $H_1: \tau_{xy.z} < 0$, reject H_0 if the computed value of $\hat{\tau}_{xy.z}$ is less than the negative of the value of $\hat{\tau}_{xy.z}$ given in Table A.24 for n and $1 - \alpha$.

For example 9.7, suppose we choose a significance level of 0.05 and test H_0: $\tau_{xy.z} = 0$ against $H_1: \tau_{xy.z} \neq 0$. Since we have a two-sided test and $n = 15$, the critical value of $\hat{\tau}_{xy.z}$ for the test is found in Table A.24 at the intersection of the row labeled 15 and the column labeled $1 - 0.05/2 = 0.975$. We find the critical value to be 0.377. Since the computed value of 0.8290 is greater than 0.377, we reject the null hypothesis in favor of the alternative. Since 0.8290 is greater than 0.570, the P value for the test is less than $2(0.001) = 0.002$.

FURTHER READING

The table of critical values of partial $\hat{\tau}$ given in Table A.24 were calculated by Maghsoodloo (T72) and Maghsoodloo and Pallos (T73). The distribution of partial $\hat{\tau}$ is the subject of an earlier paper by Moran (T74). Partial rank correlation is also the topic of papers by Hawkes (T75) and Shirahata (T76, T77).

Quade (T78) presents an alternative approach to the partial correlation problem.

Kritzer (T79) describes an asymptotic method for obtaining estimates of variance and covariance for partial rank order correlation coefficients based on contingency table data.

EXERCISES

9.18 Refer to Example 9.7. Compute $\hat{\tau}_{xz.y}$ and perform a hypothesis test to determine whether it can be concluded that $\tau_{xz.y}$ is greater than zero. Find the P value for the test. Give a verbal interpretation of $\hat{\tau}_{xz.y}$.

9.19   A real estate broker wished to know what variables are associated with the appraised value of single-family dwellings located in a certain area. A simple random sample of 6 such dwellings yielded the information shown in Table 9.34 on four variables: appraised value (W), size in square feet (X), age in years (Y), and lot size in acres (Z).

<table>
<tr><td>**TABLE 9.34**</td><td colspan="5">**Data for Exercise 9.19**</td></tr>
<tr><td></td><td>**Dwelling**</td><td>**W**</td><td>**X**</td><td>**Y**</td><td>**Z**</td></tr>
<tr><td></td><td>*1*</td><td>132</td><td>25</td><td>20</td><td>1.79</td></tr>
<tr><td></td><td>*2*</td><td>74</td><td>19</td><td>19</td><td>1.37</td></tr>
<tr><td></td><td>*3*</td><td>96</td><td>20</td><td>21</td><td>1.38</td></tr>
<tr><td></td><td>*4*</td><td>128</td><td>22</td><td>16</td><td>1.87</td></tr>
<tr><td></td><td>*5*</td><td>91</td><td>15</td><td>18</td><td>1.27</td></tr>
<tr><td></td><td>*6*</td><td>106</td><td>21</td><td>25</td><td>1.57</td></tr>
</table>

Do these data provide sufficient evidence to allow us to conclude that appraised value and age of residence are inversely correlated when size of residence is held constant? Let $\alpha = 0.05$, and find the P value.

9.20   Refer to Exercise 9.19. On the basis of these data, can one conclude that $\tau_{wz.y} > 0$? Let $\alpha = 0.01$, and find the P value for the test. Give a verbal interpretation of $\tau_{wz.y}$.

9.7

MEASURES OF ASSOCIATION FOR CONTINGENCY TABLES

Chapter 5 covered the analysis of frequency data displayed in contingency table format. We used the chi-square test of independence to determine whether we might conclude that there is an association between two categorical variables; a computed chi-square statistic that is significant allows us to conclude, at the specified level of significance, that there is an association between the two categorical variables under study. Although we may reach conclusions about the presence or absence of an association using a chi-square test of independence, the test statistic does not provide a satisfactory measure of the strength of the association between the two variables. In Section 9.6 it was noted that within the classical statistics context, the simple Pearson product–moment correlation coefficient provides a measure of the strength of the relationship between two variables. For this measure to be meaningfully interpretable, however, the two variables of interest must be measured on a continuous scale and conform to the linear model with respect to their relationship. This section presents some measures of strength of association for situations in which the Pearson correlation coefficient is not appropriate— specifically, the case in which the two variables of interest are categorical and the data consist of frequencies that may be displayed in a contingency table.

The following are some measures of association for 2 × 2 contingency tables. As previously shown, in the simplest situation yielding data that can be displayed in a contingency table, there are two dichotomous variables, each of which has only two categories. The contingency table constructed from the data collected under this model is called a 2 × 2 contingency table. The following are two measures of association that are appropriate in this case.

PHI COEFFICIENT

The phi coefficient was designed for use with dichotomous variables—that is, variables that can assume only one of two possible mutually exclusive values. Examples are gender (male, female), product quality (defective, not defective), and marital status (married, not married). In practice it is also used when values of nondichotomous variables can be meaningfully grouped into two distinct categories. Students' knowledge of a subject, for example, may be continuous and measurable through the assignment of numerical, interval-scale scores based on appropriate test performance, but it may sometimes be preferable to categorize students' performances as either pass or fail, depending on whether or not their numerical scores fall above or below some chosen value.

Suppose we have two variables, variable I and variable II. Measurements on each are obtained by noting into which of the variables' two categories a subject or object should be placed. When such determinations are made for a sample of n subjects or objects, the results may be displayed in a 2 × 2 contingency table, of which Table 9.35 is a prototype.

The phi coefficient is

$$\phi = \frac{ad - bc}{\sqrt{(a + b)(c + d)(a + c)(b + d)}} \tag{9.22}$$

The phi coefficient may assume values between -1 and $+1$. The phi coefficient is related to the chi-square statistic. The relationship is expressed by

$$\phi^2 = X^2/n \tag{9.23}$$

TABLE 9.35   **A 2 × 2 contingency table**

Variable II category	Variable I category		Total
	1	2	
1	a	b	a + b
2	c	d	c + d
Total	a + c	b + d	n

To determine whether a computed value of ϕ is significant, we may convert it to X^2 by

$$X^2 = n\phi^2 \tag{9.24}$$

We then compare the resulting X^2 with tabulated chi-square values with 1 degree of freedom to determine significance and find the P value.

We illustrate the use of the phi coefficient by means of the following example.

Example 9.8

In a study of sexual harassment in the workplace, researchers asked a sample of 125 white-collar workers in nonmanagerial positions whether they had ever been sexually harassed on the job. Table 9.36 shows the employees cross-classified by their responses to the question and gender. We wish to compute ϕ from these data to determine the strength of the relationship between the two variables. We also wish to test for significance at the 0.05 level.

TABLE 9.36

A sample of 125 employees classified by gender and sexual harassment on the job

	Sexually harassed		
Gender	Yes	No	Total
Male	15	35	50
Female	50	25	75
Total	65	60	125

By equation 9.22, we compute

$$\phi = \frac{(15)(25) - (35)(50)}{\sqrt{(50)(75)(65)(60)}} = -0.3595$$

We now have a measure of the strength of the association between gender and sexual harassment experience for our sample of 125 workers. To test for significance, we first use Equation 9.24 to compute

$$X^2 = 125(-0.3595)^2 = 16.16$$

Reference to Table A.11 with 1 degree of freedom shows that since $16.16 > 3.841$, we may reject the null hypothesis of no association between the two variables. Since $16.16 > 7.879$, the P value for the test is less than 0.005.

YULE'S Q

When measuring the strength of the association between two dichotomous variables, some researchers prefer a statistic that was introduced and called Q by

Yule (T80) in 1900. If we use the notation of Table 9.36, we may write Yule's Q as

$$Q = \frac{ad - bc}{ad + bc} \qquad \text{(9.25)}$$

Like the phi coefficient, Q may assume any value between -1 and $+1$, inclusive. We illustrate the calculation of Q by means of the following example.

Example 9.9   To illustrate the calculation of Q, let us refer to the data displayed in Table 9.36. By Equation 9.25, we have

$$Q = \frac{(15)(25) - (35)(50)}{(15)(25) + (35)(50)} = -0.647$$

The following two measures of association are for use with $r \times c$ contingency tables; that is, tables in which there are two categorical variables and one or both have more than two categories.

THE CRAMÉR STATISTIC

A statistic suggested by Cramér (T81) provides an appropriate measure of the strength of association between two categorical variables yielding data that may be displayed in a contingency table of any size. When the contingency table has two rows and two columns, the Cramér coefficient yields a value that is identical, except for a possible difference in sign, to the contingency coefficient. The Cramér coefficient is defined as

$$C = \sqrt{\frac{X^2}{n(t - 1)}} \qquad \text{(9.26)}$$

where X^2 is the chi-square statistic computed by Equation 8.1, n is the total sample size, and t is either the number of rows or the number of columns in the contingency table, whichever is smaller.

Example 9.10 illustrates the calculation of the Cramér statistic.

Example 9.10   A survey was conducted among homeowners in a certain state. One of the questions asked of respondents was "How satisfied are you with the community in which you live?" Table 9.37 classifies the respondents by their answer to the question and their place of residence.

We wish to use the Cramér statistic to measure the strength of the association between place of residence and level of satisfaction with community of residence. By Equation 8.1 we compute $X^2 = 53.178$. Since we have a sample size of 230, and since the number of rows (3) is smaller than the number of columns, we have

TABLE 9.37   **Contingency table for Example 9.10**

| Place of residence | Level of satisfaction with community of residence | | | |
	Very satisfied	Satisfied	Unsatisfied	Very unsatisfied
Rural	30	15	10	5
Suburban	40	20	15	10
Urban	10	15	20	40

$t = 3 - 1 = 2$. By Equation 9.26, then, we have

$$C = \sqrt{\frac{53.178}{230(2)}} = 0.34$$

To test C for significance, we compare X^2 in Equation 9.26 with tabulated values of chi-square in Table A.11 for $(r - 1)(c - 1)$ degrees of freedom. In other words, the significance of C depends on the significance of X^2, which is tested for significance in the manner explained in Chapter 5.

The statistic C has some interesting properties. It can assume values between 0 and 1. When, for the sample of measurements, there is no association between the two variables under study, C will be equal to zero. When the sample measurements are displayed in a square contingency table (that is, when both variables under study have the same number of categories, so that $r = c$), a computed C of 1 indicates a perfect correlation between the two variables. When r and c are unequal, a computed C of 1 does not mean that the two variables are perfectly correlated in the usual sense. It can also be shown that when $r = c = 2$, C is equal to the square of the Kendall tau statistic adjusted for ties.

An advantage of the Cramér statistic is that few assumptions are necessary for its validity. Another advantage is the fact that values of C may be used to compare contingency tables of different sizes with respect to r and c and tables based on samples of different sizes.

GOODMAN–KRUSKAL G COEFFICIENT

The Yule coefficient Q may be extended for analysis of data cast in a contingency table of size $r \times c$, in which r or c or both are greater than 2 and the categories of both variables are ordered. The statistic computed from such sample data is usually called the Goodman–Kruskal coefficient. We shall use the letter G to designate the statistic and the Greek letter γ to designate the parameter that is its population counterpart.

Suppose we have two variables of interest, variable X and variable Y, both of which are measured on an ordinal scale. We assume that X can take on the values $X_1, X_2, \ldots, X_r$ and that the values are ordered in magnitude so that $X_1 <$

$X_2 < \cdots < X_r$. Similarly, we assume that the variable Y can take on c values ordered in magnitude so that $Y_1 < Y_2 < \cdots < Y_c$. We may convert the X measurements to ranks $1, 2, \ldots, r$ and the Y measurements to ranks $1, 2, \ldots, c$ and use these ranks to label the rows and columns of a contingency table. Suppose we let the ranks of the Y measurements provide labels for the c columns and the ranks of the X measurements serve as labels for the r rows of the contingency table.

Consider now a finite population of N subjects (or objects), on whom we can take measurements of X and Y. For each subject we determine the rank of its X value and the rank of its Y value and use the results as a basis for assigning the subject to a cell of the contingency table. If a subject ranks 1 with respect to X and also 1 with respect to Y, the subject is assigned to (counted in) cell 1,1 of the contingency table. A subject ranking 2 on X and 3 on Y would be assigned to cell 2,3, and so on. The number of the N subjects assigned to cell i, j of the contingency table is designated N_{ij}. Table 9.38 is such a contingency table.

TABLE 9.38   **Contingency table for two ordered variables, X and Y, and a population of N subjects (or objects)**

X rank	Y rank						
	1	2	$\cdots$	j	$\cdots$	c	Total
1	N_{11}	N_{12}	$\cdots$	N_{1j}	$\cdots$	N_{1c}	$N_{1.}$
2	N_{21}	N_{22}	$\cdots$	N_{2j}	$\cdots$	N_{2c}	$N_{2.}$
$\vdots$							
i	N_{i1}	N_{i2}	$\cdots$	N_{ij}	$\cdots$	N_{ic}	$N_{i.}$
$\vdots$							
r	N_{r1}	N_{r2}	$\cdots$	N_{rj}	$\cdots$	N_{rc}	$N_{r.}$
Total	$N_{.1}$	$N_{.2}$	$\cdots$	$N_{.j}$	$\cdots$	$N_{.c}$	N

For the population described in Table 9.38, the coefficient γ measures the strength of the association between X and Y in the following sense: It is the difference between the probability that, for a pair of subjects, their X and Y measurements are in the same order and the probability that for a pair of subjects their X and Y measurements disagree with respect to order. We may determine whether, for a pair of subjects, their X and Y measurements agree or disagree with respect to order in exactly the same way that we did when calculating the Kendall tau statistic described in Section 9.2. Recall from that section that we let P equal the number of pairs of subjects whose X and Y measurements agree with respect to order, and we let Q equal the number of pairs of subjects whose X and Y measurements disagree with respect to order. It follows that $P + Q$ equals the number of pairs of subjects whose X and Y measurements either agree or disagree with respect to order. Consider a randomly selected pair of subjects from the population described in

Table 9.38 whose measurements on neither variable are tied. The probability that their X and Y measurements agree with respect to order is equal to $P/(P + Q)$. The probability that their X and Y measurements disagree with respect to order is equal to $Q/(P + Q)$. The coefficient γ is the difference between these two probabilities; that is,

$$\gamma = \frac{P}{P + Q} - \frac{Q}{P + Q} = \frac{P - Q}{P + Q} \tag{9.27}$$

Now consider a simple random sample of n subjects drawn from the population described by Table 9.38. From the sample we may construct a contingency table that is identical to Table 9.38 except that the symbol N is replaced by the symbol n throughout. The resulting table would be identical to Table 5.1, with the column and row headings now assumed to be ordered ranks. If we let P and Q be defined for the sample in the same way as for the population, the sample Goodman–Kruskal statistic may be described as

$$G = \frac{P - Q}{P + Q} \tag{9.28}$$

Calculation of G   When sample data are displayed in a contingency table of the type just described, P and Q are more easily and conveniently computed by the following systematic procedure. To obtain P, perform the following calculations:

1. Identify the frequency in the upper left-hand corner of the contingency table (the frequency of cell 1,1, n_{11}) as a multiplier. Call it Multiplier $P1$.
2. Add the frequencies of all remaining cells of the table that are to the right and below Multiplier $P1$ that are not in the same row or column as Multiplier $P1$. Call the result Sum $P1$.
3. Compute Product $P1 = $ (Multiplier $P1$) × (Sum $P1$).
4. Move to the next cell in the same row as Multiplier $P1$ (cell 1,2). Call its frequency Multiplier $P2$.
5. Add the frequencies of all cells of the table that are to the right of and below Multiplier $P2$, but not in the same row or column with it. Call the result Sum $P2$.
6. Product $P2 = $ (Multiplier $P2$) × (Sum $P2$).
7. Proceed as in Steps 1 through 6 until there are no more cells in row 1 that have cells below and to the right and not in the same row or column.
8. Beginning with the first cell in the row, repeat Steps 1 through 7 for each remaining row.
9. Add all products obtained in Steps 1 through 8. The result is P.

To obtain Q, do the following:

1. Identify the frequency of the cell in the upper right-hand corner of the contingency as Multiplier $Q1$.
2. Add the frequencies of all cells that are below and to the left of Multiplier $Q1$, but not in the same row or column with it. Call the result Sum $Q1$.
3. Product $Q1 = $ (Multiplier $Q1$) $\times$ (Sum $Q1$).
4. Multiplier $Q2$ is the frequency of the cell to the immediate left of the cell containing Multiplier $Q1$.
5. Sum $Q2$ is the sum of all frequencies in cells to the left of and below the cell containing Multiplier $Q2$, but are not in the same row or column with it.
6. Product $Q2 = $ (Multiplier $Q2$) $\times$ (Sum $Q2$).
7. Proceed as in Steps 1 through 6 until there are no cells in row 1 that have cells below and to the left and not in the same row or column.
8. Beginning with the last cell in the row, repeat Steps 1 through 7 for each remaining row.
9. Add the products obtained in Steps 1 through 8. The result is Q.

The calculation of G is illustrated by the following example.

Example 9.11 In a study of the relationship between price and quality of certain household products, the quality of 180 products was rated as poor, mediocre, or superior. Table 9.39 shows the 180 products cross-classified on the basis of quality rating and price. We wish to measure the strength of the association between the two variables by means of the G coefficient.

TABLE 9.39 Data for Example 9.11

Quality rating	Price category			Total
	Low	Medium	High	
Poor	20	13	12	45
Mediocre	15	45	19	79
Superior	10	17	29	56
Total	45	75	60	180

We compute P as follows:

$$20(45 + 19 + 17 + 29) = 2200$$
$$13(19 + 29) \qquad = 624$$

$$15(17 + 29) \qquad = \quad 690$$
$$45(29) \qquad\qquad = 1305$$
$$\overline{}$$
$$P = 4819$$

We next compute Q:

$$12(15 + 45 + 10 + 17) = 1044$$
$$13(15 + 10) \qquad\quad = \quad 325$$
$$19(17 + 10) \qquad\quad = \quad 513$$
$$45(10) \qquad\qquad\quad = \quad 450$$
$$\overline{}$$
$$Q = 2332$$

By Equation 9.28,

$$G = \frac{4819 - 2332}{4819 + 2332} = 0.3478$$

FURTHER READING

The Goodman–Kruskal G statistic is discussed in the papers by Goodman and Kruskal (T82, T83, T84, T85). The four papers have been published together in a book by the same authors (T86). Some of the measures of association presented in this section as well as others are discussed by Reynolds (T87). Also of interest are the papers by Freeman (T88), Janson and Vegelius (T89), Gans and Robertson (T90), and Berry and Mielke (T91).

EXERCISES

9.21 Refer to Example 5.2. Compute ϕ and test for significance at the 0.05 level. Find the P value.

9.22 Refer to Example 5.2. Compute Yule's Q and compare it with ϕ computed in Exercise 9.21.

9.23 Refer to Example 5.1. Compute Cramér's C and test for significance at the 0.05 level. Find the P value.

9.24 Refer to Example 5.1. Compute the Goodman–Kruskal G statistic.

9.25 Refer to Exercise 5.1. Compute Cramér's C and test for significance at the 0.05 level. Find the P value.

9.26 Refer to Exercise 5.2. Compute ϕ and test for significance at the 0.05 level. Find the P value.

9.27 Refer to Exercise 5.2. Compute Yule's Q.

9.28   Refer to Exercise 5.5. Compute the Goodman–Kruskal *G* statistic.

9.29   Refer to Exercise 5.6. Compute the Cramér *C* statistic and test for significance at the 0.05 level. Find the *P* value.

9.8

OTHER MEASURES OF ASSOCIATION

This section focuses on some additional measures of association. Most of these have been proposed more recently than those discussed earlier and consequently have not been as extensively used. The researcher concerned with association analysis should be aware of these procedures, since some of them may be found to be more appropriate in certain situations than the more familiar techniques.

POINT BISERIAL COEFFICIENT OF CORRELATION

Not infrequently, we may wish to assess the strength of the relationship between two variables, one of which is dichotomous and the other measured on an interval or ratio scale. For example, we might wish to measure the strength of the correlation between the gender of children and the amount of television they watch. Other pairs of dichotomous/interval-scale variables whose strength of association might be of interest to some researcher include geographic area (urban, suburban) and residential property values, product quality (defective, not defective) and amount of some ingredient in the raw material from which the product is manufactured, educational achievement among young adults (high school graduate, high school dropout) and income, and college students' fraternity/sorority membership and grade point average. Subjects in the population of interest, and consequently those in a sample drawn from the population, will have two measurements of interest, one on the dichotomous variable and one on the variable measured on the interval or ratio scale. The concept of correlation between two such variables is called biserial correlation.

Two different coefficients of biserial correlation are in use: the *biserial correlation coefficient* and the *point biserial correlation coefficient*. The discussion here will be limited to the point biserial correlation coefficient. The calculation of a point biserial correlation coefficient comes within the purview of nonparametric statistics because one of the variables involved is dichotomous. Making inferences based on a sample biserial correlation coefficient belongs to the province of classical statistics, since it is assumed for valid inferences that the two distributions of the other variable (one for each of the two values of the dichotomous variable) are normal with equal variances. Consequently the present discussion is concerned only with the calculation of the measure and its use as a descriptive statistic. Those interested in using a sample biserial correlation coefficient to make inferences will find details in other sources, such as the book by Walker and Lev (T92).

When the point biserial correlation coefficient is of interest, we use Y to designate the dichotomous variable and X to designate the other variable. To minimize the calculation burden, let one of the two possible values of Y be 1 and the other value be 0. We use the symbol r_{pb} to designate the point biserial correlation coefficient computed from sample data. The simplest computational formula for r_{pb} is

$$r_{pb} = \sqrt{\frac{n_1 n_0}{n}} \left(\frac{\bar{x}_1 - \bar{x}_0}{\sqrt{\Sigma(x - \bar{x})^2}} \right) \tag{9.29}$$

where n_1 is the number of 1's and n_0 is the number of 0's observed in a sample of n subjects or objects ($n_1 + n_0 = n$), $\bar{x}_1$ is the mean value of X for the n_1 subjects and $\bar{x}_0$ is the mean value of X for the n_0 subjects, and $\Sigma(x - \bar{x})^2$ is the numerator of the variance of the sample of all X measurements. The range of values of r_{pb} is between -1 and $+1$, inclusive.

We illustrate the calculation of r_{pb} by means of the following example.

Example 9.12   In a study of the association between income and education, the data shown in Table 9.40 were obtained on a sample of 25-year-old males who did not attend college. We use the data to calculate r_{pb}.

TABLE 9.40   **Annual income and education of a sample of 25-year-old males who did not attend college**

Subject	Income $\times$ 1,000	Completed high school (1 = yes, 0 = no)
1	$24	1
2	15	1
3	35	0
4	30	1
5	12	0
6	25	1
7	13	0
8	21	1
9	27	1
10	29	1
11	19	0
12	21	1
13	14	0
14	16	0
15	10	0

From the data in Table 9.40, we have $n_1 = 8$, $\bar{x}_1 = 24$, $n_0 = 7$, $\bar{x}_0 = 17$, and $\Sigma(x - \bar{x})^2 = 780.9333$. By Equation 9.29, we compute

$$r_{pb} = \sqrt{\frac{(8)(7)}{15}}\left(\frac{24 - 17}{\sqrt{780.9333}}\right) = 0.4840$$

MISCELLANY OF MEASURES OF ASSOCIATION

Of the numerous tests of association, correlation, dependence, and so on, let us now mention briefly only a few. Many have been developed for very special situations. The references cited will enable the reader to locate and more thoroughly investigate techniques of particular interest. In addition to these references, many of the procedures mentioned in this section are discussed in the textbooks cited in Chapter 1.

In 1965 Freeman (T93) developed a measure designed to determine the association between a nominal independent variable and an ordinal dependent variable. Exactly 20 years later Buck and Finner (T94) demonstrated that the sampling distribution of Freeman's statistic is identical to that of the Mann–Whitney U statistic.

Jolliffe (T95) proposes a runs test to detect a relationship between two variables when the form of the relationship is unknown. The author points out that although the test is not very good at detecting linear or monotonic relationships, it will detect a relationship in situations in which more specific tests are inappropriate.

The estimation of the standard errors of some measures of association using Tukey's jackknife method is discussed by Henry (T96). The method is used to test the hypothesis that a pair of partial association coefficients are equal.

Shirahata (T97) has proposed an intraclass rank test of the Spearman type to test for independence in a bivariate population. The author states that the test is useful when there is prior information that the population has the same marginal distributions. It is shown that the proposed test is asymptotically equivalent to Spearman's test under the null hypothesis of independence and its contiguous alternatives.

Measures for summarizing the strength of association between a nominal variable and an ordered categorical variable are formulated by Agresti (T98). The measures are differences or ratios of probabilities of events concerning two types of pairs of observations. They can be used to describe the degree of difference between two or more groups on an ordinal response variable.

A nonparametric measure of interclass correlation analogous to Kendall's measure of dependence is considered by Shirahata (T99), who also studies its unbiased estimator and an associated test.

Kimeldorf et al. (T100) present four procedures to measure the monotone relationship between ordinal variables. They are derived from the concept of monotone correlation. A nonlinear optimization algorithm is employed to evaluate the measures and to obtain the associated monotone scalings.

A sign test for correlation is proposed by Nelson (T101). The author states that the power of the test is probably only about half that of the most powerful test available. Its advantages are the fact that it is distribution-free and easy to use. The author provides a table of critical values for the test.

Shirahata and Araki (T102) consider two classes of rank statistics to measure the degree of association of ranked data. The association they consider is not only the monotone association, but also the departure from uniformity of the distribution of ranks. The statistics belonging to one class are based on the rank regression coefficients. The statistics in the other class are extended versions of the rank serial correlation coefficient. The asymptotic as well as the exact properties of the statistics are developed under the hypothesis of no association. Numerical examples are given.

Ledwina (T103) presents two rank tests of independence against positive quadrant dependence. The exact Bahadur slopes are given and compared in some particular cases with Bahadur slopes of Spearman's and Kendall's rank statistics. In another paper Ledwina (T104) derives the limiting Pitman efficiency of Kendall's tau, Spearman's rho, and one of the tests of independence discussed in the previously mentioned paper (T103). Bajorski (T105) derives sufficient and necessary conditions for local Bahadur optimality of the tests of independence introduced by Ledwina (T103).

For multivariate or directional data, Jupp (T106) proposes a nonparametric correlation coefficient and a two-sample statistic. The author states that both statistics are coordinate-free, resistant, and unchanged under almost all small perturbations of the data. The statistics are related to other well-known statistics. A numerical example is given.

Measures of association in contingency tables have been proposed and/or discussed by Yule and Kendall (T107), McNemar (T108), Stuart (T109), and Ives and Gibbons (T110). Blomqvist (T111) has proposed a quadrant test, and Bhuchongkul (T112) and Fieller and Pearson (T113) discuss the normal-scores test for association.

Bell and Doksum (T114) discuss a measure of correlation that uses order statistics from normally distributed random samples rather than ranks. Other articles of interest include those by Holley (T115), Hotelling and Pabst (T116), Moran (T117) and Davis (T118).

FURTHER READING

The literature on association analysis appears to be almost inexhaustible. Space limitations permit the mention here of only a few other papers that may be helpful to the researcher.

Rank correlation and concordance tests in community analyses are the subject of a paper by Jumars (T119). Turek and Suich (T120) present an exact test, developed through a reparameterization scheme, based on a statistic introduced by Goodman and Kruskal (T83). The paper also discusses the power of the test. A discussion of

rank correlation with missing data may be found in a paper by Papaioannou and Loukas (T121). Jewell (T122) investigates for the small-sample case the bias of the standard point estimators of some measures of association. The author suggests some simple alternative estimators that he claims possess superior performance in terms of bias and mean squared error. Generalizations of the class of test statistics and large deviation results derived by Ledwina (T103) are presented by Bajorski and Ledwina (T123). Somers (T124, T125, T126, T127) has written extensively on the subject of association analysis, and these papers should not be overlooked by the researcher involved in that type of analysis. Some useful publications that treat association analysis in a more general manner include those by Carroll (T128), Costner (T129), and Weisberg (T130).

EXERCISES

9.30 Each person in a sample of 16 adults was asked if he or she would be willing to participate in a product evaluation survey by a marketing research firm. Table 9.41 shows the age of each person asked to participate and whether or not he or she agreed to do so. Compute r_{pb} from these data.

TABLE 9.41 **Data for Exercise 9.30**

Age	23	35	46	30	21	30	22	55	70	63	47	41	22	28	35	54
Participated? (1 = yes, 0 = no)	0	1	1	0	0	1	0	1	1	1	1	1	0	0	0	1

9.31 The records kept by a firm that repairs small appliances include information on the type of repair, the length of time spent performing the repair, and whether or not the repair passed the final inspection process. Table 9.42 shows, for the same type of repair job, the amount of time in minutes spent on 20 repair jobs and whether or not the repair passed final inspection. Compute r_{pb} from these data.

TABLE 9.42 **Data for Exercise 9.31**

Job	Time	Passed?	Job	Time	Passed?
1	30	Yes	11	45	Yes
2	21	No	12	40	Yes
3	35	Yes	13	47	Yes
4	36	Yes	14	27	No
5	25	No	15	30	No
6	55	Yes	16	29	No
7	18	No	17	34	Yes
8	32	Yes	18	42	Yes
9	34	Yes	19	40	Yes
10	32	Yes	20	45	Yes

9.9

COMPUTER PROGRAMS

Kempi (T131) has made available a FORTRAN IV program for ranking data in ascending order. The program corrects for ties and prints the ranks and raw data upon request. After ranking, the Spearman rank correlation coefficient and the corresponding *t*-value are calcuated by a procedure that corrects for ties.

Another FORTRAN IV program described by Zar (T132) provides the one-tailed and two-tailed probabilities of given Spearman rank correlation coefficients. The program may be used alone or employed with a rank correlation program.

Berry and Mielke (T133) describe an algorithm and a FORTRAN-77 subroutine for computing Goodman and Kruskal's tau-*b* statistic (T82, T83, T84) and for the solution needed for the test described by Berry and Mielke in an earlier paper (T91).

Jupp (T106) has made available a computer program that calculates the measures discussed in the author's previously cited paper.

Kimeldorf et al. (T134, T135) describe an interactive FORTRAN program that computes various correlation measures. Data input can be finite discrete bivariate probability mass functions or ordinal contingency tables, both of which must be given in matrix form.

Brophy (T136) presents an algorithm and program for calculation of Kendall's rank correlation coefficient.

BASIC programs written by Galla (T137) calculates Kendall's tau and Kendall's partial rank correlation coefficient.

A program that computes biserial and point-biserial correlation coefficients is available for the IBM-PC from Dunlap and Kemery (T138). The program is written in FORTRAN and is designed for any computer with a FORTRAN IV or later compiler.

Most of the statistics software packages for microcomputers will compute both the Spearman and Kendall rank correlation coefficients. These include *BMDPC, CRISP, EXEC*U*STAT, MICROSTAT, NUMBER CRUNCHER STATISTICAL SYSTEM, PC STATISTICIAN, SCA, SPSS/PC, STATGRAPHICS, STATPRO,* and *SYSTAT.*

The *M/STAT-2000* package computes Kendall's partial correlation coefficients, and *SCA* calculates Kendall's coefficient of concordance.

REVIEW EXERCISES

9.32 In an evaluation of the treatment modalities provided by a psychiatric in-patient facility, five teams were asked to rank six treatment modalities on the basis of effectiveness in the treatment of patients. Each team submitted a set of ranks representing the consensus of its

members. The results are shown in Table 9.43. What can one conclude on the basis of these data?

Group rankings of six treatment modalities at an in-patient psychiatric facility

Group	Electro-convulsive therapy	Medication	Group therapy	Social activities	Occupational therapy	Recreation therapy
Patients	6	5	4	1	3	2
Social workers	6	2	1	3	5	4
Nurses	3	1	2	4	5	6
Psychologists	6	5	1	4	2	3
Psychiatric residents	3	1	2	4	6	5

9.33 In an animal-behavior study, 15 domestic animals reared together were ranked according to their relative positions in the group's pecking order and also on the basis of their "friendliness" toward humans. The results are shown in Table 9.44. Compute r_S and determine whether one can conclude that animals higher in the pecking order tend to be friendlier toward humans.

TABLE 9.44 **15 domestic animals ranked on pecking order and "friendliness" toward humans**

Animal	1	2	3	4	5	6	7	8	9	10	11	12	13	14	15
Pecking order	12	5	11	8	4	6	7	1	14	3	10	15	2	9	13
Friendliness	8	1	9	10	6	5	3	7	11	2	12	13	4	14	15

9.34 Ten seventh-grade children randomly selected from a certain public school system were ranked according to the quality of their home environment and the quality of their performance in school. The results are shown in Table 9.45. Compute r_S and determine whether one can conclude that the two variables are directly related.

TABLE 9.45 **Ten seventh-grade children ranked according to quality of home environment and quality of performance in school**

Child	1	2	3	4	5	6	7	8	9	10
Home environment	3	7	10	9	2	1	6	4	8	5
Performance in school	1	9	8	10	3	4	5	2	6	7

9.35 Four community leaders were asked to rank six community problems on the basis of priority for action. The results are shown in Table 9.46. What can one conclude from these data?

TABLE 9.46   **Community leaders' rankings of six community problems, on the basis of priority for action**

Problem	Mayor	Police chief	Public health director	School superintendent
Drug abuse	1	2	1	1
Care of elderly	5	4	2	5
Traffic	4	3	5	4
Crime	2	1	4	2
Housing	3	5	3	3
Recreation	6	6	6	6

9.36   Table 9.47 shows self-concept and academic achievement scores of 20 high school seniors. Compute $\hat{\tau}$ and determine whether one should conclude that there is a direct relationship between the two variables.

TABLE 9.47   **Self-concept and academic achievement scores of 20 high school seniors**

Subject	1	2	3	4	5	6	7	8	9	10
Academic achievement	97	83	73	88	69	70	76	60	73	99
Self-concept	66	74	11	89	6	29	59	36	53	60

Subject	11	12	13	14	15	16	17	18	19	20
Academic achievement	97	62	87	89	98	93	85	79	64	85
Self-concept	65	21	66	69	89	61	95	45	38	74

9.37   Fifteen children were given a visual-discrimination test during their first week of kindergarten and a reading-achievement test at the end of first grade. The scores on the two tests are given in Table 9.48. Compute $\hat{\tau}$ and determine whether one should conclude that the two variables are directly related.

TABLE 9.48   **Visual discrimination and reading achievement scores of 15 children**

Child	1	2	3	4	5	6	7	8	9	10	11	12	13	14	15
Visual discrimination	75	69	70	65	68	62	50	52	40	45	41	42	39	37	34
Reading achievement	95	90	82	75	70	69	60	58	55	49	42	38	35	30	20

9.38   A panel of parents, a panel of teachers, and a panel of social workers were asked to rank seven child behaviors on the basis of undesirability. The results are shown in Table 9.49. What can one conclude from these data?

TABLE 9.49   **Three panels' rankings, in order of undesirability, of seven child behaviors**

Behavior	Panel		
	Parents	Teachers	Social workers
High temper	7	6	5
Destructiveness	4	2	3
Disobedience	1	1	6
Undesirable playmates	2	7	7
Negativism	5	3	2
Anxiety	6	5	1
Lying	3	4	4

9.39 Table 9.50 shows scores made on a physical-fitness test and a self-concept scale by 10 college seniors. Compute r_S and determine whether one can conclude that there is a direct relationship between the two variables.

TABLE 9.50   **Physical fitness and self-concept scores of 10 college seniors**

Student	1	2	3	4	5	6	7	8	9	10
Physical fitness	60	70	65	72	75	77	82	84	90	95
Self-concept	55	73	62	80	81	70	83	91	87	93

9.40 Use the data of Exercise 9.37 to construct a 95% confidence interval for τ.

9.41 Refer to Exercise 5.7. Compute the Cramér statistic and test for significance at the 0.01 level. Find the P value.

9.42 Refer to Exercise 5.8. Compute the Cramér statistic and test for significance at the 0.05 level. Find the P value.

9.43 Refer to Exercise 5.9. Compute the Cramér statistic and test for significance at the 0.05 level. Find the P value.

9.44 Refer to Exercise 5.10. Compute the Goodman–Kruskal statistic.

9.45 Refer to Exercise 5.12. Compute ϕ and test for a significant association between extent of withdrawal and time of breakdown at the 0.05 level. Find the P value.

9.46 Refer to Exercise 5.16. Compute Yule's Q and test, at the 0.05 level, for a significant association between marijuana use and hard liquor use. Find the P value.

9.47 In a survey conducted by a city tourist bureau, a sample of 20 vacationing tourists completed a questionnaire designed to obtain a variety of information. Table 9.51 shows the responses of the survey respondents to questions asking for number of miles driven on the present vacation and type of overnight accommodations (economy-priced motel or hotel versus regular-price or luxury motel or hotel). Compute r_{pb} from these data.

TABLE 9.51   **Miles driven and type of accommodations reported by 20 tourists (1 = economy-priced motel or hotel, 0 = regular-priced motel or hotel)**

Tourist	Miles	Accommodations	Tourist	Miles	Accommodations
1	1100	1	11	1659	1
2	1253	0	12	1775	0
3	1713	1	13	1888	0
4	1442	0	14	1728	1
5	1739	1	15	1542	1
6	1502	1	16	858	0
7	1514	1	17	1865	1
8	862	1	18	2090	1
9	1759	1	19	1711	1
10	1550	1	20	1659	0

REFERENCES

T1 Mood, Alexander M., Franklin A. Graybill, and Duane C. Boes, *Introduction to the Theory of Statistics*, third edition, New York: McGraw-Hill, 1974.

T2 Kruskal, W. H., "Ordinal Measures of Association," *J. Amer. Statist. Assoc.*, 53 (1958), 814–861.

T3 Kendall, M. G., *Rank Correlation Methods*, fourth edition, London: Griffin, 1970.

T4 Liebetrau, Albert M., *Measures of Association*, Beverly Hills, Calif.: Sage 1983.

T5 Spearman, C., "The Proof and Measurement of Association between Two Things," *Amer. J. Psychol.*, 15 (1904), 72–101.

T6 Zar, Jerrold H., *Biostatistical Analysis*, Englewood Cliffs, N. J.: Prentice-Hall, 1974.

T7 Glasser, G. J., and R. F. Winter, "Critical Values of the Coefficient of Rank Correlation for Testing the Hypothesis of Independence," *Biometrika*, 48 (1961), 444–448.

T8 Litchfield, John T., Jr., and Frank Wilcoxon, "The Rank Correlation Method," *Analyt. Chem.* 27 (1955), 299–300.

T9 Stuart, A., "The Asymptotic Relative Efficiency of Tests and the Derivatives of Their Power Functions," *Skandinavisk Aktuarietidskrift*, 37 (1954), 163–169.

T10 Bhattacharyya, G. K., R. A. Johnson, and H. R. Neave, "Percentage Points of Non-Parametric Tests for Independence," *J. Amer. Statist. Assoc.*, 65 (1970), 976–983.

T11 Woodworth, George G., "Large Deviations and Bahadur Efficiency of Linear Rank Statistics," *Ann. Math. Statist.*, 41 (1970), 251–283.

T12 Bahadur, R. R., "Rates of Convergence of Estimates and Test Statistics," *Ann. Math. Statist.*, 38 (1967), 303–324.

T13 Daniels, H. E., "Rank Correlation and Population Models," *J. Roy. Statist. Soc., Ser. B*, 12 (1950), 171–181.

T14 Kraemer, Helena Chmurra, "The Non-Null Distribution of the Spearman Rank Correlation Coefficient," *J. Amer. Statist. Assoc.*, 69 (1974), 114–117.

T15 Zar, J. H., "Significance Testing of the Spearman Rank Correlation Coefficient," *J. Amer. Statist. Assoc.*, 67 (1972), 578–580. Addendum, Otten, A., *Ibid.*, 68 (1973), 585.

T16 Iman, Ronald L., and W. J. Conover, "Approximations of the Critical Region for Spearman's Rho with and without Ties Present," *Communic. in Statist.—Simulation and Computation*, 7 (1978), 269–282.

T17 Griffiths, D., "A Pragmatic Approach to Spearman's Rank Correlation Coefficient," *Teaching Statist.*, 2 (1980), 10–13.

T18 Kendall, M. G., "A New Measure of Rank Correlation," *Biometrika*, 30 (1938), 81–93.

T19 Gibbons, Jean Dickinson, *Nonparametric Statistical Inference*, New York: McGraw-Hill, 1971.

T20 Sillitto, G. P., "The Distribution of Kendall's Coefficient of Rank Correlation in Rankings Containing Ties," *Biometrika*, 34 (1947), 36–40.

T21 Burr, E. J., "The Distribution of Kendall's Score *S* for a Pair of Tied Rankings," *Biometrika*, 47 (1960), 151–171.

T22 Smid, L. J., "On the Distribution of the Test Statistic of Kendall and Wilcoxon When Ties Are Present," *Statistica Neerlandica*, 10 (1956), 205–214.

T23 Robillard, P., "Kendall's S Distribution with Ties in One Ranking," *J. Amer. Statist. Assoc.*, 67 (1972), 453–455.

T24 Best, D. J., "Extended Tables for Kendall's Tau," *Biometrika*, 60 (1973), 429–430.

T25 Farlie, D. J. G., "The Performance of Some Correlation Coefficients for a General Bivariate Distribution," *Biometrika*, 47 (1960), 307–323.

T26 Konijn, H. S., "On the Power of Certain Tests for Independence in Bivariate Populations," *Ann. Math. Statist.*, 27 (1956), 300–323. Errata, *Ibid.*, 29 (1958), 935.

T27 Strahan, Robert F., "Assessing Magnitude of Effect from Rank-Order Correlation Coefficients," *Educ. Psychol. Measurement*, 42 (1982), 763–765.

T28 Fieller, E. C., H. O. Hartley, and E. S. Pearson, "Tests for Rank Correlation Coefficients. I," *Biometrika*, 44 (1957), 470–481.

T29 Griffin, H. D., "Graphic Computation of Tau as a Coefficient of Disarray," *J. Amer. Statist. Assoc.*, 53 (1958), 441–447.

T30 Shah, S. M., "A Note on Griffin's Paper (Graphic Computation of Tau as a Coefficient of Disarray)," *J. Amer. Statist. Assoc.*, 56 (1961), 736.

T31 Knight, W. R., "A Computer Method for Calculating Kendall's Tau with Ungrouped Data," *J. Amer. Statist. Assoc.*, 61 (1966), 436–439.

T32 Noether, Gottfried, "Sample Size Determination for Some Common Nonparametric Tests," *J. Amer. Statist. Assoc.*, 82 (1987), 645–647.

T33 Best, D. J., M. A. Cameron, and G. K. Eagleson, "A Test for Comparing Large Sets of Tau Values," *Biometrika*, 70 (1983), 447–453.

T34 Eagleson, G. K., "A Robust Test for Multiple Comparisons of Correlation Coefficients," *Austral. J. Statist.*, 25 (1983), 256–263.

T35 Korn, Edward L., "Kendall's Tau with a Blocking Variable," *Biometrics*, 40 (1984), 209–214.

T36 Taylor, Jeremy M. G., "Kendall's and Spearman's Correlation Coefficients in the Presence of a Blocking Variable," *Biometrics*, 43 (1987), 409–416.

T37 Schumacher, E., "Kendall's Tau Used as a Coefficient of Disarray between Permutations with Unoccupied Places," in B. V. Gnedenko, M. L. Puri, and I. Vincze (eds.), *Nonparametric Statistical Inference*, Vol. II, Amsterdam: North-Holland, 1982.

T38 Wilkie, D., "Pictorial Representation of Kendall's Rank Correlation Coefficient," *Teaching Statist.*, 2 (1980), 76–78.

T39   Silverstone, H., "A Note on the Cumulants of Kendall's *S*-distribution," *Biometrika*, 37 (1950), 231–235.

T40   Noether, G. E., "Why Kendall Tau?" *Teaching Statist.*, 3 (1981), 41–43.

T41   Noether, Gottfried E., *Elements of Nonparametric Statistics*, New York: Wiley, 1967.

T42   Fligner, Michael A., and Steven W. Rust, "On the Independence Problem and Kendall's Tau," *Communic. in Statist.—Theory and Methods*, 12 (1983), 1597–1607.

T43   Samara, Basil, and Ronald H. Randles, "A Test for Correlation Based on Kendall's Tau," *Communic. in Statist.—Theory and Methods*, 17 (1988), 3191–3205.

T44   Olmstead, P. S., and John W. Tukey, "A Corner Test for Association," *Ann. Math. Statist.*, 18 (1947), 495–513.

T45   Kendall, M. G., and B. Babington-Smith, "The Problem of *m* Rankings," *Ann. Math. Statist.*, 10 (1939), 275–287.

T46   Wallis, W. A., "The Correlation Ratio for Ranked Data," *J. Amer. Statist. Assoc.*, 34 (1939), 533–538.

T47   Schucany, W. R., "A Rank Test for Two Group Concordance," (Abstract) *Ann. Math. Statist.*, 42 (1971), 1146.

T48   Schucany, W. R., and W. H. Frawley, "A Rank Test for Two Group Concordance," *Psychometrika*, 38 (1973), 249–258.

T49   Li, Loretta, and William R. Schucany, "Some Properties of a Test for Concordance of Two Groups of Rankings," *Biometrika*, 62 (1975), 417–423.

T50   Hollander, Myles, and Jayaram Sethuraman, "Testing for Agreement between Two Groups of Judges," *Biometrika*, 65 (1978), 403–411.

T51   Kraemer, Helena Chmurra, "Intergroup Concordance: Definition and Estimation," *Biometrika*, 68 (1981), 641–646.

T52   Snell, Martin C., "Recent Literature on Testing for Intergroup Concordance," *Appl. Statist.*, 32 (1983), 134–140.

T53   Baldessari, Bruno, and Francesca Gallo, "On Some Measure of Concordance," *Metron*, 35 (1977), 431–441.

T54   Shirahata, S., "Nonparametric Measures of Intraclass Correlation," *Communic. in Statist.—Theory and Methods*, 11 (1982), 1707–1721.

T55   Palachek, Albert D., and Roger A. Kerin, "Alternative Approaches to the Two-Group Concordance Problem in Brand Preference Rankings," *J. Marketing Res.*, 19 (1982), 386–389.

T56   Costello, Patricia S., and Douglas A. Wolfe, "A New Nonparametric Approach to the Problem of Agreement between Two Groups of Judges," *Communic. in Statist.—Simulation and Computation*, 14 (1985), 791–805.

T57   Ehrenberg, A. S. C., "On Sampling from a Population of Rankers," *Biometrika*, 39 (1952), 82–87.

T58   Hays, W. L., "A Note on Average Tau as a Measure of Concordance," *J. Amer. Statist. Assoc.*, 55 (1960), 331–341.

T59   Alvo, Mayer, and Paul Cabilio, "A Comparison of Approximations to the Distribution of Average Kendall Tau," *Communic. in Statist.—Theory and Methods*, 13 (1984), 3191–3216.

T60   Palachek, Albert D., and William R. Schucany, "On Approximate Confidence Intervals for Measures of Concordance," *Psychometrika*, 49 (1984), 133–141.

T61   Serlin, Ronald C., and Leonard A. Marasuilo, "Planned and Post Hoc Comparisons in Tests of Concordance and Discordance for *G* Groups of Judges," *J. Educ. Statist.*, 8 (1983), 187–205.

T62 Feigin, Paul D., and Mayer Alvo, "Intergroup Diversity and Concordance for Ranking Data: An Approach Via Metrics for Permutations," *Ann. Statist.*, 14 (1986), 691–707.

T63 Lewis, Gordon H., "Kendall's Coefficient of Concordance for Sociometric Rankings with Self Excluded," *Sociometry*, 34 (1971), 496–503.

T64 Linhart, H., "Approximate Test for *m* Rankings," *Biometrika*, 47 (1960), 476–480.

T65 Lyerly, S. B., "The Average Spearman Rank Correlation Coefficient," *Psychometrika*, 17 (1952), 421–428.

T66 Willerman, B., "The Adaptation and Use of Kendall's Coefficient of Concordance (W) to Sociometric-Type Rankings," *Psychol. Bull.*, 52 (1955), 132–133.

T67 Wood, J. T., "A Variance Stabilizing Transformation for Coefficients of Concordance and for Spearman's Rho and Kendall's Tau," *Biometrika*, 57 (1970), 619–627.

T68 Friedman, M., "A Comparison of Alternative Tests of Significance for the Problem of *m* Rankings," *Ann. Math. Statist.*, 11 (1940), 86–92.

T69 Stewart, Robert A., Graham E. Powell, Howard J. Rankin, and S. Jane Tutton, "Concordance Coefficient (W), Correction for the Inequality of Interval in the Underlying Rhos," *Perceptual and Motor Skills*, 40 (1975), 459–462.

T70 Wakimoto, Kazumasa, and Shingo Shirahata, "A Coefficient of Concordance Based on the Chart of Linked Lines," *J. Japan Statist. Soc.*, 14 (1984), 189–197.

T71 Blalock, Hubert M., Jr., *Social Statistics*, revised second edition, New York: McGraw-Hill, 1979.

T72 Maghsoodloo, S., "Estimates of the Quantiles of Kendall's Partial Rank Correlation Coefficient," *J. Statist. Comput. and Simulation*, 4 (1975), 155–164.

T73 Maghsoodloo, S., and L. Laszlo Pallos, "Asymptotic Behavior of Kendall's Partial Rank Correlation Coefficient and Additional Quantile Estimates," *J. Statist. Comput. and Simulation*, 13 (1981), 41–48.

T74 Moran, P. A. P., "Partial and Multiple Rank Correlation," *Biometrika*, 38 (1951), 26–32.

T75 Hawkes, Roland K., "The Multivariate Analysis of Ordinal Measures," *Amer. J. Sociol.*, 76 (1971), 908–926.

T76 Shirahata, S., "Tests of Partial Correlation in a Linear Model," *Biometrika*, 64 (1977), 162–164.

T77 Shirahata, Shingo, "Rank Tests of Partial Correlation," *Bull. Math. Statist., Res. Assoc. Statist. Sci.*, 19 (No. 3–4, 1981), 9–18.

T78 Quade, Dana, "Nonparametric Partial Correlation," in H. M. Blalock, Jr. (ed.), *Measurement in the Social Sciences*, Chicago: Aldine, 1974, pp. 369–398.

T79 Kritzer, Herbert M., "Comparing Partial Rank Order Correlations from Contingency Table Data," *Sociol. Methods & Research*, 8 (1980), 420–433.

T80 Yule, G. Udney, "On the Association of Attributes in Statistics: With Illustrations from the Material of the Childhood Society, &c.," *Philosoph. Transac. Roy. Soc. London, Ser. A*, 194 (1900), 257–319.

T81 Cramér, H., *Mathematical Methods of Statistics*, Princeton, N. J.: Princeton University Press, 1946.

T82 Goodman, L. A., and W. H. Kruskal, "Measures of Association for Cross-Classification," *J. Amer. Statist. Assoc.*, 49 (1954), 732–764. Errata, *Ibid.*, 52 (1957), 578.

T83 Goodman, L. A., and W. H. Kruskal, "Measures of Association for Cross-Classifications. II: Further Discussion and References," *J. Amer. Statist. Assoc.*, 54 (1959), 123–163.

T84 Goodman, L. A., and W. H. Kruskal, "Measures of Association for Cross-Classifications. III: Approximate Sample Theory," *J. Amer. Statist. Assoc.*, 58 (1963), 310–364.

T85   Goodman, Leo A., and William H. Kruskal, "Measures of Association for Cross-Classifications. IV: Simplification of Asymptotic Variances," *J. Amer. Statist. Assoc.*, 67 (1972), 415–421.

T86   Goodman, Leo A., and William H. Kruskal, *Measures of Association for Cross-Classifications*, New York: Springer-Verlag, 1979.

T87   Reynolds, H. T., *Analysis of Nominal Data*, Beverly Hills, Calif.: Sage, 1977.

T88   Freeman, Linton C., "A Further Note on Freeman's Measure of Association," *Psychometrika*, 41 (1976), 273–275.

T89   Janson, Svante, and Jan Vegelius, "The Relationship between the Phi Coefficient and the *G* Index," *Educ. Psychol. Measurement*, 40 (1980), 569–574.

T90   Gans, Lydia P., and C. A. Robertson, "Distribution of Goodman and Kruskal's Gamma and Spearman's Rho in 2 × 2 Tables for Small and Moderate Sample Sizes," *J. Amer. Statist. Assoc.*, 76 (1981), 942–946.

T91   Berry, Kenneth J., and Paul W. Mielke, Jr., "Goodman and Kruskal's TAU-B Statistic: A Nonasymptotic Test of Significance," *Sociol. Methods & Research*, 13 (1985), 543–550.

T92   Walker, Helen M., and Joseph Lev, *Statistical Inference*, New York: Henry Holt, 1953.

T93   Freeman, Linton C., *Elementary Applied Statistics: For Students in Behavioral Science*, New York: Wiley, 1965.

T94   Buck, Jane L., and Stephen L. Finner, "A Still Further Note on Freeman's Measure of Association," *Psychometrika*, 50 (1985), 365–366.

T95   Jolliffe, I. T., "Runs Tests for Detecting Dependence between Two Variables," *Statist.*, 30 (1981), 137–141.

T96   Henry, Neil W., "Jackknifing Measures of Association," *Sociol. Methods & Research*, 10 (1981), 233–240.

T97   Shirahata, S., "Intraclass Rank Tests for Independence," *Biometrika*, 68 (1981), 451–456.

T98   Agresti, Alan, "Measures of Nominal-Ordinal Association," *J. Amer. Statist. Assoc.*, 76 (1981), 524–529.

T99   Shirahata, S., "A Nonparametric Measure of Interclass Correlation," *Communic. in Statist.—Theory and Methods*, 11 (1982), 1723–1732.

T100   Kimeldorf, George, Jerrold H. May, and Allan R. Sampson, "Concordant and Discordant Monotone Correlations and Their Evaluation by Nonlinear Optimization," *TIMS/Studies in Management Sci.*, 19 (1982), 117–130.

T101   Nelson, Lloyd S., "A Sign Test for Correlation," *J. Quality Technol.*, 15 (1983), 199–200.

T102   Shirahata, Shingo, and Takaharu Araki, "Rank Statistics to Measure the Degree of Association," *J. Japan Statist. Soc.*, 14 (1984), 19–28.

T103   Ledwina, Teresa, "Large Deviations and Bahadur Slopes of Some Rank Tests of Independence," *Sankhyā*, Ser. A, Part 2, 48 (1986), 188–207.

T104   Ledwina, Teresa, "On the Limiting Pitman Efficiency of Some Rank Tests of Independence," *J. Multivariate Analysis*, 20 (1986), 265–271.

T105   Bajorski, Piotr, "Local Bahadur Optimality of Some Rank Tests of Independence," *Statist. & Probabil. Letters*, 5 (1987), 255–262.

T106   Jupp, P. E., "A Nonparametric Correlation Coefficient and a Two-Sample Test for Random Vectors or Directions," *Biometrika*, 74 (1987), 887–890.

T107   Yule, G. U., and M. G. Kendall, *An Introduction to the Theory of Statistics*, fourteenth edition, New York: Hafner, 1950.

T108   McNemar, Q., *Psychological Statistics*, third edition, New York: Wiley, 1962.

T109 Stuart, A., "The Estimation and Comparison of Strengths of Association in Contingency Tables," *Biometrika*, 40 (1953), 105–110.

T110 Ives, K. H., and J. D. Gibbons, "A Correlation Measure for Nominal Data," *Amer. Statist.*, 21 (December 1967), 16–17.

T111 Blomqvist, Nils, "On a Measure of Dependence between Two Random Variables," *Ann. Math. Statist.*, 21 (1950), 593–600.

T112 Bhuchongkul, S., "A Class of Nonparametric Tests for Independence in Bivariate Populations," *Ann. Math. Statist.*, 35 (1964), 138–149.

T113 Fieller, E. C., and E. S. Pearson, "Tests for Rank Correlation Coefficients. II," *Biometrika*, 48 (1961), 29–40.

T114 Bell, C. B., and K. A. Doksum, "Distribution-Free Tests of Independence," *Ann. Math. Statist.*, 38 (1967), 429–446.

T115 Holley, Jasper W., and Ulf Eriksson, "A Note on the Effect of Selective Sampling Procedures on the Phi Coefficient," *Multivariate Behav. Res.*, 5 (1970), 117–123.

T116 Hotelling, Harold, and Margaret Pabst, "Rank Correlation and Tests of Significance Involving No Assumption of Normality," *Ann. Math. Statist.*, 7 (1936), 29–43.

T117 Moran, P. A. P., "Recent Developments in Ranking Theory," *J. Roy. Statist. Soc., Ser. B*, 12 (1950), 153–162.

T118 Davis, J. A., "A Partial Coefficient for Goodman and Kruskal's Gamma," *J. Amer. Statist. Assoc.*, 62 (1967), 189–193.

T119 Jumars, Peter A., "Rank Correlation and Concordance Tests in Community Analyses: An Inappropriate Null Hypothesis," *Ecology*, 61 (1980), 1553–1554.

T120 Turek, Richard J., and Ronald C. Suich, "An Exact Test on the Goodman–Kruskal λ for Prediction on a Dichotomy," *J. Roy. Statist. Soc., Ser. B*, 45 (1983), 373–379.

T121 Papaioannou, Takis, and Sotiris Loukas, "Inequalities on Rank Correlation with Missing Data," *J. Roy. Statist. Soc., Ser. B*, 46 (1984), 68–71.

T122 Jewell, Nicholas P., "On the Bias of Commonly Used Measures of Association for 2×2 Tables," *Biometrics*, 42 (1986), 351–358.

T123 Bajorski, Piotr, and Teresa Ledwina, "Large Deviations and Bahadur Efficiency of Some Rank Tests of Independence," in P. Bauer, F. Konecny, and W. Wertz (eds.), *Mathematical Statistics and Probability Theory, Vol. B, Statistical Inference and Methods*, Norwell, Mass.: D. Reidel, 1987, pp. 11–23.

T124 Somers, R. H., "The Rank Analogue of Product–Moment Partial Correlation and Regression, with Application to Manifold, Ordered Contingency Tables," *Biometrika*, 46 (1959), 241–246.

T125 Somers, Robert H., "A New Asymmetric Measure of Association for Ordinal Variables," *Amer. Sociol. Rev.*, 27 (1962), 799–811.

T126 Somers, Robert H., "Simple Measures of Association for the Triple Dichotomy," *J. Roy. Statist. Soc., Ser. A*, 127 (1964), 409–415.

T127 Somers, Robert H., "Analysis of Partial Rank Correlation Measures Based on the Product–Moment Model: Part One," *Social Forces*, 53 (1974).

T128 Carroll, John B., "The Nature of the Data, or How to Choose a Correlation Coefficient," *Psychometrika*, 26 (1961), 347–372.

T129 Costner, Herbert L., "Criteria for Measures of Association," *Amer. Sociol. Rev.*, 30 (1965), 341–353.

T130 Weisberg, Herbert F., "Models of Statistical Relationship," *Amer. Polit. Sci. Rev.*, 68 (1974), 1638–1655.

T131 Kempi, Viktor, "A FORTRAN Program for Ranking and for Calculation of Spearman's Correlation Coefficient," *Comput. Methods & Programs in Biomed.*, 21 (1985), 123–125.

T132 Zar, Jerrold H., "Probabilities for Spearman Rank Correlation Coefficients," *Behav. Res. Methods and Instrumentation*, 6 (1974), 357.

T133 Berry, Kenneth J., and Paul W. Mielke, Jr., "Goodman and Kruskal's Tau-*b* Statistic: A FORTRAN-77 Subroutine," *Educ. Psychol. Measurement*, 46 (1986), 645–649.

T134 Kimeldorf, George, Jerrold H. May, and Allan R. Sampson, "User's Manual: MONCAR—A Program to Compute Concordant and Other Monotone Correlations," Working Paper Series, Graduate School of Business, University of Pittsburgh, 1980, 26 pages.

T135 Kimeldorf, George, Jerrold H. May, and Allan R. Sampson, "MONCAR—A Program to Compute Corcordant and Other Monotone Correlations," in William F. Eddy (ed.), *Computer Science and Statistics: Proceedings of the 13th Symposium on the Interface*, New York: Springer-Verlag, 1981, pp. 348–351.

T136 Brophy, Alfred L., "An Algorithm and Program for Calculation of Kendall's Rank Correlation Coefficient," *Behav. Res. Methods, Instruments, & Computers*, 18 (1986), 45–46.

T137 Galla, John P., "Kendall's tau and Kendall's Partial Correlation: Two BASIC Programs for Microcomputers," *Behav. Res. Methods, Instruments, & Computers*, 19 (1987), 55–56.

T138 Dunlap, William P., and Edward R. Kemery, "Biserial and Point-Biserial Correlation with Correction for Nonoptimal Dichotomies," *Behav. Res. Methods, Instruments, & Computers*, 20 (1988), 420–422.

E1 Pincherle, G., and D. Robinson, "Mean Blood Pressure and Its Relation to Other Factors Determined at a Routine Executive Health Examination," *J. Chronic Dis.*, 27 (1974), 245–260.

E2 Bakos, Gustav A., "Photoelectric Observations of Comet Bennett," *J. Roy. Astron. Soc. Can.*, 67 (1973), 183–189.

E3 Daniel, Jean, "A Study of the Relationship between Social Dominance and Intelligence in Albino Mice," unpublished research report, 1975.

E4 Alfrey, Allen C., Nancy L. Miller, and Donald Butkus, "Evaluation of Body Magnesium Stores," *J. Lab. Clin. Med.*, 84 (1974), 153–162.

E5 Gentry, James, and John Pike, "An Empirical Study of the Risk–Return Hypothesis Using Common Stock Portfolios of Life Insurance Companies," *J. Finan. Quant. Anal.*, 5 (1970), 179–185.

E6 Cravens, David W., and Robert B. Woodruff, "An Approach for Determining Criteria of Sales Performance," *J. Appl. Psychol.*, 57 (1973), 242–247.

E7 Krippner, Stanley, "Correlates of Reading Improvement," *J. Devel. Reading*, 7 (1963) 29–39.

E8 Johnson, Betty M., "Decision Making, Faculty Satisfaction, and the Place of the School of Nursing in the University," *Nursing Res.*, 22 (1973), 100–107.

E9 Chaiklin, Harris, and Carol Landau Frank, "Separation, Service Delivery and Family Functioning," *Public Welfare*, 31 (Winter 1973), 2–7.

E10 Pierce, E. T., "The Charge Transferred to Earth by a Lightning Flash," *J. Franklin Inst.*, 286 (1968), 353–354.

E11 Meese, A. D., and W. H. Evans, "Charge Transfer in the Lightning Stroke as Determined by the Magnetograph," *J. Franklin Inst.*, 273 (1962), 375–382.

E12 Murgatroyd, B., "An Investigation into the Practical Capacity of Roundabout Weaving Sections," *Highway Engineer*, 20 (March 1973), 6–13.

E13 de la Peña, A., V. Zarcone, and W. C. Dement, "Correlation between Measures of the Rapid Eye Movements of Wakefulness and Sleep," *Psychophysiology*, 10 (1973), 488–500.

E14 Clarke, H. Courtney, "Relationship between Whole-Blood Riboflavin Levels in the Mother and in the Prenate," *Amer. J. Obstet. Gynecol.*, 111 (1971), 43–46.

E15 Brook, C. G. D., "Determination of Body Composition of Children from Skinfold Measurements," *Arch. Dis. Child.*, 46 (1971), 182–184.

E16 Robbins, Richard, Michael Blumenthal, and Miles A. Galin, "Reduction of Vitreous Weight by Ocular Massage," *Amer. J. Ophthalmol.*, 69 (1970), 603–607.

E17 Jensen, Arthur R., "IQs of Identical Twins Reared Apart," *Behav. Genet.*, 1 (1970), 133–148.

E18 Burt, C., "The Genetic Determination of Differences in Intelligence: A Study of Monozygotic Twins Reared Together and Apart," *Br. J. Psychol.*, 57 (1966), 137–153.

E19 Bhatia, M. L., S. C. Manchanda, and Sujoy B. Roy, "Coronary Haemodynamic Studies in Chronic Severe Anaemia," *Br. Heart J.*, 31 (1969), 365–374.

E20 Poland, Alan, Donald Smith, R. Kuntzman, M. Jacobson, and A. H. Conney, "Effect of Intensive Occupational Exposure to DDT on Phenylbutazone and Cortisol Metabolism in Human Subjects," *Clin. Pharmacol. Ther.*, 11 (1970), 724–732.

E21 Goby, Marshall J., William J. Filstead, and Jean J. Rossi, "Structural Components of an Alcoholism Treatment Program," *Quart. J. Studies on Alcohol*, 35 (1974), 1266–1271.

E22 Vine, Ian, "Stereotypes in the Judgement of Personality from Handwriting," *Br. J. Soc. Clin. Psychol.*, 13 (1974), 61–64.

E23 Viney, Wayne, Ross Loomis, Jacob Hautaluoma, and Stanley Wagner, "A Comparison of Perceived Organizational Influence in Two Metropolitan Communities," *Rocky Mountain Soc. Sci. J.*, 11 (January 1974) 81–86.

E24 Robinson, Jack L., "Faculty Compensation in the Big Eight, 1971–1972," *Oklahoma Bus. Bull.*, 40 (September 1972), 10–12.

E25 Gibson, Lay James, and Richard W. Reeves, "The Roles of Hinterland Composition, Externalities, and Variable Spacing as Determinants of Economic Structure in Small Towns," *Professional Geographer*, 26 (1974), 152–158.

SIMPLE LINEAR
REGRESSION ANALYSIS

Regression analysis is one of the more widely used statistical techniques available to the researcher or decision maker. In the simple linear parametric case, the researcher usually uses the method of least squares in fitting regression lines to observed sample data, and inferences regarding population parameters are based on fairly rigid assumptions. When these assumptions are met, the familiar parametric inferential procedures are the most appropriate ones to use. If, however, the assumptions are violated, the use of inferential procedures based on them may yield misleading results.

This chapter covers some alternative nonparametric procedures that are useful in simple linear regression analysis when the investigator is not able or willing to make the assumptions necessary for the valid application of analogous parametric techniques.

10.1

FITTING THE REGRESSION LINE

Suppose that we have a sample of n pairs of observations

$$(X_1, Y_1), (X_2, Y_2), \ldots, (X_n, Y_n)$$

on the continuous variables X and Y, where each pair of observations (X_i, Y_i) is a measurement on the same (the ith) unit of association. We wish to fit to the data a regression line of the form

$$Y = a + bX \tag{10.1}$$

where a is the y intercept and b is the slope of the line obtained.

BROWN–MOOD METHOD FOR FINDING SLOPE AND y INTERCEPT

Brown and Mood (T1) and Mood (T2) describe a method for determining a and b. By this method, we first divide the Y values into two groups: (1) those occurring with X values that are less than or equal to the median of X, and (2) those occurring with values of X that are greater than the median of X. Then the desired values of a and b are those that yield a line for which the median of the deviations about the line is zero in each of the two groups.

To find a and b, we proceed as follows:

1. Prepare a scatter diagram of the sample data.
2. Draw a vertical line through the median of the X values. If one or more points fall on this median line, shift it to the left or right as necessary so that the number of points on either side of the median is as nearly equal as possible.
3. Determine the median of X and the median of Y in each of the two groups of observations formed in Step 2. That is, compute a total of four medians.
4. In the first group of observations, plot a point representing the intersection of the median of X with the median of Y. Plot a similar point for the second group of observations.
5. Draw a line connecting the two points plotted in Step 4. This line is a first approximation to the desired line.
6. If the median of the vertical deviations of the points from this line is not zero in both groups, shift the line to a new position until it is clear that the deviations in each group have a median of zero. This may be accomplished more conveniently by using a transparent ruler. If greater accuracy is needed, the iterative procedure described by Mood (T2) may be used.

7. The value of a is given by the y intercept of the final line and

$$b = \frac{Y_1 - Y_2}{X_1 - X_2}$$

(10.2)

where (X_1, Y_1) and (X_2, Y_2) are the coordinates of any two points on the line.

The following example illustrates the procedure.

Example 10.1   Clark et al. (E1) examined the fat filtration characteristics of a packed polyester-and-wool filter used in the arterial lines during clinical hemodilution. They collected data on the filter's recovery of solids for 10 patients who underwent surgery. Table 10.1 shows removal rates of lipids and cholesterol. We wish to fit a regression line to the data, where we treat cholesterol as the Y variable and lipids as the X variable.

TABLE 10.1   **Arterial line filter recovery data for 10 surgery patients**

| Patient | Removal rates, mg/kg/L $\times$ 10^{-2} | |
	Lipids (X)	Cholesterol (Y)
1	3.81	1.90
2	2.10	1.03
3	0.79	0.44
4	1.99	1.18
5	1.03	0.62
6	2.07	1.29
7	0.74	0.39
8	3.88	2.30
9	1.43	0.93
10	0.41	0.29

Source: Richard E. Clark, Harry W. Margraf, and Richard A. Beauchamp, "Fat and Solid Filtration in Clinical Perfusions," *Surgery,* 77 (1975), 216–224.

The median of the X values is 1.71, which divides the observations into two groups of five each. Below the median we have the observations $(0.79, 0.44)$, $(1.03, 0.62)$, $(0.74, 0.39)$, $(1.43, 0.93)$, and $(0.41, 0.29)$. Above the median are $(3.81, 1.90)$, $(2.10, 1.03)$, $(1.99, 1.18)$, $(2.07, 1.29)$, and $(3.88, 2.30)$. The medians of the X's and Y's below the median are 0.79 and 0.44, respectively. Above the median of X, the medians are 2.10 and 1.29 for X and Y, respectively. These data yield a first approximation to b of $b' = (1.29 - 0.44)/(2.10 - 0.79) = 0.6489$. Since the medians of the observations about the resulting line are not zero in the two groups, we adjust the

FIGURE 10.1

Relationship between the rates of cholesterol and total lipids removed in Example 10.1

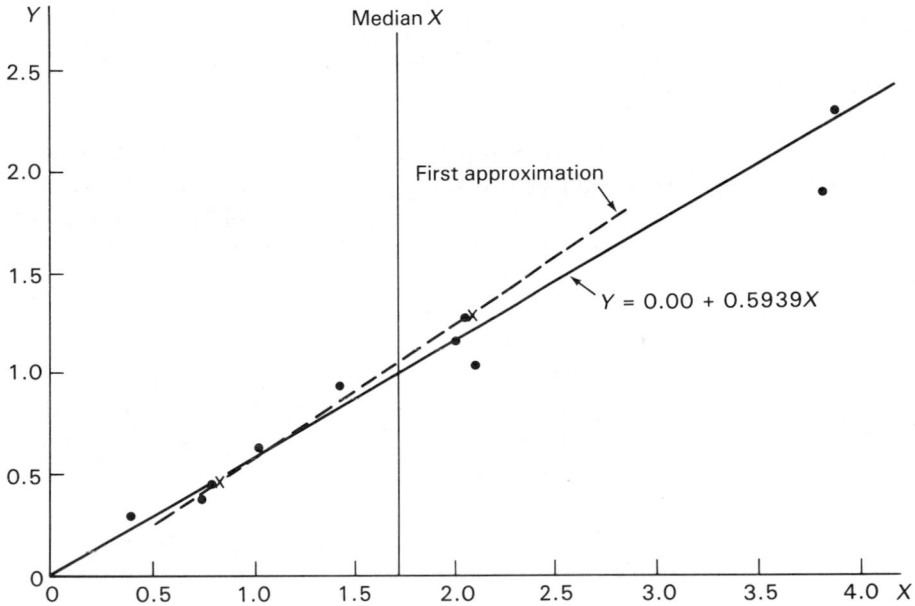

line visually. This procedure leads to a final value of $b = 0.5939$ and $a = 0.00$. Thus the equation for the line is $Y = 0.00 + 0.5939X$.

Figure 10.1 shows the scatter diagram of the original data for this example, the median of X, the first approximation to the regression line, and the final regression line.

The regression line computed from sample data in this example describes the linear relationship between the rates of cholesterol and total lipids removed in the sample. This sample line provides an estimate of the regression line that describes the linear relationship between the variables in the population from which the sample was drawn. The y intercept α and the slope β of the population regression line are estimated by a and b, respectively.

THEIL'S METHOD FOR FINDING SLOPE

The methods suggested by Brown and Mood are quick and dirty. Perhaps of more use to most researchers is the method proposed by Theil (T3) for obtaining a point estimate of the slope coefficient β. We assume that the data conform to the classical regression model

$$Y_i = \alpha + \beta X_i + e_i, \qquad i = 1,\dots,n$$

where the X_i's are known constants, α and β are unknown parameters, and Y_i is an observed value of the continuous random variable Y at the value X_i. For each value of X_i, we assume a subpopulation of Y values, and the e_i's are mutually independent. The X_i's are all distinct (no ties), and we take $X_1 < X_2 < \cdots < X_n$.

The data consist of n pairs of observations,

$$(X_1, Y_1), (X_2, Y_2), \ldots, (X_i, Y_i), \ldots, (X_n, Y_n)$$

where the ith pair represents measurements taken on the ith unit of association.

To obtain an estimator for β, we first form all possible sample slopes $S_{ij} = (Y_j - Y_i)/(X_j - X_i)$, where $i < j$. There will be $N = \binom{n}{2}$ values of S_{ij}. The estimator of β, which we designate by $\hat{\beta}$, is the median of the S_{ij} values; that is,

$$\hat{\beta} = \text{median } \{S_{ij}\} \tag{10.3}$$

Example 10.2   Yunginger and Gleich (E2) studied the sequential changes in the serum and nasal immunoglobulin E (IgE) concentrations in nonallergic adults and in adults with seasonal allergic rhinitis. They collected the data shown in Table 10.2. The table contains IgE and nasal total protein (TP) measurements taken on 10 patients with histories of typical ragweed hay fever. Because of the nature of the treatment received, these 10 patients made up the "low-dose-treated" group in the study. The data were collected during August.

TABLE 10.2   **Nasal TP and IgE determinations on 10 patients with ragweed hay fever**

Y Nasal TP ($\mu g/ml$)	206	453	141	131	203	172	153	356	297	425
X Nasal IgE/TP ($\times 10^{-7}$)	9.7	408	106	7.5	49.3	5.8	6.5	2.8	16.8	7.0

Source: John W. Yunginger and Gerald J. Gleich, "Seasonal Changes in Serum and Nasal IgE Concentrations," *J. Allergy Clin. Immunol.*, 51 (1973), 174–186.

We wish to compute $\hat{\beta}$, the estimate of the slope β of the population regression line that describes the linear relationship between the two variables. The $N = \binom{10}{2} = 45$ ordered values of S_{ij} are shown in Table 10.3. The median of these S_{ij} values is $+0.070$. Therefore the estimate of β is $\hat{\beta} = +0.070$. Discussions of the properties of this estimator can be found in the references cited in Section 10.2.

TABLE 10.3   **Ordered array of sample slopes S_{ij} for Example 10.2**

−588.000	−27.143	−5.248	−2.083	−0.121	0.399	0.747	8.718	16.562
−81.111	−24.118	−4.214	−1.749	−0.076	0.620	0.804	11.364	17.849
−61.333	−22.000	−3.290	−1.094	0.070	0.697	1.033	12.817	34.091
−54.865	−21.739	−2.892	−0.675	0.102	0.699	1.168	13.981	210.833
−47.872	−13.061	−2.869	−0.309	0.239	0.713	1.722	16.429	544.000

10.2

TESTING HYPOTHESES ABOUT α AND β

Researchers are often interested in testing hypotheses about one or both of the parameters α and β. This section presents a method for testing simultaneously the null hypothesis that $\alpha = \alpha_0$ and $\beta = \beta_0$, and two methods for testing the null hypothesis that $\beta = \beta_0$.

BROWN–MOOD METHOD

The following method for testing hypotheses about α and β is described by Brown and Mood (T1) and Mood (T2).

Assumption

The data consist of n pairs of observations $(X_1, Y_1), (X_2, Y_2), \ldots, (X_n, Y_n)$ on the continuous variables X and Y, where each pair of observations (X_i, Y_i) is a measurement on the same unit of association.

Hypotheses

$$H_0: \alpha = \alpha_0, \beta = \beta_0, \qquad H_1: \alpha \neq \alpha_0 \quad \text{and/or} \quad \beta \neq \beta_0$$

Test Statistic

To compute the test statistic, we proceed as follows:

1. Plot the data points as a scatter diagram.
2. Draw the line $Y = \alpha_0 + \beta_0 X$ on the scatter diagram.
3. Draw a vertical line on the scatter diagram through the median of the X values.
4. Let $n_1 = $ the number of data points above the hypothesized regression line and to the left of the vertical line drawn through the median of the X values. Let $n_2 = $ the number of data points above the hypothesized regression line and to the right of the vertical line drawn through the median of the X values.

Mood (T2) points out that both n_1 and n_2 have the binomial distribution with parameter 0.5, a fact that forms the basis for the test statistic.

The test statistic is

$$X^2 = \frac{8}{n}\left[\left(n_1 - \frac{n}{4}\right)^2 + \left(n_2 - \frac{n}{4}\right)^2\right]$$

(10.4)

which is distributed approximately as chi-square with two degrees of freedom when H_0 is true and n is not too small. Tate and Clelland (T4) state that the approximation tends to be good for practical work when n is about 10 or more.

Decision Rule

If the computed value of X^2 exceeds the tabulated value of chi-square for two degrees of freedom and a chosen level of significance, we can reject the null hypothesis at that level of significance.

Example 10.3

To illustrate the procedure, let us use the data of Example 10.1 and test the null hypothesis that $\alpha = 0$ and $\beta = 0.5$.

Hypotheses

$$H_0\colon \alpha = 0, \beta = 0.5, \qquad H_1\colon \alpha \neq 0, \beta \neq 0.5$$

Test Statistic

Figure 10.2 shows the scatter diagram, the hypothesized regression line, and the line drawn through the median of the X values. It shows that $n_1 = 5$ and $n_2 = 3$, so we

---

FIGURE 10.2 Scatter diagram for Example 10.3

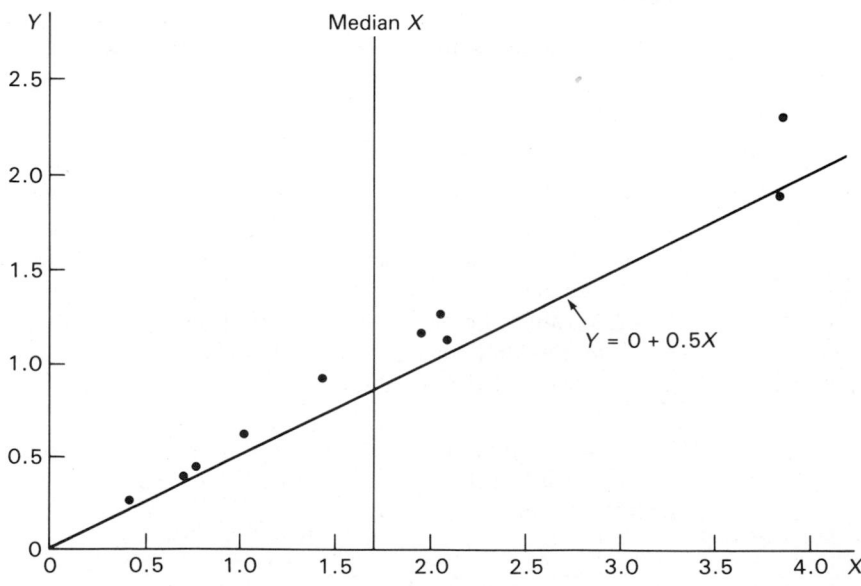

compute

$$X^2 = \frac{8}{10}\left[\left(5 - \frac{10}{4}\right)^2 + \left(3 - \frac{10}{4}\right)^2\right] = 5.2$$

Table A.11 reveals that when H_0 is true, the probability of obtaining a value of X^2 as large as or larger than 5.2 is greater than 0.05. We do not reject H_0, and we conclude that the sample may have come from a population in which the regression line has a slope of 0.5 and an intercept of 0.

In regression analysis we are usually more interested in β, the slope of the population regression line. When we wish to test a hypothesis about β only— namely, $H_0: \beta = \beta_0$ against $H_1: \beta \neq \beta_0$—we may use the following procedure described by Brown and Mood (T1) and Mood (T2).

1. Plot the data points as a scatter diagram.
2. Draw a vertical line through the median of the X values.
3. Fit the line $Y = a + \beta_0 X$ to the data, where a is the median of the deviations $Y_i - \beta_0 X$ for all observed values of Y and β_0 is the hypothesized value of β. Usually we may determine this line quite easily by plotting the line $Y = \beta_0 X$ and drawing a line parallel to $Y = \beta_0 X$, which divides the points into two equal groups.
4. Count the number of points n_1 that are above the line $Y = a + \beta_0 X$ and to the left of the median of the X values.

The test statistic is

$$X_b^2 = \frac{16}{n}\left(n_1 - \frac{n}{4}\right)^2 \qquad \textbf{(10.5)}$$

If H_0 is true, the test statistic is distributed approximately as chi-square with one degree of freedom, provided n is fairly large. Tate and Clelland (T4) recommend using the chi-square approximation for n of 20 or larger.

The following example illustrates the procedure for testing a hypothesis about β alone.

Example 10.4 Pilkey and Hower (E3) examined the changes in concentration of magnesium and strontium in the tests of a recent echinoid species collected from varying environments over much of its geographical range. They used X-ray techniques to analyze the magnesium and strontium in specimens of *Dendraster excentricus*, the common Pacific Coast sand dollar, collected from 24 localities between Vancouver Island, British Columbia, and Santa Rosalia Bay, Baja California, and calculated the percentage of calcium. Table 10.4 shows the mean summer temperature (X) at the 24 locations and the mean percent $MgCO_3$ content (Y) of the specimens

collected. We wish to test the null hypothesis that $\beta = 0$ in the regression line for the population represented by the sample.

TABLE 10.4   **Data on *Dendraster excentricus* for Example 10.4**

Location number	1	2	3	4	5	6	7	8
Summer mean temperature, °C (X)	23.0	18.7	17.5	21.0	20.0	19.0	15.3	14.0
Mean percent $MgCO_2$ (Y)	9.5	9.0	9.2	9.2	9.4	9.3	9.0	8.5
Location number	9	10	11	12	13	14	15	16
Summer mean temperature, °C (X)	14.0	13.7	13.3	13.6	13.1	13.0	13.6	14.2
Mean percent $MgCO_2$ (Y)	9.0	8.4	8.8	8.9	8.5	8.7	8.6	8.7
Location number	17	18	19	20	21	22	23	24
Summer mean temperature, °C (X)	13.9	14.8	14.2	13.0	16.1	15.9	13.0	11.7
Mean percent $MgCO_2$ (Y)	8.5	9.1	9.1	8.0	8.1	8.5	8.4	8.7

Source: Orrin H. Pilkey and John Hower, "The Effect of Environment on the Concentration of Skeletal Magnesium and Strontium in *Dendraster*," *J. Geol.*, 68 (1960), 203–214. Copyright 1960, the University of Chicago.

Figure 10.3 shows the scatter diagram and the median of the X values, which is 14.1. Since $\beta_0 = 0$, the line $Y = \beta_0 X$ is the X axis. The figure also shows the line parallel to the X axis drawn through the points in such a way as to divide them into two equal groups.

In Figure 10.3 there are $n_1 = 3$ points above the line and to the left of the median of X. Then, by Equation 10.5, we may compute

$$X_b^2 = \frac{16}{24}\left(3 - \frac{24}{4}\right)^2 = 6.00$$

Table A.11 reveals that the probability of observing a value of X_b^2 as large as 6 due to chance alone when H_0 is true is less than 0.025.

FIGURE 10.3   Scatter diagram showing relationship between mean $MgCO_3$ of *Dendraster excentricus* tests and mean summer temperature

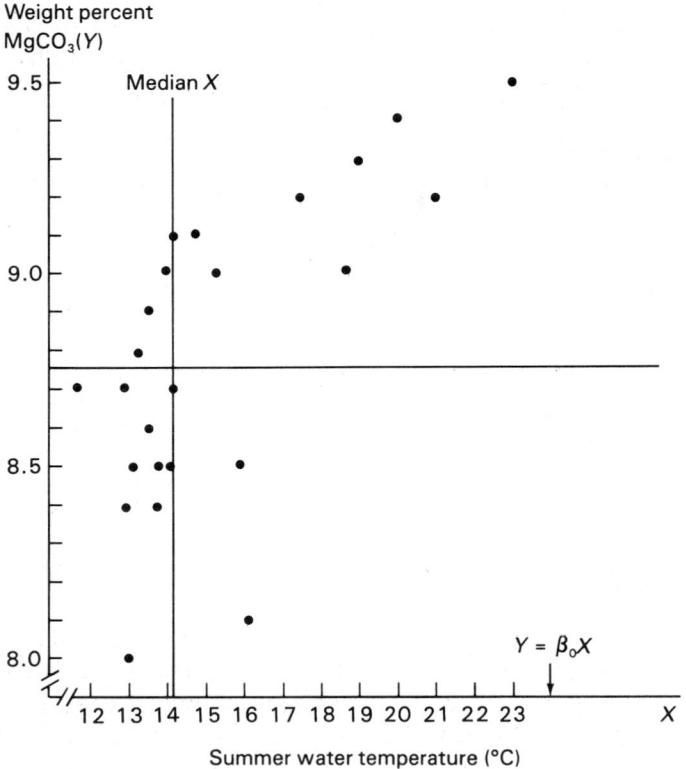

Weight percent $MgCO_3(Y)$

Summer water temperature (°C)

THEIL'S METHOD

Another method for testing $H_0: \beta = \beta_0$ has been proposed by Theil (T5). The procedure is based on the Kendall tau statistic.

Assumptions

   A. The appropriate model is

$$Y_i = \alpha + \beta X_i + e_i, \qquad i = 1,\dots,n$$

   where the X_i's are known constants and α and β are unknown parameters.

   B. For each value of X_i there is a subpopulation of Y values.

   C. Y_i is an observed value of the continuous random variable Y at the value X_i.

D. The X_i are all distinct (no ties), and we take $X_1 < X_2 < \cdots < X_n$.

E. The e_i's are mutually independent and come from the same continuous population.

The data available for analysis consist of n pairs of observations, (X_1, Y_1), $(X_2, Y_2), \ldots, (X_i, Y_i), \ldots, (X_n, Y_n)$, where the ith pair represents measurements taken on the ith unit of association.

Hypotheses

A. (Two-sided): $H_0: \beta = \beta_0$, $H_1: \beta \neq \beta_0$

B. (One-sided): $H_0: \beta \leq \beta_0$, $H_1: \beta > \beta_0$

C. (One-sided): $H_0: \beta \geq \beta_0$, $H_1: \beta < \beta_0$

Test Statistic

As mentioned earlier, the procedure described here is based on the Kendall tau statistic. Specifically we compute the Kendall statistic by comparing all possible pairs of observations of the form $(X_i, Y_i - \beta_0 X_i)$, in the same manner as described in Chapter 9. We may summarize the procedure as follows (see Chapter 9 for a more detailed description).

1. Arrange the pairs of observations $(X_i, Y_i - \beta_0 X_i)$ in a column in natural order with respect to the X values.

2. Compare each $Y_i - \beta_0 X_i$ with each $Y_j - \beta_0 X_j$ appearing below it.

3. Let P be the number of such comparisons that result in a pair $(Y_i - \beta_0 X_i, Y_j - \beta_0 X_j)$ that is in natural order, and let Q be the number of such comparisons that result in a pair that is in reverse natural order.

4. Let $S = P - Q$. The test statistic, then, is

$$\hat{\tau} = \frac{S}{n(n-1)/2} \tag{10.6}$$

Decision Rule

The decision rules for the three sets of hypotheses stated earlier are as follows:

A. (Two-sided): Refer to Table A.22. Reject H_0 at the α level of significance if the computed value of $\hat{\tau}$ is either positive and larger than the τ^* entry for n and $\alpha/2$, or negative and smaller than the negative of the τ^* entry for n and $\alpha/2$.

B. (One-sided): Refer to Table A.22. Reject H_0 at the α level of significance if the computed value of $\hat{\tau}$ is positive and larger than the τ^* entry for n and α.

C. (One-sided): Refer to Table A.22. Reject H_0 at the α level of significance if the computed value of $\hat{\tau}$ is smaller than the negative of the τ^* entry for n and α.

Example 10.5 Grundy and Metzger (E4) describe the methodology involved in a physiological procedure for estimating hepatic secretion of biliary lipids in man. The authors report the data on hourly output of biliary bile acids and biliary phospholipids shown in Table 10.5. We wish to test the null hypothesis that $\beta = 0$ in the population regression equation describing the linear relationship between the two variables.

TABLE 10.5 **Hourly output during the steady state of biliary bile acids and biliary phospholipids in 10 subjects**

Biliary bile acids (mg/hr) X	940	594	1200	1440	1112
Biliary phospholipids (mg/hr) Y	311	391	414	542	387
Biliary bile acids (mg/hr) X	625	1385	1035	931	742
Biliary phospholipids (mg/hr) Y	485	502	458	345	346

Source: Scott M. Grundy and Allan L. Metzger, "A Physiological Method for Estimation of Hepatic Secretion of Biliary Lipids in Man," *Gastroenterology*, 62 (1972), 1200–1217, Copyright 1972, Williams & Wilkins, Baltimore.

Hypotheses

$H_0: \beta = 0, \qquad H_1: \beta \neq 0$

Test Statistic

The first step in calculating $\hat{\tau}$ is to compute $Y_i - \beta_0 X_i$ for each pair of observations. Since $\beta_0 = 0$, $Y_i - \beta_0 X_i$ equals Y_i for each pair. The remainder of the steps described earlier for calculating $\hat{\tau}$ are summarized for this example in Table 10.6. From the data of this table, we compute $S = 30 - 15 = 15$. Finally, by Equation 10.6, we have

$$\hat{\tau} = \frac{15}{10(9)/2} = 0.33$$

Decision

Table A.22 reveals that when H_0 is true and $n = 10$, the probability of observing a value of $\hat{\tau}$ as large as 0.33 is approximately 0.20. This is not very persuasive evidence in favor of rejecting H_0. We conclude, therefore, that the slope of the population regression line may very well be zero.

TABLE 10.6   **Work table for computing $\hat{\tau}$ for Example 10.5**

x_i	$Y_i - \beta_0 X_i$	Number of $Y_i - \beta_0 X_i$ pairs in natural order	Number of $Y_i - \beta_0 X_i$ pairs in reverse natural order
594	391	5	4
625	485	2	6
742	346	5	2
931	345	5	1
940	311	5	0
1035	458	2	2
1112	387	3	0
1200	414	2	0
1385	502	1	0
1440	542	0	0
		$P = 30$	$Q = 15$

Ties   If there are ties in the Y variable, proceed as described. In the presence of ties, however, the results are approximate rather than exact. Sen (T6) describes a procedure based on Kendall's tau that allows for ties in the X values.

Power-Efficiency   Properties of the estimator on which this test is based, including efficiency, are discussed by Sen (T6).

FURTHER READING

For the case of simple linear regression, Hill (T7) has studied the theoretical properties of the estimators proposed by Brown and Mood (T1, T2). Adichie (T8) also comments on the Brown–Mood estimators and proposes a test of linearity against an alternative of convexity that makes use of the Brown–Mood procedure for obtaining a straight line. Kildea (T9, T10) introduced a class of modified Brown–Mood estimators incorporating weights and generalizes the procedure to the multiple-regression case.

A discussion of the extension of Theil's test to the case involving two explanatory variables may be found in the paper by Domański (T11). The author also compares the power of the extended test to that of the Fisher F test.

Comparisons among various nonparametric regression methods have been made by Gross and Tomberlin (T12) and Hussain and Sprent (T13).

Cunningham (T14) discusses four multiple monotone regression models that result from allowing the independent variables to be measured on nominal or ordinal scales and allowing the function that combines the independent variables to be either additive or merely order-preserving for each variable. The author discusses the relations among the models and the computational methods involved. By way of example, the models are applied to real data.

A brief review of robust multiple regression is contained in a paper by Andrews (T15), and additional techniques are proposed and discussed.

Hájek (T16) discusses the extension of the Kolmogorov–Smirnov test to regression alternatives.

Sign and Wilcoxon tests for linearity are proposed by Olshen (T17).

Cleveland and McGill (T18) show how to make scatter plots more powerful through the incorporation of additional graphical information.

Other papers of interest in the area of nonparametric regression analysis include those by Adichie (T19), Daniels (T20), Hogg and Randles (T21), Ghosh and Sen (T22), Jurečková (T23, T24), Konijn (T25), Koul (T26), Samanta (T27), Srivastava and Saleh (T28), Révész (T29), Wegman (T30), Cheng and Lin (T31), Brown and Maritz (T32), Gore and Rao (T33), Rao and Gore (T34), Bhattacharya (T35), Bhattacharya et al. (T36), Hettmansperger and McKean (T37), Henderson (T38), Brown (T39), Lancaster and Quade (T40), Cheng and Cheng (T41), and Whitney (T42).

A book by Eubank (T43) is devoted to the subject of spline smoothing and nonparametric regression. Rousseeuw and Leroy (T44) have written a book on robust regression and outlier detection. Nonparametric methods in general linear models by Puri and Sen (T45) emphasizes rank-based procedures. Also of interest is the doctoral thesis by Drummond (T46). Daniel (T47) and Collomb (T48) have published bibliographies on nonparametric regression procedures.

EXERCISES

10.1 Lue et al. (E5) discuss a method for determination of pulmonary blood volume in congenital heart disease. As part of their investigation, the authors collected the data on infants and children who underwent cardiac catheterization, as shown in Table 10.7 (see p. 440). Use the Brown–Mood method to test $H_0: \alpha = 5$, $\beta = -0.25$, and $H_0: \beta = -0.25$.

10.2 Chen et al. (E6) point out that lateral displacement of the nipples has long been quoted as being closely associated with Turner's syndrome. In an investigation of this phenomenon, Chen et al. (E6) measured the internipple distance and other characteristics in 698 children. Thirty-nine of their subjects were diagnosed as having Turner's syndrome. Table 10.8 (see p. 440) shows the heights and internipple (ID) distances of these 39 subjects. Use the Brown–Mood method to test $H_0: \alpha = 4$, $\beta = 0.1$ and $H_0: \beta = 0.1$

10.3 Dondero et al. (E7) point out that investigations on experimental animals have shown that testosterone stimulates the production and secretion of citric acid in the accessory glands of the male genital tract. The "citric-acid test," by which the level of citric acid provides an index of testicular secretion of androgens in humans, is based on this experiment. To establish the value of the citric-acid test for diagnostic purposes, Dondero et al. (E7) conducted a study to determine whether there is a relationship between plasma testosterone and levels of citric acid in human seminal plasma. The results are shown in Table 10.9 (see p. 441). Use the Mood–Brown method to test $H_0: \alpha = 450$, $\beta = 0$, and $H_0: \beta = 0$.

TABLE 10.7   **Derived pulmonary mean transit time, t_m (MPA–LA), and pulmonary blood flow values in infants and children who underwent cardiac catheterization**

PBF, L/min/sq M(X)	5.33	5.28	4.29	2.86	2.76	3.38	11.48	15.80
t_m (MPA–LA), sec (Y)	5.41	5.03	2.34	5.17	5.47	4.95	1.88	1.46
PBF, L/min/sq M (X)	6.18	8.99	9.76	17.26	10.03	11.99	12.34	14.15
t_m (MPA–LA), sec (Y)	3.79	1.56	2.64	1.46	2.11	1.35	1.47	1.74
PBF, L/min/sq M (X)	10.19	13.96	16.72	8.96	7.55	8.63	5.90	5.37
t_m (MPA–LA), sec (Y)	1.96	1.84	2.17	4.75	4.16	3.52	2.29	3.52
PBF, L/min/sq M (X)	5.02	3.38	3.42	3.53	3.32	10.76	6.77	3.86
t_m (MPA–LA), sec (Y)	3.49	5.12	4.70	3.98	4.46	2.42	3.29	5.83
PBF, L/min/sq M (X)	3.56	19.45	14.77	14.93	16.65	9.38	19.27	7.24
t_m (MPA–LA), sec (Y)	5.47	1.64	2.13	2.13	1.86	2.83	2.05	1.76
PBF, L/min/sq M (X)	9.56	7.39	11.57	12.35	8.96	9.17	10.15	13.90
t_m (MPA–LA), sec (Y)	2.69	3.29	2.27	1.63	2.69	2.42	3.13	2.42
PBF, L/min/sq M (X)	5.54	13.01						
t_m (MPA–LA), sec (Y)	3.39	2.47						

Source: Hung-Chi Lue, Chiung-Ming Chen, Chiung-Lin Chen, and Huoyao Wei, ''A New Approach to Pulmonary Blood Volume Determination in Congenital Heart Disease with Left-to-Right Shunts,'' *Chest*, 68 (1975), 689–696.

TABLE 10.8   **Measurements of patients with Turner's syndrome**

Height, cm (X)	44.0	50.0	50.0	50.0	49.0	49.4	65.5	62.0	76.0	73.5
ID, cm (Y)	11.0	9.0	9.9	9.0	10.2	9.0	10.5	10.0	10.0	11.8
Height, cm (X)	74.5	74.0	90.0	82.0	89.0	95.6	86.5	97.0	103.2	101.0
ID, cm (Y)	13.0	11.0	14.5	11.0	12.0	11.2	11.5	12.2	11.5	13.4
Height, cm (X)	103.0	113.5	107.0	109.2	106.5	104.5	116.0	123.1	136.5	137.0
ID, cm (Y)	13.0	13.0	14.2	12.0	14.0	12.0	12.2	15.0	17.0	16.5
Height, cm (X)	140.0	126.7	139.0	131.0	161.5	141.0	134.2	146.0	146.0	
ID, cm (Y)	15.0	18.2	14.0	16.5	16.8	21.5	22.0	20.0	28.0	

Source: Harold Chen, Ceres Espiritu, Carmelita Casquejo, Kinghan Boriboon, and Paul Woolley, Jr., ''Inter-Nipple Distance in Normal Children from Birth to 14 Years, and in Children with Turner's, Noonan's, Down's and Other Aneuploides,'' *Growth*, 38 (1974), 421–436.

TABLE 10.9   **Concentrations of plasma testosterone and seminal citric acid**

Plasma testosterone, ng/ml	110	176	178	190	217	220	236	260	276	290	297	304	357	360
Seminal citric acid, mg/ml	420	300	280	570	620	640	480	115	470	435	100	280	550	230
Plasma testosterone, ng/ml	500	520	520	526	530	531	544	552	560	560	567	569	569	577
Seminal citric acid, mg/ml	580	150	550	750	520	775	675	600	260	470	800	350	500	600
Plasma testosterone, ng/ml	360	360	368	372	373	377	390	390	415	450	463	470	488	500
Seminal citric acid, mg/ml	600	825	470	330	50	500	350	365	470	775	570	490	310	525
Plasma testosterone, ng/ml	590	597	600	637	666	670	690	706	707	755	796	800	870	1000
Seminal citric acid, mg/ml	400	225	375	430	675	475	500	525	570	700	320	415	600	775

Source: F. Dondero, F. Sciarra, and A. Isidori, "Evaluation of Relationship between Plasma Testosterone and Human Seminal Citric Acid," *Fertil. Steril.*, 23 (1972), 168–171.

10.4   Stitt et al. (E8) investigated the relationship between surface area and body weight in nine squirrel monkeys. They collected the data shown in Table 10.10. Use Theil's method (described in the preceding section) to estimate β, the slope of the population linear regression line describing the relationship between the two variables. Use body weight as the independent (x) variable.

TABLE 10.10   **Body weight and total surface area of nine squirrel monkeys**

Total area, cm²	780.6	887.6	1039.2	1040.0	1122.8
Body wt, grams	660.0	705.0	923.0	953.0	994.0
Total area, cm²	1070.4	1133.4	1125.2	1148.0	
Body wt, grams	1005.0	1018.0	1129.0	1181.0	

Source: John T. Stitt, James D. Hardy, and Ethan R. Nadel, "Surface Area of the Squirrel Monkey in Relation to Body Weight," *J. Appl. Physiol.*, 31 (1971), 140–141.

10.3

CONFIDENCE INTERVAL FOR SLOPE COEFFICIENT

Section 10.1 presented two methods for obtaining estimates of β, the slope of the linear regression line describing the relationship between two variables X and Y. Usually the investigator is more interested in a confidence interval for β. Mood (T2) suggests a trial-and-error technique for obtaining a confidence interval for β based on the Brown–Mood hypothesis-testing procedure for β. The technique is based on the fact that a $100(1 - \alpha)\%$ confidence interval consists of the values of β_0 that would not be rejected at the α level of significance. Mood (T2) also suggests an alternative approximate method based on the Brown–Mood hypothesis test for β.

The method for constructing a confidence interval for β that is presented in this section is due to Theil (T3, T49). It is based on Theil's hypothesis-testing procedure for β, which was discussed in Section 10.2. The assumptions underlying this hypothesis-testing procedure also apply to the construction of intervals described in this section. The method yields a symmetric two-sided $100(1 - \alpha)\%$ confidence interval for β in the following steps:

1. From the n pairs (X_i, Y_i), compute all possible sample slopes $S_{ij} = (Y_j - Y_i)/(X_j - X_i)$, where $i < j$. There will be $N = \binom{n}{2}$ values of S_{ij}.
2. Arrange the values of S_{ij} in order of magnitude, from smallest to largest. The lower and upper limits of the confidence interval for β consist of two values, $\hat{\beta}_L$ and $\hat{\beta}_U$, respectively, from this array.
3. Enter Table A.22 with n and $\alpha/2$ to find $S_{\alpha/2}$.
4. Subtract 2 from $S_{\alpha/2}$ to obtain $C_{\alpha/2}$—that is, $C_{\alpha/2} = S_{\alpha/2} - 2$.
5. Let

$$k = \frac{N - C_{\alpha/2}}{2} \tag{10.7}$$

6. The lower limit $\hat{\beta}_L$ of the confidence interval for β is the kth value of S_{ij}, counting from the smallest value in the ordered array prepared in Step 2. The upper limit $\hat{\beta}_U$ is the kth value of S_{ij}, counting backward from the largest value in the ordered array. In other words, we may write the confidence interval obtained for β with a confidence coefficient of $1 - \alpha$ as

$$C(\hat{\beta}_L < \beta < \hat{\beta}_U) = 1 - \alpha \tag{10.8}$$

For sample sizes not given in Table A.22, we may approximate $C_{\alpha/2}$ by

$$C_{\alpha/2} \approx z_{\alpha/2} \sqrt{\frac{n(n - 1)(2n + 5)}{18}} \tag{10.9}$$

where $z_{\alpha/2}$ is the value of z from the standard normal table that has $\alpha/2$ of the area under the curve to its right.

Example 10.6 To illustrate the construction of a confidence interval for β, let us refer to Example 10.2. We wish to construct a confidence interval for β for which the confidence coefficient is 0.95. For a 95% confidence interval, $\alpha/2 = 0.025$.

1. The original data are given in Table 10.2.
2. An ordered array of the sample slopes is given in Table 10.3.
3. When we enter Table A.22 with $n = 10$ and $\alpha/2 = 0.025$, we find that $S_{\alpha/2} = 23$.
4. Subtracting 2 from $S_{\alpha/2}$ gives $23 - 2 = 21$.
5. By Equation 10.7, we have $k = (45 - 21)/2 = 12$.
6. Thus $\hat{\beta}_L$ is the twelfth-from-the-smallest value, and $\hat{\beta}_U$ is the twelfth-from-the-largest value of the S_{ij} values given in Table 10.3. Thus $\hat{\beta}_L = -4.214$, and $\hat{\beta}_U = 1.168$. Since $1 - 2(0.025) = 0.95$, we can write the confidence interval as

$$C(-4.214 < \beta < 1.168) = 0.95$$

In other words, we say that we are 95% confident that β is between -4.214 and 1.168.

EXERCISES

10.5 Refer to Exercise 10.4. Construct a confidence interval for β with confidence coefficient 0.95.

10.6 Jowsey et al. (E9) examined the effects of dialysate calcium (Ca) and dialysate fluoride concentrations on bone resorption and mineralization, as measured histologically. Subjects were selected from a group of hemodialysis patients. They were assigned in a random fashion to groups receiving one of two dialysate Ca concentrations. The investigators took bone biopsy specimens at the beginning and at the end of the study, with an interval of at least two months between biopsies. The amounts of bone resorption were expressed as percentages of the total available surface. Table 10.11 shows the determinations of bone resorption and osteoid width at the second biopsy for five subjects who were exposed to concentrations of dialysate Ca of 6 or more milligrams per 100 ml. Construct a confidence interval for β with confidence coefficient 0.95.

10.7 Sylvester (E10) investigated the anatomical and pathological status of the cortico-spinal tracts, somato-sensory tracts, and parietal lobes of patients who had cerebral palsy and mental retardation. As part of the study, measurements were taken on the parietal lobes and medullary pyramid areas of the subjects' brains. Results for six spastic subjects are shown in Table 10.12. Construct a confidence interval for β with confidence coefficient 0.95.

TABLE 10.11	**Bone resorption and osteoid-width determinations on five hemodialysis patients exposed to dialysate Ca concentrations of 6 or more mg/100 ml**

Osteoid width, μ (Y)	Bone resorption, % (X)
8.7	9.5
9.5	18.0
15.4	20.1
15.7	18.8
16.2	30.3

Source: Jenifer Jowsey, William J. Johnson, Donald R. Taves, and Patrick J. Kelly, "Effects of Dialysate Calcium and Fluoride on Bone Disease during Hemodialysis," *J. Lab. Clin. Med.,* 79 (1972), 204–214.

TABLE 10.12	**Parietal lobe weights and pyramidal areas of six spastic subjects**

Parietal lobe weight, g (Y)	162	123	121	127	54	150
Pyramidal area, mm² (X)	10.02	6.50	9.91	10.44	7.37	9.90

Source: P. E. Sylvester, "Pyramidal, Lemniscal, and Parietal Lobe Status in Cerebral Palsy," *J. Mental Defic. Res.,* 13 (1969), 20–33.

10.4

TEST FOR PARALLELISM OF TWO REGRESSION LINES

Investigators frequently perform regression analyses on each of two sets of data that are similar. A common question in these situations is whether the two regression lines have the same slope. For example, a medical researcher may wish to know whether the slopes of the regression lines describing the linear relationship between blood pressure and age are the same in men as in women. A psychologist may want to know whether the slope of the regression line describing the relationship between the intensity of a stimulus and the speed of response is the same in two different age groups.

Such tests for equality of slopes of population regression lines are referred to as tests for homogeneity of slopes or *tests for parallelism.* In parametric statistics the tests are based on the t or the F distribution. When the assumptions underlying the use of these distributions are not met, some alternative procedure is needed. One such alternative is a nonparametric test for parallelism suggested by Hollander (T50). The test is a variation on the Wilcoxon matched-pairs signed-ranks test discussed in Chapter 4.

Assumptions

A. The data consist of two sets of observations. For each, the model

$$Y_j = \alpha + \beta X_j + e_j \tag{10.10}$$

is assumed to be appropriate. In Equation 10.10, the X_j's are known

fixed constants, α and β are unknown parameters, and Y_j is a value of the random variable Y observed at the fixed point X_j.

B. The e_j's are mutually independent. For each line, the e_j's come from the same continuous populations, but the two populations do not have to be the same.

Hypotheses

We are given two sets of data, set 1 and set 2. The underlying model for set 1 is

$$Y_{1j} = \alpha_1 + \beta_1 X_{1j} + e_{1j}, \qquad j = 1,\ldots,n_1, X_{11} \leq X_{12} \leq \cdots \leq X_{1n_1}$$

$$(10.11)$$

and the model for set 2 is

$$Y_{2j} = \alpha_2 + \beta_2 X_{2j} + e_{2j}, \qquad j = 1,\ldots,n_2, X_{21} \leq X_{22} \leq \cdots \leq X_{2n_2}$$

$$(10.12)$$

We may test the following null hypotheses H_0 against their two-sided and one-sided alternatives H_1.

A. (Two-sided): $H_0: \beta_1 = \beta_2,$ $H_1: \beta_1 \neq \beta_2$
B. (One-sided): $H_0: \beta_1 \geq \beta_2,$ $H_1: \beta_1 < \beta_2$
C. (One-sided): $H_0: \beta_1 \leq \beta_2,$ $H_1: \beta_1 > \beta_2$

Test Statistic

To calculate the test statistic, proceed as follows:

1. The test requires that the number of observation pairs in the two groups be the same and that there be an even number of observation pairs. Consequently the first step is to discard observations randomly from one or both groups as necessary to meet the requirements of the sample size.

2. Set the resulting number of observations in each group equal to $2n$.

3. From the data of set 1, form n pairs by pairing X_{1j} with $X_{1,j+n}$, where $j = 1, 2,\ldots, n$.

4. Compute n slope estimators for group 1. They have the form

$$u_{1j} = \frac{Y_{1,t+n} - Y_{1t}}{X_{1,t+n} - X_{1t}}, \qquad t = 1, 2,\ldots, n \qquad (10.13)$$

5. From the data of set 2, form n pairs as described in Step 3.

6. Compute n slope estimators for set 2. They have the form

$$u_{2t} = \frac{Y_{2,t+n} - Y_{2t}}{X_{2,t+n} - X_{2t}}, \qquad t = 1, 2,\ldots, n \qquad (10.14)$$

7. Randomly pair the u_{1j}'s with the u_{2t}'s so that each u appears in one and only one pair.

8. Compute n differences of the form

$$Z = u_{1j} - u_{2t} \tag{10.15}$$

Designate these n differences $Z_1, Z_2, \ldots, Z_n$.

The procedure from this point is identical to the procedure for computing the Wilcoxon matched-pairs signed-ranks test statistic (described in Chapter 4). In this case the Z's play the role of the D_i's of Chapter 4. The remaining steps for computing the test statistic are the following:

9. Rank the absolute values of the Z's from smallest to largest.
10. Assign to each of the resulting ranks the sign of the difference whose absolute value yielded that rank.
11. Compute

$$T_+ = \text{the sum of the ranks that have positive signs} \tag{10.16}$$

$$T_- = \text{the sum of the ranks that have negative signs} \tag{10.17}$$

The test statistic is T_+ or T_-, depending on the alternative hypothesis.

Ties   If there are ties among the $|Z|$'s, assign the tied values the mean of the ranks for which they are tied.

Decision Rule

The decision rule for this test follows the decision rule for the Wilcoxon matched-pairs signed-ranks test. See Chapter 4 for complete details and the rationale behind the decision. Briefly, the specific decision rules for the three sets of possible hypotheses are as follows:

A. (Two-sided): Let T be T_+ or T_-, whichever is smaller. Reject H_0 at the α level of significance if the calculated T is smaller than or equal to tabulated T for n and preselected $\alpha/2$ given in Table A.3.

B. (One-sided): Recall that each $Z = u_{1j} - u_{2t}$ is the difference between a slope estimator computed from sample 1 data and a slope estimator computed from sample 2 data. When $\beta_1 < \beta_2$, the estimators computed from sample 1 data tend to be smaller than those computed from sample 2 data. This situation results in more negative Z's, which in turn leads to a large value of T_-. When T_- is large, T_+ is small. Small values of T_+ (and large values of T_-) tend to support the alternative hypothesis. Sufficiently small values of T_+ cause us to reject H_0. Therefore reject H_0 at the α level of significance if the computed value of T_+ is less than or equal to tabulated T for n and preselected α (one-sided) given in Table A.3.

C. (One-sided): When $\beta_1 > \beta_2$, the estimators computed from sample 1 data tend to be larger than those computed from sample 2 data. This situation leads to positive values of Z, which in turn leads to large

values of T_+. When T_+ is large, T_- is small. Consequently, small values of T_- (and large values of T_+) tend to support the alternative hypothesis. Sufficiently small values of T_-, then, cause us to reject H_0.

Reject H_0 at the α level of significance if the computed value of T_- is less than or equal to tabulated T for n and preselected α (one-sided) given in Table A.3.

Example 10.7   When primates are used in research, it is often necessary to know the age of the animal. Reed (E11) collected measurements of skull, muzzle, long bone, and tooth development in *Papio cynocephalus* baboons. He used these data, collected over a period of five years, to develop a regression relationship between these measurements and age. Reed concluded on the basis of his results that the oral growth and development and the growth of the humerus yield an accurate tool for determining a baboon's age from birth to 60 months. Part of Reed's data are shown in Table 10.13.

TABLE 10.13   **Age, in months, and sum of skull, muzzle, and long-bone measurements, in millimeters, for male and female *Papio cynocephalus* baboons**

Male	Sum (X)	Age (Y)	Female	Sum (X)	Age (Y)
1	175.0	1.36	1	175.0	1.58
2	183.0	2.20	2	183.0	2.48
3	190.0	3.05	3	190.0	3.40
4	200.0	4.45	4	200.0	4.92
5	211.0	6.19	5	211.0	6.87
6	220.0	7.78	6	220.0	8.66
7	230.0	9.70	7	230.0	10.86
8	239.5	11.66	8	239.5	13.14
9	245.5	12.96	9	245.5	14.67
10	260.0	16.33	10	260.0	18.68
11	271.5	19.21	11	271.5	22.14
12	284.0	22.52	12	284.0	26.18
13	291.0	24.46	13	291.0	28.57
14	302.5	27.78	14	302.5	32.68
15	314.0	31.25	15	314.0	37.03
16	318.5	32.65	16	318.5	38.80
17	327.0	35.36	17	327.0	42.22
18	337.0	38.65			
19	345.5	41.52			
20	360.0	46.61			
21	375.0	52.10			
22	384.5	55.69			
23	397.0	60.55			
24	411.0	66.18			
25	419.5	69.68			
26	428.5	73.47			
27	440.0	78.41			
28	454.5	84.81			

Source: O. M. Reed, "*Papio Cynocephalus* Age Determinations," *Amer. J. Phys. Anthropol.*, 38 (1973), 309–314.

The data are from African colonies existing at the Southwest Foundation for Research and Education, San Antonio, Texas. We wish to know whether we can conclude that the slope describing the linear relationship between the sum of skull, muzzle, and longbone measurements and age for males is different from that for females.

Hypotheses

$$H_0: \beta_M = \beta_F, \qquad H_1: \beta_M \neq \beta_F$$

Test Statistic

1. To get an even number of observation pairs in each group and equal group sizes, we must randomly discard 1 observation pair from the female group and 12 observation pairs from the male group. After we do this, we can arrange the remaining 16 observations for each group in order of magnitude of the x values. These are shown in Table 10.14.

TABLE 10.14   **Age, in months, and sum of skull, muzzle, and long-bone measurements, in millimeters, for male and female *Papio cynocephalus* baboons**

Male	Sum (X)	Age (Y)	Female	Sum (X)	Age (Y)
	200.0	4.45		175.0	1.58
	211.0	6.19		183.0	2.48
	220.0	7.78		190.0	3.40
	230.0	9.70		200.0	4.92
	239.5	11.66		211.0	6.87
	260.0	16.33		220.0	8.66
	291.0	24.46		239.5	13.14
	314.0	31.25		245.5	14.67
	327.0	35.36		260.0	18.68
	345.5	41.52		271.5	22.14
	360.0	46.61		284.0	26.18
	384.5	55.69		291.0	28.57
	397.0	60.55		302.5	32.68
	411.0	66.18		314.0	37.03
	428.5	73.47		318.5	38.80
	440.0	78.41		327.0	42.22

Source: O. M. Reed, "*Papio Cynocephalus* Age Determinations," *Amer. J. Phys. Anthropol.*, 38 (1973), 309–314.

2. We now have $2n = 16$.
3. Since $n = 8$, we pair $X_{1,1}$ with $X_{1,9}$, $X_{1,2}$ with $X_{1,10}$, and so on, to $X_{1,8}$ with $X_{1,16}$.
4. We now form eight slope estimators for group 1.

$$u_{11} = \frac{Y_{1,9} - Y_{1,1}}{X_{1,9} - X_{1,1}} = \frac{35.36 - 4.45}{327.0 - 200.0} = 0.243$$

$$u_{12} = \frac{Y_{1,10} - Y_{1,2}}{X_{1,10} - X_{1,2}} = \frac{41.52 - 6.19}{345.5 - 211.0} = 0.263$$

$$u_{13} = \frac{Y_{1,11} - Y_{1,3}}{X_{1,11} - X_{1,3}} = \frac{46.61 - 7.78}{360.0 - 220.0} = 0.277$$

$$u_{14} = \frac{Y_{1,12} - Y_{1,4}}{X_{1,12} - X_{1,4}} = \frac{55.69 - 9.70}{384.5 - 230.0} = 0.298$$

$$u_{15} = \frac{Y_{1,13} - Y_{1,5}}{X_{1,13} - X_{1,5}} = \frac{60.55 - 11.66}{397.0 - 239.5} = 0.310$$

$$u_{16} = \frac{Y_{1,14} - Y_{1,6}}{X_{1,14} - X_{1,6}} = \frac{66.18 - 16.33}{411.0 - 260.0} = 0.330$$

$$u_{17} = \frac{Y_{1,15} - Y_{1,7}}{X_{1,15} - X_{1,7}} = \frac{73.47 - 24.46}{428.5 - 291.0} = 0.356$$

$$u_{18} = \frac{Y_{1,16} - Y_{1,8}}{X_{1,16} - X_{1,8}} = \frac{78.41 - 31.25}{440.0 - 314.0} = 0.374$$

5. For the data of group 2 (females), we pair $X_{2,1}$ with $X_{2,9}$, $X_{2,2}$ with $X_{2,10}$, and so on, to $X_{2,8}$ with $X_{2,16}$.

6. We then have eight slope estimators for group 2.

$$u_{21} = \frac{Y_{2,9} - Y_{2,1}}{X_{2,9} - X_{2,1}} = \frac{18.68 - 1.58}{260.0 - 175.0} = 0.201$$

$$u_{22} = \frac{Y_{2,10} - Y_{2,2}}{X_{2,10} - X_{2,2}} = \frac{22.14 - 2.48}{271.5 - 183.0} = 0.222$$

$$u_{23} = \frac{Y_{2,11} - Y_{2,3}}{X_{2,11} - X_{2,3}} = \frac{26.18 - 3.40}{284.0 - 190.0} = 0.242$$

$$u_{24} = \frac{Y_{2,12} - Y_{2,4}}{X_{2,12} - X_{2,4}} = \frac{28.57 - 4.92}{291.0 - 200.0} = 0.260$$

$$u_{25} = \frac{Y_{2,13} - Y_{2,5}}{X_{2,13} - X_{2,5}} = \frac{32.68 - 6.87}{302.5 - 211.0} = 0.282$$

$$u_{26} = \frac{Y_{2,14} - Y_{2,6}}{X_{2,14} - X_{2,6}} = \frac{37.03 - 8.66}{314.0 - 220.0} = 0.302$$

$$u_{27} = \frac{Y_{2,15} - Y_{2,7}}{X_{2,15} - X_{2,7}} = \frac{38.80 - 13.14}{318.5 - 239.5} = 0.325$$

$$u_{28} = \frac{Y_{2,16} - Y_{2,8}}{X_{2,16} - X_{2,8}} = \frac{42.22 - 14.67}{327.0 - 245.5} = 0.338$$

7. Random pairing of the u_{1j}'s and u_{2t}'s yields the following pairs.

$$u_{14}, u_{21} \qquad u_{11}, u_{24} \qquad u_{16}, u_{27} \qquad u_{15}, u_{23}$$
$$u_{18}, u_{25} \qquad u_{17}, u_{22} \qquad u_{12}, u_{28} \qquad u_{13}, u_{26}$$

8. The differences between members of these pairs give the following Z's.

$$Z_1 = 0.298 - 0.201 = 0.097 \qquad Z_5 = 0.374 - 0.282 = 0.092$$

$$Z_2 = 0.243 - 0.260 = -0.017 \qquad Z_6 = 0.356 - 0.222 = 0.134$$

$$Z_3 = 0.330 - 0.325 = 0.005 \qquad Z_7 = 0.263 - 0.338 = -0.075$$

$$Z_4 = 0.310 - 0.242 = 0.068 \qquad Z_8 = 0.277 - 0.302 = -0.025$$

9, 10. Steps 9 and 10 are summarized in Table 10.15.

TABLE 10.15   **Summary of Steps 9 and 10 for Example 10.7**

Z_i	$|Z_i|$	Rank of $|Z_i|$	Signed ranks
0.097	0.097	7	+7
−0.017	0.017	2	−2
0.005	0.005	1	+1
0.068	0.068	4	+4
0.092	0.092	6	+6
0.134	0.134	8	+8
−0.075	0.075	5	−5
−0.025	0.025	3	−3

---

11. From the entries in the last column of Table 10.15, we compute
    $T_+ = 26$ and $T_- = 10$.

Decision

Since $T_- = 10$ is the smaller of the sums computed in Step 11, we set $T = 10$. When we enter Table A.3 with $n = 8$, we find that, when H_0 is true, the probability of a value of T as extreme as 10 is $2(0.1563) = 0.3162$ (two-sided probability). We conclude that these data do not provide convincing evidence in favor of the alternative hypothesis that the two population linear regression lines have different slopes.

Large-Sample Approximation   When n is greater than 30, we cannot use Table A.3. When n is greater than 20, we may use the normal approximation given for the

Wilcoxon test in Chapter 4; that is, we compute

$$z = \frac{T - n(n + 1)/4}{\sqrt{\dfrac{n(n + 1)(2n + 1)}{24}}}$$

(10.18)

For large n, z is distributed approximately as the standard normal.

Power-Efficiency   The properties of Hollander's test for parallelism are discussed by Hollander (T50) and Hollander and Wolfe (T51). As Hollander (T50) points out, two disadvantages of his test for parallelism are its dependence on irrelevant randomizations and the requirement of equal sample sizes.

FURTHER READING

A distribution-free method for making inferences about the difference in slopes of two regression lines that lacks the disadvantage of extraneous randomization is proposed by Rao and Daley (T52). The method assumes that the data are obtained from a designed experiment with common regression constants. According to the authors, a comparison of the proposed method to its competitors indicates that it is superior in performance.

Potthoff (T53) provides a method, analogous to the two-sample Wilcoxon test, for testing the hypothesis that two simple regression lines are parallel when the two sets of error terms have two arbitrary unknown continuous distributions.

A rank score method of testing the null hypothesis that two or more slope parameters are equal has been studied by Sen (T54). The test criteria are based on the individual ranks of the different samples. In this model the intercepts are nuisance parameters. Adichie (T55) has shown that for the special case where the intercept parameters, though unknown, are all equal, suitable rank score tests for equality of slope parameters may be based on the simultaneous ranking of all the observations. The author shows that the test statistic has an asymptotic distribution that is chi-square under the hypothesis and noncentral chi-square under an appropriate sequence of alternatives. The author considers the efficiency and power of the proposed statistic.

Rao and Gore (T56) suggest some distribution-free methods for testing the hypotheses of parallelism and concurrence of two linear regressions when one can assume that the independent variable is equally spaced. The author compares the proposed procedures with nonparametric competitors and the normal theory t-test.

Salama and Quade (T57) propose a procedure for comparing two multiple regressions by means of a weighted measure of correlation.

Akritas et al. (T58) consider the problem of estimating the intercepts of several regression lines after a preliminary test on the parallelism of the lines. They propose estimators of the intercepts and study their asymptotic behavior. The authors compare their procedures with conventional estimation procedures and study their efficiency.

Also of interest are the papers by Chiang (T59) and Chiang and Puri (T60).

EXERCISES

10.8 Valentin and Olesen (E12) investigated 14 patients (4 males, 10 females) with a variety of cardiac diseases ranging in severity from mild to very severe. In another study, Nillson and Hultman (E13) used 19 healthy volunteers (10 males, 9 females). The ages of the first group ranged from 36 to 64 years, while the ages of the second group ranged from 21 to 45 years. The heights and weights of the two groups are shown in Table 10.16. Let X = height and Y = weight. Test the null hypothesis that $\beta_1 = \beta_2$.

TABLE 10.16 **Heights, centimeters, and weights, kilograms, of subjects participating in two different investigations**

Cardiac patients		Healthy volunteers	
Height	Weight	Height	Weight
177	61.3	180	79
159	60.4	190	80
154	62.8	196	90
160	63.2	177	62
157	67.3	167	62
169	63.2	170	61
160	45.0	171	66
180	87.5	163	56
158	47.9	165	52
167	58.9	164	54
154	48.2	168	58
176	63.3	185	84
155	61.3	180	83
158	59.3	185	64
		185	64
		157	52
		162	57
		165	65
		171	57

Sources: Cardiac patients: N. Valentin and K. H. Olesen, "Muscle Electrolytes and Total Exchangeable Electrolytes in Patients with Cardiac Disease," *Scand. J. Clin. Lab. Invest.* 32 (1973), 161–166. Healthy volunteers: L. H:Son Nillson and E. Hultman, "Liver Glycogen in Man—The Effect of Total Starvation or a Carbohydrate-Poor Diet Followed by Carbohydrate Refeeding," *Scand. J. Clin. Lab. Invest.*, 32 (1973), 325–330. Used by permission of Universitetsforlaget, Oslo.

10.9 Bond (E14) studied 45 patients undergoing major vaginal surgery. Twenty had epidural analgesia plus sedation, and 25 had epidural analgesia together with light halothane anaesthesia. Patients were selected at random for inclusion in either group. The purpose of the study was to examine the significance of the level of the blood pressure in blood loss in major vaginal surgery. Table 10.17 shows, for each group, the amount of blood loss (Y) and the mean operative systolic blood pressure expressed as a percentage of the pre-operative figure (X). Can we conclude on the basis of these data that the slopes of the regression lines describing the linear relationship between the two variables are different in the two groups?

TABLE 10.17   **Data on two groups of patients undergoing major vaginal surgery**

	Lumbar epidural block only			Lumbar epidural block + halothane	
	Mean operative SBP			Mean operative SBP	
Patient	Percentage of preoperative	Blood loss (ml)	Patient	Percentage of preoperative	Blood loss (ml)
1	75	260	1	54	223
2	80	179	2	71	175
3	67	338	3	50	19
4	83	302	4	72	120
5	74	179	5	56	133
6	79	52	6	67	111
7	60	347	7	57	219
8	69	105	8	56	266
9	90	307	9	53	50
10	58	263	10	63	194
11	58	61	11	62	203
12	75	181	12	62	150
13	67	38	13	80	79
14	67	201	14	41	420
15	88	370	15	73	515
16	60	178	16	83	270
17	62	61	17	65	124
18	100	95	18	58	275
19	68	273	19	63	288
20	85	204	20	54	46
			21	73	64
			22	47	89
			23	50	73
			24	62	93
			25	53	70

Source: A. G. Bond, "Conduction Anaesthesia, Blood Pressure and Haemorrhage," Br. J. Anaesthesia, 41 (1969), 942–945.

10.10   As part of a diabetes research project, Raskin et al. (E15) collected the data on age and the width of the quadriceps muscle capillary basement membrane (QCBM) for diabetic and nondiabetic subjects shown in Table 10.18. The authors point out that the morphometric analysis of QCBM width has been widely used to evaluate the presence and time of onset of diabetic microangiopathy. Determine whether one can conclude that the slope for the linear regression of QCBM width on age is greater in the diabetic group.

TABLE 10.18   **Age and basement membrane width in two groups of subjects**

Nondiabetic subjects

Case	1	2	3	4	5	6	7	8	9
Age, years	1	1	1	2	2	2	3	5	5
Basement membrane width (Å)	982	863	869	598	889	763	1159	991	793
Case	10	11	12	13	14	15	16	17	18
Age, years	6	7	7	10	12	12	12	12	12
Basement membrane width (Å)	1104	930	893	790	861	862	781	1163	960
Case	19	20	21	22	23	24	25	26	27
Age, years	12	12	13	14	14	15	15	15	16
Basement membrane width (Å)	1010	758	850	1042	841	768	1182	1178	952
Case	28	29	30	31	32	33	34	35	36
Age, years	17	19	20	20	21	21	22	23	23
Basement membrane width (Å)	1141	848	1413	1424	1116	902	797	923	1076
Case	37	38	39	40	41	42	43	44	45
Age, years	23	23	24	24	24	25	25	25	25
Basement membrane width (Å)	1167	940	1252	1111	775	772	1145	1019	1603
Case	46	47	48	49	50	51	52	53	54
Age, years	26	26	26	27	27	27	28	28	28
Basement membrane width (Å)	964	1241	1010	1278	1347	1468	1092	1065	1126
Case	55	56	57	58					
Age, years	30	30	30	30					
Basement membrane width (Å)	1286	1161	1129	967					

Diabetic patients

Case	1	2	3	4	5	6	7	8	9
Age, years	4	4	4	5	6	7	7	8	8
Basement membrane width (Å)	768	887	859	1009	969	780	1099	1152	1142
Case	10	11	12	13	14	15	16	17	18
Age, years	8	8	8	8	8	9	9	9	10
Basement membrane width (Å)	824	1297	1255	1260	694	940	880	1345	801

TABLE 10.18 (Continued)

Case	19	20	21	22	23	24	25	26	27
Age, years	10	11	11	11	11	11	12	12	12
Basement membrane width (Å)	713	1347	1459	1718	1245	1891	1619	733	995
Case	28	29	30	31	32	33	34	35	36
Age, years	12	13	14	14	14	14	15	15	15
Basement membrane width (Å)	894	1037	1139	1297	1280	1020	862	1077	1558
Case	37	38	39	40	41	42	43	44	45
Age, years	15	15	16	16	17	17	18	18	18
Basement membrane width (Å)	1852	927	1897	1210	911	1155	1774	1155	2161
Case	46	47	48	49	50	51	52	53	54
Age, years	19	20	20	21	21	21	22	22	22
Basement membrane width (Å)	2655	3339	1522	1957	3095	2779	1858	4022	1133
Case	55	56	57	58	59	60	61	62	63
Age, years	23	24	24	25	25	25	27	28	28
Basement membrane width (Å)	2455	1991	3335	3002	1447	2024	4756	1778	2620
Case	64	65	66	67					
Age, years	29	30	30	30					
Basement membrane width (Å)	1517	1771	2021	1718					

Source: Philip Raskin, James F. Marks, Henry Burns, Jr., Mary Ellen Plumer, and Marvin D. Siperstein, "Capillary Basement Membrane Width in Diabetic Children," Amer. J. Med., 58 (1975), 365–372.

10.5

ESTIMATOR AND CONFIDENCE INTERVAL FOR DIFFERENCE BETWEEN SLOPE PARAMETERS

Hollander and Wolfe (T51) have proposed a procedure for estimating the difference between the slopes of two population regression lines. We may designate this parameter $\theta = \beta_1 - \beta_2$, where β_1 and β_2 are the slopes of two population regression lines. The procedure is based on Hollander's test for parallelism described in Section 10.4; the assumptions and model for the test apply. We obtain the estimate as follows.

1. Obtain the Z values as described in Step 8 of Section 10.4.
2. Compute the $n(n + 1)/2$ values of $(Z_i + Z_j)/2$, where $i \le j$, and j takes on values between 1 and n, inclusive.
3. Compute the estimator of $\theta = \beta_1 - \beta_2$ by

$$\hat{\theta} = \text{median}\left\{\frac{Z_i + Z_j}{2}\right\}, \qquad i \le j \qquad\qquad \textbf{(10.19)}$$

This procedure is the same as that described in Chapter 4 for finding an estimator of the median difference based on the Wilcoxon test.

Example 10.8 To illustrate the calculation of $\hat{\theta}$, let us consider Example 10.7. We wish to estimate the difference between β_M and β_F, the slopes of the population linear regression lines describing the relationship between certain bone measurements and age in male and female baboons, respectively.

1. The Z's are 0.097, -0.017, 0.005, 0.068, 0.092, 0.134, -0.075, and -0.025.
2. The $n(n + 1)/2 = 36$ values of $(Z_i + Z_j)/2, i \le j$, arrayed in order of magnitude, are

-0.0750	-0.0170	0.0110	0.0365	0.0585	0.0945
-0.0500	-0.0100	0.0215	0.0375	0.0680	0.0970
-0.0460	-0.0060	0.0255	0.0400	0.0695	0.1010
-0.0350	-0.0035	0.0295	0.0485	0.0800	0.1130
-0.0250	0.0050	0.0335	0.0510	0.0825	0.1155
-0.0210	0.0085	0.0360	0.0545	0.0920	0.1340

3. The estimate of the difference between population slopes is the median of these 36 values. That is,

$$\hat{\theta} = \frac{0.0360 + 0.0365}{2} = 0.03625$$

Confidence Interval for $\theta = \beta_1 - \beta_2$ Hollander and Wolfe (T51) describe a procedure for constructing a symmetric, two-sided $100(1 - \alpha)\%$ confidence interval for $\theta = \beta_1 - \beta_2$. The procedure is based on Hollander's test for parallelism described in Section 10.4, and the model and assumptions of the test apply. The procedure consists of the following steps:

1. Obtain the values of Z as described in Step 8 of Section 10.4.
2. Compute the $n(n + 1)/2$ values of $(Z_i + Z_j)/2$, where $i \le j$, and j takes on values between 1 and n, inclusive. Arrange these averages in order of magnitude from smallest to largest.

3. Locate in Table A.3 the sample size and the appropriate value of P as determined by the desired level of confidence. When $(1 - \alpha)$ is the confidence coefficient, $P = \alpha/2$. When the exact value of $\alpha/2$ cannot be found in Table A.3, choose a neighboring value—either the closest value or one that is larger or smaller than $\alpha/2$, depending on whether a slightly wider or slightly narrower interval than desired is more acceptable.

4. The endpoints of the confidence interval are the Kth smallest and the Kth largest values of $(Z_i + Z_j)/2$. $K = T + 1$, where T is the value in the column labeled T corresponding to the value of P selected in Step 3.

This is the same procedure described in Chapter 4 for finding a confidence interval for the median difference based on the Wilcoxon test.

Example 10.9 To illustrate the construction of a confidence interval for $\beta_1 - \beta_2$, let us refer again to the data of Example 10.7. We construct an approximate 95% confidence interval for $\beta_M - \beta_F$.

1. The values of Z are given in Step 1 of Example 10.8.
2. The ordered array of $(Z_i + Z_j)/2$ values is given in Step 2 of Example 10.8.
3. We enter Table A.3 with $n = 8$. Since our confidence coefficient is $0.95 = (1 - 0.05)$, $\alpha = 0.05$ and $\alpha/2 = 0.025$. In Table A.3, for $n = 8$, the value of P closest to 0.025 is 0.0273. Therefore, for our example, $T = 4$, and $K = 4 + 1 = 5$.
4. The lower limit of our confidence interval is the fifth-smallest value of $(Z_i + Z_j)/2$, and the upper limit is the fifth-largest value. Thus the lower confidence limit is -0.0250 and the upper limit is 0.0970. We are $100[1 - 2(0.0273)] = 94.5\%$ confident that the difference in population slopes is between -0.0250 and 0.0970.

FURTHER READING

Hollander and Wolfe (T51) describe an estimator associated with the parallelism test proposed by Potthoff (T53).

EXERCISES

10.11 Refer to Exercise 10.8. Compute $\hat{\theta}$, and construct an approximate 95% confidence interval for the difference between slopes.

10.12   Refer to Exercise 10.9. Compute $\hat{\theta}$, and construct an approximate 95% confidence interval for the difference between slopes.

10.13   Refer to Exercise 10.10. Compute $\hat{\theta}$, and construct an approximate 95% confidence interval for the difference between slopes.

10.6

COMPUTER PROGRAMS

A BASIC computer program written by Smith et al. (T61) allows the user to compute a point estimate of the slope and intercept of a sample regression line, test hypotheses about the slope and intercept parameters, construct a confidence interval for a population slope, perform a test for parallelism of two regression lines, and construct a confidence interval for the difference between two slope parameters. Other computer programs available for nonparametric regression analysis include those written by Rock and Duffy (T62) and Woosley (T63).

REVIEW EXERCISES

10.14   As part of a research project, a reading specialist collected the data on IQ and reading-test scores shown in Table 10.19. Use Theil's method to construct a 95% confidence interval for β.

TABLE 10.19   **Reading test scores and IQ of seven 9th-grade students**

Student	1	2	3	4	5	6	7
Reading score	47	48	44	49	46	56	50
IQ	110	100	105	95	103	120	115

10.15   Table 10.20 shows scores measuring reading ability and intellectual ability of 12 high school juniors. Using Theil's method, test the null hypothesis that $\beta = 0$.

TABLE 10.20   **Reading ability and intellectual ability scores of 12 high school juniors**

Subject	1	2	3	4	5	6	7	8	9	10	11	12
Reading ability	223	207	231	222	221	208	215	189	201	193	211	191
Intellectual ability	105	103	106	109	101	100	99	94	91	95	98	92

10.16 Table 10.21 shows the expenditures for equipment maintenance and net income before taxes for a random sample of 10 business firms. All figures are in thousands of dollars. Use Theil's method to test the null hypothesis that $\beta = 0$.

TABLE 10.21 **Equipment maintenance expenditures (X) and net income before taxes (Y) for 10 business firms**

X	13	17	29	28	40	37	41	26	24
Y	15	21	18	14	22	23	24	16	17

10.17 Use Theil's method to construct a 95% confidence interval for β of Exercise 10.16.

10.18 Table 10.22 shows scores on tests for creativity and independence made by a sample of junior high school boys and girls. Do the data provide sufficient evidence to indicate that the slopes of the two regression lines are different?

TABLE 10.22 **Creativity and independence scores for a sample of junior high school boys and girls**

For boys

Creativity (X)	85	86	83	92	81	76	65	99	93	66	73	63
Independence (Y)	78	57	80	83	68	69	54	92	78	63	62	60

For girls

Creativity (X)	75	57	87	80	91	99	81	93	58	72	76
Independence (Y)	70	37	76	59	84	73	69	74	45	54	62

10.19 A manufacturer who wished to study the relationship between the age (X) and maintenance costs (Y) of a certain type of machine collected the data shown in Table 10.23. Use the Brown–Mood method to find the slope and intercept.

TABLE 10.23 **Data for Exercise 10.19**

X	8	6	2	1	2	6	3	1	2	5	7	5
Y	106	121	48	31	59	127	71	26	53	72	126	68
X	9	1	3	8								
Y	136	22	76	110								

10.20 Refer to Exercise 10.19. Use the Brown–Mood method to test the null hypothesis that $\alpha = 25$ and $\beta = 12$. Use a 0.05 significance level. Find the P value.

10.21 In a study of the relationship between repair costs (Y) and repair time (X) in thousandths of an hour in the repair of customer-returned items, a manufacturer collected the data shown in Table 10.24. Use Theil's method to test the null hypothesis that β is no greater than 1.

TABLE 10.24   **Data for Exercise 10.21**

X	162	166	143	159	205	211	128	164	163
Y	13.83	13.49	12.18	15.99	24.24	19.76	15.51	13.77	16.25
X	202	126	184	125	162	177			
Y	18.87	13.77	13.01	11.82	12.54	12.96			

10.22 Refer to Exercise 10.21. Construct a 95% confidence interval for β.

10.23 A retail store chain wished to study the relationship between advertising effort (X) and sales volume (Y). Two studies were conducted, one in the spring and the other in the fall of the same year. The data (coded for ease of computation) for the two studies, which lasted for 10 weeks each, are shown in Table 10.25. Can we conclude that the slopes of the two regression lines describing the linear relationship between the two variables are different in the two studies?

TABLE 10.25   **Data for Exercise 10.23**

Spring

X	160	140	40	80	160	100	80	40
Y	120	100	60	80	120	80	80	60
X	165	145						
Y	130	105						

Fall

X	30	50	60	90	50	30	80	90
Y	35	45	50	65	50	40	55	70
X	90	80						
Y	70	70						

10.24 Refer to Exercise 10.23. Construct a 95% confidence interval for the difference between population slopes.

REFERENCES

T1 Brown, G. W., and A. M. Mood, "On Median Tests for Linear Hypotheses," in Jerzy Neyman (ed.), *Proceedings of the Second Berkeley Symposium on Mathematical Statistics and Probability*, Berkeley and Los Angeles: The University of California Press, 1951, 159–166.

T2 Mood, A. M., *Introduction to the Theory of Statistics*, New York: McGraw-Hill, 1950.

T3 Theil, H., "A Rank-Invariant Method of Linear and Polynomial Regression Analysis. III," *Nederl. Akad. Wetensch. Proc., Ser. A*, 53 (1950), 1397–1412.

T4 Tate, Merle W., and Richard C. Clelland, *Nonparametric and Shortcut Statistics in the Social, Biological, and Medical Sciences*, Danville, Ill.: Interstate Printers and Publishers, 1957.

T5 Theil, H., "A Rank-Invariant Method of Linear and Polynomial Regression Analysis. I," *Nederl. Akad. Wetensch. Proc., Ser. A*, 53 (1950), 386–392.

T6 Sen, P. K., "Estimates of the Regression Coefficient Based on Kendall's Tau," *J. Amer. Statist. Assoc.*, 63 (1968), 1379–1389.

T7 Hill, Bruce Marvin, "A Test of Linearity versus Convexity of a Median Regression Curve," *Ann. Math. Statist.*, 33 (1962), 1096–1123.

T8 Adichie, J., "Estimation of Regression Parameters Based on Rank Tests," *Ann. Math. Statist.*, 38 (1967), 894–904.

T9 Kildea, Daniel G., *Median Estimators for Regression Models—The Brown–Mood Approach*, doctoral thesis, Bundoora, Victoria, Australia: Department of Mathematical Statistics, School of Physical Sciences, La Trobe University, 1978.

T10 Kildea, D. G., "Brown–Mood Type Median Estimators for Simple Regression Models," *Ann. Statist.*, 9 (1981), 438–442.

T11 Domański, C., "Notes on the Theil Test for the Hypothesis of Linearity for the Model with Two Explanatory Variables," in B. V. Gnedenko, M. L. Puri, and I. Vincze (ed.), *Nonparametric Statistical Inference*, Vol. I, Amsterdam: North-Holland, 1980, pp. 213–220.

T12 Gross, Shulamith T., and Thomas J. Tomberlin, "A Comparison of Non-Parametric Regression Methods," in *ASA Proceedings of the Social Statistics Section*, Washington, D. C.: American Statistical Association, 1983, pp. 549–550.

T13 Hussain, S. S., and P. Sprent, "Non-Parametric Regression," *J. Roy. Statist. Soc., Ser. A*, 146 (1983), 182–191.

T14 Cunningham, James P., "Multiple Monotone Regression," *Psychol. Bull.*, 92 (1982), 791–800.

T15 Andrews, D. F., "A Robust Method for Multiple Linear Regression," *Technometrics*, 16 (1974), 523–531.

T16 Hájek, J., "Extension of the Kolmogorov–Smirnov Test to Regression Alternatives," in L. LeCam and J. Neyman (eds.), *Bernoulli, Bayes, Laplace Anniversary Volume*, New York: Springer-Verlag, 1965, 45–60.

T17 Olshen, R. A., "Sign and Wilcoxon Tests for Linearity," *Ann. Math. Statist.*, 38 (1967), 1759–1769.

T18 Cleveland, William S., and Robert McGill, "The Many Faces of a Scatterplot," *J. Amer. Statist. Assoc.*, 79 (1984), 807–822.

T19 Adichie, J. N., "Asymptotic Efficiency of a Class of Nonparametric Tests for Regression Parameters," *Ann. Math. Statist.*, 38 (1967), 884–893.

T20 Daniels, H. E., "A Distribution-Free Test for Regression Parameters," *Ann. Math. Statist.*, 25 (1954), 449–513.

T21 Hogg, Robert V., and Ronald H. Randles, "Adaptive Distribution-Free Regression Methods and Their Applications," *Technometrics*, 17 (1975), 399–407.

T22 Ghosh, M., and P. Sen, "On a Class of Rank Order Tests for Regression With Partially Informed Stochastic Predictors," *Ann. Math. Statist.*, 42 (1971), 650–661.

T23 Jurečková, J., "Asymptotic Linearity of a Rank Statistic in Regression Parameter," *Ann. Math. Statist.*, 40 (1969), 1889–1900.

T24 Jurečková, Jane, "Nonparametric Estimate of Regression Coefficients," *Ann. Math. Statist.*, 42 (1971), 1328–1338.

T25 Konijn, H. S., "Non-Parametric Robust and Short-Cut Methods in Regression and Structural Analysis," *Austral. J. Statist.*, 3 (1961), 77–86.

T26 Koul, H. L., "Asymptotic Behavior of Wilcoxon Type Confidence Regions in Multiple Linear Regression," *Ann. Math. Statist.*, 40 (1969), 1950–1979.

T27 Samanta, M., "Optimum Nonparametric Inference of Partial Regression Coefficients," *Calcutta Statist. Assoc. Bull.*, 20 (1972), 109–133.

T28 Srivastava, M. S., and A. K. Md. Ehsanes Saleh, "On a Class of Nonparametric Estimates for Regression Parameters," *J. Statist. Res.*, 4 (1970), 133–139.

T29 Révész, P., "On the Nonparametric Estimation of the Regression Function," *Problems of Control and Information Theory*, 8 (1979), 297–302.

T30 Wegman, Edward J., "Two Approaches to Nonparametric Regression: Splines & Isotonic Inference," in K. Matusita (ed.), *Recent Developments in Statistical Inference and Data Analysis*, Amsterdam: North-Holland, 1980, pp. 323–334.

T31 Cheng, Kuang-Fu, and Pi-Erh Lin, "Nonparametric Estimation of a Regression Function," *Z. Wahrscheinlichkeitstheorie verw. Gebiete*, 57 (1981), 223–233.

T32 Brown, B. M., and J. S. Maritz, "Distribution-Free Methods in Regression," *Austral. J. Statist.*, 24 (1982), 318–331.

T33 Gore, A. P., and K. S. Madhava Rao, "Nonparametric Tests for Slope in Linear Regression Problems," *Biometrical J.*, 24 (1982), 229–237.

T34 Rao, K. S. Madhava, and A. P. Gore, "Nonparametric Tests for Intercept in Linear Regression Problems," *Austral. J. Statist.*, 24 (1982), 42–50.

T35 Bhattacharya, P. K., "Justification for a *K-S* Type Test for the Slope of a Truncated Regression," *Ann. Statist.*, 11 (1983), 697–701.

T36 Bhattacharya, P. K., Herman Chernoff, and S. S. Yang, "Nonparametric Estimation of the Slope of a Truncated Regression," *Ann. Statist.*, 11 (1983), 505–514.

T37 Hettmansperger, Thomas P., and Joseph W. McKean, "A Geometric Interpretation of Inferences Based on Ranks in the Linear Model," *J. Amer. Statist. Assoc.*, 78 (1983), 885–893.

T38 Henderson, Harold V., "Regression Models for an Ordered Contingency Table: An Example," *New Zealand Statist.*, 19 (November 1984), 15–21.

T39 Brown, B. M., "Grouping Problems in Distribution-Free Regression," *Austral. J. Statist.*, 27 (1985), 123–134.

T40 Lancaster, J. F., and Dana Quade, "A Nonparametric Test for Linear Regression Based on Combining Kendall's Tau with the Sign Test," *J. Amer. Statist. Assoc.*, 80 (1985), 393–397.

T41 Cheng, K. F., and P. E. Cheng, "Robust Nonparametric Estimation of a Regression Function," *Sankhyā*, 49 (1987), 9–22.

T42 Whitney, Paul, "Consistent Least Squares Nonparametric Regression," *J. Statist. Planning and Inference*, 17 (1987), 137–148.

T43 Eubank, Randall L., *Spline Smoothing and Nonparametric Regression*, New York: Marcel Dekker, 1988.

T44 Rousseeuw, Peter J., and Annick M. Leroy, *Robust Regression and Outlier Detection*, New York: Wiley, 1987.

T45 Puri, Madan Lal, and Pranab Kumar Sen, *Nonparametric Methods in General Linear Models*, New York: Wiley, 1985.

T46 Drummond, Douglas James, "The Cuts Procedure for Regression Based Solely on the Signs of the Residuals," doctoral thesis, Raleigh, N. C.: Department of Statistics, North Carolina State University, 1976.

T47   Daniel, Wayne W., *Nonparametric, Distribution-Free, and Robust Procedures in Regression Analysis: A Selected Bibliography*, Monticello, Ill.: Vance Bibliographies, June 1980.

T48   Collomb, Gérard, "Nonparametric Regression: An Up-to-Date Bibliography," *Statist.*, 16 (1985), 309–324.

T49   Theil, H., "A Rank-Invariant Method of Linear and Polynomial Regression Analysis. II," *Nederl. Akad. Wetensch. Proc., Ser. A*, 53 (1950), 521–525.

T50   Hollander, Myles, "A Distribution-Free Test for Parallelism," *J. Amer. Statist. Assoc.*, 65 (1970), 387–394.

T51   Hollander, Myles, and Douglas A. Wolfe, *Nonparametric Statistical Methods*, New York: Wiley, 1973.

T52   Rao, P. V., and R. R. Daley,"On the Use of the Scholz–Sievers Method for Distribution-Free Inference about the Difference in Slopes of Two Regression Lines," *Communic. in Statist.— Theory and Methods*, 17 (1988), 1437–1448.

T53   Potthoff, Richard F., "A Nonparametric Test of Whether Two Simple Regression Lines Are Parallel," *Ann. Statist.*, 2 (1974), 295–310.

T54   Sen, P. K., "On a Class of Rank-Order Tests for the Parallelism of Several Regression Lines," *Ann. Math. Statist.*, 40 (1969), 1668–1683.

T55   Adichie, J. N., "Rank Score Comparison of Several Regression Parameters," *Ann. Statist.*, 2 (1974), 396–402.

T56   Rao, K. S. Madhava, and A. P. Gore, "Distribution-Free Tests for Parallelism and Concurrence in Two-Sample Regression Problem," *J. Statist. Planning and Inference*, 5 (1981), 281–286.

T57   Salama, Ibrahim, and Dana Quade, "A Nonparametric Comparison of Two Multiple Regressions by Means of a Weighted Measure of Correlation," *Communic. in Statist.—Theory and Methods*, 11 (1982), 1185–1195.

T58   Akritas, M. G., A. K. Md.E. Saleh, and P. K. Sen, "Nonparametric Estimation of Intercepts After a Preliminary Test on Parallelism of Several Regression Lines," in P. K. Sen (ed.), *Biostatistics: Statistics in Biomedical, Public Health and Environmental Sciences*, New York: Elsevier, 1985.

T59   Chiang, Ching-Yuan, "Tests of the Parallelism of Several Regression Surfaces Based on Rank-Order Estimates," *J. Statist. Planning and Inference*, 14 (1986), 373–387.

T60   Chiang, Ching-Yuan, and Madan L. Puri, "Rank-Order Tests for the Parallelism of Several Surfaces," *J. Statist. Planning and Inference*, 10 (1984), 43–57.

T61   Smith, Edward C., Wayne W. Daniel, and Brian Schott, "Nonparametric Regression Analysis: A Program Package for Use on a Computer Terminal," *Behav. Res. Methods & Instrumentation*, 10 (1978), 435–436.

T62   Rock, N. M. S., and T. R. Duffy, "REGRES—A FORTRAN-77 Program to Calculate Nonparametric and Structural Parametric Solutions to Bivariate Regression Equations," *Comput. & Geosci.*, 12 (1986), 807–818.

T63   Woosley, John T., "Non-Parametric Linear Regression Analysis," *Lab. Med.*, 18 (1987), 333–334.

E1   Clark, Richard E., Harry W. Margraf, and Richard A. Beauchamp, "Fat and Solid Filtration in Clinical Perfusions," *Surgery*, 77 (1975), 216–224.

E2   Yunginger, John W., and Gerald J. Gleich, "Seasonal Changes in Serum and Nasal IgE Concentrations," *J. Allergy Clin. Immunol.*, 51 (1973), 174–186.

E3   Pilkey, Orrin H., and John Hower, "The Effect of Environment on the Concentration of Skeletal Magnesium and Strontium in *Dendraster*," *J. Geol.*, 68 (1960), 203–214.

E4   Grundy, Scott M., and Allan L. Metzger, "A Physiological Method for Estimation of Hepatic Secretion of Biliary Lipids in Man," *Gastroenterology*, 62 (1972), 1200–1217.

E5   Lue, Hung-Chi, Chiung-Ming Chen, Chiung-Lin Chen, and Huoyao Wei, "A New Approach to Pulmonary Blood Volume Determination in Congenital Heart Disease with Left-to-Right Shunts," *Chest*, 68 (1975), 689–696.

E6   Chen, Harold, Ceres Espiritu, Carmelita Casquejo, Kinghan Boriboon, and Paul Woolley, Jr., "Inter-Nipple Distance in Normal Children from Birth to 14 Years, and in Children with Turner's, Noonan's, Down's and Other Aneuploides," *Growth*, 38 (1974), 421–436.

E7   Dondero, F., F. Sciarra, and A. Isidori, "Evaluation of Relationship between Plasma Testosterone and Human Seminal Citric Acid," *Fertil. Steril.*, 23 (1972), 168–171.

E8   Stitt, John T., James D. Hardy, and Ethan R. Nadel, "Surface Area of the Squirrel Monkey in Relation to Body Weight," *J. Appl. Physiol.*, 31 (1971), 140–141.

E9   Jowsey, Jenifer, William J. Johnson, Donald R. Taves, and Patrick J. Kelly, "Effects of Dialysate Calcium and Fluoride on Bone Disease During Hemodialysis," *J. Lab. Clin. Med.*, 79 (1972), 204–214.

E10   Sylvester, P. E., "Pyramidal, Lemniscal, and Parietal Lobe Status in Cerebral Palsy," *J. Mental Defic. Res.*, 13 (1969), 20–33.

E11   Reed, O. M., "*Papio Cynocephalus* Age Determinations," *Amer. J. Phys. Anthropol.*, 38 (1973), 309–314.

E12   Valentin, N., and K. H. Olesen, "Muscle Electrolytes and Total Exchangeable Electrolytes in Patients with Cardiac Disease," *Scand. J. Clin. Lab. Invest.*, 32 (1973), 161–166.

E13   Nillson, L. H: Son, and E. Hultman, "Liver Glycogen in Man—The Effect of Total Starvation or a Carbohydrate-Poor Diet Followed by Carbohydrate Refeeding," *Scand. J. Clin. Lab. Invest.*, 32 (1973), 325–330.

E14   Bond, A. G., "Conduction Anaesthesia, Blood Pressure and Haemorrhage," *Br. J. Anaesthesia*, 41 (1969), 942–945.

E15   Raskin, Philip, James F. Marks, Henry Burns, Jr., Mary Ellen Plumer, and Marvin D. Siperstein, "Capillary Basement Membrane Width in Diabetic Children," *Amer. J. Med.*, 58 (1975), 365–372.

APPENDIX:
TABLES

TABLE A.1   **Binomial probability distribution**

$$P(r \mid n, p) = \binom{n}{r} p^r q^{n-r}$$

n = 1

r \ p	.01	.02	.03	.04	.05	.06	.07	.08	.09	.10
0	.9900	.9800	.9700	.9600	.9500	.9400	.9300	.9200	.9100	.9000
1	.0100	.0200	.0300	.0400	.0500	.0600	.0700	.0800	.0900	.1000

	.11	.12	.13	.14	.15	.16	.17	.18	.19	.20
0	.8900	.8800	.8700	.8600	.8500	.8400	.8300	.8200	.8100	.8000
1	.1100	.1200	.1300	.1400	.1500	.1600	.1700	.1800	.1900	.2000

	.21	.22	.23	.24	.25	.26	.27	.28	.29	.30
0	.7900	.7800	.7700	.7600	.7500	.7400	.7300	.7200	.7100	.7000
1	.2100	.2200	.2300	.2400	.2500	.2600	.2700	.2800	.2900	.3000

	.31	.32	.33	.34	.35	.36	.37	.38	.39	.40
0	.6900	.6800	.6700	.6600	.6500	.6400	.6300	.6200	.6100	.6000
1	.3100	.3200	.3300	.3400	.3500	.3600	.3700	.3800	.3900	.4000

	.41	.42	.43	.44	.45	.46	.47	.48	.49	.50
0	.5900	.5800	.5700	.5600	.5500	.5400	.5300	.5200	.5100	.5000
1	.4100	.4200	.4300	.4400	.4500	.4600	.4700	.4800	.4900	.5000

n = 2

r \ p	.01	.02	.03	.04	.05	.06	.07	.08	.09	.10
0	.9801	.9604	.9409	.9216	.9025	.8836	.8649	.8464	.8281	.8100
1	.0198	.0392	.0582	.0768	.0950	.1128	.1302	.1472	.1638	.1800
2	.0001	.0004	.0009	.0016	.0025	.0036	.0049	.0064	.0081	.0100

	.11	.12	.13	.14	.15	.16	.17	.18	.19	.20
0	.7921	.7744	.7569	.7396	.7225	.7056	.6889	.6724	.6561	.6400
1	.1958	.2112	.2262	.2408	.2550	.2688	.2822	.2952	.3078	.3200
2	.0121	.0144	.0169	.0196	.0225	.0256	.0289	.0324	.0361	.0400

	.21	.22	.23	.24	.25	.26	.27	.28	.29	.30
0	.6241	.6084	.5929	.5776	.5625	.5476	.5329	.5184	.5041	.4900
1	.3318	.3432	.3542	.3648	.3750	.3848	.3942	.4032	.4118	.4200
2	.0441	.0484	.0529	.0576	.0625	.0676	.0729	.0784	.0841	.0900

TABLE A.1   (continued)

n = 2 (cont.)

r \ p	.31	.32	.33	.34	.35	.36	.37	.38	.39	.40
0	.4761	.4624	.4489	.4356	.4225	.4096	.3969	.3844	.3721	.3600
1	.4278	.4352	.4422	.4488	.4550	.4608	.4662	.4712	.4758	.4800
2	.0961	.1024	.1089	.1156	.1225	.1296	.1369	.1444	.1521	.1600

r \ p	.41	.42	.43	.44	.45	.46	.47	.48	.49	.50
0	.3481	.3364	.3249	.3136	.3025	.2916	.2809	.2704	.2601	.2500
1	.4838	.4872	.4902	.4928	.4950	.4968	.4982	.4992	.4998	.5000
2	.1681	.1764	.1849	.1936	.2025	.2116	.2209	.2304	.2401	.2500

n = 3

r \ p	.01	.02	.03	.04	.05	.06	.07	.08	.09	.10
0	.9704	.9412	.9127	.8847	.8574	.8306	.8044	.7787	.7536	.7290
1	.0294	.0576	.0847	.1106	.1354	.1590	.1816	.2031	.2236	.2430
2	.0003	.0012	.0026	.0046	.0071	.0102	.0137	.0177	.0221	.0270
3	.0000	.0000	.0000	.0001	.0001	.0002	.0003	.0005	.0007	.0010

r \ p	.11	.12	.13	.14	.15	.16	.17	.18	.19	.20
0	.7050	.6815	.6585	.6361	.6141	.5927	.5718	.5514	.5314	.5120
1	.2614	.2788	.2952	.3106	.3251	.3387	.3513	.3631	.3740	.3840
2	.0323	.0380	.0441	.0506	.0574	.0645	.0720	.0797	.0877	.0960
3	.0013	.0017	.0022	.0027	.0034	.0041	.0049	.0058	.0069	.0080

r \ p	.21	.22	.23	.24	.25	.26	.27	.28	.29	.30
0	.4930	.4746	.4565	.4390	.4219	.4052	.3890	.3732	.3579	.3430
1	.3932	.4015	.4091	.4159	.4219	.4271	.4316	.4355	.4386	.4410
2	.1045	.1133	.1222	.1313	.1406	.1501	.1597	.1693	.1791	.1890
3	.0093	.0106	.0122	.0138	.0156	.0176	.0197	.0220	.0244	.0270

r \ p	.31	.32	.33	.34	.35	.36	.37	.38	.39	.40
0	.3285	.3144	.3008	.2875	.2746	.2621	.2500	.2383	.2270	.2160
1	.4428	.4439	.4444	.4443	.4436	.4424	.4406	.4382	.4354	.4320
2	.1989	.2089	.2189	.2289	.2389	.2488	.2587	.2686	.2783	.2880
3	.0298	.0328	.0359	.0393	.0429	.0467	.0507	.0549	.0593	.0640

r \ p	.41	.42	.43	.44	.45	.46	.47	.48	.49	.50
0	.2054	.1951	.1852	.1756	.1664	.1575	.1489	.1406	.1327	.1250
1	.4282	.4239	.4191	.4140	.4084	.4024	.3961	.3894	.3823	.3750
2	.2975	.3069	.3162	.3252	.3341	.3428	.3512	.3594	.3674	.3750
3	.0689	.0741	.0795	.0852	.0911	.0973	.1038	.1106	.1176	.1250

TABLE A.1 (continued)

n = 4

r \ p	.01	.02	.03	.04	.05	.06	.07	.08	.09	.10
0	.9606	.9224	.8853	.8493	.8145	.7807	.7481	.7164	.6857	.6561
1	.0388	.0753	.1095	.1416	.1715	.1993	.2252	.2492	.2713	.2916
2	.0006	.0023	.0051	.0088	.0135	.0191	.0254	.0325	.0402	.0486
3	.0000	.0000	.0001	.0002	.0005	.0008	.0013	.0019	.0027	.0036
4	.0000	.0000	.0000	.0000	.0000	.0000	.0000	.0000	.0001	.0001

r \ p	.11	.12	.13	.14	.15	.16	.17	.18	.19	.20
0	.6274	.5997	.5729	.5470	.5220	.4979	.4746	.4521	.4305	.4096
1	.3102	.3271	.3424	.3562	.3685	.3793	.3888	.3970	.4039	.4096
2	.0575	.0669	.0767	.0870	.0975	.1084	.1195	.1307	.1421	.1536
3	.0047	.0061	.0076	.0094	.0115	.0138	.0163	.0191	.0222	.0256
4	.0001	.0002	.0003	.0004	.0005	.0007	.0008	.0010	.0013	.0016

r \ p	.21	.22	.23	.24	.25	.26	.27	.28	.29	.30
0	.3895	.3702	.3515	.3336	.3164	.2999	.2840	.2687	.2541	.2401
1	.4142	.4176	.4200	.4214	.4219	.4214	.4201	.4180	.4152	.4116
2	.1651	.1767	.1882	.1996	.2109	.2221	.2331	.2439	.2544	.2646
3	.0293	.0332	.0375	.0420	.0469	.0520	.0575	.0632	.0693	.0756
4	.0019	.0023	.0028	.0033	.0039	.0046	.0053	.0061	.0071	.0081

r \ p	.31	.32	.33	.34	.35	.36	.37	.38	.39	.40
0	.2267	.2138	.2015	.1897	.1785	.1678	.1575	.1478	.1385	.1296
1	.4074	.4025	.3970	.3910	.3845	.3775	.3701	.3623	.3541	.3456
2	.2745	.2841	.2933	.3021	.3105	.3185	.3260	.3330	.3396	.3456
3	.0822	.0891	.0963	.1038	.1115	.1194	.1276	.1361	.1447	.1536
4	.0092	.0105	.0119	.0134	.0150	.0168	.0187	.0209	.0231	.0256

r \ p	.41	.42	.43	.44	.45	.46	.47	.48	.49	.50
0	.1212	.1132	.1056	.0983	.0915	.0850	.0789	.0731	.0677	.0625
1	.3368	.3278	.3185	.3091	.2995	.2897	.2799	.2700	.2600	.2500
2	.3511	.3560	.3604	.3643	.3675	.3702	.3723	.3738	.3747	.3750
3	.1627	.1719	.1813	.1908	.2005	.2102	.2201	.2300	.2400	.2500
4	.0283	.0311	.0342	.0375	.0410	.0448	.0488	.0531	.0576	.0625

n = 5

r \ p	.01	.02	.03	.04	.05	.06	.07	.08	.09	.10
0	.9510	.9039	.8587	.8154	.7738	.7339	.6957	.6591	.6240	.5905
1	.0480	.0922	.1328	.1699	.2036	.2342	.2618	.2866	.3086	.3280
2	.0010	.0038	.0082	.0142	.0214	.0299	.0394	.0498	.0610	.0729
3	.0000	.0001	.0003	.0006	.0011	.0019	.0030	.0043	.0060	.0081
4	.0000	.0000	.0000	.0000	.0000	.0001	.0001	.0002	.0003	.0004

TABLE A.1   (*continued*)

n = 5 (cont.)

p \ r	.11	.12	.13	.14	.15	.16	.17	.18	.19	.20
0	.5584	.5277	.4984	.4704	.4437	.4182	.3939	.3707	.3487	.3277
1	.3451	.3598	.3724	.3829	.3915	.3983	.4034	.4069	.4089	.4096
2	.0853	.0981	.1113	.1247	.1382	.1517	.1652	.1786	.1919	.2048
3	.0105	.0134	.0166	.0203	.0244	.0289	.0338	.0392	.0450	.0512
4	.0007	.0009	.0012	.0017	.0022	.0028	.0035	.0043	.0053	.0064
5	.0000	.0000	.0000	.0001	.0001	.0001	.0001	.0002	.0002	.0003

p \ r	.21	.22	.23	.24	.25	.26	.27	.28	.29	.30
0	.3077	.2887	.2707	.2536	.2373	.2219	.2073	.1935	.1804	.1681
1	.4090	.4072	.4043	.4003	.3955	.3898	.3834	.3762	.3685	.3602
2	.2174	.2297	.2415	.2529	.2637	.2739	.2836	.2926	.3010	.3087
3	.0578	.0648	.0721	.0798	.0879	.0962	.1049	.1138	.1229	.1323
4	.0077	.0091	.0108	.0126	.0146	.0169	.0194	.0221	.0251	.0284
5	.0004	.0005	.0006	.0008	.0010	.0012	.0014	.0017	.0021	.0024

p \ r	.31	.32	.33	.34	.35	.36	.37	.38	.39	.40
0	.1564	.1454	.1350	.1252	.1160	.1074	.0992	.0916	.0845	.0778
1	.3513	.3421	.3325	.3226	.3124	.3020	.2914	.2808	.2700	.2592
2	.3157	.3220	.3275	.3323	.3364	.3397	.3423	.3441	.3452	.3456
3	.1418	.1515	.1613	.1712	.1811	.1911	.2010	.2109	.2207	.2304
4	.0319	.0357	.0397	.0441	.0488	.0537	.0590	.0646	.0706	.0768
5	.0029	.0034	.0039	.0045	.0053	.0060	.0069	.0079	.0090	.0102

p \ r	.41	.42	.43	.44	.45	.46	.47	.48	.49	.50
0	.0715	.0656	.0602	.0551	.0503	.0459	.0418	.0380	.0345	.0312
1	.2484	.2376	.2270	.2164	.2059	.1956	.1854	.1755	.1657	.1562
2	.3452	.3442	.3424	.3400	.3369	.3332	.3289	.3240	.3185	.3125
3	.2399	.2492	.2583	.2671	.2757	.2838	.2916	.2990	.3060	.3125
4	.0834	.0902	.0974	.1049	.1128	.1209	.1293	.1380	.1470	.1562
5	.0116	.0131	.0147	.0165	.0185	.0206	.0229	.0255	.0282	.0312

n = 6

p \ r	.01	.02	.03	.04	.05	.06	.07	.08	.09	.10
0	.9415	.8858	.8330	.7828	.7351	.6899	.6470	.6064	.5679	.5314
1	.0571	.1085	.1546	.1957	.2321	.2642	.2922	.3164	.3370	.3543
2	.0014	.0055	.0120	.0204	.0305	.0422	.0550	.0688	.0833	.0984
3	.0000	.0002	.0005	.0011	.0021	.0036	.0055	.0080	.0110	.0146
4	.0000	.0000	.0000	.0000	.0001	.0002	.0003	.0005	.0008	.0012
5	.0000	.0000	.0000	.0000	.0000	.0000	.0000	.0000	.0000	.0001

TABLE A.1   (continued)

n = 6 (cont.)

p / r	.11	.12	.13	.14	.15	.16	.17	.18	.19	.20
0	.4970	.4644	.4336	.4046	.3771	.3513	.3269	.3040	.2824	.2621
1	.3685	.3800	.3888	.3952	.3993	.4015	.4018	.4004	.3975	.3932
2	.1139	.1295	.1452	.1608	.1762	.1912	.2057	.2197	.2331	.2458
3	.0188	.0236	.0289	.0349	.0415	.0486	.0562	.0643	.0729	.0819
4	.0017	.0024	.0032	.0043	.0055	.0069	.0086	.0106	.0128	.0154
5	.0001	.0001	.0002	.0003	.0004	.0005	.0007	.0009	.0012	.0015
6	.0000	.0000	.0000	.0000	.0000	.0000	.0000	.0000	.0000	.0001

p / r	.21	.22	.23	.24	.25	.26	.27	.28	.29	.30
0	.2431	.2252	.2084	.1927	.1780	.1642	.1513	.1393	.1281	.1176
1	.3877	.3811	.3735	.3651	.3560	.3462	.3358	.3251	.3139	.3025
2	.2577	.2687	.2789	.2882	.2966	.3041	.3105	.3160	.3206	.3241
3	.0913	.1011	.1111	.1214	.1318	.1424	.1531	.1639	.1746	.1852
4	.0182	.0214	.0249	.0287	.0330	.0375	.0425	.0478	.0535	.0595
5	.0019	.0024	.0030	.0036	.0044	.0053	.0063	.0074	.0087	.0102
6	.0001	.0001	.0001	.0002	.0002	.0003	.0004	.0005	.0006	.0007

p / r	.31	.32	.33	.34	.35	.36	.37	.38	.39	.40
0	.1079	.0989	.0905	.0827	.0754	.0687	.0625	.0568	.0515	.0467
1	.2909	.2792	.2673	.2555	.2437	.2319	.2203	.2089	.1976	.1866
2	.3267	.3284	.3292	.3290	.3280	.3261	.3235	.3201	.3159	.3110
3	.1957	.2061	.2162	.2260	.2355	.2446	.2533	.2616	.2693	.2765
4	.0660	.0727	.0799	.0873	.0951	.1032	.1116	.1202	.1291	.1382
5	.0119	.0137	.0157	.0180	.0205	.0232	.0262	.0295	.0330	.0369
6	.0009	.0011	.0013	.0015	.0018	.0022	.0026	.0030	.0035	.0041

p / r	.41	.42	.43	.44	.45	.46	.47	.48	.49	.50
0	.0422	.0381	.0343	.0308	.0277	.0248	.0222	.0198	.0176	.0156
1	.1759	.1654	.1552	.1454	.1359	.1267	.1179	.1095	.1014	.0938
2	.3055	.2994	.2928	.2856	.2780	.2699	.2615	.2527	.2436	.2344
3	.2831	.2891	.2945	.2992	.3032	.3065	.3091	.3110	.3121	.3125
4	.1475	.1570	.1666	.1763	.1861	.1958	.2056	.2153	.2249	.2344
5	.0410	.0455	.0503	.0554	.0609	.0667	.0729	.0795	.0864	.0938
6	.0048	.0055	.0063	.0073	.0083	.0095	.0108	.0122	.0138	.0156

n = 7

p / r	.01	.02	.03	.04	.05	.06	.07	.08	.09	.10
0	.9321	.8681	.8080	.7514	.6983	.6485	.6017	.5578	.5168	.4783
1	.0659	.1240	.1749	.2192	.2573	.2897	.3170	.3396	.3578	.3720
2	.0020	.0076	.0162	.0274	.0406	.0555	.0716	.0886	.1061	.1240
3	.0000	.0003	.0008	.0019	.0036	.0059	.0090	.0128	.0175	.0230
4	.0000	.0000	.0000	.0001	.0002	.0004	.0007	.0011	.0017	.0026
5	.0000	.0000	.0000	.0000	.0000	.0000	.0000	.0001	.0001	.0002

TABLE A.1 (*continued*)

n = 7 (cont.)

r \ p	.11	.12	.13	.14	.15	.16	.17	.18	.19	.20
0	.4423	.4087	.3773	.3479	.3206	.2951	.2714	.2493	.2288	.2097
1	.3827	.3901	.3946	.3965	.3960	.3935	.3891	.3830	.3756	.3670
2	.1419	.1596	.1769	.1936	.2097	.2248	.2391	.2523	.2643	.2753
3	.0292	.0363	.0441	.0525	.0617	.0714	.0816	.0923	.1033	.1147
4	.0036	.0049	.0066	.0086	.0109	.0136	.0167	.0203	.0242	.0287
5	.0003	.0004	.0006	.0008	.0012	.0016	.0021	.0027	.0034	.0043
6	.0000	.0000	.0000	.0000	.0001	.0001	.0001	.0002	.0003	.0004

r \ p	.21	.22	.23	.24	.25	.26	.27	.28	.29	.30
0	.1920	.1757	.1605	.1465	.1335	.1215	.1105	.1003	.0910	.0824
1	.3573	.3468	.3356	.3237	.3115	.2989	.2860	.2731	.2600	.2471
2	.2850	.2935	.3007	.3067	.3115	.3150	.3174	.3186	.3186	.3177
3	.1263	.1379	.1497	.1614	.1730	.1845	.1956	.2065	.2169	.2269
4	.0336	.0389	.0447	.0510	.0577	.0648	.0724	.0803	.0886	.0972
5	.0054	.0066	.0080	.0097	.0115	.0137	.0161	.0187	.0217	.0250
6	.0005	.0006	.0008	.0010	.0013	.0016	.0020	.0024	.0030	.0036
7	.0000	.0000	.0000	.0000	.0001	.0001	.0001	.0001	.0002	.0002

r \ p	.31	.32	.33	.34	.35	.36	.37	.38	.39	.40
0	.0745	.0672	.0606	.0546	.0490	.0440	.0394	.0352	.0314	.0280
1	.2342	.2215	.2090	.1967	.1848	.1732	.1619	.1511	.1407	.1306
2	.3156	.3127	.3088	.3040	.2985	.2922	.2853	.2778	.2698	.2613
3	.2363	.2452	.2535	.2610	.2679	.2740	.2793	.2838	.2875	.2903
4	.1062	.1154	.1248	.1345	.1442	.1541	.1640	.1739	.1838	.1935
5	.0286	.0326	.0369	.0416	.0466	.0520	.0578	.0640	.0705	.0774
6	.0043	.0051	.0061	.0071	.0084	.0098	.0113	.0131	.0150	.0172
7	.0003	.0003	.0004	.0005	.0006	.0008	.0009	.0011	.0014	.0016

r \ p	.41	.42	.43	.44	.45	.46	.47	.48	.49	.50
0	.0249	.0221	.0195	.0173	.0152	.0134	.0117	.0103	.0090	.0078
1	.1211	.1119	.1032	.0950	.0872	.0798	.0729	.0664	.0604	.0547
2	.2524	.2431	.2336	.2239	.2140	.2040	.1940	.1840	.1740	.1641
3	.2923	.2934	.2937	.2932	.2918	.2897	.2867	.2830	.2786	.2734
4	.2031	.2125	.2216	.2304	.2388	.2468	.2543	.2612	.2676	.2734
5	.0847	.0923	.1003	.1086	.1172	.1261	.1353	.1447	.1543	.1641
6	.0196	.0223	.0252	.0284	.0320	.0358	.0400	.0445	.0494	.0547
7	.0019	.0023	.0027	.0032	.0037	.0044	.0051	.0059	.0068	.0078

n = 8

r \ p	.01	.02	.03	.04	.05	.06	.07	.08	.09	.10
0	.9227	.8508	.7837	.7214	.6634	.6096	.5596	.5132	.4703	.4305
1	.0746	.1389	.1939	.2405	.2793	.3113	.3370	.3570	.3721	.3826
2	.0026	.0099	.0210	.0351	.0515	.0695	.0888	.1087	.1288	.1488
3	.0001	.0004	.0013	.0029	.0054	.0089	.0134	.0189	.0255	.0331
4	.0000	.0000	.0001	.0002	.0004	.0007	.0013	.0021	.0031	.0046
5	.0000	.0000	.0000	.0000	.0000	.0000	.0001	.0001	.0002	.0004

TABLE A.1   (continued)

n = 8 (cont.)

p r	.11	.12	.13	.14	.15	.16	.17	.18	.19	.20
0	.3937	.3596	.3282	.2992	.2725	.2479	.2252	.2044	.1853	.1678
1	.3892	.3923	.3923	.3897	.3847	.3777	.3691	.3590	.3477	.3355
2	.1684	.1872	.2052	.2220	.2376	.2518	.2646	.2758	.2855	.2936
3	.0416	.0511	.0613	.0723	.0839	.0959	.1084	.1211	.1339	.1468
4	.0064	.0087	.0115	.0147	.0185	.0228	.0277	.0332	.0393	.0459
5	.0006	.0009	.0014	.0019	.0026	.0035	.0045	.0058	.0074	.0092
6	.0000	.0001	.0001	.0002	.0002	.0003	.0005	.0006	.0009	.0011
7	.0000	.0000	.0000	.0000	.0000	.0000	.0000	.0000	.0001	.0001

p r	.21	.22	.23	.24	.25	.26	.27	.28	.29	.30
0	.1517	.1370	.1236	.1113	.1001	.0899	.0806	.0722	.0646	.0576
1	.3226	.3092	.2953	.2812	.2670	.2527	.2386	.2247	.2110	.1977
2	.3002	.3052	.3087	.3108	.3115	.3108	.3089	.3058	.3017	.2965
3	.1596	.1722	.1844	.1963	.2076	.2184	.2285	.2379	.2464	.2541
4	.0530	.0607	.0689	.0775	.0865	.0959	.1056	.1156	.1258	.1361
5	.0113	.0137	.0165	.0196	.0231	.0270	.0313	.0360	.0411	.0467
6	.0015	.0019	.0025	.0031	.0038	.0047	.0058	.0070	.0084	.0100
7	.0001	.0002	.0002	.0003	.0004	.0005	.0006	.0008	.0010	.0012
8	.0000	.0000	.0000	.0000	.0000	.0000	.0000	.0000	.0001	.0001

p r	.31	.32	.33	.34	.35	.36	.37	.38	.39	.40
0	.0514	.0457	.0406	.0360	.0319	.0281	.0248	.0218	.0192	.0168
1	.1847	.1721	.1600	.1484	.1373	.1267	.1166	.1071	.0981	.0896
2	.2904	.2835	.2758	.2675	.2587	.2494	.2397	.2297	.2194	.2090
3	.2609	.2668	.2717	.2756	.2786	.2805	.2815	.2815	.2806	.2787
4	.1465	.1569	.1673	.1775	.1875	.1973	.2067	.2157	.2242	.2322
5	.0527	.0591	.0659	.0732	.0808	.0888	.0971	.1058	.1147	.1239
6	.0118	.0139	.0162	.0188	.0217	.0250	.0285	.0324	.0367	.0413
7	.0015	.0019	.0023	.0028	.0033	.0040	.0048	.0057	.0067	.0079
8	.0001	.0001	.0001	.0002	.0002	.0003	.0004	.0004	.0005	.0007

p r	.41	.42	.43	.44	.45	.46	.47	.48	.49	.50
0	.0147	.0128	.0111	.0097	.0084	.0072	.0062	.0053	.0046	.0039
1	.0816	.0742	.0672	.0608	.0548	.0493	.0442	.0395	.0352	.0312
2	.1985	.1880	.1776	.1672	.1569	.1469	.1371	.1275	.1183	.1094
3	.2759	.2723	.2679	.2627	.2568	.2503	.2431	.2355	.2273	.2188
4	.2397	.2465	.2526	.2580	.2627	.2665	.2695	.2717	.2730	.2734
5	.1332	.1428	.1525	.1622	.1719	.1816	.1912	.2006	.2098	.2188
6	.0463	.0517	.0575	.0637	.0703	.0774	.0848	.0926	.1008	.1094
7	.0092	.0107	.0124	.0143	.0164	.0188	.0215	.0244	.0277	.0312
8	.0008	.0010	.0012	.0014	.0017	.0020	.0024	.0028	.0033	.0039

TABLE A.1   (*continued*)

n = 9

r \ p	.01	.02	.03	.04	.05	.06	.07	.08	.09	.10
0	.9135	.8337	.7602	.6925	.6302	.5730	.5204	.4722	.4279	.3874
1	.0830	.1531	.2116	.2597	.2985	.3292	.3525	.3695	.3809	.3874
2	.0034	.0125	.0262	.0433	.0629	.0840	.1061	.1285	.1507	.1722
3	.0001	.0006	.0019	.0042	.0077	.0125	.0186	.0261	.0348	.0446
4	.0000	.0000	.0001	.0003	.0006	.0012	.0021	.0034	.0052	.0074
5	.0000	.0000	.0000	.0000	.0000	.0001	.0002	.0003	.0005	.0008
6	.0000	.0000	.0000	.0000	.0000	.0000	.0000	.0000	.0000	.0001

r \ p	.11	.12	.13	.14	.15	.16	.17	.18	.19	.20
0	.3504	.3165	.2855	.2573	.2316	.2082	.1869	.1676	.1501	.1342
1	.3897	.3884	.3840	.3770	.3679	.3569	.3446	.3312	.3169	.3020
2	.1927	.2119	.2295	.2455	.2597	.2720	.2823	.2908	.2973	.3020
3	.0556	.0674	.0800	.0933	.1069	.1209	.1349	.1489	.1627	.1762
4	.0103	.0138	.0179	.0228	.0283	.0345	.0415	.0490	.0573	.0661
5	.0013	.0019	.0027	.0037	.0050	.0066	.0085	.0108	.0134	.0165
6	.0001	.0002	.0003	.0004	.0006	.0008	.0012	.0016	.0021	.0028
7	.0000	.0000	.0000	.0000	.0000	.0001	.0001	.0001	.0002	.0003

r \ p	.21	.22	.23	.24	.25	.26	.27	.28	.29	.30
0	.1199	.1069	.0952	.0846	.0751	.0665	.0589	.0520	.0458	.0404
1	.2867	.2713	.2558	.2404	.2253	.2104	.1960	.1820	.1685	.1556
2	.3049	.3061	.3056	.3037	.3003	.2957	.2899	.2831	.2754	.2668
3	.1891	.2014	.2130	.2238	.2336	.2424	.2502	.2569	.2624	.2668
4	.0754	.0852	.0954	.1060	.1168	.1278	.1388	.1499	.1608	.1715
5	.0200	.0240	.0285	.0335	.0389	.0449	.0513	.0583	.0657	.0735
6	.0036	.0045	.0057	.0070	.0087	.0105	.0127	.0151	.0179	.0210
7	.0004	.0005	.0007	.0010	.0012	.0016	.0020	.0025	.0031	.0039
8	.0000	.0000	.0001	.0001	.0001	.0001	.0002	.0002	.0003	.0004

r \ p	.31	.32	.33	.34	.35	.36	.37	.38	.39	.40
0	.0355	.0311	.0272	.0238	.0207	.0180	.0156	.0135	.0117	.0101
1	.1433	.1317	.1206	.1102	.1004	.0912	.0826	.0747	.0673	.0605
2	.2576	.2478	.2376	.2270	.2162	.2052	.1941	.1831	.1721	.1612
3	.2701	.2721	.2731	.2729	.2716	.2693	.2660	.2618	.2567	.2508
4	.1820	.1921	.2017	.2109	.2194	.2272	.2344	.2407	.2462	.2508
5	.0818	.0904	.0994	.1086	.1181	.1278	.1376	.1475	.1574	.1672
6	.0245	.0284	.0326	.0373	.0424	.0479	.0539	.0603	.0671	.0743
7	.0047	.0057	.0069	.0082	.0098	.0116	.0136	.0158	.0184	.0212
8	.0005	.0007	.0008	.0011	.0013	.0016	.0020	.0024	.0029	.0035
9	.0000	.0000	.0000	.0001	.0001	.0001	.0001	.0002	.0002	.0003

TABLE A.1   (*continued*)

n = 9 (*cont.*)

r \ p	.41	.42	.43	.44	.45	.46	.47	.48	.49	.50
0	.0087	.0074	.0064	.0054	.0046	.0039	.0033	.0028	.0023	.0020
1	.0542	.0484	.0431	.0383	.0339	.0299	.0263	.0231	.0202	.0176
2	.1506	.1402	.1301	.1204	.1110	.1020	.0934	.0853	.0776	.0703
3	.2442	.2369	.2291	.2207	.2119	.2027	.1933	.1837	.1739	.1641
4	.2545	.2573	.2592	.2601	.2600	.2590	.2571	.2543	.2506	.2461
5	.1769	.1863	.1955	.2044	.2128	.2207	.2280	.2347	.2408	.2461
6	.0819	.0900	.0983	.1070	.1160	.1253	.1348	.1445	.1542	.1641
7	.0244	.0279	.0318	.0360	.0407	.0458	.0512	.0571	.0635	.0703
8	.0042	.0051	.0060	.0071	.0083	.0097	.0114	.0132	.0153	.0176
9	.0003	.0004	.0005	.0006	.0008	.0009	.0011	.0014	.0016	.0020

n = 10

r \ p	.01	.02	.03	.04	.05	.06	.07	.08	.09	.10
0	.9044	.8171	.7374	.6648	.5987	.5386	.4840	.4344	.3894	.3487
1	.0914	.1667	.2281	.2770	.3151	.3438	.3643	.3777	.3851	.3874
2	.0042	.0153	.0317	.0519	.0746	.0988	.1234	.1478	.1714	.1937
3	.0001	.0008	.0026	.0058	.0105	.0168	.0248	.0343	.0452	.0574
4	.0000	.0000	.0001	.0004	.0010	.0019	.0033	.0052	.0078	.0112
5	.0000	.0000	.0000	.0000	.0001	.0001	.0003	.0005	.0009	.0015
6	.0000	.0000	.0000	.0000	.0000	.0000	.0000	.0000	.0001	.0001

r \ p	.11	.12	.13	.14	.15	.16	.17	.18	.19	.20
0	.3118	.2785	.2484	.2213	.1969	.1749	.1552	.1374	.1216	.1074
1	.3854	.3798	.3712	.3603	.3474	.3331	.3178	.3017	.2852	.2684
2	.2143	.2330	.2496	.2639	.2759	.2856	.2929	.2980	.3010	.3020
3	.0706	.0847	.0995	.1146	.1298	.1450	.1600	.1745	.1883	.2013
4	.0153	.0202	.0260	.0326	.0401	.0483	.0573	.0670	.0773	.0881
5	.0023	.0033	.0047	.0064	.0085	.0111	.0141	.0177	.0218	.0264
6	.0002	.0004	.0006	.0009	.0012	.0018	.0024	.0032	.0043	.0055
7	.0000	.0000	.0000	.0001	.0001	.0002	.0003	.0004	.0006	.0008
8	.0000	.0000	.0000	.0000	.0000	.0000	.0000	.0000	.0001	.0001

r \ p	.21	.22	.23	.24	.25	.26	.27	.28	.29	.30
0	.0947	.0834	.0733	.0643	.0563	.0492	.0430	.0374	.0326	.0282
1	.2517	.2351	.2188	.2030	.1877	.1730	.1590	.1456	.1330	.1211
2	.3011	.2984	.2942	.2885	.2816	.2735	.2646	.2548	.2444	.2335
3	.2134	.2244	.2343	.2429	.2503	.2563	.2609	.2642	.2662	.2668
4	.0993	.1108	.1225	.1343	.1460	.1576	.1689	.1798	.1903	.2001
5	.0317	.0375	.0439	.0509	.0584	.0664	.0750	.0839	.0933	.1029
6	.0070	.0088	.0109	.0134	.0162	.0195	.0231	.0272	.0317	.0368
7	.0011	.0014	.0019	.0024	.0031	.0039	.0049	.0060	.0074	.0090
8	.0001	.0002	.0002	.0003	.0004	.0005	.0007	.0009	.0011	.0014
9	.0000	.0000	.0000	.0000	.0000	.0000	.0001	.0001	.0001	.0001

TABLE A.1   (*continued*)

n = 10 (cont.)

r \ p	.31	.32	.33	.34	.35	.36	.37	.38	.39	.40
0	.0245	.0211	.0182	.0157	.0135	.0115	.0098	.0084	.0071	.0060
1	.1099	.0995	.0898	.0808	.0725	.0649	.0578	.0514	.0456	.0403
2	.2222	.2107	.1990	.0873	.1757	.1642	.1529	.1419	.1312	.1209
3	.2662	.2644	.2614	.2573	.2522	.2462	.2394	.2319	.2237	.2150
4	.2093	.2177	.2253	.2320	.2377	.2424	.2461	.2487	.2503	.2508
5	.1128	.1229	.1332	.1434	.1536	.1636	.1734	.1829	.1920	.2007
6	.0422	.0482	.0547	.0616	.0689	.0767	.0849	.0934	.1023	.1115
7	.0108	.0130	.0154	.0181	.0212	.0247	.0285	.0327	.0374	.0425
8	.0018	.0023	.0028	.0035	.0043	.0052	.0063	.0075	.0090	.0106
9	.0002	.0002	.0003	.0004	.0005	.0006	.0008	.0010	.0013	.0016
10	.0000	.0000	.0000	.0000	.0000	.0000	.0000	.0001	.0001	.0001

r \ p	.41	.42	.43	.44	.45	.46	.47	.48	.49	.50
0	.0051	.0043	.0036	.0030	.0025	.0021	.0017	.0014	.0012	.0010
1	.0355	.0312	.0273	.0238	.0207	.0180	.0155	.0133	.0114	.0098
2	.1111	.1017	.0927	.0843	.0763	.0688	.0619	.0554	.0494	.0439
3	.2058	.1963	.1865	.1765	.1665	.1564	.1464	.1364	.1267	.1172
4	.2503	.2488	.2462	.2427	.2384	.2331	.2271	.2204	.2130	.2051
5	.2087	.2162	.2229	.2289	.2340	.2383	.2417	.2441	.2456	.2461
6	.1209	.1304	.1401	.1499	.1596	.1692	.1786	.1878	.1966	.2051
7	.0480	.0540	.0604	.0673	.0746	.0824	.0905	.0991	.1080	.1172
8	.0125	.0147	.0171	.0198	.0229	.0263	.0301	.0343	.0389	.0439
9	.0019	.0024	.0029	.0035	.0042	.0050	.0059	.0070	.0083	.0098
10	.0001	.0002	.0002	.0003	.0003	.0004	.0005	.0006	.0008	.0010

n = 11

r \ p	.01	.02	.03	.04	.05	.06	.07	.08	.09	.10
0	.8953	.8007	.7153	.6382	.5688	.5063	.4501	.3996	.3544	.3138
1	.0995	.1798	.2433	.2925	.3293	.3555	.3727	.3823	.3855	.3835
2	.0050	.0183	.0376	.0609	.0867	.1135	.1403	.1662	.1906	.2131
3	.0002	.0011	.0035	.0076	.0137	.0217	.0317	.0434	.0566	.0710
4	.0000	.0000	.0002	.0006	.0014	.0028	.0048	.0075	.0112	.0158
5	.0000	.0000	.0000	.0000	.0001	.0002	.0005	.0009	.0015	.0025
6	.0000	.0000	.0000	.0000	.0000	.0000	.0000	.0001	.0002	.0003

r \ p	.11	.12	.13	.14	.15	.16	.17	.18	.19	.20
0	.2775	.2451	.2161	.1903	.1673	.1469	.1288	.1127	.0985	.0859
1	.3773	.3676	.3552	.3408	.3248	.3078	.2901	.2721	.2541	.2362
2	.2332	.2507	.2654	.2774	.2866	.2932	.2971	.2987	.2980	.2953
3	.0865	.1025	.1190	.1355	.1517	.1675	.1826	.1967	.2097	.2215
4	.0214	.0280	.0356	.0441	.0536	.0638	.0748	.0864	.0984	.1107
5	.0037	.0053	.0074	.0101	.0132	.0170	.0214	.0265	.0323	.0388
6	.0005	.0007	.0011	.0016	.0023	.0032	.0044	.0058	.0076	.0097
7	.0000	.0001	.0001	.0002	.0003	.0004	.0006	.0009	.0013	.0017
8	.0000	.0000	.0000	.0000	.0000	.0000	.0001	.0001	.0001	.0002

TABLE A.1 (continued)

n = 11 (cont.)

r \ p	.21	.22	.23	.24	.25	.26	.27	.28	.29	.30
0	.0748	.0650	.0564	.0489	.0422	.0364	.0314	.0270	.0231	.0198
1	.2187	.2017	.1854	.1697	.1549	.1408	.1276	.1153	.1038	.0932
2	.2907	.2845	.2768	.2680	.2581	.2474	.2360	.2242	.2121	.1998
3	.2318	.2407	.2481	.2539	.2581	.2608	.2619	.2616	.2599	.2568
4	.1232	.1358	.1482	.1603	.1721	.1832	.1937	.2035	.2123	.2201
5	.0459	.0536	.0620	.0709	.0803	.0901	.1003	.1108	.1214	.1321
6	.0122	.0151	.0185	.0224	.0268	.0317	.0371	.0431	.0496	.0566
7	.0023	.0030	.0039	.0050	.0064	.0079	.0098	.0120	.0145	.0173
8	.0003	.0004	.0006	.0008	.0011	.0014	.0018	.0023	.0030	.0037
9	.0000	.0000	.0001	.0001	.0001	.0002	.0002	.0003	.0004	.0005

r \ p	.31	.32	.33	.34	.35	.36	.37	.38	.39	.40
0	.0169	.0144	.0122	.0104	.0088	.0074	.0062	.0052	.0044	.0036
1	.0834	.0744	.0662	.0587	.0518	.0457	.0401	.0351	.0306	.0266
2	.1874	.1751	.1630	.1511	.1395	.1284	.1177	.1075	.0978	.0887
3	.2526	.2472	.2408	.2335	.2254	.2167	.2074	.1977	.1876	.1774
4	.2269	.2326	.2372	.2406	.2428	.2438	.2436	.2423	.2399	.2365
5	.1427	.1533	.1636	.1735	.1830	.1920	.2003	.2079	.2148	.2207
6	.0641	.0721	.0806	.0894	.0985	.1080	.1176	.1274	.1373	.1471
7	.0206	.0242	.0283	.0329	.0379	.0434	.0494	.0558	.0627	.0701
8	.0046	.0057	.0070	.0085	.0102	.0122	.0145	.0171	.0200	.0234
9	.0007	.0009	.0011	.0015	.0018	.0023	.0028	.0035	.0043	.0052
10	.0001	.0001	.0001	.0001	.0002	.0003	.0003	.0004	.0005	.0007

r \ p	.41	.42	.43	.44	.45	.46	.47	.48	.49	.50
0	.0030	.0025	.0021	.0017	.0014	.0011	.0009	.0008	.0006	.0005
1	.0231	.0199	.0171	.0147	.0125	.0107	.0090	.0076	.0064	.0054
2	.0801	.0721	.0646	.0577	.0513	.0454	.0401	.0352	.0308	.0269
3	.1670	.1566	.1462	.1359	.1259	.1161	.1067	.0976	.0888	.0806
4	.2321	.2267	.2206	.2136	.2060	.1978	.1892	.1801	.1707	.1611
5	.2258	.2299	.2329	.2350	.2360	.2359	.2348	.2327	.2296	.2256
6	.1569	.1664	.1757	.1846	.1931	.2010	.2083	.2148	.2206	.2256
7	.0779	.0861	.0947	.1036	.1128	.1223	.1319	.1416	.1514	.1611
8	.0271	.0312	.0357	.0407	.0462	.0521	.0585	.0654	.0727	.0806
9	.0063	.0075	.0090	.0107	.0126	.0148	.0173	.0201	.0233	.0269
10	.0009	.0011	.0014	.0017	.0021	.0025	.0031	.0037	.0045	.0054
11	.0001	.0001	.0001	.0001	.0002	.0002	.0002	.0003	.0004	.0005

TABLE A.1 (continued)

n = 12

r \ p	.01	.02	.03	.04	.05	.06	.07	.08	.09	.10
0	.8864	.7847	.6938	.6127	.5404	.4759	.4186	.3677	.3225	.2824
1	.1074	.1922	.2575	.3064	.3413	.3645	.3781	.3837	.3827	.3766
2	.0060	.0216	.0438	.0702	.0988	.1280	.1565	.1835	.2082	.2301
3	.0002	.0015	.0045	.0098	.0173	.0272	.0393	.0532	.0686	.0852
4	.0000	.0001	.0003	.0009	.0021	.0039	.0067	.0104	.0153	.0213
5	.0000	.0000	.0000	.0001	.0002	.0004	.0008	.0014	.0024	.0038
6	.0000	.0000	.0000	.0000	.0000	.0000	.0001	.0001	.0003	.0005

r \ p	.11	.12	.13	.14	.15	.16	.17	.18	.19	.20
0	.2470	.2157	.1880	.1637	.1422	.1234	.1069	.0924	.0798	.0687
1	.3663	.3529	.3372	.3197	.3012	.2821	.2627	.2434	.2245	.2062
2	.2490	.2647	.2771	.2863	.2924	.2955	.2960	.2939	.2897	.2835
3	.1026	.1203	.1380	.1553	.1720	.1876	.2021	.2151	.2265	.2362
4	.0285	.0369	.0464	.0569	.0683	.0804	.0931	.1062	.1195	.1329
5	.0056	.0081	.0111	.0148	.0193	.0245	.0305	.0373	.0449	.0532
6	.0008	.0013	.0019	.0028	.0040	.0054	.0073	.0096	.0123	.0155
7	.0001	.0001	.0002	.0004	.0006	.0009	.0013	.0018	.0025	.0033
8	.0000	.0000	.0000	.0000	.0001	.0001	.0002	.0002	.0004	.0005
9	.0000	.0000	.0000	.0000	.0000	.0000	.0000	.0000	.0000	.0001

r \ p	.21	.22	.23	.24	.25	.26	.27	.28	.29	.30
0	.0591	.0507	.0434	.0371	.0317	.0270	.0229	.0194	.0164	.0138
1	.1885	.1717	.1557	.1407	.1267	.1137	.1016	.0906	.0804	.0712
2	.2756	.2663	.2558	.2444	.2323	.2197	.2068	.1937	.1807	.1678
3	.2442	.2503	.2547	.2573	.2581	.2573	.2549	.2511	.2460	.2397
4	.1460	.1589	.1712	.1828	.1936	.2034	.2122	.2197	.2261	.2311
5	.0621	.0717	.0818	.0924	.1032	.1143	.1255	.1367	.1477	.1585
6	.0193	.0236	.0285	.0340	.0401	.0469	.0542	.0620	.0704	.0792
7	.0044	.0057	.0073	.0092	.0115	.0141	.0172	.0207	.0246	.0291
8	.0007	.0010	.0014	.0018	.0024	.0031	.0040	.0050	.0063	.0078
9	.0001	.0001	.0002	.0003	.0004	.0005	.0007	.0009	.0011	.0015
10	.0000	.0000	.0000	.0000	.0000	.0001	.0001	.0001	.0001	.0002

r \ p	.31	.32	.33	.34	.35	.36	.37	.38	.39	.40
0	.0116	.0098	.0082	.0068	.0057	.0047	.0039	.0032	.0027	.0022
1	.0628	.0552	.0484	.0422	.0368	.0319	.0276	.0237	.0204	.0174
2	.1552	.1429	.1310	.1197	.1088	.0986	.0890	.0800	.0716	.0639
3	.2324	.2241	.2151	.2055	.1954	.1849	.1742	.1634	.1526	.1419
4	.2349	.2373	.2384	.2382	.2367	.2340	.2302	.2254	.2195	.2128
5	.1688	.1787	.1879	.1963	.2039	.2106	.2163	.2210	.2246	.2270
6	.0885	.0981	.1079	.1180	.1281	.1382	.1482	.1580	.1675	.1766
7	.0341	.0396	.0456	.0521	.0591	.0666	.0746	.0830	.0918	.1009
8	.0096	.0116	.0140	.0168	.0199	.0234	.0274	.0318	.0367	.0420
9	.0019	.0024	.0031	.0038	.0048	.0059	.0071	.0087	.0104	.0125
10	.0003	.0003	.0005	.0006	.0008	.0010	.0013	.0016	.0020	.0025
11	.0000	.0000	.0000	.0001	.0001	.0001	.0001	.0002	.0002	.0003

TABLE A.1 (*continued*)

n = 12 (cont.)

r \ p	.41	.42	.43	.44	.45	.46	.47	.48	.49	.50
0	.0018	.0014	.0012	.0010	.0008	.0006	.0005	.0004	.0003	.0002
1	.0148	.0126	.0106	.0090	.0075	.0063	.0052	.0043	.0036	.0029
2	.0567	.0502	.0442	.0388	.0339	.0294	.0255	.0220	.0189	.0161
3	.1314	.1211	.1111	.1015	.0923	.0836	.0754	.0676	.0604	.0537
4	.2054	.1973	.1886	.1794	.1700	.1602	.1504	.1405	.1306	.1208
5	.2284	.2285	.2276	.2256	.2225	.2184	.2134	.2075	.2008	.1934
6	.1851	.1931	.2003	.2068	.2124	.2171	.2208	.2234	.2250	.2256
7	.1103	.1198	.1295	.1393	.1489	.1585	.1678	.1768	.1853	.1934
8	.0479	.0542	.0611	.0684	.0762	.0844	.0930	.1020	.1113	.1208
9	.0148	.0175	.0205	.0239	.0277	.0319	.0367	.0418	.0475	.0537
10	.0031	.0038	.0046	.0056	.0068	.0082	.0098	.0116	.0137	.0161
11	.0004	.0005	.0006	.0008	.0010	.0013	.0016	.0019	.0024	.0029
12	.0000	.0000	.0000	.0001	.0001	.0001	.0001	.0001	.0002	.0002

n = 13

r \ p	.01	.02	.03	.04	.05	.06	.07	.08	.09	.10
0	.8775	.7690	.6730	.5882	.5133	.4474	.3893	.3383	.2935	.2542
1	.1152	.2040	.2706	.3186	.3512	.3712	.3809	.3824	.3773	.3672
2	.0070	.0250	.0502	.0797	.1109	.1422	.1720	.1995	.2239	.2448
3	.0003	.0019	.0057	.0122	.0214	.0333	.0475	.0636	.0812	.0997
4	.0000	.0001	.0004	.0013	.0028	.0053	.0089	.0138	.0201	.0277
5	.0000	.0000	.0000	.0001	.0003	.0006	.0012	.0022	.0036	.0055
6	.0000	.0000	.0000	.0000	.0000	.0001	.0001	.0003	.0005	.0008
7	.0000	.0000	.0000	.0000	.0000	.0000	.0000	.0000	.0000	.0001

r \ p	.11	.12	.13	.14	.15	.16	.17	.18	.19	.20
0	.2198	.1898	.1636	.1408	.1209	.1037	.0887	.0758	.0646	.0550
1	.3532	.3364	.3178	.2979	.2774	.2567	.2362	.2163	.1970	.1787
2	.2619	.2753	.2849	.2910	.2937	.2934	.2903	.2848	.2773	.2680
3	.1187	.1376	.1561	.1737	.1900	.2049	.2180	.2293	.2385	.2457
4	.0367	.0469	.0583	.0707	.0838	.0976	.1116	.1258	.1399	.1535
5	.0082	.0115	.0157	.0207	.0266	.0335	.0412	.0497	.0591	.0691
6	.0013	.0021	.0031	.0045	.0063	.0085	.0112	.0145	.0185	.0230
7	.0002	.0003	.0005	.0007	.0011	.0016	.0023	.0032	.0043	.0058
8	.0000	.0000	.0001	.0001	.0001	.0002	.0004	.0005	.0008	.0011
9	.0000	.0000	.0000	.0000	.0000	.0000	.0000	.0001	.0001	.0001

TABLE A.1 (*continued*)

n = 13 (*cont.*)

p r	.21	.22	.23	.24	.25	.26	.27	.28	.29	.30
0	.0467	.0396	.0334	.0282	.0238	.0200	.0167	.0140	.0117	.0097
1	.1613	.1450	.1299	.1159	.1029	.0911	.0804	.0706	.0619	.0540
2	.2573	.2455	.2328	.2195	.2059	.1921	.1784	.1648	.1516	.1388
3	.2508	.2539	.2550	.2542	.2517	.2475	.2419	.2351	.2271	.2181
4	.1667	.1790	.1904	.2007	.2097	.2174	.2237	.2285	.2319	.2337
5	.0797	.0909	.1024	.1141	.1258	.1375	.1489	.1600	.1705	.1803
6	.0283	.0342	.0408	.0480	.0559	.0644	.0734	.0829	.0928	.1030
7	.0075	.0096	.0122	.0152	.0186	.0226	.0272	.0323	.0379	.0442
8	.0015	.0020	.0027	.0036	.0047	.0060	.0075	.0094	.0116	.0142
9	.0002	.0003	.0005	.0006	.0009	.0012	.0015	.0020	.0026	.0034
10	.0000	.0000	.0001	.0001	.0001	.0002	.0002	.0003	.0004	.0006
11	.0000	.0000	.0000	.0000	.0000	.0000	.0000	.0000	.0000	.0001

	.31	.32	.33	.34	.35	.36	.37	.38	.39	.40
0	.0080	.0066	.0055	.0045	.0037	.0030	.0025	.0020	.0016	.0013
1	.0469	.0407	.0351	.0302	.0259	.0221	.0188	.0159	.0135	.0113
2	.1265	.1148	.1037	.0933	.0836	.0746	.0663	.0586	.0516	.0453
3	.2084	.1981	.1874	.1763	.1651	.1538	.1427	.1317	.1210	.1107
4	.2341	.2331	.2307	.2270	.2222	.2163	.2095	.2018	.1934	.1845
5	.1893	.1974	.2045	.2105	.2154	.2190	.2215	.2227	.2226	.2214
6	.1134	.1239	.1343	.1446	.1546	.1643	.1734	.1820	.1898	.1968
7	.0509	.0583	.0662	.0745	.0833	.0924	.1019	.1115	.1213	.1312
8	.0172	.0206	.0244	.0288	.0336	.0390	.0449	.0513	.0582	.0656
9	.0043	.0054	.0067	.0082	.0101	.0122	.0146	.0175	.0207	.0243
10	.0008	.0010	.0013	.0017	.0022	.0027	.0034	.0043	.0053	.0065
11	.0001	.0001	.0002	.0002	.0003	.0004	.0006	.0007	.0009	.0012
12	.0000	.0000	.0000	.0000	.0000	.0000	.0001	.0001	.0001	.0001

	.41	.42	.43	.44	.45	.46	.47	.48	.49	.50
0	.0010	.0008	.0007	.0005	.0004	.0003	.0003	.0002	.0002	.0001
1	.0095	.0079	.0066	.0054	.0045	.0037	.0030	.0024	.0020	.0016
2	.0395	.0344	.0298	.0256	.0220	.0188	.0160	.0135	.0114	.0095
3	.1007	.0913	.0823	.0739	.0660	.0587	.0519	.0457	.0401	.0349
4	.1750	.1653	.1553	.1451	.1350	.1250	.1151	.1055	.0962	.0873
5	.2189	.2154	.2108	.2053	.1989	.1917	.1838	.1753	.1664	.1571
6	.2029	.2080	.2121	.2151	.2169	.2177	.2173	.2158	.2131	.2095
7	.1410	.1506	.1600	.1690	.1775	.1854	.1927	.1992	.2048	.2095
8	.0735	.0818	.0905	.0996	.1089	.1185	.1282	.1379	.1476	.1571
9	.0284	.0329	.0379	.0435	.0495	.0561	.0631	.0707	.0788	.0873
10	.0079	.0095	.0114	.0137	.0162	.0191	.0224	.0261	.0303	.0349
11	.0015	.0019	.0024	.0029	.0036	.0044	.0054	.0066	.0079	.0095
12	.0002	.0002	.0003	.0004	.0005	.0006	.0008	.0010	.0013	.0016
13	.0000	.0000	.0000	.0000	.0000	.0000	.0001	.0001	.0001	.0001

TABLE A.1   (*continued*)

n = 14

r \ p	.01	.02	.03	.04	.05	.06	.07	.08	.09	.10
0	.8687	.7536	.6528	.5647	.4877	.4205	.3620	.3112	.2670	.2288
1	.1229	.2153	.2827	.3294	.3593	.3758	.3815	.3788	.3698	.3559
2	.0081	.0286	.0568	.0892	.1229	.1559	.1867	.2141	.2377	.2570
3	.0003	.0023	.0070	.0149	.0259	.0398	.0562	.0745	.0940	.1142
4	.0000	.0001	.0006	.0017	.0037	.0070	.0116	.0178	.0256	.0349
5	.0000	.0000	.0000	.0001	.0004	.0009	.0018	.0031	.0051	.0078
6	.0000	.0000	.0000	.0000	.0000	.0001	.0002	.0004	.0008	.0013
7	.0000	.0000	.0000	.0000	.0000	.0000	.0000	.0000	.0001	.0002

r \ p	.11	.12	.13	.14	.15	.16	.17	.18	.19	.20
0	.1956	.1670	.1423	.1211	.1028	.0871	.0736	.0621	.0523	.0440
1	.3385	.3188	.2977	.2759	.2539	.2322	.2112	.1910	.1719	.1539
2	.2720	.2826	.2892	.2919	.2912	.2875	.2811	.2725	.2620	.2501
3	.1345	.1542	.1728	.1901	.2056	.2190	.2303	.2393	.2459	.2501
4	.0457	.0578	.0710	.0851	.0998	.1147	.1297	.1444	.1586	.1720
5	.0113	.0158	.0212	.0277	.0352	.0437	.0531	.0634	.0744	.0860
6	.0021	.0032	.0048	.0068	.0093	.0125	.0163	.0209	.0262	.0322
7	.0003	.0005	.0008	.0013	.0019	.0027	.0038	.0052	.0070	.0092
8	.0000	.0001	.0001	.0002	.0003	.0005	.0007	.0010	.0014	.0020
9	.0000	.0000	.0000	.0000	.0000	.0001	.0001	.0001	.0002	.0003

r \ p	.21	.22	.23	.24	.25	.26	.27	.28	.29	.30
0	.0369	.0309	.0258	.0214	.0178	.0148	.0122	.0101	.0083	.0068
1	.1372	.1218	.1077	.0948	.0832	.0726	.0632	.0548	.0473	.0407
2	.2371	.2234	.2091	.1946	.1802	.1659	.1519	.1385	.1256	.1134
3	.2521	.2520	.2499	.2459	.2402	.2331	.2248	.2154	.2052	.1943
4	.1843	.1955	.2052	.2135	.2202	.2252	.2286	.2304	.2305	.2290
5	.0980	.1103	.1226	.1348	.1468	.1583	.1691	.1792	.1883	.1963
6	.0391	.0466	.0549	.0639	.0734	.0834	.0938	.1045	.1153	.1262
7	.0119	.0150	.0188	.0231	.0280	.0335	.0397	.0464	.0538	.0618
8	.0028	.0037	.0049	.0064	.0082	.0103	.0128	.0158	.0192	.0232
9	.0005	.0007	.0010	.0013	.0018	.0024	.0032	.0041	.0052	.0066
10	.0001	.0001	.0001	.0002	.0003	.0004	.0006	.0008	.0011	.0014
11	.0000	.0000	.0000	.0000	.0000	.0001	.0001	.0001	.0002	.0002

TABLE A.1   (*continued*)

n = 14 (cont.)

p / r	.31	.32	.33	.34	.35	.36	.37	.38	.39	.40
0	.0055	.0045	.0037	.0030	.0024	.0019	.0016	.0012	.0010	.0008
1	.0349	.0298	.0253	.0215	.0181	.0152	.0128	.0106	.0088	.0073
2	.1018	.0911	.0811	.0719	.0634	.0557	.0487	.0424	.0367	.0317
3	.1830	.1715	.1598	.1481	.1366	.1253	.1144	.1039	.0940	.0845
4	.2261	.2219	.2164	.2098	.2022	.1938	.1848	.1752	.1652	.1549
5	.2032	.2088	.2132	.2161	.2178	.2181	.2170	.2147	.2112	.2066
6	.1369	.1474	.1575	.1670	.1759	.1840	.1912	.1974	.2026	.2066
7	.0703	.0793	.0886	.0983	.1082	.1183	.1283	.1383	.1480	.1574
8	.0276	.0326	.0382	.0443	.0510	.0582	.0659	.0742	.0828	.0918
9	.0083	.0102	.0125	.0152	.0183	.0218	.0258	.0303	.0353	.0408
10	.0019	.0024	.0031	.0039	.0049	.0061	.0076	.0093	.0113	.0136
11	.0003	.0004	.0006	.0007	.0010	.0013	.0016	.0021	.0026	.0033
12	.0000	.0000	.0001	.0001	.0001	.0002	.0002	.0003	.0004	.0005
13	.0000	.0000	.0000	.0000	.0000	.0000	.0000	.0000	.0000	.0001

r	.41	.42	.43	.44	.45	.46	.47	.48	.49	.50
0	.0006	.0005	.0004	.0003	.0002	.0002	.0001	.0001	.0001	.0001
1	.0060	.0049	.0040	.0033	.0027	.0021	.0017	.0014	.0011	.0009
2	.0272	.0233	.0198	.0168	.0141	.0118	.0099	.0082	.0068	.0056
3	.0757	.0674	.0597	.0527	.0462	.0403	.0350	.0303	.0260	.0222
4	.1446	.1342	.1239	.1138	.1040	.0945	.0854	.0768	.0687	.0611
5	.2009	.1943	.1869	.1788	.1701	.1610	.1515	.1418	.1320	.1222
6	.2094	.2111	.2115	.2108	.2088	.2057	.2015	.1963	.1902	.1833
7	.1663	.1747	.1824	.1892	.1952	.2003	.2043	.2071	.2089	.2095
8	.1011	.1107	.1204	.1301	.1398	.1493	.1585	.1673	.1756	.1833
9	.0469	.0534	.0605	.0682	.0762	.0848	.0937	.1030	.1125	.1222
10	.0163	.0193	.0228	.0268	.0312	.0361	.0415	.0475	.0540	.0611
11	.0041	.0051	.0063	.0076	.0093	.0112	.0134	.0160	.0189	.0222
12	.0007	.0009	.0012	.0015	.0019	.0024	.0030	.0037	.0045	.0056
13	.0001	.0001	.0001	.0002	.0002	.0003	.0004	.0005	.0007	.0009
14	.0000	.0000	.0000	.0000	.0000	.0000	.0000	.0000	.0000	.0001

n = 15

p / r	.01	.02	.03	.04	.05	.06	.07	.08	.09	.10
0	.8601	.7386	.6333	.5421	.4633	.3953	.3367	.2863	.2430	.2059
1	.1303	.2261	.2938	.3388	.3658	.3785	.3801	.3734	.3605	.3432
2	.0092	.0323	.0636	.0988	.1348	.1691	.2003	.2273	.2496	.2669
3	.0004	.0029	.0085	.0178	.0307	.0468	.0653	.0857	.1070	.1285
4	.0000	.0002	.0008	.0022	.0049	.0090	.0148	.0223	.0317	.0428
5	.0000	.0000	.0001	.0002	.0006	.0013	.0024	.0043	.0069	.0105
6	.0000	.0000	.0000	.0000	.0000	.0001	.0003	.0006	.0011	.0019
7	.0000	.0000	.0000	.0000	.0000	.0000	.0000	.0001	.0001	.0003

TABLE A.1 (continued)

n = 15 (cont.)

p r	.11	.12	.13	.14	.15	.16	.17	.18	.19	.20
0	.1741	.1470	.1238	.1041	.0874	.0731	.0611	.0510	.0424	.0352
1	.3228	.3006	.2775	.2542	.2312	.2090	.1878	.1678	.1492	.1319
2	.2793	.2870	.2903	.2897	.2856	.2787	.2692	.2578	.2449	.2309
3	.1496	.1696	.1880	.2044	.2184	.2300	.2389	.2452	.2489	.2501
4	.0555	.0694	.0843	.0998	.1156	.1314	.1468	.1615	.1752	.1876
5	.0151	.0208	.0277	.0357	.0449	.0551	.0662	.0780	.0904	.1032
6	.0031	.0047	.0069	.0097	.0132	.0175	.0226	.0285	.0353	.0430
7	.0005	.0008	.0013	.0020	.0030	.0043	.0059	.0081	.0107	.0138
8	.0001	.0001	.0002	.0003	.0005	.0008	.0012	.0018	.0025	.0035
9	.0000	.0000	.0000	.0000	.0001	.0001	.0002	.0003	.0005	.0007
10	.0000	.0000	.0000	.0000	.0000	.0000	.0000	.0000	.0001	.0001

	.21	.22	.23	.24	.25	.26	.27	.28	.29	.30
0	.0291	.0241	.0198	.0163	.0134	.0109	.0089	.0072	.0059	.0047
1	.1162	.1018	.0889	.0772	.0668	.0576	.0494	.0423	.0360	.0305
2	.2162	.2010	.1858	.1707	.1559	.1416	.1280	.1150	.1029	.0916
3	.2490	.2457	.2405	.2336	.2252	.2156	.2051	.1939	.1821	.1700
4	.1986	.2079	.2155	.2213	.2252	.2273	.2276	.2262	.2231	.2186
5	.1161	.1290	.1416	.1537	.1651	.1757	.1852	.1935	.2005	.2061
6	.0514	.0606	.0705	.0809	.0917	.1029	.1142	.1254	.1365	.1472
7	.0176	.0220	.0271	.0329	.0393	.0465	.0543	.0627	.0717	.0811
8	.0047	.0062	.0081	.0104	.0131	.0163	.0201	.0244	.0293	.0348
9	.0010	.0014	.0019	.0025	.0034	.0045	.0058	.0074	.0093	.0116
10	.0002	.0002	.0003	.0005	.0007	.0009	.0013	.0017	.0023	.0030
11	.0000	.0000	.0000	.0001	.0001	.0002	.0002	.0003	.0004	.0006
12	.0000	.0000	.0000	.0000	.0000	.0000	.0000	.0000	.0001	.0001

	.31	.32	.33	.34	.35	.36	.37	.38	.39	.40
0	.0038	.0031	.0025	.0020	.0016	.0012	.0010	.0008	.0006	.0005
1	.0258	.0217	.0182	.0152	.0126	.0104	.0086	.0071	.0058	.0047
2	.0811	.0715	.0627	.0547	.0476	.0411	.0354	.0303	.0259	.0219
3	.1579	.1457	.1338	.1222	.1110	.1002	.0901	.0805	.0716	.0634
4	.2128	.2057	.1977	.1888	.1792	.1692	.1587	.1481	.1374	.1268
5	.2103	.2130	.2142	.2140	.2123	.2093	.2051	.1997	.1933	.1859
6	.1575	.1671	.1759	.1837	.1906	.1963	.2008	.2040	.2059	.2066
7	.0910	.1011	.1114	.1217	.1319	.1419	.1516	.1608	.1693	.1771
8	.0409	.0476	.0549	.0627	.0710	.0798	.0890	.0985	.1082	.1181
9	.0143	.0174	.0210	.0251	.0298	.0349	.0407	.0470	.0538	.0612
10	.0038	.0049	.0062	.0078	.0096	.0118	.0143	.0173	.0206	.0245
11	.0008	.0011	.0014	.0018	.0024	.0030	.0038	.0048	.0060	.0074
12	.0001	.0002	.0002	.0003	.0004	.0006	.0007	.0010	.0013	.0016
13	.0000	.0000	.0000	.0000	.0001	.0001	.0001	.0001	.0002	.0003

TABLE A.1 (continued)

n = 15 (cont.)

r \ p	.41	.42	.43	.44	.45	.46	.47	.48	.49	.50
0	.0004	.0003	.0002	.0002	.0001	.0001	.0001	.0001	.0000	.0000
1	.0038	.0031	.0025	.0020	.0016	.0012	.0010	.0008	.0006	.0005
2	.0185	.0156	.0130	.0108	.0090	.0074	.0060	.0049	.0040	.0032
3	.0558	.0489	.0426	.0369	.0318	.0272	.0232	.0197	.0166	.0139
4	.1163	.1061	.0963	.0869	.0780	.0696	.0617	.0545	.0478	.0417
5	.1778	.1691	.1598	.1502	.1404	.1304	.1204	.1106	.1010	.0916
6	.2060	.2041	.2010	.1967	.1914	.1851	.1780	.1702	.1617	.1527
7	.1840	.1900	.1949	.1987	.2013	.2028	.2030	.2020	.1997	.1964
8	.1279	.1376	.1470	.1561	.1647	.1727	.1800	.1864	.1919	.1964
9	.0691	.0775	.0863	.0954	.1048	.1144	.1241	.1338	.1434	.1527
10	.0288	.0337	.0390	.0450	.0515	.0585	.0661	.0741	.0827	.0916
11	.0091	.0111	.0134	.0161	.0191	.0226	.0266	.0311	.0361	.0417
12	.0021	.0027	.0034	.0042	.0052	.0064	.0079	.0096	.0116	.0139
13	.0003	.0004	.0006	.0008	.0010	.0013	.0016	.0020	.0026	.0032
14	.0000	.0000	.0001	.0001	.0001	.0002	.0002	.0003	.0004	.0005

n = 16

r \ p	.01	.02	.03	.04	.05	.06	.07	.08	.09	.10
0	.8515	.7238	.6143	.5204	.4401	.3716	.3131	.2634	.2211	.1853
1	.1376	.2363	.3040	.3469	.3706	.3795	.3771	.3665	.3499	.3294
2	.0104	.0362	.0705	.1084	.1463	.1817	.2129	.2390	.2596	.2745
3	.0005	.0034	.0102	.0211	.0359	.0541	.0748	.0970	.1198	.1423
4	.0000	.0002	.0010	.0029	.0061	.0112	.0183	.0274	.0385	.0514
5	.0000	.0000	.0001	.0003	.0008	.0017	.0033	.0057	.0091	.0137
6	.0000	.0000	.0000	.0000	.0001	.0002	.0005	.0009	.0017	.0028
7	.0000	.0000	.0000	.0000	.0000	.0000	.0000	.0001	.0002	.0004
8	.0000	.0000	.0000	.0000	.0000	.0000	.0000	.0000	.0000	.0001

r \ p	.11	.12	.13	.14	.15	.16	.17	.18	.19	.20
0	.1550	.1293	.1077	.0895	.0743	.0614	.0507	.0418	.0343	.0281
1	.3065	.2822	.2575	.2332	.2097	.1873	.1662	.1468	.1289	.1126
2	.2841	.2886	.2886	.2847	.2775	.2675	.2554	.2416	.2267	.2111
3	.1638	.1837	.2013	.2163	.2285	.2378	.2441	.2475	.2482	.2463
4	.0658	.0814	.0977	.1144	.1311	.1472	.1625	.1766	.1892	.2001
5	.0195	.0266	.0351	.0447	.0555	.0673	.0799	.0930	.1065	.1201
6	.0044	.0067	.0096	.0133	.0180	.0235	.0300	.0374	.0458	.0550
7	.0008	.0013	.0020	.0031	.0045	.0064	.0088	.0117	.0153	.0197
8	.0001	.0002	.0003	.0006	.0009	.0014	.0020	.0029	.0041	.0055
9	.0000	.0000	.0000	.0001	.0001	.0002	.0004	.0006	.0008	.0012
10	.0000	.0000	.0000	.0000	.0000	.0000	.0001	.0001	.0001	.0002

TABLE A.1   (continued)

n = 16 (cont.)

p r	.21	.22	.23	.24	.25	.26	.27	.28	.29	.30
0	.0230	.0188	.0153	.0124	.0100	.0081	.0065	.0052	.0042	.0033
1	.0979	.0847	.0730	.0626	.0535	.0455	.0385	.0325	.0273	.0228
2	.1952	.1792	.1635	.1482	.1336	.1198	.1068	.0947	.0835	.0732
3	.2421	.2359	.2279	.2185	.2079	.1964	.1843	.1718	.1591	.1465
4	.2092	.2162	.2212	.2242	.2252	.2243	.2215	.2171	.2112	.2040
5	.1334	.1464	.1586	.1699	.1802	.1891	.1966	.2026	.2071	.2099
6	.0650	.0757	.0869	.0984	.1101	.1218	.1333	.1445	.1551	.1649
7	.0247	.0305	.0371	.0444	.0524	.0611	.0704	.0803	.0905	.1010
8	.0074	.0097	.0125	.0158	.0197	.0242	.0293	.0351	.0416	.0487
9	.0017	.0024	.0033	.0044	.0058	.0075	.0096	.0121	.0151	.0185
10	.0003	.0005	.0007	.0010	.0014	.0019	.0025	.0033	.0043	.0056
11	.0000	.0001	.0001	.0002	.0002	.0004	.0005	.0007	.0010	.0013
12	.0000	.0000	.0000	.0000	.0000	.0001	.0001	.0001	.0002	.0002

	.31	.32	.33	.34	.35	.36	.37	.38	.39	.40
0	.0026	.0021	.0016	.0013	.0010	.0008	.0006	.0005	.0004	.0003
1	.0190	.0157	.0130	.0107	.0087	.0071	.0058	.0047	.0038	.0030
2	.0639	.0555	.0480	.0413	.0353	.0301	.0255	.0215	.0180	.0150
3	.1341	.1220	.1103	.0992	.0888	.0790	.0699	.0615	.0538	.0468
4	.1958	.1865	.1766	.1662	.1553	.1444	.1333	.1224	.1118	.1014
5	.2111	.2107	.2088	.2054	.2008	.1949	.1879	.1801	.1715	.1623
6	.1739	.1818	.1885	.1940	.1982	.2010	.2024	.2024	.2010	.1983
7	.1116	.1222	.1326	.1428	.1524	.1615	.1698	.1772	.1836	.1889
8	.0564	.0647	.0735	.0827	.0923	.1022	.1122	.1222	.1320	.1417
9	.0225	.0271	.0322	.0379	.0442	.1511	.0586	.0666	.0750	.0840
10	.0071	.0089	.0111	.0137	.0167	.0201	.0241	.0286	.0336	.0392
11	.0017	.0023	.0030	.0038	.0049	.0062	.0077	.0095	.0117	.0142
12	.0003	.0004	.0006	.0008	.0011	.0014	.0019	.0024	.0031	.0040
13	.0000	.0001	.0001	.0001	.0002	.0003	.0003	.0005	.0006	.0008
14	.0000	.0000	.0000	.0000	.0000	.0000	.0000	.0001	.0001	.0001

	.41	.42	.43	.44	.45	.46	.47	.48	.49	.50
0	.0002	.0002	.0001	.0001	.0001	.0001	.0000	.0000	.0000	.0000
1	.0024	.0019	.0015	.0012	.0009	.0007	.0005	.0004	.0003	.0002
2	.0125	.0103	.0085	.0069	.0056	.0046	.0037	.0029	.0023	.0018
3	.0405	.0349	.0299	.0254	.0215	.0181	.0151	.0126	.0104	.0085
4	.0915	.0821	.0732	.0649	.0572	.0501	.0436	.0378	.0325	.0278
5	.1526	.1426	.1325	.1224	.1123	.1024	.0929	.0837	.0749	.0667
6	.1944	.1894	.1833	.1762	.1684	.1600	.1510	.1416	.1319	.1222
7	.1930	.1959	.1975	.1978	.1969	.1947	.1912	.1867	.1811	.1746
8	.1509	.1596	.1676	.1749	.1812	.1865	.1908	.1939	.1958	.1964
9	.0932	.1027	.1124	.1221	.1318	.1413	.1504	.1591	.1672	.1746
10	.0453	.0521	.0594	.0672	.0755	.0842	.0934	.1028	.1124	.1222
11	.0172	.0206	.0244	.0288	.0337	.0391	.0452	.0518	.0589	.0667
12	.0050	.0062	.0077	.0094	.0115	.0139	.0167	.0199	.0236	.0278
13	.0011	.0014	.0018	.0023	.0029	.0036	.0046	.0057	.0070	.0085
14	.0002	.0002	.0003	.0004	.0005	.0007	.0009	.0011	.0014	.0018
15	.0000	.0000	.0000	.0000	.0001	.0001	.0001	.0001	.0002	.0002

TABLE A.1 (*continued*)

n = 17

r \ p	.01	.02	.03	.04	.05	.06	.07	.08	.09	.10
0	.8429	.7093	.5958	.4996	.4181	.3493	.2912	.2423	.2012	.1668
1	.1447	.2461	.3133	.3539	.3741	.3790	.3726	.3582	.3383	.3150
2	.0117	.0402	.0775	.1180	.1575	.1935	.2244	.2492	.2677	.2800
3	.0006	.0041	.0120	.0246	.0415	.0618	.0844	.1083	.1324	.1556
4	.0000	.0003	.0013	.0036	.0076	.0138	.0222	.0330	.0458	.0605
5	.0000	.0000	.0001	.0004	.0010	.0023	.0044	.0075	.0118	.0175
6	.0000	.0000	.0000	.0000	.0001	.0003	.0007	.0013	.0023	.0039
7	.0000	.0000	.0000	.0000	.0000	.0000	.0001	.0002	.0004	.0007
8	.0000	.0000	.0000	.0000	.0000	.0000	.0000	.0000	.0000	.0001

r \ p	.11	.12	.13	.14	.15	.16	.17	.18	.19	.20
0	.1379	.1138	.0937	.0770	.0631	.0516	.0421	.0343	.0278	.0225
1	.2898	.2638	.2381	.2131	.1893	.1671	.1466	.1279	.1109	.0957
2	.2865	.2878	.2846	.2775	.2673	.2547	.2402	.2245	.2081	.1914
3	.1771	.1963	.2126	.2259	.2359	.2425	.2460	.2464	.2441	.2393
4	.0766	.0937	.1112	.1287	.1457	.1617	.1764	.1893	.2004	.2093
5	.0246	.0332	.0432	.0545	.0668	.0801	.0939	.1081	.1222	.1361
6	.0061	.0091	.0129	.0177	.0236	.0305	.0385	.0474	.0573	.0680
7	.0012	.0019	.0030	.0045	.0065	.0091	.0124	.0164	.0211	.0267
8	.0002	.0003	.0006	.0009	.0014	.0022	.0032	.0045	.0062	.0084
9	.0000	.0000	.0001	.0002	.0003	.0004	.0006	.0010	.0015	.0021
10	.0000	.0000	.0000	.0000	.0000	.0001	.0001	.0002	.0003	.0004
11	.0000	.0000	.0000	.0000	.0000	.0000	.0000	.0000	.0000	.0001

r \ p	.21	.22	.23	.24	.25	.26	.27	.28	.29	.30
0	.0182	.0146	.0118	.0094	.0075	.0060	.0047	.0038	.0030	.0023
1	.0822	.0702	.0597	.0505	.0426	.0357	.0299	.0248	.0206	.0169
2	.1747	.1584	.1427	.1277	.1136	.1005	.0883	.0772	.0672	.0581
3	.2322	.2234	.2131	.2016	.1893	.1765	.1634	.1502	.1372	.1245
4	.2161	.2205	.2228	.2228	.2209	.2170	.2115	.2044	.1961	.1868
5	.1493	.1617	.1730	.1830	.1914	.1982	.2033	.2067	.2083	.2081
6	.0794	.0912	.1034	.1156	.1276	.1393	.1504	.1608	.1701	.1784
7	.0332	.0404	.0485	.0573	.0668	.0769	.0874	.0982	.1092	.1201
8	.0110	.0143	.0181	.0226	.0279	.0338	.0404	.0478	.0558	.0644
9	.0029	.0040	.0054	.0071	.0093	.0119	.0150	.0186	.0228	.0276
10	.0006	.0009	.0013	.0018	.0025	.0033	.0044	.0058	.0074	.0095
11	.0001	.0002	.0002	.0004	.0005	.0007	.0010	.0014	.0019	.0026
12	.0000	.0000	.0000	.0001	.0001	.0001	.0002	.0003	.0004	.0006
13	.0000	.0000	.0000	.0000	.0000	.0000	.0000	.0000	.0001	.0001

TABLE A.1   (continued)

n = 17 (cont.)

r \ p	.31	.32	.33	.34	.35	.36	.37	.38	.39	.40
0	.0018	.0014	.0011	.0009	.0007	.0005	.0004	.0003	.0002	.0002
1	.0139	.0114	.0093	.0075	.0060	.0048	.0039	.0031	.0024	.0019
2	.0500	.0428	.0364	.0309	.0260	.0218	.0182	.0151	.0125	.0102
3	.1123	.1007	.0898	.0795	.0701	.0614	.0534	.0463	.0398	.0341
4	.1766	.1659	.1547	.1434	.1320	.1208	.1099	.0993	.0892	.0796
5	.2063	.2030	.1982	.1921	.1849	.1767	.1677	.1582	.1482	.1379
6	.1854	.1910	.1952	.1979	.1991	.1988	.1970	.1939	.1895	.1839
7	.1309	.1413	.1511	.1602	.1685	.1757	.1818	.1868	.1904	.1927
8	.0735	.0831	.0930	.1032	.1134	.1235	.1335	.1431	.1521	.1606
9	.0330	.0391	.0458	.0531	.0611	.0695	.0784	.0877	.0973	.1070
10	.0119	.0147	.0181	.0219	.0263	.0313	.0368	.0430	.0498	.0571
11	.0034	.0044	.0057	.0072	.0090	.0112	.0138	.0168	.0202	.0242
12	.0008	.0010	.0014	.0018	.0024	.0031	.0040	.0051	.0065	.0081
13	.0001	.0002	.0003	.0004	.0005	.0007	.0009	.0012	.0016	.0021
14	.0000	.0000	.0000	.0001	.0001	.0001	.0002	.0002	.0003	.0004
15	.0000	.0000	.0000	.0000	.0000	.0000	.0000	.0000	.0000	.0001

r \ p	.41	.42	.43	.44	.45	.46	.47	.48	.49	.50
0	.0001	.0001	.0001	.0001	.0000	.0000	.0000	.0000	.0000	.0000
1	.0015	.0012	.0009	.0007	.0005	.0004	.0003	.0002	.0002	.0001
2	.0084	.0068	.0055	.0044	.0035	.0028	.0022	.0017	.0013	.0010
3	.0290	.0246	.0207	.0173	.0144	.0119	.0097	.0079	.0064	.0052
4	.0706	.0622	.0546	.0475	.0411	.0354	.0302	.0257	.0217	.0182
5	.1276	.1172	.1070	.0971	.0875	.0784	.0697	.0616	.0541	.0472
6	.1773	.1697	.1614	.1525	.1432	.1335	.1237	.1138	.1040	.0944
7	.1936	.1932	.1914	.1883	.1841	.1787	.1723	.1650	.1570	.1484
8	.1682	.1748	.1805	.1850	.1883	.1903	.1910	.1904	.1886	.1855
9	.1169	.1266	.1361	.1453	.1540	.1621	.1694	.1758	.1812	.1855
10	.0650	.0733	.0822	.0914	.1008	.1105	.1202	.1298	.1393	.1484
11	.0287	.0338	.0394	.0457	.0525	.0599	.0678	.0763	.0851	.0944
12	.0100	.0122	.0149	.0179	.0215	.0255	.0301	.0352	.0409	.0472
13	.0027	.0034	.0043	.0054	.0068	.0084	.0103	.0125	.0151	.0182
14	.0005	.0007	.0009	.0012	.0016	.0020	.0026	.0033	.0041	.0052
15	.0001	.0001	.0001	.0002	.0003	.0003	.0005	.0006	.0008	.0010
16	.0000	.0000	.0000	.0000	.0000	.0000	.0001	.0001	.0001	.0001

n = 18

r \ p	.01	.02	.03	.04	.05	.06	.07	.08	.09	.10
0	.8345	.6951	.5780	.4796	.3972	.3283	.2708	.2229	.1831	.1501
1	.1517	.2554	.3217	.3597	.3763	.3772	.3669	.3489	.3260	.3002
2	.0130	.0443	.0846	.1274	.1683	.2047	.2348	.2579	.2741	.2835
3	.0007	.0048	.0140	.0283	.0473	.0697	.0942	.1196	.1446	.1680
4	.0000	.0004	.0016	.0044	.0093	.0167	.0266	.0390	.0536	.0700
5	.0000	.0000	.0001	.0005	.0014	.0030	.0056	.0095	.0148	.0218
6	.0000	.0000	.0000	.0000	.0002	.0004	.0009	.0018	.0032	.0052
7	.0000	.0000	.0000	.0000	.0000	.0000	.0001	.0003	.0005	.0010
8	.0000	.0000	.0000	.0000	.0000	.0000	.0000	.0000	.0001	.0002

TABLE A.1   (*continued*)

n = **18 (*cont.*)**

p r	.11	.12	.13	.14	.15	.16	.17	.18	.19	.20
0	.1227	.1002	.0815	.0662	.0536	.0434	.0349	.0281	.0225	.0180
1	.2731	.2458	.2193	.1940	.1704	.1486	.1288	.1110	.0951	.0811
2	.2869	.2850	.2785	.2685	.2556	.2407	.2243	.2071	.1897	.1723
3	.1891	.2072	.2220	.2331	.2406	.2445	.2450	.2425	.2373	.2297
4	.0877	.1060	.1244	.1423	.1592	.1746	.1882	.1996	.2087	.2153
5	.0303	.0405	.0520	.0649	.0787	.0931	.1079	.1227	.1371	.1507
6	.0081	.0120	.0168	.0229	.0301	.0384	.0479	.0584	.0697	.0816
7	.0017	.0028	.0043	.0064	.0091	.0126	.0168	.0220	.0280	.0350
8	.0003	.0005	.0009	.0014	.0022	.0033	.0047	.0066	.0090	.0120
9	.0000	.0001	.0001	.0003	.0004	.0007	.0011	.0016	.0024	.0033
10	.0000	.0000	.0000	.0000	.0001	.0001	.0002	.0003	.0005	.0008
11	.0000	.0000	.0000	.0000	.0000	.0000	.0000	.0001	.0001	.0001

	.21	.22	.23	.24	.25	.26	.27	.28	.29	.30
0	.0144	.0114	.0091	.0072	.0056	.0044	.0035	.0027	.0021	.0016
1	.0687	.0580	.0487	.0407	.0338	.0280	.0231	.0189	.0155	.0126
2	.1553	.1390	.1236	.1092	.0958	.0836	.0725	.0626	.0537	.0458
3	.2202	.2091	.1969	.1839	.1704	.1567	.1431	.1298	.1169	.1046
4	.2195	.2212	.2205	.2177	.2130	.2065	.1985	.1892	.1790	.1681
5	.1634	.1747	.1845	.1925	.1988	.2031	.2055	.2061	.2048	.2017
6	.0941	.1067	.1194	.1317	.1436	.1546	.1647	.1736	.1812	.1873
7	.0429	.0516	.0611	.0713	.0820	.0931	.1044	.1157	.1269	.1376
8	.0157	.0200	.0251	.0310	.0376	.0450	.0531	.0619	.0713	.0811
9	.0046	.0063	.0083	.0109	.0139	.0176	.0218	.0267	.0323	.0386
10	.0011	.0016	.0022	.0031	.0042	.0056	.0073	.0094	.0119	.0149
11	.0002	.0003	.0005	.0007	.0010	.0014	.0020	.0026	.0035	.0046
12	.0000	.0001	.0001	.0001	.0002	.0003	.0004	.0006	.0008	.0012
13	.0000	.0000	.0000	.0000	.0000	.0000	.0001	.0001	.0002	.0002

	.31	.32	.33	.34	.35	.36	.37	.38	.39	.40
0	.0013	.0010	.0007	.0006	.0004	.0003	.0002	.0002	.0001	.0001
1	.0102	.0082	.0066	.0052	.0042	.0033	.0026	.0020	.0016	.0012
2	.0388	.0327	.0275	.0229	.0190	.0157	.0129	.0105	.0086	.0069
3	.0930	.0822	.0722	.0630	.0547	.0471	.0404	.0344	.0292	.0246
4	.1567	.1450	.1333	.1217	.1104	.0994	.0890	.0791	.0699	.0614
5	.1971	.1911	.1838	.1755	.1664	.1566	.1463	.1358	.1252	.1146
6	.1919	.1948	.1962	.1959	.1941	.1908	.1862	.1803	.1734	.1655
7	.1478	.1572	.1656	.1730	.1792	.1840	.1875	.1895	.1900	.1892
8	.0913	.1017	.1122	.1226	.1327	.1423	.1514	.1597	.1671	.1734
9	.0456	.0532	.0614	.0701	.0794	.0890	.0988	.1087	.1187	.1284
10	.0184	.0225	.0272	.0325	.0385	.0450	.0522	.0600	.0683	.0771
11	.0060	.0077	.0097	.0122	.0151	.0184	.0223	.0267	.0318	.0374
12	.0016	.0021	.0028	.0037	.0047	.0060	.0076	.0096	.0118	.0145
13	.0003	.0005	.0006	.0009	.0012	.0016	.0021	.0027	.0035	.0045
14	.0001	.0001	.0001	.0002	.0002	.0003	.0004	.0006	.0008	.0011
15	.0000	.0000	.0000	.0000	.0000	.0000	.0001	.0001	.0001	.0002

TABLE A.1 (continued)

n = 18 (cont.)

r \ p	.41	.42	.43	.44	.45	.46	.47	.48	.49	.50
0	.0001	.0001	.0000	.0000	.0000	.0000	.0000	.0000	.0000	.0000
1	.0009	.0007	.0005	.0004	.0003	.0002	.0002	.0001	.0001	.0001
2	.0055	.0044	.0035	.0028	.0022	.0017	.0013	.0010	.0008	.0006
3	.0206	.0171	.0141	.0116	.0095	.0077	.0062	.0050	.0039	.0031
4	.0536	.0464	.0400	.0342	.0291	.0246	.0206	.0172	.0142	.0117
5	.1042	.0941	.0844	.0753	.0666	.0586	.0512	.0444	.0382	.0327
6	.1569	.1477	.1380	.1281	.1181	.1081	.0983	.0887	.0796	.0708
7	.1869	.1833	.1785	.1726	.1657	.1579	.1494	.1404	.1310	.1214
8	.1786	.1825	.1852	.1864	.1864	.1850	.1822	.1782	.1731	.1669
9	.1379	.1469	.1552	.1628	.1694	.1751	.1795	.1828	.1848	.1855
10	.0862	.0957	.1054	.1151	.1248	.1342	.1433	.1519	.1598	.1669
11	.0436	.0504	.0578	.0658	.0742	.0831	.0924	.1020	.1117	.1214
12	.0177	.0213	.0254	.0301	.0354	.0413	.0478	.1549	.0626	.0708
13	.0057	.0071	.0089	.0109	.0134	.0162	.0196	.0234	.0278	.0327
14	.0014	.0018	.0024	.0031	.0039	.0049	.0062	.0077	.0095	.0117
15	.0003	.0004	.0005	.0006	.0009	.0011	.0015	.0019	.0024	.0031
16	.0000	.0000	.0001	.0001	.0001	.0002	.0002	.0003	.0004	.0006
17	.0000	.0000	.0000	.0000	.0000	.0000	.0000	.0000	.0000	.0001

n = 19

r \ p	.01	.02	.03	.04	.05	.06	.07	.08	.09	.10
0	.8262	.6812	.5606	.4604	.3774	.3086	.2519	.2051	.1666	.1351
1	.1586	.2642	.3294	.3645	.3774	.3743	.3602	.3389	.3131	.2852
2	.0144	.0485	.0917	.1367	.1787	.2150	.2440	.2652	.2787	.2852
3	.0008	.0056	.0161	.0323	.0533	.0778	.1041	.1307	.1562	.1796
4	.0000	.0005	.0020	.0054	.0112	.0199	.0313	.0455	.0618	.0798
5	.0000	.0000	.0002	.0007	.0018	.0038	.0071	.0119	.0183	.0266
6	.0000	.0000	.0000	.0001	.0002	.0006	.0012	.0024	.0042	.0069
7	.0000	.0000	.0000	.0000	.0000	.0001	.0002	.0004	.0008	.0014
8	.0000	.0000	.0000	.0000	.0000	.0000	.0000	.0001	.0001	.0002

r \ p	.11	.12	.13	.14	.15	.16	.17	.18	.19	.20
0	.1092	.0881	.0709	.0569	.0456	.0364	.0290	.0230	.0182	.0144
1	.2565	.2284	.2014	.1761	.1529	.1318	.1129	.0961	.0813	.0685
2	.2854	.2803	.2708	.2581	.2428	.2259	.2081	.1898	.1717	.1540
3	.1999	.2166	.2293	.2381	.2428	.2439	.2415	.2361	.2282	.2182
4	.0988	.1181	.1371	.1550	.1714	.1858	.1979	.2073	.2141	.2182
5	.0366	.0483	.0614	.0757	.0907	.1062	.1216	.1365	.1507	.1636
6	.0106	.0154	.0214	.0288	.0374	.0472	.0581	.0699	.0825	.0955
7	.0024	.0039	.0059	.0087	.0122	.0167	.0221	.0285	.0359	.0443
8	.0004	.0008	.0013	.0021	.0032	.0048	.0068	.0094	.0126	.0166
9	.0001	.0001	.0002	.0004	.0007	.0011	.0017	.0025	.0036	.0051
10	.0000	.0000	.0000	.0001	.0001	.0002	.0003	.0006	.0009	.0013
11	.0000	.0000	.0000	.0000	.0000	.0000	.0001	.0001	.0002	.0003

TABLE A.1   (*continued*)

n = 19 (*cont.*)

r \ p	.21	.22	.23	.24	.25	.26	.27	.28	.29	.30
0	.0113	.0089	.0070	.0054	.0042	.0033	.0025	.0019	.0015	.0011
1	.0573	.0477	.0396	.0326	.0268	.0219	.0178	.0144	.0116	.0093
2	.1371	.1212	.1064	.0927	.0803	.0692	.0592	.0503	.0426	.0358
3	.2065	.1937	.1800	.1659	.1517	.1377	.1240	.1109	.0985	.0869
4	.2196	.2185	.2151	.2096	.2023	.1935	.1835	.1726	.1610	.1491
5	.1751	.1849	.1928	.1986	.2023	.2040	.2036	.2013	.1973	.1916
6	.1086	.1217	.1343	.1463	.1574	.1672	.1757	.1827	.1880	.1916
7	.0536	.0637	.0745	.0858	.0974	.1091	.1207	.1320	.1426	.1525
8	.0214	.0270	.0334	.0406	.0487	.0575	.0670	.0770	.0874	.0981
9	.0069	.0093	.0122	.0157	.0198	.0247	.0303	.0366	.0436	.0514
10	.0018	.0026	.0036	.0050	.0066	.0087	.0112	.0142	.0178	.0220
11	.0004	.0006	.0009	.0013	.0018	.0025	.0034	.0045	.0060	.0077
12	.0001	.0001	.0002	.0003	.0004	.0006	.0008	.0012	.0016	.0022
13	.0000	.0000	.0000	.0000	.0001	.0001	.0002	.0002	.0004	.0005
14	.0000	.0000	.0000	.0000	.0000	.0000	.0000	.0000	.0001	.0001

r \ p	.31	.32	.33	.34	.35	.36	.37	.38	.39	.40
0	.0009	.0007	.0005	.0004	.0003	.0002	.0002	.0001	.0001	.0001
1	.0074	.0059	.0046	.0036	.0029	.0022	.0017	.0013	.0010	.0008
2	.0299	.0249	.0206	.0169	.0138	.0112	.0091	.0073	.0058	.0046
3	.0762	.0664	.0574	.0494	.0422	.0358	.0302	.0253	.0211	.0175
4	.1370	.1249	.1131	.1017	.0909	.0806	.0710	.0621	.0540	.0467
5	.1846	.1764	.1672	.1572	.1468	.1360	.1251	.1143	.1036	.0933
6	.1935	.1936	.1921	.1890	.1844	.1785	.1714	.1634	.1546	.1451
7	.1615	.1692	.1757	.1808	.1844	.1865	.1870	.1860	.1835	.1797
8	.1088	.1195	.1298	.1397	.1489	.1573	.1647	.1710	.1760	.1797
9	.0597	.0687	.0782	.0880	.0980	.1082	.1182	.1281	.1375	.1464
10	.0268	.0323	.0385	.0453	.0528	.0608	.0694	.0785	.0879	.0976
11	.0099	.0124	.0155	.0191	.0233	.0280	.0334	.0394	.0460	.0532
12	.0030	.0039	.0051	.0066	.0083	.0105	.0131	.0161	.0196	.0237
13	.0007	.0010	.0014	.0018	.0024	.0032	.0041	.0053	.0067	.0085
14	.0001	.0002	.0003	.0004	.0006	.0008	.0010	.0014	.0018	.0024
15	.0000	.0000	.0000	.0001	.0001	.0001	.0002	.0003	.0004	.0005
16	.0000	.0000	.0000	.0000	.0000	.0000	.0000	.0000	.0001	.0001

TABLE A.1   (continued)

n = 19 (cont.)

r \ p	.41	.42	.43	.44	.45	.46	.47	.48	.49	.50
0	.0000	.0000	.0000	.0000	.0000	.0000	.0000	.0000	.0000	.0000
1	.0006	.0004	.0003	.0002	.0002	.0001	.0001	.0001	.0001	.0000
2	.0037	.0029	.0022	.0017	.0013	.0010	.0008	.0006	.0004	.0003
3	.0144	.0118	.0096	.0077	.0062	.0049	.0039	.0031	.0024	.0018
4	.0400	.0341	.0289	.0243	.0203	.0168	.0138	.0113	.0092	.0074
5	.0834	.0741	.0653	.0572	.0497	.0429	.0368	.0313	.0265	.0222
6	.1353	.1252	.1150	.1049	.0949	.0853	.0761	.0674	.0593	.0518
7	.1746	.1683	.1611	.1530	.1443	.1350	.1254	.1156	.1058	.0961
8	.1820	.1829	.1823	.1803	.1771	.1725	.1668	.1601	.1525	.1442
9	.1546	.1618	.1681	.1732	.1771	.1796	.1808	.1806	.1791	.1762
10	.1074	.1172	.1268	.1361	.1449	.1530	.1603	.1667	.1721	.1762
11	.0611	.0694	.0783	.0875	.0970	.1066	.1163	.1259	.1352	.1442
12	.0283	.0335	.0394	.0458	.0529	.0606	.0688	.0775	.0866	.0961
13	.0106	.0131	.0160	.0194	.0233	.0278	.0328	.0385	.0448	.0518
14	.0032	.0041	.0052	.0065	.0082	.0101	.0125	.0152	.0185	.0222
15	.0007	.0010	.0013	.0017	.0022	.0029	.0037	.0047	.0059	.0074
16	.0001	.0002	.0002	.0003	.0005	.0006	.0008	.0011	.0014	.0018
17	.0000	.0000	.0000	.0000	.0001	.0001	.0001	.0002	.0002	.0003

n = 20

r \ p	.01	.02	.03	.04	.05	.06	.07	.08	.09	.10
0	.8179	.6676	.5438	.4420	.3585	.2901	.2342	.1887	.1516	.1216
1	.1652	.2725	.3364	.3683	.3774	.3703	.3526	.3282	.3000	.2702
2	.0159	.0528	.0988	.1458	.1887	.2246	.2521	.2711	.2818	.2852
3	.0010	.0065	.0183	.0364	.0596	.0860	.1139	.1414	.1672	.1901
4	.0000	.0006	.0024	.0065	.0133	.0233	.0364	.0523	.0703	.0898
5	.0000	.0000	.0002	.0009	.0022	.0048	.0088	.0145	.0222	.0319
6	.0000	.0000	.0000	.0001	.0003	.0008	.0017	.0032	.0055	.0089
7	.0000	.0000	.0000	.0000	.0000	.0001	.0002	.0005	.0011	.0020
8	.0000	.0000	.0000	.0000	.0000	.0000	.0000	.0001	.0002	.0004
9	.0000	.0000	.0000	.0000	.0000	.0000	.0000	.0000	.0000	.0001

r \ p	.11	.12	.13	.14	.15	.16	.17	.18	.19	.20
0	.0972	.0776	.0617	.0490	.0388	.0306	.0241	.0189	.0148	.0115
1	.2403	.2115	.1844	.1595	.1368	.1165	.0986	.0829	.0693	.0576
2	.2822	.2740	.2618	.2466	.2293	.2109	.1919	.1730	.1545	.1369
3	.2093	.2242	.2347	.2409	.2428	.2410	.2358	.2278	.2175	.2054
4	.1099	.1299	.1491	.1666	.1821	.1951	.2053	.2125	.2168	.2182
5	.0435	.0567	.0713	.0868	.1028	.1189	.1345	.1493	.1627	.1746
6	.0134	.0193	.0266	.0353	.0454	.0566	.0689	.0819	.0954	.1091
7	.0033	.0053	.0080	.0115	.0160	.0216	.0282	.0360	.0448	.0545
8	.0007	.0012	.0019	.0030	.0046	.0067	.0094	.0128	.0171	.0222
9	.0001	.0002	.0004	.0007	.0011	.0017	.0026	.0038	.0053	.0074
10	.0000	.0000	.0001	.0001	.0002	.0004	.0006	.0009	.0014	.0020
11	.0000	.0000	.0000	.0000	.0000	.0001	.0001	.0002	.0003	.0005
12	.0000	.0000	.0000	.0000	.0000	.0000	.0000	.0000	.0001	.0001

TABLE A.1   *(continued)*

n = 20 (cont.)

r \ p	.21	.22	.23	.24	.25	.26	.27	.28	.29	.30
0	.0090	.0069	.0054	.0041	.0032	.0024	.0018	.0014	.0011	.0008
1	.0477	.0392	.0321	.0261	.0211	.0170	.0137	.0109	.0087	.0068
2	.1204	.1050	.0910	.0783	.0669	.0569	.0480	.0403	.0336	.0278
3	.1920	.1777	.1631	.1484	.1339	.1199	.1065	.0940	.0823	.0716
4	.2169	.2131	.2070	.1991	.1897	.1790	.1675	.1553	.1429	.1304
5	.1845	.1923	.1979	.2012	.2023	.2013	.1982	.1933	.1868	.1789
6	.1226	.1356	.1478	.1589	.1686	.1768	.1833	.1879	.1907	.1916
7	.0652	.0765	.0883	.1003	.1124	.1242	.1356	.1462	.1558	.1643
8	.0282	.0351	.0429	.0515	.0609	.0709	.0815	.0924	.1034	.1144
9	.0100	.0132	.0171	.0217	.0271	.0332	.0402	.0479	.0563	.0654
10	.0029	.0041	.0056	.0075	.0099	.0128	.0163	.0205	.0253	.0308
11	.0007	.0010	.0015	.0022	.0030	.0041	.0055	.0072	.0094	.0120
12	.0001	.0002	.0003	.0005	.0008	.0011	.0015	.0021	.0029	.0039
13	.0000	.0000	.0001	.0001	.0002	.0002	.0003	.0005	.0007	.0010
14	.0000	.0000	.0000	.0000	.0000	.0000	.0001	.0001	.0001	.0002

r \ p	.31	.32	.33	.34	.35	.36	.37	.38	.39	.40
0	.0006	.0004	.0003	.0002	.0002	.0001	.0001	.0001	.0001	.0000
1	.0054	.0042	.0033	.0025	.0020	.0015	.0011	.0009	.0007	.0005
2	.0229	.0188	.0153	.0124	.0100	.0080	.0064	.0050	.0040	.0031
3	.0619	.0531	.0453	.0383	.0323	.0270	.0224	.0185	.0152	.0123
4	.1181	.1062	.0947	.0839	.0738	.0645	.0559	.0482	.0412	.0350
5	.1698	.1599	.1493	.1384	.1272	.1161	.1051	.0945	.0843	.0746
6	.1907	.1881	.1839	.1782	.1712	.1632	.1543	.1447	.1347	.1244
7	.1714	.1770	.1811	.1836	.1844	.1836	.1812	.1774	.1722	.1659
8	.1251	.1354	.1450	.1537	.1614	.1678	.1730	.1767	.1790	.1797
9	.0750	.0849	.0952	.1056	.1158	.1259	.1354	.1444	.1526	.1597
10	.0370	.0440	.0516	.0598	.0686	.0779	.0875	.0974	.1073	.1171
11	.0151	.0188	.0231	.0280	.0336	.0398	.0467	.0542	.0624	.0710
12	.0051	.0066	.0085	.0108	.0136	.0168	.0206	.0249	.0299	.0355
13	.0014	.0019	.0026	.0034	.0045	.0058	.0074	.0094	.0118	.0146
14	.0003	.0005	.0006	.0009	.0012	.0016	.0022	.0029	.0038	.0049
15	.0001	.0001	.0001	.0002	.0003	.0004	.0005	.0007	.0010	.0013
16	.0000	.0000	.0000	.0000	.0000	.0001	.0001	.0001	.0002	.0003

TABLE A.1   (continued)

n = 20 (cont.)

r \ p	.41	.42	.43	.44	.45	.46	.47	.48	.49	.50
0	.0000	.0000	.0000	.0000	.0000	.0000	.0000	.0000	.0000	.0000
1	.0004	.0003	.0002	.0001	.0001	.0001	.0001	.0000	.0000	.0000
2	.0024	.0018	.0014	.0011	.0008	.0006	.0005	.0003	.0002	.0002
3	.0100	.0080	.0064	.0051	.0040	.0031	.0024	.0019	.0014	.0011
4	.0295	.0247	.0206	.0170	.0139	.0113	.0092	.0074	.0059	.0046
5	.0656	.0573	.0496	.0427	.0365	.0309	.0260	.0217	.0180	.0148
6	.1140	.1037	.0936	.0839	.0746	.0658	.0577	.0501	.0432	.0370
7	.1585	.1502	.1413	.1318	.1221	.1122	.1023	.0925	.0830	.0739
8	.1790	.1768	.1732	.1683	.1623	.1553	.1474	.1388	.1296	.1201
9	.1658	.1707	.1742	.1763	.1771	.1763	.1742	.1708	.1661	.1602
10	.1268	.1359	.1446	.1524	.1593	.1652	.1700	.1734	.1755	.1762
11	.0801	.0895	.0991	.1089	.1185	.1280	.1370	.1455	.1533	.1602
12	.0417	.0486	.0561	.0642	.0727	.0818	.0911	.1007	.1105	.1201
13	.0178	.0217	.0260	.0310	.0366	.0429	.0497	.0572	.0653	.0739
14	.0062	.0078	.0098	.0122	.0150	.0183	.0221	.0264	.0314	.0370
15	.0017	.0023	.0030	.0038	.0049	.0062	.0078	.0098	.0121	.0148
16	.0004	.0005	.0007	.0009	.0013	.0017	.0022	.0028	.0036	.0046
17	.0001	.0001	.0001	.0002	.0002	.0003	.0005	.0006	.0008	.0011
18	.0000	.0000	.0000	.0000	.0000	.0000	.0001	.0001	.0001	.0002

n = 25

r \ p	.01	.02	.03	.04	.05	.06	.07	.08	.09	.10
0	.7778	.6035	.4670	.3604	.2774	.2129	.1630	.1244	.0946	.0718
1	.1964	.3079	.3611	.3754	.3650	.3398	.3066	.2704	.2340	.1994
2	.0238	.0754	.1340	.1877	.2305	.2602	.2770	.2821	.2777	.2659
3	.0018	.0118	.0318	.0600	.0930	.1273	.1598	.1881	.2106	.2265
4	.0001	.0013	.0054	.0137	.0269	.0447	.0662	.0899	.1145	.1384
5	.0000	.0001	.0007	.0024	.0060	.0120	.0209	.0329	.0476	.0646
6	.0000	.0000	.0001	.0003	.0010	.0026	.0052	.0095	.0157	.0239
7	.0000	.0000	.0000	.0000	.0001	.0004	.0011	.0022	.0042	.0072
8	.0000	.0000	.0000	.0000	.0000	.0001	.0002	.0004	.0009	.0018
9	.0000	.0000	.0000	.0000	.0000	.0000	.0000	.0001	.0002	.0004
10	.0000	.0000	.0000	.0000	.0000	.0000	.0000	.0000	.0000	.0001

TABLE A.1   (*continued*)

n = 25 (cont.)

p r	.11	.12	.13	.14	.15	.16	.17	.18	.19	.20
0	.0543	.0409	.0308	.0230	.0172	.0128	.0095	.0070	.0052	.0038
1	.1678	.1395	.1149	.0938	.0759	.0609	.0486	.0384	.0302	.0236
2	.2488	.2283	.2060	.1832	.1607	.1392	.1193	.1012	.0851	.0708
3	.2358	.2387	.2360	.2286	.2174	.2033	.1874	.1704	.1530	.1358
4	.1603	.1790	.1940	.2047	.2110	.2130	.2111	.2057	.1974	.1867
5	.0832	.1025	.1217	.1399	.1564	.1704	.1816	.1897	.1945	.1960
6	.0343	.0466	.0606	.0759	.0920	.1082	.1240	.1388	.1520	.1633
7	.0115	.0173	.0246	.0336	.0441	.0559	.0689	.0827	.0968	.1108
8	.0032	.0053	.0083	.0123	.0175	.0240	.0318	.0408	.0511	.0623
9	.0007	.0014	.0023	.0038	.0058	.0086	.0123	.0169	.0226	.0294
10	.0001	.0003	.0006	.0010	.0016	.0026	.0040	.0059	.0085	.0118
11	.0000	.0001	.0001	.0002	.0004	.0007	.0011	.0018	.0027	.0040
12	.0000	.0000	.0000	.0000	.0001	.0002	.0003	.0005	.0007	.0012
13	.0000	.0000	.0000	.0000	.0000	.0000	.0001	.0001	.0002	.0003
14	.0000	.0000	.0000	.0000	.0000	.0000	.0000	.0000	.0000	.0001

p r	.21	.22	.23	.24	.25	.26	.27	.28	.29	.30
0	.0028	.0020	.0015	.0010	.0008	.0005	.0004	.0003	.0002	.0001
1	.0183	.0141	.0109	.0083	.0063	.0047	.0035	.0026	.0020	.0014
2	.0585	.0479	.0389	.0314	.0251	.0199	.0157	.0123	.0096	.0074
3	.1192	.1035	.0891	.0759	.0641	.0537	.0446	.0367	.0300	.0243
4	.1742	.1606	.1463	.1318	.1175	.1037	.0906	.0785	.0673	.0572
5	.1945	.1903	.1836	.1749	.1645	.1531	.1408	.1282	.1155	.1030
6	.1724	.1789	.1828	.1841	.1828	.1793	.1736	.1661	.1572	.1472
7	.1244	.1369	.1482	.1578	.1654	.1709	.1743	.1754	.1743	.1712
8	.0744	.0869	.0996	.1121	.1241	.1351	.1450	.1535	.1602	.1651
9	.0373	.0463	.0562	.0669	.0781	.0897	.1013	.1127	.1236	.1336
10	.0159	.0209	.0269	.0338	.0417	.0504	.0600	.0701	.0808	.0916
11	.0058	.0080	.0109	.0145	.0189	.0242	.0302	.0372	.0450	.0536
12	.0018	.0026	.0038	.0054	.0074	.0099	.0130	.0169	.0214	.0268
13	.0005	.0007	.0011	.0017	.0025	.0035	.0048	.0066	.0088	.0115
14	.0001	.0002	.0003	.0005	.0007	.0010	.0015	.0022	.0031	.0042
15	.0000	.0000	.0001	.0001	.0002	.0003	.0004	.0006	.0009	.0013
16	.0000	.0000	.0000	.0000	.0000	.0001	.0001	.0002	.0002	.0004
17	.0000	.0000	.0000	.0000	.0000	.0000	.0000	.0000	.0001	.0001

TABLE A.1   (continued)

n = 25 (cont.)

p r	.31	.32	.33	.34	.35	.36	.37	.38	.39	.40
0	.0001	.0001	.0000	.0000	.0000	.0000	.0000	.0000	.0000	.0000
1	.0011	.0008	.0006	.0004	.0003	.0002	.0001	.0001	.0001	.0000
2	.0057	.0043	.0033	.0025	.0018	.0014	.0010	.0007	.0005	.0004
3	.0195	.0156	.0123	.0097	.0076	.0058	.0045	.0034	.0026	.0019
4	.0482	.0403	.0334	.0274	.0224	.0181	.0145	.0115	.0091	.0071
5	.0910	.0797	.0691	.0594	.0506	.0427	.0357	.0297	.0244	.0199
6	.1363	.1250	.1134	.1020	.0908	.0801	.0700	.0606	.0520	.0442
7	.1662	.1596	.1516	.1426	.1327	.1222	.1115	.1008	.0902	.0800
8	.1680	.1690	.1681	.1652	.1607	.1547	.1474	.1390	.1298	.1200
9	.1426	.1502	.1563	.1608	.1635	.1644	.1635	.1609	.1567	.1511
10	.1025	.1131	.1232	.1325	.1409	.1479	.1536	.1578	.1603	.1612
11	.0628	.0726	.0828	.0931	.1034	.1135	.1230	.1319	.1398	.1465
12	.0329	.0399	.0476	.0560	.0650	.0745	.0843	.0943	.1043	.1140
13	.0148	.0188	.0234	.0288	.0350	.0419	.0495	.0578	.0667	.0760
14	.0057	.0076	.0099	.0127	.0161	.0202	.0249	.0304	.0365	.0434
15	.0019	.0026	.0036	.0048	.0064	.0083	.0107	.0136	.0171	.0212
16	.0005	.0008	.0011	.0015	.0021	.0029	.0039	.0052	.0068	.0088
17	.0001	.0002	.0003	.0004	.0006	.0009	.0012	.0017	.0023	.0031
18	.0000	.0000	.0001	.0001	.0001	.0002	.0003	.0005	.0007	.0009
19	.0000	.0000	.0000	.0000	.0000	.0000	.0001	.0001	.0002	.0002

	.41	.42	.43	.44	.45	.46	.47	.48	.49	.50
0	.0000	.0000	.0000	.0000	.0000	.0000	.0000	.0000	.0000	.0000
1	.0000	.0000	.0000	.0000	.0000	.0000	.0000	.0000	.0000	.0000
2	.0003	.0002	.0001	.0001	.0001	.0000	.0000	.0000	.0000	.0000
3	.0014	.0011	.0008	.0006	.0004	.0003	.0002	.0001	.0001	.0001
4	.0055	.0042	.0032	.0024	.0018	.0014	.0010	.0007	.0005	.0004
5	.0161	.0129	.0102	.0081	.0063	.0049	.0037	.0028	.0021	.0016
6	.0372	.0311	.0257	.0211	.0172	.0138	.0110	.0087	.0068	.0053
7	.0703	.0611	.0527	.0450	.0381	.0319	.0265	.0218	.0178	.0143
8	.1099	.0996	.0895	.0796	.0701	.0612	.0529	.0453	.0384	.0322
9	.1442	.1363	.1275	.1181	.1084	.0985	.0886	.0790	.0697	.0609
10	.1603	.1579	.1539	.1485	.1419	.1342	.1257	.1166	.1071	.0974
11	.1519	.1559	.1583	.1591	.1583	.1559	.1521	.1468	.1404	.1328
12	.1232	.1317	.1393	.1458	.1511	.1550	.1573	.1581	.1573	.1550
13	.0856	.0954	.1051	.1146	.1236	.1320	.1395	.1460	.1512	.1550
14	.0510	.0592	.0680	.0772	.0867	.0964	.1060	.1155	.1245	.1328
15	.0260	.0314	.0376	.0445	.0520	.0602	.0690	.0782	.0877	.0974
16	.0113	.0142	.0177	.0218	.0266	.0321	.0382	.0451	.0527	.0609
17	.0042	.0055	.0071	.0091	.0115	.0145	.0179	.0220	.0268	.0322
18	.0013	.0018	.0024	.0032	.0042	.0055	.0071	.0090	.0114	.0143
19	.0003	.0005	.0007	.0009	.0013	.0017	.0023	.0031	.0040	.0053
20	.0001	.0001	.0001	.0002	.0003	.0004	.0006	.0009	.0012	.0016
21	.0000	.0000	.0000	.0000	.0001	.0001	.0001	.0002	.0003	.0004
22	.0000	.0000	.0000	.0000	.0000	.0000	.0000	.0000	.0000	.0001

TABLE A.2   **Standard normal curve areas (entries in the body of the table give the area under the standard normal curve from 0 to z)**

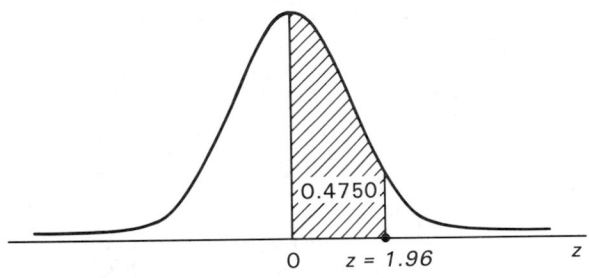

z	.00	.01	.02	.03	.04	.05	.06	.07	.08	.09
0.0	.0000	.0040	.0080	.0120	.0160	.0199	.0239	.0279	.0319	.0359
0.1	.0398	.0438	.0478	.0517	.0557	.0596	.0636	.0675	.0714	.0753
0.2	.0793	.0832	.0871	.0910	.0948	.0987	.1026	.1064	.1103	.1141
0.3	.1179	.1217	.1255	.1293	.1331	.1368	.1406	.1443	.1480	.1517
0.4	.1554	.1591	.1628	.1664	.1700	.1736	.1772	.1808	.1844	.1879
0.5	.1915	.1950	.1985	.2019	.2054	.2088	.2123	.2157	.2190	.2224
0.6	.2257	.2291	.2324	.2357	.2389	.2422	.2454	.2486	.2517	.2549
0.7	.2580	.2611	.2642	.2673	.2704	.2734	.2764	.2794	.2823	.2852
0.8	.2881	.2910	.2939	.2967	.2995	.3023	.3051	.3078	.3106	.3133
0.9	.3159	.3186	.3212	.3238	.3264	.3289	.3315	.3340	.3365	.3389
1.0	.3413	.3438	.3461	.3485	.3508	.3531	.3554	.3577	.3599	.3621
1.1	.3643	.3665	.3686	.3708	.3729	.3749	.3770	.3790	.3810	.3830
1.2	.3849	.3869	.3888	.3907	.3925	.3944	.3962	.3980	.3997	.4015
1.3	.4032	.4049	.4066	.4082	.4099	.4115	.4131	.4147	.4162	.4177
1.4	.4192	.4207	.4222	.4236	.4251	.4265	.4279	.4292	.4306	.4319
1.5	.4332	.4345	.4357	.4370	.4382	.4394	.4406	.4418	.4429	.4441
1.6	.4452	.4463	.4474	.4484	.4495	.4505	.4515	.4525	.4535	.4545
1.7	.4554	.4564	.4573	.4582	.4591	.4599	.4608	.4616	.4625	.4633
1.8	.4641	.4649	.4656	.4664	.4671	.4678	.4686	.4693	.4699	.4706
1.9	.4713	.4719	.4726	.4732	.4738	.4744	.4750	.4756	.4761	.4767
2.0	.4772	.4778	.4783	.4788	.4793	.4798	.4803	.4808	.4812	.4817
2.1	.4821	.4826	.4830	.4834	.4838	.4842	.4846	.4850	.4854	.4857
2.2	.4861	.4864	.4868	.4871	.4875	.4878	.4881	.4884	.4887	.4890
2.3	.4893	.4896	.4898	.4901	.4904	.4906	.4909	.4911	.4913	.4916
2.4	.4918	.4920	.4922	.4925	.4927	.4929	.4931	.4932	.4934	.4936
2.5	.4938	.4940	.4941	.4943	.4945	.4946	.4948	.4949	.4951	.4952
2.6	.4953	.4955	.4956	.4957	.4959	.4960	.4961	.4962	.4963	.4964
2.7	.4965	.4966	.4967	.4968	.4969	.4970	.4971	.4972	.4973	.4974
2.8	.4974	.4975	.4976	.4977	.4977	.4978	.4979	.4979	.4980	.4981
2.9	.4981	.4982	.4982	.4983	.4984	.4984	.4985	.4985	.4986	.4986
3.0	.4987	.4987	.4987	.4988	.4988	.4989	.4989	.4989	.4990	.4990

APPENDIX: TABLES

TABLE A.3   **Probability levels for the Wilcoxon signed-rank test**

T	P	T	P	T	P	T	P	T	P	T	P
n = 5		**n = 8**		**n = 10**		**n = 11**		**n = 12**		**n = 13**	
*0	.0313	0	.0039	0	.0010	0	.0005	0	.0002	0	.0001
1	.0625	1	.0078	1	.0020	1	.0010	1	.0005	1	.0002
2	.0938	2	.0117	2	.0029	2	.0015	2	.0007	2	.0004
3	.1563	3	.0195	3	.0049	3	.0024	3	.0012	3	.0006
4	.2188	4	.0273	4	.0068	4	.0034	4	.0017	4	.0009
5	.3125	*5	.0391	5	.0098	5	.0049	5	.0024	5	.0012
6	.4063	6	.0547	6	.0137	6	.0068	6	.0034	6	.0017
7	.5000	7	.0742	7	.0186	7	.0093	7	.0046	7	.0023
		8	.0977	8	.0244	8	.0122	8	.0061	8	.0031
n = 6		9	.1250	9	.0322	9	.0161	9	.0081	9	.0040
0	.0156	10	.1563	*10	.0420	10	.0210	10	.0105	10	.0052
1	.0313	11	.1914	11	.0527	11	.0269	11	.0134	11	.0067
*2	.0469	12	.2305	12	.0654	12	.0337	12	.0171	12	.0085
3	.0781	13	.2734	13	.0801	*13	.0415	13	.0212	13	.0107
4	.1094	14	.3203	14	.0967	14	.0508	14	.0261	14	.0133
5	.1563	15	.3711	15	.1162	15	.0615	15	.0320	15	.0164
6	.2188	16	.4219	16	.1377	16	.0737	16	.0386	16	.0199
7	.2813	17	.4727	17	.1611	17	.0874	*17	.0461	17	.0239
8	.3438	18	.5273	18	.1875	18	.1030	18	.0549	18	.0287
9	.4219	**n = 9**		19	.2158	19	.1201	19	.0647	19	.0341
10	.5000	0	.0020	20	.2461	20	.1392	20	.0757	20	.0402
		1	.0039	21	.2783	21	.1602	21	.0881	*21	.0471
n = 7		2	.0059	22	.3125	22	.1826	22	.1018	22	.0549
0	.0078	3	.0098	23	.3477	23	.2065	23	.1167	23	.0636
1	.0156	4	.0137	24	.3848	24	.2324	24	.1331	24	.0732
2	.0234	5	.0195	25	.4229	25	.2598	25	.1506	25	.0839
*3	.0391	6	.0273	26	.4609	26	.2886	26	.1697	26	.0955
4	.0547	7	.0371	27	.5000	27	.3188	27	.1902	27	.1082
5	.0781	*8	.0488			28	.3501	28	.2119	28	.1219
6	.1094	9	.0645			29	.3823	29	.2349	29	.1367
7	.1484	10	.0820			30	.4155	30	.2593	30	.1527
8	.1875	11	.1016			31	.4492	31	.2847	31	.1698
9	.2344	12	.1250			32	.4829	32	.3110	32	.1879
10	.2891	13	.1504			33	.5171	33	.3386	33	.2072
11	.3438	14	.1797					34	.3667	34	.2274
12	.4063	15	.2129					35	.3955	35	.2487
13	.4688	16	.2480					36	.4250	36	.2709
14	.5313	17	.2852					37	.4548	37	.2939
		18	.3262					38	.4849	38	.3177
		19	.3672					39	.5151	39	.3424
		20	.4102							40	.3677
		21	.4551							41	.3934
		22	.5000							42	.4197
										43	.4463
										44	.4730
										45	.5000

* For given n, the smallest rank total for which the probability level is equal to or less than 0.0500.

TABLE A.3   (*continued*)

T	P	T	P	T	P	T	P	T	P	T	P
n = 14		**n = 14**		**n = 15**		**n = 16**		**n = 17**		**n = 17**	
0	.0001	50	.4516	47	.2444	39	.0719	25	.0064	74	.4633
2	.0002	51	.4758	48	.2622	40	.0795	26	.0075	75	.4816
3	.0003	52	.5000	49	.2807	41	.0877	27	.0087	76	.5000
4	.0004			50	.2997	42	.0964	28	.0101		
5	.0006	**n = 15**		51	.3193	43	.1057	29	.0116	**n = 18**	
6	.0009	1	.0001	52	.3394	44	.1156	30	.0133	6	.0001
7	.0012	3	.0002	53	.3599	45	.1261	31	.0153	10	.0002
8	.0015	5	.0003	54	.3808	46	.1372	32	.0174	12	.0003
9	.0020	6	.0004	55	.4020	47	.1489	33	.0198	14	.0004
10	.0026	7	.0006	56	.4235	48	.1613	34	.0224	15	.0005
11	.0034	8	.0008	57	.4452	49	.1742	35	.0253	16	.0006
12	.0043	9	.0010	58	.4670	50	.1877	36	.0284	17	.0008
13	.0054	10	.0013	59	.4890	51	.2019	37	.0319	18	.0010
14	.0067	11	.0017	60	.5110	52	.2166	38	.0357	19	.0012
15	.0083	12	.0021	**n = 16**		53	.2319	39	.0398	20	.0014
16	.0101	13	.0027	3	.0001	54	.2477	40	.0443	21	.0017
17	.0123	14	.0034	5	.0002	55	.2641	*41	.0492	22	.0020
18	.0148	15	.0042	7	.0003	56	.2809	42	.0544	23	.0024
19	.0176	16	.0051	8	.0004	57	.2983	43	.0601	24	.0028
20	.0209	17	.0062	9	.0005	58	.3161	44	.0662	25	.0033
21	.0247	18	.0075	10	.0007	59	.3343	45	.0727	26	.0038
22	.0290	19	.0090	11	.0008	60	.3529	46	.0797	27	.0045
23	.0338	20	.0108	12	.0011	61	.3718	47	.0871	28	.0052
24	.0392	21	.0128	13	.0013	62	.3910	48	.0950	29	.0060
*25	.0453	22	.0151	14	.0017	63	.4104	49	.1034	30	.0069
26	.0520	23	.0177	15	.0021	64	.4301	50	.1123	31	.0080
27	.0594	24	.0206	16	.0026	65	.4500	51	.1218	32	.0091
28	.0676	25	.0240	17	.0031	66	.4699	52	.1317	33	.0104
29	.0765	26	.0277	18	.0038	67	.4900	53	.1421	34	.0118
30	.0863	27	.0319	19	.0046	68	.5100	54	.1530	35	.0134
31	.0969	28	.0365	20	.0055			55	.1645	36	.0152
32	.1083	29	.0416	21	.0065	**n = 17**		56	.1764	37	.0171
33	.1206	*30	.0473	22	.0078	4	.0001	57	.1889	38	.0192
34	.1338	31	.0535	23	.0091	8	.0002	58	.2019	39	.0216
35	.1479	32	.0603	24	.0107	9	.0003	59	.2153	40	.0241
36	.1629	33	.0677	25	.0125	11	.0004	60	.2293	41	.0269
37	.1788	34	.0757	26	.0145	12	.0005	61	.2437	42	.0300
38	.1955	35	.0844	27	.0168	13	.0007	62	.2585	43	.0333
39	.2131	36	.0938	28	.0193	14	.0008	63	.2738	44	.0368
40	.2316	37	.1039	29	.0222	15	.0010	64	.2895	45	.0407
41	.2508	38	.1147	30	.0253	16	.0013	65	.3056	46	.0449
42	.2708	39	.1262	31	.0288	17	.0016	66	.3221	*47	.0494
43	.2915	40	.1384	32	.0327	18	.0019	67	.3389	48	.0542
44	.3129	41	.1514	33	.0370	19	.0023	68	.3559	49	.0594
45	.3349	42	.1651	34	.0416	20	.0028	69	.3733	50	.0649
46	.3574	43	.1796	*35	.0467	21	.0033	70	.3910	51	.0708
47	.3804	44	.1947	36	.0523	22	.0040	71	.4088	52	.0770
48	.4039	45	.2106	37	.0583	23	.0047	72	.4268	53	.0837
49	.4276	46	.2271	38	.0649	24	.0055	73	.4450	54	.0907

TABLE A.3   (continued)

T	P	T	P	T	P	T	P	T	P	T	P
n = 18		*n* = 19		*n* = 19		*n* = 20		*n* = 20		*n* = 21	
55	.0982	30	.0036	79	.2706	48	.0164	97	.3921	61	.0298
56	.1061	31	.0041	80	.2839	49	.0181	98	.4062	62	.0323
57	.1144	32	.0047	81	.2974	50	.0200	99	.4204	63	.0351
58	.1231	33	.0054	82	.3113	51	.0220	100	.4347	64	.0380
59	.1323	34	.0062	83	.3254	52	.0242	101	.4492	65	.0411
60	.1419	35	.0070	84	.3397	53	.0266	102	.4636	66	.0444
61	.1519	36	.0080	85	.3543	54	.0291	103	.4782	*67	.0479
62	.1624	37	.0090	86	.3690	55	.0319	104	.4927	68	.0516
63	.1733	38	.0102	87	.3840	56	.0348	105	.5073	69	.0555
64	.1846	39	.0115	88	.3991	57	.0379	*n* = 21		70	.0597
65	.1964	40	.0129	89	.4144	58	.0413	14	.0001	71	.0640
66	.2086	41	.0145	90	.4298	59	.0448	20	.0002	72	.0686
67	.2211	42	.0162	91	.4453	*60	.0487	22	.0003	73	.0735
68	.2341	43	.0180	92	.4609	61	.0527	24	.0004	74	.0786
69	.2475	44	.0201	93	.4765	62	.0570	26	.0005	75	.0839
70	.2613	45	.0223	94	.4922	63	.0615	27	.0006	76	.0895
71	.2754	46	.0247	95	.5078	64	.0664	28	.0007	77	.0953
72	.2899	47	.0273	*n* = 20		65	.0715	29	.0008	78	.1015
73	.3047	48	.0301	11	.0001	66	.0768	30	.0009	79	.1078
74	.3198	49	.0331	16	.0002	67	.0825	31	.0011	80	.1145
75	.3353	50	.0364	19	.0003	68	.0884	32	.0012	81	.1214
76	.3509	51	.0399	20	.0004	69	.0947	33	.0014	82	.1286
77	.3669	52	.0437	22	.0005	70	.1012	34	.0016	83	.1361
78	.3830	*53	.0478	23	.0006	71	.1081	35	.0019	84	.1439
79	.3994	54	.0521	24	.0007	72	.1153	36	.0021	85	.1519
80	.4159	55	.0567	25	.0008	73	.1227	37	.0024	86	.1602
81	.4325	56	.0616	26	.0010	74	.1305	38	.0028	87	.1688
82	.4493	57	.0668	27	.0012	75	.1387	39	.0031	88	.1777
83	.4661	58	.0723	28	.0014	76	.1471	40	.0036	89	.1869
84	.4831	59	.0782	29	.0016	77	.1559	41	.0040	90	.1963
85	.5000	60	.0844	30	.0018	78	.1650	42	.0045	91	.2060
		61	.0909	31	.0021	79	.1744	43	.0051	92	.2160
n = 19		62	.0978	32	.0024	80	.1841	44	.0057	93	.2262
9	.0001	63	.1051	33	.0028	81	.1942	45	.0063	94	.2367
13	.0002	64	.1127	34	.0032	82	.2045	46	.0071	95	.2474
15	.0003	65	.1206	35	.0036	83	.2152	47	.0079	96	.2584
17	.0004	66	.1290	36	.0042	84	.2262	48	.0088	97	.2696
18	.0005	67	.1377	37	.0047	85	.2375	49	.0097	98	.2810
19	.0006	68	.1467	38	.0053	86	.2490	50	.0108	99	.2927
20	.0007	69	.1562	39	.0060	87	.2608	51	.0119	100	.3046
21	.0008	70	.1660	40	.0068	88	.2729	52	.0132	101	.3166
22	.0010	71	.1762	41	.0077	89	.2853	53	.0145	102	.3289
23	.0012	72	.1868	42	.0086	90	.2979	54	.0160	103	.3414
24	.0014	73	.1977	43	.0096	91	.3108	55	.0175	104	.3540
25	.0017	74	.2090	44	.0107	92	.3238	56	.0192	105	.3667
26	.0020	75	.2207	45	.0120	93	.3371	57	.0210	106	.3796
27	.0023	76	.2327	46	.0133	94	.3506	58	.0230	107	.3927
28	.0027	77	.2450	47	.0148	95	.3643	59	.0251	108	.4058
29	.0031	78	.2576			96	.3781	60	.0273	109	.4191

TABLE A.3 (*continued*)

T	P	T	P	T	P	T	P	T	P	T	P
n = 21		**n = 22**		**n = 22**		**n = 23**		**n = 23**		**n = 24**	
110	.4324	67	.0271	116	.3751	68	.0163	117	.2700	62	.0053
111	.4459	68	.0293	117	.3873	69	.0177	118	.2800	63	.0058
112	.4593	69	.0317	118	.3995	70	.0192	119	.2902	64	.0063
113	.4729	70	.0342	119	.4119	71	.0208	120	.3005	65	.0069
114	.4864	71	.0369	120	.4243	72	.0224	121	.3110	66	.0075
115	.5000	72	.0397	121	.4368	73	.0242	122	.3217	67	.0082
		73	.0427	122	.4494	74	.0261	123	.3325	68	.0089
		74	.0459	123	.4620	75	.0281	124	.3434	69	.0097
n = 22		*75	.0492	124	.4746	76	.0303	125	.3545	70	.0106
18	.0001	76	.0527	125	.4873	77	.0325	126	.3657	71	.0115
23	.0002	77	.0564	126	.5000	78	.0349	127	.3770	72	.0124
26	.0003	78	.0603			79	.0274	128	.3884	73	.0135
29	.0004	79	.0644	**n = 23**		80	.0401	129	.3999	74	.0146
30	.0005	80	.0687	21	.0001	81	.0429	130	.4115	75	.0157
32	.0006	81	.0733	28	.0002	82	.0459	131	.4231	76	.0170
33	.0007	82	.0780	31	.0003	*83	.0490	132	.4348	77	.0183
34	.0008	83	.0829	33	.0004	84	.0523	133	.4466	78	.0197
35	.0010	84	.0881	35	.0005	85	.0557	134	.4584	79	.0212
36	.0011	85	.0935	36	.0006	86	.0593	135	.4703	80	.0228
37	.0013	86	.0991	38	.0007	87	.0631	136	.4822	81	.0245
38	.0014	87	.1050	39	.0008	88	.0671	137	.4941	82	.0263
39	.0016	88	.1111	40	.0009	89	.0712	138	.5060	83	.0282
40	.0018	89	.1174	41	.0011	90	.0755			84	.0302
41	.0021	90	.1240	42	.0012	91	.0801	**n = 24**		85	.0323
42	.0023	91	.1308	43	.0014	92	.0848	25	.0001	86	.0346
43	.0026	92	.1378	44	.0015	93	.0897	32	.0002	87	.0369
44	.0030	93	.1451	45	.0017	94	.0948	36	.0003	88	.0394
45	.0033	94	.1527	46	.0019	95	.1001	38	.0004	89	.0420
46	.0037	95	.1604	47	.0022	96	.1056	40	.0005	90	.0447
47	.0042	96	.1685	48	.0024	97	.1113	42	.0006	*91	.0475
48	.0046	97	.1767	49	.0027	98	.1172	43	.0007	92	.0505
49	.0052	98	.1853	50	.0030	99	.1234	44	.0008	93	.0537
50	.0057	99	.1940	51	.0034	100	.1297	45	.0009	94	.0570
51	.0064	100	.2030	52	.0037	101	.1363	46	.0010	95	.0604
52	.0070	101	.2122	53	.0041	102	.1431	47	.0011	96	.0640
53	.0078	102	.2217	54	.0046	103	.1501	48	.0013	97	.0678
54	.0086	103	.2314	55	.0051	104	.1573	49	.0014	98	.0717
55	.0095	104	.2413	56	.0056	105	.1647	50	.0016	99	.0758
56	.0104	105	.2514	57	.0061	106	.1723	51	.0018	100	.0800
57	.0115	106	.2618	58	.0068	107	.1802	52	.0020	101	.0844
58	.0126	107	.2723	59	.0074	108	.1883	53	.0022	102	.0890
59	.0138	108	.2830	60	.0082	109	.1965	54	.0024	103	.0938
60	.0151	109	.2940	61	.0089	110	.2050	55	.0027	104	.0987
61	.0164	110	.3051	62	.0098	111	.2137	56	.0029	105	.1038
62	.0179	111	.3164	63	.0107	112	.2226	57	.0033	106	.1091
63	.0195	112	.3278	64	.0117	113	.2317	58	.0036	107	.1146
64	.0212	113	.3394	65	.0127	114	.2410	59	.0040	108	.1203
65	.0231	114	.3512	66	.0138	115	.2505	60	.0044	109	.1261
66	.0250	115	.3631	67	.0150	116	.2601	61	.0048	110	.1322

TABLE A.3   (continued)

T	P	T	P	T	P	T	P	T	P	T	P
n = 24		**n = 25**		**n = 25**		**n = 25**		**n = 26**		**n = 26**	
111	.1384	50	.0008	99	.0452	148	.3556	81	.0076	130	.1289
112	.1448	51	.0009	*100	.0479	149	.3655	82	.0082	131	.1344
113	.1515	52	.0010	101	.0507	150	.3755	83	.0088	132	.1399
114	.1583	53	.0011	102	.0537	151	.3856	84	.0095	133	.1457
115	.1653	54	.0013	103	.0567	152	.3957	85	.0102	134	.1516
116	.1724	55	.0014	104	.0600	153	.4060	86	.0110	135	.1576
117	.1798	56	.0015	105	.0633	154	.4163	87	.0118	136	.1638
118	.1874	57	.0017	106	.0668	155	.4266	88	.0127	137	.1702
119	.1951	58	.0019	107	.0705	156	.4370	89	.0136	138	.1767
120	.2031	59	.0021	108	.0742	157	.4474	90	.0146	139	.1833
121	.2112	60	.0023	109	.0782	158	.4579	91	.0156	140	.1901
122	.2195	61	.0025	110	.0822	159	.4684	92	.0167	141	.1970
123	.2279	62	.0028	111	.0865	160	.4789	93	.0179	142	.2041
124	.2366	63	.0031	112	.0909	161	.4895	94	.0191	143	.2114
125	.2454	64	.0034	113	.0954	162	.5000	95	.0204	144	.2187
126	.2544	65	.0037	114	.1001			96	.0217	145	.2262
127	.2635	66	.0040	115	.1050	**n = 26**		97	.0232	146	.2339
128	.2728	67	.0044	116	.1100	34	.0001	98	.0247	147	.2417
129	.2823	68	.0048	117	.1152	42	.0002	99	.0263	148	.2496
130	.2919	69	.0053	118	.1205	46	.0003	100	.0279	149	.2577
131	.3017	70	.0057	119	.1261	49	.0004	101	.0297	150	.2658
132	.3115	71	.0062	120	.1317	51	.0005	102	.0315	151	.2741
133	.3216	72	.0068	121	.1376	53	.0006	103	.0334	152	.2826
134	.3317	73	.0074	122	.1436	55	.0007	104	.0355	153	.2911
135	.3420	74	.0080	123	.1498	56	.0008	105	.0376	154	.2998
136	.3524	75	.0087	124	.1562	57	.0009	106	.0398	155	.3085
137	.3629	76	.0094	125	.1627	58	.0010	107	.0421	156	.3174
138	.3735	77	.0101	126	.1694	59	.0011	108	.0445	157	.3264
139	.3841	78	.0110	127	.1763	60	.0012	109	.0470	158	.3355
140	.3949	79	.0118	128	.1833	61	.0013	*110	.0497	159	.3447
141	.4058	80	.0128	129	.1905	62	.0015	111	.0524	160	.3539
142	.4167	81	.0137	130	.1979	63	.0016	112	.0553	161	.3633
143	.4277	82	.0148	131	.2054	64	.0018	113	.0582	162	.3727
144	.4387	83	.0159	132	.2131	65	.0020	114	.0613	163	.3822
145	.4498	84	.0171	133	.2209	66	.0021	115	.0646	164	.3918
146	.4609	85	.0183	134	.2289	67	.0023	116	.0679	165	.4014
147	.4721	86	.0197	135	.2371	68	.0026	117	.0714	166	.4111
148	.4832	87	.0211	136	.2454	69	.0028	118	.0750	167	.4208
149	.4944	88	.0226	137	.2539	70	.0031	119	.0787	168	.4306
150	.5056	89	.0241	138	.2625	71	.0033	120	.0825	169	.4405
		90	.0258	139	.2712	72	.0036	121	.0865	170	.4503
n = 25		91	.0275	140	.2801	73	.0040	122	.0907	171	.4602
29	.0001	92	.0294	141	.2891	74	.0043	123	.0950	172	.4702
37	.0002	93	.0313	142	.2983	75	.0047	124	.0994	173	.4801
41	.0003	94	.0334	143	.3075	76	.0051	125	.1039	174	.4900
43	.0004	95	.0355	144	.3169	77	.0055	126	.1086	175	.5000
45	.0005	96	.0377	145	.3264	78	.0060	127	.1135		
47	.0006	97	.0401	146	.3360	79	.0065	128	.1185		
48	.0007	98	.0426	147	.3458	80	.0070	129	.1236		

TABLE A.3 (continued)

T	P	T	P	T	P	T	P	T	P	T	P
n = 27		*n = 27*		*n = 27*		*n = 28*		*n = 28*		*n = 28*	
39	.0001	105	.0218	154	.2066	74	.0012	123	.0349	172	.2466
47	.0002	106	.0231	155	.2135	75	.0013	124	.0368	173	.2538
52	.0003	107	.0246	156	.2205	76	.0015	125	.0387	174	.2611
55	.0004	108	.0260	157	.2277	77	.0016	126	.0407	175	.2685
57	.0005	109	.0276	158	.2349	78	.0017	127	.0428	176	.2759
59	.0006	110	.0292	159	.2423	79	.0019	128	.0450	177	.2835
61	.0007	111	.0309	160	.2498	80	.0020	129	.0473	178	.2912
62	.0008	112	.0327	161	.2574	81	.0022	*130	.0496	179	.2990
64	.0009	113	.0346	162	.2652	82	.0024	131	.0521	180	.3068
65	.0010	114	.0366	163	.2730	83	.0026	132	.0546	181	.3148
66	.0011	115	.0386	164	.2810	84	.0028	133	.0573	182	.3228
67	.0012	116	.0407	165	.2890	85	.0030	134	.0600	183	.3309
68	.0014	117	.0430	166	.2972	86	.0033	135	.0628	184	.3391
69	.0015	118	.0453	167	.3055	87	.0035	136	.0657	185	.3474
70	.0016	*119	.0477	168	.3138	88	.0038	137	.0688	186	.3557
71	.0018	120	.0502	169	.3223	89	.0041	138	.0719	187	.3641
72	.0019	121	.0528	170	.3308	90	.0044	139	.0751	188	.3725
73	.0021	122	.0555	171	.3395	91	.0048	140	.0785	189	.3811
74	.0023	123	.0583	172	.3482	92	.0051	141	.0819	190	.3896
75	.0025	124	.0613	173	.3570	93	.0055	142	.0855	191	.3983
76	.0027	125	.0643	174	.3659	94	.0059	143	.0891	192	.4070
77	.0030	126	.0674	175	.3748	95	.0064	144	.0929	193	.4157
78	.0032	127	.0707	176	.3838	96	.0068	145	.0968	194	.4245
79	.0035	128	.0741	177	.3929	97	.0073	146	.1008	195	.4333
80	.0038	129	.0776	178	.4020	98	.0078	147	.1049	196	.4421
81	.0041	130	.0812	179	.4112	99	.0084	148	.1091	197	.4510
82	.0044	131	.0849	180	.4204	100	.0089	149	.1135	198	.4598
83	.0048	132	.0888	181	.4297	101	.0096	150	.1180	199	.4687
84	.0052	133	.0927	182	.4390	102	.0102	151	.1225	200	.4777
85	.0056	134	.0968	183	.4483	103	.0109	152	.1273	201	.4866
86	.0060	135	.1010	184	.4577	104	.0116	153	.1321	202	.4955
87	.0065	136	.1054	185	.4670	105	.0124	154	.1370	203	.5045
88	.0070	137	.1099	186	.4764	106	.0132	155	.1421		
89	.0075	138	.1145	187	.4859	107	.0140	156	.1473	*n = 29*	
90	.0081	139	.1193	188	.4953	108	.0149	157	.1526	50	.0001
91	.0087	140	.1242	189	.5047	109	.0159	158	.1580	59	.0002
92	.0093	141	.1292			110	.0168	159	.1636	65	.0003
93	.0100	142	.1343	*n = 28*		111	.0179	160	.1693	68	.0004
94	.0107	143	.1396	44	.0001	112	.0190	161	.1751	71	.0005
95	.0115	144	.1450	53	.0002	113	.0201	162	.1810	73	.0006
96	.0123	145	.1506	58	.0003	114	.0213	163	.1870	75	.0007
97	.0131	146	.1563	61	.0004	115	.0226	164	.1932	76	.0008
98	.0140	147	.1621	64	.0005	116	.0239	165	.1995	78	.0009
99	.0150	148	.1681	66	.0006	117	.0252	166	.2059	79	.0010
100	.0159	149	.1742	68	.0007	118	.0267	167	.2124	80	.0011
101	.0170	150	.1804	69	.0008	119	.0282	168	.2190	81	.0012
102	.0181	151	.1868	70	.0009	120	.0298	169	.2257	82	.0013
103	.0193	152	.1932	72	.0010	121	.0314	170	.2326	83	.0014
104	.0205	153	.1999	73	.0011	122	.0331	171	.2395	84	.0015

T	P	T	P	T	P	T	P	T	P	T	P
n = 29		*n* = 29		*n* = 29		*n* = 30		*n* = 30		*n* = 30	
85	.0016	134	.0362	183	.2340	90	.0013	139	.0275	188	.1854
86	.0018	135	.0380	184	.2406	91	.0014	140	.0288	189	.1909
87	.0019	136	.0399	185	.2473	92	.0015	141	.0303	190	.1965
88	.0021	137	.0418	186	.2541	93	.0016	142	.0318	191	.2022
89	.0022	138	.0439	187	.2611	94	.0017	143	.0333	192	.2081
90	.0024	139	.0460	188	.2681	95	.0019	144	.0349	193	.2140
91	.0026	*140	.0482	189	.2752	96	.0020	145	.0366	194	.2200
92	.0028	141	.0504	190	.2824	97	.0022	146	.0384	195	.2261
93	.0030	142	.0528	191	.2896	98	.0023	147	.0402	196	.2323
94	.0032	143	.0552	192	.2970	99	.0025	148	.0420	197	.2386
95	.0035	144	.0577	193	.3044	100	.0027	149	.0440	198	.2449
96	.0037	145	.0603	194	.3120	101	.0029	150	.0460	199	.2514
97	.0040	146	.0630	195	.3196	102	.0031	*151	.0481	200	.2579
98	.0043	147	.0658	196	.3272	103	.0033	152	.0502	201	.2646
99	.0046	148	.0687	197	.3350	104	.0036	153	.0524	202	.2713
100	.0049	149	.0716	198	.3428	105	.0038	154	.0547	203	.2781
101	.0053	150	.0747	199	.3507	106	.0041	155	.0571	204	.2849
102	.0057	151	.0778	200	.3586	107	.0044	156	.0595	205	.2919
103	.0061	152	.0811	201	.3666	108	.0047	157	.0621	206	.2989
104	.0065	153	.0844	202	.3747	109	.0050	158	.0647	207	.3060
105	.0069	154	.0879	203	.3828	110	.0053	159	.0674	208	.3132
106	.0074	155	.0914	204	.3909	111	.0057	160	.0701	209	.3204
107	.0079	156	.0951	205	.3991	112	.0060	161	.0730	210	.3277
108	.0084	157	.0988	206	.4074	113	.0064	162	.0759	211	.3351
109	.0089	158	.1027	207	.4157	114	.0068	163	.0790	212	.3425
110	.0095	159	.1066	208	.4240	115	.0073	164	.0821	213	.3500
111	.0101	160	.1107	209	.4324	116	.0077	165	.0853	214	.3576
112	.0108	161	.1149	210	.4408	117	.0082	166	.0886	215	.3652
113	.0115	162	.1191	211	.4492	118	.0087	167	.0920	216	.3728
114	.0122	163	.1235	212	.4576	119	.0093	168	.0955	217	.3805
115	.0129	164	.1280	213	.4661	120	.0098	169	.0990	218	.3883
116	.0137	165	.1326	214	.4745	121	.0104	170	.1027	219	.3961
117	.0145	166	.1373	215	.4830	122	.0110	171	.1065	220	.4039
118	.0154	167	.1421	216	.4915	123	.0117	172	.1103	221	.4118
119	.0163	168	.1471	217	.5000	124	.0124	173	.1143	222	.4197
120	.0173	169	.1521			125	.0131	174	.1183	223	.4276
121	.0183	170	.1572	*n* = 30		126	.0139	175	.1225	224	.4356
122	.0193	171	.1625	55	.0001	127	.0147	176	.1267	225	.4436
123	.0204	172	.1679	66	.0002	128	.0155	177	.1311	226	.4516
124	.0216	173	.1733	71	.0003	129	.0164	178	.1355	227	.4596
125	.0228	174	.1789	75	.0004	130	.0173	179	.1400	228	.4677
126	.0240	175	.1846	78	.0005	131	.0182	180	.1447	229	.4758
127	.0253	176	.1904	80	.0006	132	.0192	181	.1494	230	.4838
128	.0267	177	.1963	82	.0007	133	.0202	182	.1543	231	.4919
129	.0281	178	.2023	84	.0008	134	.0213	183	.1592	232	.5000
130	.0296	179	.2085	85	.0009	135	.0225	184	.1642		
131	.0311	180	.2147	87	.0010	136	.0236	185	.1694		
132	.0328	181	.2210	88	.0011	137	.0249	186	.1746		
133	.0344	182	.2274	89	.0012	138	.0261	187	.1799		

Source: Frank Wilcoxon, S. K. Katti, and Roberta A. Wilcox, "Critical Values and Probability Levels for the Wilcoxon Rank Sum Test and the Wilcoxon Signed Rank Test." Originally prepared and distributed by Lederle Laboratories Division, American Cyanamid Company, Pearl River, New York, in cooperation with the Department of Statistics, The Florida State University, Tallahassee, Florida. Revised October 1968. Copyright 1963 by the American Cyanamid Company and The Florida State University. Reproduced by permission of S. K. Katti.

TABLE A.4 — Table of confidence limits for a proportion*

n	r	Confidence coefficient, % 90	95	99	n	r	Confidence coefficient, % 90	95	99	n	r	Confidence coefficient, % 90	95	99
1	0	0.000	0.000	0.000	9	3	0.129	0.098	0.053	13	6	0.246	0.224	0.159
	1	.100	.050	.010		4	.210	.169	.105+		7	.276	.260	.213
2	0	0.000	0.000	0.000		5	.232	.251	.171		8	.379	.327	.273
	1	.051	.025+	.005+		6	.390	.289	.250		9	.455+	.413	.302
	2	.316	.224	.100		7	.485-	.442	.344		10	.530	.480	.406
3	0	0.000	0.000	0.000		8	.609	.557	.402		11	.621	.566	.477
	1	.035-	.017	.003		9	.768	.711	.598		12	.724	.673	.571
	2	.196	.135+	.059	10	0	0.000	0.000	0.000		13	.827	.775-	.698
	3	.464	.368	.215+		1	.010	.005+	.001	14	0	0.000	0.000	0.000
4	0	0.000	0.000	0.000		2	.055-	.037	.016		1	.007	.004	.001
	1	.026	.013	.003		3	.116	.087	.048		2	.039	.026	.011
	2	.143	.098	.042		4	.188	.150	.093		3	.081	.061	.033
	3	.320	.249	.141		5	.222	.222	.150		4	.131	.104	.064
	4	.500	.473	.316		6	.341	.267	.218		5	.163	.153	.102
5	0	0.000	0.000	0.000		7	.352	.381	.297		6	.224	.206	.146
	1	.021	.010	.002		8	.500	.397	.376		7	.261	.206	.195
	2	.112	.076	.033		9	.648	.603	.488		8	.355-	.312	.249
	3	.247	.189	.106		10	.778	.733	.624		9	.406	.371	.286
	4	.379	.343	.222	11	0	0.000	0.000	0.000		10	.422	.389	.364
	5	.621	.500	.398		1	.010	.005-	.001		11	.578	.500	.392
6	0	0.000	0.000	0.000		2	.049	.033	.014		12	.635-	.611	.500
	1	.017	.009	.002		3	.105-	.079	.043		13	.739	.688	.608
	2	.093	.063	.027		4	.169	.135+	.084		14	.837	.794	.714
	3	.201	.153	.085-		5	.197	.200	.134	15	0	0.000	0.000	0.000
	4	.333	.271	.173		6	.302	.250	.194		1	.007	.003	.001
	5	.458	.402	.294		7	.315+	.333	.262		2	.036	.024	.010
	6	.655+	.598	.464		8	.423	.369	.340		3	.076	.057	.031
7	0	0.000	0.000	0.000		9	.577	.500	.407		4	.122	.097	.059
	1	.015-	.007	.001		10	.685-	.631	.500		5	.154	.142	.094
	2	.079	.053	.023		11	.803	.750	.641		6	.205+	.191	.135
	3	.170	.129	.071	12	0	0.000	0.000	0.000		7	.247	.191	.179
	4	.279	.225+	.142		1	.009	.004	.001		8	.325+	.294	.229
	5	.316	.341	.236		2	.045+	.030	.013		9	.325+	.332	.273
	6	.500	.446	.357		3	.096	.072	.039		10	.400	.369	.328
	7	.684	.623	.500		4	.154	.123	.076		11	.500	.448	.373
8	0	0.000	0.000	0.000		5	.184	.181	.121		12	.600	.552	.461
	1	.013	.006	.001		6	.271	.236	.175		13	.675-	.631	.539
	2	.069	.046	.020		7	.294	.294	.235		14	.753	.698	.627
	3	.147	.111	.061		8	.398	.346	.302		15	.846	.809	.727
	4	.240	.193	.121		9	.500	.450	.321	16	0	0.000	0.000	0.000
	5	.255-	.289	.198		10	.602	.550	.445		1	.007	.003	.001
	6	.418	.315+	.293		11	.706	.654	.555		2	.034	.023	.010
	7	.582	.500	.410		12	.816	.764	.679		3	.071	.053	.029
	8	.745+	.685-	.549	13	0	0.000	0.000	0.000		4	.114	.090	.055+
9	0	0.000	0.000	0.000		1	.008	.004	.001		5	.147	.132	.088
	1	.012	.006	.001		2	.042	.028	.012		6	.189	.178	.125+
	2	.061	.041	.017		3	.088	.066	.036		7	.235+	.178	.166
						4	.142	.113	.069		8	.299	.272	.212
						5	.173	.166	.111		9	.305+	.272	.261
											10	.381	.352	.295+

* Calculated by Edwin L. Crow, Eleanor G. Crow, and Robert S. Gardner according to a modification of a proposal of Theodore E. Sterne.

503

TABLE A.4   (*continued*)

n	r	Confidence coefficient, % 90	95	99	n	r	Confidence coefficient, % 90	95	99	n	r	Confidence coefficient, % 90	95	99
16	11	0.450	0.429	0.357	19	6	0.151	0.147	0.103	21	15	0.542	0.494	0.409
	12	.550	.500	.421		7	.209	.150	.137		16	.593	.545$^-$	.466
	13	.619	.571	.475$^+$		8	.238	.222	.173		17	.647	.602	.534
	14	.695$^-$	.648	.549		9	.265$^+$	.232	.212		18	.694	.662	.591
	15	.765$^-$	.728	.643		10	.337	.312	.218		19	.755$^+$	.724	.658
	16	.853	.822	.736		11	.386	.345$^-$	.293		20	.809	.787	.717
						12	.386	.365$^-$	.305$^+$		21	.877	.863	.799
17	0	0.000	0.000	0.000		13	.440	.426	.383					
	1	.006	.003	.001		14	.560	.500	.436	22	0	0.000	0.000	0.000
	2	.032	.021	.009		15	.614	.574	.485$^-$		1	.005$^-$	.002	.000
	3	.067	.050	.027		16	.663	.635$^+$	.545$^-$		2	.024	.016	.007
	4	.107	.085$^-$	.052		17	.735$^-$	.684	.617		3	.051	.038	.021
	5	.140	.124	.082		18	.791	.768	.695$^-$		4	.082	.065$^-$	.039
	6	.175$^+$	.166	.117		19	.870	.850	.782		5	.115$^-$	.094	.062
	7	.225$^+$	.166	.155$^+$							6	.115$^-$	.126	.088
	8	.277	.253	.197	20	0	0.000	0.000	0.000		7	.181	.132	.116
	9	.290	.254	.242		1	.005$^+$	.003	.001		8	.181	.187	.147
	10	.364	.337	.242		2	.027	.018	.008		9	.236	.205$^+$	.179
	11	.432	.406	.338		3	.056	.042	.023		10	.289	.260	.194
	12	.500	.456	.380		4	.090	.071	.044		11	.289	.264	.242
	13	.568	.511	.413		5	.126	.104	.069		12	.340	.326	.273
	14	.636	.583	.500		6	.141	.140	.098		13	.393	.383	.318
	15	.710	.663	.587		7	.201	.143	.129		14	.444	.418	.334
	16	.775$^-$	.746	.654		8	.221	.209	.163		15	.500	.424	.396
	17	.860	.834	.758		9	.255$^-$	.222	.200		16	.556	.500	.450
						10	.325$^-$	.293	.209		17	.607	.576	.495
18	0	0.000	0.000	0.000		11	.358	.293	.274		18	.660	.611	.546
	1	.006	.003	.001		12	.367	.351	.293		19	.711	.674	.604
	2	.030	.020	.008		13	.422	.411	.363		20	.764	.736	.666
	3	.063	.047	.025$^+$		14	.500	.467	.399		21	.819	.795$^-$	.727
	4	.101	.080	.049		15	.578	.533	.424		22	.885$^+$	.868	.806
	5	.135$^-$	.116	.077		16	.633	.589	.500					
	6	.163	.156	.110		17	.672	.649	.576	23	0	0.000	0.000	0.000
	7	.216	.157	.145$^+$		18	.745$^+$	.707	.625$^+$		1	.005$^-$	.002	.000
	8	.257	.236	.184		19	.797	.778	.707		2	.023	.016	.007
	9	.277	.242	.226		20	.874	.857	.791		3	.049	.037	.020
	10	.349	.325$^-$	.228							4	.078	.062	.038
	11	.416	.375$^-$	.314	21	0	0.000	0.000	0.000		5	.110	.090	.059
	12	.464	.381	.318		1	.005$^+$	.002	.000		6	.110	.120	.084
	13	.518	.444	.397		2	.026	.017	.007		7	.173	.127	.111
	14	.581	.556	.466		3	.054	.040	.022		8	.173	.178	.140
	15	.651	.619	.534		4	.086	.068	.041		9	.228	.198	.171
	16	.723	.675$^+$	.603		5	.121	.099	.065$^+$		10	.273	.247	.187
	17	.784	.758	.682		6	.130	.132	.092		11	.274	.255$^-$	.229
	18	.865$^+$	.843	.772		7	.191	.137	.122		12	.328	.317	.265$^+$
						8	.191	.197	.155$^-$		13	.381	.360	.298
19	0	0.000	0.000	0.000		9	.245$^-$	.213	.189		14	.431	.360	.323
	1	.006	.003	.001		10	.306	.276	.201		15	.478	.409	.384
	2	.028	.019	.008		11	.306	.276	.257		16	.521	.457	.420
	3	.059	.044	.024		12	.353	.338	.283		17	.569	.543	.429
	4	.095$^+$	.075$^+$	.046		13	.407	.398	.339		18	.619	.591	.500
	5	.130	.110	.073		14	.458	.449	.347		19	.672	.640	.571

The observed proportion in a random sample of size *n* is *r/n*. The table gives the lower confidence limit for the population proportion π, as a function of *n* and *r*. The upper confidence limit = 1 − (lower confidence limit, entered with *n* − *r* instead of *r*).

TABLE A.4   (continued)

n	r	90	95	99	n	r	90	95	99	n	r	90	95	99
		Confidence coefficient, %					Confidence coefficient, %					Confidence coefficient, %		
23	20	0.726	0.683	0.614	25	21	0.693	0.664	0.597	27	17	0.447	0.430	0.383
	21	.772	.745$^+$	.677		22	.745$^+$	.697	.648		18	.500	.437	.413
	22	.827	.802	.735$^-$		23	.786	.762	.695$^+$		19	.553	.500	.419
	23	.890	.873	.813		24	.842	.815$^-$	.755$^-$		20	.593	.563	.461
24	0	0.000	0.000	0.000		25	.899	.882	.825$^-$		21	.635$^-$	.585$^+$	.539
	1	.004	.002	.000							22	.674	.636	.581
	2	.022	.015$^+$	.006	26	0	0.000	0.000	0.000		23	.709	.684	.616
	3	.047	.035$^-$	.019		1	.004	.002	.000		24	.761	.731	.668
	4	.075$^-$	.059	.036		2	.021	.014	.006		25	.796	.777	.703
	5	.105$^-$	.086	.057		3	.043	.032	.017		26	.855$^-$	.825$^+$	.775$^+$
	6	.105$^-$	.115$^-$	.080		4	.069	.054	.033		27	.907	.890	.834
	7	.165$^-$	.122	.106		5	.097	.079	.052	28	0	0.000	0.000	0.000
	8	.165$^-$	.169	.133		6	.097	.106	.073		1	.004	.002	.000
	9	.221	.191	.163		7	.151	.114	.097		2	.019	.013	.005$^+$
	10	.259	.234	.181		8	.151	.154	.122		3	.040	.030	.016
	11	.264	.246	.216		9	.209	.180	.149		4	.064	.050	.031
	12	.317	.308	.257		10	.233	.212	.170		5	.089	.073	.048
	13	.370	.339	.280		11	.247	.230	.195$^-$		6	.089	.098	.068
	14	.413	.347	.313		12	.299	.282	.234		7	.139	.106	.089
	15	.447	.396	.362		13	.342	.282	.234		8	.139	.142	.112
	16	.447	.443	.364		14	.342	.325$^+$	.298		9	.197	.170	.137
	17	.553	.500	.416		15	.377	.374	.322		10	.208	.192	.162
	18	.577	.557	.464		16	.419	.421	.342		11	.232	.217	.175$^+$
	19	.630	.604	.536		17	.460	.458	.393		12	.284	.258	.214
	20	.683	.653	.584		18	.540	.494	.438		13	.310	.259	.218
	21	.736	.692	.636		19	.581	.535$^-$	.474		14	.312	.307	.272
	22	.779	.754	.687		20	.623	.579	.513		15	.355$^-$	.355$^-$	.272
	23	.835$^+$	.809	.741		21	.658	.626	.558		16	.396	.381	.323
	24	.895$^+$	.878	.819		22	.701	.675$^-$	.607		17	.435$^+$	.384	.364
25	0	0.000	0.000	0.000		23	.753	.718	.658		18	.473	.424	.364
	1	.004	.002	.000		24	.791	.770	.702		19	.527	.463	.408
	2	.021	.014	.006		25	.849	.820	.766		20	.565$^-$	.537	.449
	3	.045$^-$	.034	.018		26	.903	.886	.830		21	.604	.576	.500
	4	.072	.057	.034	27	0	0.000	0.000	0.000		22	.645$^+$	.616	.551
	5	.101	.082	.054		1	.004	.002	.000		23	.688	.643	.592
	6	.101	.110	.077		2	.020	.013	.006		24	.716	.693	.636
	7	.158	.118	.101		3	.042	.031	.017		25	.768	.741	.677
	8	.158	.161	.127		4	.066	.052	.032		26	.799	.783	.728
	9	.214	.185$^+$	.155		5	.093	.076	.050		27	.861	.830	.782
	10	.246	.222	.175$^+$		6	.093	.101	.070		28	.911	.894	.838
	11	.255$^-$	.238	.205$^+$		7	.145$^+$	.110	.093	29	0	0.000	0.000	0.000
	12	.307	.296	.245$^+$		8	.145$^+$	.148	.117		1	.004	.002	.000
	13	.360	.317	.245$^+$		9	.204	.175$^-$	.143		2	.018	.012	.005$^+$
	14	.389	.336	.305$^-$		10	.221	.202	.166		3	.039	.029	.015$^+$
	15	.389	.384	.342		11	.239	.223	.185$^-$		4	.062	.049	.030
	16	.432	.431	.352		12	.291	.269	.224		5	.086	.070	.046
	17	.500	.475$^-$	.403		13	.326	.269	.225$^-$		6	.086	.094	.065$^+$
	18	.568	.525$^+$	.451		14	.326	.316	.284		7	.134	.103	.086
	19	.611	.569	.500		15	.365$^+$	.364	.298		8	.134	.136	.108
	20	.638	.616	.549		16	.407	.402	.332		9	.189	.166	.132

TABLE A.4   *(continued)*

n	r	Confidence coefficient, %			n	r	Confidence coefficient, %			n	r	Confidence coefficient, %		
		90	95	99			90	95	99			90	95	99
29	10	0.189	0.184	0.157	29	28	0.866	0.834	0.789	30	15	0.336	0.324	0.256
	11	.225⁻	.211	.165⁺		29	.914	.897	.840		16	.376	.324	.308
	12	.276	.247	.206							17	.416	.364	.329
	13	.294	.251	.211	30	0	0.000	0.000	0.000		18	.446	.403	.345⁻
	14	.303	.299	.260		1	.004	.002	.000		19	.476	.440	.388
	15	.345⁻	.339	.263		2	.018	.012	.005⁺		20	.508	.476	.430
	16	.385⁺	.339	.316		3	.037	.028	.015⁻		21	.545⁻	.524	.462
	17	.425⁻	.374	.346		4	.059	.047	.028		22	.584	.560	.495⁻
	18	.463	.413	.354		5	.083	.068	.045⁻		23	.624	.597	.531
	19	.500	.451	.397		6	.083	.091	.063		24	.664	.636	.570
	20	.537	.500	.438		7	.129	.100	.083		25	.705⁺	.676	.612
	21	.575⁺	.549	.477		8	.129	.131	.104		26	.735⁺	.708	.655⁺
	22	.615⁻	.587	.523		9	.182	.163	.127		27	.781	.756	.690
	23	.655⁺	.626	.562		10	.182	.175⁺	.151		28	.818	.795⁻	.744
	24	.697	.661	.603		11	.219	.205⁺	.151		29	.871	.837	.794
	25	.721	.701	.646		12	.265⁻	.236	.198		30	.917	.900	.849
	26	.775⁺	.749	.684		13	.265⁻	.244	.206					
	27	.811	.789	.737		14	.295⁻	.292	.249					

Source: E. L. Crow, "Confidence Intervals for a Proportion," *Biometrika*, 43 (1956), 423–435; reprinted by permission of the Biometrika Trustees.

TABLE A.5 Lower critical values of *r* in the runs test

n_1 \ n_2	2	3	4	5	6	7	8	9	10	11	12	13	14	15	16	17	18	19	20
2											2	2	2	2	2	2	2	2	2
3					2	2	2	2	2	2	2	2	2	3	3	3	3	3	3
4				2	2	2	3	3	3	3	3	3	3	3	4	4	4	4	4
5			2	2	3	3	3	3	3	4	4	4	4	4	4	4	5	5	5
6		2	2	3	3	3	3	4	4	4	4	5	5	5	5	5	5	6	6
7		2	2	3	3	3	4	4	5	5	5	5	5	6	6	6	6	6	6
8		2	3	3	3	4	4	5	5	5	6	6	6	6	6	7	7	7	7
9		2	3	3	4	4	5	5	5	6	6	6	7	7	7	7	8	8	8
10		2	3	3	4	5	5	5	6	6	7	7	7	7	8	8	8	8	9
11		2	3	4	4	5	5	6	6	7	7	7	8	8	8	9	9	9	9
12	2	2	3	4	4	5	6	6	7	7	7	8	8	8	9	9	9	10	10
13	2	2	3	4	5	5	6	6	7	7	8	8	9	9	9	10	10	10	10
14	2	2	3	4	5	5	6	7	7	8	8	9	9	9	10	10	10	11	11
15	2	3	3	4	5	6	6	7	7	8	8	9	9	10	10	11	11	11	12
16	2	3	4	4	5	6	6	7	8	8	9	9	10	10	11	11	11	12	12
17	2	3	4	4	5	6	7	7	8	9	9	10	10	11	11	11	12	12	13
18	2	3	4	5	5	6	7	8	8	9	9	10	10	11	11	12	12	13	13
19	2	3	4	5	6	6	7	8	8	9	10	10	11	11	12	12	13	13	13
20	2	3	4	5	6	6	7	8	9	9	10	10	11	12	12	13	13	13	14

Source: Frieda S. Swed and C. Eisenhart, "Tables for Testing Randomness of Grouping in a Sequence of Alternatives," *Ann. Math. Statist.*, 14 (1943), 66–87.

Note: For the one-sample runs test, any value of *r* that is equal to or smaller than that shown in the body of this table for given value of n_1 and n_2 is significant at the 0.05 level.

TABLE A.6 Upper critical values of *r* in the runs test

n_1 \ n_2	2	3	4	5	6	7	8	9	10	11	12	13	14	15	16	17	18	19	20
2																			
3																			
4				9	9														
5			9	10	10	11	11												
6			9	10	11	12	12	13	13	13	13								
7				11	12	13	13	14	14	14	14	15	15	15					
8				11	12	13	14	14	15	15	16	16	16	16	17	17	17	17	17
9						13	14	14	15	16	16	16	17	17	18	18	18	18	18
10						13	14	15	16	16	17	17	18	18	18	19	19	20	20
11						13	14	15	16	17	17	18	19	19	19	20	20	21	21
12						13	14	16	16	17	18	19	19	20	20	21	21	22	22
13							15	16	17	18	19	19	20	20	21	21	22	22	23
14							15	16	17	18	19	20	20	21	22	22	23	23	24
15							15	16	18	18	19	20	21	22	22	23	23	24	25
16							17	18	19	20	21	21	22	23	23	24	24	25	25
17							17	18	19	20	21	22	23	23	24	25	25	26	26
18							17	18	19	20	21	22	23	24	25	25	26	26	27
19							17	18	20	21	22	23	23	24	25	26	26	27	27
20							17	18	20	21	22	23	24	25	25	26	27	27	28

Source: Frieda S. Swed and C. Eisenhart, "Tables for Testing Randomness of Grouping in a Sequence of Alternatives," *Ann. Math. Statist.*, 14 (1943), 66–87.

Note: For the one-sample runs test, any value of *r* that is equal to or larger than that shown in the body of this table for given values of n_1 and n_2 is significant at the 0.05 level.

TABLE A.7   Quantiles of the Mann−Whitney test statistic

n_1	p	$n_2 = 2$	3	4	5	6	7	8	9	10	11	12	13	14	15	16	17	18	19	20
2	.001	0	0	0	0	0	0	0	0	0	0	0	0	0	0	0	0	0	0	0
	.005	0	0	0	0	0	0	0	0	0	0	0	0	0	0	0	0	0	1	1
	.01	0	0	0	0	0	0	0	0	0	0	0	1	1	1	1	1	1	2	2
	.025	0	0	0	0	0	0	1	1	1	1	2	2	2	2	2	3	3	3	3
	.05	0	0	0	1	1	1	2	2	2	2	3	3	4	4	4	4	5	5	5
	.10	0	1	1	2	2	2	3	3	4	4	5	5	5	6	6	7	7	8	8
3	.001	0	0	0	0	0	0	0	0	0	0	0	0	0	0	0	1	1	1	1
	.005	0	0	0	0	0	0	0	1	1	1	2	2	2	3	3	3	3	4	4
	.01	0	0	0	0	0	1	1	2	2	2	3	3	3	4	4	5	5	5	6
	.025	0	0	0	1	2	2	3	3	4	4	5	5	6	6	7	7	8	8	9
	.05	0	1	1	2	3	3	4	5	5	6	6	7	8	8	9	10	10	11	12
	.10	1	2	2	3	4	5	6	6	7	8	9	10	11	11	12	13	14	15	16
4	.001	0	0	0	0	0	0	0	0	1	1	1	2	2	2	3	3	4	4	4
	.005	0	0	0	0	1	1	2	2	3	3	4	4	5	6	6	7	7	8	9
	.01	0	0	0	1	2	2	3	4	4	5	6	6	7	9	8	9	10	10	11
	.025	0	0	1	2	3	4	5	5	6	7	8	9	10	11	12	12	13	14	15
	.05	0	1	2	3	4	5	6	7	8	9	10	11	12	13	15	16	17	18	19
	.10	1	2	4	5	6	7	8	10	11	12	13	14	16	17	18	19	21	22	23
5	.001	0	0	0	0	0	0	1	2	2	3	3	4	4	5	6	6	7	8	8
	.005	0	0	0	1	2	2	3	4	5	6	7	8	8	9	10	11	12	13	14
	.01	0	0	1	2	3	4	5	6	7	8	9	10	11	12	13	14	15	16	17
	.025	0	1	2	3	4	6	7	8	9	10	12	13	14	15	16	18	19	20	21
	.05	1	2	3	5	6	7	9	10	12	13	14	16	17	19	20	21	23	24	26
	.10	2	3	5	6	8	9	11	13	14	16	18	19	21	23	24	26	28	29	31
6	.001	0	0	0	0	0	0	2	3	4	5	5	6	7	8	9	10	11	12	13
	.005	0	0	1	2	3	4	5	6	7	8	10	11	12	13	14	16	17	18	19
	.01	0	0	2	3	4	5	7	8	9	10	12	13	14	16	17	19	20	21	23
	.025	0	2	3	4	6	7	9	11	12	14	15	17	18	20	22	23	25	26	28
	.05	1	3	4	6	8	9	11	13	15	17	18	20	22	24	26	27	29	31	33
	.10	2	4	6	8	10	12	14	16	18	20	22	24	26	28	30	32	35	37	39
7	.001	0	0	0	1	2	3	4	5	6	7	8	9	10	11	12	14	15	16	17
	.005	0	0	1	2	4	5	7	8	10	11	13	14	16	17	19	20	22	23	25
	.01	0	1	2	4	5	7	8	10	12	13	15	17	18	20	22	24	25	27	29
	.025	0	2	4	6	7	9	11	13	15	17	19	21	23	25	27	29	31	33	35
	.05	1	3	5	7	9	12	14	16	18	20	22	25	27	29	31	34	36	38	40
	.10	2	5	7	9	12	14	17	19	22	24	27	29	32	34	37	39	42	44	47
8	.001	0	0	0	1	2	3	5	6	7	9	10	12	13	15	16	18	19	21	22
	.005	0	0	2	3	5	7	8	10	12	14	16	18	19	21	23	25	27	29	31
	.01	0	1	3	5	7	8	10	12	14	16	18	21	23	25	27	29	31	33	35
	.025	1	3	5	7	9	11	14	16	18	20	23	25	27	30	32	35	37	39	42
	.05	2	4	6	9	11	14	16	19	21	24	27	29	32	34	37	40	42	45	48
	.10	3	6	8	11	14	17	20	23	25	28	31	34	37	40	43	46	49	52	55
9	.001	0	0	0	2	3	4	6	8	9	11	13	15	16	18	20	22	24	26	27
	.005	0	1	2	4	6	8	10	12	14	17	19	21	23	25	28	30	32	34	37
	.01	0	2	4	6	8	10	12	15	17	19	22	24	27	29	32	34	37	39	41
	.025	1	3	5	8	11	13	16	18	21	24	27	29	32	35	38	40	43	46	49
	.05	2	5	7	10	13	16	19	22	25	28	31	34	37	40	43	46	49	52	55
	.10	3	6	10	13	16	19	23	26	29	32	36	39	42	46	49	53	56	59	63

TABLE A.7   *(continued)*

n_1	p	$n_2=2$	3	4	5	6	7	8	9	10	11	12	13	14	15	16	17	18	19	20
	.001	0	0	1	2	4	6	7	9	11	13	15	18	20	22	24	26	28	30	33
	.005	0	1	3	5	7	10	12	14	17	19	22	25	27	30	32	35	38	40	43
10	.01	0	2	4	7	9	12	14	17	20	23	25	28	31	34	37	39	42	45	48
	.025	1	4	6	9	12	15	18	21	24	27	30	34	37	40	43	46	49	53	56
	.05	2	5	8	12	15	18	21	25	28	32	35	38	42	45	49	52	56	59	63
	.10	4	7	11	14	18	22	25	29	33	37	40	44	48	52	55	59	63	67	71
	.001	0	0	1	3	5	7	9	11	13	16	18	21	23	25	28	30	33	35	38
	.005	0	1	3	6	8	11	14	17	19	22	25	28	31	34	37	40	43	46	49
11	.01	0	2	5	8	10	13	16	19	23	26	29	32	35	38	42	45	48	51	54
	.025	1	4	7	10	14	17	20	24	27	31	34	38	41	45	48	52	56	59	63
	.05	2	6	9	13	17	20	24	28	32	35	39	43	47	51	55	58	62	66	70
	.10	4	8	12	16	20	24	28	32	37	41	45	49	53	58	62	66	70	74	79
	.001	0	0	1	3	5	8	10	13	15	18	21	24	26	29	32	35	38	41	43
	.005	0	2	4	7	10	13	16	19	22	25	28	32	35	38	42	45	48	52	55
12	.01	0	3	6	9	12	15	18	22	25	29	32	36	39	43	47	50	54	57	61
	.025	2	5	8	12	15	19	23	27	30	34	38	42	46	50	54	58	62	66	70
	.05	3	6	10	14	18	22	27	31	35	39	43	48	52	56	61	65	69	73	78
	.10	5	9	13	18	22	27	31	36	40	45	50	54	59	64	68	73	78	82	87
	.001	0	0	2	4	6	9	12	15	18	21	24	27	30	33	36	39	43	46	49
	.005	0	2	4	8	11	14	18	21	25	28	32	35	39	43	46	50	54	58	61
13	.01	1	3	6	10	13	17	21	24	28	32	36	40	44	48	52	56	60	64	68
	.025	2	5	9	13	17	21	25	29	34	38	42	46	51	55	60	64	68	73	77
	.05	3	7	11	16	20	25	29	34	38	43	48	52	57	62	66	71	76	81	85
	.10	5	10	14	19	24	29	34	39	44	49	54	59	64	69	75	80	85	90	95
	.001	0	0	2	4	7	10	13	16	20	23	26	30	33	37	40	44	47	51	55
	.005	0	2	5	8	12	16	19	23	27	31	35	39	43	47	51	55	59	64	68
14	.01	1	3	7	11	14	18	23	27	31	35	39	44	48	52	57	61	66	70	74
	.025	2	6	10	14	18	23	27	32	37	41	46	51	56	60	65	70	75	79	84
	.05	4	8	12	17	22	27	32	37	42	47	52	57	62	67	72	78	83	88	93
	.10	5	11	16	21	26	32	37	42	48	53	59	64	70	75	81	86	92	98	103
	.001	0	0	2	5	8	11	15	18	22	25	29	33	37	41	44	48	52	56	60
	.005	0	3	6	9	13	17	21	25	30	34	38	43	47	52	56	61	65	70	74
15	.01	1	4	8	12	16	20	25	29	34	38	43	48	52	57	62	67	71	76	81
	.025	2	6	11	15	20	25	30	35	40	45	50	55	60	65	71	76	81	86	91
	.05	4	8	13	19	24	29	34	40	45	51	56	62	67	73	78	84	89	95	101
	.10	6	11	17	23	28	34	40	46	52	58	64	69	75	81	87	93	99	105	111
	.001	0	0	3	6	9	12	16	20	24	28	32	36	40	44	49	53	57	61	66
	.005	0	3	6	10	14	19	23	28	32	37	42	46	51	56	61	66	71	75	80
16	.01	1	4	8	13	17	22	27	32	37	42	47	52	57	62	67	72	77	83	88
	.025	2	7	12	16	22	27	32	38	43	48	54	60	65	71	76	82	87	93	99
	.05	4	9	15	20	26	31	37	43	49	55	61	66	72	78	84	90	96	102	108
	.10	6	12	18	24	30	37	43	49	55	62	68	75	81	87	94	100	107	113	120
	.001	0	1	3	6	10	14	18	22	26	30	35	39	44	48	53	58	62	67	71
	.005	0	3	7	11	16	20	25	30	35	40	45	50	55	61	66	71	76	82	87
17	.01	1	5	9	14	19	24	29	34	39	45	50	56	61	67	72	78	83	89	94
	.025	3	7	12	18	23	29	35	40	46	52	58	64	70	76	82	88	94	100	106
	.05	4	10	16	21	27	34	40	46	52	58	65	71	78	84	90	97	103	110	116
	.10	7	13	19	26	32	39	46	53	59	66	73	80	86	93	100	107	114	121	128
	.001	0	1	4	7	11	15	19	24	28	33	38	43	47	52	57	62	67	72	77
	.005	0	3	7	12	17	22	27	32	38	43	48	54	59	65	71	76	82	88	93
18	.01	1	5	10	15	20	25	31	37	42	48	54	60	66	71	77	83	89	95	101
	.025	3	8	13	19	25	31	37	43	49	56	62	68	75	81	87	94	100	107	113
	.05	5	10	17	23	29	36	42	49	56	62	69	76	83	89	96	103	110	117	124
	.10	7	14	21	28	35	42	49	56	63	70	78	85	92	99	107	114	121	129	136

TABLE A.7 (continued)

n_1	p	$n_2=2$	3	4	5	6	7	8	9	10	11	12	13	14	15	16	17	18	19	20
	.001	0	1	4	8	12	16	21	26	30	35	41	46	51	56	61	67	72	78	83
	.005	1	4	8	13	18	23	29	34	40	46	52	58	64	70	75	82	88	94	100
19	.01	2	5	10	16	21	27	33	39	45	51	57	64	70	76	83	89	95	102	108
	.025	3	8	14	20	26	33	39	46	53	59	66	73	79	86	93	100	107	114	120
	.05	5	11	18	24	31	38	45	52	59	66	73	81	88	95	102	110	117	124	131
	.10	8	15	22	29	37	44	52	59	67	74	82	90	98	105	113	121	129	136	144
	.001	0	1	4	8	13	17	22	27	33	38	43	49	55	60	66	71	77	83	89
	.005	1	4	9	14	19	25	31	37	43	49	55	61	68	74	80	87	93	100	106
20	.01	2	6	11	17	23	29	35	41	48	54	61	68	74	81	88	94	101	108	115
	.025	3	9	15	21	28	35	42	49	56	63	70	77	84	91	99	106	113	120	128
	.05	5	12	19	26	33	40	48	55	63	70	78	85	93	101	108	116	124	131	139
	.10	8	16	23	31	39	47	55	63	71	79	87	95	103	111	120	128	136	144	152

Source: Adapted from L. R. Verdooren, "Extended Tables of Critical Values for Wilcoxon's Test Statistic," *Biometrika*, 50 (1963), 177–186; used by permission of the Biometrika Trustees. The adaptation is due to W. J. Conover, *Practical Nonparametric Statistics*, New York: Wiley, 1971, 384–388.

TABLE A.8

Upper tail probabilities for the null distribution of the Ansari–Bradley W statistic: $2 \leq n_1 \leq n_2$, $(n_1 + n_2) \leq 20$

$n_1 = 2$

x	$n_2 = 2$	$n_2 = 3$	$n_2 = 4$	$n_2 = 5$	$n_2 = 6$	$n_2 = 7$	$n_2 = 8$	$n_2 = 9$	$n_2 = 10$
2	1.0000	1.0000	1.0000	1.0000	1.0000	1.0000	1.0000	1.0000	1.0000
3	.8333	.9000	.9333	.9524	.9643	.9722	.9778	.9818	.9848
4	.1667	.5000	.6667	.7619	.8214	.8611	.8889	.9091	.9242
5		.2000	.3333	.5238	.6429	.7222	.7778	.8182	.8485
6			.0667	.2381	.3571	.5000	.6000	.6727	.7273
7				.0952	.1786	.3056	.4000	.5091	.5909
8					.0357	.1389	.2222	.3273	.4091
9						.0556	.1111	.2000	.2727
10							.0222	.0909	.1515
11								.0364	.0758
12									.0152

$n_1 = 2$

x	$n_2 = 11$	$n_2 = 12$	$n_2 = 13$	$n_2 = 14$	$n_2 = 15$	$n_2 = 16$	$n_2 = 17$	$n_2 = 18$
2	1.0000	1.0000	1.0000	1.0000	1.0000	1.0000	1.0000	1.0000
3	.9872	.9890	.9905	.9917	.9926	.9935	.9942	.9947
4	.9359	.9451	.9524	.9583	.9632	.9673	.9708	.9737
5	.8718	.8901	.9048	.9167	.9265	.9346	.9415	.9474
6	.7692	.8022	.8286	.8500	.8676	.8824	.8947	.9053
7	.6538	.7033	.7429	.7750	.8015	.8235	.8421	.8579
8	.5000	.5714	.6286	.6750	.7132	.7451	.7719	.7947
9	.3590	.4286	.5048	.5667	.6176	.6601	.6959	.7263
10	.2308	.2967	.3714	.4333	.5000	.5556	.6023	.6421
11	.1410	.1978	.2667	.3250	.3897	.4444	.5029	.5526
12	.0641	.1099	.1714	.2250	.2868	.3399	.3977	.4474
13	.0256	.0549	.1048	.1500	.2059	.2549	.3099	.3579
14		.0110	.0476	.0833	.1324	.1765	.2281	.2737
15			.0190	.0417	.0809	.1176	.1637	.2053
16				.0083	.0368	.0654	.1053	.1421
17					.0147	.0327	.0643	.0947
18						.0065	.0292	.0526
19							.0117	.0263
20								.0053

Source: Myles Hollander and Douglas A. Wolfe, *Nonparametric Statistical Methods, Copyright* © 1973 John Wiley & Sons, Inc. Reprinted by permission of John Wiley & Sons, Inc.

TABLE A.8   (continued)

$n_1 = 3$

x	$n_2 = 3$	$n_2 = 4$	$n_2 = 5$	$n_2 = 6$	$n_2 = 7$	$n_2 = 8$	$n_2 = 9$	$n_2 = 10$	$n_2 = 11$
4	1.0000	1.0000	1.0000	1.0000	1.0000	1.0000	1.0000	1.0000	1.0000
5	.9000	.9429	.9643	.9762	.9833	.9879	.9909	.9930	.9945
6	.7000	.8286	.8929	.9286	.9500	.9636	.9727	.9790	.9835
7	.3000	.5714	.7143	.8095	.8667	.9030	.9273	.9441	.9560
8	.1000	.3429	.5000	.6548	.7500	.8182	.8636	.8951	.9176
9		.1429	.2857	.4643	.5833	.6909	.7636	.8182	.8571
10		.0286	.1071	.2857	.4167	.5455	.6364	.7168	.7747
11			.0357	.1429	.2500	.3939	.5000	.5979	.6703
12				.0595	.1333	.2606	.3636	.4755	.5604
13				.0119	.0500	.1455	.2364	.3497	.4396
14					.0167	.0727	.1364	.2413	.3297
15						.0303	.0727	.1503	.2253
16						.0061	.0273	.0839	.1429
17							.0091	.0420	.0824
18								.0175	.0440
19								.0035	.0165
20									.0055

$n_1 = 3$

x	$n_2 = 12$	$n_2 = 13$	$n_2 = 14$	$n_2 = 15$	$n_2 = 16$	$n_2 = 17$
4	1.0000	1.0000	1.0000	1.0000	1.0000	1.0000
5	.9956	.9964	.9971	.9975	.9979	.9982
6	.9868	.9893	.9912	.9926	.9938	.9947
7	.9648	.9714	.9765	.9804	.9835	.9860
8	.9341	.9464	.9559	.9632	.9690	.9737
9	.8857	.9071	.9235	.9363	.9463	.9544
10	.8198	.8536	.8794	.8995	.9154	.9281
11	.7341	.7821	.8206	.8505	.8741	.8930
12	.6374	.6964	.7485	.7892	.8225	.8491
13	.5297	.6000	.6632	.7132	.7575	.7930
14	.4242	.5000	.5735	.6324	.6852	.7281
15	.3209	.4000	.4794	.5441	.6058	.6561
16	.2286	.3036	.3868	.4559	.5232	.5789
17	.1516	.2179	.2985	.3676	.4396	.5000
18	.0945	.1464	.2206	.2868	.3591	.4211
19	.0527	.0929	.1529	.2108	.2817	.3439
20	.0264	.0536	.1015	.1495	.2136	.2719
21	.0110	.0286	.0632	.1005	.1548	.2070
22	.0022	.0107	.0353	.0637	.1073	.1509
23		.0036	.0176	.0368	.0712	.1070
24			.0074	.0196	.0444	.0719
25			.0015	.0074	.0248	.0456
26				.0025	.0124	.0263
27					.0052	.0140
28					.0010	.0053
29						.0018

TABLE A.8   (continued)

$n_1 = 4$

x	$n_2 = 4$	$n_2 = 5$	$n_2 = 6$	$n_2 = 7$	$n_2 = 8$	$n_2 = 9$	$n_2 = 10$	$n_2 = 11$	$n_2 = 12$
6	1.0000	1.0000	1.0000	1.0000	1.0000	1.0000	1.0000	1.0000	1.0000
7	.9857	.9921	.9952	.9970	.9980	.9986	.9990	.9993	.9995
8	.9286	.9603	.9762	.9848	.9899	.9930	.9950	.9963	.9973
9	.8000	.8889	.9333	.9576	.9717	.9804	.9860	.9897	.9923
10	.6286	.7778	.8571	.9091	.9394	.9580	.9700	.9780	.9835
11	.3714	.6032	.7333	.8242	.8788	.9161	.9401	.9560	.9670
12	.2000	.4286	.5810	.7152	.7980	.8573	.8961	.9238	.9429
13	.0714	.2619	.4190	.5818	.6889	.7762	.8342	.8769	.9066
14	.0143	.1349	.2667	.4424	.5677	.6783	.7542	.8154	.8582
15		.0476	.1429	.3030	.4323	.5650	.6593	.7385	.7951
16		.0159	.0667	.1939	.3111	.4503	.5554	.6520	.7225
17			.0238	.1061	.2020	.3357	.4446	.5546	.6374
18			.0048	.0515	.1212	.2378	.3407	.4564	.5473
19				.0182	.0606	.1538	.2458	.3590	.4527
20				.0061	.0283	.0923	.1658	.2711	.3626
21					.0101	.0490	.1039	.1934	.2775
22					.0020	.0238	.0599	.1319	.2049
23						.0084	.0300	.0821	.1418
24						.0028	.0140	.0484	.0934
25							.0050	.0256	.0571
26							.0010	.0125	.0330
27								.0044	.0165
28								.0015	.0077
29									.0027
30									.0005

$n_1 = 4$

x	$n_2 = 13$	$n_2 = 14$	$n_2 = 15$	$n_2 = 16$
6	1.0000	1.0000	1.0000	1.0000
7	.9996	.9997	.9997	.9998
8	.9979	.9984	.9987	.9990
9	.9941	.9954	.9964	.9971
10	.9874	.9902	.9923	.9938
11	.9748	.9804	.9845	.9876
12	.9563	.9660	.9732	.9785
13	.9286	.9444	.9561	.9649
14	.8908	.9144	.9324	.9459
15	.8408	.8742	.9002	.9197
16	.7811	.8245	.8599	.8867
17	.7101	.7647	.8101	.8448
18	.6319	.6967	.7528	.7961
19	.5471	.6209	.6873	.7391
20	.4613	.5412	.6166	.6764
21	.3761	.4588	.5413	.6078
22	.2979	.3791	.4654	.5368
23	.2261	.3033	.3896	.4632
24	.1655	.2353	.3189	.3922

TABLE A.8   (continued)

$n_1 = 4$

x	$n_2 = 13$	$n_2 = 14$	$n_2 = 15$	$n_2 = 16$
25	.1151	.1755	.2531	.3236
26	.0765	.1258	.1953	.2609
27	.0471	.0856	.1450	.2039
28	.0277	.0556	.1042	.1552
29	.0147	.0340	.0712	.1133
30	.0071	.0196	.0470	.0803
31	.0025	.0098	.0289	.0541
32	.0008	.0046	.0170	.0351
33	.	.0016	.0090	.0215
34		.0003	.0044	.0124
35			.0015	.0062
36			.0005	.0029
37				.0010
38				.0002

$n_1 = 5$

x	$n_2 = 5$	$n_2 = 6$	$n_2 = 7$	$n_2 = 8$	$n_2 = 9$	$n_2 = 10$	$n_2 = 11$
9	1.0000	1.0000	1.0000	1.0000	1.0000	1.0000	1.0000
10	.9921	.9957	.9975	.9984	.9990	.9993	.9995
11	.9762	.9870	.9924	.9953	.9970	.9980	.9986
12	.9286	.9610	.9773	.9860	.9910	.9940	.9959
13	.8492	.9156	.9495	.9689	.9800	.9867	.9908
14	.7302	.8420	.9015	.9386	.9600	.9734	.9817
15	.5873	.7446	.8333	.8936	.9291	.9524	.9670
16	.4127	.6147	.7374	.8275	.8821	.9197	.9437
17	.2698	.4805	.6237	.7451	.8212	.8761	.9116
18	.1508	.3463	.5000	.6457	.7423	.8182	.8681
19	.0714	.2294	.3763	.5385	.6523	.7483	.8132
20	.0238	.1342	.2626	.4266	.5514	.6663	.7468
21	.0079	.0693	.1667	.3209	.4486	.5771	.6708
22		.0303	.0985	.2269	.3477	.4832	.5870
23		.0108	.0505	.1507	.2577	.3916	.5000
24		.0022	.0227	.0917	.1788	.3044	.4130
25			.0076	.0513	.1179	.2268	.3292
26			.0025	.0249	.0709	.1608	.2532
27				.0109	.0400	.1086	.1868
28				.0039	.0200	.0686	.1319
29				.0008	.0090	.0406	.0884
30					.0030	.0220	.0563
31					.0010	.0107	.0330
32						.0047	.0183
33						.0017	.0092
34						.0003	.0041
35							.0014
36							.0005

TABLE A.8   (*continued*)

$n_1 = 5$

x	$n_2 = 12$	$n_2 = 13$	$n_2 = 14$	$n_2 = 15$
9	1.0000	1.0000	1.0000	1.0000
10	.9997	.9998	.9998	.9999
11	.9990	.9993	.9995	.9996
12	.9971	.9979	.9985	.9988
13	.9935	.9953	.9966	.9974
14	.9871	.9907	.9931	.9948
15	.9767	.9832	.9876	.9907
16	.9601	.9711	.9787	.9840
17	.9368	.9538	.9659	.9743
18	.9047	.9295	.9476	.9604
19	.8633	.8978	.9235	.9417
20	.8116	.8569	.8920	.9171
21	.7508	.8079	.8533	.8861
22	.6810	.7498	.8067	.8483
23	.6054	.6846	.7530	.8038
24	.5254	.6130	.6923	.7523
25	.4449	.5383	.6267	.6950
26	.3662	.4617	.5572	.6329
27	.2928	.3870	.4864	.5673
28	.2262	.3154	.4157	.5000
29	.1690	.2502	.3478	.4327
30	.1214	.1921	.2840	.3671
31	.0835	.1431	.2262	.3050
32	.0546	.1022	.1751	.2477
33	.0339	.0705	.1318	.1962
34	.0197	.0462	.0960	.1517
35	.0107	.0289	.0675	.1139
36	.0052	.0168	.0455	.0829
37	.0023	.0093	.0294	.0583
38	.0008	.0047	.0181	.0396
39	.0002	.0021	.0105	.0257
40		.0007	.0057	.0160
41		.0002	.0028	.0093
42			.0012	.0052
43			.0004	.0026
44			.0001	.0012
45				.0004
46				.0001

TABLE A.8   (continued)

$n_1 = 6$

x	$n_2 = 6$	$n_2 = 7$	$n_2 = 8$	$n_2 = 9$	$n_2 = 10$	$n_2 = 11$	$n_2 = 12$	$n_2 = 13$	$n_2 = 14$
12	1.0000	1.0000	1.0000	1.0000	1.0000	1.0000	1.0000	1.0000	1.0000
13	.9989	.9994	.9997	.9998	.9999	.9999	.9999	1.0000	1.0000
14	.9946	.9971	.9983	.9990	.9994	.9996	.9997	.9998	.9999
15	.9848	.9918	.9953	.9972	.9983	.9989	.9992	.9995	.9996
16	.9632	.9802	.9887	.9932	.9958	.9973	.9982	.9987	.9991
17	.9264	.9592	.9760	.9856	.9910	.9942	.9961	.9973	.9981
18	.8658	.9242	.9547	.9724	.9825	.9887	.9925	.9948	.9964
19	.7846	.8735	.9217	.9518	.9692	.9799	.9865	.9907	.9935
20	.6807	.8048	.8751	.9215	.9487	.9663	.9772	.9843	.9890
21	.5649	.7203	.8139	.8803	.9202	.9469	.9636	.9749	.9823
22	.4351	.6189	.7366	.8260	.8812	.9199	.9445	.9613	.9725
23	.3193	.5122	.6474	.7600	.8322	.8849	.9190	.9431	.9591
24	.2154	.4038	.5501	.6829	.7717	.8407	.8860	.9191	.9413
25	.1342	.3030	.4499	.5984	.7025	.7877	.8451	.8887	.9184
26	.0736	.2133	.3526	.5085	.6246	.7259	.7962	.8514	.8896
27	.0368	.1410	.2634	.4190	.5425	.6574	.7398	.8074	.8549
28	.0152	.0851	.1861	.3323	.4575	.5831	.6765	.7564	.8138
29	.0054	.0484	.1249	.2543	.3754	.5065	.6082	.6996	.7668
30	.0011	.0239	.0783	.1860	.2975	.4292	.5364	.6376	.7139
31		.0105	.0453	.1303	.2283	.3549	.4636	.5723	.6566
32		.0035	.0240	.0859	.1678	.2851	.3918	.5049	.5954
33		.0012	.0113	.0539	.1188	.2226	.3235	.4376	.5322
34			.0047	.0312	.0798	.1678	.2602	.3716	.4678
35			.0017	.0170	.0513	.1226	.2038	.3094	.4046
36			.0003	.0082	.0308	.0859	.1549	.2518	.3434
37				.0036	.0175	.0579	.1140	.2002	.2861
38				.0012	.0090	.0370	.0810	.1550	.2332
39				.0004	.0042	.0226	.0555	.1170	.1862
40					.0017	.0128	.0364	.0855	.1451
41					.0006	.0069	.0228	.0608	.1104
42					.0001	.0033	.0135	.0415	.0816
43						.0015	.0075	.0274	.0587
44						.0005	.0039	.0172	.0409
45						.0002	.0018	.0104	.0275
46							.0008	.0058	.0177
47							.0003	.0031	.0110
48							.0001	.0015	.0065
49								.0007	.0036
50								.0002	.0019
51								.0001	.0009
52									.0004
53									.0001
54									.0000

TABLE A.8　*(continued)*

$n_1 = 7$

x	$n_2 = 7$	$n_2 = 8$	$n_2 = 9$	$n_2 = 10$	$n_2 = 11$	$n_2 = 12$	$n_2 = 13$
16	1.0000	1.0000	1.0000	1.0000	1.0000	1.0000	1.0000
17	.9994	.9997	.9998	.9999	1.0000	1.0000	1.0000
18	.9983	.9991	.9995	.9997	.9998	.9999	.9999
19	.9948	.9972	.9984	.9991	.9994	.9996	.9998
20	.9878	.9935	.9963	.9978	.9987	.9992	.9995
21	.9744	.9862	.9921	.9954	.9972	.9982	.9988
22	.9534	.9744	.9851	.9912	.9946	.9966	.9978
23	.9196	.9549	.9734	.9841	.9901	.9937	.9959
24	.8730	.9270	.9559	.9734	.9833	.9893	.9930
25	.8106	.8878	.9306	.9574	.9729	.9826	.9885
26	.7348	.8375	.8965	.9354	.9583	.9730	.9820
27	.6463	.7748	.8523	.9059	.9381	.9595	.9727
28	.5507	.7021	.7981	.8685	.9118	.9415	.9602
29	.4493	.6194	.7336	.8221	.8782	.9181	.9435
30	.3537	.5324	.6608	.7676	.8374	.8889	.9223
31	.2652	.4435	.5820	.7052	.7887	.8532	.8958
32	.1894	.3577	.5000	.6368	.7333	.8111	.8637
33	.1270	.2777	.4180	.5637	.6714	.7626	.8258
34	.0804	.2075	.3392	.4888	.6050	.7085	.7822
35	.0466	.1478	.2664	.4139	.5353	.6494	.7332
36	.0256	.1005	.2019	.3421	.4647	.5869	.6795
37	.0122	.0648	.1477	.2753	.3950	.5220	.6219
38	.0052	.0393	.1035	.2154	.3286	.4568	.5616
39	.0017	.0221	.0694	.1633	.2667	.3925	.5000
40	.0006	.0115	.0441	.1199	.2113	.3311	.4384
41		.0053	.0266	.0847	.1626	.2735	.3781
42		.0022	.0149	.0576	.1218	.2213	.3205
43		.0008	.0079	.0375	.0882	.1749	.2668
44		.0002	.0037	.0233	.0619	.1350	.2178
45			.0016	.0136	.0417	.1014	.1742
46			.0005	.0075	.0271	.0742	.1363
47			.0002	.0038	.0167	.0526	.1042
48				.0017	.0099	.0361	.0777
49				.0007	.0054	.0239	.0565
50				.0003	.0028	.0152	.0398
51				.0001	.0013	.0092	.0273
52					.0006	.0053	.0180
53					.0002	.0029	.0115
54					.0001	.0015	.0070
55						.0007	.0041
56						.0003	.0022
57						.0001	.0012
58						.0000	.0005
59							.0002
60							.0001
61							.0000

TABLE A.8   (*continued*)

$n_1 = 8$

x	$n_2 = 8$	$n_2 = 9$	$n_2 = 10$	$n_2 = 11$	$n_2 = 12$
20	1.0000	1.0000	1.0000	1.0000	1.0000
21	.9999	1.0000	1.0000	1.0000	1.0000
22	.9996	.9998	.9999	.9999	1.0000
23	.9989	.9994	.9997	.9998	.9999
24	.9974	.9986	.9992	.9996	.9997
25	.9941	.9969	.9983	.9990	.9994
26	.9885	.9938	.9965	.9980	.9988
27	.9789	.9886	.9935	.9962	.9977
28	.9643	.9804	.9887	.9934	.9960
29	.9428	.9680	.9813	.9889	.9932
30	.9133	.9504	.9704	.9823	.9890
31	.8737	.9262	.9551	.9728	.9830
32	.8246	.8947	.9344	.9598	.9745
33	.7650	.8549	.9075	.9423	.9629
34	.6970	.8069	.8738	.9199	.9477
35	.6212	.7508	.8328	.8918	.9281
36	.5413	.6877	.7847	.8578	.9038
37	.4587	.6184	.7296	.8174	.8742
38	.3788	.5457	.6686	.7710	.8392
39	.3030	.4714	.6031	.7189	.7986
40	.2350	.3983	.5347	.6621	.7528
41	.1754	.3281	.4653	.6015	.7022
42	.1263	.2636	.3969	.5386	.6476
43	.0867	.2055	.3314	.4746	.5898
44	.0572	.1557	.2704	.4113	.5302
45	.0357	.1139	.2153	.3500	.4698
46	.0211	.0807	.1672	.2925	.4102
47	.0115	.0548	.1262	.2394	.3524
48	.0059	.0358	.0925	.1919	.2978
49	.0026	.0221	.0656	.1503	.2472
50	.0011	.0131	.0449	.1150	.2014
51	.0004	.0072	.0296	.0856	.1608
52	.0001	.0037	.0187	.0621	.1258
53		.0017	.0113	.0437	.0962
54		.0007	.0065	.0298	.0719
55		.0002	.0035	.0196	.0523
56		.0001	.0017	.0124	.0371
57			.0008	.0075	.0255
58			.0003	.0043	.0170
59			.0001	.0023	.0110
60			.0000	.0012	.0068
61				.0006	.0040
62				.0002	.0023
63				.0001	.0012
64				.0000	.0006
65					.0003
66					.0001
67					.0000
68					.0000

TABLE A.8   (*continued*)

$n_1 = 9$				$n_1 = 9$			
x	$n_2 = 9$	$n_2 = 10$	$n_2 = 11$	x	$n_2 = 9$	$n_2 = 10$	$n_2 = 11$
25	1.0000	1.0000	1.0000	50	.2167	.3673	.5000
26	1.0000	1.0000	1.0000	51	.1687	.3092	.4407
27	.9999	.9999	1.0000	52	.1276	.2552	.3827
28	.9996	.9998	.9999	53	.0938	.2064	.3271
29	.9991	.9995	.9997	54	.0668	.1632	.2749
30	.9981	.9990	.9995	55	.0460	.1262	.2269
31	.9963	.9980	.9989	56	.0305	.0952	.1840
32	.9932	.9964	.9980	57	.0195	.0700	.1462
33	.9882	.9937	.9964	58	.0118	.0500	.1138
34	.9805	.9894	.9940	59	.0068	.0347	.0867
35	.9695	.9831	.9903	60	.0037	.0232	.0645
36	.9540	.9741	.9849	61	.0019	.0150	.0468
37	.9332	.9618	.9773	62	.0009	.0093	.0331
38	.9062	.9453	.9669	63	.0004	.0056	.0227
39	.8724	.9240	.9532	64	.0001	.0031	.0151
40	.8313	.8972	.9355	65	.0000	.0017	.0097
41	.7833	.8646	.9133	66		.0008	.0060
42	.7283	.8259	.8862	67		.0004	.0036
43	.6677	.7813	.8538	68		.0002	.0020
44	.6025	.7310	.8160	69		.0001	.0011
45	.5346	.6759	.7731	70		.0000	.0005
46	.4654	.6166	.7251	71			.0003
47	.3975	.5548	.6729	72			.0001
48	.3323	.4916	.6173	73			.0000
49	.2717	.4287	.5593	74			.0000

$n_1 = 10$		$n_1 = 10$		$n_1 = 10$	
x	$n_2 = 10$	x	$n_2 = 10$	x	$n_2 = 10$
30	1.0000	47	.8993	64	.1007
31	1.0000	48	.8694	65	.0761
32	1.0000	49	.8344	66	.0560
33	.9999	50	.7940	67	.0403
34	.9998	51	.7486	68	.0282
35	.9996	52	.6986	69	.0192
36	.9992	53	.6449	70	.0126
37	.9984	54	.5881	71	.0080
38	.9971	55	.5296	72	.0049
39	.9951	56	.4704	73	.0029
40	.9920	57	.4119	74	.0016
41	.9874	58	.3551	75	.0008
42	.9808	59	.3014	76	.0004
43	.9718	60	.2514	77	.0002
44	.9597	61	.2060	78	.0001
45	.9440	62	.1656	79	.0000
46	.9239	63	.1306	80	.0000

Computed by G. A. Mack on the Ohio State University IBM 370/165.

TABLE A.9   **Approximate critical values C_α for the Hollander test of extreme reactions**

	n_1								
N	4	5	6	7	8	9	10	11	12
8	5.00	5.00							
	—	—							
9	5.00	10.00	17.50						
	—	—	—						
10	5.00	14.80	23.33	28.00	42.00				
	5.00	10.00	17.50	28.00	—				
11	8.75	14.80	26.83	39.43	49.88	60.00			
	5.00	10.00	17.50	28.00	42.00	—			
12	8.75	20.00	29.50	42.00	58.00	68.89	82.50		
	5.00	10.00	17.50	28.00	42.00	60.00	—		
13	10.00	21.20	34.00	49.42	63.88	80.00	100.10	110.00	
	5.00	10.00	23.33	34.86	49.88	68.89	82.50	—	
14	13.00	23.20	38.83	54.86	73.50	90.22	110.00	132.00	154.90
	5.00	14.80	23.33	39.43	55.50	75.56	92.40	110.00	—
15	14.00	26.80	42.83	61.43	79.50	101.60	122.10	144.90	168.70
	5.00	14.80	28.00	42.00	59.50	79.56	102.50	120.90	143.00
16	14.00	29.20	47.50	67.71	89.88	112.20	135.60	161.60	187.00
	5.00	17.20	30.83	47.71	67.88	88.89	110.40	136.20	164.70
17	17.00	33.20	53.33	74.86	98.88	124.00	150.00	176.90	205.70
	8.75	17.20	34.00	52.00	73.88	96.00	120.90	148.20	177.70
18	18.75	36.80	58.00	82.86	108.00	135.60	164.40	194.20	224.90
	8.75	20.00	37.33	56.00	79.50	104.00	131.60	161.60	190.90
19	20.00	40.00	64.00	90.86	118.90	148.90	180.10	212.60	246.00
	8.75	21.20	39.33	60.86	85.50	112.89	142.10	174.00	206.30
20	21.00	44.80	70.00	98.86	129.90	162.20	196.40	231.60	267.70
	8.75	23.20	42.00	66.86	91.88	122.20	153.60	188.00	222.90

Source: M. Hollander, "A Nonparametric Test for the Two-Sample Problem," *Psychometrika*, 28 (1963), 395–403.

Note: The top number opposite each value of *N* is the critical value for $\alpha \approx 0.05$; the bottom number is for $\alpha \approx 0.01$.

TABLE A.10   **Significance tests in a 2 × 2 contingency table**

		Probability			
	a	0.05	0.025	0.01	0.005
A = 3 B = 3	3	**0**.050	—	—	—
A = 4 B = 4	4	**0**.014	**0**.014	—	—
3	4	**0**.029	—	—	—
A = 5 B = 5	5	**1**.024	**1**.024	**0**.004	**0**.004
	4	**0**.024	**0**.024	—	—
4	5	**1**.048	**0**.008	**0**.008	—
	4	**0**.040	—	—	—
3	5	**0**.018	**0**.018	—	—
2	5	**0**.048	—	—	—
A = 6 B = 6	6	**2**.030	**1**.008	**1**.008	**0**.001
	5	**1**.040	**0**.008	**0**.008	—
	4	**0**.030	—	—	—
5	6	**1**.015+	**1**.015+	**0**.002	**0**.002
	5	**0**.013	**0**.013	—	—
	4	**0**.045+	—	—	—
4	6	**1**.033	**0**.005⁻	**0**.005⁻	**0**.005⁻
	5	**0**.024	**0**.024	—	—
3	6	**0**.012	**0**.012	—	—
	5	**0**.048	—	—	—
2	6	**0**.036	—	—	—
A = 7 B = 7	7	**3**.035⁻	**2**.010+	**1**.002	**1**.002
	6	**1**.015⁻	**1**.015⁻	**0**.002	**0**.002
	5	**0**.010+	**0**.010+	—	—
	4	**0**.035⁻	—	—	—
6	7	**2**.021	**2**.021	**1**.005⁻	**1**.005⁻
	6	**1**.025+	**0**.004	**0**.004	**0**.004
	5	**0**.016	**0**.016	—	—
	4	**0**.049	—	—	—
5	7	**2**.045+	**1**.010+	**0**.001	**0**.001
	6	**1**.045+	**0**.008	**0**.008	—
	5	**0**.027	—	—	—
4	7	**1**.024	**1**.024	**0**.003	**0**.003
	6	**0**.015+	**0**.015+	—	—
	5	**0**.045+	—	—	—
3	7	**0**.008	**0**.008	**0**.008	—
	6	**0**.033	—	—	—
2	7	**0**.028	—	—	—
A = 8 B = 8	8	**4**.038	**3**.013	**2**.003	**2**.003
	7	**2**.020	**2**.020	**1**.005+	**0**.001
	6	**1**.020	**1**.020	**0**.003	**0**.003
	5	**0**.013	**0**.013	—	—
	4	**0**.038	—	—	—

1. Bold type, for given a, A and B, shows the value of b (< a), which is just significant at the probability level quoted (single-tail test).
2. Small type, for given A, B and r = a + b, shows the exact probability (if there is independence) that b is equal to or less than the integer shown in bold type.

TABLE A.10   (*continued*)

	a	Probability			
		0.05	0.025	0.01	0.005
$A = 8$ $B = 7$	8	**3** .026	**2** .007	**2** .007	**1** .001
	7	**2** .035⁻	**1** .009	**1** .009	**0** .001
	6	**1** .032	**0** .006	**0** .006	—
	5	**0** .019	**0** .019	—	—
6	8	**2** .015⁻	**2** .015⁻	**1** .003	**1** .003
	7	**1** .016	**1** .016	**0** .002	**0** .002
	6	**0** .009	**0** .009	**0** .009	—
	5	**0** .028	—	—	—
5	8	**2** .035⁻	**1** .007	**1** .007	**0** .001
	7	**1** .032	**0** .005⁻	**0** .005⁻	**0** .005⁻
	6	**0** .016	**0** .016	—	—
	5	**0** .044	—	—	—
4	8	**1** .018	**1** .018	**0** .002	**0** .002
	7	**0** .010⁺	**0** .010⁺	—	—
	6	**0** .030	—	—	—
3	8	**0** .006	**0** .006	**0** .006	—
	7	**0** .024	**0** .024	—	—
2	8	**0** .022	**0** .022	—	—
$A = 9$ $B = 9$	9	**5** .041	**4** .015⁻	**3** .005⁻	**3** .005⁻
	8	**3** .025⁻	**3** .025⁻	**2** .008	**1** .002
	7	**2** .028	**1** .008	**1** .008	**0** .001
	6	**1** .025⁻	**1** .025⁻	**0** .005⁻	**0** .005⁻
	5	**0** .015⁻	**0** .015⁻	—	—
	4	**0** .041	—	—	—
8	9	**4** .029	**3** .009	**3** .009	**2** .002
	8	**3** .043	**2** .013	**1** .003	**1** .003
	7	**2** .044	**1** .012	**0** .002	**0** .002
	6	**1** .036	**0** .007	**0** .007	—
	5	**0** .020	**0** .020	—	—
7	9	**3** .019	**3** .019	**2** .005⁻	**2** .005⁻
	8	**2** .024	**2** .024	**1** .006	**0** .001
	7	**1** .020	**1** .020	**0** .003	**0** .003
	6	**0** .010⁺	**0** .010⁺	—	—
	5	**0** .029	—	—	—
6	9	**3** .044	**2** .011	**1** .002	**1** .002
	8	**2** .047	**1** .011	**0** .001	**0** .001
	7	**1** .035⁻	**0** .006	**0** .006	—
	6	**0** .017	**0** .017	—	—
	5	**0** .042	—	—	—
5	9	**2** .027	**1** .005⁻	**1** .005⁻	**1** .005⁻
	8	**1** .023	**1** .023	**0** .003	**0** .003
	7	**0** .010⁺	**0** .010⁺	—	—
	6	**0** .028	—	—	—
4	9	**1** .014	**1** .014	**0** .001	**0** .001
	8	**0** .007	**0** .007	**0** .007	—
	7	**0** .021	**0** .021	—	—
	6	**0** .049	—	—	—
3	9	**1** .045⁺	**0** .005⁻	**0** .005⁻	**0** .005⁻
	8	**0** .018	**0** .018	—	—
	7	**0** .045⁺	—	—	—
2	9	**0** .018	**0** .018	—	—

TABLE A.10   (continued)

	a	Probability			
		0.05	**0.025**	**0.01**	**0.005**
$A = 10\ B = 10$	10	**6** .043	**5** .016	**4** .005$^+$	**3** .002
	9	**4** .029	**3** .010$^-$	**3** .010$^-$	**2** .003
	8	**3** .035$^-$	**2** .012	**1** .003	**1** .003
	7	**2** .035$^-$	**1** .010$^-$	**1** .010$^-$	**0** .002
	6	**1** .029	**0** .005$^+$	**0** .005$^+$	—
	5	**0** .016	**0** .016	—	—
	4	**0** .043	—	—	—
9	10	**5** .033	**4** .011	**3** .003	**3** .003
	9	**4** .050$^-$	**3** .017	**2** .005$^-$	**2** .005$^-$
	8	**2** .019	**2** .019	**1** .004	**1** .004
	7	**1** .015$^-$	**1** .015$^-$	**0** .002	**0** .002
	6	**1** .040	**0** .008	**0** .008	—
	5	**0** .022	**0** .022	—	—
8	10	**4** .023	**4** .023	**3** .007	**2** .002
	9	**3** .032	**2** .009	**2** .009	**1** .002
	8	**2** .031	**1** .008	**1** .008	**0** .001
	7	**1** .023	**1** .023	**0** .004	**0** .004
	6	**0** .011	**0** .011	—	—
	5	**0** .029	—	—	—
7	10	**3** .015$^-$	**3** .015$^-$	**2** .003	**2** .003
	9	**2** .018	**2** .018	**1** .004	**1** .004
	8	**1** .013	**1** .013	**0** .002	**0** .002
	7	**1** .036	**0** .006	**0** .006	—
	6	**0** .017	**0** .017	—	—
	5	**0** .041	—	—	—
6	10	**3** .036	**2** .008	**2** .008	**1** .001
	9	**2** .036	**1** .008	**1** .008	**0** .001
	8	**1** .024	**1** .024	**0** .003	**0** .003
	7	**0** .010$^+$	**0** .010$^+$	—	—
	6	**0** .026	—	—	—
5	10	**2** .022	**2** .022	**1** .004	**1** .004
	9	**1** .017	**1** .017	**0** .002	**0** .002
	8	**1** .047	**0** .007	**0** .007	—
	7	**0** .019	**0** .019	—	—
	6	**0** .042	—	—	—
4	10	**1** .011	**1** .011	**0** .001	**0** .001
	9	**1** .041	**0** .005$^-$	**0** .005$^-$	**0** .005$^-$
	8	**0** .015$^-$	**0** .015$^-$	—	—
	7	**0** .035$^-$	—	—	—
3	10	**1** .038	**0** .003	**0** .003	**0** .003
	9	**0** .014	**0** .014	—	—
	8	**0** .035$^-$	—	—	—
2	10	**0** .015$^+$	**0** .015$^+$	—	—
	9	**0** .045$^+$	—	—	—
$A = 11\ B = 11$	11	**7** .045$^+$	**6** .018	**5** .006	**4** .002
	10	**5** .032	**4** .012	**3** .004	**3** .004
	9	**4** .040	**3** .015$^-$	**2** .004	**2** .004
	8	**3** .043	**2** .015$^-$	**1** .004	**1** .004
	7	**2** .040	**1** .012	**0** .002	**0** .002
	6	**1** .032	**0** .006	**0** .006	—
	5	**0** .018	**0** .018	—	—
	4	**0** .045$^+$	—	—	—

TABLE A.10   (*continued*)

		Probability			
	a	0.05	0.025	0.01	0.005
A = 11 *B* = 10	11	**6** .035⁺	**5** .012	**4** .004	**4** .004
	10	**4** .021	**4** .021	**3** .007	**2** .002
	9	**3** .024	**3** .024	**2** .007	**1** .002
	8	**2** .023	**2** .023	**1** .006	**0** .001
	7	**1** .017	**1** .017	**0** .003	**0** .003
	6	**1** .043	**0** .009	**0** .009	—
	5	**0** .023	**0** .023	—	—
9	11	**5** .026	**4** .008	**4** .008	**3** .002
	10	**4** .038	**3** .012	**2** .003	**2** .003
	9	**3** .040	**2** .012	**1** .003	**1** .003
	8	**2** .035⁻	**1** .009	**1** .009	**0** .001
	7	**1** .025⁻	**1** .025⁻	**0** .004	**0** .004
	6	**0** .012	**0** .012	—	—
	5	**0** .030	—	—	—
8	11	**4** .018	**4** .018	**3** .005⁻	**3** .005⁻
	10	**3** .024	**3** .024	**2** .006	**1** .001
	9	**2** .022	**2** .022	**1** .005⁻	**1** .005⁻
	8	**1** .015⁻	**1** .015⁻	**0** .002	**0** .002
	7	**1** .037	**0** .007	**0** .007	—
	6	**0** .017	**0** .017	—	—
	5	**0** .040	—	—	—
7	11	**4** .043	**3** .011	**2** .002	**2** .002
	10	**3** .047	**2** .013	**1** .002	**1** .002
	9	**2** .039	**1** .009	**1** .009	**0** .001
	8	**1** .025⁻	**1** .025⁻	**0** .004	**0** .004
	7	**0** .010⁺	**0** .010⁺	—	—
	6	**0** .025⁻	**0** .025⁻	—	—
6	11	**3** .029	**2** .006	**2** .006	**1** .001
	10	**2** .028	**1** .005⁺	**1** .005⁺	**0** .001
	9	**1** .018	**1** .018	**0** .002	**0** .002
	8	**1** .043	**0** .007	**0** .007	—
	7	**0** .017	**0** .017	—	—
	6	**0** .037	—	—	—
5	11	**2** .018	**2** .018	**1** .003	**1** .003
	10	**1** .013	**1** .013	**0** .001	**0** .001
	9	**1** .036	**0** .005⁻	**0** .005⁻	**0** .005⁻
	8	**0** .013	**0** .013	—	—
	7	**0** .029	—	—	—
4	11	**1** .009	**1** .009	**1** .009	**0** .001
	10	**1** .033	**0** .004	**0** .004	**0** .004
	9	**0** .011	**0** .011	—	—
	8	**0** .026	—	—	—
3	11	**1** .003	**0** .003	**0** .003	**0** .003
	10	**0** .011	**0** .011	—	—
	9	**0** .027	—	—	—
2	11	**0** .013	**0** .013	—	—
	10	**0** .038	—	—	—

TABLE A.10 (*continued*)

			Probability			
		a	0.05	0.025	0.01	0.005
A = 12 B = 12		12	**8** .047	**7** .019	**6** .007	**5** .002
		11	**6** .034	**5** .014	**4** .005⁻	**4** .005⁻
		10	**5** .045⁻	**4** .018	**3** .006	**2** .002
		9	**4** .050⁻	**3** .020	**2** .006	**1** .001
		8	**3** .050⁻	**2** .018	**1** .005⁻	**1** .005⁻
		7	**2** .045⁻	**1** .014	**0** .002	**0** .002
		6	**1** .034	**0** .007	**0** .007	—
		5	**0** .019	**0** .019	—	—
		4	**0** .047	—	—	—
	11	12	**7** .037	**6** .014	**5** .005⁻	**5** .005⁻
		11	**5** .024	**5** .024	**4** .008	**3** .002
		10	**4** .029	**3** .010⁺	**2** .003	**2** .003
		9	**3** .030	**2** .009	**2** .009	**1** .002
		8	**2** .026	**1** .007	**1** .007	**0** .001
		7	**1** .019	**1** .019	**0** .003	**0** .003
		6	**1** .045⁻	**0** .009	**0** .009	—
		5	**0** .024	**0** .024	—	—
	10	12	**6** .029	**5** .010⁻	**5** .010⁻	**4** .003
		11	**5** .043	**4** .015⁺	**3** .005⁻	**3** .005⁻
		10	**4** .048	**3** .017	**2** .005⁻	**2** .005⁻
		9	**3** .046	**2** .015	**1** .004	**1** .004
		8	**2** .038	**1** .010⁺	**0** .002	**0** .002
		7	**1** .026	**0** .005⁻	**0** .005⁻	**0** .005⁻
		6	**0** .012	**0** .012	—	—
		5	**0** .030	—	—	—
	9	12	**5** .021	**5** .021	**4** .006	**3** .002
		11	**4** .029	**3** .009	**3** .009	**2** .002
		10	**3** .029	**2** .008	**2** .008	**1** .002
		9	**2** .024	**2** .024	**1** .006	**0** .001
		8	**1** .016	**1** .016	**0** .002	**0** .002
		7	**1** .037	**0** .007	**0** .007	—
		6	**0** .017	**0** .017	—	—
		5	**0** .039	—	—	—
	8	12	**5** .049	**4** .014	**3** .004	**3** .004
		11	**3** .018	**3** .018	**2** .004	**2** .004
		10	**2** .015⁺	**2** .015⁺	**1** .003	**1** .003
		9	**2** .040	**1** .010⁻	**1** .010⁻	**0** .001
		8	**1** .025⁻	**1** .025⁻	**0** .004	**0** .004
		7	**0** .010⁺	**0** .010⁺	—	—
		6	**0** .024	**0** .024	—	—
	7	12	**4** .036	**3** .009	**3** .009	**2** .002
		11	**3** .038	**2** .010⁻	**2** .010⁻	**1** .002
		10	**2** .029	**1** .006	**1** .006	**0** .001
		9	**1** .017	**1** .017	**0** .002	**0** .002
		8	**1** .040	**0** .007	**0** .007	—
		7	**0** .016	**0** .016	—	—
		6	**0** .034	—	—	—

TABLE A.10   (continued)

	a	Probability 0.05	0.025	0.01	0.005
A = 12 B = 6	12	**3** .025⁻	**3** .025⁻	**2** .005⁻	**2** .005⁻
	11	**2** .022	**2** .022	**1** .004	**1** .004
	10	**1** .013	**1** .013	**0** .002	**0** .002
	9	**1** .032	**0** .005⁻	**0** .005⁻	**0** .005⁻
	8	**0** .011	**0** .011	—	—
	7	**0** .025⁻	**0** .025⁻	—	—
	6	**0** .050⁻	—	—	—
5	12	**2** .015⁻	**2** .015⁻	**1** .002	**1** .002
	11	**1** .010⁻	**1** .010⁻	**1** .010⁻	**0** .001
	10	**1** .028	**0** .003	**0** .003	**0** .003
	9	**0** .009	**0** .009	**0** .009	—
	8	**0** .020	**0** .020	—	—
	7	**0** .041	—	—	—
4	12	**2** .050	**1** .007	**1** .007	**0** .001
	11	**1** .027	**0** .003	**0** .003	**0** .003
	10	**0** .008	**0** .008	**0** .008	—
	9	**0** .019	**0** .019	—	—
	8	**0** .038	—	—	—
3	12	**1** .029	**0** .002	**0** .002	**0** .002
	11	**0** .009	**0** .009	**0** .009	—
	10	**0** .022	**0** .022	—	—
	9	**0** .044	—	—	—
2	12	**0** .011	**0** .011	—	—
	11	**0** .033	—	—	—
A = 13 B = 13	13	**9** .048	**8** .020	**7** .007	**6** .003
	12	**7** .037	**6** .015⁺	**5** .006	**4** .002
	11	**6** .048	**5** .021	**4** .008	**3** .002
	10	**4** .024	**4** .024	**3** .008	**2** .002
	9	**3** .024	**3** .024	**2** .008	**1** .002
	8	**2** .021	**2** .021	**1** .006	**0** .001
	7	**2** .048	**1** .015⁺	**0** .003	**0** .003
	6	**1** .037	**0** .007	**0** .007	—
	5	**0** .020	**0** .020	—	—
	4	**0** .048	—	—	—
12	13	**8** .039	**7** .015⁻	**6** .005⁺	**5** .002
	12	**6** .027	**5** .010⁻	**5** .010⁻	**4** .003
	11	**5** .033	**4** .013	**3** .004	**3** .004
	10	**4** .036	**3** .013	**2** .004	**2** .004
	9	**3** .034	**2** .011	**1** .003	**1** .003
	8	**2** .029	**1** .008	**1** .008	**0** .001
	7	**1** .020	**1** .020	**0** .004	**0** .004
	6	**1** .046	**0** .010⁻	**0** .010⁻	—
	5	**0** .024	**0** .024	—	—
11	13	**7** .031	**6** .011	**5** .003	**5** .003
	12	**6** .048	**5** .018	**4** .006	**3** .002
	11	**4** .021	**4** .021	**3** .007	**2** .002
	10	**3** .021	**3** .021	**2** .006	**1** .001
	9	**3** .050⁻	**2** .017	**1** .004	**1** .004
	8	**2** .040	**1** .011	**0** .002	**0** .002
	7	**1** .027	**0** .005⁻	**0** .005⁻	**0** .005⁻
	6	**0** .013	**0** .013	—	—
	5	**0** .030	—	—	—

TABLE A.10   (continued)

		Probability			
	a	0.05	0.025	0.01	0.005
$A = 13\ B = 10$	13	**6** .024	**6** .024	**5** .007	**4** .002
	12	**5** .035⁻	**4** .012	**3** .003	**3** .003
	11	**4** .037	**3** .012	**2** .003	**2** .003
	10	**3** .033	**2** .010⁺	**1** .002	**1** .002
	9	**2** .026	**1** .006	**1** .006	**0** .001
	8	**1** .017	**1** .017	**0** .003	**0** .003
	7	**1** .038	**0** .007	**0** .007	—
	6	**0** .017	**0** .017	—	—
	5	**0** .038	—	—	—
9	13	**5** .017	**5** .017	**4** .005⁻	**4** .005⁻
	12	**4** .023	**4** .023	**3** .007	**2** .001
	11	**3** .022	**3** .022	**2** .006	**1** .001
	10	**2** .017	**2** .017	**1** .004	**1** .004
	9	**2** .040	**1** .010⁺	**0** .001	**0** .001
	8	**1** .025⁻	**1** .025⁻	**0** .004	**0** .004
	7	**0** .010⁺	**0** .010⁺	—	—
	6	**0** .023	**0** .023	—	—
	5	**0** .049	—	—	—
8	13	**5** .042	**4** .012	**3** .003	**3** .003
	12	**4** .047	**3** .014	**2** .003	**2** .003
	11	**3** .041	**2** .011	**1** .002	**1** .002
	10	**2** .029	**1** .007	**1** .007	**0** .001
	9	**1** .017	**1** .017	**0** .002	**0** .002
	8	**1** .037	**0** .006	**0** .006	—
	7	**0** .015⁻	**0** .015⁻	—	—
	6	**0** .032	—	—	—
7	13	**4** .031	**3** .007	**3** .007	**2** .001
	12	**3** .031	**2** .007	**2** .007	**1** .001
	11	**2** .022	**2** .022	**1** .004	**1** .004
	10	**1** .012	**1** .012	**0** .002	**0** .002
	9	**1** .029	**0** .004	**0** .004	**0** .004
	8	**0** .010⁺	**0** .010⁺	—	—
	7	**0** .022	**0** .022	—	—
	6	**0** .044	—	—	—
6	13	**3** .021	**3** .021	**2** .004	**2** .004
	12	**2** .017	**2** .017	**1** .003	**1** .003
	11	**2** .046	**1** .010⁻	**1** .010⁻	**0** .001
	10	**1** .024	**1** .024	**0** .003	**0** .003
	9	**1** .050⁻	**0** .008	**0** .008	—
	8	**0** .017	**0** .017	—	—
	7	**0** .034	—	—	—
5	13	**2** .012	**2** .012	**1** .002	**1** .002
	12	**2** .044	**1** .008	**1** .008	**0** .001
	11	**1** .022	**1** .022	**0** .002	**0** .002
	10	**1** .047	**0** .007	**0** .007	—
	9	**0** .015⁻	**0** .015⁻	—	—
	8	**0** .029	—	—	—
4	13	**2** .044	**1** .006	**1** .006	**0** .000
	12	**1** .022	**1** .022	**0** .002	**0** .002
	11	**0** .006	**0** .006	**0** .006	—
	10	**0** .015⁻	**0** .015⁻	—	—
	9	**0** .029	—	—	—

TABLE A.10   *(continued)*

		Probability				
		a	0.05	0.025	0.01	0.005
$A = 13$ $B = 3$		13	**1** .025	**1** .025	**0** .002	**0** .002
		12	**0** .007	**0** .007	**0** .007	—
		11	**0** .018	**0** .018	—	—
		10	**0** .036	—	—	—
	2	13	**0** .010$^-$	**0** .010$^-$	**0** .010$^-$	—
		12	**0** .029	—	—	—
$A = 14$ $B = 14$		14	**10** .049	**9** .020	**8** .008	**7** .003
		13	**8** .038	**7** .016	**6** .006	**5** .002
		12	**6** .023	**6** .023	**5** .009	**4** .003
		11	**5** .027	**4** .011	**3** .004	**3** .004
		10	**4** .028	**3** .011	**2** .003	**2** .003
		9	**3** .027	**2** .009	**2** .009	**1** .002
		8	**2** .023	**2** .023	**1** .006	**0** .001
		7	**1** .016	**1** .016	**0** .003	**0** .003
		6	**1** .038	**0** .008	**0** .008	—
		5	**0** .020	**0** .020	—	—
		4	**0** .049	—	—	—
	13	14	**9** .041	**8** .016	**7** .006	**6** .002
		13	**7** .029	**6** .011	**5** .004	**5** .004
		12	**6** .037	**5** .015$^+$	**4** .005$^+$	**3** .002
		11	**5** .041	**4** .017	**3** .006	**2** .001
		10	**4** .041	**3** .016	**2** .005$^-$	**2** .005$^-$
		9	**3** .038	**2** .013	**1** .003	**1** .003
		8	**2** .031	**1** .009	**1** .009	**0** .001
		7	**1** .021	**1** .021	**0** .004	**0** .004
		6	**1** .048	**0** .010$^+$	—	—
		5	**0** .025$^-$	**0** .025$^-$	—	—
	12	14	**8** .033	**7** .012	**6** .004	**6** .004
		13	**6** .021	**6** .021	**5** .007	**4** .002
		12	**5** .025$^+$	**4** .009	**4** .009	**3** .003
		11	**4** .026	**3** .009	**3** .009	**2** .002
		10	**3** .024	**3** .024	**2** .007	**1** .002
		9	**2** .019	**2** .019	**1** .005$^-$	**1** .005$^-$
		8	**2** .042	**1** .012	**0** .002	**0** .002
		7	**1** .028	**0** .005$^+$	**0** .005$^+$	—
		6	**0** .013	**0** .013	—	—
		5	**0** .030	—	—	—
	11	14	**7** .026	**6** .009	**6** .009	**5** .003
		13	**6** .039	**5** .014	**4** .004	**4** .004
		12	**5** .043	**4** .016	**3** .005$^-$	**3** .005$^-$
		11	**4** .042	**3** .015$^-$	**2** .004	**2** .004
		10	**3** .036	**2** .011	**1** .003	**1** .003
		9	**2** .027	**1** .007	**1** .007	**0** .001
		8	**1** .017	**1** .017	**0** .003	**0** .003
		7	**1** .038	**0** .007	**0** .007	—
		6	**0** .017	**0** .017	—	—
		5	**0** .038	—	—	—

TABLE A.10   (continued)

		Probability			
	a	0.05	0.025	0.01	0.005
$A = 14$ $B = 10$	14	**6** .020	**6** .020	**5** .006	**4** .002
	13	**5** .028	**4** .009	**4** .009	**3** .002
	12	**4** .028	**3** .009	**3** .009	**2** .002
	11	**3** .024	**3** .024	**2** .007	**1** .001
	10	**2** .018	**2** .018	**1** .004	**1** .004
	9	**2** .040	**1** .011	**0** .002	**0** .002
	8	**1** .024	**1** .024	**0** .004	**0** .004
	7	**0** .010⁻	**0** .010⁻	**0** .010⁻	—
	6	**0** .022	**0** .022	—	—
	5	**0** .047	—	—	—
9	14	**6** .047	**5** .014	**4** .004	**4** .004
	13	**4** .018	**4** .018	**3** .005⁻	**3** .005⁻
	12	**3** .017	**3** .017	**2** .004	**2** .004
	11	**3** .042	**2** .012	**1** .002	**1** .002
	10	**2** .029	**1** .007	**1** .007	**0** .001
	9	**1** .017	**1** .017	**0** .002	**0** .002
	8	**1** .036	**0** .006	**0** .006	—
	7	**0** .014	**0** .014	—	—
	6	**0** .030	—	—	—
8	14	**5** .036	**4** .010⁻	**4** .010⁻	**3** .002
	13	**4** .039	**3** .011	**2** .002	**2** .002
	12	**3** .032	**2** .008	**2** .008	**1** .001
	11	**2** .022	**2** .022	**1** .005⁻	**1** .005⁻
	10	**2** .048	**1** .012	**0** .002	**0** .002
	9	**1** .026	**0** .004	**0** .004	**0** .004
	8	**0** .009	**0** .009	**0** .009	—
	7	**0** .020	**0** .020	—	—
	6	**0** .040	—	—	—
7	14	**4** .026	**3** .006	**3** .006	**2** .001
	13	**3** .025	**2** .006	**2** .006	**1** .001
	12	**2** .017	**2** .017	**1** .003	**1** .003
	11	**2** .041	**1** .009	**1** .009	**0** .001
	10	**1** .021	**1** .021	**0** .003	**0** .003
	9	**1** .043	**0** .007	**0** .007	—
	8	**0** .015⁻	**0** .015⁻	—	—
	7	**0** .030	—	—	—
6	14	**3** .018	**3** .018	**2** .003	**2** .003
	13	**2** .014	**2** .014	**1** .002	**1** .002
	12	**2** .037	**1** .007	**1** .007	**0** .001
	11	**1** .018	**1** .018	**0** .002	**0** .002
	10	**1** .038	**0** .005⁺	**0** .005⁺	—
	9	**0** .012	**0** .012	—	—
	8	**0** .024	**0** .024	—	—
	7	**0** .044	—	—	—
5	14	**2** .010⁺	**2** .010⁺	**1** .001	**1** .001
	13	**2** .037	**1** .006	**1** .006	**0** .001
	12	**1** .017	**1** .017	**0** .002	**0** .002
	11	**1** .038	**0** .005⁻	**0** .005⁻	**0** .005⁻
	10	**0** .011	**0** .011	—	—
	9	**0** .022	**0** .022	—	—
	8	**0** .040	—	—	—

TABLE A.10   (continued)

	a	Probability			
		0.05	0.025	0.01	0.005
A = 14 B = 4	14	**2** .039	**1** .005⁻	**1** .005⁻	**1** .005⁻
	13	**1** .019	**1** .019	**0** .002	**0** .002
	12	**1** .044	**0** .005⁻	**0** .005⁻	**0** .005⁻
	11	**0** .011	**0** .011	—	—
	10	**0** .023	**0** .023	—	—
	9	**0** .041	—	—	—
3	14	**1** .022	**1** .022	**0** .001	**0** .001
	13	**0** .006	**0** .006	**0** .006	—
	12	**0** .015⁻	**0** .015⁻	—	—
	11	**0** .029	—	—	—
2	14	**0** .008	**0** .008	**0** .008	—
	13	**0** .025	**0** .025	—	—
	12	**0** .050	—	—	—
A = 15 B = 15	15	**11** .050⁻	**10** .021	**9** .008	**8** .003
	14	**9** .040	**8** .018	**7** .007	**6** .003
	13	**7** .025⁺	**6** .010⁺	**5** .004	**5** .004
	12	**6** .030	**5** .013	**4** .005⁻	**4** .005⁻
	11	**5** .033	**4** .013	**3** .005⁻	**3** .005⁻
	10	**4** .033	**3** .013	**2** .004	**2** .004
	9	**3** .030	**2** .010⁺	**1** .003	**1** .003
	8	**2** .025⁺	**1** .007	**1** .007	**0** .001
	7	**1** .018	**1** .018	**0** .003	**0** .003
	6	**1** .040	**0** .008	**0** .008	—
	5	**0** .021	**0** .021	—	—
	4	**0** .050⁻	—	—	—
14	15	**10** .042	**9** .017	**8** .006	**7** .002
	14	**8** .031	**7** .013	**6** .005⁻	**6** .005⁻
	13	**7** .041	**6** .017	**5** .007	**4** .002
	12	**6** .046	**5** .020	**4** .007	**3** .002
	11	**5** .048	**4** .020	**3** .007	**2** .002
	10	**4** .046	**3** .018	**2** .006	**1** .001
	9	**3** .041	**2** .014	**1** .004	**1** .004
	8	**2** .033	**1** .009	**1** .009	**0** .001
	7	**1** .022	**1** .022	**0** .004	**0** .004
	6	**1** .049	**0** .011	—	—
	5	**0** .025⁺	—	—	—
13	15	**9** .035⁻	**8** .013	**7** .035⁻	**7** .005⁻
	14	**7** .023	**7** .023	**6** .009	**5** .003
	13	**6** .029	**5** .011	**4** .004	**4** .004
	12	**5** .031	**4** .012	**3** .004	**3** .004
	11	**4** .030	**3** .011	**2** .003	**2** .003
	10	**3** .026	**2** .008	**2** .008	**1** .002
	9	**2** .020	**2** .020	**1** .005⁺	**0** .001
	8	**2** .043	**1** .013	**0** .002	**0** .002
	7	**1** .029	**0** .005⁺	**0** .005⁺	—
	6	**0** .013	**0** .013	—	—
	5	**0** .031	—	—	—

TABLE A.10   *(continued)*

	a	0.05	0.025	0.01	0.005
		Probability			
$A = 15\ B = 12$	15	**8** .028	**7** .010⁻	**7** .010⁻	**6** .003
	14	**7** .043	**6** .016	**5** .006	**4** .002
	13	**6** .049	**5** .019	**4** .007	**3** .002
	12	**5** .049	**4** .019	**3** .006	**2** .002
	11	**4** .045⁺	**3** .017	**2** .005⁻	**2** .005⁻
	10	**3** .038	**2** .012	**1** .003	**1** .003
	9	**2** .028	**1** .007	**1** .007	**0** .001
	8	**1** .018	**1** .018	**0** .003	**0** .003
	7	**1** .038	**0** .007	**0** .007	—
	6	**0** .017	**0** .017	—	—
	5	**0** .037	—	—	—
11	15	**7** .022	**7** .022	**6** .007	**5** .002
	14	**6** .032	**5** .011	**4** .003	**4** .003
	13	**5** .034	**4** .012	**3** .003	**3** .003
	12	**4** .032	**3** .010⁺	**2** .003	**2** .003
	11	**3** .026	**2** .008	**2** .008	**1** .002
	10	**2** .019	**2** .019	**1** .004	**1** .004
	9	**2** .040	**1** .011	**0** .002	**0** .002
	8	**1** .024	**1** .024	**0** .004	**0** .004
	7	**1** .049	**0** .010⁻	**0** .010⁻	—
	6	**0** .022	**0** .022	—	—
	5	**0** .046	—	—	—
10	15	**6** .017	**6** .017	**5** .005⁻	**5** .005⁻
	14	**5** .023	**5** .023	**4** .007	**3** .002
	13	**4** .022	**4** .022	**3** .007	**2** .001
	12	**3** .018	**3** .018	**2** .005⁻	**2** .005⁻
	11	**3** .042	**2** .013	**1** .003	**1** .003
	10	**2** .029	**1** .007	**1** .007	**0** .001
	9	**1** .016	**1** .016	**0** .002	**0** .002
	8	**1** .034	**0** .006	**0** .006	—
	7	**0** .013	**0** .013	—	—
	6	**0** .028	—	—	—
9	15	**6** .042	**5** .012	**4** .003	**4** .003
	14	**5** .047	**4** .015⁻	**3** .004	**3** .004
	13	**4** .042	**3** .013	**2** .003	**2** .003
	12	**3** .032	**2** .009	**2** .009	**1** .002
	11	**2** .021	**2** .021	**1** .005⁻	**1** .005⁻
	10	**2** .045⁻	**1** .011	**0** .002	**0** .002
	9	**1** .024	**1** .024	**0** .004	**0** .004
	8	**1** .048	**0** .009	**0** .009	—
	7	**0** .019	**0** .019	—	—
	6	**0** .037	—	—	—
8	15	**5** .032	**4** .008	**4** .008	**3** .002
	14	**4** .033	**3** .009	**3** .009	**2** .002
	13	**3** .026	**2** .006	**2** .006	**1** .001
	12	**2** .017	**2** .017	**1** .003	**1** .003
	11	**2** .037	**1** .008	**1** .008	**0** .001
	10	**1** .019	**1** .019	**0** .003	**0** .003
	9	**1** .038	**0** .006	**0** .006	—
	8	**0** .013	**0** .013	—	—
	7	**0** .026	—	—	—
	6	**0** .050⁻	—	—	—

TABLE A.10 (continued)

		Probability			
	a	0.05	0.025	0.01	0.005
A = 15 B = 7	15	**4** .023	**4** .023	**3** .005⁻	**3** .005⁻
	14	**3** .021	**3** .021	**2** .004	**2** .004
	13	**2** .014	**2** .014	**1** .002	**1** .002
	12	**2** .032	**1** .007	**1** .007	**0** .001
	11	**1** .015⁺	**1** .015⁺	**0** .002	**0** .002
	10	**1** .032	**0** .005⁻	**0** .005⁻	**0** .005⁻
	9	**0** .010⁺	**0** .010⁺	—	—
	8	**0** .020	**0** .020	—	—
	7	**0** .038	—	—	—
6	15	**3** .015⁺	**3** .015⁺	**2** .003	**2** .003
	14	**2** .011	**2** .011	**1** .002	**1** .002
	13	**2** .031	**1** .006	**1** .006	**0** .001
	12	**1** .014	**1** .014	**0** .002	**0** .002
	11	**1** .029	**0** .004	**0** .004	**0** .004
	10	**0** .009	**0** .009	**0** .009	—
	9	**0** .017	**0** .017	—	—
	8	**0** .032	—	—	—
5	15	**2** .009	**2** .009	**2** .009	**1** .001
	14	**2** .032	**1** .005⁻	**1** .005⁻	**1** .005⁻
	13	**1** .014	**1** .014	**0** .001	**0** .001
	12	**1** .031	**0** .004	**0** .004	**0** .004
	11	**0** .008	**0** .008	**0** .008	—
	10	**0** .016	**0** .016	—	—
	9	**0** .030	—	—	—
4	15	**2** .035⁺	**1** .004	**1** .004	**1** .004
	14	**1** .016	**1** .016	**0** .001	**0** .001
	13	**1** .037	**0** .004	**0** .004	**0** .004
	12	**0** .009	**0** .009	**0** .009	—
	11	**0** .018	**0** .018	—	—
	10	**0** .033	—	—	—
3	15	**1** .020	**1** .020	**0** .001	**0** .001
	14	**0** .005⁻	**0** .005⁻	**0** .005⁻	**0** .005⁻
	13	**0** .012	**0** .012	—	—
	12	**0** .025⁻	**0** .025⁻	—	—
	11	**0** .043	—	—	—
2	15	**0** .007	**0** .007	**0** .007	—
	14	**0** .022	**0** .022	—	—
	13	**0** .044	—	—	—
A = 16 B = 16	16	**11** .022	**11** .022	**10** .009	**9** .003
	15	**10** .041	**9** .019	**8** .008	**7** .003
	14	**8** .027	**7** .012	**6** .005⁻	**6** .005⁻
	13	**7** .033	**6** .015⁻	**5** .006	**4** .002
	12	**6** .037	**5** .016	**4** .006	**3** .002
	11	**5** .038	**4** .016	**3** .006	**2** .002
	10	**4** .037	**3** .015⁻	**2** .005⁻	**2** .005⁻
	9	**3** .033	**2** .012	**1** .003	**1** .003
	8	**2** .027	**1** .008	**1** .008	**0** .001
	7	**1** .019	**1** .019	**0** .003	**0** .003
	6	**1** .041	**0** .009	**0** .009	—
	5	**0** .022	**0** .022	—	—

TABLE A.10   (*continued*)

		Probability			
	a	**0.05**	**0.025**	**0.01**	**0.005**
A = 16 *B* = 15	16	**11** .043	**10** .018	**9** .007	**8** .002
	15	**9** .033	**8** .014	**7** .005⁺	**6** .002
	14	**8** .044	**7** .019	**6** .008	**5** .003
	13	**6** .023	**6** .023	**5** .009	**4** .003
	12	**5** .024	**5** .024	**4** .009	**3** .003
	11	**4** .023	**4** .023	**3** .008	**2** .002
	10	**4** .049	**3** .020	**2** .006	**1** .001
	9	**3** .043	**2** .016	**1** .004	**1** .004
	8	**2** .035⁻	**1** .010⁺	**0** .002	**0** .002
	7	**1** .023	**1** .023	**0** .004	**0** .004
	6	**0** .011	**0** .011	—	—
	5	**0** .026	—		
	14	16	**10** .037	**9** .014	**8** .005⁺	**7** .002
	15	**8** .025⁺	**7** .010⁻	**7** .010⁻	**6** .003
	14	**7** .032	**6** .013	**5** .005⁻	**5** .005⁻
	13	**6** .035⁺	**5** .014	**4** .005⁺	**3** .001
	12	**5** .035⁺	**4** .014	**3** .005⁻	**3** .005⁻
	11	**4** .033	**3** .012	**2** .004	**2** .004
	10	**3** .028	**2** .009	**2** .009	**1** .002
	9	**2** .021	**2** .021	**1** .006	**0** .001
	8	**2** .045⁻	**1** .013	**0** .002	**0** .002
	7	**1** .030	**0** .006	**0** .006	—
	6	**0** .013	**0** .013	—	—
	5	**0** .031	—	—	—
	13	16	**9** .030	**8** .011	**7** .004	**7** .004
	15	**8** .047	**7** .019	**6** .007	**5** .002
	14	**6** .023	**6** .023	**5** .008	**4** .003
	13	**5** .023	**5** .023	**4** .008	**3** .003
	12	**4** .022	**4** .022	**3** .007	**2** .002
	11	**4** .048	**3** .018	**2** .005⁺	**1** .001
	10	**3** .039	**2** .013	**1** .003	**1** .003
	9	**2** .029	**1** .008	**1** .008	**0** .001
	8	**1** .018	**1** .018	**0** .003	**0** .003
	7	**1** .038	**0** .007	**0** .007	—
	6	**0** .017	**0** .017	—	—
	5	**0** .037	—	—	—
	12	16	**8** .024	**8** .024	**7** .008	**6** .002
	15	**7** .036	**6** .013	**5** .004	**5** .004
	14	**6** .040	**5** .015⁻	**4** .005⁻	**4** .005⁻
	13	**5** .039	**4** .014	**3** .004	**3** .004
	12	**4** .034	**3** .012	**2** .003	**2** .003
	11	**3** .027	**2** .008	**2** .008	**1** .002
	10	**2** .019	**2** .019	**1** .005⁻	**1** .005⁻
	9	**2** .040	**1** .011	**0** .002	**0** .002
	8	**1** .024	**1** .024	**0** .004	**0** .004
	7	**1** .048	**0** .010⁻	**0** .010⁻	—
	6	**0** .021	**0** .021	—	—
	5	**0** .044	—	—	—

TABLE A.10   (continued)

	a	Probability			
		0.05	0.025	0.01	0.005
$A = 16\ B = 11$	16	7 .019	7 .019	6 .006	5 .002
	15	6 .027	5 .009	5 .009	4 .002
	14	5 .027	4 .009	4 .009	3 .002
	13	4 .024	4 .024	3 .008	2 .002
	12	3 .019	3 .019	2 .005$^+$	1 .001
	11	3 .041	2 .013	1 .003	1 .003
	10	2 .028	1 .007	1 .007	0 .001
	9	1 .016	1 .016	0 .002	0 .002
	8	1 .033	0 .006	0 .006	—
	7	0 .013	0 .013	—	—
	6	0 .027	—	—	—
10	16	7 .046	6 .014	5 .004	5 .004
	15	5 .018	5 .018	4 .005$^+$	3 .001
	14	4 .018	4 .018	3 .005$^-$	3 .005$^-$
	13	4 .042	3 .014	2 .003	2 .003
	12	3 .032	2 .009	2 .009	1 .002
	11	2 .021	2 .021	1 .005$^-$	1 .005$^-$
	10	2 .042	1 .011	0 .002	0 .002
	9	1 .023	1 .023	0 .004	0 .004
	8	1 .045$^-$	0 .008	0 .008	—
	7	0 .017	0 .017	—	—
	6	0 .035$^-$	—	—	—
9	16	6 .037	5 .010$^-$	5 .010$^-$	4 .002
	15	5 .040	4 .012	3 .003	3 .003
	14	4 .034	3 .010$^-$	3 .010$^-$	2 .002
	13	3 .025$^+$	2 .007	2 .007	1 .001
	12	2 .016	2 .016	1 .003	1 .003
	11	2 .033	1 .008	1 .008	0 .001
	10	1 .017	1 .017	0 .002	0 .002
	9	1 .034	0 .006	0 .006	—
	8	0 .012	0 .012	—	—
	7	0 .024	0 .024	—	—
	6	0 .045$^+$	—	—	—
8	16	5 .028	4 .007	4 .007	3 .001
	15	4 .028	3 .007	3 .007	2 .001
	14	3 .021	3 .021	2 .005$^-$	2 .005$^-$
	13	3 .047	2 .013	1 .002	1 .002
	12	2 .028	1 .006	1 .006	0 .001
	11	1 .014	1 .014	0 .002	0 .002
	10	1 .027	0 .004	0 .004	0 .004
	9	0 .009	0 .009	0 .009	—
	8	0 .017	0 .017	—	—
	7	0 .033	—	—	—
7	16	4 .020	4 .020	3 .004	3 .004
	15	3 .017	3 .017	2 .003	2 .003
	14	3 .045$^+$	2 .011	1 .002	1 .002
	13	2 .026	1 .005$^-$	1 .005$^-$	1 .005$^-$
	12	1 .012	1 .012	0 .001	0 .001
	11	1 .024	1 .024	0 .003	0 .003
	10	1 .045$^-$	0 .007	0 .007	—
	9	0 .014	0 .014	—	—
	8	0 .026	—	—	—
	7	0 .047	—	—	—

TABLE A.10   (continued)

	a	Probability 0.05	0.025	0.01	0.005
$A = 16$ $B = 6$	16	**3** .013	**3** .013	**2** .002	**2** .002
	15	**3** .046	**2** .009	**2** .009	**1** .001
	14	**2** .025+	**1** .004	**1** .004	**1** .004
	13	**1** .011	**1** .011	**0** .001	**0** .001
	12	**1** .023	**1** .023	**0** .003	**0** .003
	11	**1** .043	**0** .006	**0** .006	—
	10	**0** .012	**0** .012	—	—
	9	**0** .023	**0** .023	—	—
	8	**0** .040	—	—	—
5	16	**3** .048	**2** .008	**2** .008	**1** .001
	15	**2** .028	**1** .004	**1** .004	**1** .004
	14	**1** .011	**1** .011	**0** .001	**0** .001
	13	**1** .025+	**0** .003	**0** .003	**0** .003
	12	**1** .047	**0** .006	**0** .006	—
	11	**0** .012	**0** .012	—	—
	10	**0** .023	**0** .023	—	—
	9	**0** .039	—	—	—
4	16	**2** .032	**1** .004	**1** .004	**1** .004
	15	**1** .013	**1** .013	**0** .001	**1** .001
	14	**1** .032	**0** .003	**0** .003	**0** .003
	13	**0** .007	**0** .007	**0** .007	—
	12	**0** .014	**0** .014	—	—
	11	**0** .026	—	—	—
	10	**0** .043	—	—	—
3	16	**1** .018	**1** .018	**0** .001	**0** .001
	15	**0** .004	**0** .004	**0** .004	**0** .004
	14	**0** .010+	**0** .010+	—	—
	13	**0** .021	**0** .021	—	—
	12	**0** .036	—	—	—
2	16	**0** .007	**0** .007	**0** .007	—
	15	**0** .020	**0** .020	—	—
	14	**0** .039	—	—	—
$A = 17$ $B = 17$	17	**12** .022	**12** .022	**11** .009	**10** .004
	16	**11** .043	**10** .020	**9** .008	**8** .003
	15	**9** .029	**8** .013	**7** .005+	**6** .002
	14	**8** .035+	**7** .016	**6** .007	**5** .002
	13	**7** .040	**6** .018	**5** .007	**4** .003
	12	**6** .042	**5** .019	**4** .007	**3** .002
	11	**5** .042	**4** .018	**3** .007	**2** .002
	10	**4** .040	**3** .016	**2** .005+	**1** .001
	9	**3** .035+	**2** .013	**1** .003	**1** .003
	8	**2** .029	**1** .008	**1** .008	**0** .001
	7	**1** .020	**1** .020	**0** .004	**0** .004
	6	**1** .043	**0** .009	**0** .009	—
	5	**0** .022	**0** .022	—	—

TABLE A.10   (continued)

	a	Probability 0.05	0.025	0.01	0.005
A = 17 B = 16	17	**12**.044	**11**.018	**10**.007	**9**.003
	16	**10**.035⁻	**9**.015⁻	**8**.006	**7**.002
	15	**9**.046	**8**.021	**7**.009	**6**.003
	14	**7**.025⁺	**6**.011	**5**.004	**5**.004
	13	**6**.027	**5**.011	**4**.004	**4**.004
	12	**5**.027	**4**.011	**3**.004	**3**.004
	11	**4**.025⁺	**3**.009	**3**.009	**2**.003
	10	**3**.022	**3**.022	**2**.007	**1**.002
	9	**3**.046	**2**.017	**1**.004	**1**.004
	8	**2**.036	**1**.011	**0**.002	**0**.002
	7	**1**.024	**1**.024	**0**.005⁻	**0**.005⁻
	6	**0**.011	**0**.011	—	—
	5	**0**.026	—	—	—
15	17	**11**.038	**10**.015⁻	**9**.006	**8**.002
	16	**9**.027	**8**.011	**7**.004	**7**.004
	15	**8**.035⁺	**7**.015⁻	**6**.006	**5**.002
	14	**7**.040	**6**.017	**5**.006	**4**.002
	13	**6**.041	**5**.017	**4**.006	**3**.002
	12	**5**.039	**4**.016	**3**.005⁺	**2**.001
	11	**4**.035⁺	**3**.013	**2**.004	**2**.004
	10	**3**.029	**2**.010⁻	**2**.010⁻	**1**.002
	9	**2**.022	**2**.022	**1**.006	**0**.001
	8	**2**.046	**1**.014	**0**.002	**0**.002
	7	**1**.030	**0**.006	**0**.006	—
	6	**0**.014	**0**.014	—	—
	5	**0**.031	—	—	—
14	17	**10**.032	**9**.012	**8**.004	**8**.004
	16	**8**.021	**8**.021	**7**.008	**6**.003
	15	**7**.026	**6**.010⁻	**6**.010⁻	**5**.003
	14	**6**.028	**5**.011	**4**.004	**4**.004
	13	**5**.027	**4**.010⁻	**4**.010⁻	**3**.003
	12	**4**.024	**4**.024	**3**.008	**2**.002
	11	**4**.049	**3**.019	**2**.006	**1**.001
	10	**3**.040	**2**.014	**1**.003	**1**.003
	9	**2**.029	**1**.008	**1**.008	**0**.001
	8	**1**.018	**1**.018	**0**.003	**0**.003
	7	**1**.038	**0**.007	**0**.007	—
	6	**0**.017	**0**.017	—	—
	5	**0**.036	—	—	—
13	17	**9**.026	**8**.009	**8**.009	**7**.003
	16	**8**.040	**7**.015⁺	**6**.005⁺	**5**.002
	15	**7**.045⁺	**6**.018	**5**.006	**4**.002
	14	**6**.045⁺	**5**.018	**4**.006	**3**.002
	13	**5**.042	**4**.016	**3**.005⁺	**2**.001
	12	**4**.035⁺	**3**.013	**2**.004	**2**.004
	11	**3**.028	**2**.009	**2**.009	**1**.002
	10	**2**.019	**2**.019	**1**.005⁻	**1**.005⁻
	9	**2**.040	**1**.011	**0**.002	**0**.002
	8	**1**.024	**1**.024	**0**.004	**0**.004
	7	**1**.047	**0**.010⁻	**0**.010⁻	—
	6	**0**.021	**0**.021	—	—
	5	**0**.043	—	—	—

TABLE A.10   (continued)

		Probability			
		a	0.05	0.025	0.01	0.005
A = 17 *B* = 12		17	**8** .021	**8** .021	**7** .007	**6** .002
		16	**7** .030	**6** .011	**5** .003	**5** .003
		15	**6** .033	**5** .012	**4** .004	**4** .004
		14	**5** .030	**4** .011	**3** .003	**3** .003
		13	**4** .026	**3** .008	**3** .008	**2** .002
		12	**3** .020	**3** .020	**2** .006	**1** .001
		11	**3** .041	**2** .013	**1** .003	**1** .003
		10	**2** .028	**1** .007	**1** .007	**0** .001
		9	**1** .016	**1** .016	**0** .002	**0** .002
		8	**1** .032	**0** .006	**0** .006	—
		7	**0** .012	**0** .012	—	—
		6	**0** .026	—	—	—
	11	17	**7** .016	**7** .016	**6** .005⁻	**6** .005⁻
		16	**6** .022	**6** .022	**5** .007	**4** .002
		15	**5** .022	**5** .022	**4** .007	**3** .002
		14	**4** .019	**4** .019	**3** .006	**2** .001
		13	**4** .042	**3** .014	**2** .004	**2** .004
		12	**3** .031	**2** .009	**2** .009	**1** .002
		11	**2** .020	**2** .020	**1** .005⁻	**1** .005⁻
		10	**2** .040	**1** .011	**0** .001	**0** .001
		9	**1** .022	**1** .022	**0** .004	**0** .004
		8	**1** .042	**0** .008	**0** .008	—
		7	**0** .016	**0** .016	—	—
		6	**0** .033	—	—	—
	10	17	**7** .041	**6** .012	**5** .003	**5** .003
		16	**6** .047	**5** .015⁺	**4** .004	**4** .004
		15	**5** .043	**4** .014	**3** .004	**3** .004
		14	**4** .034	**3** .010⁺	**2** .002	**2** .002
		13	**3** .024	**3** .024	**2** .007	**1** .001
		12	**3** .049	**2** .015⁺	**1** .003	**1** .003
		11	**2** .031	**1** .007	**1** .007	**0** .001
		10	**1** .016	**1** .016	**0** .002	**0** .002
		9	**1** .031	**0** .005⁺	**0** .005⁺	—
		8	**0** .011	**0** .011	—	—
		7	**0** .022	**0** .022	—	—
		6	**0** .042	—	—	—
	9	17	**6** .032	**5** .008	**5** .008	**4** .002
		16	**5** .034	**4** .010⁻	**4** .010⁻	**3** .002
		15	**4** .028	**3** .008	**3** .008	**2** .002
		14	**3** .020	**3** .020	**2** .005⁻	**2** .005⁻
		13	**3** .042	**2** .012	**1** .002	**1** .002
		12	**2** .025⁺	**1** .006	**1** .006	**0** .001
		11	**2** .048	**1** .012	**0** .002	**0** .002
		10	**1** .024	**1** .024	**0** .004	**0** .004
		9	**1** .045⁻	**0** .008	**0** .008	—
		8	**0** .016	**0** .016	—	—
		7	**0** .030	—	—	—

TABLE A.10   (continued)

		Probability			
	a	0.05	0.025	0.01	0.005
A = 17 B = 8	17	**5** .024	**5** .024	**4** .006	**3** .001
	16	**4** .023	**4** .023	**3** .006	**2** .001
	15	**3** .017	**3** .017	**2** .004	**2** .004
	14	**3** .039	**2** .010⁻	**2** .010⁻	**1** .002
	13	**2** .022	**2** .022	**1** .004	**1** .004
	12	**2** .043	**1** .010⁻	**1** .010⁻	**0** .001
	11	**1** .020	**1** .020	**0** .003	**0** .003
	10	**1** .038	**0** .006	**0** .006	—
	9	**0** .012	**0** .012	—	—
	8	**0** .022	**0** .022	—	—
	7	**0** .040	—	—	—
7	17	**4** .017	**4** .017	**3** .003	**3** .003
	16	**3** .014	**3** .014	**2** .003	**2** .003
	15	**3** .038	**2** .009	**2** .009	**1** .001
	14	**2** .021	**2** .021	**1** .004	**1** .004
	13	**2** .042	**1** .009	**1** .009	**0** .001
	12	**1** .018	**1** .018	**0** .002	**0** .002
	11	**1** .034	**0** .005⁻	**0** .005⁻	**0** .005⁻
	10	**0** .010⁻	**0** .010⁻	**0** .010⁻	—
	9	**0** .019	**0** .019	—	—
	8	**0** .033	—	—	—
6	17	**3** .011	**3** .011	**2** .002	**2** .002
	16	**3** .040	**2** .008	**2** .008	**1** .001
	15	**2** .021	**2** .021	**1** .003	**1** .003
	14	**2** .045⁺	**1** .009	**1** .009	**0** .001
	13	**1** .018	**1** .018	**0** .002	**0** .002
	12	**1** .035⁻	**0** .005⁻	**0** .005⁻	**0** .005⁻
	11	**0** .009	**0** .009	**0** .009	—
	10	**0** .017	**0** .017	—	—
	9	**0** .030	—	—	—
	8	**0** .050⁻	—	—	—
5	17	**3** .043	**2** .006	**2** .006	**1** .001
	16	**2** .024	**2** .024	**1** .003	**1** .003
	15	**1** .009	**1** .009	**1** .009	**0** .001
	14	**1** .021	**1** .021	**0** .002	**0** .002
	13	**1** .039	**0** .005⁻	**0** .005⁻	**0** .005⁻
	12	**0** .010⁻	**0** .010⁻	**0** .010⁻	—
	11	**0** .018	**0** .018	—	—
	10	**0** .030	—	—	—
	9	**0** .049	—	—	—
4	17	**2** .029	**1** .003	**1** .003	**1** .003
	16	**1** .011	**1** .011	**0** .001	**0** .001
	15	**1** .028	**0** .003	**0** .003	**0** .003
	14	**0** .006	**0** .006	**0** .006	—
	13	**0** .012	**0** .012	—	—
	12	**0** .021	**0** .021	—	—
	11	**0** .035⁺	—	—	—

TABLE A.10   (*continued*)

			Probability			
		a	**0.05**	**0.025**	**0.01**	**0.005**
A = 17 *B* = 3		17	**1** .016	**1** .016	**0** .001	**0** .001
		16	**1** .046	**0** .004	**0** .004	**0** .004
		15	**0** .009	**0** .009	**0** .009	—
		14	**0** .018	**0** .018	—	—
		13	**0** .031	—	—	—
		12	**0** .049	—	—	—
	2	17	**0** .006	**0** .006	**0** .006	—
		16	**0** .018	**0** .018	—	—
		15	**0** .035$^+$	—	—	—
A = 18 *B* = 18		18	**13** .023	**13** .023	**12** .010$^-$	**11** .004
		17	**12** .044	**11** .020	**10** .009	**9** .004
		16	**10** .030	**9** .014	**8** .006	**7** .002
		15	**9** .038	**8** .018	**7** .008	**6** .003
		14	**8** .043	**7** .020	**6** .009	**5** .003
		13	**7** .046	**6** .022	**5** .009	**4** .003
		12	**6** .047	**5** .022	**4** .009	**3** .003
		11	**5** .046	**4** .020	**3** .008	**2** .002
		10	**4** .043	**3** .018	**2** .006	**1** .001
		9	**3** .038	**2** .014	**1** .004	**1** .004
		8	**2** .030	**1** .009	**1** .009	**0** .001
		7	**1** .020	**1** .020	**0** .004	**0** .004
		6	**1** .044	**0** .010$^-$	**0** .010$^-$	—
		5	**0** .023	**0** .023	—	—
	17	18	**13** .045$^+$	**12** .019	**11** .008	**10** .003
		17	**11** .036	**10** .016	**9** .007	**8** .002
		16	**10** .049	**9** .023	**8** .010$^-$	**7** .004
		15	**8** .028	**7** .012	**6** .005$^-$	**6** .005$^-$
		14	**7** .030	**6** .013	**5** .005$^+$	**4** .002
		13	**6** .031	**5** .013	**4** .005$^-$	**4** .005$^-$
		12	**5** .030	**4** .012	**3** .004	**3** .004
		11	**4** .028	**3** .010$^+$	**2** .003	**2** .003
		10	**3** .023	**3** .023	**2** .008	**1** .002
		9	**3** .047	**2** .018	**1** .005$^-$	**1** .005$^-$
		8	**2** .037	**1** .011	**0** .002	**0** .002
		7	**1** .025$^-$	**1** .025$^-$	**0** .005$^-$	**0** .005$^-$
		6	**0** .011	**0** .011	—	—
		5	**0** .026	—	—	—
	16	18	**12** .039	**11** .016	**10** .006	**9** .002
		17	**10** .029	**9** .012	**8** .005$^-$	**8** .005$^-$
		16	**9** .038	**8** .017	**7** .007	**6** .002
		15	**8** .043	**7** .019	**6** .008	**5** .003
		14	**7** .046	**6** .020	**5** .008	**4** .003
		13	**6** .045$^+$	**5** .020	**4** .007	**3** .002
		12	**5** .042	**4** .018	**3** .006	**2** .002
		11	**4** .037	**3** .015$^-$	**2** .004	**2** .004
		10	**3** .031	**2** .011	**1** .003	**1** .003
		9	**2** .023	**2** .023	**1** .006	**0** .001
		8	**2** .046	**1** .014	**0** .002	**0** .002
		7	**1** .030	**0** .006	**0** .006	—
		6	**0** .014	**0** .014	—	—
		5	**0** .031	—	—	—

TABLE A.10   (continued)

		Probability			
	a	0.05	0.025	0.01	0.005
A = 18 B = 15	18	**11** .033	**10** .013	**9** .005⁻	**9** .005⁻
	17	**9** .023	**9** .023	**8** .009	**7** .003
	16	**8** .029	**7** .012	**6** .004	**6** .004
	15	**7** .031	**6** .013	**5** .005⁻	**5** .005⁻
	14	**6** .031	**5** .013	**4** .004	**4** .004
	13	**5** .029	**4** .011	**3** .004	**3** .004
	12	**4** .025⁺	**3** .009	**3** .009	**2** .003
	11	**3** .020	**3** .020	**2** .006	**1** .001
	10	**3** .041	**2** .014	**1** .004	**1** .004
	9	**2** .030	**1** .008	**1** .008	**0** .001
	8	**1** .018	**1** .018	**0** .003	**0** .003
	7	**1** .038	**0** .007	**0** .007	—
	6	**0** .017	**0** .017	—	—
	5	**0** .036	—	—	—
14	18	**10** .028	**9** .010⁻	**9** .010⁻	**8** .003
	17	**9** .043	**8** .017	**7** .006	**6** .002
	16	**8** .050⁻	**7** .021	**6** .008	**5** .003
	15	**6** .022	**6** .022	**5** .008	**4** .003
	14	**6** .049	**5** .020	**4** .007	**3** .002
	13	**5** .044	**4** .017	**3** .006	**2** .001
	12	**4** .037	**3** .013	**2** .004	**2** .004
	11	**3** .028	**2** .009	**2** .009	**1** .002
	10	**2** .020	**2** .020	**1** .005⁻	**1** .005⁻
	9	**2** .039	**1** .011	**0** .002	**0** .002
	8	**1** .024	**1** .024	**0** .004	**0** .004
	7	**1** .047	**0** .009	**0** .009	—
	6	**0** .020	**0** .020	—	—
	5	**0** .043	—	—	—
13	18	**9** .023	**9** .023	**8** .008	**7** .002
	17	**8** .034	**7** .012	**6** .004	**6** .004
	16	**7** .037	**6** .014	**5** .005⁻	**5** .005⁻
	15	**6** .036	**5** .014	**4** .004	**4** .004
	14	**5** .032	**4** .012	**3** .004	**3** .004
	13	**4** .027	**3** .009	**3** .009	**2** .002
	12	**3** .020	**3** .020	**2** .006	**1** .001
	11	**3** .040	**2** .013	**1** .003	**1** .003
	10	**2** .027	**1** .007	**1** .007	**0** .001
	9	**1** .015⁺	**1** .015⁺	**0** .002	**0** .002
	8	**1** .031	**0** .006	**0** .006	—
	7	**0** .012	**0** .012	—	—
	6	**0** .025⁺	—	—	—

TABLE A.10   (continued)

		Probability			
	a	0.05	0.025	0.01	0.005
A = 18 B = 12	18	8 .018	8 .018	7 .006	6 .002
	17	7 .026	6 .009	6 .009	5 .003
	16	6 .027	5 .009	5 .009	4 .003
	15	5 .024	5 .024	4 .008	3 .002
	14	4 .020	4 .020	3 .006	2 .001
	13	4 .042	3 .014	2 .004	2 .004
	12	3 .030	2 .009	2 .009	1 .002
	11	2 .019	2 .019	1 .005$^-$	1 .005$^-$
	10	2 .038	1 .010$^+$	0 .001	0 .001
	9	1 .021	1 .021	0 .003	0 .003
	8	1 .040	0 .007	0 .007	—
	7	0 .016	0 .016	—	—
	6	0 .031	—	—	—
11	18	8 .045$^+$	7 .014	6 .004	6 .004
	17	6 .018	6 .018	5 .006	4 .001
	16	5 .018	5 .018	4 .005$^+$	3 .001
	15	5 .043	4 .015$^-$	3 .004	3 .004
	14	4 .033	3 .011	2 .003	2 .003
	13	3 .023	3 .023	2 .007	1 .001
	12	3 .046	2 .014	1 .003	1 .003
	11	2 .029	1 .007	1 .007	0 .001
	10	1 .015$^-$	1 .015$^-$	0 .002	0 .002
	9	1 .029	0 .005$^-$	0 .005$^-$	0 .005$^-$
	8	0 .010$^+$	0 .010$^+$	—	—
	7	0 .020	0 .020	—	—
	6	0 .039	—	—	—
10	18	7 .037	6 .010$^+$	5 .003	5 .003
	17	6 .041	5 .013	4 .003	4 .003
	16	5 .036	4 .011	3 .003	3 .003
	15	4 .028	3 .008	3 .008	2 .002
	14	3 .019	3 .019	2 .005$^-$	2 .005$^-$
	13	3 .039	2 .011	1 .002	1 .002
	12	2 .023	2 .023	1 .005$^+$	0 .001
	11	2 .043	1 .011	0 .001	0 .001
	10	1 .022	1 .022	0 .003	0 .003
	9	1 .040	0 .007	0 .007	—
	8	0 .014	0 .014	—	—
	7	0 .027	—	—	—
	6	0 .049	—	—	—
9	18	6 .029	5 .007	5 .007	4 .002
	17	5 .030	4 .008	4 .008	3 .002
	16	4 .023	4 .023	3 .006	2 .001
	15	3 .016	3 .016	2 .004	2 .004
	14	3 .034	2 .009	2 .009	1 .002
	13	2 .019	2 .019	1 .004	1 .004
	12	2 .037	1 .009	1 .009	0 .001
	11	1 .018	1 .018	0 .002	0 .002
	10	1 .033	0 .005$^+$	0 .005$^+$	—
	9	0 .010$^+$	0 .010$^+$	—	—
	8	0 .020	0 .020	—	—
	7	0 .036	—	—	—

TABLE A.10   (continued)

		Probability			
	a	0.05	0.025	0.01	0.005
$A = 18$ $B = 8$	18	**5**.022	**5**.022	**4**.005⁻	**4**.005⁻
	17	**4**.020	**4**.020	**3**.004	**3**.004
	16	**3**.014	**3**.014	**2**.003	**2**.003
	15	**3**.032	**2**.008	**2**.008	**1**.001
	14	**2**.017	**2**.017	**1**.003	**1**.003
	13	**2**.034	**1**.007	**1**.007	**0**.001
	12	**1**.015⁺	**1**.015⁺	**0**.002	**0**.002
	11	**1**.028	**0**.004	**0**.004	**0**.004
	10	**1**.049	**0**.008	**0**.008	—
	9	**0**.016	**0**.016	—	—
	8	**0**.028	—	—	—
	7	**0**.048	—	—	—
7	18	**4**.015⁺	**4**.015⁺	**3**.003	**3**.003
	17	**3**.012	**3**.012	**2**.002	**2**.002
	16	**3**.032	**2**.007	**2**.007	**1**.001
	15	**2**.017	**2**.017	**1**.003	**1**.003
	14	**2**.034	**1**.007	**1**.007	**0**.001
	13	**1**.014	**1**.014	**0**.002	**0**.002
	12	**1**.027	**0**.004	**0**.004	**0**.004
	11	**1**.046	**0**.007	**0**.007	—
	10	**0**.013	**0**.013	—	—
	9	**0**.024	**0**.024	—	—
	8	**0**.040	—	—	—
6	18	**3**.010⁻	**3**.010⁻	**3**.010⁻	**2**.001
	17	**3**.035⁺	**2**.006	**2**.006	**1**.001
	16	**2**.018	**2**.018	**1**.003	**1**.003
	15	**2**.038	**1**.007	**1**.007	**0**.001
	14	**1**.015⁻	**1**.015⁻	**0**.002	**0**.002
	13	**1**.028	**0**.003	**0**.003	**0**.003
	12	**1**.048	**0**.007	**0**.007	—
	11	**0**.013	**0**.013	—	—
	10	**0**.022	**0**.022	—	—
	9	**0**.037	—	—	—
5	18	**3**.040	**2**.006	**2**.006	**1**.001
	17	**2**.021	**2**.021	**1**.003	**1**.003
	16	**2**.048	**1**.008	**1**.008	**0**.001
	15	**1**.017	**1**.017	**0**.002	**0**.002
	14	**1**.033	**0**.004	**0**.004	**0**.004
	13	**0**.007	**0**.007	**0**.007	—
	12	**0**.014	**0**.014	—	—
	11	**0**.024	**0**.024	—	—
	10	**0**.038	—	—	—
4	18	**2**.026	**1**.003	**1**.003	**1**.003
	17	**1**.010⁻	**1**.010⁻	**1**.010⁻	**0**.001
	16	**1**.024	**1**.024	**0**.002	**0**.002
	15	**1**.046	**0**.005⁻	**0**.005⁻	**0**.005⁻
	14	**0**.010⁻	**0**.010⁻	**0**.010⁻	—
	13	**0**.017	**0**.017	—	—
	12	**0**.029	—	—	—
	11	**0**.045⁺	—	—	—

TABLE A.10 (continued)

		Probability			
	a	0.05	0.025	0.01	0.005
A = 18 B = 3	18	**1** .014	**1** .014	**0** .001	**0** .001
	17	**1** .041	**0** .003	**0** .003	**0** .003
	16	**0** .008	**0** .008	**0** .008	—
	15	**0** .015⁺	**0** .015⁺	—	—
	14	**0** .026	—	—	—
	13	**0** .042	—	—	—
2	18	**0** .005⁺	**0** .005⁺	**0** .005⁺	—
	17	**0** .016	**0** .016	—	—
	16	**0** .032	—	—	—
A = 19 B = 19	19	**14** .023	**14** .023	**13** .010⁻	**12** .004
	18	**13** .045⁻	**12** .021	**11** .009	**10** .004
	17	**11** .031	**10** .015⁻	**9** .006	**8** .003
	16	**10** .039	**9** .019	**8** .009	**7** .003
	15	**9** .046	**8** .022	**6** .004	**6** .004
	14	**8** .050⁻	**7** .024	**5** .004	**5** .004
	13	**6** .025⁺	**5** .011	**4** .004	**4** .004
	12	**5** .024	**5** .024	**3** .003	**3** .003
	11	**5** .050⁻	**4** .022	**3** .009	**2** .003
	10	**4** .046	**3** .019	**2** .006	**1** .002
	9	**3** .039	**2** .015⁻	**1** .004	**1** .004
	8	**2** .031	**1** .009	**1** .009	**0** .002
	7	**1** .021	**1** .021	**0** .004	**0** .004
	6	**1** .045⁻	**0** .010⁻	**0** .010⁻	—
	5	**0** .023	**0** .023	—	—
18	19	**14** .046	**13** .020	**12** .008	**11** .003
	18	**12** .037	**11** .017	**10** .007	**9** .003
	17	**10** .024	**10** .024	**8** .004	**8** .004
	16	**9** .030	**8** .014	**7** .006	**6** .002
	15	**8** .033	**7** .015⁺	**6** .006	**5** .002
	14	**7** .035⁺	**6** .016	**5** .006	**4** .002
	13	**6** .035⁻	**5** .015⁺	**4** .006	**3** .002
	12	**5** .033	**4** .014	**3** .005⁻	**3** .005⁻
	11	**4** .030	**3** .011	**2** .004	**2** .004
	10	**3** .025⁻	**3** .025⁻	**2** .008	**1** .002
	9	**3** .049	**2** .019	**1** .005⁺	**0** .001
	8	**2** .038	**1** .012	**0** .002	**0** .002
	7	**1** .025⁺	**0** .005⁻	**0** .005⁻	**0** .005⁻
	6	**0** .012	**0** .012	—	—
	5	**0** .027	—	—	—
17	19	**13** .040	**12** .016	**11** .006	**10** .002
	18	**11** .030	**10** .013	**9** .005⁺	**8** .002
	17	**10** .040	**9** .018	**8** .008	**7** .003
	16	**9** .047	**8** .022	**7** .009	**6** .003
	15	**8** .050⁻	**7** .023	**6** .010⁻	**5** .004
	14	**6** .023	**6** .023	**5** .010⁻	**4** .003
	13	**6** .049	**5** .022	**4** .008	**3** .003

TABLE A.10   (continued)

		Probability			
	a	0.05	0.025	0.01	0.005
A = 19 B = 17	12	5 .045⁻	4 .019	3 .007	2 .002
	11	4 .039	3 .015⁺	2 .005⁻	2 .005⁻
	10	3 .032	2 .011	1 .003	1 .003
	9	2 .024	2 .024	1 .007	0 .001
	8	2 .047	1 .015⁻	0 .002	0 .002
	7	1 .031	0 .006	0 .006	—
	6	0 .014	0 .014	—	—
	5	0 .031	—	—	—
16	19	12 .035⁻	11 .013	10 .005⁻	10 .005⁻
	18	10 .024	10 .024	9 .010⁻	8 .004
	17	9 .031	8 .013	7 .005⁺	6 .002
	16	8 .035⁻	7 .015⁺	6 .006	5 .002
	15	7 .036	6 .015⁺	5 .006	4 .002
	14	6 .034	5 .014	4 .005⁺	3 .002
	13	5 .031	4 .013	3 .004	3 .004
	12	4 .027	3 .010⁻	3 .010⁻	2 .003
	11	3 .021	3 .021	2 .007	1 .002
	10	3 .042	2 .015⁻	1 .004	1 .004
	9	2 .030	1 .009	1 .009	0 .001
	8	1 .018	1 .018	0 .003	0 .003
	7	1 .037	0 .007	0 .007	—
	6	0 .017	0 .017	—	—
	5	0 .036	—	—	—
15	19	11 .029	10 .011	9 .004	9 .004
	18	10 .046	9 .019	8 .007	7 .002
	17	8 .023	8 .023	7 .009	6 .003
	16	7 .025⁻	7 .025⁻	6 .010⁻	5 .003
	15	6 .024	6 .024	5 .009	4 .003
	14	5 .022	5 .022	4 .008	3 .002
	13	5 .045⁺	4 .018	3 .006	2 .002
	12	4 .037	3 .014	2 .004	2 .004
	11	3 .029	2 .009	2 .009	1 .002
	10	2 .020	2 .020	1 .005ˡ	0 .001
	9	2 .039	1 .011	0 .002	0 .002
	8	1 .023	1 .023	0 .004	0 .004
	7	1 .046	0 .009	0 .009	—
	6	0 .020	0 .020	—	—
	5	0 .042	—	—	—
14	19	10 .024	10 .024	9 .008	8 .003
	18	9 .037	8 .014	7 .005⁻	7 .005⁻
	17	8 .042	7 .017	6 .006	5 .002
	16	7 .042	6 .017	5 .006	4 .002
	15	6 .039	5 .015⁺	4 .005⁺	3 .001
	14	5 .034	4 .013	3 .004	3 .004
	13	4 .027	3 .009	3 .009	2 .003
	12	3 .020	3 .020	2 .006	1 .001
	11	3 .040	2 .013	1 .003	1 .003
	10	2 .027	1 .007	1 .007	0 .001
	9	1 .015⁻	1 .015⁻	0 .002	0 .002
	8	1 .030	0 .005⁺	0 .005⁺	—
	7	0 .012	0 .012	—	—
	6	0 .024	0 .024	—	—
	5	0 .049	—	—	—

TABLE A.10 (continued)

		Probability			
	a	0.05	0.025	0.01	0.005
A = 19 B = 13	19	**9** .020	**9** .020	**8** .006	**7** .002
	18	**8** .029	**7** .010⁺	**6** .003	**6** .003
	17	**7** .031	**6** .011	**5** .004	**5** .004
	16	**6** .029	**5** .011	**4** .003	**4** .003
	15	**5** .025⁺	**4** .009	**4** .009	**3** .003
	14	**4** .020	**4** .020	**3** .006	**2** .002
	13	**4** .041	**3** .015⁻	**2** .004	**2** .004
	12	**3** .029	**2** .009	**2** .009	**1** .002
	11	**2** .019	**2** .019	**1** .005⁻	**1** .005⁻
	10	**2** .036	**1** .010⁻	**1** .010⁻	**0** .001
	9	**1** .020	**1** .020	**0** .003	**0** .003
	8	**1** .038	**0** .007	**0** .007	—
	7	**0** .015⁻	**0** .015⁻	—	—
	6	**0** .030	—	—	—
12	19	**9** .049	**8** .016	**7** .005⁻	**7** .005⁻
	18	**7** .022	**7** .022	**6** .007	**5** .002
	17	**6** .022	**6** .022	**5** .007	**4** .002
	16	**5** .019	**5** .019	**4** .006	**3** .002
	15	**5** .042	**4** .015⁺	**3** .004	**3** .004
	14	**4** .032	**3** .011	**2** .003	**2** .003
	13	**3** .023	**3** .023	**2** .006	**1** .001
	12	**3** .043	**2** .014	**1** .003	**1** .003
	11	**2** .027	**1** .007	**1** .007	**0** .001
	10	**2** .050⁻	**1** .014	**0** .002	**0** .002
	9	**1** .027	**0** .005⁻	**0** .005⁻	**0** .005⁻
	8	**1** .050⁻	**0** .010⁻	**0** .010⁻	—
	7	**0** .019	**0** .019	—	—
	6	**0** .037	—	—	—
11	19	**8** .041	**7** .012	**6** .003	**6** .003
	18	**7** .047	**6** .016	**5** .004	**5** .004
	17	**6** .043	**5** .015⁻	**4** .004	**4** .004
	16	**5** .035⁺	**4** .012	**3** .003	**3** .003
	15	**4** .027	**3** .008	**3** .008	**2** .002
	14	**3** .018	**3** .018	**2** .005⁻	**2** .005⁻
	13	**3** .035⁺	**2** .010⁺	**1** .002	**1** .002
	12	**2** .021	**2** .021	**1** .005⁻	**1** .005⁻
	11	**2** .040	**1** .010⁺	**0** .001	**0** .001
	10	**1** .020	**1** .020	**0** .003	**0** .003
	9	**1** .037	**0** .006	**0** .006	—
	8	**0** .013	**0** .013	—	—
	7	**0** .025⁻	**0** .025⁻	—	—
	6	**0** .046	—	—	—

TABLE A.10 (continued)

	a	0.05	0.025	0.01	0.005
A = 19 B = 10	19	7 .033	6 .009	6 .009	5 .002
	18	6 .036	5 .011	4 .003	4 .003
	17	5 .030	4 .009	4 .009	3 .002
	16	4 .022	4 .022	3 .006	2 .001
	15	4 .047	3 .015$^-$	2 .004	2 .004
	14	3 .030	2 .008	2 .008	1 .002
	13	2 .017	2 .017	1 .004	1 .004
	12	2 .033	1 .008	1 .008	0 .001
	11	1 .016	1 .016	0 .002	0 .002
	10	1 .029	0 .005$^-$	0 .005$^-$	0 .005$^-$
	9	0 .009	0 .009	0 .009	—
	8	0 .018	0 .018	—	—
	7	0 .032	—	—	—
9	19	6 .026	5 .006	5 .006	4 .001
	18	5 .026	4 .007	4 .007	3 .001
	17	4 .020	4 .020	3 .005$^-$	3 .005$^-$
	16	4 .044	3 .013	2 .003	2 .003
	15	3 .028	2 .007	2 .007	1 .001
	14	2 .015$^-$	2 .015$^-$	1 .003	1 .003
	13	2 .029	1 .006	1 .006	0 .001
	12	1 .013	1 .013	0 .002	0 .002
	11	1 .024	1 .024	0 .004	0 .004
	10	1 .042	0 .007	0 .007	—
	9	0 .013	0 .013	—	—
	8	0 .024	0 .024	—	—
	7	0 .043	—	—	—
8	19	5 .019	5 .019	4 .004	4 .004
	18	4 .017	4 .017	3 .004	3 .004
	17	4 .044	3 .011	2 .002	2 .002
	16	3 .027	2 .006	2 .006	1 .001
	15	2 .013	2 .013	1 .002	1 .002
	14	2 .027	1 .006	1 .006	0 .001
	13	2 .049	1 .011	0 .001	0 .001
	12	1 .021	1 .021	0 .003	0 .003
	11	1 .038	0 .006	0 .006	—
	10	0 .011	0 .011	—	—
	9	0 .020	0 .020	—	—
	8	0 .034	—	—	—
7	19	4 .013	4 .013	3 .002	3 .002
	18	4 .047	3 .010$^+$	2 .002	2 .002
	17	3 .028	2 .006	2 .006	1 .001
	16	2 .014	2 .014	1 .002	1 .002
	15	2 .028	1 .005$^+$	1 .005$^+$	0 .001
	14	1 .011	1 .011	0 .001	0 .001
	13	1 .021	1 .021	0 .003	0 .003
	12	1 .037	0 .005$^+$	0 .005$^+$	—
	11	0 .010$^-$	0 .010$^-$	0 .010$^-$	—
	10	0 .017	0 .017	—	—
	9	0 .030	—	—	—
	8	0 .048	—	—	—

TABLE A.10   (*continued*)

		Probability			
	a	0.05	0.025	0.01	0.005
A = 19 B = 6	19	**4**.050⁻	**3**.009	**3**.009	**2**.001
	18	**3**.031	**2**.005⁺	**2**.005⁺	**1**.001
	17	**2**.015⁺	**2**.015⁺	**1**.002	**1**.002
	16	**2**.032	**1**.006	**1**.006	**0**.000
	15	**1**.012	**1**.012	**0**.001	**0**.001
	14	**1**.023	**1**.023	**0**.003	**0**.003
	13	**1**.039	**0**.005⁺	**0**.005⁺	—
	12	**0**.010⁻	**0**.010⁻	**0**.010⁻	—
	11	**0**.017	**0**.017	—	—
	10	**0**.028	—	—	—
	9	**0**.045⁺	—	—	—
5	19	**3**.036	**2**.005⁻	**2**.005⁻	**2**.005⁻
	18	**2**.018	**2**.018	**1**.002	**1**.002
	17	**2**.042	**1**.006	**1**.006	**0**.000
	16	**1**.014	**1**.014	**0**.001	**0**.001
	15	**1**.028	**0**.003	**0**.003	**0**.003
	14	**1**.047	**0**.006	**0**.006	—
	13	**0**.011	**0**.011	—	—
	12	**0**.019	**0**.019	—	—
	11	**0**.030	—	—	—
	10	**0**.047	—	—	—
4	19	**2**.024	**2**.024	**1**.002	**1**.002
	18	**1**.009	**1**.009	**1**.009	**0**.001
	17	**1**.021	**1**.021	**0**.002	**0**.002
	16	**1**.040	**0**.004	**0**.004	**0**.004
	15	**0**.008	**0**.008	**0**.008	—
	14	**0**.014	**0**.014	—	—
	13	**0**.024	**0**.024	—	—
	12	**0**.037	—	—	—
3	19	**1**.013	**1**.013	**0**.001	**0**.001
	18	**1**.038	**0**.003	**0**.003	**0**.003
	17	**0**.006	**0**.006	**0**.006	—
	16	**0**.013	**0**.013	—	—
	15	**0**.023	**0**.023	—	—
	14	**0**.036	—	—	—
2	19	**0**.005⁻	**0**.005⁻	**0**.005⁻	**0**.005⁻
	18	**0**.014	**0**.014	—	—
	17	**0**.029	—	—	—
	16	**0**.048	—	—	—

TABLE A.10   (continued)

		Probability			
	a	0.05	0.025	0.01	0.005
$A = 20\ B = 20$	20	15 .024	15 .024	13 .004	13 .004
	19	14 .046	13 .022	12 .010⁻	11 .004
	18	12 .032	11 .015⁺	10 .007	9 .003
	17	11 .041	10 .020	9 .009	8 .004
	16	10 .048	9 .024	7 .005⁻	7 .005⁻
	15	8 .027	7 .012	6 .005⁺	5 .002
	14	7 .028	6 .013	5 .005⁺	4 .002
	13	6 .028	5 .012	4 .005⁻	4 .005⁻
	12	5 .027	4 .011	3 .004	3 .004
	11	4 .024	4 .024	3 .009	2 .003
	10	4 .048	3 .020	2 .007	1 .002
	9	3 .041	2 .015⁺	1 .004	1 .004
	8	2 .032	1 .010⁻	1 .010⁻	0 .002
	7	1 .022	1 .022	0 .004	0 .004
	6	1 .046	0 .010⁺	—	—
	5	0 .024	0 .024	—	—
19	20	15 .047	14 .020	13 .008	12 .003
	19	13 .039	12 .018	11 .008	10 .003
	18	11 .026	10 .012	9 .005⁻	9 .005⁻
	17	10 .032	9 .015⁻	8 .006	7 .002
	16	9 .036	8 .017	7 .007	6 .003
	15	8 .038	7 .018	6 .008	5 .003
	14	7 .039	6 .018	5 .007	4 .003
	13	6 .038	5 .017	4 .007	3 .002
	12	5 .035⁺	4 .015⁺	3 .005⁺	2 .002
	11	4 .031	3 .012	2 .004	2 .004
	10	3 .026	2 .009	2 .009	1 .002
	9	2 .019	2 .019	1 .005⁺	0 .001
	8	2 .039	1 .012	0 .002	0 .002
	7	1 .026	0 .005⁺	0 .005⁺	—
	6	0 .012	0 .012	—	—
	5	0 .027	—	—	—
18	20	14 .041	13 .017	12 .007	11 .003
	19	12 .032	11 .014	10 .006	9 .002
	18	11 .043	10 .020	9 .008	8 .003
	17	10 .050⁻	9 .024	7 .004	7 .004
	16	8 .026	7 .011	6 .005⁻	6 .005⁻
	15	7 .027	6 .012	5 .004	5 .004
	14	6 .026	5 .011	4 .004	4 .004
	13	5 .024	5 .024	4 .009	3 .003
	12	5 .047	4 .020	3 .007	2 .002
	11	4 .041	3 .016	2 .005⁺	1 .001
	10	3 .033	2 .012	1 .003	1 .003
	9	2 .024	2 .024	1 .007	0 .001
	8	2 .048	1 .015⁻	0 .003	0 .003
	7	1 .031	0 .006	0 .006	—
	6	0 .014	0 .014	—	—
	5	0 .031	—	—	—

TABLE A.10   (*continued*)

		Probability			
	a	0.05	0.025	0.01	0.005
A = 20 B = 17	20	**13** .036	**12** .014	**11** .005 $^+$	**10** .002
	19	**11** .026	**10** .011	**9** .004	**9** .004
	18	**10** .034	**9** .015 $^-$	**8** .006	**7** .002
	17	**9** .038	**8** .017	**7** .007	**6** .003
	16	**8** .040	**7** .018	**6** .007	**5** .003
	15	**7** .039	**6** .017	**5** .007	**4** .002
	14	**6** .037	**5** .016	**4** .006	**3** .002
	13	**5** .033	**4** .013	**3** .005 $^-$	**3** .005 $^-$
	12	**4** .028	**3** .010 $^+$	**2** .003	**2** .003
	11	**3** .022	**3** .022	**2** .007	**1** .002
	10	**3** .042	**2** .015 $^+$	**1** .004	**1** .004
	9	**2** .031	**1** .009	**1** .009	**0** .001
	8	**1** .019	**1** .019	**0** .003	**0** .003
	7	**1** .037	**0** .008	**0** .008	—
	6	**0** .017	**0** .017	—	—
	5	**0** .036	—	—	—
16	20	**12** .031	**11** .012	**10** .004	**10** .004
	19	**11** .049	**10** .021	**9** .008	**8** .003
	18	**9** .026	**8** .011	**7** .004	**7** .004
	17	**8** .028	**7** .012	**6** .004	**6** .004
	16	**7** .028	**6** .012	**5** .004	**5** .004
	15	**6** .026	**5** .011	**4** .004	**4** .004
	14	**5** .023	**5** .023	**4** .009	**3** .003
	13	**5** .046	**4** .019	**3** .007	**2** .002
	12	**4** .038	**3** .014	**2** .004	**2** .004
	11	**3** .029	**2** .010 $^-$	**2** .010 $^-$	**1** .002
	10	**2** .020	**2** .020	**1** .005 $^+$	**0** .001
	9	**2** .039	**1** .011	**0** .002	**0** .002
	8	**1** .023	**1** .023	**0** .004	**0** .004
	7	**1** .045 $^+$	**0** .009	**0** .009	—
	6	**0** .020	**0** .020	—	—
	5	**0** .041	—	—	—
15	20	**11** .026	**10** .009	**10** .009	**9** .003
	19	**10** .040	**9** .016	**8** .006	**7** .002
	18	**9** .046	**8** .019	**7** .007	**6** .002
	17	**8** .047	**7** .020	**6** .008	**5** .002
	16	**7** .045 $^-$	**6** .019	**5** .007	**4** .002
	15	**6** .040	**5** .017	**4** .006	**3** .002
	14	**5** .034	**4** .013	**3** .004	**3** .004
	13	**4** .028	**3** .010 $^-$	**3** .010 $^-$	**2** .003
	12	**3** .020	**3** .020	**2** .006	**1** .001
	11	**3** .039	**2** .013	**1** .003	**1** .003
	10	**2** .026	**1** .007	**1** .007	**0** .001
	9	**2** .049	**1** .015 $^-$	**0** .002	**0** .002
	8	**1** .029	**0** .005 $^+$	**0** .005 $^+$	—
	7	**0** .012	**0** .012	—	—
	6	**0** .024	**0** .024	—	—
	5	**0** .048	—	—	—

TABLE A.10   (*continued*)

	a	0.05	0.025	0.01	0.005
		Probability			
A = 20 *B* = 14	20	**10** .022	**10** .022	**9** .007	**8** .002
	19	**9** .032	**8** .012	**7** .004	**7** .004
	18	**8** .035⁺	**7** .014	**6** .005⁻	**6** .005⁻
	17	**7** .035⁻	**6** .013	**5** .005⁻	**5** .005⁻
	16	**6** .031	**5** .012	**4** .004	**4** .004
	15	**5** .026	**4** .009	**4** .009	**3** .003
	14	**4** .020	**4** .020	**3** .007	**2** .002
	13	**4** .040	**3** .015⁻	**2** .004	**2** .004
	12	**3** .029	**2** .009	**2** .009	**1** .002
	11	**2** .018	**2** .018	**1** .005⁻	**1** .005⁻
	10	**2** .035⁺	**1** .010⁻	**1** .010⁻	**0** .001
	9	**1** .019	**1** .019	**0** .003	**0** .003
	8	**1** .037	**0** .007	**0** .007	—
	7	**0** .014	**0** .014	—	—
	6	**0** .029	—	—	—
13	20	**9** .017	**9** .017	**8** .005⁺	**7** .002
	19	**8** .025⁻	**8** .025⁻	**7** .008	**6** .003
	18	**7** .026	**6** .009	**6** .009	**5** .003
	17	**6** .024	**6** .024	**5** .008	**4** .002
	16	**5** .020	**5** .020	**4** .007	**3** .002
	15	**5** .041	**4** .015⁺	**3** .005⁻	**3** .005⁻
	14	**4** .031	**3** .011	**2** .003	**2** .003
	13	**3** .022	**3** .022	**2** .006	**1** .001
	12	**3** .041	**2** .013	**1** .003	**1** .003
	11	**2** .026	**1** .007	**1** .007	**0** .001
	10	**2** .047	**1** .013	**0** .002	**0** .002
	9	**1** .026	**0** .004	**0** .004	**0** .004
	8	**1** .047	**0** .009	**0** .009	—
	7	**0** .018	**0** .018	—	—
	6	**0** .035⁻	—	—	—
12	20	**9** .044	**8** .014	**7** .004	**7** .004
	19	**7** .019	**7** .019	**6** .006	**5** .002
	18	**6** .018	**6** .018	**5** .006	**4** .002
	17	**6** .043	**5** .016	**4** .005⁻	**4** .005⁻
	16	**5** .034	**4** .012	**3** .003	**3** .003
	15	**4** .025⁺	**3** .008	**3** .008	**2** .002
	14	**4** .049	**3** .017	**2** .005⁻	**2** .005⁻
	13	**3** .033	**2** .010⁻	**2** .010⁻	**1** .002
	12	**2** .020	**2** .020	**1** .005⁻	**1** .005⁻
	11	**2** .036	**1** .009	**1** .009	**0** .001
	10	**1** .018	**1** .018	**0** .003	**0** .003
	9	**1** .034	**0** .006	**0** .006	—
	8	**0** .012	**0** .012	—	—
	7	**0** .023	**0** .023	—	—
	6	**0** .043	—	—	—

TABLE A.10   (*continued*)

		Probability			
	a	**0.05**	**0.025**	**0.01**	**0.005**
A = 20 B = 11	20	8 .037	7 .010+	6 .003	6 .003
	19	7 .042	6 .013	5 .004	5 .004
	18	6 .037	5 .012	4 .003	4 .003
	17	5 .029	4 .009	4 .009	3 .002
	16	4 .021	4 .021	3 .006	2 .001
	15	4 .042	3 .014	2 .003	2 .003
	14	3 .028	2 .008	2 .008	1 .001
	13	2 .016	2 .016	1 .003	1 .003
	12	2 .029	1 .007	1 .007	0 .001
	11	1 .014	1 .014	0 .002	0 .002
	10	1 .026	0 .004	0 .004	0 .004
	9	1 .046	0 .008	0 .008	—
	8	0 .016	0 .016	—	—
	7	0 .029	—	—	—
10	20	7 .030	6 .008	6 .008	5 .002
	19	6 .031	5 .009	5 .009	4 .002
	18	5 .026	4 .007	4 .007	3 .002
	17	4 .018	4 .018	3 .005−	3 .005−
	16	4 .039	3 .012	2 .003	2 .003
	15	3 .024	3 .024	2 .006	1 .001
	14	3 .045+	2 .013	1 .003	1 .003
	13	2 .025+	1 .006	1 .006	0 .001
	12	2 .045−	1 .011	0 .001	0 .001
	11	1 .021	1 .021	0 .003	0 .003
	10	1 .037	0 .006	0 .006	—
	9	0 .012	0 .012	—	—
	8	0 .022	0 .022	—	—
	7	0 .038	—	—	—
9	20	6 .023	6 .023	5 .005+	4 .001
	19	5 .022	5 .022	4 .005+	3 .001
	18	4 .016	4 .016	3 .004	3 .004
	17	4 .037	3 .010+	2 .002	2 .002
	16	3 .022	3 .022	2 .005+	1 .001
	15	3 .043	2 .012	1 .002	1 .002
	14	2 .023	2 .023	1 .005−	1 .005−
	13	2 .041	1 .009	1 .009	0 .001
	12	1 .018	1 .018	0 .002	0 .002
	11	1 .032	0 .005−	0 .005−	0 .005−
	10	0 .009	0 .009	0 .009	—
	9	0 .017	0 .017	—	—
	8	0 .029	—	—	—
	7	0 .050−	—	—	—

TABLE A.10   (continued)

	a	Probability			
		0.05	0.025	0.01	0.005
A = 20 B = 8	20	**5** .017	**5** .017	**4** .003	**4** .003
	19	**4** .015⁻	**4** .015⁻	**3** .003	**3** .003
	18	**4** .038	**3** .009	**3** .009	**2** .002
	17	**3** .022	**3** .022	**2** .005⁻	**2** .005⁻
	16	**3** .044	**2** .011	**1** .002	**1** .002
	15	**2** .022	**2** .022	**1** .004	**1** .004
	14	**2** .040	**1** .009	**1** .009	**0** .001
	13	**1** .016	**1** .016	**0** .002	**0** .002
	12	**1** .029	**0** .004	**0** .004	**0** .004
	11	**1** .048	**0** .008	**0** .008	—
	10	**0** .014	**0** .014	—	—
	9	**0** .024	**0** .024	—	—
	8	**0** .041	—	—	—
7	20	**4** .012	**4** .012	**3** .002	**3** .002
	19	**4** .042	**3** .009	**3** .009	**2** .001
	18	**3** .024	**3** .024	**2** .005⁻	**2** .005⁻
	17	**3** .050⁻	**2** .011	**1** .002	**1** .002
	16	**2** .023	**2** .023	**1** .004	**1** .004
	15	**2** .043	**1** .009	**1** .009	**0** .001
	14	**1** .016	**1** .016	**0** .002	**0** .002
	13	**1** .029	**0** .004	**0** .004	**0** .004
	12	**1** .048	**0** .007	**0** .007	—
	11	**0** .013	**0** .013	—	—
	10	**0** .022	**0** .022	—	—
	9	**0** .036	—	—	—
6	20	**4** .046	**3** .008	**3** .008	**2** .001
	19	**3** .028	**2** .005⁻	**2** .005⁻	**2** .005⁻
	18	**2** .013	**2** .013	**1** .002	**1** .002
	17	**2** .028	**1** .004	**1** .004	**1** .004
	16	**1** .010⁻	**1** .010⁻	**1** .010⁻	**0** .001
	15	**1** .018	**1** .018	**0** .002	**0** .002
	14	**1** .032	**0** .004	**0** .004	**0** .004
	13	**0** .007	**0** .007	**0** .007	—
	12	**0** .013	**0** .013	—	—
	11	**0** .022	**0** .022	—	—
	10	**0** .035⁻	—	—	—
5	20	**3** .033	**2** .004	**2** .004	**2** .004
	19	**2** .016	**2** .016	**1** .002	**1** .002
	18	**2** .038	**1** .005⁺	**1** .005⁺	**0** .000
	17	**1** .012	**1** .012	**0** .001	**0** .001
	16	**1** .023	**1** .023	**0** .002	**0** .002
	15	**1** .040	**0** .005⁻	**0** .005⁻	**0** .005⁻
	14	**0** .009	**0** .009	**0** .009	—
	13	**0** .015⁻	**0** .015⁻	—	—
	12	**0** .024	**0** .024	—	—
	11	**0** .038	—	—	—

TABLE A.10 (*continued*)

	a	**0.05**	**0.025**	**0.01**	**0.005**
		Probability			
A = 20 *B* = 4	20	**2** .022	**2** .022	**1** .002	**1** .002
	19	**1** .008	**1** .008	**1** .008	**0** .000
	18	**1** .018	**1** .018	**0** .001	**0** .001
	17	**1** .035⁺	**0** .003	**0** .003	**0** .003
	16	**0** .007	**0** .007	**0** .007	—
	15	**0** .012	**0** .012	—	—
	14	**0** .020	**0** .020	—	—
	13	**0** .031	—	—	—
	12	**0** .047	—	—	—
3	20	**1** .012	**1** .012	**0** .001	**0** .001
	19	**1** .034	**0** .002	**0** .002	**0** .002
	18	**0** .006	**0** .006	**0** .006	—
	17	**0** .011	**0** .011	—	—
	16	**0** .020	**0** .020	—	—
	15	**0** .032	—	—	—
	14	**0** .047	—	—	—
2	20	**0** .004	**0** .004	**0** .004	**0** .004
	19	**0** .013	**0** .013	—	—
	18	**0** .026	—	—	—
	17	**0** .043	—	—	—
1	20	**0** .048	—	—	—

Source: Entries for *A* = 3, *B* = 3 through *A* = 15, *B* = 2 from Table 38 of E. S. Pearson and H. O. Hartley, *Biometrika Tables for Statisticians*, Volume 1, third edition, London: The Syndics of the Cambridge University Press, 1966. Entries for *A* = 16, *B* = 16 through *A* = 20, *B* = 1 from R. Latscha, "Tests of Significance in a 2 × 2 Contingency Table: Extension of Finney's Table," *Biometrika*, 40 (1953), 74–86; used by permission of the Biometrika Trustees.

TABLE A.11 **Percentiles of the chi-square distribution, $P(X^2 \leq \chi^2)$**

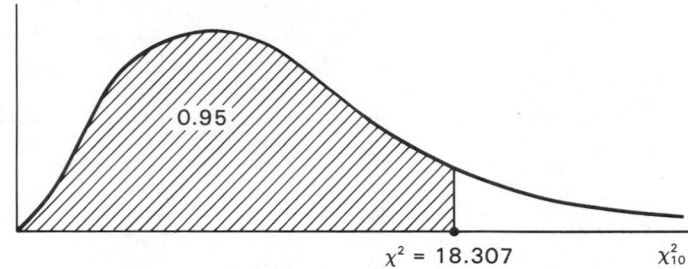

$\chi^2 = 18.307$ χ^2_{10}

df	$\chi^2_{0.005}$	$\chi^2_{0.025}$	$\chi^2_{0.05}$	$\chi^2_{0.90}$	$\chi^2_{0.95}$	$\chi^2_{0.975}$	$\chi^2_{0.99}$	$\chi^2_{0.995}$
1	0.0000393	0.000982	0.00393	2.706	3.841	5.024	6.635	7.879
2	0.0100	0.0506	0.103	4.605	5.991	7.378	9.210	10.597
3	0.0717	0.216	0.352	6.251	7.815	9.348	11.345	12.838
4	0.207	0.484	0.711	7.779	9.488	11.143	13.277	14.860
5	0.412	0.831	1.145	9.236	11.070	12.832	15.086	16.750
6	0.676	1.237	1.635	10.645	12.592	14.449	16.812	18.548
7	0.989	1.690	2.167	12.017	14.067	16.013	18.475	20.278
8	1.344	2.180	2.733	13.362	15.507	17.535	20.090	21.955
9	1.735	2.700	3.325	14.684	16.919	19.023	21.666	23.589
10	2.156	3.247	3.940	15.987	18.307	20.483	23.209	25.188
11	2.603	3.816	4.575	17.275	19.675	21.920	24.725	26.757
12	3.074	4.404	5.226	18.549	21.026	23.336	26.217	28.300
13	3.565	5.009	5.892	19.812	22.362	24.736	27.688	29.819
14	4.075	5.629	6.571	21.064	23.685	26.119	29.141	31.319
15	4.601	6.262	7.261	22.307	24.996	27.488	30.578	32.801
16	5.142	6.908	7.962	23.542	26.296	28.845	32.000	34.267
17	5.697	7.564	8.672	24.769	27.587	30.191	33.409	35.718
18	6.265	8.231	9.390	25.989	28.869	31.526	34.805	37.156
19	6.844	8.907	10.117	27.204	30.144	32.852	36.191	38.582
20	7.434	9.591	10.851	28.412	31.410	34.170	37.566	39.997
21	8.034	10.283	11.591	29.615	32.671	35.479	38.932	41.401
22	8.643	10.982	12.338	30.813	33.924	36.781	40.289	42.796
23	9.260	11.688	13.091	32.007	35.172	38.076	41.638	44.181
24	9.886	12.401	13.848	33.196	36.415	39.364	42.980	45.558
25	10.520	13.120	14.611	34.382	37.652	40.646	44.314	46.928
26	11.160	13.844	15.379	35.563	38.885	41.923	45.642	48.290
27	11.808	14.573	16.151	36.741	40.113	43.194	46.963	49.645
28	12.461	15.308	16.928	37.916	41.337	44.461	48.278	50.993
29	13.121	16.047	17.708	39.087	42.557	45.722	49.588	52.336
30	13.787	16.791	18.493	40.256	43.773	46.979	50.892	53.672
35	17.192	20.569	22.465	46.059	49.802	53.203	57.342	60.275
40	20.707	24.433	26.509	51.805	55.758	59.342	63.691	66.766
45	24.311	28.366	30.612	57.505	61.656	65.410	69.957	73.166
50	27.991	32.357	34.764	63.167	67.505	71.420	76.154	79.490
60	35.535	40.482	43.188	74.397	79.082	83.298	88.379	91.952
70	43.275	48.758	51.739	85.527	90.531	95.023	100.425	104.215
80	51.172	57.153	60.391	96.578	101.879	106.629	112.329	116.321
90	59.196	65.647	69.126	107.565	113.145	118.136	124.116	128.299
100	67.328	74.222	77.929	118.498	124.342	129.561	135.807	140.169

Source: A. Hald and S. A. Sinkbaek, "A Table of Percentage Points of the χ^2 Distribution," *Skandinavisk Aktuarietidskrift*, 33 (1950), 168–175. Used by permission.

TABLE A.12 **Critical values of the Kruskal–Wallis test statistic**

Sample sizes					Sample sizes				
n_1	n_2	n_3	Critical value	α	n_1	n_2	n_3	Critical value	α
2	1	1	2.7000	0.500				4.7000	0.101
2	2	1	3.6000	0.200	4	4	1	6.6667	0.010
2	2	2	4.5714	0.067				6.1667	0.022
			3.7143	0.200				4.9667	0.048
3	1	1	3.2000	0.300				4.8667	0.054
3	2	1	4.2857	0.100				4.1667	0.082
			3.8571	0.133				4.0667	0.102
3	2	2	5.3572	0.029	4	4	2	7.0364	0.006
			4.7143	0.048				6.8727	0.011
			4.5000	0.067				5.4545	0.046
			4.4643	0.105				5.2364	0.052
3	3	1	5.1429	0.043				4.5545	0.098
			4.5714	0.100				4.4455	0.103
			4.0000	0.129	4	4	3	7.1439	0.010
3	3	2	6.2500	0.011				7.1364	0.011
			5.3611	0.032				5.5985	0.049
			5.1389	0.061				5.5758	0.051
			4.5556	0.100				4.5455	0.099
			4.2500	0.121				4.4773	0.102
3	3	3	7.2000	0.004	4	4	4	7.6538	0.008
			6.4889	0.011				7.5385	0.011
			5.6889	0.029				5.6923	0.049
			5.6000	0.050				5.6538	0.054
			5.0667	0.086				4.6539	0.097
			4.6222	0.100				4.5001	0.104
4	1	1	3.5714	0.200	5	1	1	3.8571	0.143
4	2	1	4.8214	0.057	5	2	1	5.2500	0.036
			4.5000	0.076				5.0000	0.048
			4.0179	0.114				4.4500	0.071
4	2	2	6.0000	0.014				4.2000	0.095
			5.3333	0.033				4.0500	0.119
			5.1250	0.052	5	2	2	6.5333	0.008
			4.4583	0.100				6.1333	0.013
			4.1667	0.105				5.1600	0.034
4	3	1	5.8333	0.021				5.0400	0.056
			5.2083	0.050				4.3733	0.090
			5.0000	0.057				4.2933	0.122
			4.0556	0.093	5	3	1	6.4000	0.012
			3.8889	0.129				4.9600	0.048
4	3	2	6.4444	0.008				4.8711	0.052
			6.3000	0.011				4.0178	0.095
			5.4444	0.046				3.8400	0.123
			5.4000	0.051	5	3	2	6.9091	0.009
			4.5111	0.098				6.8218	0.010
			4.4444	0.102				5.2509	0.049
4	3	3	6.7455	0.010				5.1055	0.052
			6.7091	0.013				4.6509	0.091
			5.7909	0.046				4.4945	0.101
			5.7273	0.050	5	3	3	7.0788	0.009
			4.7091	0.092				6.9818	0.011

TABLE A.12   (continued)

Sample sizes					Sample sizes				
n_1	n_2	n_3	Critical value	α	n_1	n_2	n_3	Critical value	α
5	3	3	5.6485	0.049	5	5	1	6.8364	0.011
			5.5152	0.051				5.1273	0.046
			4.5333	0.097				4.9091	0.053
			4.4121	0.109				4.1091	0.086
5	4	1	6.9545	0.008				4.0364	0.105
			6.8400	0.011	5	5	2	7.3385	0.010
			4.9855	0.044				7.2692	0.010
			4.8600	0.056				5.3385	0.047
			3.9873	0.098				5.2462	0.051
			3.9600	0.102				4.6231	0.097
5	4	2	7.2045	0.009				4.5077	0.100
			7.1182	0.010	5	5	3	7.5780	0.010
			5.2727	0.049				7.5429	0.010
			5.2682	0.050				5.7055	0.046
			4.5409	0.098				5.6264	0.051
			4.5182	0.101				4.5451	0.100
5	4	3	7.4449	0.010				4.5363	0.102
			7.3949	0.011	5	5	4	7.8229	0.010
			5.6564	0.049				7.7914	0.010
			5.6308	0.050				5.6657	0.049
			4.5487	0.099				5.6429	0.050
			4.5231	0.103				4.5229	0.099
5	4	4	7.7604	0.009				4.5200	0.101
			7.7440	0.011	5	5	5	8.0000	0.009
			5.6571	0.049				7.9800	0.010
			5.6176	0.050				5.7800	0.049
			4.6187	0.100				5.6600	0.051
			4.5527	0.102				4.5600	0.100
5	5	1	7.3091	0.009				4.5000	0.102

Source: W. H. Kruskal and W. A. Wallis, "Use of Ranks in One-Criterion Analysis of Variance," *J. Amer. Statist. Assoc.*, 47 (1952), 583–621, Addendum, *Ibid.*, 48 (1953), 907–911.

TABLE A.13(a)

Critical values of J, the Jonckheere–Terpstra test statistic (for nominal values of α shown); exact significance levels in parentheses

n_1	n_2	n_3	$\alpha = 0.5$	$\alpha = 0.2$	$\alpha = 0.1$	$\alpha = 0.05$	$\alpha = 0.025$	$\alpha = 0.01$	$\alpha = 0.005$
2	2	2	6 (.57778)	8 (.28889)	9 (.16667)	10 (.08889)	11 (0.3333)	12 (.01111)	12 (.01111)
			7 (.42222)	9 (.16667)	10 (.08889)	11 (.03333)	12 (.01111)	—	—
2	2	3	8 (.56190)	11 (.21905)	12 (.13810)	13 (.07619)	14 (.03810)	15 (.01429)	15 (.01429)
			9 (.43810)	12 (.13810)	13 (.07619)	14 (.03810)	15 (.01429)	16 (.00476)	16 (.00476)
2	2	4	10 (.55238)	13 (.25714)	15 (.11667)	16 (.07143)	17 (.03810)	18 (.01905)	19 (.00714)
			11 (.44762)	14 (.18095)	16 (.07143)	17 (.03810)	18 (.01905)	19 (.00714)	20 (.00238)
2	2	5	12 (.54497)	16 (.21561)	18 (.10450)	19 (.06614)	20 (.03968)	22 (.01058)	22 (.01058)
			13 (.45503)	17 (.15344)	19 (.06614)	20 (.03968)	21 (.02116)	23 (.00397)	23 (.00397)
2	2	6	14 (.53968)	18 (.24444)	20 (.13571)	22 (.06349)	23 (.03968)	25 (.01270)	26 (.00635)
			15 (.46032)	19 (.18492)	21 (.09444)	23 (.03968)	24 (.02381)	26 (.00635)	27 (.00238)
2	2	7	16 (.53535)	21 (.21212)	23 (.12172)	25 (.06061)	27 (.02525)	28 (.01515)	29 (.00808)
			17 (.46465)	22 (.16364)	24 (.08788)	26 (.04040)	28 (.01515)	29 (.00808)	30 (.00404)
2	2	8	18 (.53199)	23 (.23535)	26 (.11178)	28 (.05892)	30 (.02694)	32 (.01010)	33 (.00539)
			19 (.46801)	24 (.18855)	27 (.08215)	29 (.04040)	31 (.01684)	33 (.00539)	34 (.00269)
2	3	3	11 (.50000)	14 (.22143)	15 (.15179)	17 (.05714)	18 (.03036)	19 (.01429)	20 (.00536)
			12 (.40000)	15 (.15179)	16 (.09643)	18 (.03036)	19 (.01429)	20 (.00536)	21 (.00179)
2	3	4	13 (.54286)	17 (.22222)	19 (.11190)	20 (.07381)	22 (.02619)	23 (.01349)	24 (.00635)
			14 (.45714)	18 (.16190)	20 (.07381)	21 (.04524)	23 (.01349)	24 (.00635)	25 (.00238)
2	3	5	16 (.50000)	20 (.22302)	22 (.12421)	24 (.05913)	25 (.03810)	27 (.01310)	28 (.00675)
			17 (.42500)	21 (.16944)	23 (.08770)	25 (.03810)	26 (.02302)	28 (.00675)	29 (.00317)
2	3	6	18 (.53355)	23 (.22338)	25 (.13398)	27 (.07143)	29 (.03290)	31 (.01255)	32 (.00714)
			19 (.46645)	24 (.17554)	26 (.09957)	28 (.04957)	30 (.02100)	32 (.00714)	33 (.00368)
2	3	7	21 (.50000)	26 (.22374)	29 (.10960)	31 (.06023)	33 (.02929)	35 (.01225)	36 (.00732)
			22 (.44003)	27 (.18030)	30 (.08232)	32 (.04268)	34 (.01032)	36 (.00732)	37 (.00417)
2	3	8	23 (.52727)	29 (.22393)	32 (.11826)	35 (.05198)	37 (.02650)	39 (.01189)	40 (.00754)
			24 (.47273)	30 (.18430)	33 (.09192)	36 (.03768)	38 (.01810)	40 (.00754)	41 (.00451)
2	4	4	16 (.53746)	20 (.25587)	23 (.10794)	25 (.05016)	26 (.03206)	28 (.01079)	29 (.00540)
			17 (.46254)	21 (.19810)	24 (.07556)	26 (.03206)	27 (.01905)	29 (.00540)	30 (.00254)
2	4	5	19 (.53261)	24 (.22872)	27 (.10491)	29 (.05397)	30 (.03680)	32 (.01501)	33 (.00880)
			20 (.46739)	25 (.18095)	28 (.07662)	30 (.03680)	31 (.02395)	33 (.00880)	34 (.00491)
2	4	6	22 (.52929)	28 (.20859)	31 (.10245)	33 (.05685)	35 (.02821)	37 (.01219)	38 (.00758)
			23 (.47071)	29 (.16797)	32 (.07742)	34 (.04076)	36 (.01898)	38 (.00758)	39 (.00440)
2	4	7	25 (.52634)	31 (.23209)	35 (.10047)	37 (.05921)	39 (.03193)	42 (.01033)	43 (.00660)
			26 (.47366)	32 (.19305)	36 (.07797)	38 (.04406)	40 (.02261)	43 (.00660)	44 (.00408)
2	4	8	28 (.52410)	35 (.21496)	38 (.12266)	41 (.06112)	44 (.02593)	46 (.01310)	48 (.00593)
			29 (.47590)	36 (.18077)	39 (.09879)	42 (.04686)	45 (.01863)	47 (.00892)	49 (.00377)
2	5	5	23 (.50000)	28 (.23274)	31 (.11935)	34 (.05014)	35 (.03565)	38 (.01046)	39 (.00643)
			24 (.44228)	29 (.19000)	32 (.09157)	35 (.03565)	36 (.02453)	39 (.00643)	40 (.00373)
2	5	6	26 (.52597)	32 (.23596)	36 (.10462)	38 (.06277)	40 (.03469)	43 (.01179)	44 (.00777)
			27 (.47403)	33 (.19708)	37 (.08178)	39 (.04715)	41 (.02486)	44 (.00777)	45 (.00491)
2	5	7	30 (.50000)	37 (.20292)	40 (.11588)	43 (.05821)	46 (.02507)	48 (.01290)	50 (.00601)
			31 (.45303)	38 (.17057)	41 (.09355)	44 (.04477)	47 (.01820)	49 (.00894)	51 (.00393)
2	5	8	33 (.52151)	41 (.20773)	45 (.10400)	48 (.05459)	51 (.02519)	53 (.01383)	55 (.00701)
			34 (.47849)	42 (.17764)	46 (.08500)	49 (.04283)	52 (.01885)	54 (.00996)	56 (.00482)
2	6	6	30 (.52338)	37 (.22198)	41 (.10607)	44 (.05260)	46 (.03031)	49 (.01139)	51 (.00526)
			31 (.47662)	38 (.18816)	42 (.08528)	45 (.04027)	47 (.02235)	50 (.00786)	52 (.00343)

TABLE A.13(a) *(continued)*

n_1	n_2	n_3	$\alpha = 0.5$	$\alpha = 0.2$	$\alpha = 0.1$	$\alpha = 0.05$	$\alpha = 0.025$	$\alpha = 0.01$	$\alpha = 0.005$
2	6	7	34 (.52125)	42 (.21088)	46 (.10721)	49 (.05720)	52 (.02703)	55 (.01103)	57 (.00551)
			35 (.47875)	43 (.18087)	47 (.08803)	50 (.04521)	53 (.02040)	56 (.00789)	58 (.00376)
2	6	8	38 (.51949)	47 (.20176)	51 (.10804)	54 (.06118)	57 (.03135)	61 (.01070)	63 (.00569)
			39 (.48051)	48 (.17491)	52 (.09031)	55 (.04953)	58 (.02449)	62 (.00788)	64 (.00404)
2	7	7	39 (.50000)	47 (.21740)	52 (.10029)	55 (.05628)	58 (.02858)	61 (.01293)	64 (.00509)
			40 (.46130)	48 (.18948)	53 (.08358)	56 (.04543)	59 (.02225)	62 (.00964)	65 (.00360)
2	7	8	43 (.51781)	52 (.22285)	57 (.11128)	61 (.05543)	64 (.02987)	68 (.01127)	70 (.00642)
			44 (.48219)	53 (.19675)	58 (.09468)	62 (.04555)	65 (.02381)	69 (.00857)	71 (.00474)
2	8	8	48 (.51641)	58 (.21616)	63 (.11392)	68 (.05085)	71 (.02858)	75 (.01170)	78 (.00537)
			49 (.48359)	59 (.19248)	64 (.09833)	69 (.04231)	72 (.02319)	76 (.00913)	79 (.00404)
3	3	3	14 (.50000)	17 (.25952)	19 (.13869)	21 (.06131)	22 (.03690)	24 (.01071)	24 (.01071)
			15 (.41548)	18 (.19405)	20 (.09464)	22 (.03690)	23 (.02083)	25 (.00476)	25 (.00476)
3	3	4	17 (.50000)	21 (.22833)	23 (.13000)	25 (.06405)	27 (.02643)	28 (.01548)	29 (.00857)
			18 (.42667)	22 (.17500)	24 (.09310)	26 (.04214)	28 (.01548)	29 (.00857)	30 (.00429)
3	3	5	20 (.50000)	25 (.20584)	27 (.12348)	29 (.06623)	31 (.03106)	33 (.01234)	34 (.00714)
			21 (.43528)	26 (.16147)	28 (.09177)	30 (.04621)	32 (.02002)	34 (.00714)	35 (.00390)
3	3	6	23 (.50000)	28 (.23193)	31 (.11845)	33 (.06791)	35 (.03506)	38 (.01017)	39 (.00622)
			24 (.44210)	29 (.18912)	32 (.09075)	34 (.04946)	36 (.02408)	39 (.00622)	40 (.00357)
3	3	7	26 (.50000)	32 (.21323)	35 (.11451)	38 (.05219)	40 (.02768)	42 (.01320)	44 (.00551)
			27 (.44761)	33 (.17619)	36 (.08989)	39 (.03849)	41 (.01941)	43 (.00868)	45 (.00335)
3	3	8	29 (.50000)	35 (.23428)	39 (.11131)	42 (.05446)	44 (.03092)	47 (.01122)	48 (.00759)
			30 (.45216)	36 (.19843)	40 (.08914)	43 (.04144)	45 (.02259)	48 (.00759)	49 (.00498)
3	4	4	20 (.53221)	25 (.23247)	28 (.10926)	30 (.05758)	32 (.02649)	34 (.01030)	35 (.00589)
			21 (.46779)	26 (.18528)	29 (.08043)	31 (.03974)	33 (.01688)	35 (.00589)	36 (.00320)
3	4	5	24 (.50000)	29 (.23579)	32 (.12269)	35 (.05281)	37 (.02648)	39 (.01169)	40 (.00732)
			25 (.44304)	30 (.19325)	33 (.09481)	36 (.03791)	38 (.01789)	40 (.00732)	41 (.00440)
3	4	6	27 (.52566)	33 (.23834)	37 (.10723)	39 (.06505)	42 (.02642)	44 (.01284)	46 (.00553)
			28 (.47434)	34 (.19973)	38 (.08432)	40 (.04923)	43 (.01865)	45 (.00856)	47 (.00343)
3	4	7	31 (.50000)	38 (.20504)	41 (.11810)	44 (.06003)	47 (.02633)	49 (.01379)	51 (.00657)
			32 (.45344)	39 (.17279)	42 (.09566)	45 (.04644)	48 (.01926)	50 (.00963)	52 (.00435)
3	4	8	34 (.52137)	42 (.20952)	46 (.10583)	49 (.05607)	52 (.02624)	55 (.01058)	57 (.00522)
			35 (.47863)	43 (.17947)	47 (.08672)	50 (.04419)	53 (.01974)	56 (.00752)	58 (.00354)
3	5	5	28 (.50000)	34 (.22029)	37 (.12200)	40 (.05823)	42 (.03227)	45 (.01116)	46 (.00740)
			29 (.44913)	35 (.18365)	38 (.09706)	41 (.04382)	43 (.02324)	46 (.00740)	47 (.00475)
3	5	6	32 (.50000)	39 (.20820)	42 (.12137)	45 (.06278)	48 (.02822)	51 (.01071)	53 (.00500)
			33 (.45405)	40 (.17607)	48 (.09882)	46 (.04890)	49 (.02085)	52 (.00741)	54 (.00328)
3	5	7	36 (.50000)	43 (.22963)	48 (.10022)	51 (.05332)	54 (.02518)	57 (.01033)	59 (.00519)
			37 (.45809)	44 (.19851)	49 (.08220)	52 (.04211)	55 (.01903)	58 (.00740)	60 (.00356)
3	5	8	40 (.50000)	48 (.21844)	53 (.10138)	56 (.05718)	59 (.02926)	63 (.01001)	65 (.00534)
			41 (.46147)	49 (.19057)	54 (.08461)	57 (.04627)	60 (.02284)	64 (.00737)	66 (.00380)
3	6	6	36 (.52087)	44 (.21513)	48 (.11162)	51 (.06089)	54 (.02965)	57 (.01264)	59 (.00656)
			37 (.47913)	45 (.18533)	49 (.09226)	52 (.04855)	55 (.02267)	58 (.00919)	60 (.00459)
3	6	7	41 (.50000)	49 (.22091)	54 (.10392)	57 (.05931)	60 (.03081)	64 (.01085)	66 (.00590)
			42 (.46187)	50 (.19315)	55 (.08704)	58 (.04821)	61 (.02420)	65 (.00807)	67 (.00425)
3	6	8	45 (.51759)	54 (.22580)	59 (.11440)	63 (.05797)	67 (.02551)	70 (.01238)	73 (.00539)
			46 (.48241)	55 (.19983)	60 (.09770)	64 (.04788)	68 (.02027)	71 (.00950)	74 (.00397)
3	7	7	46 (.50000)	55 (.21371)	60 (.10697)	64 (.05366)	67 (.02919)	71 (.01125)	73 (.00651)
			47 (.46502)	56 (.18868)	61 (.09112)	65 (.04421)	68 (.02337)	72 (.00861)	74 (.00486)

TABLE A.13(a) *(continued)*

n_1	n_2	n_3	$\alpha = 0.5$	$\alpha = 0.2$	$\alpha = 0.1$	$\alpha = 0.05$	$\alpha = 0.025$	$\alpha = 0.01$	$\alpha = 0.005$
3	7	8	51 (.50000)	61 (.20768)	66 (.10953)	70 (.05853)	74 (.02783)	78 (.01156)	81 (.00540)
			52 (.46769)	62 (.18490)	67 (.09460)	71 (.04917)	75 (.02265)	79 (.00907)	82 (.00410)
3	8	8	56 (.51497)	67 (.21440)	73 (.10544)	78 (.05022)	81 (.02986)	86 (.01089)	89 (.00539)
			57 (.48503)	68 (.19289)	74 (.09197)	79 (.04251)	82 (.02477)	87 (.00869)	90 (.00418)
4	4	4	24 (.52840)	30 (.21573)	33 (.10993)	35 (.06323)	37 (.03296)	39 (.01530)	41 (.00615)
			25 (.47160)	31 (.17558)	34 (.08439)	36 (0.4632)	38 (.02286)	40 (.00993)	42 (.00367)
4	4	5	28 (.52535)	35 (.20291)	38 (.11051)	41 (.05178)	43 (.02833)	45 (.01412)	47 (.00630)
			29 (.47465)	36 (.16825)	39 (.08738)	42 (.03873)	44 (.02027)	46 (.00959)	48 (.00402)
4	4	6	32 (.52292)	39 (.22651)	43 (.11087)	46 (.05649)	48 (.03336)	51 (.01321)	53 (.00639)
			33 (.47708)	40 (.19294)	44 (.08984)	47 (.04376)	49 (.02497)	52 (.00931)	54 (.00429)
4	4	7	36 (.52091)	44 (.21471)	48 (.11118)	51 (.06052)	54 (.02939)	57 (.01248)	59 (.00645)
			37 (.47909)	45 (.18488)	49 (.09184)	52 (.04822)	55 (.02244)	58 (.00906)	60 (.00450)
4	4	8	40 (.51923)	49 (.20504)	53 (.11139)	57 (.05216)	60 (.02636)	63 (.01188)	65 (.00649)
			41 (.48077)	50 (.17830)	54 (.09353)	58 (.04204)	61 (.02049)	64 (.00885)	66 (.00468)
4	5	5	33 (.50000)	40 (.21074)	44 (.10139)	47 (.05094)	49 (.02980)	52 (.01162)	54 (.00557)
			34 (.45453)	41 (.17872)	45 (.08177)	48 (.03928)	50 (.02220)	53 (.00815)	55 (.00371)
4	5	6	37 (.52068)	45 (.21719)	49 (.11377)	53 (.05021)	55 (.03096)	58 (.01346)	61 (.00502)
			38 (.47932)	46 (.18750)	50 (.09435)	54 (.03970)	56 (.02382)	59 (.00987)	62 (.00347)
4	5	7	42 (.50000)	50 (.22261)	55 (.10570)	58 (.06081)	62 (.02519)	65 (.01147)	67 (.00633)
			43 (.46215)	51 (.19494)	56 (.08875)	59 (.04959)	63 (.01963)	66 (.00858)	68 (.00459)
4	5	8	46 (.51748)	56 (.20134)	60 (.11594)	64 (.05923)	68 (.02636)	71 (.01294)	74 (.00572)
			47 (.48252)	57 (.17722)	61 (.09919)	65 (.04905)	69 (.02102)	72 (.00998)	75 (.00425)
4	6	6	42 (.51886)	51 (.20965)	55 (.11612)	59 (.05592)	62 (.02909)	66 (.01031)	68 (.00565)
			43 (.48114)	52 (.18307)	56 (.09810)	60 (.04546)	63 (.02287)	67 (.00769)	69 (.00408)
4	6	7	47 (.51733)	57 (.20342)	62 (.10126)	66 (.05067)	69 (.02756)	73 (.01066)	75 (.00619)
			48 (.48267)	58 (.17938)	63 (.08619)	67 (.04174)	70 (.02208)	74 (.00818)	76 (.00463)
4	6	8	52 (.51603)	62 (.22166)	68 (.10397)	72 (.05539)	76 (.02631)	80 (.01095)	83 (.00513)
			53 (.48397)	63 (.19820)	69 (.08972)	73 (.04651)	77 (.02141)	81 (.00859)	84 (.00390)
4	7	7	53 (.50000)	63 (.21068)	68 (.11261)	73 (.05154)	76 (.02963)	81 (.01000)	83 (.00607)
			54 (.46809)	64 (.18800)	69 (.09759)	74 (.04318)	77 (.02426)	82 (.00783)	84 (.00465)
4	7	8	58 (.51481)	69 (.21695)	75 (.10806)	80 (.05226)	84 (.02621)	88 (.01177)	91 (.00595)
			59 (.48519)	70 (.19552)	76 (.09450)	81 (.04441)	85 (.02170)	89 (.00946)	92 (.00466)
4	8	8	64 (.51376)	76 (.21292)	82 (.11160)	87 (.05754)	92 (.02610)	97 (.01023)	100 (.00538)
			65 (.48624)	77 (.19320)	83 (.09869)	88 (.04966)	93 (.02191)	98 (.00831)	101 (.00428)
5	5	5	38 (.50000)	46 (.20318)	50 (.10490)	53 (.05715)	56 (.02788)	59 (.01196)	61 (.00626)
			39 (.45888)	47 (.17478)	51 (.08666)	54 (.04558)	57 (.02136)	60 (.00873)	62 (.00440)
5	5	6	43 (.50000)	51 (.22463)	56 (.10781)	60 (.05124)	63 (.02637)	66 (.01222)	69 (.00501)
			44 (.46248)	52 (.19706)	57 (.09078)	61 (.04151)	64 (.02066)	67 (.00921)	70 (.00360)
5	5	7	48 (.50000)	57 (.21690)	62 (.11026)	66 (.05631)	70 (.02514)	73 (.01241)	76 (.00554)
			49 (.46549)	58 (.19200)	63 (.09430)	67 (.04665)	71 (.02008)	74 (.00960)	77 (.00413)
5	5	8	53 (.50000)	63 (.21043)	68 (.11235)	73 (.05135)	76 (.02948)	80 (.01256)	83 (.00601)
			54 (.46806)	64 (.18774)	69 (.09734)	74 (.04300)	77 (.02413)	81 (.00992)	84 (.00461)
5	6	6	48 (.51720)	58 (.20518)	63 (.10301)	67 (.05205)	70 (.02859)	74 (.01125)	76 (.00661)
			49 (.48280)	59 (.18118)	64 (.08787)	68 (.04301)	71 (.02299)	75 (.00868)	77 (.00498)
5	6	7	54 (.50000)	64 (.21215)	69 (.11412)	74 (.05272)	78 (.02507)	82 (.01048)	84 (.00641)
			55 (.46829)	65 (.18952)	70 (.09906)	75 (.04427)	79 (.02042)	83 (.00824)	85 (.00494)
5	6	8	59 (.51473)	70 (.21820)	76 (.10935)	81 (.05328)	85 (.02694)	89 (.01223)	92 (.00624)
			60 (.48527)	71 (.19681)	77 (.09575)	82 (.04535)	86 (.02235)	90 (.00985)	93 (.00490)
5	7	7	60 (.50000)	71 (.20814)	77 (.10319)	81 (.05828)	85 (.02998)	90 (.01125)	93 (.00571)
			61 (.47066)	72 (.18741)	78 (.09019)	82 (.04981)	86 (.02499)	91 (.00904)	94 (.00447)

TABLE A.13(a)　*(continued)*

n_1	n_2	n_3	$\alpha = 0.5$	$\alpha = 0.2$	$\alpha = 0.1$	$\alpha = 0.05$	$\alpha = 0.025$	$\alpha = 0.01$	$\alpha = 0.005$
5	7	8	66 (.50000)	78 (.20471)	84 (.10680)	89 (.05493)	93 (.02948)	98 (.01193)	102 (.00515)
			67 (.47271)	79 (.18559)	85 (.09438)	90 (.04739)	94 (.02489)	99 (.00077)	103 (.00410)
5	8	8	72 (.51273)	85 (.21165)	92 (.10461)	97 (.05635)	102 (.02724)	107 (.01166)	111 (.00536)
			73 (.48727)	86 (.19344)	93 (.09319)	98 (.04917)	103 (.02322)	108 (.00968)	112 (.00434)
6	6	6	54 (.51582)	65 (.20145)	70 (.10721)	74 (.05805)	78 (.02816)	82 (.01206)	85 (.00581)
			55 (.48418)	66 (.17959)	71 (.09285)	75 (.04897)	79 (.02306)	83 (.00954)	86 (.00447)
6	6	7	60 (.51464)	71 (.21964)	77 (.11084)	82 (.05446)	86 (.02778)	91 (.01031)	94 (.00520)
			61 (.48536)	72 (.19831)	78 (.09721)	83 (.04645)	87 (.02311)	92 (.00827)	95 (.00406)
6	6	8	66 (.51362)	78 (.21527)	85 (.10104)	90 (.05151)	94 (.02745)	99 (.01100)	102 (.00588)
			67 (.48638)	79 (.19561)	86 (.08914)	91 (.04436)	95 (.02313)	100 (.00899)	103 (.00471)
6	7	7	67 (.50000)	79 (.20598)	85 (.10808)	90 (.05595)	95 (.02559)	100 (.01016)	103 (.00541)
			68 (.47285)	80 (.18689)	86 (.09563)	91 (.04835)	96 (.02153)	101 (.00829)	104 (.00432)
6	7	8	73 (.51267)	86 (.21274)	93 (.10571)	99 (.05002)	103 (.02787)	109 (.01002)	112 (.00558)
			74 (.48733)	87 (.19456)	94 (.09426)	100 (.04351)	104 (.02380)	110 (.00829)	113 (.00454)
6	8	8	80 (.51184)	94 (.21055)	101 (.10987)	107 (.05532)	112 (.02822)	118 (.01098)	122 (.00533)
			81 (.48816)	95 (.19364)	102 (.09885)	108 (.04873)	113 (.02437)	119 (.00923)	123 (.00439)
7	7	7	74 (.50000)	87 (.20413)	94 (.10045)	99 (.05401)	104 (.02609)	109 (.01118)	113 (.00515)
			75 (.47473)	88 (.18643)	95 (.08944)	100 (.04711)	105 (.02225)	110 (.00929)	114 (.00418)
7	7	8	81 (.50000)	95 (.20251)	102 (.10477)	108 (.05235)	113 (.02653)	119 (.01023)	122 (.00597)
			82 (.47637)	96 (.18602)	103 (.09414)	109 (.04605)	114 (.02288)	120 (.00859)	123 (.00494)
7	8	8	88 (.51108)	103 (.20959)	111 (.10393)	117 (.05443)	123 (.02539)	129 (.01041)	133 (.00530)
			89 (.48892)	104 (.19380)	112 (.09402)	118 (.04834)	124 (.02209)	130 (.00885)	134 (.00443)
8	8	8	96 (.51040)	112 (.20874)	120 (.10852)	127 (.05365)	133 (.02629)	139 (.01152)	144 (.00527)
			97 (.48960)	113 (.19393)	121 (.09891)	128 (.04798)	134 (.02310)	140 (.00992)	145 (.00445)

Source: Robert E. Odeh, ''On Jonckheere's *k*-Sample Test against Ordered Alternatives,'' *Technometrics*, 13 (1971). 912–918.

TABLE A.13(b)

Critical values of *J*, the Jonckheere–Terpstra test statistic (for nominal values of α shown and *k* samples all of size *n*); exact significance levels in parentheses

	k	α = 0.5	α = 0.2	α = 0.1	α = 0.05	α = 0.025	α = 0.01	α = 0.005
n = 2	4	12 (.54921) 13 (.45079)	15 (.26825) 16 (.19286)	17 (.13016) 18 (.08294)	18 (.08294) 19 (.04841)	20 (.02619) 21 (.01230)	21 (.01230) 22 (.00516)	22 (.00516) 23 (.00159)
	5	20 (.53534) 21 (.46466)	25 (.21102) 26 (.16246)	27 (.12133) 28 (.08779)	29 (.06126) 30 (.04116)	31 (.02646) 32 (.01623)	32 (.01623) 33 (.00939)	34 (.00511) 35 (.00257)
	6	30 (.52707) 31 (.47293)	36 (.22650) 37 (.18713)	39 (.12151) 40 (.09533)	42 (.05533) 43 (.04083)	44 (.02944) 45 (.02071)	46 (.01418) 47 (.00944)	48 (.00608) 49 (.00379)
n = 3	4	27 (.52760) 28 (.47240)	33 (.22197) 34 (.18229)	36 (.11663) 37 (.09067)	39 (.05145) 40 (.03744)	41 (.02657) 42 (.01834)	43 (.01229) 44 (.00797)	44 (.00797) 45 (.00498)
	5	45 (.51980) 46 (.48020)	53 (.22740) 54 (.19822)	58 (.10487) 59 (.08738)	61 (.05884) 62 (.04752)	64 (.02995) 65 (.02335)	68 (.01023) 69 (.00755)	70 (.00549) 71 (.00392)
	6	68 (.50000) 69 (.46981)	79 (.20145) 80 (.18058)	84 (.11087) 85 (.09686)	89 (.05331) 90 (.04524)	93 (.02262) 94 (.02201)	97 (.01193) 98 (.00958)	100 (.00604) 101 (.00473)
n = 4	4	48 (.51826) 49 (.48174)	57 (.21724) 58 (.19096)	62 (.10581) 63 (.08950)	66 (.05142) 67 (.04198)	69 (.02715) 70 (.02150)	72 (.01304) 73 (.00998)	75 (.00562) 76 (.00414)
	5	80 (.51305) 81 (.48695)	93 (.20589) 94 (.18756)	99 (.11129) 100 (.09910)	105 (.05211) 106 (.04523)	109 (.02876) 110 (.02450)	115 (.01016) 116 (.00839)	118 (.00561) 119 (.00455)
	6	120 (.50994) 121 (.49006)	137 (.20490) 138 (.19092)	146 (.10048) 147 (.09181)	153 (.05084) 154 (.04567)	159 (.02572) 160 (.02274)	166 (.01025) 167 (.00888)	170 (.00568) 171 (.00486)
n = 5	4	75 (.51321) 76 (.48679)	88 (.20295) 89 (.18455)	94 (.10832) 95 (.09621)	99 (.05735) 100 (.04983)	104 (.02708) 105 (.02296)	109 (.01125) 110 (.00928)	113 (.00502) 114 (.00404)
	5	125 (.50942) 126 (.49058)	143 (.20345) 144 (.19032)	152 (.10385) 153 (.09542)	159 (.05492) 160 (.04970)	166 (.02603) 167 (.02318)	173 (.01095) 174 (.00958)	178 (.00545) 179 (.00470)
	6	188 (.50000) 189 (.48567)	211 (.20386) 212 (.19377)	223 (.10319) 224 (.09679)	233 (.05153) 234 (.04775)	241 (.02701) 242 (.02477)	251 (.01067) 252 (.00964)	258 (.00510) 259 (.00456)
n = 6	4	108 (.51013) 109 (.48987)	125 (.20037) 126 (.18631)	133 (.10521) 134 (.09607)	140 (.05287) 141 (.04743)	146 (.02647) 147 (.02336)	153 (.01035) 154 (.00894)	157 (.00565) 158 (.00481)
	5	180 (.50721) 181 (.49279)	203 (.20745) 204 (.19719)	215 (.10494) 216 (.09842)	225 (.05229) 226 (.04844)	234 (.02505) 235 (.02292)	243 (.01072) 244 (.00969)	250 (.00510) 251 (.00456)
	6	270 (.50548) 271 (.49452)	301 (.20070) 302 (.19304)	316 (.10478) 317 (.09982)	329 (.05285) 330 (.04990)	341 (.02523) 342 (.02361)	353 (.01078) 354 (.00999)	362 (.00527) 363 (.00485)

TABLE A.14 **Kendall's coefficient of concordance**

***k* = 3**

***b* = 2**

W	*P*
.000	1.000
.250	.833
.750	.500
1.000	.167

***b* = 3**

W	*P*
.000	1.000
.111	.944
.333	.528
.444	.361
.778	.194
1.000	.028

***b* = 4**

W	*P*
.000	1.000
.062	.931
.188	.653
.250	.431
.438	.273
.562	.125
.750	.069
.812	.042
1.000	.005

***b* = 5**

W	*P*
.000	1.000
.040	.954
.120	.691
.160	.522
.280	.367
.360	.182
.480	.124
.520	.093
.640	.039
.760	.024
.840	.008
1.000	.001

***b* = 6**

W	*P*
.000	1.000
.028	.956
.083	.740
.111	.570
.194	.430

***b* = 6 (cont.)**

W	*P*
.250	.252
.333	.184
.361	.142
.444	.072
.528	.052
.583	.029
.694	.012
.750	.008
.778	.006
.861	.002
1.000	.000

***b* = 7**

W	*P*
.000	1.000
.020	.964
.061	.768
.082	.620
.143	.486
.184	.305
.245	.237
.265	.192
.326	.112
.388	.085
.429	.051
.510	.027
.551	.021
.571	.016
.633	.008
.735	.004
.755	.003
.796	.001
.878	.000
1.000	.000

***b* = 8**

W	*P*
.000	1.000
.016	.967
.047	.794
.062	.654
.109	.531
.141	.355
.188	.285
.203	.236
.250	.149
.297	.120
.328	.079

***b* = 8 (cont.)**

W	*P*
.391	.047
.422	.038
.438	.030
.484	.018
.562	.010
.578	.008
.609	.005
.672	.002
.750	.001
.766	.001
.812	.000
.891	.000
1.000	.000

***b* = 9**

W	*P*
.000	1.000
.012	.971
.037	.814
.049	.685
.086	.569
.111	.398
.148	.328
.160	.278
.198	.187
.235	.154
.259	.107
.309	.069
.333	.057
.346	.048
.383	.031
.444	.019
.457	.016
.482	.010
.531	.006
.593	.004
.605	.003
.642	.001
.704	.001
.753	.000
⋮	⋮
1.000	.000

***b* = 10**

W	*P*
.000	1.000

***b* = 10 (cont.)**

W	*P*
.010	.974
.030	.830
.040	.710
.070	.601
.090	.436
.120	.368
.130	.316
.160	.222
.190	.187
.210	.135
.250	.092
.270	.078
.280	.066
.310	.046
.360	.030
.370	.026
.390	.018
.430	.012
.480	.007
.490	.006
.520	.003
.570	.002
.610	.001
.630	.001
.640	.001
.670	.000
⋮	⋮
1.000	.000

***b* = 11**

W	*P*
.000	1.000
.008	.976
.025	.844
.033	.732
.058	.629
.074	.470
.099	.403
.107	.351
.132	.256
.157	.219
.174	.163
.207	.116
.223	.100
.231	.087
.256	.062

***b* = 11 (cont.)**

W	*P*
.298	.043
.306	.037
.322	.027
.355	.019
.397	.013
.405	.011
.430	.007
.471	.005
.504	.003
.521	.002
.529	.002
.554	.001
.603	.001
.620	.000
⋮	⋮
1.000	.000

***b* = 12**

W	*P*
.000	1.000
.007	.978
.021	.856
.028	.751
.049	.654
.062	.500
.083	.434
.090	.383
.111	.287
.132	.249
.146	.191
.174	.141
.188	.123
.194	.108
.215	.080
.250	.058
.257	.050
.271	.038
.299	.028
.333	.019
.340	.017
.361	.011
.396	.008
.424	.005
.438	.004
.444	.004
.465	.002

TABLE A.14   (*continued*)

k = 3

| b = 12 (cont.) | | b = 13 (cont.) | | b = 14 (cont.) | | b = 14 (cont.) | | b = 15 (cont.) | |
W	P	W	P	W	P	W	P	W	P
.507	.002	.219	.064	.020	.781	.429	.002	.191	.059
.521	.001	.231	.050	.036	.694	.464	.001	.213	.047
.528	.001	.254	.038	.046	.551	.474	.001	.218	.043
.549	.001	.284	.027	.061	.489	.495	.000	.231	.030
.562	.001	.290	.025	.066	.438	.	.	.253	.022
.583	.000	.308	.016	.082	.344	.	.	.271	.018
.	.	.337	.012	.097	.305	.	.	.280	.015
.	.	.361	.008	.107	.242	1.000	.000	.284	.011
.	.	.373	.007	.128	.188	**b = 15**		.298	.010
1.000	.000	.379	.006	.138	.167	**W**	**P**	.324	.007
		.396	.004	.143	.150	.000	1.000	.333	.005
b = 13		.432	.003	.158	.117	.004	.982	.338	.005
W	**P**	.444	.002	.184	.089	.013	.882	.351	.004
.000	1.000	.450	.002	.189	.079	.018	.794	.360	.004
.006	.980	.467	.001	.199	.063	.031	.711	.373	.003
.018	.866	.479	.001	.219	.049	.040	.573	.404	.002
.024	.767	.497	.001	.245	.036	.053	.513	.413	.001
.041	.657	.538	.001	.250	.033	.058	.463	.431	.001
.053	.527	.550	.000	.265	.023	.071	.369	.444	.001
.071	.463	.	.	.291	.018	.084	.330	.458	.001
.077	.412	.	.	.311	.011	.093	.267	.480	.000
.095	.316	.	.	.321	.010	.111	.211	.	.
.112	.278	1.000	.000	.327	.009	.120	.189	.	.
.124	.217	**b = 14**		.342	.007	.124	.170	.	.
.148	.165	**W**	**P**	.372	.005	.138	.136	1.000	.000
.160	.145	.000	1.000	.383	.003	.160	.106		
.166	.129	.005	.981	.388	.003	.164	.096		
.183	.098	.015	.874	.403	.003	.173	.077		
.213	.073			.413	.002				

k = 4

| b = 2 | | b = 3 | | b = 3 (cont.) | | b = 4 (cont.) | | b = 4 (cont.) | |
W	P	W	P	W	P	W	P	W	P
.000	1.000	.022	1.000	.644	.161	.050	.930	.325	.321
.100	.958	.067	.958	.733	.075	.075	.898	.375	.237
.200	.833	.111	.910	.778	.054	.100	.794	.400	.199
.300	.792	.200	.727	.822	.026	.125	.753	.425	.188
.400	.625	.244	.615	.911	.017	.150	.680	.450	.159
.500	.542	.289	.524	1.000	.002	.175	.651	.475	.141
.600	.458	.378	.446			.200	.528	.500	.106
.700	.375	.422	.328	**b = 4**		.225	.513	.525	.093
.800	.208	.467	.293	**W**	**P**	.250	.432	.550	.077
.900	.167	.556	.207	.000	1.000	.275	.390	.575	.069
1.000	.042	.600	.182	.025	.992	.300	.352	.600	.058

TABLE A.14   (continued)

k = 4

b = 4 (cont.)

W	P
.625	.054
.650	.036
.675	.035
.700	.020
.725	.013
.775	.011
.800	.006
.825	.005
.850	.002
.900	.002
.925	.001
1.000	.000

b = 5

W	P
.008	1.000
.024	.974
.040	.944
.072	.857
.088	.769
.104	.710
.136	.652
.152	.563
.168	.520
.200	.443
.216	.406
.232	.368
.264	.301
.280	.266
.296	.232
.328	.213
.344	.162
.360	.151
.392	.119
.408	.102
.424	.089
.456	.071
.472	.067
.488	.057
.520	.049
.536	.033
.552	.032
.584	.024
.600	.021
.616	.015
.648	.011
.664	.009
.680	.008
.712	.006
.728	.003
.744	.002

b = 5 (cont.)

W	P
.776	.002
.792	.001
.808	.001
.840	.000
.	.
.	.
.	.
1.000	.000

b = 6

W	P
.000	1.000
.011	.996
.022	.952
.033	.938
.044	.878
.056	.843
.067	.797
.078	.779
.089	.676
.100	.666
.111	.608
.122	.566
.133	.541
.144	.517
.167	.427
.178	.385
.189	.374
.200	.337
.211	.321
.222	.274
.233	.259
.244	.232
.256	.221
.267	.193
.278	.190
.289	.162
.300	.154
.311	.127
.322	.113
.344	.109
.356	.088
.367	.087
.378	.073
.389	.067
.400	.063
.411	.058
.422	.043
.433	.041
.444	.036

b = 6 (cont.)

W	P
.456	.033
.467	.031
.478	.027
.489	.021
.500	.021
.522	.017
.533	.015
.544	.015
.556	.011
.567	.010
.578	.009
.589	.008
.600	.006
.611	.006
.633	.004
.644	.003
.656	.003
.667	.002
.678	.002
.700	.001
.711	.001
.722	.001
.733	.001
.744	.001
.756	.000
.	.
.	.
.	.
1.000	.000

b = 7

W	P
.004	1.000
.012	.984
.020	.964
.037	.905
.045	.846
.053	.795
.069	.754
.078	.678
.086	.652
.102	.596
.110	.564
.118	.533
.135	.460
.143	.420
.151	.378
.167	.358
.176	.306
.184	.300
.200	.264

b = 7 (cont.)

W	P
.208	.239
.216	.216
.233	.188
.241	.182
.249	.163
.265	.150
.273	.122
.282	.118
.298	.101
.306	.093
.314	.081
.331	.073
.339	.062
.347	.058
.363	.051
.371	.040
.380	.037
.396	.034
.404	.032
.412	.030
.429	.024
.437	.021
.445	.018
.461	.016
.469	.014
.478	.013
.494	.009
.502	.008
.510	.008
.527	.007
.535	.006
.543	.004
.559	.004
.567	.003
.576	.003
.592	.003
.600	.002
.608	.002
.624	.001
.633	.001
.641	.001
.657	.001
.665	.001
.673	.001
.690	.000
.	.
.	.
.	.
1.000	.000

b = 8

W	P
.000	1.000
.006	.998
.012	.967
.019	.957
.025	.914
.031	.890
.038	.853
.044	.842
.050	.764
.056	.754
.062	.709
.069	.677
.075	.660
.081	.637
.094	.557
.100	.509
.106	.500
.112	.471
.119	.453
.125	.404
.131	.390
.137	.364
.144	.348
.156	.325
.162	.297
.169	.283
.175	.247
.181	.231
.194	.217
.200	.185
.206	.182
.212	.162
.219	.155
.225	.153
.231	.144
.238	.122
.244	.120
.250	.112
.256	.106
.262	.098
.269	.091
.281	.077
.294	.067
.300	.062
.306	.061
.312	.052
.319	.049
.325	.046
.331	.043
.338	.038

TABLE A.14 (*continued*)

k = 4				k = 5					
b = 8 (cont.)		b = 8 (cont.)		b = 3		b = 3 (cont.)		b = 3 (cont.)	
W	P	W	P	W	P	W	P	W	P
.344	.037	.500	.004	.000	1.000	.333	.475	.667	.063
.356	.031	.506	.004	.022	1.000	.356	.432	.689	.056
.362	.028	.512	.003	.044	.988	.378	.406	.711	.045
.369	.026	.519	.003	.067	.972	.400	.347	.733	.038
.375	.023	.525	.002	.089	.941	.422	.326	.756	.028
.381	.021	.531	.002	.111	.914	.444	.291	.778	.026
.394	.019	.538	.002	.133	.845	.467	.253	.800	.017
.400	.015	.544	.002	.156	.831	.489	.236	.822	.015
.406	.015	.550	.002	.178	.768	.511	.213	.844	.008
.412	.013	.556	.002	.200	.720	.533	.172	.867	.005
.419	.013	.562	.001	.222	.682	.556	.163	.889	.004
.425	.011	.569	.001	.244	.649	.578	.127	.911	.003
.431	.010	.575	.001	.267	.595	.600	.117	.956	.001
.438	.009	.581	.001	.289	.559	.622	.096	1.000	.000
.444	.008	.594	.001	.311	.493	.644	.080		
.450	.008	.606	.001						
.456	.008	.612	.000						
.462	.007	.	.						
.469	.007	.	.						
.475	.006	.	.						
.481	.005	1.000	.000						
.494	.004								

Source: Donald B. Owen, *Handbook of Statistical Tables*, Reading, Mass: Addison-Wesley, 1962, and Maurice G. Kendall, *Rank Correlation Methods*, fourth edition, Charles Griffin & Company, Ltd, High Wycombe, Bucks., England; reprinted by permission.

Note: **P** = the probability of a computed value of W greater than or equal to the tabulated value.

TABLE A.15   **Critical values of minimum r_j for comparison of k treatments against one control in b sets of observations: a one-tailed critical region with an experimentwise error rate**

b	Level of significance for min r_j	2	3	4	5	6	7	8	9
		k = number of treatments (excluding control)							
4	.15	0 (.113)*	—	—	—	—	—	—	—
	.10	—	—	—	—	—	—	—	—
	.05	—	—	—	—	—	—	—	—
5	.15	0 (.058)	0 (.082)	0 (.104)	0	—	—	—	—
	.10	0 (.058)	0 (.082)	—	—	—	—	—	—
	.05	—	—	—	—	—	—	—	—
6	.15	0 (.030)	0 (.043)	0 (.055)	0	0	0	0	0
	.10	0 (.030)	0 (.043)	0 (.055)	0	0	—	—	—
	.05	0 (.030)	0 (.043)	—	—	—	—	—	—
7	.15	1 (.113)	0 (.022)	0 (.029)	0	0	0	0	0
	.10	0 (.015)	0 (.022)	0 (.029)	0	0	0	0	0
	.05	0 (.015)	0 (.022)	0 (.029)	0	—	—	—	—
8	.15	1 (.066)	1 (.092)	1	0	0	0	0	0
	.10	1 (.066)	1 (.092)	0	0	0	0	0	0
	.05	0 (.008)	0 (.011)	0	0	0	0	0	0
9	.15	1 (.037)	1 (.053)	1	1	1	1	1	1
	.10	1 (.037)	1 (.053)	1	1	0	0	0	0
	.05	1 (.037)	0 (.006)	0	0	0	0	0	0
10	.15	2 (.100)	2 (.139)	1	1	1	1	1	1
	.10	1 (.021)	1 (.030)	1	1	1	1	1	1
	.05	1 (.021)	1 (.030)	1	0	0	0	0	0
11	.15	2 (.061)	2 (.087)	2	2	1	1	1	1
	.10	2 (.061)	2 (.087)	1	1	1	1	1	1
	.05	1 (.011)	1 (.017)	1	1	1	1	0	0
12	.15	3 (.131)	2 (.053)	2	2	2	2	2	2
	.10	2 (.037)	2 (.053)	2	2	1	1	1	1
	.05	2 (.037)	1 (.009)	1	1	1	1	1	1
13	.15	3 (.085)	3 (.119)	2	2	2	2	2	2
	.10	3 (.085)	2 (.031)	2	2	2	2	2	2
	.05	2 (.022)	2 (.031)	2	1	1	1	1	1
14	.15	3 (.054)	3 (.077)	3	3	3	2	2	2
	.10	3 (.054)	3 (.077)	2	2	2	2	2	2
	.05	2 (.013)	2 (.018)	2	2	2	2	1	1
15	.15	4 (.108)	4 (.149)	3	3	3	3	3	3
	.10	3 (.034)	3 (.048)	3	3	3	2	2	2
	.05	3 (.034)	3 (.034)	2	2	2	2	2	2
16	.15	4 (.072)	4	4	3	3	3	3	3
	.10	4 (.072)	3	3	3	3	3	3	3
	.05	3 (.021)	3	3	3	2	2	2	2

*( ) Exact cumulative probability.

TABLE A.15   (*continued*)

b	Level of significance for min r_j	k = number of treatments (excluding control)							
		2	3	4	5	6	7	8	9
17	.15	5 (.129)	4	4	4	4	4	3	3
	.10	4 (.046)	4	4	3	3	3	3	3
	.05	4 (.046)	3	3	3	3	3	2	2
18	.15	5 (.089)	5	4	4	4	4	4	4
	.10	5 (.089)	4	4	4	4	4	3	3
	.05	4 (.029)	4	3	3	3	3	3	3
19	.15	6 (.149)	5	5	5	4	4	4	4
	.10	5 (.060)	5	4	4	4	4	4	4
	.05	4 (.019)	4	4	4	3	3	3	3
20	.15	6 (.105)	6	5	5	5	5	5	5
	.10	5 (.039)	5	5	5	4	4	4	4
	.05	5 (.039)	4	4	4	4	4	3	3
21	.15	6 (.073)	6	6	5	5	5	5	5
	.10	6 (.073)	5	5	5	5	5	5	5
	.05	5 (.026)	5	5	4	4	4	4	4
22	.15	7 (.121)	6	6	6	6	6	5	5
	.10	6 (.050)	6	6	5	5	5	5	5
	.05	6 (.050)	5	5	5	4	4	4	4
23	.15	7 (.086)	7	6	6	6	6	6	6
	.10	7 (.086)	6	6	6	6	5	5	5
	.05	6 (.033)	6	5	5	5	5	5	5
24	.15	8 (.136)	7	7	7	6	6	6	6
	.10	7 (.060)	7	6	6	6	6	6	6
	.05	6 (.022)	6	6	5	5	5	5	5
25	.15	8	8	7	7	7	7	7	7
	.10	7	7	7	7	6	6	6	6
	.05	7	6	6	6	6	6	5	5
30	.15	10	10	9	9	9	9	9	9
	.10	10	9	9	9	8	8	8	8
	.05	9	8	8	8	8	8	7	7
35	.15	12	12	12	11	11	11	11	11
	.10	12	11	11	11	10	10	10	10
	.05	11	10	10	10	10	9	9	9
40	.15	15	14	14	13	13	13	13	13
	.10	14	13	13	13	13	12	12	12
	.05	13	12	12	12	12	11	11	11
45	.15	17	16	16	16	15	15	15	15
	.10	16	16	15	15	15	14	14	14
	.05	15	14	14	14	14	13	13	13
50	.15	19	18	18	18	17	17	17	17
	.10	18	18	17	17	17	17	16	16
	.05	17	17	16	16	16	16	15	15

Source: A. L. Rhyne, Jr., and R. G. D. Steel. "Tables for a Treatments versus Control Multiple Comparisons Sign Test," *Technometrics*, Vol. 7, No. 3 (Aug. 1965), pp. 297–298; reprinted by permission.

TABLE A.16 Critical values of minimum$_j$[min(r_j, $b - r_j$)] for comparison of k treatments against one control in b sets of observations: a two-tailed critical region with an experimentwise error rate

b	Level of significance for min$_j$[min(r_j, $b - r_j$)]	k = number of treatments (excluding control) 2	3	4	5	6	7	8	9
6	.10	0 (.060)*	0 (.085)	—	—	—	—	—	—
	.05	—	—	—	—	—	—	—	—
	.01	—	—	—	—	—	—	—	—
7	.10	0 (.030)	0 (.044)	0 (.057)	—	—	—	—	—
	.05	0 (.030)	—	—	—	—	—	—	—
	.01	—	—	—	—	—	—	—	—
8	.10	0 (.015)	0 (.023)	0	0	0	0	0	0
	.05	0 (.015)	0 (.023)	0	—	—	—	—	—
	.01	—	—	—	—	—	—	—	—
9	.10	1 (.074)	0 (.011)	0	0	0	0	0	0
	.05	0 (.008)	0 (.011)	0	0	0	0	—	—
	.01	—	—	—	—	—	—	—	—
10	.10	1 (.041)	1 (.060)	1	0	0	0	0	0
	.05	1 (.041)	0 (.006)	0	0	0	0	0	0
	.01	0 (.004)	0 (.006)	—	—	—	—	—	—
11	.10	1 (.023)	1 (.034)	1	1	1	1	0	0
	.05	1 (.023)	1 (.034)	0	0	0	0	0	0
	.01	0 (.002)	0 (.003)	0	—	—	—	—	—
12	.10	2 (.073)	1 (.018)	1	1	1	1	1	1
	.05	1 (.012)	1 (.018)	1	1	0	0	0	0
	.01	0 (.001)	0 (.001)	0	0	0	0	—	—
13	.10	2 (.043)	2 (.063)	2	1	1	1	1	1
	.05	2 (.043)	1 (.011)	1	1	1	1	1	1
	.01	1 (.007)	0 (.001)	0	0	0	0	0	0
14	.10	2 (.025)	2 (.037)	2	2	2	2	1	1
	.05	2 (.025)	2 (.037)	1	1	1	1	1	1
	.01	1 (.004)	1 (.005)	0	0	0	0	0	0
15	.10	3 (.067)	3 (.096)	2	2	2	2	2	2
	.05	2 (.014)	2 (.021)	2	2	1	1	1	1
	.01	1 (.002)	1 (.003)	1	1	0	0	0	0
16	.10	3 (.041)	3	3	2	2	2	2	2
	.05	3 (.041)	2	2	2	2	2	2	2
	.01	2 (.008)	1	1	1	1	1	1	0
17	.10	4 (.093)	3	3	3	3	3	2	2
	.05	3 (.024)	3	2	2	2	2	2	2
	.01	2 (.005)	1	1	1	1	1	1	1

TABLE A.16   *(continued)*

b	Level of significance for $\min_j[\min(r_j, b - r_j)]$	k = number of treatments (excluding control) 2	3	4	5	6	7	8	9
18	.10	4 (.059)	4	3	3	3	3	3	3
	.05	3 (.015)	3	3	3	2	2	2	2
	.01	2 (.003)	2	2	1	1	1	1	1
19	.10	4 (.037)	4	4	4	3	3	3	3
	.05	4 (.037)	3	3	3	3	3	3	3
	.01	3 (.009)	2	2	2	2	2	2	1
20	.10	5 (.079)	4	4	4	4	4	4	3
	.05	4 (.023)	4	3	3	3	3	3	3
	.01	3 (.005)	2	2	2	2	2	2	2
21	.10	5 (.051)	5	4	4	4	4	4	4
	.05	4 (.014)	4	4	4	4	3	3	3
	.01	3 (.003)	3	3	2	2	2	2	2
22	.10	6 (.099)	5	5	5	5	4	4	4
	.05	5 (.033)	4	4	4	4	4	4	4
	.01	4 (.009)	3	3	3	3	3	2	2
23	.10	6 (.066)	5	5	5	5	5	5	5
	.05	5 (.021)	5	5	4	4	4	4	4
	.01	4 (.005)	3	3	3	3	3	3	3
24	.10	6 (.043)	6	6	5	5	5	5	5
	.05	6 (.043)	5	5	5	5	5	4	4
	.01	4 (.003)	4	4	3	3	3	3	3
25	.10	7	6	6	6	6	6	5	5
	.05	6	6	5	5	5	5	5	5
	.01	4	4	4	4	4	4	3	3
30	.10	9	8	8	8	8	7	7	7
	.05	8	8	7	7	7	7	7	7
	.01	6	6	6	6	5	5	5	5
35	.10	11	10	10	10	10	9	9	9
	.05	10	9	9	9	9	9	9	8
	.01	8	8	8	7	7	7	7	7
40	.10	13	12	12	12	12	11	11	11
	.05	12	12	11	11	11	11	11	10
	.01	10	10	9	9	9	9	9	9
45	.10	15	14	14	14	14	13	13	13
	.05	14	14	13	13	13	13	12	12
	.01	12	12	11	11	11	11	11	11
50	.10	17	17	16	16	16	16	15	15
	.05	16	16	15	15	15	15	14	14
	.01	14	14	13	13	13	13	13	12

Source: A. L. Rhyne, Jr., and R. G. D. Steel, "Tables for a Treatments versus Control Multiple Comparisons Sign Test," *Technometrics*, Vol. 7, No. 3 (Aug. 1965), p. 299; reprinted by permission.

Selected critical values of _L_, for Page's ordered alternatives test

	3 α			4 α		
_b_	0.001	0.01	0.05	0.001	0.01	0.05
2			28		60	58
3		42	41	89	87	84
4	56	55	54	117	114	111
5	70	68	66	145	141	137
6	83	81	79	172	167	163
7	96	93	91	198	193	189
8	109	106	104	225	220	214
9	121	119	116	252	246	240
10	134	131	128	278	272	266
11	147	144	141	305	298	292
12	160	156	153	331	324	317
13	172	169	165			
14	185	181	178			
15	197	194	190			
16	210	206	202			
17	223	218	215			
18	235	231	227			
19	248	243	239			
20	260	256	251			

	5 α			6 α		
_b_	0.001	0.01	0.05	0.001	0.01	0.05
2	109	106	103	178	173	166
3	160	155	150	260	252	244
4	210	204	197	341	331	321
5	259	251	244	420	409	397
6	307	299	291	499	486	474
7	355	346	338	577	563	550
8	403	393	384	655	640	625
9	451	441	431	733	717	701
10	499	487	477	811	793	777
11	546	534	523	888	869	852
12	593	581	570	965	946	928

	7 α			8 α		
_b_	0.001	0.01	0.05	0.001	0.01	0.05
2	269	261	252	388	376	362
3	394	382	370	567	549	532
4	516	501	487	743	722	701
5	637	620	603	917	893	869
6	757	737	719	1090	1063	1037
7	876	855	835	1262	1232	1204
8	994	972	950	1433	1401	1371
9	1113	1088	1065	1603	1569	1537
10	1230	1205	1180	1773	1736	1703
11	1348	1321	1295	1943	1905	1868
12	1465	1437	1410	2112	2072	2035

Source: E. B. Page, "Ordered Hypotheses for Multiple Treatments: A Significance Test for Linear Ranks," _J. Amer. Statist. Assoc._, 58 (1963), 216–230.

TABLE A.18   **Quantiles of the Kolmogorov test statistic**

One-sided test Two-sided test	$p = 0.90$ $p = 0.80$	0.95 0.90	0.975 0.95	0.99 0.98	0.995 0.99
$n = 1$	.900	.950	.975	.990	.995
2	.684	.776	.842	.900	.929
3	.565	.636	.708	.785	.829
4	.493	.565	.624	.689	.734
5	.447	.509	.563	.627	.669
6	.410	.468	.519	.577	.617
7	.381	.436	.483	.538	.576
8	.358	.410	.454	.507	.542
9	.339	.387	.430	.480	.513
10	.323	.369	.409	.457	.489
11	.308	.352	.391	.437	.468
12	.296	.338	.375	.419	.449
13	.285	.325	.361	.404	.432
14	.275	.314	.349	.390	.418
15	.266	.304	.338	.377	.404
16	.258	.295	.327	.366	.392
17	.250	.286	.318	.355	.381
18	.244	.279	.309	.346	.371
19	.237	.271	.301	.337	.361
20	.232	.265	.294	.329	.352
21	.226	.259	.287	.321	.344
22	.221	.253	.281	.314	.337
23	.216	.247	.275	.307	.330
24	.212	.242	.269	.301	.323
25	.208	.238	.264	.295	.317
26	.204	.233	.259	.290	.311
27	.200	.229	.254	.284	.305
28	.197	.225	.250	.279	.300
29	.193	.221	.246	.275	.295
30	.190	.218	.242	.270	.290
31	.187	.214	.238	.266	.285
32	.184	.211	.234	.262	.281
33	.182	.208	.231	.258	.277
34	.179	.205	.227	.254	.273
35	.177	.202	.224	.251	.269
36	.174	.199	.221	.247	.265
37	.172	.196	.218	.244	.262
38	.170	.194	.215	.241	.258
39	.168	.191	.213	.238	.255
40	.165	.189	.210	.235	.252
Approximation for $n > 40$:	$\dfrac{1.07}{\sqrt{n}}$	$\dfrac{1.22}{\sqrt{n}}$	$\dfrac{1.36}{\sqrt{n}}$	$\dfrac{1.52}{\sqrt{n}}$	$\dfrac{1.63}{\sqrt{n}}$

Source: L. H. Miller, "Table of Percentage Points of Kolmogorov Statistics,"
J. Amer. Statist. Assoc., 51 (1956), 111–121.

TABLE A.19(a)   Critical values for Lilliefors test, normal case 1 (μ unknown, σ^2 known)

	α				
n	0.20	0.15	0.10	0.05	0.01
3	.392	.308	.428	.453	.495
4	.351	.366	.384	.410	.455
5	.318	.333	.350	.376	.423
6	.294	.307	.324	.348	.396
7	.276	.288	.305	.328	.374
8	.260	.272	.288	.311	.353
9	.246	.258	.272	.294	.334
10	.234	.245	.259	.280	.323
11	.225	.235	.249	.269	.309
12	.216	.226	.238	.259	.300
13	.209	.218	.230	.249	.285
14	.202	.211	.224	.242	.280
15	.195	.205	.217	.235	.270
16	.189	.197	.209	.227	.261
17	.184	.192	.203	.220	.256
18	.179	.187	.198	.215	.246
19	.174	.182	.194	.210	.242
20	.170	.178	.189	.205	.235
21	.166	.174	.184	.199	.230
22	.163	.171	.180	.195	.227
23	.160	.167	.177	.193	.221
24	.156	.164	.173	.188	.217
25	.154	.160	.170	.185	.214
26	.151	.158	.167	.181	.209
27	.147	.154	.163	.177	.205
28	.146	.153	.161	.174	.202
29	.143	.149	.158	.172	.198
30	.141	.147	.155	.169	.193

TABLE A.19(b)   Critical values for Lilliefors test, normal case 2 (μ known, σ^2 unknown)

	α				
n	0.20	0.15	0.10	0.05	0.01
2	.739	.770	.797	.820	.837
3	.551	.599	.657	.722	.798
4	.499	.529	.565	.621	.734
5	.440	.470	.507	.567	.660
6	.400	.429	.464	.514	.607
7	.375	.395	.429	.477	.566
8	.451	.374	.405	.450	.534
9	.332	.353	.382	.425	.505
10	.315	.335	.361	.401	.477
11	.300	.320	.346	.387	.466
12	.289	.307	.332	.371	.444
13	.277	.296	.320	.358	.428
14	.266	.284	.307	.341	.410
15	.259	.275	.297	.331	.397
16	.251	.257	.288	.322	.387
17	.244	.260	.282	.313	.377
18	.236	.251	.271	.302	.369
19	.231	.246	.266	.297	.357
20	.226	.241	.260	.290	.348
21	.219	.233	.252	.282	.337

	α				
n	**0.20**	**0.15**	**0.10**	**0.05**	**0.01**
22	.214	.228	.247	.278	.334
23	.210	.223	.242	.270	.319
24	.205	.218	.236	.263	.317
25	.202	.214	.231	.256	.308
26	.197	.210	.227	.255	.305
27	.194	.208	.224	.250	.302
28	.191	.203	.219	.244	.292
29	.188	.200	.217	.242	.290
30	.185	.198	.212	.236	.284
50	$1.02/\sqrt{n}$	$1.080/\sqrt{n}$	$1.170/\sqrt{n}$	$1.310/\sqrt{n}$	$1.595/\sqrt{n}$
100	$1.04/\sqrt{n}$	$1.100/\sqrt{n}$	$1.180/\sqrt{n}$	$1.320/\sqrt{n}$	$1.610/\sqrt{n}$
≥101	$1.06/\sqrt{n}$	$1.120/\sqrt{n}$	$1.190/\sqrt{n}$	$1.333/\sqrt{n}$	$1.625/\sqrt{n}$

TABLE A.19(c)   **Critical values for Lilliefors test, normal case 3 (μ, σ^2 both unknown)**

	α				
n	**0.20**	**0.15**	**0.10**	**0.05**	**0.01**
4	.303	.320	.344	.374	.414
5	.290	.302	.319	.344	.398
6	.268	.280	.295	.321	.371
7	.252	.264	.280	.304	.353
8	.239	.251	.266	.290	.333
9	.227	.239	.253	.275	.319
10	.217	.228	.241	.262	.303
11	.209	.219	.232	.252	.291
12	.201	.210	.223	.243	.281
13	.193	.203	.215	.233	.270
14	.187	.196	.209	.227	.264
15	.181	.190	.202	.219	.256
16	.176	.184	.195	.212	.248
17	.170	.179	.190	.207	.241
18	.166	.174	.185	.201	.234
19	.162	.171	.181	.197	.230
20	.159	.167	.177	.192	.223
21	.155	.163	.173	.188	.219
22	.152	.160	.170	.185	.214
23	.149	.156	.165	.181	.210
24	.145	.153	.162	.177	.205
25	.144	.151	.159	.173	.202
26	.141	.147	.156	.170	.198
27	.138	.145	.153	.166	.193
28	.136	.142	.151	.165	.191
29	.134	.140	.149	.162	.188
30	.132	.138	.146	.159	.183
31	$\dfrac{0.741}{d_n}$	$\dfrac{0.775}{d_n}$	$\dfrac{0.819}{d_n}$	$\dfrac{0.895}{d_n}$	$\dfrac{1.035}{d_n}$

$$d_n = (\sqrt{n} - 0.01 + 0.83/\sqrt{n})$$

Source: Andrew L. Mason and C. B. Bell, ''New Lilliefors and Srinivasan Tables with Applications,'' *Communic. Statist.—Simul.*, Vol. 15, No. 2 (1986), pp. 457–459. Copyright (c) 1986 by Marcel Dekker, Inc.; reprinted by permission.

TABLE A.20(a)   **Quantiles of the Smirnov test statistic for two samples of equal size n**

One-sided test	$p = 0.90$	0.95	0.975	0.99	0.995
Two-sided test	$p = 0.80$	0.90	0.95	0.98	0.99
$n = 3$	2/3	2/3			
4	3/4	3/4	3/4		
5	3/5	3/5	4/5	4/5	4/5
6	3/6	4/6	4/6	5/6	5/6
7	4/7	4/7	5/7	5/7	5/7
8	4/8	4/8	5/8	5/8	6/8
9	4/9	5/9	5/9	6/9	6/9
10	4/10	5/10	6/10	6/10	7/10
11	5/11	5/11	6/11	7/11	7/11
12	5/12	5/12	6/12	7/12	7/12
13	5/13	6/13	6/13	7/13	8/13
14	5/14	6/14	7/14	7/14	8/14
15	5/15	6/15	7/15	8/15	8/15
16	6/16	6/16	7/16	8/16	9/16
17	6/17	7/17	7/17	8/17	9/17
18	6/18	7/18	8/18	9/18	9/18
19	6/19	7/19	8/19	9/19	9/19
20	6/20	7/20	8/20	9/20	10/20
21	6/21	7/21	8/21	9/21	10/21
22	7/22	8/22	8/22	10/22	10/22
23	7/23	8/23	9/23	10/23	10/23
24	7/24	8/24	9/24	10/24	11/24
25	7/25	8/25	9/25	10/25	11/25
26	7/26	8/26	9/26	10/26	11/26
27	7/27	8/27	9/27	11/27	11/27
28	8/28	9/28	10/28	11/28	12/28
29	8/29	9/29	10/29	11/29	12/29
30	8/30	9/30	10/30	11/30	12/30
31	8/31	9/31	10/31	11/31	12/31
32	8/32	9/32	10/32	12/32	12/32
33	8/33	9/33	11/33	12/33	13/33
34	8/34	10/34	11/34	12/34	13/34
35	8/35	10/35	11/35	12/35	13/35
36	9/36	10/36	11/36	12/36	13/36
37	9/37	10/37	11/37	13/37	13/37
38	9/38	10/38	11/38	13/38	14/38
39	9/39	10/39	11/39	13/39	14/39
40	9/40	10/40	12/40	13/40	14/40
Approximation for $n > 40$:	$\dfrac{1.52}{\sqrt{n}}$	$\dfrac{1.73}{\sqrt{n}}$	$\dfrac{1.92}{\sqrt{n}}$	$\dfrac{2.15}{\sqrt{n}}$	$\dfrac{2.30}{\sqrt{n}}$

Source: Z. W. Birnbaum and R. A. Hall, "Small-Sample Distribution for Multi-Sample Statistics of the Smirnov Type," *Ann. Math. Statist.*, 31 (1960), 710–720.

TABLE A.20(b) — **Quantiles of the Smirnov test statistic for two samples of different size**

One-sided test Two-sided test	$p = 0.90$ $p = 0.80$	0.95 0.90	0.975 0.95	0.99 0.98	0.995 0.99
$N_1 = 1$ $N_2 = 9$	17/18				
10	9/10				
$N_1 = 2$ $N_2 = 3$	5/6				
4	3/4				
5	4/5	4/5			
6	5/6	5/6			
7	5/7	6/7			
8	3/4	7/8	7/8		
9	7/9	8/9	8/9		
10	7/10	4/5	9/10		
$N_1 = 3$ $N_2 = 4$	3/4	3/4			
5	2/3	4/5	4/5		
6	2/3	2/3	5/6		
7	2/3	5/7	6/7	6/7	
8	5/8	3/4	3/4	7/8	
9	2/3	2/3	7/9	8/9	8/9
10	3/5	7/10	4/5	9/10	9/10
12	7/12	2/3	3/4	5/6	11/12
$N_1 = 4$ $N_2 = 5$	3/5	3/4	4/5	4/5	
6	7/12	2/3	3/4	5/6	5/6
7	17/28	5/7	3/4	6/7	6/7
8	5/8	5/8	3/4	7/8	7/8
9	5/9	2/3	3/4	7/9	8/9
10	11/20	13/20	7/10	4/5	4/5
12	7/12	2/3	2/3	3/4	5/6
16	9/16	5/8	11/16	3/4	13/16
$N_1 = 5$ $N_2 = 6$	3/5	2/3	2/3	5/6	5/6
7	4/7	23/35	5/7	29/35	6/7
8	11/20	5/8	27/40	4/5	4/5
9	5/9	3/5	31/45	7/9	4/5
10	1/2	3/5	7/10	7/10	4/5
15	8/15	3/5	2/3	11/15	11/15
20	1/2	11/20	3/5	7/10	3/4
$N_1 = 6$ $N_2 = 7$	23/42	4/7	29/42	5/7	5/6
8	1/2	7/12	2/3	3/4	3/4
9	1/2	5/9	2/3	13/18	7/9
10	1/2	17/30	19/30	7/10	11/15
12	1/2	7/12	7/12	2/3	3/4
18	4/9	5/9	11/18	2/3	13/18
24	11/24	1/2	7/12	5/8	2/3
$N_1 = 7$ $N_2 = 8$	27/56	33/56	5/8	41/56	3/4
9	31/63	5/9	40/63	5/7	47/63
10	33/70	39/70	43/70	7/10	5/7
14	3/7	1/2	4/7	9/14	5/7
28	3/7	13/28	15/28	17/28	9/14
$N_1 = 8$ $N_2 = 9$	4/9	13/24	5/8	2/3	3/4
10	19/40	21/40	23/40	27/40	7/10
12	11/24	1/2	7/12	5/8	2/3
16	7/16	1/2	9/16	5/8	5/8
32	13/32	7/16	1/2	9/16	19/32

TABLE A.20(b)   (*continued*)

One-sided test	$p = 0.90$	0.95	0.975	0.99	0.995
Two-sided test	$p = 0.80$	0.90	0.95	0.98	0.99
$N_1 = 9$ $N_2 = 10$	7/15	1/2	26/45	2/3	31/45
12	4/9	1/2	5/9	11/18	2/3
15	19/45	22/45	8/15	3/5	29/45
18	7/18	4/9	1/2	5/9	11/18
36	13/36	5/12	17/36	19/36	5/9
$N_1 = 10$ $N_2 = 15$	2/5	7/15	1/2	17/30	19/30
20	2/5	9/20	1/2	11/20	3/5
40	7/20	2/5	9/20	1/2	
$N_1 = 12$ $N_2 = 15$	23/60	9/20	1/2	11/20	7/12
16	3/8	7/16	23/48	13/24	7/12
18	13/36	5/12	17/36	19/36	5/9
20	11/30	5/12	7/15	31/60	17/30
$N_1 = 15$ $N_2 = 20$	7/20	2/5	13/30	29/60	31/60
$N_1 = 16$ $N_2 = 20$	27/80	31/80	17/40	19/40	41/80
Large-sample approximation:	$1.07\sqrt{\dfrac{m+n}{mn}}$	$1.22\sqrt{\dfrac{m+n}{mn}}$	$1.36\sqrt{\dfrac{m+n}{mn}}$	$1.52\sqrt{\dfrac{m+n}{mn}}$	$1.63\sqrt{\dfrac{m+n}{mn}}$

Source: Frank J. Massey, Jr., "Distribution Table for the Deviation between Two Sample Cumulatives," *Ann. Math. Statist.*, 23 (1952), 435–441. This table incorporates the corrections reported in Louis S. Davis, "Table Errata 266," *Math. of Computation*, 12 (1958), 262–263

TABLE A.21

Critical values of Spearman's rank correlation coefficient

$\alpha(2):$ $\alpha(1):$ n	0.50 0.25	0.20 0.10	0.10 0.05	0.05 0.025	0.02 0.01	0.01 0.005	0.005 0.0025	0.002 0.001	0.001 0.0005
4	0.600	1.000	1.000						
5	0.500	0.800	0.900	1.000	1.000				
6	0.371	0.657	0.829	0.886	0.943	1.000	1.000		
7	0.321	0.571	0.714	0.786	0.893	0.929	0.964	1.000	1.000
8	0.310	0.524	0.643	0.738	0.833	0.881	0.905	0.952	0.976
9	0.267	0.483	0.600	0.700	0.783	0.833	0.867	0.917	0.933
10	0.248	0.455	0.564	0.648	0.745	0.794	0.830	0.879	0.903
11	0.236	0.427	0.536	0.618	0.709	0.755	0.800	0.845	0.873
12	0.217	0.406	0.503	0.587	0.678	0.727	0.769	0.818	0.846
13	0.209	0.385	0.484	0.560	0.648	0.703	0.747	0.791	0.824
14	0.200	0.367	0.464	0.538	0.626	0.679	0.723	0.771	0.802
15	0.189	0.354	0.446	0.521	0.604	0.654	0.700	0.750	0.779
16	0.182	0.341	0.429	0.503	0.582	0.635	0.679	0.729	0.762
17	0.176	0.328	0.414	0.485	0.566	0.615	0.662	0.713	0.748
18	0.170	0.317	0.401	0.472	0.550	0.600	0.643	0.695	0.728
19	0.165	0.309	0.391	0.460	0.535	0.584	0.628	0.677	0.712
20	0.161	0.299	0.380	0.447	0.520	0.570	0.612	0.662	0.696
21	0.156	0.292	0.370	0.435	0.508	0.556	0.599	0.648	0.681
22	0.152	0.284	0.361	0.425	0.496	0.544	0.586	0.634	0.667
23	0.148	0.278	0.353	0.415	0.486	0.532	0.573	0.622	0.654
24	0.144	0.271	0.344	0.406	0.476	0.521	0.562	0.610	0.642
25	0.142	0.265	0.337	0.398	0.466	0.511	0.551	0.598	0.630
26	0.138	0.259	0.331	0.390	0.457	0.501	0.541	0.587	0.619
27	0.136	0.255	0.324	0.382	0.448	0.491	0.531	0.577	0.608
28	0.133	0.250	0.317	0.375	0.440	0.483	0.522	0.567	0.598
29	0.130	0.245	0.312	0.368	0.433	0.475	0.513	0.558	0.589
30	0.128	0.240	0.306	0.362	0.425	0.467	0.504	0.549	0.580
31	0.126	0.236	0.301	0.356	0.418	0.459	0.496	0.541	0.571
32	0.124	0.232	0.296	0.350	0.412	0.452	0.489	0.533	0.563
33	0.121	0.229	0.291	0.345	0.405	0.446	0.482	0.525	0.554
34	0.120	0.225	0.287	0.340	0.399	0.439	0.475	0.517	0.547
35	0.118	0.222	0.283	0.335	0.394	0.433	0.468	0.510	0.539
36	0.116	0.219	0.279	0.330	0.388	0.427	0.462	0.504	0.533
37	0.114	0.216	0.275	0.325	0.383	0.421	0.456	0.497	0.526
38	0.113	0.212	0.271	0.321	0.378	0.415	0.450	0.491	0.519
39	0.111	0.210	0.267	0.317	0.373	0.410	0.444	0.485	0.513
40	0.110	0.207	0.264	0.313	0.368	0.405	0.439	0.479	0.507
41	0.108	0.204	0.261	0.309	0.364	0.400	0.433	0.473	0.501
42	0.107	0.202	0.257	0.305	0.359	0.395	0.428	0.468	0.495
43	0.105	0.199	0.254	0.301	0.355	0.391	0.423	0.463	0.490
44	0.104	0.197	0.251	0.298	0.351	0.386	0.419	0.458	0.484
45	0.103	0.194	0.248	0.294	0.347	0.382	0.414	0.453	0.479
46	0.102	0.192	0.246	0.291	0.343	0.378	0.410	0.448	0.474
47	0.101	0.190	0.243	0.288	0.340	0.374	0.405	0.443	0.469
48	0.100	0.188	0.240	0.285	0.336	0.370	0.401	0.439	0.465
49	0.098	0.186	0.238	0.282	0.333	0.366	0.397	0.434	0.460
50	0.097	0.184	0.235	0.279	0.329	0.363	0.393	0.430	0.456

α(2):	0.50	0.20	0.10	0.05	0.02	0.01	0.005	0.002	0.001
α(1):	0.25	0.10	0.05	0.025	0.01	0.005	0.0025	0.001	0.0005
n									
51	0.096	0.182	0.233	0.276	0.326	0.359	0.390	0.426	0.451
52	0.095	0.180	0.231	0.274	0.323	0.356	0.386	0.422	0.447
53	0.095	0.179	0.228	0.271	0.320	0.352	0.382	0.418	0.443
54	0.094	0.177	0.226	0.268	0.317	0.349	0.379	0.414	0.439
55	0.093	0.175	0.224	0.266	0.314	0.346	0.375	0.411	0.435
56	0.092	0.174	0.222	0.264	0.311	0.343	0.372	0.407	0.432
57	0.091	0.172	0.220	0.261	0.308	0.340	0.369	0.404	0.428
58	0.090	0.171	0.218	0.259	0.306	0.337	0.366	0.400	0.424
59	0.089	0.169	0.216	0.257	0.303	0.334	0.363	0.397	0.421
60	0.089	0.168	0.214	0.255	0.300	0.331	0.360	0.394	0.418
61	0.088	0.166	0.213	0.252	0.298	0.329	0.357	0.391	0.414
62	0.087	0.165	0.211	0.250	0.296	0.326	0.354	0.388	0.411
63	0.086	0.163	0.209	0.248	0.293	0.323	0.351	0.385	0.408
64	0.086	0.162	0.207	0.246	0.291	0.321	0.348	0.382	0.405
65	0.085	0.161	0.206	0.244	0.289	0.318	0.346	0.379	0.402
66	0.084	0.160	0.204	0.243	0.287	0.316	0.343	0.376	0.399
67	0.084	0.158	0.203	0.241	0.284	0.314	0.341	0.373	0.396
68	0.083	0.157	0.201	0.239	0.282	0.311	0.338	0.370	0.393
69	0.082	0.156	0.200	0.237	0.280	0.309	0.336	0.368	0.390
70	0.082	0.155	0.198	0.235	0.278	0.307	0.333	0.365	0.388
71	0.081	0.154	0.197	0.234	0.276	0.305	0.331	0.363	0.385
72	0.081	0.153	0.195	0.232	0.274	0.303	0.329	0.360	0.382
73	0.080	0.152	0.194	0.230	0.272	0.301	0.327	0.358	0.380
74	0.080	0.151	0.193	0.229	0.271	0.299	0.324	0.355	0.377
75	0.079	0.150	0.191	0.227	0.269	0.297	0.322	0.353	0.375
76	0.078	0.149	0.190	0.226	0.267	0.295	0.320	0.351	0.372
77	0.078	0.148	0.189	0.224	0.265	0.293	0.318	0.349	0.370
78	0.077	0.147	0.188	0.223	0.264	0.291	0.316	0.346	0.368
79	0.077	0.146	0.186	0.221	0.262	0.289	0.314	0.344	0.365
80	0.076	0.145	0.185	0.220	0.260	0.287	0.312	0.342	0.363
81	0.076	0.144	0.184	0.219	0.259	0.285	0.310	0.340	0.361
82	0.075	0.143	0.183	0.217	0.257	0.284	0.308	0.338	0.359
83	0.075	0.142	0.182	0.216	0.255	0.282	0.306	0.336	0.357
84	0.074	0.141	0.181	0.215	0.254	0.280	0.305	0.334	0.355
85	0.074	0.140	0.180	0.213	0.252	0.279	0.303	0.332	0.353
86	0.074	0.139	0.179	0.212	0.251	0.277	0.301	0.330	0.351
87	0.073	0.139	0.177	0.211	0.250	0.276	0.299	0.328	0.349
88	0.073	0.138	0.176	0.210	0.248	0.274	0.298	0.327	0.347
89	0.072	0.137	0.175	0.209	0.247	0.272	0.296	0.325	0.345
90	0.072	0.136	0.174	0.207	0.245	0.271	0.294	0.323	0.343
91	0.072	0.135	0.173	0.206	0.244	0.269	0.293	0.321	0.341
92	0.071	0.135	0.173	0.205	0.243	0.268	0.291	0.319	0.339
93	0.071	0.134	0.172	0.204	0.241	0.267	0.290	0.318	0.338
94	0.070	0.133	0.171	0.203	0.240	0.265	0.288	0.316	0.336
95	0.070	0.133	0.170	0.202	0.239	0.264	0.287	0.314	0.334
96	0.070	0.132	0.169	0.201	0.238	0.262	0.285	0.313	0.332
97	0.069	0.131	0.168	0.200	0.236	0.261	0.284	0.311	0.331
98	0.069	0.130	0.167	0.199	0.235	0.260	0.282	0.310	0.329
99	0.068	0.130	0.166	0.198	0.234	0.258	0.281	0.308	0.327
100	0.068	0.129	0.165	0.197	0.233	0.257	0.279	0.307	0.326

Source: Jerrold H. Zar, *Biostatistical Analysis*, 2e, © 1984, pp. 577–578. Reprinted by permission of Prentice Hall, Inc., Englewood Cliffs, New Jersey.

TABLE A.22   **Critical values for use with the Kendall tau statistic**

α	0.005		0.010		0.025		0.050		0.100	
n	S	τ*	S	τ*	S	τ*	S	τ*	S	τ*
4	8	1.000	8	1.000	8	1.000	6	1.000	6	1.000
5	12	1.000	10	1.000	10	1.000	8	.800	8	.800
6	15	1.000	13	.867	13	.867	11	.733	9	.600
7	19	.905	17	.810	15	.714	13	.619	11	.524
8	22	.786	20	.714	18	.643	16	.571	12	.429
9	26	.722	24	.667	20	.556	18	.500	14	.389
10	29	.644	27	.600	23	.511	21	.467	17	.378
11	33	.600	31	.564	27	.491	23	.418	19	.345
12	38	.576	36	.545	30	.455	26	.394	20	.303
13	44	.564	40	.513	34	.436	28	.359	24	.308
14	47	.516	43	.473	37	.407	33	.363	25	.275
15	53	.505	49	.467	41	.390	35	.333	29	.276
16	58	.483	52	.433	46	.383	38	.317	30	.250
17	64	.471	58	.426	50	.368	42	.309	34	.250
18	69	.451	63	.412	53	.346	45	.294	37	.242
19	75	.439	67	.392	57	.333	49	.287	39	.228
20	80	.421	72	.379	62	.326	52	.274	42	.221
21	86	.410	78	.371	66	.314	56	.267	44	.210
22	91	.394	83	.359	71	.307	61	.264	47	.203
23	99	.391	89	.352	75	.296	65	.257	51	.202
24	104	.377	94	.341	80	.290	68	.246	54	.196
25	110	.367	100	.333	86	.287	72	.240	58	.193
26	117	.360	107	.329	91	.280	77	.237	61	.188
27	125	.356	113	.322	95	.271	81	.231	63	.179
28	130	.344	118	.312	100	.265	86	.228	68	.180
29	138	.340	126	.310	106	.261	90	.222	70	.172
30	145	.333	131	.301	111	.255	95	.218	75	.172
31	151	.325	137	.295	117	.252	99	.213	77	.166
32	160	.323	144	.290	122	.246	104	.210	82	.165
33	166	.314	152	.288	128	.242	108	.205	86	.163
34	175	.312	157	.280	133	.237	113	.201	89	.159
35	181	.304	165	.277	139	.234	117	.197	93	.156
36	190	.302	172	.273	146	.232	122	.194	96	.152
37	198	.297	178	.267	152	.228	128	.192	100	.150
38	205	.292	185	.263	157	.223	133	.189	105	.149
39	213	.287	193	.260	163	.220	139	.188	109	.147
40	222	.285	200	.256	170	.218	144	.185	112	.144

Source: L. Kaarsemaker and A. van Wijngaarden, "Tables for Use in Rank Correlation," *Statistica Neerlandica*, 7 (1953), 41–54.

* The column labeled S contains, for each n, the smallest value of S for which $P(S \geq S) \leq \alpha$. The column labeled τ* contains, for each n, the smallest value of τ* for which $P(\tau \geq \tau^*) \leq \alpha$.

TABLE A.23   **Probability of a sum of absolute value equal to or greater than S when a sample of n is drawn from an unassociated population**

S	n 2	4	6	8	10	12	14	∞
0	1.0000	1.0000	1.0000	1.0000	1.0000		1.0000	1.000000
1	1.0000	0.7500	0.9333	0.9036	0.9106		0.9115	0.912037
2	1.0000	0.7500	0.7556	0.7544	0.7567		0.7580	0.754630
3	1.0000	0.4167	0.6000	0.6000	0.6008		0.6039	0.599537
4	1.0000	0.4167	0.4667	0.4619	0.4662		0.4690	0.462963
5	0.0000	0.3333	0.3111	0.3508	0.3519		0.3547	0.346933
6	0.0000	0.3333	0.2222	0.2619	0.2589		0.2611	0.252025
7	0.0000	0.3333	0.1556	0.1821	0.1867		0.1876	0.177662
8	0.0000	0.3333	0.1111	0.1258	0.1333		0.1322	0.121817
9	0.0000	0.0000	0.1000	0.0839	0.0928		0.0918	0.081471
10	0.0000	0.0000	0.1000	0.0554	0.0642		0.0632	0.053295
11	0.0000	0.0000	0.1000	0.0375	0.0436		0.0432	0.034189
12	0.0000	0.0000	0.1000	0.0304	0.0290		0.0296	0.021557
13	0.0000	0.0000	0.0000	0.0286	0.0190		0.0202	0.013386
14	0.0000	0.0000	0.0000	0.0286	0.0127		0.0139	0.008200
15	0.0000	0.0000	0.0000	0.0286	0.0095		0.0096	0.004963
16	0.0000	0.0000	0.0000	0.0286	0.0083		0.0066	0.002972
17	0.0000	0.0000	0.0000	0.0000	0.0079		0.0045	0.001762
18	0.0000	0.0000	0.0000	0.0000	0.0079		0.0031	0.001036
19	0.0000	0.0000	0.0000	0.0000	0.0079		0.0021	0.000604
20	0.0000	0.0000	0.0000	0.0000	0.0079		0.0014	0.000350
21	0.0000	0.0000	0.0000	0.0000	0.0000		0.0010	0.000201
22	0.0000	0.0000	0.0000	0.0000	0.0000		0.0008	0.000115
23	0.0000	0.0000	0.0000	0.0000	0.0000		0.0006	0.000065
24	0.0000	0.0000	0.0000	0.0000	0.0000		0.0006	0.000036
25	0.0000	0.0000	0.0000	0.0000	0.0000		0.0006	0.000020
26	0.0000	0.0000	0.0000	0.0000	0.0000		0.0006	0.000011
27	0.0000	0.0000	0.0000	0.0000	0.0000		0.0006	0.000006
28	0.0000	0.0000	0.0000	0.0000	0.0000		0.0006	0.000003
29	0.0000	0.0000	0.0000	0.0000	0.0000		0.0000	0.000002
30	0.0000	0.0000	0.0000	0.0000	0.0000		0.0000	0.000001
31 or over	0.0000	0.0000	0.0000	0.0000	0.0000		0.0000	0.000000

Source: P. S. Olmstead and John W. Tukey, "A Corner Test for Association," *Ann. Math. Statist.*, 18 (1947), 495–513.

TABLE A.24

Estimates of the quantiles of Kendall's partial rank correlation coefficient

1 − α n	Quantiles 0.75	0.80	0.85	0.90	0.925	0.950	0.975	0.98	0.99	0.995	0.999
3	0.50	1	—	1	1	1	1	1	1	1	1
4	0.4472	0.5000	—	0.7071	0.7071	0.7071	1	1	1	1	1
5	0.3333	0.4082	0.4286	0.5345	0.6124	0.6667	0.8018	0.8165	0.8165	1.0	1.0
6	0.2773	0.3273	0.3889	0.4725	0.5330	0.6001	0.6667	0.7222	0.7638	0.8660	1.0
7	0.233	0.282	—	0.421	0.475	0.527	0.617	0.632	0.712	0.761	0.901
8	0.206	0.254	—	0.382	0.430	0.484	0.565	0.580	0.648	0.713	0.807
9	0.187	0.230	—	0.347	0.391	0.443	0.515	0.542	0.602	0.660	0.757
10	0.170	0.215	—	0.325	0.365	0.413	0.480	0.504	0.562	0.614	0.718
11	0.162	0.202	—	0.305	0.343	0.387	0.453	0.475	0.530	0.581	0.677
12	0.153	0.190	—	0.288	0.322	0.465	0.430	0.451	0.505	0.548	0.643
13	0.145	0.180	—	0.273	0.305	0.347	0.410	0.428	0.481	0.527	0.616
14	0.137	0.172	—	0.260	0.293	0.331	0.391	0.408	0.458	0.503	0.590
15	0.133	0.166	0.204	0.251	0.280	0.319	0.377	0.394	0.442	0.485	0.570
16	0.125	0.157	—	0.240	0.267	0.305	0.361	0.377	0.423	0.466	0.549
17	0.121	0.151	—	0.231	0.258	0.294	0.348	0.363	0.410	0.450	0.532
18	0.117	0.147	—	0.222	0.250	0.284	0.336	0.351	0.395	0.434	0.514
19	0.114	0.141	—	0.215	0.241	0.275	0.326	0.340	0.382	0.421	0.498
20	0.111	0.139	0.170	0.210	0.236	0.268	0.318	0.332	0.374	0.412	0.488
25	0.098	0.122	0.149	0.185	0.207	0.236	0.279	0.293	0.329	0.363	0.430
30	0.088	0.110	0.135	0.167	0.187	0.213	0.253	0.264	0.298	0.329	0.390
35	0.081	0.101	0.124	0.153	0.171	0.196	0.232	0.243	0.274	0.303	0.361
40	0.075	0.094	0.115	0.142	0.159	0.182	0.216	0.226	0.255	0.282	0.335
45	0.071	0.088	0.108	0.133	0.150	0.171	0.203	0.212	0.240	0.265	0.316
50	0.067	0.083	0.102	0.126	0.141	0.161	0.192	0.201	0.225	0.250	0.298
60	0.060	0.075	0.093	0.114	0.128	0.147	0.174	0.182	0.206	0.227	0.270
70	0.056	0.070	0.086	0.106	0.119	0.135	0.160	0.168	0.190	0.210	0.251
80	0.052	0.065	0.080	0.098	0.110	0.126	0.150	0.157	0.178	0.197	0.235
90	0.049	0.061	0.075	0.092	0.104	0.119	0.141	0.148	0.167	0.185	0.221

Source: S. Maghsoodloo and L. Laszlo Pallos, "Asymptotic Behavior of Kendall's Partial Rank Correlation Coefficient and Additional Quantile Estimates," *J. Statist. Comput. Simul.* Vol. 13 (1981), pp. 41–48; and S. Maghsoodloo, "Estimates of the Quantiles of Kendall's Partial Rank Correlation Coefficient," *J. Statist. Comput. Simul.* Vol. 4 (1975), pp. 155–164; reprinted by permission of the copyright holder, Gordon and Breach Science Publishers, Inc.

SOLUTIONS TO ODD-NUMBERED EXERCISES

CHAPTER 2

2.1 Calculation of $X_i - M_0$ yields nine negative differences, two positive differences, and one zero difference.

$$k = 2, \quad P(k \leq 2 \,|\, 11, 0.50) = 0.0328$$

Since $0.0328 < 0.05$, reject H_0 and conclude that H_1 is true.

$$k' = 2, P \text{ value} = 0.0328.$$

2.3

$D_i = X_i - M_0$	+24.5	+47.7	+7.3	+48.9	−6.6
Signed rank of $|D_i|$	+6	+9	+3	+10	−2
$D_i = X_i - M_0$	+59.6	+72.4	+20.4	+50.9	+57.5
Signed rank of $|D_i|$	+14	+15	+5	+11	+12
$D_i = X_i - M_0$	−1.5	+59.3	+10.6	+46.8	+31.7
Signed rank of $|D_i|$	−1	+13	+4	+8	+7

$T_+ = 117, T_- = 3$ (Test statistic)

Since $3 < 16$, reject H_0 at the 0.004 level of significance. P value < 0.004.

2.5

$D_i = X_i - M_0$	+17	−16	−28	+3	−23	+26	+21	+19
Signed rank of $|D_i|$	+6	−5	−14.5	+2	−11.5	+13	+10	+8
$D_i = X_i - M_0$	+7	+1	−20	+6	−28	+23	0	+18
Signed rank of $|D_i|$	+4	+1	−9	+3	−14.5	+11.5	—	+7

$T_+ = 65.5, T_- = 54.5$ (Test statistic)

Since $55 > 54.5 > 54$, then $2(0.4020) > P$ value $> 2(0.3808)$; $0.8040 > P$ value > 0.7616. Do not reject H_0.

2.7 Sample median $= 10$. The ordered sample: 6, 6, 7, 7, 8, 8, 9, 10, 11, 12, 12, 13, 14, 15, 16.

$M_L = 7, M_U = 13$.

The confidence coefficient is 0.9648.

2.9 The ordered sample: 0.64, 0.65, 0.70, 0.71, 0.78, 0.82, 0.85, 0.86, 0.86, 1.00, 1.10

When $n = 11$ and $1 - \alpha = 0.9462$, $K = 12$. There are $[(11)(10)/2] + 11 = 66$ averages. The $(66/2) + 1 = 34$ smallest averages are as follows.

0.640	0.680	0.730	0.750	0.775	0.785
0.645	0.700	0.735	0.750	0.780	0.800
0.650	0.705	0.740	0.755	0.780	0.815
0.670	0.710	0.745	0.755	0.780	0.820
0.675	0.710	0.745	0.760	0.780	
0.675	0.715	0.750	0.765	0.785	

The 12 largest averages are as follows: 1.10, 1.05, 1.00, 0.98, 0.98, 0.975, 0.960, 0.940, 0.930, 0.930, 0.925, 0.910. The estimate of the population median is $(0.815 + 0.820)/2 = 0.8175$. The lower and upper limits of the 94.62% confidence interval for the population median are 0.715 and 0.910.

2.11 The ordered sample: 4290, 5280, 5280, 5555, 5610. With $n = 5$ and $1 - \alpha = 0.9374$, $K = 2$. There are $[5(4)/2] + 5 = 15$ averages. The eight smallest averages are 4290, 4785, 4785, 4922.5, 4950, 5280, 5280, 5280. The point estimate of the population median is 5280. The two largest averages are 5610.0 and 5582.5. The lower and upper limits of the 93.74% confidence interval are 4290 and 5582.5.

2.13 $S = 6, s = 6$. Since P value $= 0.0690$, we cannot reject H_0.

2.15 $S = 4, s = 5$ when $\alpha = 0.0065$. Since $4 < 5$, we cannot reject H_0. P value $= 0.0936$.

2.17 From Table A.4, the lower limit $= 0.236$ and the upper limit $= 0.675$.

2.19 $n_1 =$ number of days with 50% or less $= 10$
$n_2 =$ number of days with more than 50% $= 20$
$r = 14$
Critical values of r are 9 and 20.
Since $9 < 14 < 20$, we cannot reject the hypothesis of randomness.

2.21 $n_1 =$ number of observations above $1435 = 17$
$n_2 =$ number of observations below $1435 = 15$
$r = 19$
Critical values of r are 11 and 23.
Since $11 < 19 < 23$, we cannot reject the hypothesis of randomness.

2.23 $n' = 25$; $C = (25 + 1)/2 = 13$; $n = 12$. There are 9 plus signs and 3 minus signs. $P(K \leq 3 \mid 12, 0.50) = 0.0729$. We cannot reject H_0.

2.25 The differences $X_i - 20$ are: 2, 4, 17, 8, -5, -6, 2, -4, -2, -3, 3, -4, 0, -2, -5. $P(K \leq 6 \mid 14, 0.50) = 0.3954 = P$ value. The population median may be 20.

2.27 $P(r \geq 6 \mid 20, 0.25) = 0.3829 = P$ value. We cannot reject $H_0: p \leq 0.25$.

2.29

$D_i = X_i - 70$	+10	−2	−40	−3	0	−8	−1
Signed rank of $\|D_i\|$	+12	−4	−22	−6	Delete	−9.5	−1.5
$D_i = X_i - 70$	−5	−17	−41	−5	−2	−8	
Signed rank of $\|D_i\|$	−7.5	−15.5	−23	−7.5	−4	−9.5	
$D_i = X_i - 70$	−14	−24	−22	−31	+2	−34	−1
Signed rank of $\|D_i\|$	−13	−18	−17	−20	+4	−21	−1.5
$D_i = X_i - 70$	−30	−9	−16	−17	−45		
Signed rank of $\|D_i\|$	−19	−11	−14	−15.5	−24		

$T_+ = 16$, P value < 0.0001

2.31 $r = 10$, $n_1 = 12$, $n_2 = 12$. Since $7 < 10 < 19$, we cannot reject the null hypothesis of randomness.

2.33 $P(r \geq 11 \mid 25, 0.5) = 0.7878$. The data do not support ESP.

2.35 $n' = 30$, $\quad C = 30/2 = 15$
There are 15 minus signs and 0 plus signs.
$P(K \leq 0 \mid 15, 0.50) = 0$
Since $0 < 0.025$, reject H_0. There is a trend.

2.37 $S = 9$, $\quad s = 10$
Since $9 < 10$, we cannot reject H_0.
P value $= 0.0593$.

2.39 The Wilcoxon signed-ranks test used the most information contained in the data.

$D_i = X_i - M_0$	Signed rank of $\|D_i\|$
−24	−19.5
−23	−17.5
−22	−16
−21	−14.5
−12	−8
−11	−7
−6	−4.5
−5	−3
1	1
2	2
6	4.5
10	6
14	9
15	10
16	11.5
16	11.5
17	13
21	14.5
23	17.5
24	19.5

$T_+ = 120$
$T_- = 90$ (Test statistic)

Since $90 > 43$, we cannot reject H_0 at 0.01 level of significance.
P value $= 0.2979$

2.41 The ordered sample: 27,500, 28,900, 29,300, 29,900, 30,100, 30,300, 31,200, 32,200, 32,500, 33,900, 34,300, 34,900, 35,000, 35,100, 35,300, 36,200, 37,200, 38,000, 43,000.
The estimate of the population median is 33,900.

$$P(K \leq 4 \mid 19, 0.50) = 0.0095$$

$$P(K \leq 5 \mid 19, 0.50) = 0.0317$$

$$K' = 4, \quad K' + 1 = 5$$

$$M_L = 30,100$$

$$M_U = 35,300$$

2.43 $\hat{p} = 2900/4000 = 0.725$

$$0.725 \pm 1.96 \sqrt{\frac{(0.725)(0.275)}{4000}}$$

$$0.725 \pm 0.0138$$

$$C(0.711 \leq p \leq 0.739) = 0.95$$

CHAPTER 3

3.1 The sample median is 14.
Since 4 of the 116 observations are equal to the sample median, a method of handling ties must be employed. For demonstration purposes, the first two methods mentioned in the text will be illustrated.

a. Eliminate from the analysis the 4 observations that are equal to the median. The resulting contingency table is as follows.

	Less than three months	Three months or more	Total
Above 14	33	23	56
Below 14	38	18	56
Total	71	41	112

$$\hat{p} = \frac{33 + 23}{71 + 41} = 0.50$$

$$T = \frac{(33/71) - (23/41)}{\sqrt{(0.50)(1 - 0.50)(1/71 + 1/41)}} = -0.98$$

Since $-1.96 < -0.98 < 1.96$, H_0 is not rejected.
P value $= 2(0.1635) = 0.3270$

b. Categorize the observations as exceeding the median or not exceeding the median. The resulting contingency table is as follows.

	Less than three months	Three months or more	Total
Above 14	33	23	56
14 or below	40	20	60
Total	$\overline{73}$	$\overline{43}$	$\overline{116}$

$$\hat{p} = \frac{33 + 23}{116} = 0.48$$

$$T = \frac{(33/73) - (23/43)}{\sqrt{(0.48)(1 - 0.48)(1/73 + 1/43)}} = -0.86$$

P value $= 2(0.1949) = 0.3898$

3.3

X	Rank	Y	Rank
		23	1
		40	2
42	3		
		45	4
		50	5
58	6		
62	7		
		68	8
73	9		
90	10		
Total	35		

$$w_\alpha = 5, \qquad w_{1-\alpha} = (5)(5) - 5 = 20$$

H_0 cannot be rejected by the Mann–Whitney test.
P value $> 0.10, \qquad T = 35 - 5(6)/2 = 20$

3.5 **a.** Graphic solution

$$w_{0.025} = 32, \qquad L \approx -16, \qquad U \approx 4$$

Since the horizontal and vertical scales are the same, and since the constructed lines are each at a 45-degree angle to the horizontal axis, the magnitudes of the endpoints of the confidence interval, except for sign, are determined once the constructed lines cross either axis. The point where a line crosses the vertical axis gives an endpoint value of the confidence interval, but the sign is changed. In the present example, the top line crosses the vertical axis at approximately $+16$. The sign is changed to give an endpoint of -16.

Graph for Exercise 3.5

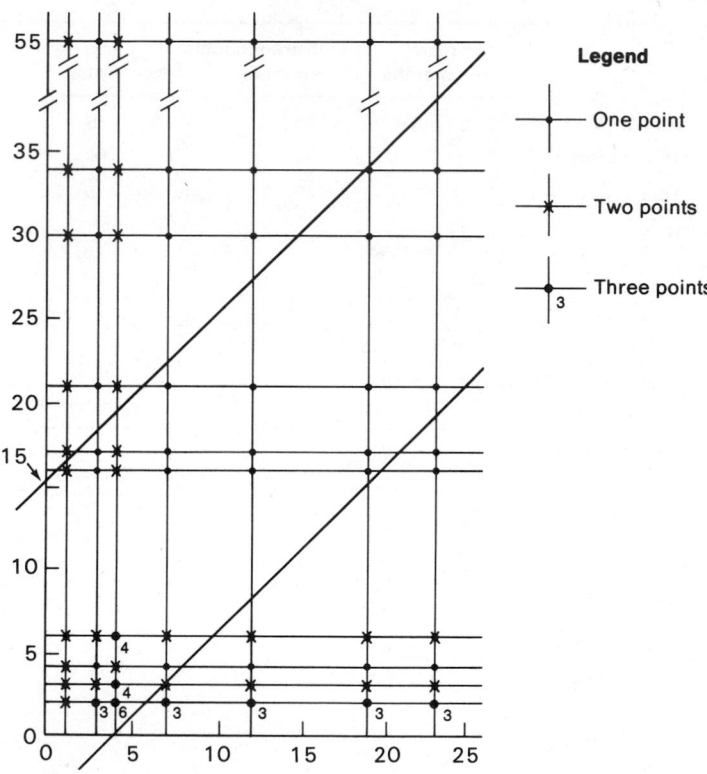

b. Arithmetic solution
Ordered samples are:
X: 1, 1, 3, 4, 4, 7, 12, 19, 23
Y: 2, 2, 2, 3, 3, 4, 6, 6, 16, 17, 21, 30, 34, 53
$w_{\alpha/2} = 32$

X	1	1	3	4	4	7	12	19	23
Y									
2	−1	−1	1	2	2	5	10	17	21
2	−1	−1	1	2	2	5	10	17	21
2	−1	−1	1	2	2	5	10	17	21
3	−2	−2	0	1	1	4	9	16	20
3	−2	−2	0	1	1	4	9	16	20
4	−3	−3	−1	0	0	3	8	15	19
6	−5	−5	−3	−2	−2	1	6	13	17
6	−5	−5	−3	−2	−2	1	6	13	17
16	−15	−15	−13	−12	−12	−9	−4	3	7
17	−16	−16	−14	−13	−13	−10	−5	2	6
21	−20	−20	−18	−17	−17	−14	−9	−2	2
30	−29	−29	−27	−26	−26	−23	−18	−11	−7
34	−33	−33	−31	−30	−30	−27	−22	−15	−11
53	−52	−52	−50	−49	−49	−46	−41	−34	−30

$$L = -15, \quad U = 3$$

3.7

X	Rank	Y	Rank
328	1		
		336	2
347	3		
		372	4
		425	5
		428	5
433	4		
		434	3
478	2		
607	1		
Total	11		

$$n_1 = 5, \quad n_2 = 5$$

Upper critical value is 20 for $\alpha/2 = 0.0238$, and lower critical value is 11 for $1 - \alpha/2 = 0.9762$. Reject H_0.

3.11

$$\begin{array}{ccccccccc} 3 & 4 & \cdots & 12.2 & 12.8 & 15 & 17 & 40 & \cdots & 1100 \\ X & X & \cdots & X & Y & X & X & Y & \cdots & Y \end{array}$$

$$r = 4$$

$$z = \frac{4 - \left[\dfrac{2(16)(27)}{16 + 27} + 1 \right]}{\sqrt{\dfrac{2(16)(27)[2(16)(27) - 16 - 27)]}{(16 + 27)^2(16 + 27 - 1)}}} = \frac{-17.09}{3.02} = -5.66$$

Since $-5.66 < -1.96$, reject H_0 and conclude that the population distribution functions are different.

P value < 0.001

3.13

Observation	Group	Rank
35	Y	1
38	Y	2
40	X	3
41	X	4
42	X	5
43	Y	6
48	X	7
50	X	8
51	X	9
53	X	10
54	Y	11
55	Y	12
56	X	13
57	X	14
58	X	15
65	Y	16
69	Y	17
75	Y	18
85	Y	19
88	Y	20

$$\bar{r} = 8.8$$

$$G = (3 - 8.8)^2 + (4 - 8.8)^2 + \cdots + (15 - 8.8)^2 = 159.60$$

Since $159.60 < 196.4$, reject H_0.

3.15 After rearranging the data, $A = 12$, $B = 10$, and $a = 9$.
From Table A.10, the critical value of b is found to be 3. Since the observed value of $b = 6$ (after rearrangement) is greater than 3, H_0 cannot be rejected.
P value > 0.05

3.17

X	Rank	Y	Rank
206	1		
211	2		
213	3		
229	4		
258	5		
267	6		
281	7		
281	8		
		290	9
		290	9
317	8		
321	7		
		360	6
		400	5
		403	4
		420	3
		460	2
		660	1
Total	51		

$$n_1 = 8, \qquad n_2 = 10$$

Upper critical value is 52 for $\alpha/2 = 0.0187$, and lower critical value is 29 for $1 - \alpha/2 = 0.9813$. We cannot reject H_0.

3.19

X	Rank	Y	Rank
18	1		
		21	2
		24	3
		28	4
124	5		
163	6		
	$\overline{12}$		$\overline{9}$

$T = 12 - 3(4)/2 = 6$

α	$W_{1-\alpha/2}$
0.001	$(3)(3) - 0 = 9$
0.005	$(3)(3) - 0 = 9$
0.01	$(3)(3) - 0 = 9$
0.025	$(3)(3) - 0 = 9$
0.05	$(3)(3) - 1 = 8$
0.10	$(3)(3) - 1 = 7$

Since $6 > 7$, P value > 0.20.

3.21

X	Rank	Y	Rank
0.75	1		
1.76	2		
2.48	3		
4.88	4		
5.10	5		
		5.68	6
		5.68	7
6.01	8		
7.13	9		
		11.63	10
		16.30	11
		21.46	12
		33.30	13
		44.20	14
	$\overline{32}$		$\overline{73}$

$T = 32 - \dfrac{7(8)}{2} = 4$

Since $4 < 5$, reject H_0.
$0.002 < P$ value < 0.01

3.23 $w_{\alpha/2} = 24$

B	A 32.0	32.5	34.0	40.0	49.0	51.5	52.0	52.0	65.0	74.0
27.0	5	5.5	7	13	22	24.5	25	25	38	47
28.0	4	4.5	6	12	21	23.5	24	24	37	46
30.0	2	2.5	4	10	19	21.5	22	22	35	44
30.5	1.5	2	3.5	9.5	18.5	21	21.5	21.5	34.5	43.5
31.0	1	1.5	3	9	18	20.5	21	21	34	43
31.0	1	1.5	3	9	18	20.5	21	21	34	43
31.5	0.5	1	2.5	8.5	17.5	20	20.5	20.5	33.5	42.5
32.5	−0.5	0	1.5	7.5	16.5	19	19.5	19.5	32.5	41.5
33.0	−1	−0.5	1	7	16	18.5	19	19	32	41
34.5	−2.5	−2	−0.5	5.5	14.5	17	17.5	17.5	30.5	39.5

$L = 4, \quad U = 24$

3.25

	EH	EMR	Total
> 72	13	4	17
≤ 72	12	16	28
Total	25	20	45

$$\hat{p} = \frac{13 + 4}{25 + 20} = 0.38$$

$$T = \frac{(13/25) - (4/20)}{\sqrt{(0.38)(0.62)(1/25 + 1/20)}} = 2.20$$

Since $2.20 > 1.96$, H_0 is rejected.
P value $= 2(0.0217) = 0.0434$

3.27

X	Rank	Y	Rank
14.3	1		
14.6	2.5		
14.6	2.5		
14.7	4		
14.9	5		
		15.5	6.5
		15.5	6.5
15.8	8		
16.1	9		
16.4	10		
		16.7	11.5
		16.7	11.5
16.8	14		
		16.8	14
		16.8	14
17.2	16.5		
		17.2	16.5
		17.6	18
		17.9	19
		18	20
Total	72.5		

$$T = 72.5 - \frac{(10)(11)}{2} = 17.5$$

$$w_\alpha = 28$$

Since $17.5 < 28$, reject H_0.

3.29

X	Rank	Y	Rank	X	Rank	Y	Rank
		20	1	32	13.5		
		22	3			33	15
		22	3			34	17
		22	3	34	17		
		23	5	34	17		
		26	6	35	19		
		27	7	37	20		
		29	8	39	21.5		
		30	9.5	39	21.5		
30	9.5			40	23		
		31	11.5	43	24.5		
31	11.5			43	24.5		
32	13.5			45	26		
				Total	262		

$$T = 262 - \frac{14(15)}{2} = 157$$

Reject H_0, since $157 > [(12)(14) - 46]$; $157 > 122$.

3.31

X	Rank	Y	Rank	X	Rank	Y	Rank
		8	1			16.5	19
		8.75	2	17	20		
		9.1	3			17.25	21
10	4					17.75	22
11	5.5					18	23
		11	5.5	18.2	24		
		11.2	7	19	25.5		
12.5	8.5					19	25.5
12.5	8.5					20	27.5
		13	10.5	20	27.5		
		13	10.5			20.3	29
		13.25	12	21	30		
		13.5	13			21.2	31
		15	14.5	21.5	32		
15	14.5			22.25	33		
15.4	16			22.5	34		
		16.1	17	23	35		
		16.3	18	**Total**	318		

$$w_{\alpha/2} = 91, \qquad w_{1-\alpha/2} = (15)(20) - 91 = 209$$

$$T = 318 - \frac{(15)(16)}{2} = 198$$

We cannot reject H_0 by the Mann–Whitney test.

3.33

A	Rank	B	Rank	A	Rank	B	Rank
60	1.5					70	18.5
60	1.5					70	18.5
62	4			70	18.5		
62	4			70	18.5		
62	4			71	22.5		
		63	6	71	22.5		
65	7.5					71	22.5
65	7.5					71	22.5
66	9					72	27
		67	10			72	27
		68	13			72	27
		68	13			72	27
		68	13	72	27		
68	13					74	30
68	13					75	31
		69	16		174		

$$w_\alpha = 78, \qquad w_{1-\alpha} = (15)(16) - 78 = 162$$

$$T = 174 - \frac{(15)(16)}{2} = 54$$

Since $54 < 162$, we cannot reject H_0.

3.35 $T = 1 + 2 + 3 + 4 + 5 + 6 + 7 + 8 + 13 + 17 + 18 + 17 + 16 + 13 + 10 + 9 + 8 + 7 + 3 = 167$

$n_1 = 19, \qquad n_2 = 16$

Since $19 + 16 = 35$ is odd, we compute the large sample approximation as:

$$T^* = \frac{167 - [19(19 + 16 + 1)^2/4(19 + 16)]}{(19)(16)[16(19 + 16)(624.5) - (19 + 16 + 1)^4]/16(16 + 19)^2(19 + 16 - 1)}$$

$$T^* = 0.01$$

Since $0.01 < 1.96$, we cannot reject H_0.

3.37 Answers vary, depending on results of randomization. For example, let $k = 5$; then $m_1 = 2$.
Subdivision of group A:

Subsample	Observations	Sums of squares
1	9, 11, 9, 13, 10	11.2
2	8, 7, 12, 11, 9	12.4

Subdivision of group B:

Subsample	Observations	Sums of squares
1	12, 10, 13, 11, 11	5.2
2	15, 15, 14, 15, 11	12
3	14, 14, 13, 13, 9	17.2

Group A	Rank	Group B	Rank
		5.2	1
11.2	2		
		12	3
12.4	4		
		17.2	5
Total	6		

$$w_\alpha = 0, \qquad w_{1-\alpha} = 6$$

$$T = 6 - \frac{(2)(3)}{2} = 3$$

We cannot reject H_0 for this randomization.

3.39

Sample	Remember	Does not remember	Total
A	9	6	15
B	4	8	12
Total	13	14	

$$A = 15, \qquad B = 12, \qquad a = 9, \qquad b = 4$$

We cannot reject H_0.

CHAPTER 4

4.1 The signs of the differences in observed values $(X_i - Y_i)$ are

$$-, +, +, +, +, +, +, +, +, +$$

P value $= P(k \le 1|10, 0.50) = 0.0108$

4.3 The signs of the observed differences $(X_i - Y_i)$ are

$$+, +, +, +, +, +, +, +, +, +, +, +$$

P value $= P(K = 0|12, 0.50) = 0.0002$

4.5

$D_i = Y_i = X_i$	-53	-13	2	-17	-6	-26	-2	-24
Signed rank of $|D_i|$	-8	-4	$+1.5$	-5	-3	-7	-1.5	-6

$T_+ = 1.5$ (test statistic); $T_- = 34.5$; $0.0078 < P$ value < 0.0117

4.7

Ordered D_i	-0.75	1.00	3.00	3.50	3.50	5.75
	6.25	7.25	7.75	8.75	9.25	10.50
	11.00	12.25	12.50	13.75	16.00	17.00

$M_L = 3.50;$ $\quad M_U = 12.25$
$K' = 4;$ $\quad C(3.50 \le M \le 12.25) = 0.969$

4.9 $D_i = $ (after $-$ before) $= +0.1, -0.2, +0.3, +0.3, -0.3, +1.1, +0.3, +0.6, -0.5, +1.3,$ $+0.5, +0.7, +1.4, +1.3, +0.6;$ $K = 17$
  Ordered $U_{ij} = -0.50, -0.35, -0.30, -0.25, -0.20, -0.20, -0.10, -0.10, -0.10, -0.05,$ $-0.01, 0.00, 0.00, 0.00, 0.05, 0.05, 0.05,^* \ldots, 0.95,^{**} 0.95, 1.00, 1.00, 1.00, 1.00, 1.05, 1.10, 1.20,$ $1.20, 1.25, 1.30, 1.30, 1.30, 1.35, 1.35, 1.40.$
  *Lower limit; **Upper limit

4.11 $z = \dfrac{24 - 18}{\sqrt{24 + 18}} = 0.93;$ P value $= 0.1762$

4.13 The differences $(X - Y)$ yield four negative answers, two positive answers, and one zero answer. $P(K \le 2|6, 0.5) = 0.3438 = P$ value

4.15 The ordered differences (after–before) are $-32, -20, -5, 6, 7$

$$K' = 0; \qquad M_L = -32; \qquad M_U = 7$$

Confidence coefficient $= 1 - 2(0.0312) = 0.9376$

4.17 The differences (before–after) yield eight positive answers, one negative answer, and one zero answer. $P(K \leq 1 \mid 9, 0.5) = 0.0196 = P$ value

4.19

$D_i = Y_i - X_i$	$+5$	$+5$	$+15$	$+3$	$+4$	$+1$	$+4$
Signed rank of $\mid D_i \mid$	$+12$	$+12$	$+17.5$	$+5.5$	$+8.5$	$+1.5$	$+8.5$
$D_i = Y_i - X_i$	-5	$+3$	-2	$+2$	0	$+8$	$+4$
Signed rank of $\mid D_i \mid$	-12	$+5.5$	-3.5	$+3.5$	Omit	$+15$	$+8.5$
$D_i = Y_i - X_i$	$+4$	-1	0	-7	$+10$	$+15$	
Signed rank of $\mid D_i \mid$	$+8.5$	-1.5	Omit	-14	$+16$	$+17.5$	

$T_+ = 140, T_- = 31.$ $X = $ before, $Y = $ after

P value $= 2(0.0080) = 0.0160$

Reject H_0.

4.21 The differences $(A-E)$ yield one zero, one positive, and ten negative differences. $P(K \leq 1 \mid 11, 0.5) = 0.0059 = P$ value

4.23 The signs of the differences in the observed values $(X_i - Y_i)$ are

$$+, -, -, +, -, -, -, -, -, 0, -, -$$

Eliminate the zero difference from analysis. Effective $n = 11$.

P value $= P(k \leq 2 \mid 11, 0.50) = 0.0328$

We cannot reject H_0 at the 0.05 level.

4.25

$D_i = Y_i - X_i$	Signed rank of $\mid D_i \mid$
12	$+5.5$
16	$+8$
15	$+7$
12	$+5.5$
8	$+4$
-1	-1
0	omit
7	$+3$
5	$+2$

$T_+ = 35,$ $T_- = 1$ (test statistic)

P value $= 0.0078$

Reject H_0, since $0.0078 < 0.01$.

4.27

$D_i = Y_i - X_i$	Signed rank of $|D_i|$
-10	-7
-2	-1.5
-5	-4.5
$+3$	$+3$
$+2$	$+1.5$
-5	-4.5
-10	-7
-18	-10
-15	-9
-10	-7

$T_+ = 4.5$ (test statistic), $T_- = -50.5$

$0.0068 < P$ value < 0.0098
Reject H_0.

CHAPTER 5

5.1 H_0: There is no association between hospital size and willingness to return survey questionnaires
H_1: The two variables are related
$X^2 = 56.8856$; df $= 4$

Expected frequencies (E)	152.338	289.662	84.4406	160.559
$(O - E)^2/E$	12.9044	6.78669	1.08222	0.569152
Expected frequencies (E)	59.97	114.03	41.3586	
$(O - E)^2/E$	0.0687149	0.0361376	15.897	
Expected frequencies (E)	78.6414	6.89311	13.1069	
$(O - E)^2/E$	8.36047	7.32731	3.85354	

P value < 0.005

5.3 H_0: There is no relationship between the two variables
H_1: There is a relationship between the two variables
$X^2 = 12.3705$; df $= 4$

Expected frequencies (E)	10.8308	23.3846	21.4154	10.0923	14.2769
$(O - E)^2/E$	4.30804	0.243168	2.02459	2.38652	0.36313
Expected frequencies (E)	33.1692	71.6154	65.5846	30.9077	43.7231
$(O - E)^2/E$	1.40671	0.0794026	0.661086	0.779268	0.118574

$0.01 < P$ value < 0.025

5.5 H_0: The two variables are independent
     H_1: The two variables are not independent
     $X^2 = 29.5767$;   df = 9

Expected frequencies (E)	6.92641	7.61905	10.8052	6.64935
$(O - E)^2/E$	0.535782	2.8003	0.0600024	8.12591
Expected frequencies (E)	14.7186	16.1905	22.961	14.1299
$(O - E)^2/E$	0.939497	0.00224088	0.402218	0.0535834
Expected frequencies (E)	15.368	16.9048	23.974	14.7532
$(O - E)^2/E$	0.364867	6.02872	0.658746	0.954831
Expected frequencies (E)	12.987	14.2857	20.2597	12.4675
$(O - E)^2/E$	4.94401	1.95571	0.149485	1.60086

P value < 0.005

5.7 H_0: The two variables are independent
     H_1: The two variables are not independent
     $X^2 = 15.3219$;   df = 5

Expected frequencies (E)	107.25	36.4	62.4	44.2
$(O - E)^2/E$	1.07751	0.531867	0.656409	0.00090495
Expected frequencies (E)	29.9	5.85	57.75	19.6
$(O - E)^2/E$	0.562209	2.53376	2.00108	0.987759
Expected frequencies (E)	33.6	23.8	16.1	3.15
$(O - E)^2/E$	1.21905	0.0016808	1.0441	4.70556

$0.005 < P$ value < 0.01

5.9 H_0: Social class and political interests are not related
     H_1: The two variables are related
     $X^2 = 102.105$;   df = 4

Expected frequencies (E)	211.183	146.77	34.0469	123.908
$(O - E)^2/E$	14.2289	9.21189	9.56597	2.04246
Expected frequencies (E)	86.115	19.9765	123.908	86.115
$(O - E)^2/E$	0.764716	28.8903	12.2176	23.395
Expected frequencies (E)	19.9765			
$(O - E)^2/E$	1.78804			

P value < 0.005

5.11 H_0: Sex and age when contact lenses first worn are independent
H_1: The two variables are not independent
$X^2 = 12.8131;$ df $= 2$

Expected frequencies (E)	3.48315	45.6292	12.8876
$(O - E)^2/E$	0.631533	1.27561	6.443
Expected frequencies (E)	6.51685	85.3708	24.1124
$(O - E)^2/E$	0.337544	0.681792	3.44367

P value < 0.005

5.13 H_0: The two groups do not differ with respect to rate of recovery
H_1: The two groups do differ with respect to rate of recovery
$X^2 = 5.1948;$ df $= 1$

Expected frequencies (E)	14	14	11	11
$(O - E)^2/E$	1.14286	1.14286	1.45455	1.45455

$0.01 < P$ value < 0.025

5.15 H_0: The two groups do not differ
H_1: The two groups do differ
$X^2 = 41.195;$ df $= 1$

Expected frequencies (E)	728.432	122.568	733.568	123.432
$(O - E)^2/E$	2.97707	17.6928	2.95619	17.5689

P value < 0.005

5.17 H_0: The three groups are homogeneous
H_1: The three groups are not homogeneous
$X^2 = 10.1083;$ df $= 6$

Expected frequencies (E)	24.8	14.7396	28.0755	56.3849
$(O - E)^2/E$	0.0016131	0.107774	0.0411974	0.00262748
Expected frequencies (E)	46.8	27.8151	52.9811	106.404
$(O - E)^2/E$	2.97521	0.833544	0.167741	3.60901
Expected frequencies (E)	140.4	83.4453	158.943	319.211
$(O - E)^2/E$	0.958406	0.151429	0.103534	1.15619

P value > 0.10

5.19 H_0: The two populations represented are homogeneous
H_1: The two populations are not homogeneous
$X^2 = 15.7942;$ df $= 2$

Expected frequencies (E)	11.8788	16.1212		
$(O - E)^2/E$	5.22572	3.85055		
Expected frequencies (E)	8.90909	12.0909	7.21212	9.78788
$(O - E)^2/E$	1.87848	1.38414	1.98943	1.4659

P value < 0.005

5.21 H_0: There is no difference in choice of major between the two groups
H_1: There is a difference
$X^2 = 0.811311;$ df $= 4$

Expected frequencies (E)	22.1878	24.8122	37.2944
$(O - E)^2/E$	0.215726	0.192912	0.196283
Expected frequencies (E)	41.7056	8.96954	10.0305
$(O - E)^2/E$	0.17552	0.000103423	0.0000924784
Expected frequencies (E)	14.1624	15.8376	10.3858
$(O - E)^2/E$	0.00186307	0.00166603	0.0143302
Expected frequencies (E)	11.6142		
$(O - E)^2/E$	0.0128146		

P value > 0.10

5.23 Independence

5.25 Independence

5.27 Homogeneity

5.29 H_0: The functional areas respond to recommendations in the same way
H_1: The functional areas respond differently
$X^2 = 9.25;$ df $= 4$

Expected frequencies (E)	$(O - E)^2/E$
20.77	0.86
6.23	2.87
20.77	0.68
6.23	2.28
8.46	0.03
2.54	0.08
6.15	0.22
1.85	0.72
3.85	0.35
1.15	1.15

$0.1 > P$ value > 0.05

5.31 H_0: The type of artist does not depend on gender
H_1: The type of artist does depend on gender
df = 6, $X^2_{6,0.95} = 12.59$
Reject H_0 because 41.13 > 12.59.
P value < 0.005

5.33 H_0: Response rates are the same or more for white than for colored paper
H_1: Response rates for colored paper are greater than for white paper
$$X^2 = \frac{400((50)(67) - (133)(150))^2}{(50 + 133)(150 + 67)(133 + 67)(50 + 200)} = 69.39$$
df = 1
Reject H_0.
P value < 0.005

5.35 H_0: Executives do not differ in their opinions on inflation
H_1: Executives' opinions do differ
$X^2 = 59.29;$ df = 4

Expected frequencies (E)	$(O - E)^2/E$
52.00	14.02
48.00	15.19
52.00	10.17
48.00	11.02
52.00	0.08
48.00	0.08
52.00	0.94
48.00	1.02
52.00	3.25
48.00	3.25

P value < 0.005

CHAPTER 6

6.1

	Group			
	I	**II**	**III**	**Total**
> 168.05	4	5	2	11
≤ 168.05	8	0	3	11
	12	5	5	22

$X^2 = 6.53333;$ df = 2

Expected frequencies (E)	6	6	2.5
$(O - E)^2/E$	0.666667	0.666667	2.5
Expected frequencies (E)	2.5	2.5	2.5
$(O - E)^2/E$	2.5	0.1	0.1

$0.025 < P$ value < 0.05

6.3

	Controls	MD	Polio	Total
> 1.3	0	45	1	46
≤ 1.3	16	16	19	51
	16	61	20	97

$X^2 = 45.851$;　df = 2

Expected frequencies (E)	7.58763	8.41237	28.9278
$(O - E)^2/E$	7.58763	6.84374	8.92963
Expected frequencies (E)	32.0722	9.48454	10.5155
$(O - E)^2/E$	8.05416	7.58997	6.84586

P value < 0.005

6.5

	Normal	Moderate	Severe	Total
> 0.135	10	1	0	11
≤ 0.135	0	5	6	11
	10	6	6	22

$X^2 = 18.6667$;　df = 2

Expected frequencies (E)	5	5	3	3	3	3
$(O - E)^2/E$	5	5	1.33333	1.33333	3	3

P value < 0.005

6.7

	Ranks						
Controls	7.5	7.5	11	17.5	17.5	20.5	
LSD	2	4.5	4.5	7.5	11	14	
UML	1	3	7.5	11	14	14	
Controls	20.5	23.5	25	26	27		$R_1 = 203.5$
LSD	16	20.5					$R_2 = 80.0$
UML	20.5	23.5					$R_3 = 94.5$

$$H = \frac{12}{27(27 + 1)}\left[\frac{203.5^2}{11} + \frac{80^2}{8} + \frac{94.5^2}{8}\right] - 3(27 + 1)$$
$$= 6.175, 0.025 < P \text{ value} < 0.05$$

Correcting for ties:

$$\Sigma T = 3(2^3 - 2) + 2(3^3 - 3) + 2(4^3 - 4) = 186$$

$$H_C = \frac{6.175}{1 - (186/19656)} = 6.234, \qquad 0.025 < P \text{ value} < 0.05$$

6.9

Ranks by method

1	2	3	4	5	6
25	30	9	24	8	12
23	1	6	20	13	17
22	19	2	10	29	21
26	5	15	11	4	
32	31	27	16		
18			14		
3					
28					
7					

$R_1 = \overline{184}$ $R_2 = \overline{86}$ $R_3 = \overline{59}$ $R_4 = \overline{95}$ $R_5 = \overline{54}$ $R_6 = \overline{50}$

$$H = \frac{12}{32(32+1)}\left[\frac{184^2}{9} + \frac{86^2}{5} + \frac{59^2}{5} + \frac{95^2}{6} + \frac{54^2}{4} + \frac{50^2}{3}\right] - 3(32+1)$$

$$= 3.31$$

(These intermediate calculations have probably been carried to more decimals than the original author's calculations.) P value > 0.10

6.11 $U_{6,7} = 23.5,$ $U_{6,8} = 52,$ $U_{7,8} = 42.5$

$$J = 23.5 + 52 + 42.5 = 118$$

$$N = 6 + 6 + 12 = 24, \quad \Sigma n_i^2 = 6^2 + 6^2 + 12^2 = 216$$

$$\Sigma n_i^2(2n_i + 3) = 6^2(2 \cdot 6 + 3) + 6^2(2 \cdot 6 + 3) + 12^2(2 \cdot 12 + 3)$$
$$= 4968$$

$$z = \frac{118 - [(24^2 - 216)/4]}{\sqrt{\dfrac{24^2(2 \cdot 24 + 3) - 4968}{72}}} = 1.52$$

P value $= 0.0643$

6.13 $\bar{R}_1 = 203.5/11 = 18.5,\ \bar{R}_2 = 80/8 = 10,\ \bar{R}_3 = 94.5/8 = 11.81$

$$0.20/(3)(3-1) = 0.033,\ z \approx 1.84$$

$$1.84\sqrt{\frac{[27(27^2 - 1) - 186][1/11 + 1/8]}{12(26)}} = 6.75$$

Since $|18.5 - 10| = 8.5 > 6.75$, conclude that the medians of populations 1 and 2 are not equal. Since $|18.5 - 11.81| = 6.69 < 6.75$, conclude that the medians of populations 1 and 3 may be equal.

$$1.84\sqrt{\frac{3[27(27^2 - 1) - 186]}{6(27)(26)}} = 6.85$$

Since $|10 - 11.81| = 1.81 < 6.85$, conclude that the medians of populations 2 and 3 may be equal.

6.15

Table of ranks									Total
Controls	24	23	25	27	26				125
Neonatal hepatitis	17	22	18	19	20	21			117
Biliary	5	12	14.5	14.5	2	16	3	9	
atresia	8	13	1	6	10	4	11	7	136

$$H = \frac{12}{27(28)}\left[\frac{125^2}{5} + \frac{117^2}{6} + \frac{136^2}{16}\right] - 3(28) = 20.17$$

Since $20.17 > \chi_2^2 = 10.597$, P value < 0.005.
$\bar{R}_1 = 125/5 = 25$, $\bar{R}_2 = 117/6 = 19.5$, $\bar{R}_3 = 136/16 = 8.5$
$0.15/3(2) = 0.025$, $z = 1.96$

$$|25 - 19.5| < 1.96\sqrt{\frac{27(28)}{12}\left(\frac{1}{5} + \frac{1}{6}\right)},$$

$|5.5| < 9.42$, not significant

$$|25 - 8.5| > 1.96\sqrt{\frac{(27)(28)}{12}\left(\frac{1}{5} + \frac{1}{16}\right)},$$

$|16.5| > 7.97$, significant

$$|19.5 - 8.5| > 1.96\sqrt{\frac{(27)(28)}{12}\left(\frac{1}{6} + \frac{1}{16}\right)},$$

$|11.0| > 7.45$, significant

6.17

Table of ranks									Total
Carriers	1	9	4	3	2				19
HB-antibody positive	5	12	13	6	7	11	10	8	72
Controls	17	16	18	15	19	14			99

$$H = \frac{12}{19(20)}\left[\frac{19^2}{5} + \frac{72^2}{8} + \frac{99^2}{6}\right] - 3(20) = 14.327$$

Since $14.327 > \chi_2^2 = 10.597$, P value < 0.005.
$\bar{R}_1 = 19/5 = 3.8$, $\bar{R}_2 = 72/8 = 9.0$, $\bar{R}_3 = 99/6 = 16.5$
$0.15/3(2) = 0.025$, $z = 1.96$

$$|3.8 - 9.0| < 1.96\sqrt{\frac{19(20)}{12}\left(\frac{1}{5} + \frac{1}{8}\right)},$$

$|-5.7| < 6.29$, not significant

$$|9.0 - 16.5| > 1.96\sqrt{\frac{19(20)}{12}\left(\frac{1}{8} + \frac{1}{6}\right)},$$

$|-7.5| > 5.96$, significant

$$|3.8 - 16.5| > 1.96\sqrt{\frac{(19)(20)}{12}\left(\frac{1}{5} + \frac{1}{6}\right)},$$

$|-12.7| > 6.68$, significant

6.19

Table of ranks						Total
Nonsmokers	5.5	7	3.5	8	5.5	
	2	1	10.5	3.5	10.5	57
Light	24	14	13	10.5	18.5	
smokers	18.5	18.5	15			132
Medium	10.5	22.5	21	18.5	25	
smokers	22.5	16				136
Heavy	32.5	26	30	30	32.5	
smokers	28	27	30			236

$$H = \frac{12}{33(34)}\left[\frac{57^2}{10} + \frac{132^2}{8} + \frac{136^2}{7} + \frac{236^2}{8}\right] - 3(34) = 27.489$$

Since $27.489 > \chi_3^2 = 12.838$, P value < 0.005.

6.21

	Very isolated	Moderately isolated	Rural nonisolated	Urban ghetto	Total
> 33	2	10	15	18	45
≤ 33	21	13	8	5	47
Total	23	23	23	23	92

Expected frequencies:
First row: $(23)(45)/92 = 11.25$
Second row: $(23)(47)/92 = 11.75$

$$X^2 = \frac{(2 - 11.25)^2}{11.25} + \cdots + \frac{(5 - 11.75)^2}{11.75} = 25.533$$

Since $25.533 > \chi_3^2 = 12.838$, P value < 0.005.

6.23 $U_{N-MR} = 1$, $\quad U_{N-SR} = 0$, $\quad U_{MR-SR} = 5.5$

$\qquad J = 1 + 0 + 5.5 = 6.5$

$$z = \frac{6.5 - \{[23^2 - (11^2 + 6^2 + 6^2)]/4\}}{\sqrt{\{23^2[2\cdot 23 + 3] - [(11^2(2\cdot 11 + 3) + 6^2(2\cdot 6 + 3) + 6^2(2\cdot 6 + 3)]\}/72}}$$
$$= -4.452$$

P value < 0.001

6.25 $U_{I-II} = 57$, $\quad U_{I-III} = 49$, $\quad U_{II-III} = 24$

$\qquad J = 57 + 49 + 24 = 130$

$$z = \frac{130 - \{[22^2 - (10^2 + 6^2 + 6^2)]/4\}}{\sqrt{\{22^2(2\cdot 22 + 3) - [(10^2(2\cdot 10 + 3) + 6^2(2\cdot 6 + 3) + 6^2(2\cdot 6 + 3)]\}/72}}$$
$$= 3.1705$$

P value < 0.001

6.27

Ranks

A	B	C	D
16	13	19.5	2
5	7.5	19.5	9.5
7.5	19.5	23	5
2	13	13	2
23	16	19.5	11
5	23	16	9.5

$$R_1 = 58.5, \qquad R_2 = 92, \qquad R_3 = 110.5, \qquad R_4 = 39$$

$$H = \frac{12}{24(24+1)} \left[\frac{58.5^2}{6} + \frac{92^2}{6} + \frac{110.5^2}{6} + \frac{39^2}{6} \right] - 3(24+1)$$
$$= 10.39$$

$0.005 < P$ value < 0.01

6.29

$$R_A = \frac{11}{4} = 2.75, \qquad R_B = \frac{18}{3} = 6, \qquad R_C = \frac{26}{3} = 8.67$$

$$\frac{0.15}{3(2)} = 0.025, \qquad z = 1.96$$

$$1.96 \sqrt{\frac{10(10+1)}{12} \left(\frac{1}{4} + \frac{1}{3} \right)} = 4.53$$

Since $|2.75 - 8.67| > 4.53$, conclude that the medians of populations A and C are different. Since $|2.75 - 6| < 4.53$, conclude that the medians of populations B and C may not be different.

CHAPTER 7

7.1

Table of ranks

Patients	1	2	3	4	5	6	7	8	9	10	Total
A	1.5	1	1	1	1	2.5	2	1	1.5	1	$R_1 = 13.5$
B	1.5	2	3	3	3	1	1	2	1.5	2	$R_2 = 20.0$
C	3	3	2	2	2	2.5	3	3	3	3	$R_3 = 26.5$

$$\chi_r^2 = \frac{12}{10(3)(3+1)} (13.5^2 + 20.0^2 + 26.5^2) - 3(10)(3+1)$$

$$= 8.45; \qquad W = \frac{8.45}{10(2)} = 0.422$$

Since $0.390 < 0.422 < 0.430$, $0.012 < P$ value < 0.018.

7.3

Table of ranks									Total
Subject	1	2	3	4	5	6	7	8	
0	1	1	1	1	1	1	1	1	$R_1 = 8$
24	2	2	2.5	3	2	3	2	2.5	$R_2 = 19$
72	3	3	2.5	2	3	2	3	2.5	$R_3 = 21$

$$\chi_r^2 = \frac{12}{8(3)(3+1)}(8^2 + 19^2 + 21^2) - 3(8)(3+1) = 12.25; \; W = \frac{12.25}{8(2)} = 0.766$$

P value $= 0.001$

7.5

Table of ranks									Total
Patient	1	2	3	5	6	7	9	10	
C1	2	1.5	3	3	3	3	4	4	
C2	4	4	2	4	4	4	3	3	
Placebo	3	3	4	2	2	2	2	1	
Test drug	1	1.5	1	1	1	1	1	2	
Patient	12	13	14	15	18	19	22		
C1	2	3	4	4	1	4	4		$R_1 = 45.5$
C2	4	4	3	1	2	2	2		$R_2 = 46.0$
Placebo	3	2	2	3	3	3	3		$R_3 = 38.0$
Test drug	1	1	1	2	4	1	1		$R_4 = 20.5$

$$\chi_r^2 = \frac{12}{15(4)(4+1)}(45.5^2 + 46.0^2 + 38.0^2 + 20.5^2) - 3(15)(4+1)$$
$$= 17.02$$

P value < 0.005

7.7 $0.05/(3)(3-1) = 0.0083 \approx 0.01$

$$z = 2.33; \; 2.33\sqrt{\frac{8(3)(3+1)}{6}} = 9.32$$

$|8 - 19| = 11$, reject H_0; $\quad |8 - 21| = 13$, reject H_0; $\quad |19 - 21| = 2$, do not reject H_0

7.9

Dog	A	B	C	D	E	Total
Before	1	1	2	1	1	$R_1 = 6$
15	2	2	1	2	3	$R_2 = 10$
60	3	3.5	3	3	2	$R_3 = 14.5$
120	4	3.5	4	4	4	$R_4 = 19.5$

$$L = 6 + 2(10) + 3(14.5) + 4(19.5) = 147.5$$

Since $147.5 > 145, p < 0.001$.

7.11

Table of ranks

Case	4	5	6	7	8	9	10
1	7	6	3	4.5	4.5	2	1
2	6	5	7	4	3	2	1
3	4	6.5	6.5	5	2	3	1
4	7	6	5	4	3	1.5	1.5
5	7	6	5	4	3	2	1
6	7	6	5	4	2	3	1
7	6	7	4	5	1	2	3
8	7	6	4	3	5	2	1
9	6	7	5	3	2	4	1
10	7	6	5	4	3	2	1
Rank sums:	64	61.5	49.5	40.5	28.5	23.5	12.5

$$L = 12.5 + 2(23.5) + 3(28.5) + 4(40.5) + 5(49.5) + 6(61.5) + 7(64)$$
$$= 1371.5$$

Since $1371.5 > 1230$, P value < 0.001.

7.13

Ranks by treatment						Rank sums
0	1	1	1	1	1	5
1	2	1	1	1	2	7
2	2	2	1	2	1	8
4	1	1	2	2	1	7
9	2	2	2	2	2	10
18	1	2	2	1	2	8

$$T = \frac{12(6-1)}{5(6)(2-1)(2+1)}(5^2 + 7^2 + 8^2 + 7^2 + 10^2 + 8^2)$$
$$-\frac{3(5)(6-1)(2+1)}{2-1} = 9$$

Since $9 < \chi_5^2 = 9.236$, P value > 0.10.

7.15 The C_j's are 6, 5, 1, 3, and the R_i's are 2, 3, 1, 3, 3, 3

$$Q = \frac{4(4-1)(6^2 + 5^2 + 1^2 + 3^2) - (4-1)(15)^2}{4(15) - (2^2 + 3^2 + 1^2 + 3^2 + 3^2 + 3^2)} = 9.3158$$

$0.025 < P$ value < 0.050.

7.17 The C_j's are 6, 8, 5, 3, and the R_i's are 3, 3, 2, 2, 2, 3, 3, 1, 2, 1

$$Q = \frac{4(4-1)(6^2 + 8^2 + 5^2 + 3^2) - (4-1)(22)^2}{4(22) - (3^2 + 3^2 + 2^2 + \cdots + 1^2)} = 4.5882$$

Since $4.5882 < \chi_3^2 = 6.251$, P value > 0.10.

7.19 $R_1 = 27.5, R_2 = 53.5, R_3 = 55, R_4 = 44$

$$0.18/(4)(3) = 0.015, z = 2.17$$

$$2.17\sqrt{\frac{18(4)(5)}{6}} = 16.809$$

$$|27.5 - 53.5| = 26, \text{ significant}$$

$$|27.5 - 55| = 27.5, \text{significant}$$
$$|27.5 - 44| = 16.5, \text{not significant}$$
$$|53.5 - 55| = 1.5, \text{not significant}$$
$$|53.5 - 44| = 9.5, \text{not significant}$$
$$|55 - 44| = 11, \text{not significant}$$

7.21 $R_1 = 38, R_2 = 32, R_3 = 20$

$$0.15/(3)(2) = 0.025, z = 1.96$$

$$1.96\sqrt{\frac{15(3)(4)}{6}} = 10.735$$

$$|38 - 32| = 6, \text{not significant}$$
$$|38 - 20| = 18, \text{significant}$$
$$|32 - 20| = 12, \text{significant}$$

7.23

Table of ranks

	A	B	C	D
1	3		1	2
2		1	2	3
3	3	2	1	
4	3	2		1
	9	5	4	6

$$T = \frac{12(3)}{3(4)(2)(4)}(9^2 + 5^2 + 4^2 + 6^2) - \frac{3(3)(3)(4)}{2} = 5.25$$

Since $5.25 < \chi_3^2 = 6.251$, P value > 0.10.

7.25 After eliminating the row of all 1's, we have:

$$C_1 = 4, \quad C_2 = 8, \quad C_3 = 7, \quad C_4 = 4,$$
$$R_1 = 3, \quad R_2 = 2, \quad R_3 = 2, \quad R_4 = 2, \quad R_5 = 3, \quad R_6 = 1,$$
$$R_7 = 3, \quad R_8 = 3, \quad R_{10} = 2, \quad R_{11} = 1, \quad R_{12} = 1, \quad N = 23$$

$$Q = \frac{4(3)(4^2 + 8^2 + 7^2 + 4^2) - 3(23)^2}{4(23) - (3^2 + 2^2 + \cdots + 1^2)} = 4.135$$

Since $4.135 < 6.251$, P value > 0.10.

7.27

		Ranks by treatment		
	A	B	C	Total
	9	9	28	46
	1	16.5	30	47.5
	9	24	13.5	46.5
	7	22.5	16.5	46
	6	25.5	15	46.5
	20.5	3.5	20.5	44.5
	13.5	9	28	46.5
	2	18	28	48
	11.5	11.5	22.5	45.5
	3.5	19	25.5	48
Rank sums	83	154.5	227.5	

$$\Sigma\hat{R}^2_{\cdot j} = 82{,}515.5, \qquad \Sigma\hat{R}^2_{i\cdot} = 21{,}633.5, \qquad \Sigma\Sigma\hat{R}^2_{ij} = 9{,}447.5$$

$$T = \frac{(2)\{82515.5 - [(3)(10^2)/4][(3)(10) + 1]^2\}}{9447.5 - (1/3)(21633.5)}$$

$$T = 9.3392$$

Reject H_0. $0.005 < P$ value < 0.01.

7.29

		Ranks by treatment		
	1	2	3	Total
	4	9	24	37
	2	12	21	35
	3	18.5	18.5	40
	7	20	14	41
	1	11	23	35
	8	17	13	38
	5	10	22	37
	6	15.5	15.5	37
Rank sums	36	113	151	

$$\Sigma\hat{R}^2_{\cdot j} = 36{,}866, \qquad \Sigma\hat{R}^2_{i\cdot} = 11{,}282, \qquad \Sigma\Sigma\hat{R}^2_{ij} = 4{,}899$$

$$T = \frac{(2)\{36866 - [(3)(8)^2/4][(3)(8) + 1]^2\}}{4899 - (1/3)(11282)}$$

$$T = 12.06$$

Reject H_0. P value < 0.005.

7.31

Differences		
D_{iIH}	D_{iIC}	D_{iHC}
−.02	−.79	−.77
.12	−.84	−.96
.37	.31	−.06
.01	.11	.1

Differences

D_{iIH}	D_{iIC}	D_{iHC}
.08	−.95	−1.03
−.52	−1.17	−.65
−.07	−.17	−.1
.18	.13	−.05
−.15	−.85	−.7
−.09	.03	.12

$$z_{IH} = -.005, \quad z_{IC} = -.48, \quad z_{HC} = -.355$$

$$m_A = \frac{0 + -.005 + -.48}{3} = -.1617$$

$$m_B = \frac{.005 + 0 + -.355}{3} = -.117$$

$$m_C = \frac{.48 + .355 + 0}{3} = .278$$

$$m_I - m_C = -0.44$$

7.33

Differences

	24	72	24	72
	2.4	10.4	5.3	11.5
	4.4	8.9	3.9	2.2
	6.5	6.5	3.6	8.6
	6.1	4.3	5.4	5.4
Number of minuses, r_j	0	0		
Number of pluses, $b - r_j$	8	8		

Reject at 0.05 level of significance.

7.35

Ranks by treatment

I	H	C
1	2	3
2	1	3
3	1	2
3	2	1
2	1	3
1	2	3
1	2	3
3	1	2
1	2	3
2	3	1
2	3	1
Rank sums 19	17	24

$$L = 19 + 2(17) + 3(24) = 125$$

We cannot reject H_0.

7.37

Table of ranks

	A	B	C	D	E	F
1			1	2		3
2	1			2		3
3		1		2	3	
4	2	1		3		
5			3	2	1	
6		1			2	3
7		1	2			3
8	1				2	3
9	1	2	3			
10	1		3		2	
	6	6	12	11	10	15

$$t = 6, \quad k = 3, \quad r = 5$$

$$T = \frac{12(6 - 1)}{(5)(6)(3 - 1)(3 + 1)} (6^2 + 6^2 + 12^2 + 11^2 + 10^2 + 15^2)$$

$$- \frac{(3)(5)(6 - 1)(3 + 1)}{(3 - 1)}$$

$$T = 15.5$$

Reject H_0 at the $\alpha = 0.01$ level.

CHAPTER 8

8.1

CDQ

z_{L_i}	Expected relative frequency	Expected frequency	$\dfrac{(O - E)^2}{E}$
2.30	0.0107	2.14	2.1400
2.06	0.0090	1.80	0.8000
1.70	0.0249	4.98	20.1607
1.33	0.0472	9.44	0.6307
0.97	0.0742	14.84	0.3144
0.61	0.1049	20.98	10.6959
0.25	0.1304	26.08	3.7733
−0.11	0.1425	28.50	1.0614
−0.47	0.1370	27.40	1.5898
−0.83	0.1159	23.18	2.2240
−1.19	0.0863	17.26	3.4709
−1.55	0.0564	11.28	3.4963
−1.91	0.0325	6.50	6.5000
	0.0281	5.62	5.6200
			62.4774

Since $62.4774 > \chi^2_{11} = 26.757$, P value < 0.005.

z_{L_i}	BPS Expected relative frequency	Expected frequency	$\dfrac{(O-E)^2}{E}$
2.23	0.0129	2.58	2.5800
1.97	0.0115	2.30	0.2130
1.67	0.0231	4.62	0.0313
1.36	0.0394	7.88	0.4485
1.06	0.0577	11.54	6.2020
0.76	0.0790	15.80	1.1165
0.46	0.0992	19.84	0.1706
0.16	0.1136	22.72	1.7358
−0.14	0.1193	23.86	1.9723
−0.44	0.1143	22.86	3.4339
−0.74	0.1004	20.08	2.4963
−1.04	0.0804	16.08	1.0352
−1.35	0.0607	12.14	0.0609
−1.65	0.0390	7.80	63.1846
	0.0495	9.90	9.9000
			94.5809

Since $94.5809 > \chi^2_{12} = 28.300$, P value < 0.005.

8.3 Expected frequency under H_0 is $(7 + 12 + 6 + 10 + 12 + 6)/6 = 8.83$

$$X^2 = \frac{(7 - 8.83)^2}{8.83} + \frac{(12 - 8.83)^2}{8.83} + \cdots + \frac{(6 - 8.83)^2}{8.83} = 4.6243$$

Since $4.6243 < \chi^2_5 = 9.236$, P value > 0.10.

8.5

Expected relative frequency	0.0068	0.0494	0.1543	0.2676		
Expected frequency	9.07	65.90	205.84	356.98		
$(O - E)^2/E$	1.0391	1.2020	0.0001	0.0706		
Expected relative frequency	0.2786	0.1740	0.0604	0.0090		
Expected frequency	371.65	232.12	80.57	12.01		
$(O - E)^2/E$	0.1190	2.4567	1.6615	0.0816	**Total**	6.6306

Since $6.6306 < \chi^2_7 = 12.017$, P value > 0.10.

8.7 $\hat{\lambda} = \dfrac{0(24) + 1(16) + \cdots + 12(1)}{120} = \dfrac{380}{120} = 3.17$

$$f(0) = 0.0420 \qquad f(1) = 0.1332 \qquad f(2) = 0.2110 \qquad f(12) = 0.0001$$
$$f(3) = 0.2230 \qquad f(4) = 0.1767 \qquad f(5) = 0.1120$$
$$f(6) = 0.0592 \qquad f(7) = 0.0268 \qquad f(8) = 0.0106$$
$$f(9) = 0.0037 \qquad f(10) = 0.0012 \qquad f(11) = 0.0003$$

Expected frequencies	$\dfrac{(O - E)^2}{E}$
5.04	71.3257
15.98	0.0000
25.32	3.4306
26.76	2.8676
21.20	1.8132
13.44	1.4668
7.10	0.1704
3.22	0.9840
1.27 ⎫	
0.44 ⎪	
0.14 ⎬ 1.90	43.5842
0.04 ⎪	
0.01 ⎭	
	125.6425

Since $125.6425 > \chi^2_7 = 20.278$, P value < 0.005.

8.9

$S(x)$	$z = \dfrac{(x_i - 388.89)}{55.95}$	$F_0(x)$	$\lvert S(x_i) - F_0(x_i) \rvert$	$\lvert S(x_{i-1}) - F_0(x_i) \rvert$
0.0066	−3.38	0.0000(approx)	0.0066	0.0000
0.0132	−2.57	0.0051	0.0081	0.0015
0.0265	−2.12	0.0170	0.0095	0.0038
0.0397	−1.86	0.0314	0.0083	0.0049
0.0927	−1.59	0.0559	0.0368	0.0162
0.1126	−1.41	0.0793	0.0333	0.0134
0.1457	−1.23	0.1093	0.0364	0.0033
0.1523	−1.14	0.1271	0.0252	0.0186
0.1722	−1.05	0.1469	0.0253	0.0054
0.1788	−0.96	0.1685	0.0103	0.0037
0.2119	−0.87	0.1922	0.0197	0.0134
0.2185	−0.78	0.2177	0.0008	0.0058
0.2583	−0.70	0.2420	0.0163	0.0235
0.2649	−0.61	0.2709	0.0060	0.0126
0.2980	−0.52	0.3015	0.0035	0.0366
0.3377	−0.43	0.3336	0.0041	0.0356
0.3642	−0.34	0.3669	0.0027	0.0292
0.3841	−0.25	0.4013	0.0172	0.0371
0.4371	−0.16	0.4364	0.0007	0.0523
0.4437	−0.07	0.4721	0.0284	0.0350
0.4901	0.02	0.5080	0.0179	0.0643
0.4967	0.11	0.5438	0.0471	0.0537
0.5563	0.20	0.5793	0.0230	0.0826
0.5762	0.29	0.6141	0.0379	0.0578
0.6556	0.38	0.6480	0.0076	0.0718
0.6622	0.43	0.6664	0.0042	0.0108
0.7218	0.56	0.7123	0.0095	0.0501
0.7881	0.73	0.7673	0.0208	0.0455
0.8146	0.82	0.7939	0.0207	0.0058
0.8411	0.91	0.8186	0.0225	0.0040
0.8477	1.00	0.8413	0.0064	0.0002
0.8874	1.09	0.8621	0.0253	0.0144
0.9272	1.27	0.8980	0.0292	0.0106

$S(x)$	$z = \dfrac{(x_i - 388.89)}{55.95}$	$F_0(x)$	$|S(x_i) - F_0(x_i)|$	$|S(x_{i-1}) - F_0(x_i)|$
0.9338	1.36	0.9131	0.0207	0.0141
0.9404	1.45	0.9265	0.0139	0.0073
0.9603	1.63	0.9484	0.0119	0.0080
0.9801	1.72	0.9573	0.0228	0.0030
0.9934	1.81	0.9649	0.0285	0.0152
1.0000	1.99	0.9767	0.0233	0.0167

$\bar{x} = 388.89$; $s = 55.95$; $D = 0.0826$

Since $0.0826 < 1.07/\sqrt{151} = 0.0871$, P value > 0.20.

8.11

$F_0(x)$	0.2671	0.6197	0.8524	0.9548
$S(x)$	0.3000	0.6000	0.8333	0.9500
$|S(x) - F_0(x)|$	0.0329	0.0197	0.0191	0.0048
$F_0(x)$	0.9886	0.9975	0.9995	0.9999
$S(x)$	1.0000	1.0000	1.0000	1.0000
$|S(x) - F_0(x)|$	0.0114	0.0025	0.0005	0.0001

$D = 0.0329$

Since $0.0329 < 1.07/\sqrt{60} = 0.1381$, P value > 0.20.

8.13

Ordered sample values

X	Y	$S_1(x) - S_2(x)$
	9.3	$0 - 1/10 = -1/10$
10.2		$1/10 - 1/10 = 0$
	10.4	$1/10 - 2/10 = -1/10$
10.9		$2/10 - 2/10 = 0$
11.1		$3/10 - 2/10 = 1/10$
11.2		$4/10 - 2/10 = 2/10$
	11.2 } Tie	$4/10 - 3/10 = 1/10$
or		or
	11.2 } Tie	$3/10 - 3/10 = 0$
11.2		$4/10 - 3/10 = 1/10$
11.3		$5/10 - 3/10 = 2/10$
11.3		$6/10 - 3/10 = 3/10$
11.4		$7/10 - 3/10 = 4/10$
	11.4 } Tie	$7/10 - 4/10 = 3/10$
or		or
	11.4 } Tie	$6/10 - 4/10 = 2/10$
11.4		$7/10 - 4/10 = 3/10$
	11.6	$7/10 - 5/10 = 2/10$

Ordered sample values

X	Y	$S_1(x) - S_2(x)$
11.8		$8/10 - 5/10 = 3/10$
11.9		$9/10 - 5/10 = 4/10$
	11.9 } Tie	$9/10 - 6/10 = 3/10$
or		or
	11.9 } Tie	$8/10 - 6/10 = 2/10$
11.9		$9/10 - 6/10 = 3/10$
	12.2	$10/10 - 6/10 = 4/10$
	13.2	$10/10 - 7/10 = 3/10$
	13.8	$10/10 - 8/10 = 2/10$
	14.2	$10/10 - 9/10 = 1/10$
	14.5	$10/10 - 10/10 = 0$

$D = 4/10$, P value $= 0.20$

8.15

Ordered sample values

X	Y	$S_1(x) - S_2(x)$
118.5		$1/9 - 0 = 1/9$
135.0		$2/9 - 0 = 2/9$
	182.7	$2/9 - 1/8 = 7/72$
	204.0	$2/9 - 2/8 = -2/72$
261.7		$3/9 - 2/8 = 6/72$
	273.0	$3/9 - 3/8 = -3/72$
	297.7	$3/9 - 4/8 = -12/72$
	304.0	$3/9 - 5/8 = -21/72$
314.0		$4/9 - 5/8 = -13/72$
315.7		$5/9 - 5/8 = -5/72$
	324.3	$5/9 - 6/8 = -14/72$
336.9		$6/9 - 6/8 = -6/72$
	347.9	$6/9 - 7/8 = -15/72$
	351.2	$6/9 - 8/8 = -24/72$
352.7		$7/9 - 8/8 = -16/72$
565.5		$8/9 - 8/8 = -8/72$
797.5		$9/9 - 8/8 = 0$

$D = 24/72 = 0.33$

Since $0.33 < 4/9 = 0.44$, P value > 0.20.

8.17

$S_1(x)$	$S_2(x)$	$S_1(x) - S_2(x)$
0.1064	0.0741	0.0323
0.2128	0.1605	0.0523
0.4043	0.2840	0.1203
0.5745	0.4444	0.1301
0.8298	0.5556	0.2742
0.9574	0.7037	0.2537
1.0000	0.7778	0.2222
1.0000	0.8642	0.1358
1.0000	0.9136	0.0864
1.0000	0.9506	0.0494
1.0000	1.0000	0.0000

$D = 0.2742$

Since

$$0.2742 > 1.36\sqrt{\frac{47 + 81}{(47)(81)}} = 0.2494 \quad \text{and}$$

$$< 1.52\sqrt{(47 + 81)/(47)(81)} = 0.2787$$

$0.05 > P \text{ value} > 0.02$

8.19 $w_{1-\alpha} = \dfrac{1.22}{\sqrt{151}} = 0.0993$

S(x)	S(x) − 0.0993	S(x) + 0.0993	L(x)	U(x)
0.0066	−0.0927	0.1059	0	0.1059
0.0132	−0.0861	0.1125	0	0.1125
0.0265	−0.0728	0.1258	0	0.1258
0.0397	−0.0596	0.1390	0	0.1390
0.0927	−0.0066	0.1920	0	0.1920
0.1126	0.0133	0.2119	0.0133	0.2119
0.1457	0.0464	0.2450	0.0464	0.2450
0.1523	0.0530	0.2516	0.0530	0.2516
0.1722	0.0729	0.2715	0.0729	0.2715
0.1788	0.0795	0.2781	0.0795	0.2781
0.2119	0.1126	0.3112	0.1126	0.3112
0.2185	0.1192	0.3178	0.1192	0.3178
0.2583	0.1590	0.3576	0.1590	0.3576
0.2649	0.1656	0.3642	0.1656	0.3642
0.2980	0.1987	0.3973	0.1987	0.3973
0.3377	0.2384	0.4370	0.2384	0.4370
0.3642	0.2649	0.4635	0.2649	0.4635
0.3841	0.2848	0.4834	0.2848	0.4834
0.4371	0.3378	0.5364	0.3378	0.5364
0.4437	0.3444	0.5430	0.3444	0.5430
0.4901	0.3908	0.5894	0.3908	0.5894
0.4967	0.3974	0.5960	0.3974	0.5960
0.5563	0.4570	0.6556	0.4570	0.6556
0.5762	0.4769	0.6755	0.4769	0.6755
0.6556	0.5563	0.7549	0.5563	0.7549
0.6622	0.5629	0.7615	0.5629	0.7615
0.7218	0.6225	0.8211	0.6225	0.8211
0.7881	0.6888	0.8874	0.6888	0.8874
0.8146	0.7153	0.9139	0.7153	0.9139
0.8411	0.7418	0.9404	0.7418	0.9404
0.8477	0.7484	0.9470	0.7484	0.9470
0.8874	0.7881	0.9867	0.7881	0.9867
0.9272	0.8279	1.0265		1
0.9338	0.8345	1.0331		1
0.9404	0.8411	1.0397		1
0.9603	0.8610	1.0596		1
0.9801	0.8808	1.0794		1
0.9934	0.8941	1.0927		1
1.0000	0.9007	1.0993		1

8.21 $\hat{\mu} = 105,$ $\hat{\sigma}^2 = 365.97,$ $\hat{\sigma} = 19.13$

Score	z_{L_i}	Expected relative frequency	Expected frequency	$(O - E)^2/E$
< 50		0.0020	0.60 ⎫	
50–59	−2.88	0.0074	2.22 ⎬ 2.82	0.2384
60–69	−2.35	0.0242	7.26	0.2187
70–79	−1.83	0.0615	18.45	0.0164
80–89	−1.31	0.1226	36.78	0.2819
90–99	−0.78	0.1797	53.91	0.4472
100–109	−0.26	0.2052	61.56	0.0031
110–119	0.26	0.1797	53.91	0.2836
120–129	0.78	0.1226	35.78	0.7408
130–139	1.31	0.0615	18.45	0.3524
140–149	1.83	0.0242	7.26	0.0093
150–159	2.35	0.0074	2.22 ⎫	0.3005
> 159	2.88	0.0024	0.72 ⎬ 2.94	
				2.8923

Since $2.8923 < \chi_8^2 = 13.362,$ P value > 0.10.

8.23

Expected relative frequency	Expected frequency	$(O - E)^2/E$
$f(0) = 0.0010$	0.10 ⎫	
$f(1) = 0.0098$	0.98 ⎬ 1.08	0.0059
$f(2) = 0.0439$	4.39	0.0848
$f(3) = 0.1172$	11.72	1.9009
$f(4) = 0.2051$	20.51	0.0117
$f(5) = 0.2461$	24.61	0.2321
$f(6) = 0.2051$	20.51	0.0127
$f(7) = 0.1172$	11.72	0.0067
$f(8) = 0.0439$	4.39	0.0346
$f(9) = 0.0098$	0.98 ⎫	
$f(10) = 0.0010$	0.10 ⎬ 1.08	3.4133
		5.7027

Since $5.7027 < \chi_8^2 = 13.362,$ P value > 0.10.

8.25

$S_2(x) - S_1(x)$
$0.0933 - 0.0200 = 0.0733$
$0.2867 - 0.0533 = 0.2334$
$0.5533 - 0.1200 = 0.4333$
$0.6733 - 0.2667 = 0.4066$
$0.8133 - 0.3467 = 0.4666$
$0.9667 - 0.5667 = 0.3000$
$0.9067 - 0.7133 = 0.1934$
$0.9400 - 0.8133 = 0.1267$
$0.9733 - 0.9200 = 0.0533$
$0.9867 - 0.9800 = 0.0067$
$1.0000 - 1.0000 = 0.0000$

$$D^- = 0.4666$$

$$1.07 \sqrt{\frac{150 + 150}{(150)(150)}} = (1.07)(0.11547005) = 0.1236$$

$$1.22(0.11547005) = 0.1409$$

$$1.36(0.11547005) = 0.1570$$

$$1.52(0.11547005) = 0.1755$$

$$1.63(0.11547005) = 0.1882$$

P value < 0.005

8.27 Sample statistics:

Mean $= 108.0190$

Std. Dev. $= 1.3018$

Melting points (sorted)	Z_i	$P(Z \leq Z_i)$	$F_0(x)$	Cum. Freq.	$S(x)$	$S(x) - F_0(x)$	$S(x_{i-1}) - F_0(x)$
105.5	−1.94	0.0262	0.0262	1	0.048	0.0214	−0.0262
106.2	−1.40	0.0808	0.0546	2	0.095	0.0144	−0.0332
106.6	−1.09	0.1379	0.0833	3	0.143	0.0050	−0.0427
107.0	−0.78	0.2177	0.1344	4	0.190	−0.0272	−0.0748
107.1	−0.71	0.2389	0.1045	6	0.286	0.0468	−0.0484
107.1	−0.71	0.2389	0.1344	6	0.286	0.0468	0.0468
107.3	−0.55	0.2912	0.1568	7	0.333	0.0421	−0.0055
107.6	−0.32	0.3745	0.2177	8	0.381	0.0065	−0.0412
107.7	−0.25	0.4013	0.1836	9	0.429	0.0273	−0.0203
108.0	−0.01	0.496	0.3124	10	0.476	−0.0198	−0.0674
108.0	−0.01	0.496	0.1836	10	0.476	−0.0198	−0.0198
108.2	0.14	0.5557	0.3721	12	0.571	0.0157	−0.0795
108.3	0.22	0.5871	0.215	13	0.619	0.0319	−0.0157
108.3	0.22	0.5871	0.3721	14	0.667	0.0796	0.0319
108.4	0.29	0.6141	0.242	15	0.714	0.1002*	0.0526
108.8	0.60	0.7257	0.4837	16	0.762	0.0362	−0.0114
109.1	0.83	0.7967	0.313	18	0.857	0.0604	−0.0348
109.1	0.83	0.7967	0.4837	18	0.857	0.0604	0.0604
109.2	0.91	0.8186	0.3349	19	0.905	0.0862	0.0385
109.4	1.06	0.8554	0.5205	20	0.952	0.0970	0.0494
111.5	2.67	0.9962	0.4757	21	1.000	0.0038	−0.0438

$D = 0.1002$

Since $0.1002 < 0.188$, do not reject H_0.

8.29

Ordered sample values

Women	Men	$S_1(x) - S_2(x)$
21		$1/10 - 0 = 1/10$
	25	$1/10 - 1/10 = 0$
26		$2/10 - 1/10 = 1/10$
27		$3/10 - 1/10 = 2/10$
31		$4/10 - 1/10 = 3/10$
33		$5/10 - 1/10 = 4/10$
	33 } Tie	$5/10 - 2/10 = 3/10$
or		
	33 } Tie	$4/10 - 2/10 = 2/10$
33		$5/10 - 2/10 = 3/10$
	34	$5/10 - 3/10 = 2/10$
35		$6/10 - 3/10 = 3/10$
	35 } Tie	$6/10 - 4/10 = 2/10$
or		
	35 } Tie	$5/10 - 4/10 = 1/10$
35		$6/10 - 4/10 = 2/10$
37		$7/10 - 4/10 = 3/10$
	39	$7/10 - 5/10 = 2/10$
42		$8/10 - 5/10 = 3/10$
	42 } Tie	$8/10 - 6/10 = 2/10$
or		
	42 } Tie	$7/10 - 6/10 = 1/10$
42		$8/10 - 6/10 = 2/10$
	43	$8/10 - 7/10 = 1/10$
47		$9/10 - 7/10 = 2/10$
	47 } Tie	$9/10 - 8/10 = 1/10$
or		
	47 } Tie	$8/10 - 8/10 = 0/10$
47		$9/10 - 8/10 = 1/10$
54		$10/10 - 8/10 = 2/10$
	54 } Tie	$10/10 - 9/10 = 1/10$
or		
	54 } Tie	$9/10 - 9/10 = 0/10$
54		$10/10 - 9/10 = 1/10$
	56	$10/10 - 10/10 = 0$

$D = 4/10$; do not reject.

CHAPTER 9

9.1

Rank of								
H	1	2	4	3	6	7	5	9
r	1	2	3	4	5	6	7	8
d_i^2	0	0	1	1	1	1	4	1
Rank of								
H	8	10	12	11	13	15	14	16
r	9	10	11	12	13	14	15	16
d_i^2	1	0	1	1	0	1	1	0

Total $d_i^2 = 14$

$$r_S = 1 - \frac{6(14)}{16(16^2 - 1)} = 0.9794$$

Since $0.9794 > 0.762$, P value < 0.001 for a two-sided test.

9.3

Rank of Serum Mg	14	13	12	11	10	8.5	8.5
Bone Mg	14	10	12	8	9	13	11
d_i^2	0	9	0	9	1	20.25	6.25
Rank of Serum Mg	7	6	5	4	2	3	1
Bone Mg	7	6	5	2	3	4	1
d_i^2	0	0	0	4	1	1	0

Total $d_i^2 = 51.5$

$$r_S = 1 - \frac{6(51.5)}{14(14^2 - 1)} = 0.8868$$

Since $0.8868 > 0.802$, P value < 0.001 for a two-sided test.

9.5

(X, Y) rankings	(1, 12)	(2, 9)	(3, 1)	(4, 7)	(5, 3)	(6, 10)
Y pairs in natural order	0	2	9	3	6	1
Y pairs in reverse natural order	11	8	0	5	1	5
(X, Y) rankings	(7, 11)	(8, 8)	(9, 2)	(10, 6)	(11, 4)	(12, 5)
Y pairs in natural order	0	0	3	0	1	0
Y pairs in reverse natural order	5	4	0	2	0	0

$P = 25,$ $Q = 41$

$$S = 25 - 41 = -16, \qquad \hat{\tau} = \frac{-16}{12(12 - 1)/2} = -0.24$$

Since $-0.24 > -0.303$, P value is greater than 0.10.

9.7

Number of Y pairs in natural order	11	11	11	11	9	10	7	6
Number of Y pairs in reverse natural order	0	0	0	0	2	0	2	0
Number of Y pairs in natural order	6	3	3	2	2	2	1	0
Number of Y pairs in reverse natural order	0	3	2	2	0	0	0	0

$P = 95,$ $Q = 11$

$$S = 86; \qquad T_x = \frac{5(4) + 3(2) + 2}{2} = 14; \qquad T_y = \frac{2(1)}{2} = 1$$

$$\hat{\tau} = \frac{84}{\sqrt{16(15)/2 - 14}\sqrt{16(15)/2 - 1}} = 0.7479$$

Since $0.7479 > 0.483$, P value < 0.01, for a two-sided test.

9.9

Ordered values of eye track	561.7	702.4	772.6	777.2	854.6
Fixation rate	3.43	3.68	3.97	3.81	4.33
No. of Y pairs in natural order	8	7	4	4	3
No. of Y pairs in reverse natural order	0	0	2	1	1
Ordered values of eye track	870.2	892.9	926.4	980.8	
Fixation rate	4.53	3.80	4.41	4.85	
No. of Y pairs in natural order	1	2	1	0	
No. of Y pairs in reverse natural order	1	0	0	0	

$P = 30$, $Q = 6$

$$S = 30 - 6 = 24, \qquad \hat{\tau} = \frac{24}{9(8)/2} = 0.667$$

Since $\hat{\tau} = 0.667 = \tau^* = 0.677$, P value $= 0.01$ for a one-sided test.

9.11

Ordered values of left eye	1.315	1.340	1.374	1.540	1.635
Right eye	1.310	1.345	1.379	1.540	1.625
No. of Y pairs in natural order	9	8	7	6	5
No. of Y pairs in reverse natural order	0	0	0	0	0
Ordered values of left eye	1.735	1.850	1.910	1.915	2.040
Right eye	1.750	1.840	1.915	1.905	2.032
No. of Y pairs in natural order	4	3	1	1	0
No. of Y pairs in reverse natural order	0	0	1	0	0

$P = 44$, $Q = 1$

$$S = 44 - 1 = 43, \qquad \hat{\tau} = \frac{43}{10(9)/2} = 0.9556$$

$$C_1 = 9 \qquad C_2 = 9 \qquad C_3 = 9 \qquad C_4 = 9 \qquad C_5 = 9$$
$$C_6 = 9 \qquad C_7 = 9 \qquad C_8 = 8 \qquad C_9 = 8 \qquad C_{10} = 9$$
$$\Sigma C_i = 8(9) + 2(8) = 88, \qquad \Sigma C_i^2 = 8(81) + 2(64) = 776$$

$$\hat{\sigma}^2 = 4(776) - 2(88) - \frac{2(17)(88)^2}{10(9)} = 2.4889$$

$$C\left[0.9556 - \frac{2}{10(9)}(1.645)\sqrt{2.4889} \leq \tau \leq 0.9556\right.$$
$$\left. + \frac{2}{10(9)}(1.645)\sqrt{2.4889}\right] = 0.90$$

$$C(0.9556 - 0.0577 \leq \tau \leq 0.9556 + 0.0577) = 0.90$$

$$C(0.8979 \leq \tau \leq 1.0000) = 0.90$$

9.13 Median Y (Total DDT) = 1092.5
Median X $(6 - \beta - \text{hydroxycortisol}) = 251$
Counting from top: $+2$ Counting from right: $+5$
Counting from bottom: $+4$ Counting from left: $+4$
$S = |2 + 5 + 4 + 4| = 15$, P value $= 0.004963$

9.15
$$W = \frac{12(12^2 + 10^2 + 8^2 + 28.5^2 + 24^2 + 25^2 + 35.5^2 + 22^2 + 30^2 + 25^2) - 3(4)^2(10)(10 + 1)^2}{4^2(10)(10^2 - 1)}$$

$$= 0.568$$
$$X^2 = 4(10 - 1)(0.57) = 20.52$$

$0.01 < P$ value < 0.025

9.17

Table of ranks

1970 population	No. of establishments	No. of functions	No. of primary functions	No. of functional units	R_i
37	26	26	27.5	23.5	140.0
36	28.5	28.5	27.5	28	148.5
35	22	17	16.5	21.5	112.0
34	30	30	29	30	153.0
33	14	19	19.5	12.5	98.0
32	37	37	37	35	178.0
31	35	32.5	32.5	36	167.0
30	33	32.5	31	33	159.5
29	36	34.5	35.5	37	172.0
28	4	1	1	5	39.0
27	31	34.5	32.5	32	157.0
26	12	9	13.5	11	71.5
25	32	31	34	31	153.0
24	20	10.5	12	19	85.5
23	24	25	25.5	25	122.5
22	13	12	15	14	76.0
21	18	18	19.5	16.5	93.0
20	17	14.5	16.5	15	83.0
19	34	36	35.5	34	158.5
18	27	27	30	27	129.0
17	21	22	21	21.5	102.5
16	28.5	28.5	25.5	29	127.5
15	19	23	23	20	100.0
14	11	14.5	10	12.5	62.0

Table of ranks

1970 population	No. of establishments	No. of functions	No. of primary functions	No. of functional units	R_i
13	2	7	5	6.5	33.5
12	25	21	22	23.5	103.5
11	6	2.5	3	4	26.5
10	8	13	10	10	51.0
9	23	24	24	26	106.0
8	10	8	7	9	42.0
7	16	20	18	18	79.0
6	9	10.5	10	8	43.5
5	1	5.5	3	1	15.5
4	3	4	6	2	19.0
3	15	16	13.5	16.5	64.0
2	7	5.5	8	6.5	29.0
1	5	2.5	3	3	14.5

$$W = \frac{12(140.0^2 + 148.5^2 + \cdots + 14.5^2) - 3(5)^2(37)(37+1)^2}{5^2(37)(37^2-1)}$$

$$= 0.8591$$

$$X^2 = 5(37-1)(0.8591) = 154.638$$

Since $154.638 > 66.766$, P value < 0.005.

9.19

W	X	Natural order	Reverse natural order
74	19	4	1
91	15	4	0
96	20	3	0
106	21	2	0
128	22	1	0
132	25	0	0

$$P = 14, \quad Q = 1, \quad S = 14 - 1 = 13$$

$$\hat{\tau}_{WX} = \frac{13}{(6)(5)/2} = 0.867$$

W	X	Natural order	Reverse natural order
74	19	3	2
91	18	3	1
96	21	2	1
106	25	0	2
128	16	1	0
132	20	0	0

$$P = 9, \quad Q = 6, \quad S = 9 - 6 = 3$$

$$\hat{\tau}_{WY} = \frac{3}{(6)(5)/2} = 0.20$$

Y	X	Natural order	Reverse natural order
16	22	1	4
18	15	4	0
19	19	3	0
20	25	0	2
21	20	1	0
25	21	0	0

$$P = 9, \qquad Q = 6, \qquad S = 9 - 6 = 3$$

$$\hat{\tau}_{YX} = \frac{3}{(6)(5)/2} = 0.2$$

$$\hat{\tau}_{WY.X} = \frac{0.20 - (0.867)(0.2)}{\sqrt{(1 - 0.867^2)(1 - 0.2^2)}} = 0.054$$

We cannot reject H_0, P value > 0.25

9.21
$$\phi = \frac{(20)(14) - (16)(6)}{\sqrt{(20 + 16)(6 + 14)(20 + 6)(16 + 14)}} = 0.25$$

$$X^2 = (56)(0.25^2) = 3.5$$

$0.05 < P$ value < 0.10

9.23
$$C = \sqrt{\frac{47.9}{2746(3 - 1)}} = 0.093$$

Reject H_0. P value < 0.005

9.25
$$C = \sqrt{\frac{56.89}{1001(2 - 1)}} = 0.238$$

Reject H_0. $0.005 < P$ value < 0.01

9.27
$$Q = \frac{(45)(43) - (19)(6)}{(45)(43) + (19)(6)} = 0.89$$

9.29
$$C = \sqrt{\frac{11.56}{256(2 - 1)}} = 0.213$$

Reject H_0. $0.005 < P$ value < 0.01

9.31
$$n_1 = 14, \qquad x_1 = 39$$
$$n_2 = 6, \qquad x_2 = 25$$

$$r_{pb} = \sqrt{\frac{(14)(6)}{20}} \left[\frac{39 - 25}{\sqrt{1599}} \right] = 0.718$$

9.33

d_i	4	4	2	−2	−2	1	4	−6
d^2	16	16	4	4	4	1	16	36
d_i	3	1	−2	2	−2	5	−2	
d_i^2	9	1	4	4	4	25	4	

$$r_S = 1 - \frac{6(148)}{15(224)} = 0.7357, \qquad 0.0025 > P \text{ value} > 0.001$$

9.35

$$W = \frac{12(5^2 + 16^2 + 16^2 + 9^2 + 14^2 + 24^2) - 3(16)(6)(49)}{16(6)(35)} = 0.7643$$

$$X^2 = 4(5)(0.7643)$$
$$= 15.286$$

$0.005 < P \text{ value} < 0.01$

9.37

Y pairs in natural order	14	13	12	8	9	9	8	6
Y pairs in reverse natural order	0	0	0	3	1	0	0	1
Y pairs in natural order	6	5	3	3	1	1	0	
Y pairs in reverse natural order	0	0	1	0	1	0	0	

$$P = 98, \quad Q = 7$$

$$S = 98 - 7 = 91, \quad \hat{t} = \frac{91}{15(14)/2} = 0.8667, \quad P \text{ value} < 0.005$$

9.39

d_i	0	−1	0	−1	−1	3	0	−1	1	0
d_i^2	0	1	0	1	1	9	0	1	1	0

$$r_S = 1 - \frac{6(14)}{10(99)} = 0.9152, \quad P \text{ value} < 0.001$$

9.41

$$C = \sqrt{\frac{15.3}{440(2 - 1)}} = 0.186$$

Reject H_0; $0.005 < P \text{ value} < 0.01$

9.43

$$C = \sqrt{\frac{102.105}{852(3 - 1)}} = 0.245$$

Reject H_0; $0.025 > P \text{ value} > 0.005$

9.45

$$\phi = \frac{(10)(12) - (2)(2)}{\sqrt{(10 + 2)(2 + 12)(10 + 2)(2 + 12)}} = 0.69$$

$$X^2 = (26)(0.69^2) = 12.37$$

Reject H_0; $P \text{ value} < 0.005$

9.47

$$n_1 = 14, \bar{x}_1 = 1595$$
$$n_2 = 6, \bar{x}_2 = 1479$$

$$r_{pb} = \sqrt{\frac{(14)(6)}{20}}\left[\frac{1595 - 1479}{\sqrt{1999092}}\right] = 0.168$$

CHAPTER 10

10.1 $Y = 5 - 0.25X$, $\quad n_1 = 15$, $\quad n_2 = 19$

$$X^2 = \frac{8}{50}\left[\left(15 - \frac{50}{4}\right)^2 + \left(19 - \frac{50}{4}\right)^2\right] = 7.76$$

Since $7.378 < 7.76 < 9.210, 0.025 > P$ value > 0.01.

$\quad H_0: \beta = -0.25;$ $\quad H_1: \beta \neq -0.25;$ $\quad Y = 5.31 - 0.25X;$ $\quad n_1 = 12$

$$X_b^2 = \frac{16}{50}\left(12 - \frac{50}{4}\right)^2 = 0.08$$

Since $0.08 < 2.706, P$ value > 0.10.

$Y_i - (-0.25X)$				
6.74	5.80	5.08	5.28	6.05
6.35	4.75	5.78	4.51	5.68
3.41	5.41	4.62	5.33	3.76
5.88	5.34	4.35	6.35	4.86
6.16	3.81	4.56	6.99	4.74

Median $= (5.29 + 5.33)/2 = 5.31$

10.3 $H_0: \alpha = 450, \beta = 0$ $\quad$ Median $X = 500$

$\quad H_1: \alpha \neq 450, \beta \neq 0$ $\quad n_1 = 14, n_2 = 20$

$$X^2 = \frac{8}{56}\left[\left(14 - \frac{56}{4}\right)^2 + \left(20 - \frac{56}{4}\right)^2\right] = 5.14$$

Since $4.605 < 5.14 < 5.991, 0.10 > P$ value > 0.05.

$\quad H_0: \beta = 0;$ $\quad H_1: \beta \neq 0;$ $\quad n_1 = 10$

$$X_b^2 = \frac{16}{56}\left(10 - \frac{56}{4}\right)^2 = 4.57$$

Since $3.841 < 4.57 < 5.024, 0.025 < P$ value < 0.05.

10.5 The ordered array of S_{ij} values is as follows.

−4.7636	0.3805	0.4419	0.6093	0.8138	1.0246
−0.0739	0.4175	0.4737	0.6145	0.8400	1.1775
0.0178	0.4217	0.4841	0.6954	0.8853	1.4369
0.0267	0.4385	0.5471	0.7052	0.9833	2.0195
0.0896	0.4409	0.5604	0.7348	0.9855	2.3778
0.1348	0.4417	0.5846	0.7853	0.9916	4.8462

$$S_{\alpha/2} = 20, \quad C_{\alpha/2} = 20 - 2 = 18, \quad k = (36 - 18)/2 = 9$$

$$\hat{\beta}_L = 0.4217, \quad \hat{\beta}_U = 0.9833$$

10.7   Ordered array of S_{ij}

-2900.0000	1.0152	26.3780
-83.3333	7.9412	37.9447
-79.3103	11.0795	40.7547
-42.5926	11.3208	100.0000
-0.5865	23.7785	372.7273

$$\hat{\beta} = 11.0795, \quad S_{\alpha/2} = 13, \quad C_{\alpha/2} = 13 - 2 = 11$$

$$k = \frac{15 - 11}{2} = 2, \quad \hat{\beta}_L = -83.3333, \quad \hat{\beta}_U = 100.000$$

10.9   Solution varies, depending on which values are randomly eliminated from the data set.

10.11   Solution varies, depending on which values are randomly eliminated from the data set.

10.13   Solution varies, depending on which values are eliminated from the data set.

10.15

X	189	191	193	201	207	208
Y_i	94	92	95	91	103	100
Number in natural order	9	9	8	8	3	4
Number in reverse natural order	2	1	1	0	4	2
X	211	215	221	222	223	231
Y_i	98	99	101	109	105	106
Number in natural order	5	4	3	0	1	0
Number in reverse natural order	0	0	0	2	0	0

$$P = 54; \quad Q = 12; \quad S = 54 - 12 = 42; \quad \hat{t} = \frac{42}{12(11)/2} = 0.6364$$

P value < 0.005

10.17

Ordered array of S_{ij}		Ordered array of S_{ij}	
-1.000	0.313	0.043	0.625
-0.750	0.321	0.077	0.636
-0.636	0.333	0.100	0.667
-0.571	0.364	0.125	0.667
-0.556	0.412	0.182	0.769
-0.500	0.429	0.188	1.000
-0.333	0.462	0.200	1.500
-0.250	0.500	0.250	2.000
-0.067	0.533	0.259	4.000

$$S_{\alpha/2} = 20, \quad C_{\alpha/2} = 18, \quad k = \frac{36 - 18}{2} = 9$$

$$\hat{\beta}_L = -0.067, \quad \hat{\beta}_U = 0.625$$

10.19 $Y = +22.5 + 13.1X$

10.21

X	$Y - \beta_0 X$	Number of pairs in natural order	Number of pairs in reverse natural order
125	−113.18	12	2
126	−112.23	13	0
128	−112.49	12	0
143	−130.82	11	0
159	−143.01	10	0
162	−149.46	7	2
162	−148.17	7	1
163	−146.75	7	0
164	−150.23	6	0
166	−152.51	5	0
177	−164.04	4	0
184	−170.99	3	0
202	−183.13	1	1
205	−180.76	1	0
211	−191.24	0	0

$P = 99$ $Q = 6$

$$S = 99 - 6 = 93$$

$$\hat{\tau} = \frac{93}{15(14)} = 0.8857$$

Reject H_0 ; P value < 0.005

INDEX